JOURNEYMAN'S GUIDE TO THE NATIONAL ELECTRICAL CODE®

JOURNEYMAN'S GUIDE TO THE NATIONAL ELECTRICAL CODE

1996 Edition

F. Marco Gotshaw

Prentice Hall PTR
Upper Saddle River, NJ 07458

Library of Congress Cataloging-in-Publication Data
Gotshaw, F. Marco
Journeyman's guide to the National Electrical Code, 1996 / F. Marco Gotshaw.
p. cm.
Includes index.
ISBN 0-13-234659-1
1. National Fire Protection Association. National Electrical Code (1996) 2. Electric engineering—United States—Insurance requirements. I. National Fire Protection Association. National Electrical Code (1996) II. Title.
TK260.G68 1996
621.319'24'0218—dc20 96-15712
CIP

Editorial/production supervision: *Rainbow Graphics, Inc.*
Cover designer: *Karen Marsilio*
Manufacturing manager: *Alexis Heydt*

Prentice-Hall, Inc.
A Simon & Schuster Company
Upper Saddle River, NJ 07458

The publisher offers discounts on this book when ordered in bulk quantities. For more information, contact:

Corporate Sales Department
Prentice Hall PTR
One Lake Street
Upper Saddle River, NJ 07458

Phone: 800–382–3419
FAX: 201–236–7141
E-Mail: corpsales@prenhall.com

Note: The material in this book may be subject to local codes and regulations. Check with local building officials when planning your project.

Printed in the United States of America

10 9 8 7 6 5 4 3 2 1

ISBN 0-13-234659-1

Prentice-Hall International (UK) Limited, *London*
Prentice-Hall of Australia Pty. Limited, *Sydney*
Prentice-Hall Canada Inc., *Toronto*
Prentice-Hall Hispanoamericana, S.A., *Mexico*
Prentice-Hall of India Private Limited, *New Delhi*
Prentice-Hall of Japan, Inc., *Tokyo*
Simon & Schuster Asia Pte. Ltd., *Singapore*
Editora Prentice-Hall do Brasil, Ltda., *Rio de Janeiro*

In memory of my father
C. D. Gotshaw
whose guidance, inspiration, and faith
in me helped to make this book possible.

Contents

Preface

"How to Effectively Use the National Electrical Code" can still be used as a subtitle for this third edition of *The Journeyman's Guide to the National Electrical Code—1996 Edition.* In addition to addressing many of the changes in the 1996 edition of the NE Code, the 1996 edition of the text offers detailed pictures and drawings of the material being discussed. The outline of this book was restructured and expanded in comparison to the 1993 edition. More emphasis was placed on grounding and bonding of the structural distribution, motors and transformers.

The book's instruction is directed at electrical personnel in all phases of the electrical industry—design, manufacture, installation, repair, and service—who must ensure that the electrical function or installation at least meets NE Code standards (the safety minimum).

The book discusses the NE Code requirements from two points of view: practical and safe. For the electrical function or installation to be practical, the electrical equipment must be installed and used within the guidelines of the NE Code. Electrical safety actually has two facets: personnel safety and fire safety. Personnel safety indicates protection from electrocution or mechanical hazards involved with electrical equipment. Fire safety indicates protection of property and persons from the adverse effects of fire within the structure or equipment served. Both fall within the guidelines of the NE Code.

Various parts or excerpts from the National Electrical Code are covered in handbooks, workbooks, self-pace textbooks, etc., but this book can function as both a reference book for the electrical person who needs to take an in-depth look at the Code, and as a textbook in a formal setting (of instruction) for training electrical persons.

General electrical equipment installations (Chapters 1 through 4 in the NE Code) are investigated in a logical sequence with the central theme that good design and function starts with the proper use, installation, and protection of the electrical distribution conductors. This theme is developed in such a manner that NE Code rules governing special occupancies, special equipment, or special conditions (Chapters 5 through 7 in the NE Code) can be understood although not investigated in the text.

The text begins with the structure of the NE Code, which is a formal outline of chapters, articles, and sections. Code reference, and how to go about it, is explained with detailed reference, discussion, and worked examples. Key electrical terms are defined as presented, such that the information presented will not fall to the way-side because of unfamiliar terminology.

The material is presented in a logical manner starting with the ampacity of conductors, which is a critical function of electrical distribution. Conductor installation and protection against physical damage require sizing different types of conductor raceways, junction and outlet boxes, and other types of conductor enclosures.

The conductors must be protected from the effects of overcurrent. The different types of overcurrent and overload protection, where within the power distribution it must be located, and how it must be sized are investigated. The discussion also includes what constitutes a connected load, and the different types of power distribution, whether branch circuit, feeder, or service.

The conductors must have a source of power. The power distribution system starts at the service of the building or other structure, and continues on to the connected load. The service equipment is a key point of discussion within the text, because at that particular point, both the overcurrent protection for the conductors (and indirectly the connected load), the central tie point for the equipment grounding/bonding system, and the reference grounding of the distribution (neutral conductor) must be established or located. The discussion also includes working clearances about electrical equipment.

The lowest level of protected supply to equipment or other connected loads is the branch circuit. Lighting circuits are investigated first. Detailed drawings of the various types of lighting circuits, including two-, three-, and four-way switching, are investigated.

The different types of branch circuits, and how the circuits are sized according to overcurrent protection, wire size, and intended load, are investigated and explained. Appliances, motors, and transformers are key topics in these investigations.

Because motor control is a contributing factor as to how the motor(s) should be supplied (conductors) and protected (fuses, circuit breakers, and other, separate overload devices), simple-motor control circuits, ways to draw them, amd methods for protecting them are introduced.

The more common power distribution transformer connection drawings are also introduced and explained. Transformer connections mandate the type of power distribution within a building or other structure, and the way it must be installed and protected (against overcurrent).

The branch circuits are supplied from either feeders, panels, or other load centers including service equipment. These topics are discussed as individual chapters in the text.

Information contained in each chapter is examined with review questions located at the end of the respective chapter.

Some of the topics discussed require solving math problems with guidelines, set values, or maximum limits specified in the NE Code. These problems are introduced with step-by-step examples that include both the math requirements and the NE Code reference that requires the computation. Each example is followed by a Code challenge problem that is correlated to the example just discussed. Answers to these Code challenge problems are offered as an appendix at the end of the text. In addition to review questions, the information contained in these chapters is also examined with a set of review problems, correlated to both the example and the parallel code challenge, at the end of the chapter.

Because the required math introduced in the text with various electrical formulas may be a problem to some, the last two chapters of the text are offered as a self-help alternative. Chapter 18 investigates the basic math skills with worked examples. Chapter 19 investigates the more common electrical formulas involved with Code computations by detailed drawings and worked examples.

Acknowledgments

Thanks to my family and friends, and all the people in electrical technology, including students, electricians, contractors, engineers, and governing authorities, whose interest and support continue to make this book a good source for understanding the National Electrical Code.

Special thanks to Marion Davis Inc., Mayfair mills, and Brunswick mills for help with the photos and artwork in this 1996 edition.

About the Author

Marco Gotshaw, a native of South Carolina, is an instructor of electricity and electronics at Greenville Technical College located in Greenville, South Carolina. His specialty is instruction of the NE Code and electrical control. He has over eighteen years of experience as an electrician in both industry and construction. He is a master electrician licensed by the state of South Carolina.

Development of this text stems from formal classroom lecture, and study of the NE Code. The information contained herein has been proven by course repetition to be one of the better sources for answers to questions concerning the National Electrical Code.

Mr. Gotshaw is a graduate of Greenville Technical College with associate degrees in industrial electricity, industrial electronics, and electronic engineering technology.

Chapter 1

Introduction to the National Electrical Code

The National Electrical Code (NE Code) is a book of rules and regulations whose prime objective is (1) the safeguard of human life and (2) the supply and protection of electrical equipment.

The NE Code is sponsored by the National Fire Protection Association (NFPA) in the interest of public safety. Other associations that either amend or support the NE Code include the National Electrical Manufacturer's Association (NEMA), the American National Standards Institute (ANSI), the Institute of Electrical and Electronic Engineers (IEEE), the International Association of Electrical Inspectors (IAEI), and Underwriter's Laboratory (UL).

NEMA offers standards in terms of size or capability for electrical equipment. For instance, manufactured motor controllers (starters) are sized in accordance with full-load current and/or horsepower ratings, but are referenced in terms of *NEMA sizes:* NEMA size 1, NEMA size 2, and so on.

ANSI offers standards in terms of measurement, functional testing, design, symbols, and related manufacturing, installation, and testing data.

IEEE offers update on the latest trends or developments in the electrical world. Information is dispersed to its members through journals, at regional or national conventions, and in local chapter meetings.

IAEI is concerned with interpretation or understanding of the NE Code. IAEI is not a legal enforcer of the NE Code. Their interpretations cannot be considered as the final decision on NE Code matters.

UL operates testing laboratories and offers a listing service to equipment manufacturers. UL investigates all phases of manufacturing, installation, and consumer use of electrical products. Their evaluation often detects hazards or faults in equipment that might endanger or otherwise affect life and property.

UL is not the only listing and testing authority, but it is one of the more predominant authorities in the United States. Equipment listed with UL and bearing the UL logo is not necessarily "approved" for the intended electrical installation by the NE Code. The UL logo or labeling only indicates that the equipment passed the UL tests.

1.1 NE CODE STRUCTURE AND REFERENCE

As outlined in Section 90-3, the National Electrical Code book is structured as a formal outline which has main headings of the introduction and nine chapters. The nine chapters fall into four basic categories: Chapters 1 through 4 deal with general equipment installations. Chapters 5 through 7 deal with special installations. Chapter 8 deals with communications. Chapter 9 offers tables and examples. Chapters 5 through 9 either amend, explain, or restrict the first four chapters.

Specifically, the individual chapters offer either rules, regulations, or recommendations in the following areas of electrical installation:

Article 90: Introduction
Chapter 1: Definitions and Working Clearances
Chapter 2: Wiring Design and Protection
Chapter 3: Wiring Methods and Materials
Chapter 4: Equipment for General Use
Chapter 5: Special Occupancies
Chapter 6: Special Equipment
Chapter 7: Special Conditions
Chapter 8: Communications
Chapter 9: Tables and Examples

The NE Code book offers two means of reference for individual or specific rules, regulations, or recommendations: The *Table of Contents* (located at the front of the book) and the *Index* (located at the back of the book). Both are useful in terms of road maps as to "where to find what."

The chapters, as outlined above, are listed in the *Table of Contents* in numerical order. The chapters are divided into topics referred to as *Articles.* The Articles, which are also listed in numerical order, are divided into *Parts,* which are listed in alphabetical order. Both the Article and the Parts of the Article are listed with corresponding page numbers.

The *Table of Contents* is a good place to look if you don't know the NE Code name for a particular topic, subject, or term. For example, suppose you have a question about lighting fixture wiring. Lighting fixtures would be considered as equipment for general use. Scanning the contents of Chapter 4, Art. 410 governs the installation and supply wiring of lighting fixtures. Further scanning, Part F governs the wiring of lighting fixtures.

The *Index* deals with keys words which relate to specific subject matter. The term lighting fixtures is listed in the index in boldface print. Reference is made to Article (Art.) 410. Subtopics or subject matter pertaining to lighting fixtures are indented and listed in alphabetical order following the key term. Wiring (of lighting fixtures) is listed as 410-F (Art. 410, Part F). Notice in either reference the actual page listing was not given. Only the *Article* reference.

Without the page number, the text of the NE Code can be scanned for Art. 410 by leafing through the pages of the text. The Article number and its subject matter are listed as a center heading at the top of each successive right-hand page. Once Art. 410 is found, Part F is found by further leafing through the pages.

Part F will also have a center heading found in the body of the page (either right-hand or left-hand, so both pages must be scanned). The heading "F. Wiring of Fixtures" will be printed in boldface type.

When the student electrician or other electrical person makes reference to NE Code material, it is imperative that the road map used to find the specific rule or recommendation be listed as the reference for the answer of a question, or for the discussion of the subject matter in question.

1.2 ARTICLES

As outlined in the Table of Contents, each chapter of the NE Code is broken down into topic matter called *articles.* The article deals with a specific topic which falls under the general heading of the chapter. With the exception of Article 90, which contains the introduction to the NE Code, when reference is made to a specific Article, the chapter is indicated by the first digit in the article "call number," such that further reference to the chapter is not required.

Example: Chapter 4, Article 430
Reference: Art. 430

As found in the Table of Contents, the general heading of Chapter 4 is "Equipment for General Use." Article 430's heading reads "Motors, Motor Circuits, and Controllers." In other words, motors are equipment for general use. As found in the index, the subject "motors" is listed in boldface type with the reference: "Art. 430."

1.3 PARTS

Most articles are further subdivided into *parts.* Parts of the article zero in on a specific subject matter that falls within the scope of both the article and the chapter in which they are contained.

Example: Chapter 1, Article 110, Part A

Reference: Art. 110, Part A, or 110-A

As found in the Table of Contents, the heading of Chapter 1 reads "General." Article 110's heading reads "Requirements for Electrical Installations." Article 100 is subdivided into parts A and B, whose headings read "General" and "Over 600 Volts, Nominal," respectively. This indicates that general requirements for electrical installations, regardless of voltage rating, are contained in Part A of Art. 110, and additional general requirements for electrical installations rated over 600 volts, nominal are contained in Part B of Art. 110.

Section 110-30 (first paragraph of Part B) stipulates the rules listed in Part B of Art. 110 either supplement or amend Part A for the higher voltages.

Not all articles are divided into parts. In the Table of Contents, Articles 200, 215, and 225 are all contained in Chapter 2, but none contain parts in their listings.

1.4 SECTIONS

With the exception of Art. 100, all articles or parts are further subdivded into *sections*. Sections are individual paragraphs that discuss specific subject matter within the part or article. Sections, which are the stated rules of the NE Code, are not listed in the Table of Contents because they are too numerous.

A section is listed under the article or part with the article number suffixed with a dash and the section number. This section number and the heading of the subject matter discussed will appear in boldface print at the start of the first paragraph contained in the referenced section.

Example: Chapter 2, Article 240, Part A, Section 240-3

Reference: Sec. 240-3 or just 240-3

The *call number* used to list the section refers to both the chapter and article such that further reference to either is not required. In the above example, the listing of Sec. 240-3 identifies the third paragraph under Art. 240. The 2 in the call number of both the article and the section (240) identify Chapter 2 of the NE Code.

Subsections are the final divisions of the outline. Subsections are additional subject matter that support the sections they follow. They are indicated in reference by the section number followed by parentheses ( ) that contain a lowercase letter.

To follow with formal outline practice, a section will contain at least two subsections. If a section contains three subsections, the subsections would be identified as subsection (a), (b), and (c). The lowercase letter contained in the parentheses will appear with the heading of the subject matter discussed in boldface print at the start of the first paragraph contained in the referenced subsection.

Example: Chapter 2, Article 210, Part A, Section 210-5, Subsection (a).

Reference: Sec. 210-5(a) or just 210-5(a)

Investigating the example, Sec. 210-5 defines the "Color Code (requirements) for Branch Circuits," subsection (a) lists the color code requirements for the "Grounded (or neutral) Conductor." Subsection (b) lists the color code requirements for the "Equipment Grounding (bonding) Conductor." Notice there is no subsection (c) contained in Sec. 210-5.

A subsection can be further divided into additional subsections that are referenced by parentheses ( ) containing consecutive numbers. If a subsection contains two "inner" subsections, they are identified as subsections (1) and (2). The number contained in the parentheses will appear with the heading of the subject matter discussed in boldface print at the start of the first paragraph contained in the referenced (inner) subsection.

Example: Chapter 2, Article 220, Part A, Section 220-3, Subsection (d), Subsection (2)

Reference: Sec. 220-3(d)(2) or just 220-3(d)(2)

Investigating the example, Sec. 220-3(d) subsection (1) is printed with the heading "Dwelling Units" in boldface type. Subsection (2) is printed with the heading "Other than Dwelling Units" in boldface type. Both subsections occur under subsection (d) whose heading reads "Loads for Additions to Existing Installations." Subsection (d) occurs under Section 220-3 whose heading reads "Computation of Branch Circuits." In other words, the guidelines and/or requirements for computations of branch-circuit loads to be added to an existing dwelling installation are found in Sec. 220-3(d)(1). The guidelines and/or requirements for computations of branch-circuit loads to be added to an existing nondwelling installation are found in Sec. 220-3(d)(2).

1.5 EXCEPTIONS

The section or subsection states the *golden rule* of the NE Code, which either allows or forbids a particular method of installation, or places restrictions on devices or other electrical equipment. When extenuating circumstances or actual practices in the field have proved safe in certain situations, *exceptions* are given that either reverse or in some manner alter the golden rule.

If there is only one exception, the paragraph following the given section will start with "Exception:". To distinguish the exception from the section, the exception paragraph will be printed in italics.

Example: Chapter 3, Article 310, Section 310-2, Subsection (a), Exception

Reference: Sec. 310-2, Excp., or Sec. 310-2, Ex.

Investigating the example, Sec. 310-2 requires all conductors used in the distribution of electrical power within a building or other structure to be insulated. The exception to Sec. 310-2 will allow conductors either just "covered" (no insulation or dielectric rating), or bare conductors to be installed when permitted as specified elsewhere within the NE Code.

If two or more exceptions are listed under a given section or subsection, each exception will be listed as "Exception No. —" in numerical order as they occur.

Example: Chapter 2, Part B, Article 240, Section 240-22, Exception No. 1

Reference: Sec. 240-22, Excp. No. 1; Sec. 240-22, Ex. 1; 240-22, Excp. 1; or 240-22, Ex. 1

Investigating the example, Sec. 240-22's golden rule will not allow an overcurrent protection device to be connected in series with any circuit (load or current-carrying) conductor that is intentionally grounded. Exception No. 1 will allow a circuit breaker to protect a grounded conductor because it will disconnect all the circuit (load supply) conductors simultaneously.

Exception No. 2 will allow the "heater" element of a thermal motor overload relay to be installed in series with the grounded supply conductor of the motor because its response to overheating causes all the contacts within the motor-starter contactor to open simultaneously.

1.6 FINE-PRINT NOTES

In many instances a section will be followed by a *fine-print note.* Fine-print notes will occur as "(FPN):" at the start of a new paragraph under the section to which they pertain.

Fine-print notes are footnotes that further explain requirements of the section, exception, or table they follow. In some instances, a fine-print note will direct you to another section, article, or chapter that will further clarify or enhance the section in question.

Example: Chapter 3, Article 380, Part A, Section 380-13, Subsection (d), Fine-Print Note

Reference: Sec. 380-13(d), FPN, or just 380-13(d), FPN

Investigating the example, Sec. 380-13(d) will allow a knife switch to be used as a motor-circuit switch. The definition of a motor-circuit switch, as indicated by Sec. 380-13(d), FPN, is located in Art. 100.

Fine-print notes may also list recommended practices, actions, or calculations that will help in meeting the requirements of the NE Code regulation they follow.

If more than one fine-print note is listed for a given section, each note will be listed in numerical order with "(FPN No. . . .):".

Example: Chapter 2, Article 215, Section 215-2, Subsection (b), Fine-Print Note No. 2

Reference: Sec. 215-2(b), FPN No. 2, or just 215-2(b), FPN No. 2

Investigating the example, Sec. 215-2(b) requires the rated ampacity of feeder conductors to be at least equal to the rated ampacity of the service-entrance conductors when the load ampacity is 55 amps or less, and the feeder conductors convey the total load supplied by the service-entrance conductors.

The first FPN under this section refers the student electrician to the examples worked out in Chapter 9 of the NE Code. The second FPN discusses the voltage drop of feeder conductors and spells out recommended maximum percentages for feeder voltage drop. The third FPN refers the student electrician to Sec. 210-19(a) for a discussion of voltage drop considerations for branch circuits.

Section 90-5 cites a fine-print note as an NE Code recommendation as compared to an NE Code ruling. Recommendations are *explanatory material* contained in FPNs, whereas, mandatory NE Code rules are preceded or referenced by the word "shall."

1.7 TABLES

Tables are used in the NE Code when a composite of similar information and/or NE Code requirements is given. Tables are listed and used in many different ways throughout the NE Code. In some instances the table will be identified by the section it supports.

Example: Chapter 2, Article 220, Part A, Section 220-3(b), Table 220-3(b)
Reference: Table 220-3(b), or T220-3(b)

Investigating the example, Sec. 220-3(b) lists the guidelines for computing the branch-circuit lighting loads for certain occupancies. Table 220-3(b) lists the unit wattage or volt-amp load to be used in the computations according to the square footage floor area and type occupancy.

In some instances a group of tables will be referenced by a single section.

Example: Chapter 3, Article 310, Section 310-15, Subsection (a), Table 310-16
Reference: Table 310-16 or T310-16

Investigating the example, the allowable (rated) ampacity of individual insulated circuit conductors within a structural power distribution (branch-circuit, feeder, or service conductors) that is rated at 2,000 volts or less, is determined by Tables 310-16 through 310-19 in accordance with Sec. 310-15(a). Table 310-16 [following 310-15(d) in the NE Code] deals specifically with the rated ampacity of up to three single, individually insulated, current-carrying (circuit) conductors installed (contained) in a single raceway (conduit), multiconductor cable, or trench (direct burial).

Tables may be listed in a group in one area (section or part) of the NE Code, and referenced by several sections of different articles throughout the NE Code.

Example: Chapter 3, Article 346, Part A, Section 346-6; Chapter 9, Table 1
Reference: Sec. 346-6; Ch. 9, Table 1, or just 346-6; Ch. 9, T1
If information was determined by the specific table: Ch. 9, Table 1 or Ch. 9, T-1.

Investigating the example, Articles 345 through 351 lists the installation requirements for different types of conduit or tubing raceways. When determining the number of conductors allowed in a single conduit, most of these sections make reference to the tables in Chapter 9. Article 346 governs the installation and use of rigid metal conduit. Section 346-6 refers to Table 1 in Chapter 9 for conduit sizing.

Chapter 9 is titled "Tables and Examples." None of the tables found in either Part A of Chapter 9, or the additional tables listed in Appendix C have a sectional reference in their heading.

A table found in one part of the NE Code may be referenced by a section in another part of the NE Code.

Example: Chapter 4, Article 430, Part D, Section 430-52, Subsection (c)(1), Table 430-152

Reference: Table 430-152, or T430-152

Investigating the example, Part D of Art. 430 governs the "Branch-Circuit, Short-Circuit, and Ground-Fault (Overcurrent) Protection" for motors. Section 430-52(b) stipulates the branch-circuit overcurrent protection device (fuses or circuit breakers) for the motor must carry (convey) the high-value in-rush starting current of the motor without opening (failing).

Section 430-52(c)(1) will not allow either the rating of the branch-circuit fuses or the setting of the circuit breaker to exceed the current rating computed according to the overrate percentages listed in T430-152.

Table 430-152 is found in Part M, which is located near the end of Art. 430, several pages beyond Sec. 430-52 in the NE Code text.

1.8 NOTES TO TABLES

Notes normally give support material for the table in question, although in some instances notes restrict the values listed in tables. (As a rule, notes are only used with tables in the NE Code.) Unlike fine-print notes (FPNs) previously discussed, table notes are mandatory NE Code rules that must be abided by when using the table(s). The notes may be listed in paragraph form as an introduction to the table.

Example: Chapter 4, Article 430, Part A, Section 430-6, Subsection (a), Table 430-148, Note.

Reference: Table 430-148, Note, or T430-148, Note

Investigating the example, Sec. 430-6(a) requires the NE Code table-value of motor full-load current (FLA or FLC) to be used instead of the actual motor nameplate full-load current when sizing the power distribution and control equipment for a motor. This would include supply conductors, disconnects, overcurrent protection devices (fuses or circuit breakers), and electromechanical starters or other types of controllers for the motor. Everything in the motor circuit except running overload overcurrent protection (starter heaters or other overload detection devices).

The NE Code table-value motor full-load currents are listed in Tables 430-147 through 430-150 (found at the end of Art. 430).

The paragraph notes listed at the start of Table 430-148, describe the types of motors the table listing of motor full-load currents should be used for. The notes also describe the circuit or system voltage ranges that can be used with the motor voltage ratings.

Sometimes the notes are footnoted in the table with an obelus (dagger) or an asterisk (*). The notes will be listed immediately after the table.

> Example: Chapter 7, Article 710, Part A, Section 710-4, Subsection (b), Table 710-4(b), Asterisk (*) Note.
>
> Reference: Table 710-4(b), Note, or T710-4(b), Note

Investigating the example, Chapter 7 lists special conditions that must be considered for some electrical installations or equipment. Article 710 deals specifically with distribution voltages in excess of 600 volts nominal. Part A lists the general requirements for such installations. Section 710-4 lists the wiring methods that must be used. Subsection 710-4(b) lists restrictions placed on underground installations. Table 710-4(b) lists minimum burial depth/cover requirements for underground cables. The single asterisk note beside the column heading "Rigid Nonmetallic Conduit Approved for Direct Burial" further defines "approved," and lists alternative methods of installation and protection for conduits that are not approved.

In some instances the note will appear as part of the table. Tables 310-16 through 310-19 were mentioned previously as the "ampacity tables" for circuit conductors operating within a voltage range of 0 to 2,000 volts.

> Example: Chapter 3, Article 310, Section 310-15, Subsection (a), Table 310-16, Correction Factors
>
> Reference: Table 310-16, CF

Investigating the example, the allowable ampacities listed in T310-16, as stipulated in the table heading, are based on an ambient (surrounding air or earth) temperature of 30°C (86°F). When the expected ambient temperature will exceed this ambient base, the conductor ampacity, listed according to wire size, conductor material, and type of conductor insulation, must be derated according to the *"CORRECTION FACTORS"* listed in the lower part of the table.

When several notes list procedures or calculations that must be used to modify table values, the notes will be listed in numerical order following the table.

> Example: Chapter 2, Article 220, Part B, Section 220-19, Table 220-19, Note 4
>
> Reference: Table 220-19, Note 4; T220-19, Note 4; or T220-19, N-4

Investigating the example, Art. 220 governs minimum load computations for branch circuits, feeders, or services. Section 220-19 governs the load requirements of dwelling cooking equipment. Table 220-19 allows a derate of the demand load that is less than nameplate rating. The headings of Columns A, B, and C list a range of ratings in parentheses that the cooking equipment nameplate must fall within for the respective columns to be used. Notes 1 through 4 are used to modify the requirements of the table.

In Chapter 9, five notes are listed under Part A which apply to all the tables in Chapter 9. They are labeled as "Notes to Tables." These notes are listed in numerical order with the number listed at the start of the paragraph containing the note. Reference to these notes must contain both the chapter and part reference.

Example: Chapter 9, Part A, Note 4
Reference: Ch. 9, Pt. A, N-4, or Ch. 9-A, N-4

1.9 DEFINITIONS

Definitions of key words or electrical terms that are used throughout the NE Code are listed in Art. 100. The words or terms are listed in alphabetical order.

Example: The definition of "Approved" is found in Chapter 1, Article 100, Part A
Reference: Art. 100, Def.; Art. 100; or simply "Def"

More than one definition may be listed under a key word. When this occurs, the key word will appear in boldface print with a colon (:). The definitions will follow with another keyword or phrase in regular print ended in a colon at the start of the paragraph (of the definition). For example, there are three definitions contained in Art. 100 for a conductor. The "Conductor:" can be either "Bare:", "Covered:", or "Insulated:".

If a definition pertains only to a discrete subject matter that is discussed in a section, exception, or table, and is not used throughout the NE Code, the definition will be listed as a note following the NE Code regulation to which it pertains.

Example: Chapter 4, Article 430, Part F, Section 430-71, Definition (of a motor control circuit)
Reference: Sec. 430-71, Def, or 430-71, Def

The definition of a motor control circuit is only required in Part F of Art. 430 which governs specifically motor control circuits. This definition is not a general definition that would be used throughout the NE Code.

1.10 INTENT OF THE NATIONAL ELECTRICAL CODE

Article 90, the introduction, gives insight as to the purpose and scope of the National Electrical Code.

The purpose of the NE Code, as described by Sec. 90-1(a), is to provide necessary information for the safeguarding of people and property (in that order) against electrical hazards. Section 90-1(b) states that compliance with the NE Code will make a building or other structure hazard free (from the standpoint of fire or electrical shock), but not necessarily efficient, convenient (to the owner or user), or adequate for the intended use or for future expansion.

Section 90-1(c) stipulates the NE Code is not intended to be a design specification book, nor is it intended to be an instruction book for the untrained electrician. Clarity at this point is required, but as you can see, is not offered. Actually the NE Code book is a safety checklist that should be used by electrical engineers and/or contractors to ensure their design of an intended installation will comply with the practical safeguards as set forth by the National Electrical Code.

Many design or engineering handbooks describe in detail how an electrical feat can be accomplished. These books offer substantiating theory and often many examples. The theory of electricity and its application is not within the scope of the National Electrical Code.

An instruction manual or textbook, such as the one you are now reading, offers insight and some theory, as well as formulas that relate to NE Code rules and regulations. The NE Code cannot explain the origin of each rule or how it can be applied to the numerous electrical installations that do or can exist, therefore the NE Code book is not an instruction manual for untrained persons.

Section 90-2 lists first the electrical installations that are covered by the NE Code [subsection (a)], and then the electrical installations that are not covered by the NE Code [subsection (b)].

The NE Code or its rules and regulations are not mandatory for any given electrical installation unless the authority having jurisdiction accepts its intent. Section 90-4 indicates the authority having jurisdiction could be the electrical inspector at the city, county, or state level, an appointed insurance inspector, or the Fire Chief or Fire Marshal assigned to the locale in which the intended electrical installation will occur [remember the National Electrical Code is sponsored by the National Fire Protection Association (NFPA)].

In other words, instances will occur where your installation will be NE Code correct, but will still not pass the electrical inspection. This is rare, and normally will happen only when prevailing conditions, such as high humidity or chemical concentrations in the air of a particular locale (salt and moisture in coastal areas), prohibit installations that are generally allowed by the NE Code.

In other instances, due to mechanical failures or safety hazards to personnel, general NE Code requirements may be prohibited by local ordinance. As a rule, NE Code-correct installations will pass any inspection, because most authorities having jurisdiction adopt the National Electrical Code as their local ordinance.

Section 90-4 and the AHJ (Authority Having Jurisdiction) will be referred to many times in the development of this text.

1.11 MARGIN LINES AND BULLETS

The National Electrical Code is edited or up-dated on a three-year cycle. Changes from the previous 1993 edition to the present 1996 edition are indicated as a vertical line (margin line) or bullet (solid dot) in the outside margin of the text on each page as they occur.

A vertical line adjacent to the text indicates added, revised, or in some form edited material. A bullet indicates deleted material.

In some instances, deleted material is done away with entirely (deleted). In other instances, text of the previous edition was only moved to another article or section within a given article. Unfortunately, relocation is not indicated by the bullet. However, when relocated, the added text, at its new location, is indicated with a vertical line.

1.12 NE CODE TERMS

Many electrical terms are essential for understanding the National Electrical Code. The student electrician should understand the following:

Approved (Art. 100 and Sec. 110-2): Judged as suitable for the intended use by the governing authority (AHJ).

Authority having jurisdiction (*AHJ*) (Sec. 90-4): An office, organization, or individual responsible for approving or granting permission as to equipment, procedure, or installation requirements on electrical implementation.

Identified (Art. 100): Indicates that either equipment or material has been judged (normally by a testing agency) suitable for a particular purpose as spelled out by NE Code rules.

Labeled (Art. 100): Indicates the equipment was manufactured under controlled standards, or the equipment was tested and can be expected to perform in a prescribed manner.

Listed (Art. 100): States that either the equipment or material in question meets appropriate standards, or will perform in a specified manner.

Shall (Sec. 90-5): Spells out a mandatory NE Code rule that must be followed.

REVIEW QUESTIONS

1. Article ____________ gives insight as to the purpose and scope of the NE Code.
2. In terms of safeguarding human life from the affects of fire and electrocution (electrical hazards), the NE Code sets minimum requirements for electrical installations in residential, industrial, institutional, as well as commercial settings.

   True or False
3. Rules governing electrical equipment for general use are found in Chapter ____________ of the NE Code.
4. Rules governing either wiring design or protection from overcurrent are found in Chapter ____________ of the NE Code.
5. Rules concerning special equipment installations are found in Chapter ____________ of the NE Code.
6. Rules concerning communication equipment installations are found in Chapter ____________ of the NE Code.
7. Rules concerning definitions of electrical terms or working clearances about electrical equipment are found in Chapter ____________ of the NE Code.

8. Rules concerning special operating conditions, such as over 600 volts, are found in Chapter _____________ of the NE Code.

9. Rules concerning special occupancy conditions, such as aircraft hangars, are found in Chapter _____________ of the NE Code.

10. Rules governing either wiring methods or type of materials that can be used are found in Chapter _____________ of the NE Code.

11. With the exception of Chapter 9, each chapter in the NE Code is broken down into articles. Each article may be broken down into parts. Each article or part is further subdivided into sections. And, each section may be further divided into subsections.

    True or False

12. If the section states the mandatory rule (golden rule) of the NE Code that must be followed, _____________ either amend or negate the golden rule when special conditions or alternate requirements are met.

13. _____________ are used in the NE Code when a composite of similar information and/or NE Code requirements are given.

14. FPNs are used in the NE Code to add clarity or explanation to a section. Fine-Print Notes are a mandatory requirement.

    True or False

15. _____________ is (definition) the current in amperes a conductor can carry continuously under the conditions of use, without exceeding the temperature rating of its insulation.

16. A _____________ conductor is (definition) a conductor that has no insulation or cover whatsoever along its length.

17. Installation requirements for underground (direct burial) feeder or branch-circuit cables operating at 600 volts or less are found in Sec. _____________ and Table _____________.

18. Restrictions placed on electrical installations in hospitals (health care facilities) are found in Art. _____________.

19. If circuit or load conductors must be paralleled due to the high-magnitude of the full-load current, their installation must meet the requirements of Sec. _____________.

20. The rated ampacity of conductors rated 0 through 2000 volts is determined from Tables _____________ through _____________ and their accompanying notes.

21. Standard ratings of fuses or circuit breakers installed as conductor overcurrent protection are listed in Sec. _____________.

22. One article in the National Electrical Code that pertains specifically to branch circuits is Art. _____________.

23. "Nonmetallic-sheathed cable" is the NE Code name for Romex cable. Installation and protection requirements for this type of cable wiring method are found in Art. _____________.

24. A vertical line drawn in the outside margin beside a portion of NE Code text indicates material in the previous NE Code edition has been deleted.

    True or False

25. Mandatory NE Code rules are characterized by the use of the word _____________.

Chapter 2

Conductors and Cables

The supply power for a building or other structure, which is normally in the form of service panels, switchgear, or other load centers, is normally located in one central location. Connected loads can be lighting loads, heating loads, motor loads, or combination-equipment loads (combination of these three principle loads). Individually insulated conductors, installed either in some form of raceway, or bound in a multiconductor cable, are used to convey the electrical power to and from the various load locations.

The measure of the ability of a conductor to convey electrical current is referred to as its *ampacity* (Art. 100). Current flowing along the length of the conductor generates heat due to the minor resistance of the conductor. The interaction of the current flow to the resistance appears as both heat and voltage drop along the length of the conductor.

The *rated ampacity* of a conductor is determined according to its resistance per unit length, which reflects the amount of heat generated, and the ambient air (or earth) temperature along the run of the raceway or cable (length of wire), which reflects the amount of heat dissipated.

Article 100 of the NE Code defines the rated ampacity of a conductor as the amount of current it can convey continuously without exceeding its temperature rating. The temperature rating of a circuit conductor is determined by the type of insulation employed. Some types of insulation have higher operating temperatures than others. Regardless of the type, if too much heat is generated due to either an increase in current magnitude, or to a loss of heat dissipation (higher ambient temperature), or both, the insulation will lose its dielectric strength (the ability to block voltage) and short-circuit the conductors.

Correctly sizing circuit conductors for their rated ampacity and expected operating temperature is of prime importance for efficiency and longevity in an electrical power distribution system.

2.1 CONDUCTOR AMPACITIES

Article 310 deals with "Conductors for General Wiring." Section 310-15(a) refers the student electrician to the appropriate ampacity tables that are to be used according to the rated voltage of the conductors. The voltage ratings are grouped as 0 to 2,000 volts and 2001 to 35,000 volts. Our discussion will be restricted to Tables 310-16 through 310-19, which govern conductor ampacities with insulations rated at 2,000 volts or less.

The headings of the tables reflect the types of conductors or cables, the installation methods, and the ambient temperature rating for which the respective table-listed ampacities apply. Table 310-16 [follows Sec. 310-15(d) in the NE Code] deals with up to three individually insulated conductors installed either in raceway, or as a multiconductor cable wiring method (conduit or other raceway not required). Note that no distinction is made between metallic and nonmetallic conduit. The ambient temperature rating of this table is 30°C (degrees Celsius) or 86°F (degrees Fahrenheit). Table 310-16 is the workhorse for most general wiring installations up to 600 volts.

Table 310-17 lists ampacities of single insulated conductors installed in free air. The ambient temperature rating for this table is also based on 30°C or 86°F. Table 310-17 is used to determine the rated ampacity by wire size for service drops to various buildings or other structures, or for feeder or branch circuit conductors routed on supports along the surface of a building or along poles from one building to another.

Table 310-18 parallels the installation requirements of T310-16 (not more than three single insulated conductors in raceway or cable), but with higher temperature ratings for the conductor insulations. The insulation temperature ratings (column headings) in T310-16 range between 60°C and 90°C as compared to T310-18 whose range extends from 150°C to 250°C. The conductor ampacities listed in T310-16 were determined at an ambient (environmental) air temperature of 30°C (86°F). The conductor ampacities listed in T310-18 were determined at an ambient air temperature of 40°C (104°F). Note that higher-temperature cables or conductors for wire sizes larger than No. 4/0 (four-ought) AWG (American Wire Gage) are not available (indicated by the four dots).

Table 310-19 parallels T310-17 with higher temperature ratings for the conductor insulations. Both tables reflect the ampacities of single insulated conductors installed in free air. Table 310-17 is based on an ambient temperature of 30°C, with conductor insulations ranging between 60° and 90°C, as compared to T310-19, which is based on an ambient temperature of 40°C, with conductor insulations ranging between 150° and 250°C.

The notes that follow T310-19 ("Notes to Ampacity Tables of 0 to 2,000 Volts") relate to particular circumstances that either restrict or in some way modify the table-value ampacities listed in the tables.

Tables 310-16 through 310-19 are all structured and used in the same basic manner. The leftmost column and rightmost column list the size of the wire or conductor in terms of AWG (American Wire Gage) up through wire size No. 4/0 (0000, referred to as "four-ought"), and *kcmil* [thousand's (k) of circular (c) mils] from wire size 250 up to 2,000. In previous NE Code editions, kcmil was represented as MCM, with M designating "thousands of." The k used in this edition of the NE Code is part of the metric conversion that was begun several years ago for the United States.

Both systems of wire sizing indicate the cross-sectional area of the conductor. (Conversion of circular mils from the American Wire Gage is done for you, and can be read directly from the wire gage in Table 8 of Chapter 9.) The wire size, in Tables 310-16 through 310-19, indicates the row in which the listed ampacity will be found for a particular type of insulation and conductor material.

Analysis of the conductor ampacity tables will be based on T310-16. The tables are divided in half by a bold vertical line according to the conductive material used. The left half of the table contains an inner-column heading labeled "copper." The right half of the table contains an inner-column heading labeled "aluminum or copper-clad aluminum." The columns of ampacities listed under these inner headings identify ampacities of the respective conductive material of individual circuit conductors that are insulated with or by the respective insulation material identified in the individual columns.

The headings of the individual columns identify both the type of insulation material and its temperature rating. These headings and respective individual columns are repeated twice, once under each type of conductive material in each table. The temperature ratings and other characteristics of the particular type of insulation listed are further described in Table 310-13 for most of the insulations listed.

The table-rated ampacity of a conductor will be found where the row, identified by wire size, crosses or intersects the column, which is identifed by the type of conductor material, its insulation, and its associated temperature rating.

Example 2–1. Determine the table-rated ampacity of three No. 14 AWG THW copper conductors installed in conduit raceway.

Rated ampacity: 20 amps Ref. T310-16

The table-rated ampacity of up to three No. 14 AWG THW copper conductors (Fig. 2-1) installed in raceway would be found in T310-16. The wire size No. 14 identifies the third row from top to bottom in the leftmost column. THW copper identifies the second column from left to right under the heading "Temperature Rating of Conductor. See Table 310-13.", or the third column from left to right if the "Size" column is counted. By

Figure 2-1 THW is the letter designation of thermoplastic insulation that is both heat (H) and moisture resistant (W—wet). (*Courtesy of Southwire Company*)

this intersection of column and row, the table-rated ampacity of a No. 14 AWG THW copper conductor is 20 amps.

Code challenge 2–1. Determine the table-rated ampacity of three No. 4 AWG RHW copper conductors contained in a single raceway.

Rated ampacity: (A) ___________ amps* Ref. (B) ___________________

Example 2–2. Determine the table-rated ampacity of three 250 kcmil THHN aluminum conductors contained in rigid metal conduit, and buried as a feeder between two buildings.

Rated ampacity: 230 amps Ref. T310-16

The table-rated ampacity of 250 kcmil THHN (Fig. 2-2) aluminum conductors is found at the intersection of the sixteenth row and the third column on the aluminum side of Table 310-16.

Figure 2-2 THHN/THWN is the letter designation of a thermoplastic insulation that is nylon jacketed. (*Courtesy of Southwire Company*)

Code challenge 2–2. Determine the table-rated ampacity of three 600 kcmil THWN-2 copper-clad aluminum conductors contained in rigid nonmetallic conduit, and buried as a feeder to a remote pumphouse.

Rated ampacity: (A) ___________ amps Ref. (B) ___________________

How is the required wire size determined if the load amperage is known? For branch-circuit conductors, Sec. 210-19(a) requires the conductors to have a rated ampacity at least equal to the maximum amperage or current load to be served. For feeder conductors, Sec. 215-2 also requires the conductors to have an ampacity at least equal to the maximum amperage or current load served. With these two sections in mind, the table-rated ampacity of the intended circuit conductors would have to be greater than or equal to (≥) either the nameplate full-load current of the supplied equipment, or the computed rated-load current of an individual circuit or load center.

The rated ampacity of the circuit conductors is also affected or controlled by the rating or setting of the overcurrent protection device (OCPD) of the circuit. Section 240-3 requires current-carrying circuit conductors to be protected against the affects of overcurrent (current in excess of the rated-load current, Art. 100) in accordance with their rated ampacities as referenced by Sec. 310-15. Standard ratings of circuit fuses and standard settings of circuit breakers (both are OCPDs) are listed in Sec. 240-6(a) of the NE Code.

*Answers to Code challenge problems are listed in the Appendix at the end of the text.

For some installations, Sec. 240-3(b) will allow the circuit conductors to be sized with an ampacity at least equal to the rated-load current, and when the conductor's rated ampacity does not correspond to a standard rating of OCPD, the next higher standard size can be used without further upgrade or increase in the circuit conductor ampacity.

In general, the overcurrent protection device should be sized first at a value greater than or equal to the rated-load current computation. The conductors should then be sized for a rated ampacity greater than or equal to the rating or setting of the OCPD.

Example 2–3. If the full-load current rating of a single unit of equipment is 62 amps, determine the wire size of TW-insulated copper conductors required to supply the intended load.

Rating or setting of the OCPD: 70 amps — Ref. 240-6(a)
Table-rated ampacity: 70 amps — Ref. 240-3
Wire size: No. 4 AWG TW copper — Ref. T310-16

For a 62-amp rated-load current, the next higher standard size of fuse or circuit breaker is 70 amps. The table-rated ampacity of No. 4 AWG TW copper is listed in T310-16 as 70 amps. If the table-rated ampacity had been only 65 amps, Sec. 240-3(b) would still have allowed the No. 4 AWG TW copper conductors to be used.

Code challenge 2–3. If the full-load current rating of a single unit of equipment is 116 amps, determine the wire size of TW aluminum conductors required to supply the intended load.

Rating or setting of the OCPD: (A) ___________ amps Ref. (B) _______________
Table-rated ampacity: (C) ___________ amps Ref. (D) _______________
Wire size: (E) ___________ TW aluminum Ref. (F)_______________

2.2 WIRE TERMINAL TEMPERATURE LIMITATIONS

Circuit conductors will have a minimum of two terminations: At the point of power supply and at the point of connected load. Instead of termination to a single connected load, sometimes the circuit conductors will be terminated in a splice or tap to other circuit conductors, which may extend to several connected loads on the same circuit.

At the supply end, the circuit conductors are connected to the load terminal of a fuseholder or circuit breaker, which may be a screw terminal or pressure lug. In-line splices or taps of circuit conductors between the power supply and the connected load are made with wire nuts, crimp sleeves, or other types of pressure connectors. At the load end, the circuit conductors are normally connected to screw terminals on the base of a lighting fixture, on the side or back of a receptacle, or to a terminal strip located on the inside of a motor or other equipment terminal housing. In some instances the circuit conductors are terminated by a splice to lose "field leads" located in an equipment enclosure.

As a general rule, materials that will conduct electricity will also conduct heat. If the wire terminal on a fuseholder or circuit breaker is operating at a temperature higher than the connected conductor temperature, the heat will try to establish equilibrium along the length of the conductor. The circuit conductor is functioning as a heat sink, lowering the operating temperature of the fuse or circuit breaker.

The same is true for the load-end connections. If the wire terminals in a motor housing are operating at a higher temperature than the supply conductors, some of the heat will be transferred to the conductors. This is the way the electrical circuit should be designed.

The temperature rating of a conductor is an indication of how well the dielectric of the insulation can hold up at higher ambient (surrounding air) temperatures. A higher ambient temperature rating, as indicated in the ampacity tables, yields a higher value of rated ampacity. The increase in allowable load current (rated ampacity) increases the rate of heat generation along the length of the conductor.

Like insulated conductors, wire terminals also have an ambient temperature rating. All wire terminals have an ambient temperature rating of at least 60°C. Some wire terminals, especially those designed for large wire size or high-value load currents, have temperature ratings of 75°C. Other than special-rated devices, wire terminals do not have a 90°C temperature rating.

If the operating temperature of the circuit conductor is higher than the operating temperature of the wire terminal, the heat will travel in reverse, causing the wire terminals to overheat and prematurely fail. Heat tranfer from the circuit conductors can cause the supply fuses to open, or a thermal circuit breaker to open at a current value lower than the protection rating.

Heat transfer from the circuit conductors can cause a convenience receptacle to overheat. The excess heat can bend or distort the pressure clips inside the receptacle which maintains tension on the blades of the plug. The loss of tension may lead to low-voltage problems in the cord-supplied equipment due to the loose connection.

To prevent the wire terminals from heat-sinking higher-temperature rated circuit conductors, the NE Code restricts the operating temperature of the circuit conductors by restricting their rated ampacity.

Section 110-14 lists the rules that affect electrical connections (wire terminals). Subsection (c) stipulates that the temperature rating associated with the rated ampacity of a conductor shall be based on the *lowest* temperature rating of any wire terminal, conductor, or other device within the circuit (connected to that conductor).

In other words, if THHN copper is selected for the circuit conductors, the temperature rating for ampacity selection and resulting wire size can initially be taken from the 90°C temperature column of T310-16. If the wire terminals on the supply circuit breaker are rated at only 75°C, then the THHN copper conductors must be sized according to the ampacities listed in the 75°C column for insulated copper conductors. If the circuit is to supply convenience receptacles which have a terminal temperature rating of only 60°C, then the THHN copper conductors must be sized according to the ampacities listed in the 60°C copper column of T310-16 (rated ampacity determined according to the lowest temperature rating).

Lowering the temperature rating of the circuit conductors has the affect of increasing the wire size in most applications. An increase in wire size means a decrease in conductor resistance and a decrease in generated heat. As a result, the circuit conductors have a lower operating temperature which will heat sink the wire terminal.

When the temperature rating of the wire terminal or other device cannot be determined, subsections (1) and (2) to Sec. 110-14(c) are used as guidelines for wire terminal temperature compensation.

When the circuit is rated at 100 amps or less [normally determined by either the current rating of the connected load (full-load or rated-load current), or by the rating or setting of the circuit overcurrent protection device], or if the circuit consists of conductors wire size No. 1 AWG or smaller, the rated ampacity must be selected according to wire size and conductor material from the 60°C temperature column in the ampacity tables.

As the section reads, only conductors rated at 60°C, which are listed as either TW conductors, or UF cables in T310-16, can be used for circuits within this rating. Exception No. 1 to Sec. 110-14(c)(1) will allow the use of higher temperature-rated conductors (other insulations), provided they are sized according to the ampacities listed in the 60°C temperature column of the tables.

Example 2–4. Determine the NE Code-minimum wire size of XHHW-2 copper conductors required to supply a connected load of 40 amperes.

Wire terminal temperature limitations: 60°C Ref. 110-14(c)(1), Ex. 1
Table-rated ampacity: 40 amps
Wire size: No. 8 AWG XHHW-2 copper Ref. T310-16

In terms of rated ampacity, No. 10 AWG XHHW-2 copper (Fig. 2-3) is rated at 40 amperes, but the circuit conductors must be sized according to the 60°C column of T310-16. In the 60°C column, No. 10 AWG copper is rated at a maximum of only 30 amperes: No. 8 AWG copper is required.

Figure 2-3 XHHW is the letter designation of a crosslinked polyethylene insulation that is both high-heat and moisture resistant. (*Courtesy of Southwire Company*)

Code challenge 2–4. Determine the NE Code-minimum wire size of XHHW-2 aluminum conductors required to supply a connected load of 50 amperes.

Wire terminal temperature limitations: (A) ___________ °C Ref. (B) ___________
Table-rated ampacity: (C) ___________ amps
Wire size: (D) ___________ XHHW-2 aluminum Ref. (E) ___________

When the circuit is rated over 100 amps, or will consist of conductors wire size No. 1/0 AWG or larger, the rated ampacity can be selected according to wire size and conductor material from the 75°C temperature column in the ampacity tables, provided the conductor insulation temperature rating is at least 75°C. (If TW insulation was chosen, the wire size for the higher-amperage circuit would still be mandated to the 60°C column by the conductor insulation temperature rating.)

As the section reads, only conductors rated at 75°C can be used for circuits within this higher rating. Exception No. 1 to Sec. 110-14(c)(2) will allow the use of 90°C temperature-rated conductors, provided they are sized according to the ampacities listed in the 75°C temperature column of the tables.

Example 2–5. Determine the NE Code-minimum wire size of RHH aluminum conductors required to supply a single-phase, 3-wire, 200-amp, 120/240-volt power panel.

Wire terminal temperature limitations: 75°C Ref. 110-14(c)(2), Ex. 1
Table-rated ampacity: 205 amps
Wire size: 250 kcmil RHH aluminum Ref. T310-16

In terms of rated ampacity, No. 4/0 AWG RHH aluminum (Fig. 2-4) is rated at 205 amperes, but the circuit conductors must be sized according to the 75°C column of T310-16. In the 75°C column, No. 4/0 AWG aluminum is rated at a maximum of only 180 amperes: 250 kcmil aluminum conductors are required.

Figure 2-4 RHH, RHW, or USE is the letter designation of a vulcanized interlinked polyethylene (VIP) insulation. (*Courtesy of Southwire Company*)

Code challenge 2–5. Determine the NE Code-minimum wire size RHH copper conductors required to supply a 3-phase, 4-wire, 400-amp, 480Y/277-volt power panel.

Wire terminal temperature limitations: (A) __________ °C Ref. (B) __________
Table-rated ampacity: (C) __________ amps
Wire size: (D) __________ RHH copper Ref. (E) __________

2.3 AMPACITY DERATES DUE TO PREVAILING CONDITIONS

The ampacities of the circuit conductors just discussed are referred to as *table-rated ampacities*. Each table is based on an ambient (surrounding or environmental) air or earth (direct burial) temperature. Tables 310-16 and 310-18 are also based on not more than three circuit conductors in a given raceway.

When the expected ambient temperature will exceed the base temperature of the table, or if more than three circuit (current-carrying) conductors will be installed in a single raceway or contained in a single multiconductor cable, then the table-rated ampacity must be derated (rate of heat generation lowered) to compensate for the loss of heat dissipation.

Prevailing conditions that warrant conductor ampacity derates include:

1. Higher ambient (air or earth) temperatures.
2. More than three current-carrying conductors.
3. Continuous loads.
4. Voltage-drop compensation.

As defined in Art. 100, a load is considered to be continuous if the rated-load current will be drawn without interruption for *three hours or more.* Continuous loads are normally limited to nondwelling lighting loads, as referenced by Sec. 210-22(c) to store lighting.

In addition to heat generation, load current flowing against the wire resistance along the length of the circuit conductors also generates some voltage drop. All electrical equipment is designed with a voltage rating, which is normally matched to the voltage rating of the power supply to the building or other structure. If conductor voltage drop is not kept to a minimum, the equipment may fail prematurely. Low voltage can cause motors to overheat, or suffer from low torque. Low voltage can cause incandescent lighting lamps to glow dim. Electric-discharge lighting tubes may also be dim, but the contained ballasts normally overheat and start cycling on and off when supplied low voltage, because they are thermally protected.

All of these prevailing conditions represent situations in which the table-rated ampacity of the conductors, as were found in the previous examples, will not be suited (correctly rated) to supply the intended current load.

In the development of this text, conductor-ampacity derates for higher ambient temperatures, more than three current-carrying conductors in a common raceway, and continuous load rating will all be grouped together since they all directly affect the table-rated conductor ampacity, and are all mandatory NE Code requirements. The minimum conductor size for the maximum allowed voltage drop will then be computed. The largest, in comparison, will then be taken as NE Code minimum.

2.4 HIGHER AMBIENT TEMPERATURES

The ampacities as listed in the tables are based on a specific ambient (air or earth) temperature. Tables 310-16 and 310-17 are based on an ambient air temperature of 30°C (86°F). Tables 310-18 and 310-19 are based on an ambient air temperature of 40°C (104°F). When the ambient air or earth (for underground installations) temperature is higher than the table-based ambient temperature, the table-rated ampacity of the conductors must be lowered to a "working" or rated-load ampacity.

What value of derate percentage to use is determined by the expected ambient temperature and the type of conductor insulation. The derate percentages (decimal fractions) are listed in the lower third of the ampacity tables under the heading of *"Correction Factors."*

The higher ambient temperatures are listed as the row headings with ranges of temperature in "degrees Celsius" on the left, and "degrees Fahrenheit" on the right. If a decimal value is not listed (replaced with four dots), the insulation is not rated for a current-carrying conductor at this or any higher elevated temperature.

To derate the conductor's table-rated ampacity to a maximum rated-load ampacity for a given higher temperature, the following formula is given:

$$\text{table-rated ampacity} \times \text{higher temperature derate} = \text{rated-load ampacity}$$

Example 2–6. If three No. 1/0 (one-ought) AWG THHN aluminum conductors are installed in EMT conduit that passes through an ambient air temperature of 100°F, determine the maximum rated-load ampacity of the circuit conductors for this temperature.

Rated-load ampacity determination:

Table-rated ampacity: 135 amps — Ref. T310-16
Higher-temperature derate: 0.91 — Ref. T310-16, CF
Derate computation:

$$135 \text{ amps} \times 0.91 = 122.8 \text{ amps}$$

Wire terminal temperature limitations: 75°C — Ref. 110-14(c)(2), Ex. 1
Table-rated ampacity: 120 amps
Maximum rated-load ampacity: 120 amps — Ref. 310-15(c)

The table-rated ampacity of the No. 1/0 AWG THHN aluminum conductors was determined by the intersection of row twelve and column three under the aluminum heading in T310-16. The required derate for the higher ambient temperature of 100°F was determined by the intersection of row five in the correction factors table and the same insulation column of the ampacity table.

The wording of Secs. 110-14(c)(1) and (2) in the 1993 NE Code restricted the use of the higher-value rated ampacities found in the 90°C column of the ampacity tables, such that the derate found in this column had to be applied to the rated ampacity of the same conductor material (copper or aluminum) found in either the 60° or 75°C column according to either wire size, or circuit rating.

The only advantage of using a higher temperature-rated circuit conductor in the previous edition was the less severe derate for the higher temperature. For an expected ambient temperature of 100°F, a circuit conductor with a 60°C temperature was restricted to 82% of the table-rated ampacity found in the 60°C temperature column. The same conductor with a 90°C temperature was restricted to only 91% of the table-rated ampacity found in the 60°C temperature column.

Revision of these sections in the 1996 NE Code allows the higher values of rated ampacity (in the higher-temperature columns) to be used for derating purposes. After derate for higher ambient temperature, the computed rated-load ampacity cannot exceed the table-rated ampacity listed in either the 60° or 75°C temperature column, again according to wire size or circuit rating.

Section 310-15(c) was introduced with this example to justify the lower rated ampacity as becoming the NE Code maximum. After derate for the higher ambient temperature, the rated-load ampacity of the 90°C wire exceeded the table-rated ampacity for the same wire size (No. 1/0 AWG aluminum) listed in the 75°C column. The 75°C listing became the NE Code maximum. The 75°C column was referenced by Sec. 110-14(c)(2) because the wire size was greater than No. 1 AWG.

Code challenge 2–6. If three No. 4/0 (four-ought) AWG THWN-2 copper conductors are installed in rigid metal conduit that passes through an ambient air temperature of 125°F, determine the maximum rated-load ampacity of the circuit conductors for this temperature.

Rated-load ampacity determination:
Table-rated ampacity: (A) ____________ amps Ref. (B) ________________
Higher-temperature derate: (C) ____________ Ref. (D) ________________
Derate computation:

____________ amps × ____________ = (E) ____________ amps

Wire terminal temperature limitations: (F) ____________ °C Ref. (G) ____________
Table-rated ampacity: (H) ____________ amps
Maximum rated-load ampacity: (I) ____________ amps Ref. (J) ____________

As a rule, either the circuit full-load current rating, or the rating of the circuit overcurrent protection device is determined, and the conductors must be sized for this current. When these conductors, or sections of these conductors, will either pass through or be installed in, higher ambient temperatures, the derate formula can be worked in reverse:

Rated-load ampacity ÷ higher temperature derate = table-rated ampacity

When circuit conductors are sized in this manner, the computed table-rated ampacity is the NE Code-minimum ampacity for the given conditions. The actual table-rated ampacity, found according to wire size in the respective ampacity table, must be greater than or equal to the computed table-rated ampacity.

For higher temperature-rated conductors, the original full-load current rating or setting of the circuit OCPD without temperature compensation cannot exceed the table-rated ampacity of the circuit conductors for the same wire size (same row) listed in either the 60° or 75°C columns, according to the circuit rating or determined wire size. If the original full-load current does exceed the lower-temperature ampacity rating, then the wire size must be increased until the lower-temperature rating either equals or exceeds the original full-load current rating.

Example 2–7. Size the underground service lateral for a structure that is supplied with a 100-amp service. Use triplexed USE-RHW (75°C, Fig. 2-5) aluminum wire installed in conduit, which extends from below ground (finished grade) up to the meter

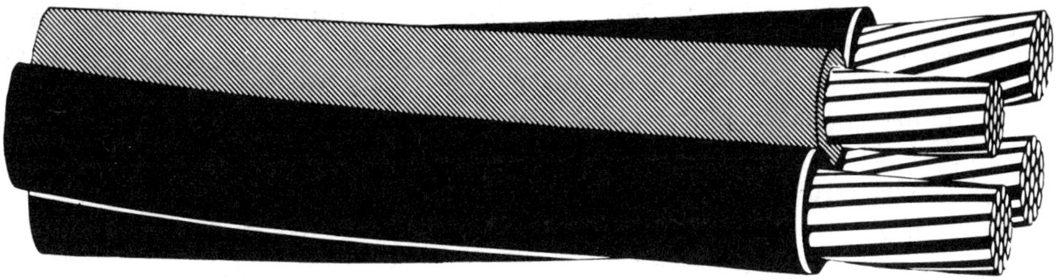

Figure 2-5 Quadraplexed mobile home feeder conductors must also provide a separate equipment-grounding conductor. (*Courtesy of Southwire Company*)

enclosure (can) on the side wall of the building. The maximum expected air/surface temperature on the side of the building is 110°F.

Table-rated ampacity determination:
Higher-temperature derate: 0.82 — Ref. T310-16, CF
Overrate computation:

$$100 \text{ amps} \div 0.82 = 122.0 \text{ amps}$$

Table-rated ampacity: 135 amps
Wire size: No. 2/0 AWG USE-RHW aluminum — Ref. T310-16
Wire terminal temperature limitations: 60°C — Ref. 110-14(c)(1), Ex. 1
Table-rated ampacity: 100 amps
Wire size: No. 1/0 AWG USE-RHW aluminum — Ref. T310-16
NE Code-minimum: Compare for larger — Ref. 310-15(c)
Wire size: No. 2/0 AWG USE/RHW aluminum

The required derate for the higher ambient temperature of 110°F was determined at the intersection of row five in the correction factors table and the second insulation temperature column (RHW) on the aluminum side of the ampacity table.

To size the wire, the table-rated ampacities listed in the RHW-aluminum column were scanned for an ampacity value that was greater than or equal to 122 amps. The 135-amp listing corresponded by row to wire size No. 2/0 AWG USE/RHW aluminum conductors.

To honor the wire terminal temperature limitations, the wire was technically sized a second time using only the full-load current rating of the service and the 60°C column (100 amps or less).

The requirements of Sec. 310-15(c) were applied in reverse to what the section reads. The worst case or lowest ampacity rating for the computed wire sizes is reflected in the largest wire size in comparison, hence the No. 2/0 AWG aluminum became NE Code minimum.

Code challenge 2–7. Size the underground service lateral for a structure supplied with a 60-amp service. Use triplexed USE-RHW (75°C) copper cable installed in conduit, which extends up the pole to a transformer. The expected air/surface temperature of the conduit is 125°F.

Table-rated ampacity determination:
Higher-temperature derate: (A) ____________ Ref. (B) ____________
Overrate computation:

____________ amps ÷ ____________ = (C) ____________ amps

Table-rated ampacity: (D) ____________ amps
Wire size: (E) ____________ USE-RHW copper Ref. (F) ____________
Wire terminal temperature limitations: (G) ____________ °C Ref. (H) ____________
Table-rated ampacity: (I) ____________ amps
Wire size: (J) ____________ USE-RHW copper Ref. (K) ____________
NE Code-minimum: Compare for (L) ____________ Ref. (M) ____________
Wire size: (N) ____________ USE-RHW copper

Where ambient air or earth temperatures are less than the table base of ambient temperature, the ampacity correction factors, which are shown as a multiplier (percentage) greater than 1, can be used to overrate the conductor table-rated ampacity. The formulas would be worked in the same manner. Both methods would either increase the rated ampacity of a selected wire size, or decrease the wire size for a given rated-load current.

2.5 MORE THAN THREE CIRCUIT CONDUCTORS

Conductors installed in a closed raceway fall victim to their own heat generation (I^2R Losses). The heat generated by the flow of current in the conductors cannot be readily dissipated into the surrounding air. Dead air, which the air inside the conduit (raceway) or cable represents, is a good insulator. The generated heat of the conductors must migrate through this dead air to the wall surface of the raceway, and then be dissipated through the walls of the conduit to the outside air (or dirt, or concrete, or whatever) that surrounds the conduit. Restricting the total count of current-carrying conductors to a maximum of three reduces the heat generated to a rate that can be dissipated.

When more than three conductors are to be installed in a common (single) raceway or cable, the table found under Note 8(a) to the ampacity tables (following T310-19) must be used to derate the table-rated ampacity of the conductors.

The "Number of Current-Carrying Conductors" in the Note 8(a) Table are established as ranges that require a percentage (less than one) derate. To derate the table-rated ampacity of a conductor to a rated-load ampacity when more than three current-carrying conductors will be installed in a common (single) raceway, the same formula as offered for derate of higher ambient temperature is offered:

table-rated ampacity × more-than-three derate = rated-load ampacity

Example 2–8. Determine the maximum rated-load ampacity of 20 No. 14 AWG THHN copper conductors installed in a single raceway.

Rated-load ampacity determination:
Table-rated ampacity: 25 amps — Ref. T310-16
More-than-three derate: 50% — Ref. T310-16, N-8(a) Tbl
Derate computation:

$$25 \text{ amps} \times 0.50 = 12.5 \text{ amps}$$

Wire terminal temperature limitations: 60°C — Ref. 110-14(c)(1), Ex. 1
Table-rated ampacity: 20 amps
Maximum rated-load ampacity: 12.5 amps — Ref. 310-15(c)

The maximum full-load current load of any one of the twenty No. 14 AWG THWN copper conductors contained in a common raceway or cable is restricted to a maximum of 12.5 amperes.

The obelus (dagger) footnote to T310-16 does not restrict the table-rated ampacity of the No. 14 AWG THHN copper conductors in this computation. The footnote only restricts the rating of the branch-circuit fuses or circuit breaker protecting the No. 14 AWG copper conductors to a maximum rating or setting of 15 amperes.

To be protected at the 15-amp rating of the OCPD, none of these No. 14 AWG copper conductors could supply convenience receptacles on multioutlet branch circuits. The load must be maintained at a maximum of 12.5 amperes. Since convenience receptacle outlets (common duplex receptacle) represent unknown load values of circuit current, there would be no way of ensuring the 12.5 amp maximum: The wire size would have to be increased to No. 12 AWG copper to convey the available 15 amperes.

Code challenge 2–8. Determine the maximum rated-load ampacity of eight No. 12 THW copper conductors installed in a single conduit run.

Rated-load ampacity determination:
Table-rated ampacity: (A) ____________ amps Ref. (B) ____________
More-than-three derate: (C) ____________ % Ref. (D) ____________
Derate computation:

____________ amps × ____________ = (E) ____________ amps

Wire terminal temperature limitations: (F) ____________ °C Ref. (G) ____________
Table-rated ampacity: (H) ____________ amps
Maximum rated-load ampacity: (I) ____________ amps Ref. (J) ____________

As a rule, either the full-load current rating, or the rating of the circuit overcurrent protection device is determined, and the conductors must be sized for this current. When these conductors or sections of these conductors will be installed in a raceway that will contain more than three current-carrying conductors, the derate formula can be worked in reverse:

Rated-load ampacity ÷ more-than-three derate = table-rated ampacity

When conductors are sized in this manner, the computed table-rated ampacity is NE Code-minimum. As with higher ambient temperature compensation, the actual table-rated ampacity, determined by wire size, must be greater than or equal to the computed table-rated ampacity.

Example 2–9. If the circuit conductors will be installed in a single conduit that contains a total of six current-carrying conductors, determine the NE Code-minimum wire size of XHHW-2 copper conductors required to supply a 250-amp feeder load.

Table-rated ampacity determination:
More-than-three derate: 80% — Ref. T310-16, N-8(a) Tbl
Overrate computation:

250 amps ÷ 0.80 = 312.5 amps

Table-rated ampacity: 320 amps
Wire size: 300 kcmil XHHW-2 copper — Ref. T310-16
Wire terminal temperature limitations: 75°C — Ref. 110-14(c)(2), Ex. 1
Table-rated ampacity: 255 amps
Wire size: 250 kcmil XHHW-2 copper — Ref. T310-16
NE Code-minimum: compare for larger — Ref. 310-15(c)
Wire size: 300 kcmil XHHW-2 copper — Ref. T310-16

As with higher ambient temperatures, more than three circuit conductors in a single raceway reduces the rate of heat dissipation. Section 110-14(c)(2) allows the rated ampacity listed in the 90°C column for these XHHW-2 copper conductors to be derated for the temperature compensation. Technically the 75°C column ampacity for the 300 kcmil circuit conductors could have been evaluated for at least 250 amps (the rated-load ampacity without compensation). As long as the 75°C column ampacity equaled or exceeded the 250-amp current requirement, the last two steps could have been left off.

Code challenge 2–9. If the circuit conductors will be installed in a single conduit that contains a total of nine current-carrying conductors, determine the NE Code-minimum wire size of THHN aluminum conductors required to supply a 200-amp feeder load.

Table-rated ampacity determination:
More-than-three derate: (A) ____________ % Ref. (B) ____________________
Overrate computation:

____________ amps ÷ ____________ = (C) ____________ amps

Table-rated ampacity: (D) ____________ amps
Wire size: (E) ____________ THHN aluminum Ref.(F) ______________________
Wire terminal temperature limitations: (G) ____________ °C Ref. (H) ____________
Table-rated ampacity: (I) ____________ amps
Wire size: (J) ____________ THHN aluminum Ref. (K)______________________
NE Code-minimum: compare for (L) ____________ Ref. (M)______________________
Wire size: (N) ____________ THHN aluminum

2.6 EQUIPMENT-GROUNDING CONDUCTORS

Which derate percentage of Note 8(a) Table to use is determined by the number of current-carrying conductors that will be installed in a single raceway. These circuit conductors generate heat under normal circuit operation. When separate equipment-grounding conductors are also installed, Note 11 to the ampacity tables exempts their count when applying the provisions of Note 8(a).

In other words, when one or more equipment-grounding conductors are pulled in the raceway with the various circuit conductors, these grounding conductors, regardless of their number, are omitted when determining the total number or count of circuit (current-carrying) conductors.

As will be discussed in Chapter 9 (Grounding and Bonding), all the noncurrent-carrying metal parts used to house the electrical equipment in the power distribution (raceways, boxes, and other metal enclosures) must be effectively bonded together (made electrically continuous) and connected to both earth ground (through the service grounding-electrode system) and the grounded service-neutral conductor.

Bonding of these noncurrent-carrying parts to form a grid, protects persons from stray voltages that can occur from static electricity. Grounding of this bonding grid ensures that any noncurrent-carrying metal parts of the power distribution system will be held at ground potential (zero volts, reference to earth ground) in the event of an accidental or intentional ground fault.

High-magnitude fault overcurrent is caused to flow by way of the grounded service-neutral conductor when an ungrounded circuit conductor is either accidentally or purposely grounded by coming into contact with the metal enclosure. The separate equipment-grounding conductor provides a low-impedance fault-return path back to the service neutral conductor for further conveyance of the fault overcurrent back to the transformer or other power supply.

Because separate equipment-grounding conductors do not generate heat, their presence in the raceway actually lowers the operating temperature of the other circuit conductors by acting as a heat sink. Equipment-grounding conductors or bonding jumpers are not counted when determining the number of current-carrying conductors.

Example 2–10. A single conduit will contain three 2-wire branch circuits consisting of No. 8 AWG RHH copper conductors. If three No. 10 AWG RHH copper equipment-grounding conductors will also be pulled into the raceway, determine the NE Code-maximum load ampacity of the circuit conductors.

The equipment-grounding conductors do not have to be counted.

Ref. T310-16, N-11

Total number of circuit conductors computation:

3 circuits × 2 conductors per circuit = 6 conductors

Rated-load ampacity determination:

Table-rated ampacity: 55 amps — Ref. T310-16

More-than-three derate: 80% — Ref. T310-16, N-8(a) Tbl

Derate computation:

55 amps × 0.80 = 44 amps

Wire terminal temperature limitations: 60°C — Ref. 110-14(c)(1), Ex. 1

Table-rated ampacity: 40 amps

Maximum rated-load ampacity: 40 amps — Ref. 310-15(c)

In this example, if the three equipment-grounding conductors are counted, the derate of Note 8(a) table restricts the rated-load ampacity of the No. 8 AWG RHH copper conductors to 70% of the higher-temperature rating of 55 amps. The maximum-load current is further restricted to only 38.5 amps. This would be the lower rated-load current in comparison to the wire terminal temperature restricted rated-load current.

Not counting the three equipment-grounding conductors, the temperature limitations of the wire terminals was the governing section. Section 310-15(c) restricts the rated-load ampacity to a maximum of 40 amperes.

Code challenge 2–10. A single conduit will contain four 2-wire branch circuits consisting of No. 6 AWG XHH aluminum conductors. If four No. 8 AWG XHH aluminum equipment-grounding conductors will also be pulled into the raceway, determine the NE Code-maximum load ampacity of the circuit conductors.

The equipment-grounding conductors (A) ____________ (do or do not) have to be counted.

Ref. (B)________________________

Total number of circuit conductors computation:

____________ circuits × ____________ conductors/circuit = (C) ____________ conductors

Rated-load ampacity determination:

Table-rated ampacity: (D) ____________ amps Ref. (E)____________________

More-than-three derate: (F) ____________ % Ref. (G)____________________

Derate computation:

____________ amps × ____________ = (H) ____________ amps

Wire terminal temperature limitations: (I) ____________ °C Ref. (J) ____________
Table-rated ampacity: (K) ____________ amps
Maximum rated-load ampacity: (L) ____________ amps Ref. (M) ______________

When the rated-load current of the respective circuit conductors is known and more than three circuit conductors will be installed in a single raceway with associated equipment-grounding conductors, the formula would be worked in reverse as shown in previous examples. The equipment-grounding conductors would not be counted in determining the total number of current-carrying conductors.

2.7 MULTIWIRE CIRCUIT NEUTRAL CONDUCTORS

When a two-wire branch circuit operates with one circuit conductor grounded, the grounded-circuit conductor is not a neutral conductor, only the circuit return. As such, the grounded-circuit conductor must be full size or equal in size to the ungrounded-circuit conductor.

A single-phase, 3-wire system, such as the 120/240-volt service to the average single-family dwelling, operates with the center wire (midpoint connection between the two 120-volt windings in the secondary of the transformer) grounded, such that 2-wire, 120-volt circuits can be supplied from either one of the ungrounded phase conductors to the common circuit conductor. The common circuit conductor is the grounded neutral conductor of the power supply.

Although not a multiphase system, the two windings of the transformer in this 3-wire supply are wired in such a manner that the other two ends of the transformer windings (the ungrounded supply conductors) are electrically 180° out of phase (electrical timing) with each other. A load connected between the two ungrounded-circuit conductors will operate at 240 volts.

For 120-volt loads, the lower-magnitude current from one of the ungrounded-circuit conductors will cancel or subtract from the higher-magnitude current of the other ungrounded-circuit conductor, such that the common neutral conductor only conveys the difference of the two currents.

Because the ungrounded phase conductors supply 240-volt loads in addition to 120-volt loads, and because the grounded neutral conductor conveys only the imbalance of the 120-volt load currents, the common neutral conductor is normally reduced in wire size as compared to the phase conductors. (Reference Sec. 220-22 in the NE Code.)

If the 120-volt loads are balanced between the two phase conductors, and all loads are connected, the neutral conductor load current should equal zero amps. The neutral, in this balanced load condition, is functioning as a voltage reference.

In a 3-phase, 4-wire wye distribution system, which is commonly used in commercial or industrial buildings, the grounded neutral conductor also carries the imbalanced load-current of the lower-voltage loads. The more common wye distribution voltages include 208Y/120, 460Y/265, and 480Y/277 volts.

In a wye configuration, three single-phase sources are generated at 120 electrical degrees out of time with each other. One end of each of the three single-phase windings of the generator (alternator for ac) or transformer (either the primary or secondary circuit) are brought out as the three ungrounded phase conductors of the power distribution. The other end of the three windings are tied common, referenced to earth ground (through a grounding-electrode system), and brought out as the fourth-wire grounded neutral conductor of the system.

As with the single-phase, 3-wire distribution system, the fourth-wire neutral conductor carries a reduced current load as compared to the ungrounded phase conductors.

Reduced current means a reduction in heat generation when the circuit conductors are operating under full-load conditions. According to the type of load supplied, Note 10 to the ampacity tables will sometimes allow the neutral conductor to be omitted when applying the provisions of Note 8(a).

Note 10(a) will allow the feeder neutral conductor to be omitted when it conveys only the unbalanced load current from the ungrounded-circuit conductors of the same circuit.

Example 2–11. Determine the NE-Code minimum wire size of the ungrounded-circuit conductors, if three single-phase, 3-wire, 100-amp, 120/240-volt feeder circuits will be installed in a single raceway. Use THW copper conductors.

The grounded feeder-neutral conductors do not have to be counted as current-carrying conductors.

Ref. T310-16, N-10(a)

Total number of circuit conductors computation:

$$3 \text{ circuits} \times 2 \text{ conductors per circuit} = 6 \text{ conductors}$$

Table-rated ampacity determination:
More-than-three derate: 80% — Ref. T310-16, N-8(a) Tbl
Overrate computation:

$$100 \text{ amps} \div 0.80 = 125 \text{ amps}$$

Table-rated ampacity: 130 amps
Wire size: No. 1 AWG THW copper — Ref. T310-16
Wire terminal temperature limitations: 60°C — Ref. 110-14(c)(1), Ex. 1
Table-rated ampacity: 110 amps
Wire size: No. 1 AWG THW copper — Ref. T310-16
NE Code-minimum: No comparison; equal — Ref. 310-15(c)
Wire size: No. 1 AWG THW copper

Code challenge 2–11. Determine the NE-Code minimum wire size of the ungrounded-circuit conductors, if four single-phase, 3-wire, 150-amp, 120/240-volt feeder circuits will be installed in a single raceway. Use RHH aluminum conductors.

The grounded feeder-neutral conductors (A) ____________ (do or do not) have to be counted as current-carrying conductors.

Ref. (B)______________________

Total number of circuit conductors computation:

___________ circuits × ___________ conductors/circuit = (C) ___________ conductors

Table-rated ampacity determination:

More-than-three derate: (D) ____________ % Ref. (E)______________________

Overrate computation:

____________ amps ÷ ____________ = (F) ____________ amps

Table-rated ampacity: (G) ____________ amps

Wire size: (H) ____________ RHH aluminum Ref. (I) ______________________

Wire terminal temperature limitations: (J) ____________ °C Ref. (K) ____________

Table-rated ampacity: (L) ____________ amps

Wire size: (M) ____________ RHH aluminum Ref. (N)______________________

NE Code-minimum: Compare for (O) ____________ Ref. (P)____________________

Wire size: (Q) ____________ RHH aluminum

The circuits in this example and the accompanying code challenge problem were assumed to supply resistive, or what is presently described as "linear" loads. Linear indicates that the power factor of the circuits is near unity (the respective line current is in phase [electrical timing] with its line voltage).

When multiwire circuits supply nonlinear loads, such as electromagnetic ballasts in electric-discharge lighting, there is a phase shift between the supply voltage applied to the ballast and the line current drawn by the ballast. The phase shift occurs on each leg or phase of a 3-phase, 4-wire wye distribution, such that the current components on the neutral conductor cannot effectively cancel for only imbalanced current as with linear (resistive) loads. Because of the phase shift, the neutral conductor conveys load current even when the phase-to-neutral currents appear to be balanced.

When the 3-phase, 4-wire wye system supplies electronic ballasts for electric-discharge lighting or phase converters for ac variable-speed motor drives, the nonlinear load also produces harmonic currents on the neutral conductor that have a tendency to load and sometimes overload the neutral conductor.

In the United States, electrical power is generated at 60 hertz. Harmonics are multiples of this 60-hertz base frequency. If the neutral component from the ballast or phase converter is 120 hertz, this is an even harmonic of the base frequency (times 2). If the neutral component is operating at 180 hertz, this would be an odd harmonic (times 3).

When applying the provisions of Note 8(a), Note 10(c) requires the grounded neutral conductor of a multiwire, multiphase circuit to be counted as a current-carrying conductor if the circuit supplies nonlinear loads.

Example 2–12. Determine the NE Code-maximum load-current rating of No. 12 AWG THW copper conductors if a total of four 3-phase, 4-wire branch circuits will be installed in a common conduit. The branch circuits will supply nonlinear loads.

The grounded neutral conductor of each multiwire branch circuit must be counted as a current-carrying conductor.

Ref. T310-16, N-10(c)

Total number of circuit conductors computation:

4 circuits × 4 conductors per circuit = 16 conductors

Rated-load ampacity determination:

Table-rated ampacity: 25 amps — Ref. T310-16

More-than-three derate: 50% — Ref. T310-16, N-8(a) Tbl

Derate computation:

25 amps × 0.50 = 12.5 amps

Wire terminal temperature limitations: 60°C — Ref. 110-14(c)(1), Ex. 1

Table-rated ampacity: 25 amps

Maximum rated-load ampacity: 12.5 amps — Ref. 310-15(c)

Code challenge 2–12. Determine the NE Code-maximum load current of No. 14 AWG THWN copper conductors if a total of three 3-phase, 4-wire branch circuits will be installed in a common conduit. The branch circuits will supply nonlinear loads.

The grounded neutral conductor of each multiwire branch circuit (A) ________ (does, does not) have to be counted as a current-carrying conductor.

Ref. (B)________________

Total number of circuit conductors computation:

________ circuits × ________ conductors/circuit = (C) ________ conductors

Rated-load ampacity determination:

Table-rated ampacity: (D) ________ amps Ref. (E)________________

More-than-three derate: (F) ________ % Ref. (G)________________

Derate computation:

________ amps × ________ = (H) ________ amps

Wire terminal temperature limitations: (I) ________ °C Ref. (J) ________

Table-rated ampacity: (K) ________ amps

Maximum rated-load ampacity: (L) ________ amps Ref. (M) ________

2.8 CONTINUOUS LOAD

As defined in Art. 100, a circuit load is rated as continuous if it draws uninterrupted current for three hours or more. Continuous current draw means continuous heat generation along the length of the circuit conductors. To reduce this rate of heat generation, continuous-load currents are restricted to a *maximum* of 80% of the table-rated ampacity of the circuit conductors.

As indicated in Sec. 210-22(c), continuous-operation (load) circuits are normally in the form of retail store or other nondwelling structure lighting circuits.

For the load computations of branch circuits, Sec. 220-3(a) requires an *overrate* of 125% to be applied to that part of the circuit load that is rated continuous. To overrate the continuous load to 125% to determine the branch-circuit rating is equivalent to restricting the continuous-load current rating to 80% of a selected branch-circuit rating. The two are reciprocal of each other.

To say it another way, a continuous load of 16 amps would require a branch circuit rated at a minimum of 20 amps: 125% of 16 amps = 20 amps. Or, a 20-amp branch circuit would be restricted to a maximum continuous-load current of only 16 amps: 80% of 20 amps = 16 amps.

Section 220-10(b) makes the same requirement of the feeder circuit. In the 1993 NE Code, the 125% overrate was applied only to the circuit OCPD rating, because the rated ampacity of the circuit conductor (as defined in Art. 100 and listed in the ampacity tables) was (and technically still is—1996 edition) the magnitude of current the conductor can convey "continuously" without exceeding its temperature rating.

The wording of Secs. 220-3(a) and 220-10(b) in the 1996 edition of the NE Code was changed to require the 125% overrate (80% derate) for continuous-load current applies to *both* the rating or setting of the OCPD and to the rated ampacity of the circuit conductors.

For 15-, 20-, and 30-amp multioutlet lighting branch circuits, the 80% restriction for continuous-load current must be applied to the rating of the branch circuit OCPD, even though the table-listed ampacity of the circuit conductors is a higher rating.

The formulas offered for derate due to higher ambient temperature or more than three circuit conductors are modified somewhat for continuous load derate computations. To determine the maximum load current for a given wire size, the table-rated ampacity of the conductor must be multiplied by 80% to apply the derate for continuous load:

$$0.80 \text{ (continuous-load derate)} \times \text{table-rated ampacity} = \text{maximum-load ampacity}$$

Example 2–13. Determine the NE Code-maximum continuous-load current of No. 14 AWG THHN copper conductors.

Continuous-load derate: 80% — Ref. 220-3(a)

Rated-load ampacity determination:

Table-rated ampacity: 20 amps — Ref. T310-16

OCPD amperage restriction: 15 amps — Ref. T310-16, Footnote

Maximum continuous-load ampacity computation:

$$0.80 \times 15 \text{ amps} = 12 \text{ amps}$$

Code challenge 2–13. Determine the NE Code-maximum continuous-load current of No. 12 AWG THHN copper conductors.

Continuous-load derate: (A) ____________ % Ref. (B) ____________________

Rated-load ampacity determination:

Table-rated ampacity: (C) ____________ amps Ref. (D) ____________________

OCPD amperage restriction: (E) ____________ amps Ref. (F) ________________

Maximum continuous-load ampacity computation:

____________ × ____________ amps = (G) ____________ amps

As with higher ambient temperatures or more than three current-carrying conductors installed in a single raceway, the rated-load current is normally computed first, and the 125% overrate for continuous load is applied to determine the table-rated ampacity according to wire size.

1.25 (continuous-load overrate) × rated-load ampacity = table-rated ampacity

When circuit conductors are sized in this manner, the computed table-rated ampacity is NE Code-minimum. The actual table-rated ampacity, determined by wire size, must be greater than or equal to the computed table-rated ampacity.

Example 2–14. Determine the NE Code-minimum wire size of XHHW-2 copper conductors required to supply a 320-amp feeder load that is rated as continuous.

Table-rated ampacity determination:

Continuous-load overrate: 125% Ref. 220-10(b)

Overrate computation:

$$1.25 \times 320 \text{ amps} = 400 \text{ amps}$$

Wire terminal temperature limitations: 75°C Ref. 110-14(c)(2), Ex. 1

Table-rated ampacity: 420 amps

Wire size: 600 kcmil XHHW-2 copper Ref. T310-16

In comparison to higher ambient temperatures or more than three circuit conductors in a single raceway, which reduce the rate of heat dissipation from the circuit conductors, continuous-load current actually increases the rate of heat generation.

The derates or overrates for higher ambient or more than three affect only the rated ampacity of the circuit conductors, not the rating or setting of the overcurrent protection device (fuses or circuit breaker). The overrate or derate for continuous load (according to how it is applied) affects both the rated ampacity of the circuit conductors and the rating of circuit overcurrent protection device.

Referring back to the examples of ampacity derates for either higher ambient temperature, or more than three circuit conductors, two computed full-load currents were used to size the circuit conductors: The overrated full-load current, which compensated for the loss of heat dissipation, was used in the higher-temperature column to size the wire. The original full-load current was used with the lower temperature column, to honor the wire-terminal temperature restrictions. Section 310-15(c) required the larger wire size in comparison to be NE Code-minimum.

In the previous example, the circuit conductors, rated at 420 amps, will be protected by fuses or a circuit breaker rated at 400 amps. The ampacity listed in the higher-temperature column was not considered for sizing the wire since a higher ambient temperature was not specified. The wire terminal temperature restrictions were used to size the circuit conductors according to the rated ampacities listed in the 75°C temperature column.

Code challenge 2–14. Determine the NE Code-minimum wire size of THHN aluminum conductors required to supply a 160-amp feeder load that is rated as continuous.

Table-rated ampacity determination:
Continuous-load overrate: (A) ___________ % Ref. (B) ____________________
Overrate computation:

___________ × ___________ amps = (C) ___________ amps

Wire terminal temperature limitations: (D) ___________ °C Ref. (E) __________
Table-rated ampacity: (F) ___________ amps
Wire size: (G) ___________ THHN aluminum Ref. (H) ____________________

2.9 MORE THAN ONE PREVAILING CONDITION

If the ambient temperature is higher than the base temperature of the respective ampacity table and the total number of current-carrying conductors in a single raceway exceeds three, the column heading of percentages in the Note 8(a) table requires the table-rated ampacity of the conductors to be double derated.

To apply the double derate, the conductor's table-rated ampacity is first multiplied by the derate for higher ambient temperature. The resulting ampacity is then multiplied by the derate for more than three circuit conductors.

Example 2–15. Determine the maximum-load ampacity of 12 No. 4 AWG XHHW copper conductors contained in a single conduit, which passes through an environmental air temperature of 52°C.

Table-rated ampacity: 95 amps Ref. T310-16
Higher-temperature derate: 0.76 Ref. T310-16, CF
More-than-three derate: 50% Ref. T310-16, N-8(a) Tbl
Rated-load ampacity determination:
Derate computation for higher ambient temperature:

$$95 \text{ amps} \times 0.76 = 72.2 \text{ amps}$$

Derate computation for more-than-three circuit conductors:

$$72.2 \text{ amps} \times 0.50 = 36.1 \text{ amps}$$

Wire terminal temperature limitations: 60°C — Ref. 110-14(c)(1), Ex. 1
Table-rated ampacity: 70 amps
NE Code-maximum: Compare for smaller — Ref. 310-15(c)
Table-rated ampacity: 36.1 amps

Code challenge 2–15. Determine the maximum-load ampacity of six No. 1 AWG THW copper conductors contained in a single conduit, which passes through an environmental air temperature of 124°F.

Table-rated ampacity: (A) ____________ amps — Ref. (B) ____________
Higher-temperature derate: (C) ____________ — Ref. (D) ____________
More-than-three derate: (E) ____________ % — Ref. (F) ____________
Rated-load ampacity determination:
Derate computation for higher ambient temperature:

____________ amps × ____________ = (G) ____________ amps

Derate computation for more-than-three circuit conductors:

____________ amps × ____________ = (H) ____________ amps

Wire terminal temperature limitations: (I) ____________ °C Ref. (J) ____________
Table-rated ampacity: (K) ____________ amps
NE Code-maximum: Compare for (L) ____________ Ref. (M) ____________
Table-rated ampacity: (N) ____________ amps

For most electrical installations, the rated- or full-load current of the respective circuit (either the branch circuit, the feeder circuit, or the service to the building) is determined by the summation of either nameplate load ratings, or by minimum values assigned by the NE Code. This computed full-load current is then divided by the correction factors for either higher ambient temperatures or more-than-three circuit conductors, or both. In other words, the computed rated-load current is overrated (increased) to increase the circuit wire size. The increase in wire size, which decreases the rate of heat generation along the length of the conductor, compensates for the loss in heat dissipation due to higher operating temperatures.

The double derate is applied to the computed rated-load current by first dividing by the derate for higher temperature, and then dividing again by the derate for more-than-three circuit conductors.

When computed in this manner, the actual table-rated ampacity, listed in the respective ampacity table according to the type of insulation and conductor material, must be greater than or equal to the computed table-rated ampacity.

As with single overrates (or derates) and wire terminal temperature limitations, the actual full-load current rating without the double overrate cannot exceed the rated-ampacity listed in either the 60° or 75°C temperature column according to the wire size selected by the double overrate.

Example 2–16. A total equipment full-load current of 70 amperes will be supplied with THW aluminum conductors. If the conductors will be installed in a single raceway, which contains a total of 6 current-carrying conductors, and passes through an ambient temperature of 35°C, determine the NE Code-minimum conductor wire size allowed for the given load conditions.

Higher-temperature derate: 0.94 — Ref. T310-16, CF

More-than-three derate: 80% — Ref. T310-16, N-8(a) Tbl

Table-rated ampacity determination:

Overrate computation for higher ambient temperature:

$$70 \text{ amps} \div 0.94 = 74.5 \text{ amps}$$

Overrate computation for more-than-three circuit conductors:

$$74.5 \text{ amps} \div 0.80 = 93.1 \text{ amps}$$

Table-rated ampacity: 100 amps

Wire size: No. 1 AWG THW aluminum — Ref. T310-16

Wire terminal temperature limitations: 60°C — Ref. 110-14(c)(1), Ex. 1

Table-rated ampacity: 75 amps

Wire size: No. 2 AWG THW aluminum — Ref. T310-16

NE Code-minimum: Compare for larger — Ref. 310-15(c)

Wire size: No. 1 AWG THW aluminum

Code challenge 2–16. A total equipment full-load current of 115 amperes will be supplied with THW-2 copper conductors. If the conductors will be installed in a single raceway, which contains a total of 9 current-carrying conductors, and passes through an ambient temperature of 105°F, determine the NE Code-minimum conductor wire size allowed for the given load conditions.

Higher-temperature derate: (A) ____________ Ref. (B) ______________________

More-than-three derate: (C) ____________ % Ref. (D) ______________________

Table-rated ampacity determination:

Overrate computation for higher ambient temperature:

____________ amps ÷ ____________ = (E) ____________ amps

Overrate computation for more-than-three circuit conductors:

____________ amps ÷ ____________ = (F) ____________ amps

Table-rated ampacity: (G) ____________ amps
Wire size: (H) ____________ THW-2 copper Ref. (I) ______________________

Wire terminal temperature limitations: (J) ____________ °C Ref. (K) ____________
Table-rated ampacity: (L) ____________ amps
Wire size: (M) ____________ THW-2 copper Ref. (N)______________________

NE Code-minimum: Compare for (O) ____________ Ref. (P) ____________________
Wire size: (Q) ____________ THW-2 copper

When the raceway for circuit conductors supplying a continuous load will pass through an ambient temperature higher than the table base, Secs. 220-3(a) and 220-10(b) exempts the 125% overrate for temperature compensation computations. Using the higher-temperature listed table-rated ampacities and the adjusted full-load current, the circuit conductors are sized in the same manner as shown in the previous examples.

The 125% overrate for continuous-load *only affects* the wire size as selected based on the wire terminal temperature limitations of Secs. 110-14(c)(1) and (2). Instead of using the full-load current without temperature compensation to size the wire based on terminal temperature limitations, the continuous-load/full-load current must be first overrated to 125% to meet the NE Code requirements.

If the raceway will contain more than three circuit conductors, the continuous-load/full-load current is overrated according to the percentages found in Note 8(a) Table. The wire size is selected according to the table-rated ampacities listed in the higher-temperature column.

For wire terminal temperature limitations, the continuous-load/full-load current is overrated by 125%, the wire is sized again by the 60°C column for original full-load currents up to 100 amps, and by the 75°C temperature column for original full-load currents rated greater than 100 amps.

The largest in wire size, as referenced by Sec. 310-15(c), becomes the NE Code-minimum wire size for the circuit.

When more than three circuit conductors which supply continuous-load current will be routed in a raceway through an area of higher ambient temperature, the NE Code does not require a triple derate to size the conductors. Instead the continuous-load current is only double overrated to compensate for the double loss of effective heat dissipation. The original full-load current, overrated to 125% to adjust for continuous load is used to size the circuit conductors for the wire terminal temperature limitations. The largest wire size in comparison becomes NE Code-minimum.

Example 2–17. A continuous full-load current of 40 amperes will be supplied with THW copper conductors. If the conductors will be installed in a single raceway, which contains a total of 6 current-carrying conductors and passes through an ambient temperature of 110°F, determine the NE Code-minimum conductor wire size allowed for the given load conditions.

Higher-temperature derate: 0.82 Ref. T310-16, CF

More-than-three derate: 80% Ref. T310-16, N-8(a) Tbl

Continuous-load overrate: 125% Ref. 220-3(a)

Temperature-compensated table-rated ampacity determination:
Overrate computation for higher ambient temperature:

$$40 \text{ amps} \div 0.82 = 48.8 \text{ amps}$$

Overrate computation for more-than-three circuit conductors:

$$48.8 \text{ amps} \div 0.80 = 61.0 \text{ amps}$$

Table-rated ampacity: 65 amps
Wire size: No. 6 AWG THW copper   Ref. T310-16
Wire terminal temperature limitations: 60°C   Ref. 110-14(c)(1), Ex. 1
Overrate computation for continuous load:

$$1.25 \times 40 \text{ amps} = 50 \text{ amps}$$

Table-rated ampacity: 55 amps
Wire size: No. 6 AWG THW copper   Ref. T310-16
NE Code-minimum: No comparison; equal   Ref. 310-15(c)
Wire size: No. 6 AWG THW copper

Code challenge 2–17. A continuous full-load current of 100 amperes will be supplied with THHN copper conductors. If the conductors will be installed in EMT conduit, which passes through an ambient temperature of 47°C, and contains a total of nine current-carrying conductors, determine the NE Code-minimum conductor wire size allowed for the given load conditions.

Higher-temperature derate: (A) ____________   Ref. (B) ____________
More-than-three derate: (C) ____________ %   Ref. (D) ____________
Continuous-load overrate: (E) ____________ %   Ref. (F) ____________
Temperature-compensated table-rated ampacity determination:
Overrate computation for higher ambient temperature:

____________ amps ÷ ____________ = (G) ____________ amps

Overrate computation for more-than-three circuit conductors:

____________ amps ÷ ____________ = (H) ____________ amps

Table-rated ampacity: (I) ____________ amps
Wire size: (J) ____________ THHN copper   Ref. (K) ____________
Wire terminal temperature limitations: (L) ____________ °C   Ref. (M) ____________
Overrate computation for continuous load:

____________ × ____________ amps = (N) ____________ amps

Table-rated ampacity: (O) ___________ amps
Wire size: (P) ___________ THHN copper Ref. (Q)___________________
NE Code-minimum: Compare for (R) ___________ Ref. (S) ___________________
Wire size: (T) ___________ THHN copper

When the ambient air or earth temperature is less than the base temperature of the respective ampacity table and more than three conductors are contained in the given raceway or trench (underground installations), the conductors can be overrated for the lower temperature and derated for "more than three" current-carrying conductors. The formulas, whether solved for minimum table-rated ampacity, or maximum rated-load ampacity, are worked in the same manner as shown in the previous examples. In either application, the overrate of 125%, or the derate to 80% for continuous load is applied only for sizing the circuit conductors for wire terminal temperature limitations.

Derates or corrections for the prevailing conditions cannot be allowed for if they are not mentioned on an NE Code test or licensing exam, nor should they be allowed for in the real electrical installation world if they do not exist.

2.10 VOLTAGE-DROP COMPENSATION

The concern for increasing the wire size in these previous sections has been to increase the cross-sectional area of the circuit conductors, which decreases the conductor resistance. A decrease in conductor resistance decreases the wattage losses or the amount of heat it will generate. The decrease in conductor resistance also compensates for or lowers the amount of voltage drop along the length of the conductor. The wattage loss is a result of this voltage drop.

If higher ambient temperatures are not encountered, or by electrical design not more than three circuit conductors will be installed in a single raceway, then an increase in wire size is not warranted. In other words, the table-rated ampacity after adjustment for wire terminal temperature compensation is taken as is to size the circuit conductors. The fine-print note to Sec. 310-15 stipulates that sized in this manner, no consideration has been given to voltage-drop compensation.

There is no magic-cutoff in terms of wire length for the conductors which will tell the electrician when or when not to check the electrical installation requirements for voltage drop. Circuit conductors of relative short length can suffer from excessive voltage drop due to high-current rated loads.

The NE Code does not make it mandatory to adjust the conductor's rated ampacity for a larger wire size to allow for voltage drop. As indicated by Sec. 90-5 and the fine-print notes (FPNs) following Secs. 210-19(a) (FPN No. 4—branch circuits) and 215-2(b) (FPN No. 2—feeders), adjusting wire size for voltage drop compensation is an NE Code recommendation.

The two fine-print notes say almost identically the same thing. The NE Code recommends a *maximum* voltage drop of 5% from point of supply (service equipment) to the point of the connected load (load end of the branch circuit).

If the service equipment panel supplies only branch circuit loads, which is typical of a dwelling structure, then technically each branch circuit could be rated at a maximum of 5% in terms of voltage drop and meet the NE Code guidelines. Applied in this manner, equipment operating at 15 amps and supplied with No. 14 AWG copper conductors from a 15-amp, 120-volt branch circuit would receive a minimum of 114 volts if the circuit conductors were sized at the 5% maximum.

Often feeders from the service power panel of a structure supply lower ampere-rated panels or other load centers which in turn supply branch circuits for loads located in the immediate area of the particular load center. Given this application, "sized to prevent a voltage drop of 3% to the farthest outlet . . ." (found in both fine print notes) indicates a 3% maximum allowance should be applied to the branch circuit in the distribution, and 2% should be applied to the feeder circuit. The overall voltage drop, as shown in Figure 2-6, is still limited to a maximum of 5%.

In terms of minimum voltage drop, the resistance instead of ampacity is used to size the wire. A branch circuit consists of a minimum of a supply and return conductor (two conductors) which are connected in series with the load. All conductor materials have some degree of voltage drop due to their resistance. In a series circuit (refer to Sec. 18.2 of this text), the voltage drop across each series resistance, the supply conductor, the load, and the return conductor, will be proportional to the magnitude of each of the three individual resistances and the total circuit current conveyed.

In an ideal branch or feeder circuit, the wire resistance would be zero. The total circuit resistance would be limited to only the load resistance, and the available supply voltage would be applied to the connected load.

In a real world branch or feeder circuit, a percentage of the available supply voltage is dropped along the length of both the supply and return conductors. What's left of the supply voltage is applied to the connected load. If the wire resistance is increased, the load receives even less voltage. If the wire resistance is decreased, the load voltage approaches the supply voltage rating.

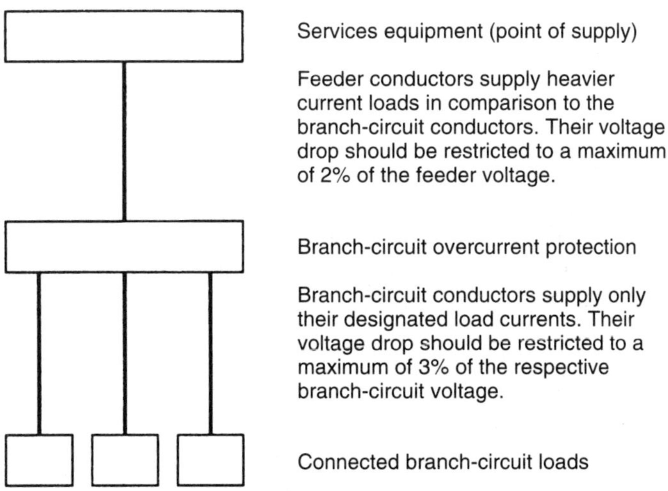

Figure 2-6 Voltage drop adjustment of a conductor's ampacity (wire size) is a Code recommendation versus a Code requirement.

The conductor resistance of manufactured wire is fixed or maintained (relatively) constant according to the conductor material, its cross-sectional area (wire size), and its cut length. The amount of voltage drop along the conductor is governed by the current flowing through it and its fixed resistance.

The circuit conductor resistance is a factor of both its cross-sectional area and its overall length. An increase in conductor length, which will yield an increase in voltage drop, can be compensated for by an increase in wire size. The decrease in resistance by an increase in wire size offsets the increase in resistance due to the increase in overall length.

An increase in circuit current for a given wire size will also increase the voltage drop along its length. An increase in wire size, which again lowers the resistance, can be used to compensate or offset the increase in voltage drop in this situation as well.

Table 8 of Chapter 9 of the NE Code, titled "Conductor Properties," gives a list of conductor resistances according to conductive material and wire size. The wire sizes are listed in the first column on the left of the table; the corresponding resistances are listed according to conductor material in the last three columns on the right of the table. The resistances are listed in ohms per thousand (k) feet because the resistance of drawn wire is relatively small when compared to its length. Aluminum (the right-most column) has only one value of resistance according to wire size. Copper has two columns identified as "Uncoated ohm/kFT" and "Coated ohm/kFT."

To compute the actual voltage drop of a set of circuit conductors, Ohm's law (V = IR) can be modified to take into account the length of the conductors, and the fact that the resistance, listed in Table 8, is per 1,000 feet. In formula,

$$VD = \frac{2 \times L \times I \times R \times (0.866)}{1{,}000}$$

Where:

VD = calculated or actual voltage drop

L = one-way length of the conductor in feet

I = (or A) load current in amps

R = conductor resistance in ohms per thousand feet as read from Table 8, Chapter 9

The formula, without the 0.866 factor, is used to determine the voltage drop of a 2-wire dc circuit, a 2-wire ac circuit, or a 3-wire (phase-to-phase and phase-to-neutral), single-phase ac circuit.

The factor of 0.866, which is equal to the square root of 3 divided by 2, is used only when the voltage drop being determined is for a 3-phase, 3-wire, ac load.

For the same voltage rating, the difference in currents between 3-phase and single-phase load currents for the same wattage or apparent power rating is the square root of 3. Since this is a single-phase voltage-drop formula, the 3-phase current can only be substituted for the single-phase current if it is multiplied by the square root of 3.

The plug-in length in the formula is a one-way length. The 2 was added to account for the voltage drop along the return conductor (twice the length). When the load is rated

3-phase and 3-wire, the current (assuming an alternation peak), actually flows to the load on one conductor, divides, and returns along the other two conductors. To compensate for the two return conductors instead of one, the formula is divided by 2.

Since both of these factors apply only for a 3-phase circuit, the square root of 3 is first divided by 2, and the result of 0.866 is then applied to the formula.

The factor of resistance, instead of impedance, assumes the circuit is supplying a balanced load at 100% power factor. Instead of pure resistance, ac circuits will normally have a vector sum of resistance and reactance called impedance. The formula assumes negligible reactance at unity power factor (allowing the resistance to equal the impedance), so the dc resistance of Table 8 is used.

When the conductor material is copper, whether to use coated or uncoated copper must be determined. Coated indicates a thin film of tin on the copper wire. Some types of insulations that contain rubber compounds will chemically react with copper. The tin coating prevents the reaction because tin will not react with rubber. Other uses for tinned copper conductors include splice or terminal make-up that will include a solder make-up. For most general wiring applications, the voltage-drop computations are based on the resistance of uncoated copper wire.

To determine the maximum voltage drop of a circuit, the nominal supply voltage and whether the calculation will be for a feeder (2%) or a branch circuit (3%) must be known. The maximum voltage drop (percentage) is set (fixed) by the supply voltage, the current is set by the connected load, and the length of the conductor is set by the location of the load. The only variable in the ohm's law voltage-drop formula is the value of resistance, which can only be changed by changing the size of the wire.

The voltage-drop formula can be worked in one of two ways. One way, the wire size is given, and with this value of fixed resistance, the actual voltage drop can be calculated and then compared to the allowed maximum voltage drop of the circuit. The 2%, 3%, or 5% allowance is applied to the circuit voltage rating, according to the type circuit in question.

Example 2–18. Determine the voltage drop of a stranded No. 10 AWG THW copper conductor supplying a 20-amp load rated at 120 volts. The load is located 175 feet from the branch-circuit overcurrent device.

$$\text{actual } VD = \frac{2 \times 175 \ \text{feet} \times 20 \ \text{amps} \times 1.24 \ \text{ohms}^*}{1{,}000 \ \text{feet}}$$

$$\text{actual } VD = 8.68 \ \text{volts}$$

*Ref. Table 8, ch. 9

On a test or other exam, where the actual voltage drop of a given circuit conductor is in question, this formula is ideal, simply because the known values of one-way length, full-load current, and resistance are known.

To determine if the actual voltage drop in this example fell within the 3% allowed (branch circuit), the percentage would have to be taken of the assigned circuit voltage: 3% of 120 volts is only 3.6 volts. The use of the No. 10 AWG copper circuit conductors, which yields a voltage drop of 8.68 volts, does not keep the circuit within the NE Code-recommended voltage-drop range.

To determine the required wire size for the maximum of 3% voltage drop, the formula will have to be reworked, using the wire resistance of No. 8 AWG copper. If the actual voltage drop of the No. 8 AWG copper conductor does not fall within the range, the formula will have to be reworked again, using the resistance of No. 6 AWG copper. No. 6 AWG copper conductors will bring the circuit in the example within the 3% range. For some circuits, the formula may have to be reworked several times to determine the NE Code-minimum wire size in terms of voltage-drop compensation.

Code challenge 2–18. Determine the actual voltage drop of a No. 4 RHW copper conductor supplying a 55-amp load rated at 240 volts. The load is located 400 feet from the branch-circuit overcurrent device.

$$\text{actual } VD = \frac{2 \times \text{(A) \_\_\_\_\_ feet} \times \text{(B) \_\_\_\_\_ amps} \times \text{(C) \_\_\_\_\_ ohms *}}{1{,}000 \text{ feet}}$$

actual VD = (D) _____ volts *Ref: (E) _____

A better method of sizing the wire, which requires the formula to be worked (or solved) only one time, is to determine the maximum allowed voltage drop, plug it into the formula, and solve for a maximum value of resistance (R). By formula:

$$R_{\text{max}} = \frac{1000 \times VD_{\text{max}}}{2 \times L \times I \times (0.866)}$$

The values of the unknowns are the same as were offered with the actual voltage-drop formula; the equation was only rearranged to solve for R_{max}.

The use of VD_{max} in the formula requires a second computation, which can be plugged directly into the formula: The maximum voltage-drop percentage, according to whether it is a branch circuit (3%) or a feeder circuit (2%) is applied to the nominal voltage rating of the circuit.

With this method, the calculated value of R can be taken into Table 8 of Chapter 9 and compared to the table-listed values of R. The table-listed value of R must be less than or equal to the calculated value of R. (Since the calculated R-value is a numerator in the actual voltage-drop formula, any table-listed R-value larger than the calculated R-value would make the calculated voltage drop larger.) The table-listed value of R will correspond by row to the correct wire size for the maximum allowed voltage drop.

Example 2–19. Using THW aluminum, determine the size of conductor required to supply a 150-amp panel located 700 feet (conductor length) from the switchgear. The distribution voltage is 208Y/120, 3-phase, 4-wire.

$$R_{\text{max}} = \frac{1{,}000 \text{ feet} \times (0.02 \times 208 \text{ volts})}{2 \times 700 \text{ feet} \times 150 \text{ amps} \times 0.866}$$

$$R_{\text{max}} = \frac{4{,}160 \text{ volts}}{181{,}860 \text{ amps}}$$

$$R_{\text{max}} = 0.0229 \text{ ohms}$$

Table-listed value of R: 0.0212 ohms

Wire size: 1,000 kcmil THW aluminum Ref. Ch. 9, T-8

Code challenge 2–19. Using THHN copper, determine the size of conductor required to supply a 200-amp panel, located 1,200 feet (conductor length) from the switchgear. The distribution voltage is 480Y/277, 3-phase, 4-wire.

$$R_{\text{max}} = \frac{1{,}000 \text{ feet} \times ([A] \_\_\_\_\_ \times [B] \_\_\_\_\_ \text{ volts})}{2 \times \_\_\_\_\_ \text{ feet} \times \_\_\_\_\_ \text{ amps} \times (C) \_\_\_\_\_}$$

$$R_{\text{max}} = \frac{\_\_\_\_\_ \text{ volts}}{\_\_\_\_\_ \text{ amps}}$$

$$R_{\text{max}} = (D) \_\_\_\_\_ \text{ ohms}$$

Table-rated value of R: (E) ___________ ohms

Wire size: (F) ___________ THHN copper Ref. (G)____________________

When other prevailing conditions such as continuous load, higher or lower ambient temperatures, or more than three current-carrying conductors in a single raceway are also present in a long distance wire run (wire length), these prevailing conditions, which require mandatory conductor ampacity adjustment by the NE Code, should be considered first, and the correct conductor size determined.

The correct (minimum) conductor wire size for the maximum voltage drop allowed (for feeder, branch circuit, or overall system drop) should then be determined, using only the original full-load or continuous-load current (as compared to the adjusted full-load current for the other prevailing conditions), and the two calculated conductor sizes compared. Using the largest conductor wire size of the two for the example or actual installation will allow for all the prevailing conditions as required by Sec. 310-15(c).

Example 2–20. Using THWN copper, determine the wire size required to supply a 64-amp, 480-volt, 3-phase continuous load located 800 feet (conductor length) from the panel. The panel is rated as 480/277-volt, 3-phase, 4-wire. The THWN (single insulated) copper conductors will be run in rigid metal conduit, which contains a total of 9 current-carrying conductors. The conduit will pass through an ambient temperature of 115°F.

Higher-temperature derate: 0.75 Ref. T310-16, CF

More-than-three derate: 70% Ref. T310-16, N-8(a) Tbl

Continuous-load overrate: 125% Ref. 220-3(a)

Temperature-compensated table-rated ampacity determination:

Overrate computation for higher ambient temperature:

$$64 \text{ amps} \div 0.75 = 85.3 \text{ amps}$$

Overrate computation for more-than-three circuit conductors:

$$85.3 \text{ amps} \div 0.70 = 121.9 \text{ amps}$$

Table-rated ampacity: 130 amps

Wire size: No. 1 AWG THWN copper — Ref. T310-16

Wire terminal temperature limitations: 60°C — Ref. 110-14(c)(1), Ex. 1

Overrate computation for continuous load:

$$1.25 \times 64 \text{ amps} = 80 \text{ amps}$$

Table-rated ampacity: 85 amps

Wire size: No. 3 AWG THWN copper — Ref. T310-16

Voltage-drop compensation computations: — Ref. 210-19(a), FPN No. 4

$$R_{\max} = \frac{1{,}000 \text{ feet} \times (0.03 \times 480 \text{ volts})}{2 \times 800 \text{ feet} \times 80 \text{ amps} \times 0.866}$$

$$R_{\max} = \frac{14{,}400 \text{ volts}}{110{,}848 \text{ amps}}$$

$$R_{\max} = 0.1299 \text{ ohms}$$

Table-listed value of R: 0.122 ohms

Wire size: No. 1/0 AWG THWN copper — Ref. Ch. 9, T-8

NE Code-minimum: Compare for larger — Ref. 310-15(c)

Wire size: No. 1/0 AWG THWN copper

The overrated continuous-load current of 80 amps was used in the voltage-drop formula instead of the 64-amp full-load rating, simply because of the lack of information in this example. The load could be increased to any value up to 80 amps without opening the circuit overcurrent protection device(s). The circuit conductors would be overload, which would contribute to a higher voltage drop. Sized in the manner shown, the maximum voltage drop will occur only when the load current equals 80 amps. At the indicated load of only 64 amps, the voltage drop will be less than the allowed 3%.

Code challenge 2–20. Using THHN copper, determine the wire size required to supply a 48-amp, 3-phase, continuous load located 800 feet from the panel. The panel is rated as 208Y/120-volt, 3-phase, 4-wire. The THHN (single insulated) copper conductors will be run in EMT conduit, which contains a total of 6 current-carrying conductors. The conduit will pass through an ambient air temperature of 94°F.

Higher-temperature derate: (A) ____________ Ref. (B) ____________

More-than-three derate: (C) ____________ % Ref. (D) ____________

Continuous-load overrate: (E) ____________ % Ref. (F) ____________

Temperature-compensated table-rated ampacity determination:

Overrate computation for higher ambient temperature:

____________ amps ÷ ____________ = (G) ____________ amps

Overrate computation for more-than-three circuit conductors:

____________ amps ÷ ____________ = (H) ____________ amps

Table-rated ampacity: (I) ____________ amps
Wire size: (J) ____________ THHN copper Ref. (K)____________________
Wire terminal temperature limitations: (L) ____________ °C Ref. (M) ____________
Overrate computation for continuous load:

____________ × ____________ amps = (N) ____________ amps

Table-rated ampacity: (O) ____________ amps
Wire size: (P) ____________ THHN copper Ref. (Q)____________________
Voltage-drop compensation computations: Ref. (R)____________________

$$R_{\text{max}} = \frac{1{,}000 \text{ feet} \times ([S] \_\_\_\_\_\_ \times [T] \_\_\_\_\_\_ \text{ volts})}{2 \times \_\_\_\_\_\_ \text{ feet} \times \_\_\_\_\_\_ \text{ amps} \times (U) \_\_\_\_\_\_}$$

$$R_{\text{max}} = \frac{\_\_\_\_\_\_ \text{ volts}}{\_\_\_\_\_\_ \text{ amps}}$$

$$R_{\text{max}} = (V) \_\_\_\_\_\_ \text{ ohms}$$

Table-rated value of R: (W) ____________ ohms
Wire size: (X) ____________ THHN copper Ref. (Y)____________________
NE Code-minimum: Compare for (Z) ____________ Ref. (1A) ____________________
Wire size: (1B) ____________ THHN copper

2.11 CABLE AMPACITIES

Ampacity corrections discussed up to now dealt with individual insulated conductors (wires) installed in some type of raceway. The raceway is used to provide protection against physical damage for both the conductor and its insulation.

Many circuits, whether service, feeder, or branch circuit, can be supplied by various forms (types) of multiconductor cable. Multiconductor cables, such as USE cable (underground service-entrance cables—previously shown in Fig. 2-5), consists of two or more single insulated conductors wrapped or twisted over and around each other to hold or bind them together as a single cable.

Other cables, such as two-conductor Romex (Fig. 2-7), consists of two single insulated conductors that are drawn side-by-side through an extrusion process which forms an outersheath of vinyl (or some other type of nonmetallic sheath) over the conductors. The outersheath binds the two conductors together. The 2-conductor cable is referred to as a flat cable. If Romex cable is manufactured as a 3-conductor cable, the three individual conductors are first spiral wrapped and then coated with the extrusion process.

Figure 2-7 Type NM cable (nonmetallic-sheathed cable) can be used for both exposed and concealed work in normally dry locations. (*Courtesy of Southwire Company*)

Romex cable is designed to be routed through the hollow chases of wooden structure walls, ceiling joists, and floor joists. The location (discussed in Chapter 6 of this text), concealed within these spaces, protect them from physical damage. The outer sheath aids in the installation through either bored holes or notches in the wood members.

USE cable is designed for direct burial. The covered trench provides the required physical protection. When this underground service cable emerges from the ground it must be installed in conduit to provide equivalent physical protection. An outersheath is not necessary for this type of cable.

The outersheath of Romex cable provides only limited physical protection. When Romex cable is installed exposed, such as in an open basement of a dwelling, it must be installed to follow the surface of the above floor joists. When Romex cable leaves the joist area in the form of a drop down to a surface-mounted receptacle outlet, or as a supply to a water heater, it must be installed in protective conduit.

The rated ampacity of the individual insulated conductors that make up these multiconductor cables, which include Romex (nonmetallic sheathed cable—Art. 336), BX (type AC or armored cable—Art. 333), UF (underground feeder or branch circuit cable—Art. 339), and SE (type SE or USE service-entrance cable—Art. 338), as well as many other types found in Chapter 3 of the NE Code, is determined in the same manner as the individual conductors installed in conduit or other raceway.

2.12 ROMEX AND BX CABLE

Article 336 governs the installation and use of Romex or nonmetallic-sheathed cable. Part C lists the construction specifications. As outlined in Sec. 336-30(b), the individual insulated conductors within the multiconductor cable must have a 90°C temperature rating. Romex cable is manufactured in three principal types: NM, NMC, and NMS.

According to Sec. 336-30(a), the acronyms NM and NMS identify a nonmetallic sheath covering which is both flame retardant and moisture resistant. Type NMC [Sec. 336-30(a)(2)] is also fungus and corrosion resistant. Types NMS [Sec. 336-30(a)(3)] is a new type of nonmetallic sheath cable designed specifically for signaling circuits (smart house wiring—Art. 780].

In terms of rated ampacity, the general rule of Sec. 336-30(b) limits the rated-load ampacity of Romex cables, according to wire size, to the ampacities found in the 60°C column of T310-16.

Although in past years, Romex cable was manufactured in both copper and aluminum, new Romex cable is only available in copper. In the description (definition) of Romex in Sec. 336-1, reference to conductor material is not made. Section 110-5 stipulates that when the conductive material is not specified, the material and sizes given shall apply to copper conductors.

Romex cable is also referenced according to the number of circuit conductors contained. For example, the term 12/2 w/GND Romex cable refers to a flat nonmetallic sheath cable which contains two insulated copper conductors and a bare copper equipment-grounding conductor. The bare conductor is located between the two other insulated conductors.

A 12/3 Romex cable contains three insulated circuit conductors but no ground (no equipment-grounding conductor or other grounding means). A 10/3 w/GND Romex cable contains three insulated No. 10 AWG copper conductors and a fourth-wire No. 10 AWG bare copper equipment-grounding conductor.

Romex multiconductor cable is available in wire sizes No. 14 through No. 2 AWG copper, as either a 2-conductor or 3-conductor cable with or without ground (separate equipment-grounding conductor).

For general installation requirements, the rated ampacity of Romex multiconductor cable is taken directly from the 60°C copper column in T310-16, according to wire size. This temperature limitation or ampacity derate is in agreement with the wire terminal temperature limitations of Sec. 110-14(c)(1), since Romex cable is not available in wires even up to No. 1 AWG.

Example 2–21. Determine the full-load current rating of a No. 10/3 w/GND Romex cable.

Cable temperature limitation: 60°C	Ref. 336-30(b)
Base rated ampacity on copper conductors.	Ref. 110-5
Rated-load ampacity: 30 amps	Ref. T310-16

In reference to Note 11 to T310-16, the equipment-grounding conductor in a single Romex cable is not counted when determining the number of current-carrying (circuit) conductors for application of Note 8(a): The No. 10/3 w/GND Romex cable in this example contains only three circuit conductors, although it actually contains four conductors in its nonmetallic sheath.

Code challenge 2–21. Determine the rated-load ampacity of a No. 12/3 AWG w/GND Romex cable.

Cable temperature limitation: (A) ____________ °C	Ref. (B) ________________
Base rated ampacity on (C) ____________ **conductors.**	Ref. (D) ________________
Rated-load ampacity: (E) ____________ amps	Ref. (F) ________________

When Romex cable will be subjected to ambient temperatures in excess of 86°F (30°C), the table-listed ampacities must be derated according to the correction factors listed at the bottom of T310-16 in the same manner as single insulated conductors installed in raceway were in previous sections of this text.

When Romex is routed through the attic of the average dwelling structure (home), higher ambient temperatures can be expected. Exactly what temperature to expect is somewhat dependent on what part of the country the dwelling will be built. Summer weather near the equator, such as a Florida residence, will normally have hotter days than a dwelling built on the shores of Lake Michigan.

For energy efficiency, most new or remodeled homes are equipped with an attic (roof) ventilating fan. If the thermostat of the attic fan is set for the fan to come on at 120°F, then you can expect at least 120°F. True?

The exception to Sec. 336-30(b) will allow the derate found in the 90°C column to be applied to the ampacity listed in the 90°C column to compensate for these higher attic (ambient) temperatures. As with higher temperature-rated, single insulated circuit conductors installed in raceway, the original rated-load current cannot exceed the table-rated current, without derate, found in the 60°C temperature column.

Example 2–22. Determine the maximum-load ampacity of a No. 14/2 w/GND Romex cable installed in an ambient temperature of 120°F.

Table-rated ampacity: 25 amps Ref. T310-16 — Ref. 336-30(b)
Higher-temperature derate: 0.82 — Ref. T310-16, CF
Rated-load ampacity computation:

$$25 \text{ amps} \times 0.82 = 20.5 \text{ amps}$$

Conductor temperature limitation: 60°C — Ref. 336-30(b)
Rated-load ampacity: 15 amps — Ref. T310-16, footnote
NE Code-minimum: Compare for smaller — Ref. 310-15(c)
Rated-load ampacity: 15 amps

In a dwelling structure, 15- and 20-amp, 120-volt, multioutlet branch circuits can supply both general (overhead) lighting loads and convenience receptacle loads. The receptacle outlets represent unknown current loads, such that the obelisk footnote to T310-16 restricts the rating of the branch-circuit fuse or circuit breaker for the No. 14 AWG Romex cable to a maximum rating or setting of 15 amps. Hence the rated-load ampacity of only 15 amps for the No. 14 AWG copper wire when the table listing in the 60°C column was 20 amps.

This poses yet another problem. Instead of the 120°F setting of the attic ventilating fan, the installing electrician could elect to evaluate the attic temperature at 130°F. This would lower the rated ampacity determined from the 90°C column even further.

If the temperature-compensated current is less than 15 amperes, then technically the No. 14 AWG Romex cable could not be installed in the dwelling attic, because of the unknown receptacle current loads. Instead, a No. 12 AWG Romex cable would have to be installed, protected by a 15-amp overcurrent protection device to meet the temperature limitations of the ampacity table.

In the same light, a 20-amp, 120-volt, multioutlet branch circuit often requires a No. 10 AWG Romex cable to compensate for the higher ambient temperatures found in the dwelling attics.

Code challenge 2–22. Determine the maximum-load ampacity of a No. 12/2 w/GND Romex cable installed in a dwelling attic if the expected ambient temperature is 130°F.

Table-rated ampacity: (A) ___________ amps
Ref. (B) ________________ Ref. (C) ___________________
Higher-temperature derate: (D) ____________ Ref. (E) ___________________
Rated-load ampacity computation:

___________ amps × ___________ = (F) ___________ amps

Conductor temperature limitation: (G) ___________ °C Ref. (H) _____________
Rated-load ampacity: (I) ___________ amps Ref. (J) ___________________
NE Code-minimum: Compare for (K) ___________ Ref. (L) _________________
Rated-load ampacity: (M) ___________ amps

Note 8(a) to the ampacity tables rated 0 to 2,000 volts also applies to multiconductor cables that contain more than three circuit (current-carrying) conductors. At first, it would seem that this table does not apply to Romex cable, since the maximum number of insulated conductors available in Romex cable is only three.

When a 40-circuit (40 poles or breaker slots) service panel is set in a dwelling or small business (small-scale commercial or industrial) structure, several Romex or other multiconductor cables are extended from the single enclosure to supply feeder and branch circuits.

If the electrician elects to install large conduit risers in the chase above and below the panel to convey the cables into the attic or under the floor into a crawl space or a basement, then the individual circuit conductors in the multiconductor cables must be derated for more than three circuit conductors in a common raceway according to the total number of circuit conductors in the cables and the derate percentages listed in N-8(a) Table. If the cables exit the panel enclosure in individual knockouts and cable clamps and penetrate the wall plate or seal in individual holes, but share a common chase (which contains the recessed panel), then the more-than-three derate is not required.

If the cables exit the panel through individual knockouts and cable clamps, but exit the side of the chase through a common hole, and this hole is extended as additional holes across two or more wall cavities (other chases formed by the wall studs for a lateral run of the cables), then the more-than-three derate does apply.

The second paragraph to Note 8(a), which follows the table, requires multiconductor cables which are bundled or otherwise grouped together for a span (length or distance) of 24 inches or more (stud walls are normally built 16 inches on center) to be derated as if the cables were installed in conduit.

Example 2–23. Determine the maximum full-load ampacity of six No. 10/2 w/GND Romex copper cables that are installed in a common conduit riser that extends from the top of an outside service panel into the structure attic through the overhang. The conduit is 4 feet long.

Table-rated ampacity: 40 amps Ref. T310-16 Ref. 336-30(b), Ex.
Total number of circuit conductors determination: ________ Ref. T310-16, N-11

6 cables × 2 conductors per cable = 12 conductors

More-than-three derate: 50% Ref. T310-16, N-8(a) Tbl
Rated-load ampacity computation:

40 amps × 0.50 = 20 amps

Conductor temperature limitation: 60°C Ref. 336-30(b)
Rated-load ampacity: 30 amps Ref. T310-16
NE Code-minimum: Compare for smaller Ref. 310-15(c)
Rated-load ampacity: 20 amps

In reference to Note 11 to T310-16, the equipment-grounding conductor in each Romex cable was not counted in determining the total count of circuit (current-carrying) conductors. It will be shown in Chapter 9 of this text that grounding or bonding conductors (by definition, Art. 100) are not current-carrying conductors under normal load conditions.

Although the obelisk footnote to T310-16 restricts the rating of the branch-circuit fuses or circuit breaker for No. 10 Romex cable to a maximum of 30 amps, the derated ampacity in this example would restrict the OCPD rating to a maximum of 20 amps.

Code challenge 2–23. Determine the maximum full-load current of eight No. 12/2 w/GND Romex copper cables that pass horizontally through common holes bored in three adjacent wall studs.

Table-rated ampacity: (A) __________ amps
Ref. (B) ________________ Ref. (C)______________
Total number of circuit conductors determination: Ref. (D)______________

__________ cables × __________ conductors per cable = (E) __________ conductors

More-than-three derate: (F) __________ % Ref. (G)______________
Rated-load ampacity computation:

__________ amps × __________ = (H) __________ amps

Conductor temperature limitation: (I) __________ °C Ref. (J) ______________
Rated-load ampacity: (K) __________ amps Ref. (L) ______________
NE Code-minimum: Compare for (M) __________ Ref. (N)______________
Rated-load ampacity: (O) __________ amps

The discussion in the section seems to be slanted towards Romex or other nonmetallic-sheathed multiconductor cables. The same rules, in terms of rated ampacity and insulated wooden structures, applies to BX multiconductor cables.

Article 333 in the NE Code governs the use and installation of BX cable (Fig. 2-8). BX is a tradename for type AC or armored cable. The exception to Sec. 333-20 requires BX cables installed in direct contact with thermal insulations, which cables installed in the outer walls or attic of a wooden structure such as the average home would be, must contain conductors, installed at the point of manufacture, rated for 90°C. Their rated-load ampacity, after any applicable derates for higher ambient temperature, cannot exceed the table-listed ampacity found in the 60°C temperature column of T310-16. These are the same requirements placed on Romex cable by Sec. 336-30(b).

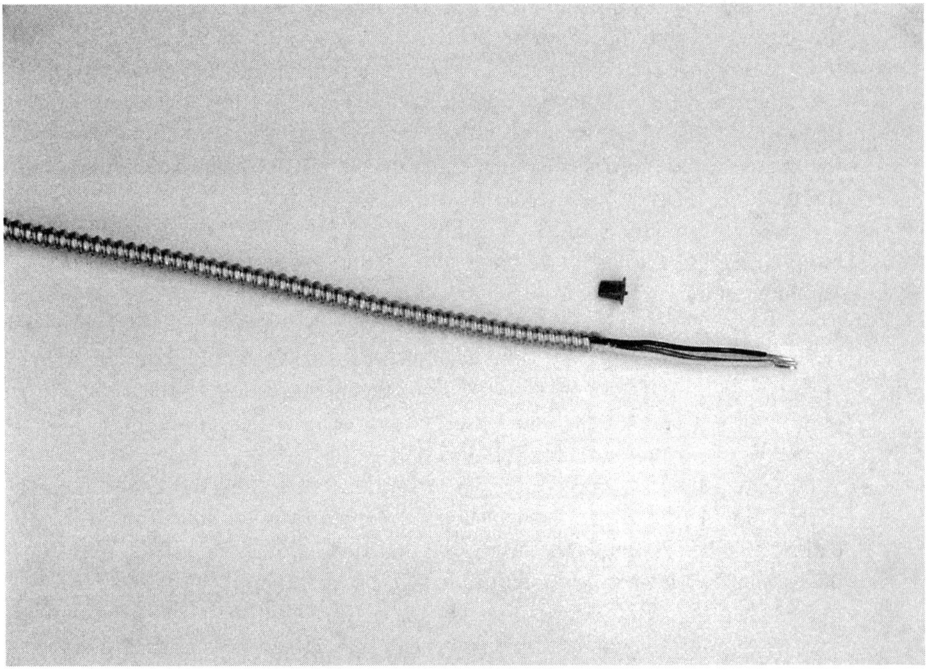

Figure 2-8 BX cable contains the same type of insulated circuit conductors as Romex, but installed in flexible conduit or armor.

If BX cables are either installed in protective raceway, bundled, or otherwise grouped or stacked, as described for Romex cables, then the more-than-three derate of N-8(a) table also applies.

2.13 UF CABLE

Rules governing the installation and use of UF cable are found in Art. 339. Type UF cable, which gets its acronym (abbreviated name) from "underground feeder," is suitable for use as either a branch-circuit or feeder-circuit wiring method.

Figure 2-9 Type UF cable is generallly used as a feeder to outside loads fed from an existing building. (*Courtesy of Southwire Company*)

Type UF cable (Fig. 2-9) is similar in appearance to Romex multiconductor cable with a thicker outer sheath. Technically, UF cable is more or less a souped-up version of Romex cable that can be installed either inside or outside a structure in either wet or corrosive locations. As the name implies, UF cable is also suitable for direct trench burial in the ground without the use of protective conduit.

In comparison Romex cable is not approved either for direct burial, or for use in wet or corrosive locations. For interior wet installations, UF cable can be installed instead of Romex cable. Section 339-3(a)(4) would require the UF cable to be installed within the guidelines set forth for Romex cable in Art. 336. (Cable installation requirements are discussed in Chapter 6 of this text.)

As required by Sec. 339-5 and listed in T310-16, the rated-load ampacity of UF cable is limited to the table-rated ampacity, found according to wire size, in the 60°C temperature column of T310-16.

Type UF cable is used for branch circuits that supply walkway, driveway, or other types of outside lighting. UF cable can also be used to supply either branch or feeder circuits (which require a load-end panel or other load center) in remote well/pump houses, boat houses, small utility buildings, garages, or other structures from a service or other subfed panel of the main building on a property.

Like Romex or any other type of nonmetallic sheathed cable, UF cable must be protected from physical damage when it emerges from the trench in a direct burial application, or when it leaves the protective floorline of overhead joists, seals, or plates in an underfloor crawl space or open (unfinished) basement application.

When a conduit riser is used to run UF cable up the side of the building for entry into the interior, the conduit is subject to higher ambient temperature as the result of summer weather. In some instances the protective raceway (conduit) is subject to direct sunlight. As a result, the table-rated ampacity of type UF cable must be derated for the expected temperature.

Example 2–24. Determine the maximum-load ampacity of a No. 10/2 w/GND UF cable installed in protective conduit along the wall of a structure where the expected surface temperature is 120°F.

Table-rated ampacity: 30 amps Ref. T310-16
Higher-temperature derate: 0.58 Ref. T310-16, CF
Rated-load ampacity computation:

$$30 \text{ amps} \times 0.58 = 17.4 \text{ amps}$$

Conductor temperature limitation: 60°C — Ref. 339-5
Rated-load ampacity: 30 amps — Ref. T310-16, footnote
NE Code-minimum: Compare for smaller — Ref. 310-15(c)
Rated-load ampacity: 17.4 amps

If the UF cable in this example will be used to supply a 30-amp load center located in a remote garage or other outbuilding (separate from the main structure), the footnote to T310-16 restricts the rating of the feeder fuses or circuit breaker to a maximum of 30 amps. The rating of the load center and the overcurrent protection device agree.

Due to the expected surface temperature of the wall where the protective conduit will be installed, the combined full-load current of all the branch circuits (normally two) supplied from the load center in the remote building is restricted to a maximum of only 17.4 amps.

If the load cannot be restricted, or if the load center supplies convenience receptacle loads where the value of expected load current is technically unknown, then the feeder OCPD rating will have to be restricted to a maximum of 15 amperes, or the wire size will have to be increased.

If the 30-amp circuit is required, then 30 amps divided by 0.58 requires a wire size that has a listed ampacity in the 60°C copper temperature column greater than or equal to 51.7 amps: A No. 6 UF cable is the minimum wire size allowed by the NE Code.

Code challenge 2–24. Determine the maximum-load ampacity of a No. 12/2 w/GND UF cable installed in protective conduit along the wall of a structure where the expected surface temperature is 110°F.

Table-rated ampacity: (A) ____________ amps — Ref. (B)________________
Higher-temperature derate: (C) ____________ — Ref. (D)________________
Rated-load ampacity computation:

____________ amps × ____________ = (E) ____________ amps

Conductor temperature limitation: (F) ____________ °C — Ref. (G)________________
Rated-load ampacity: (H) ____________ amps — Ref. (I) ________________
NE Code-minimum: Compare for (J) ____________ — Ref. (K)________________
Rated-load ampacity: (L) ____________ amps

2.14 SE CABLE

The discussion of rated-load ampacities for cable wiring methods would not be complete if type SE or service-entrance cable was not addressed. The installation and use of service-entrance cable, both type SE and USE (underground service-entrance cable), are governed by Art. 338 of the NE Code.

Like UF cable, type SE multiconductor cable actually has a dual role in feeder or branch-circuit applications. Its primary use is, as the acronym implies: Service-entrance cable. As defined in Art. 100, service-entrance conductors extend from the point of connection to the building of an aerial service-drop installation, or from the point of entry into the building for an underground service lateral.

When installed as the service-entrance conductors the requirements of both Arts. 338 and 230 apply. In terms of rated ampacity, the exact temperature column to use in T310-16 is not spelled out in Art. 338. In keeping with NE Code practice, conductors installed outside of a structure are subject to weather, which means a wet location. As found in T310-13, which lists the properties of conductor insulations, insulations suitable for wet locations are normally limited to the 75°C column in T310-16.

Type SE cable, as shown in Fig. 2-10, is available in both copper and aluminum, as well as copper-clad aluminum. The rated ampacity would be determined by wire size and conductor material in the 75°C temperature column.

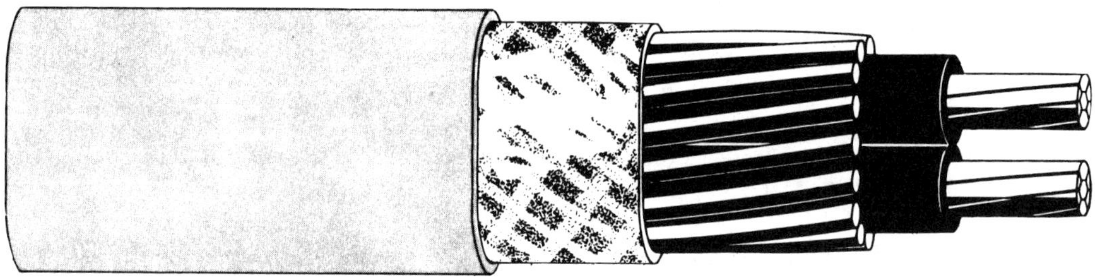

Figure 2-10 Type SE cable can be used between the service drop and the service equipment of a building. (*Courtesy of Southwire Company*)

For a single-phase, 120/240-volt, 3-wire service rated up to 400 amps, type SE cable is extended from the point of attachment to the structure to the metering base (where used) and from the metering base to the line side of the service disconnecting means for the building power distribution.

Installed to follow the outside surface of the building, the multiconductor cable is protected from physical damage by its location. Installed as such, it is exposed to warmer summer temperatures and in some instances (on the east side or west side of the structure) to direct sunlight. Such exposure requires a derate for higher surface temperatures.

Example 2–25. Determine the NE Code-minimum wire size of copper SE cable for the service-entrance to a structure supplied by a 200-amp, 120/240-volt, single-phase, 3-wire, service drop. The SE cable will be run down the exterior wall to the meter enclosure and exposed to direct sunlight. The surface temperature is not expected to exceed 125°F.

Table-rated ampacity determination:

Higher-temperature derate: 0.67 Ref. T310-16, CF

Overrate computation:

$$200 \text{ amps} \div 0.67 = 298.5 \text{ amps}$$

Table-rated ampacity: 310 amps
Wire size: 350 kcmil SE copper — Ref. T310-16
Wire terminal temperature limitations: 75°C — Ref. 110-14(c)(2)
Table-rated ampacity: 200 amps
Wire size: No. 3/0 AWG SE copper — Ref. T310-16
NE Code-minimum: Compare for larger — Ref. 310-15(c)
Wire size: 350 kcmil SE copper

The wire terminal temperature limitations imposed by Sec. 110-14(c)(2) coincided with the insulation temperature rating of the SE cable in this example, such that compensation for the higher ambient temperature required the larger wire size.

Code challenge 2–25. Determine the NE Code-minimum wire size of aluminum SE cable for the service-entrance to a structure supplied by a 125-amp, 120/240-volt, single-phase, 3-wire service drop. The SE cable will be run down the exterior wall to the meter enclosure where the surface temperature is not expected to exceed 115°F.

Table-rated ampacity determination:
Higher-temperature derate: (A) ____________ Ref. (B)____________
Overrate computation:

____________ amps ÷ ____________ = (C) ____________ amps

Table-rated ampacity: (D) ____________ amps
Wire size: (E) ____________ SE aluminum — Ref. (F)____________
Wire terminal temperature limitations: (G) ____________ °C Ref. (H) ____________
Table-rated ampacity: (I) ____________ amps
Wire size: (J) ____________ SE aluminum — Ref. (K) ____________
NE Code-minimum: No comparison; (L) ____________ Ref. (M) ____________
Wire size: (N) ____________ USE aluminum

Due to the diversity of loading in the average dwelling structure (home), Note 3 of the ampacity tables (following T310-19 in the NE Code) will allow a reduction in wire size of the SE cable for either the service-entrance or main feeder into the dwelling structure.

The higher ampacities found in the Note 3 table are limited to dwelling structures that are supplied specifically with a 120/240-volt, single-phase, 3-wire service rated between 100 and 400 amps. When using the Note 3 table, neither the wire terminal temperature restrictions of Sec. 110-14(c) nor the derates for higher ambient temperatures, listed as correction factors in the footnote to T310-16, apply.

The required wire size of the SE cable or the main feeder circuit conductors is taken directly from the Note 3 table according to the service amperage rating.

Example 2–26. Determine the NE Code-minimum wire size of copper SE cable for the service entrance to a dwelling structure supplied by a 200-amp, 120/240-volt, single-phase, 3-wire service drop. The SE cable will be run down the exterior wall to the meter enclosure and exposed to direct sunlight where the surface temperature is not expected to exceed 125°F.

Wire size: No. 2/0 AWG SE copper Ref. T310-16, N-3 Tbl

Code challenge 2–26. Determine the NE Code-minimum wire size of aluminum SE cable for the service entrance to a dwelling structure supplied by a 125-amp, 120/240-volt, single-phase, 3-wire service drop. The SE cable will be run down the exterior wall to the meter enclosure where the surface temperature is not expected to exceed 115°F.

Wire size: (A) ___________ SE aluminum Ref. (B)____________________

Because Romex cable is limited in wire size by the manufacturer to No. 2 AWG copper, SE cable is often used as the feeder supply to a lower amperage panel or other load center from the service equipment in a wooden structure. In addition, SE cable is often used as the branch-circuit supply to air conditioning equipment or appliances such as cooking equipment in the average home, when the rated amperage of Romex wire is exceeded.

There are two basic types of SE cable. One type is a flat cable which consists of two individual insulated conductors and a bare conductor that is braided around the other two conductors. The assembly is encased (bound) in a nonmetallic sheath (shown previously in Fig. 2-10).

The other type of SE cable is often referred to as SER cable (Fig. 2-11). The R indicates a round cable. This cable consists of three insulated conductors and one bare conductor. The bare conductor is round and laid beside, instead of braided around, the other three insulated conductors. The four conductors are spiral wrapped and then bound in an outer nonmetallic sheath.

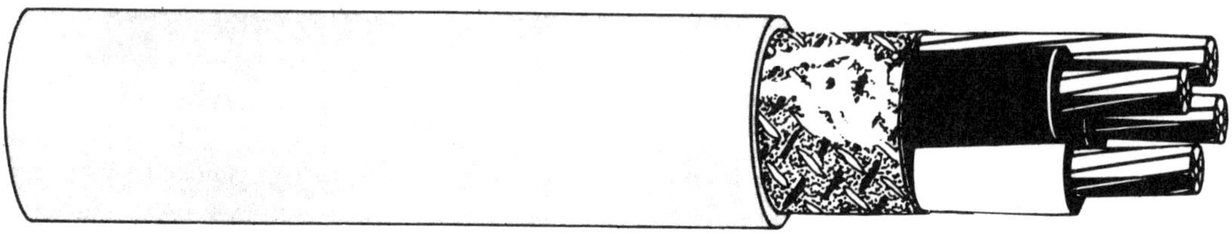

Figure 2-11 This type of SE cable is primarily used as a panel feeder in a multiple-unit dwelling (fourth-wire equipment ground). (*Courtesy of Southwire Company*)

Because the equipment-grounding means and the grounded neutral or grounded circuit conductor (2-wire branch circuit) must be kept separate (isolated) on the load side of the service disconnecting means for a building or other structure (discussed in detail in Chapter 9 of this text), the use of the two-conductor flat cable is limited to 2-wire branch circuits.

Two-wire branch circuits are used to supply equipment rated at either 120 or 240 volts. The SER cable must be used when the branch or feeder circuit is rated multiwire (three-wire—both voltages on the single-phase system, 120 and 240 volts, are supplied). Type SER cable would be used in the same manner as 3-conductor w/GND Romex cable.

When used instead of Romex, Sec. 338-4 requires the SE cable to be sized and installed in the same manner as Romex cable (by reference to Art. 336). This means the rated-load ampacity of the SE cable, which is normally found in the 75°C temperature column is restricted to the ampacities found in the 60°C.

Example 2–27. Determine the contained conductor wire size of the aluminum SE cable used to supply a 50-amp household cooking range receptacle from an individual 50-amp, 120/240-volt, 3-wire, multiwire branch circuit.

Cable temperature limitation: 60°C Ref. 338-4 Ref. 336-30(b)
Table-rated ampacity: 55 amps
Wire size: No. 4 AWG SE aluminum Ref. T310-16

Code challenge 2–27. Determine the required wire size of copper service-entrance cable used to supply a built-in oven with an individual at 40-amp, 120/240-volt, 3-wire, multiwire branch circuit.

Cable temperature limitation: (A) ____________ °C.
Ref. (B) ________________ Ref. (C) ____________________
Table-rated ampacity: (D) ____________ amps
Wire size: (E) ____________ SE copper Ref. (F) ____________________

Where multiconductor cables have extended length (a UF cable run to a remote building, Romex run along the rafters of a long warehouse, or SER cable run to a second floor panel in a dwelling structure from the basement service panel), they should also be sized for voltage drop compensation.

Except for the provisions of the Note 3 Table to T310-16, all multiconductor wiring methods found in different Articles within Chapter 3 to the NE Code are subject to the derates required by all the prevailing installation conditions discussed in this chapter.

REVIEW QUESTIONS

1. The measure of the ability of a conductor to convey electrical current is referred to as its ____________.
2. Interaction of load current flowing against the resistance of a conductor along its length appears as both ____________ and ____________ drop.
3. The rated-load ampacity of circuit conductors, whether individual insulated conductors installed in a common raceway or contained in multiconductor cables, is governed by the requirements of Article ____________.
4. The table-rated ampacity of circuit conductors with insulation voltage ratings between 0 and 2,000 volts is found in Tables ____________ through ____________.

5. Before the table-rated ampacity of circuit conductors with insulation voltage ratings between 0 and 2,000 volts can be used as the rated-load ampacity, the requirements of the notes that follow Table 310-19 must be satisfied.

   True or False

6. Temperature ratings and other characteristics of conductor insulations as listed in T310-16 are found in Table ____________.
7. In general, power or load conductors (circuit conductors, other than fixture wires or flexible cords) must be protected against overcurrent in accordance with their rated ampacities as determined by Sec. ____________.
8. Overcurrent of a conductor (or connected equipment) is defined as current in excess of rated or normal current that can be the result of either ____________, ____________ and/or ____________.
9. The standard ratings of circuit fuses, or the standard settings of circuit breakers overcurrent protection devices are listed in the NE Code in Sec. ____________.

Questions 10 and 11: Which temperature column to use to select the rated ampacity of a circuit conductor is determined by the lowest temperature rating of any connected termination, conductor, or device. For general applications (where terminal or device temperature ratings are not known),

10. The table-rated ampacity for circuit conductors protected or rated at 100 amps or less, or rated wire size up to No. 1 AWG shall be selected from the ____________ °C temperature column of T310-16 (or other ampacity tables if they apply).
11. The table-rated ampacity for circuit conductors protected or rated over 100 amps, or rated wire size larger than No. 1 AWG shall be selected from the ____________ °C temperature column of T310-16 (or other ampacity tables if they apply).
12. When the ambient air temperature will exceed 86°F, the table-rated ampacity, as listed in T310-16 according to wire size, must be derated according to the ____________ ____________ listed in the lower third of the table.
13. When a defined higher temperature and a rated-load or circuit current are given, the rated-load current must be ____________ (multiplied or divided) by the derate factor to determine the NE Code-minimum table-rated ampacity of the circuit conductors in question.
14. When more than three circuit conductors will be installed in a single raceway, the table-rated ampacity, as listed in T310-16 according to wire size, must be derated according to the percentages listed in ____________ table.
15. When more than three circuit conductors will be installed in a single raceway, and a rated-load or circuit current is given, the rated-load current must be ____________ (multiplied or divided) by the derate factor to determine the NE Code-minimum table-rated ampacity of the circuit conductors in question.
16. When determining the number of current-carrying conductors contained in a single raceway, equipment-grounding conductors or bonding jumpers must be counted when establishing the required derate percentage for more-than-three circuit conductors.

    True or False

17. When determining the number of current-carrying conductors when multiwire branch circuits are installed in a single raceway, neutral conductors supplying linear loads must be counted when establishing the required derate percentage for more-than-three conductors.

    True or False

18. When determining the number of current-carrying conductors when multiwire branch circuits are installed in a single raceway, neutral conductors supplying nonlinear loads must be counted when establishing the required derate percentage for more-than-three conductors.

    True or False

19. Load current must be considered as continuous when the current is conveyed without interrupt for ____________ hours or more.
20. The connected load is restricted to a maximum of ____________ % of either the branch-circuit or feeder OCPD rating, or the table-rated ampacity of the circuit conductors when the load is specified as continuous.
21. When a rated load is specified as continuous, its value or magnitude must be increased by a factor of ____________ % to determine either the rating of the branch-circuit or feeder OCPD, or the table-rated ampacity of the circuit conductors.
22. When more than three circuit conductors will be installed in a single raceway that will pass through a higher ambient, the table-rated ampacity, as listed in T310-16 according to wire size, must be derated for each type of loss of heat dissipation.

    True or False

23. When more than three circuit conductors contained in a single raceway that passes through higher ambient temperatures supplies continuous loads, their rated ampacity, as listed in T310-16, must be derated three times to compensate for the losses of heat dissipation.

    True or False

24. Adjusting power or load conductor ampacities by increasing wire size to compensate for excess voltage drop is an NE Code requirement as specified in the fine-print notes of Secs. 210-19(a), 215-2(b), and 310-15.

    True or False

25. The factor of 0.866 in the voltage-drop formula must be used on all circuit voltage computations, whether single-phase or 3-phase, ac or dc.

    True or False

26. The table-rated ampacity of Romex (nonmetallic-sheathed) cable must be determined according to wire size and conductor material as referenced in the ____________ °C temperature column of Table 310-16.
27. When SE cable is used for interior wiring of a structure, whether feeder or branch circuit, it must be installed as if it were Romex cable.

    True or False

REVIEW PROBLEMS

1. Determine the table-rated ampacity of three No. 8 AWG RHH copper conductors installed in conduit raceway.
2. Determine the table-rated ampacity of three 500 kcmil XHHW-2 copper-clad aluminum conductors contained in rigid metal conduit and buried as a feeder between two buildings.

3. If the full-load current rating of a single unit equipment is 86 amps, determine the wire size of TW-insulated copper conductors required to supply the intended load.
4. Determine the NE Code-minimum wire size of THHN copper conductors required to supply a connected load of 45 amperes.
5. Determine the NE Code-minimum wire size of THWN copper-clad aluminum conductors required to supply a single-phase, 3-wire, 150-amp, 120/240-volt power panel.
6. If three No. 2/0 AWG THHN copper conductors will be installed in IMC conduit that passes through an ambient air temperature of 104°F, determine the maximum rated-load ampacity of the circuit conductors for this temperature.
7. Size the underground service lateral for a structure that is supplied with a 200-amp service. Use triplexed USE-RHW (75°C) copper-clad aluminum wire installed in conduit, which extends from below ground (finished grade) up to the meter enclosure (can) on the side wall of the building. The maximum expected air/surface temperature on the side of the building is 115°F.
8. Determine the maximum rated-load ampacity of 12 No. 10 AWG THHN copper conductors installed in a single raceway.
9. If the circuit conductors will be installed in a single conduit that contains a total of eight current-carrying conductors, determine the NE Code-minimum wire size of THHN copper conductors required to supply a 150-amp feeder load.
10. A single conduit will contain two 2-wire branch circuits consisting of No. 10 AWG RHW copper conductors. If two No. 10 AWG RHW copper equipment-grounding conductors will also be pulled into the raceway, determine the NE Code-maximum load ampacity of the circuit conductors.
11. Determine the NE Code-minimum wire size of the ungrounded circuit conductors, if two single-phase, 3-wire, 125-amp, 120/240-volt feeder circuits will be installed in a single raceway. Use THHN copper conductors.
12. Determine the NE Code-maximum load current of No. 10 AWG THW copper-clad aluminum conductors, if a total of two 3-phase, 4-wire branch circuits will be installed in a common conduit. The branch circuits will supply nonlinear loads.
13. Determined the NE Code-maximum continuous-load current of No. 10 AWG THHW copper conductors.
14. Determine the NE Code-minimum wire size of THHN copper conductors required to supply a 120-amp feeder load that is rated as continuous.
15. Determine the maximum-load ampacity of six No. 4 AWG THW copper conductors contained in a single conduit that passes through an environmental air temperature of 37°C.
16. A total equipment full-load current rating of 45 amperes will be supplied with THW aluminum conductors. If the conductors will be installed in a single raceway that contains a total of 7 current-carrying conductors, and passes through an ambient temperature of 40°C, determine the NE Code-minimum conductor wire size allowed for the given load conditions.
17. A continuous full-load current of 56 amperes will be supplied with THW copper conductors. If the conductors will be installed in a single raceway that contains a total of 9 current-carrying conductors, and passes through an ambient temperature of 90°F, determine the NE Code-minimum conductor wire size allowed for the given load conditions.
18. Determine the voltage drop of stranded No. 6 AWG RHW copper circuit conductors supplying a 50-amp load rated at 240 volts. The load is located 290 feet from the branch-circuit overcurrent protection device.

19. Using RHW copper, determine the wire size of circuit conductors required to serve a 100-amp panel located 1,400 feet (conductor length) from the switchgear. The distribution voltage is 575, 3-phase, 3-wire delta.
20. Using THWN copper, determine the wire size of circuit conductors required to supply a 72-amp, 480-volt, 3-phase continuous load located 1,000 feet (conductor length) from the panel. The panel is rated as 460/265-volt, 3-phase, 4-wire. The THHN (single insulated) copper conductors are run in rigid metal conduit, which contains a total of 8 current-carrying conductors. The conduit passes through an ambient temperature of 104°F.
21. Determine the full-load current rating of a No. 8/2 w/GND Romex cable.
22. Determine the maximum-load ampacity of a No. 10/3 w/GND Romex cable installed in an ambient temperature of 38°C.
23. Determine the maximum full-load ampacity of six No. 14/2 w/GND Romex copper cables installed in a common conduit riser which extends from the top of an outside service panel into the structure attic through the overhang. The conduit is 8 feet long.
24. Determine the maximum-load ampacity of a No. 14/2 w/GND UF cable installed in protective conduit along the wall of a structure where the expected surface temperature is 114°F.
25. By the ampacity tables, determine the NE Code-minimum wire size of aluminum SE cable for the service entrance to a dwelling structure supplied by a 100-amp, 120/240-volt, single-phase, 3-wire service drop. The SE cable will be run down the exterior wall to the meter enclosure and exposed to direct sunlight. The surface temperature is not expected to exceed 120°F.
26. By the notes to the ampacity tables, determine the NE Code-minimum wire size of aluminum SE cable for the service entrance to a dwelling structure supplied by a 100-amp, 120/240-volt, single-phase, 3-wire service drop. The SE cable will be run down the exterior wall to the meter enclosure and exposed to direct sunlight. The surface temperature is not expected to exceed 120°F.
27. Determine the contained conductor wire size of the copper SE cable used to supply a 30-amp household electric dryer receptacle from an individual 30-amp, 120/240-volt, 3-wire multiwire branch circuit.

Chapter 3

Raceways

Article 100 defines a *raceway* as a closed channel designed expressly for holding wires, cables, busbars, and the like. Raceway may be either metal or nonmetal (insulating material). The list of NE Code-recognized raceways includes:

Busways	Art. 364
Cellular concrete floor raceways	Art. 358
Cellular metal floor raceways	Art. 356
Electrical metallic tubing (EMT)	Art. 348
Electrical nonmetallic tubing (ENT)	Art. 331
Flexible metal conduit (flex or "greenfield")	Art. 350
Flexible metallic tubing	Art. 349
Intermediate metal conduit (IMC)	Art. 345
Liquidtight flexible metal conduit (sealtight)	Art. 351-A
Liquidtight flexible nonmetallic conduit	Art. 351-B
Rigid metal conduit (rigid)	Art. 346
Rigid nonmetallic conduit (including PVC)	Art. 347
Surface metallic raceways	Art. 352-A
Surface nonmetallic raceways	Art. 352-B

Underfloor raceways	Art. 354
Wireways	Art. 362

By definition, Auxiliary Gutters (Art. 374), Cable Bus (Art. 365), and Cable Trays (Art. 318) are not listed as raceways. An auxiliary gutter is normally required where a concentration of electrical equipment, such as fused disconnects and motor controllers for several motors, is located in a confined area of a room or other area of the structure. Auxiliary gutters (refer to Chapter 5 of this text) can be considered as extra long junction boxes—not raceways.

As compared to raceway, cable bus and cable trays are rigid means of open support for individual cable runs between supply and load centers within a building or other structure. Neither are required to be enclosed (exceptions noted), and neither are required to be mechanically or electrically continuous between the supply and load points. Raceway, as shown in Figure 3-1, is required to be mechanically continuous between the supply and load points (Sec. 300-12), and when metal raceway is used, electrically continuous. Raceway must always totally enclose the contained conductors.

Figure 3-1 Conduit, tubing, or other types of piping raceway are closed channel used to protect circuit conductors from exposure to physical damage.

The concerns in this chapter are limited to sizing conduit and tubings, or other piping-type raceways, and cable tray. The sizing of other types of raceway, as outlined in the specific article, is normally by design of either the intended structure, or by the design of the raceway itself.

3.1 SIZING CONDUIT AND TUBING

The types of conduit and tubing (raceway) as listed above are governed by the individual articles contained in Chapter 3.

The type of conduit or tubing which can be used or installed is governed by the environment of the intended installation. Is it a wet or dry location? What is the expected temperature range? Is it a hazardous (due to easily ignitable fumes or vapors, dust, or flying particles) location? To what extent of physical damage will the conduit or tubing be exposed? The answers to these questions will prohibit the use of some conduit or tubings in favor of others.

The size of the conduit or tubing is governed by the type of conduit or tubing, the individual wire sizes of the circuit or circuits, and the count or number of conductors to be installed in the conduit or tubing.

Conduit and tubings are manufactured in nominal sizes that range from ½ inch up to 6 inches in some types. The actual diameters of the conduits vary around the nominal diameter according to the wall thickness. The diameter of a rigid (thick wall) metal conduit is less than the diameter of intermediate (medium wall) metal conduit. The diameter of thick wall PVC (schedule 80) is less than the diameter of medium wall PVC (schedule 40). (Analysis based on internal diameter values is found in Table 4 of Chapter 9 in the NE Code.) The number of conductors that can be installed must be based on the type of conduit to be installed.

Standard wire sizes are listed by the American Wire Gage (AWG) in Table 8 of Chapter 9 of the NE Code from No. 18 through No. 4/0 (four-ought). Wire sizes greater than No. 4/0 AWG are listed in circular-mil area. These range from 250 to 2,000 kcmil [thousands of (k) circular (c)-mils].

The concentric size or cross-sectional area (c.s.a.) of the conductive material within the cover or insulation is governed by these standard wire sizes. In other words, the c.s.a. of a No. 4 AWG bare copper conductor is physically equivalent to the c.s.a. of either a No. 4 AWG bare copper-clad aluminum conductor, or a No. 4 AWG bare aluminum conductor.

This is true except for compact-stranded conductors. Compact-stranded conductors are compressed after the individual (concentric-stranded) bare conductors are spiral-wrapped to yield a standard wire size. A No. 4 AWG compact-stranded bare aluminum building wire will be smaller in size (both in c.s.a. and diameter) than a No. 4 AWG bare aluminum conductor that is not compact-stranded (analysis based on comparison of listed diameters for wire size No. 4 AWG in Tables 5A and 8: 0.213 inch versus 0.232 inch).

The overall cross-sectional area (c.s.a.) of a conductor, which is used to correctly size the conduit or tubing, is governed both by the wire gage (wire size) and by the type of insulation on the conductor. According to the composition of the insulation, some compounds will be thicker than others for the same dielectric strength or voltage rating. (Dielectric strength reflects the highest voltage the insulator can withstand without allowing the currrent to arc through it.) As a rule, rubber-based compounds will have a heavier or wider cross-sectional area than either vinyl or plastic compounds for the same voltage ratings.

The total conductor c.s.a. and resulting conduit or tubing size (determined by the percent-fill allowance of the conduit c.s.a.) is determined by the tables and guidelines

(Notes) listed in Part A of Chapter 9. Chapter 9 is referred to under the various raceway articles of Chapter 3, usually under the heading of "Number of Conductors in Conduit (or Tubing)."

All the conduit or tubing tables are based on the percent-fill allowances listed in Table 1. The c.s.a. of a single insulated conductor divided by the available c.s.a. of the conduit or tubing cannot exceed a ratio of 0.53; in other words, a 53% fill allowance. For two conductors the ratio is reduced to 0.31, or 31% fill. For more than 2 conductors, the percent-fill is increased to 40%.

As illustrated in Figure 3-2, the difference in the percent-fill allowances comes from the pressure points or total contact area between the side length of the conductor(s) and the conduit or tubing wall. For a single conductor, the cross-sectional area of the conduit and the conductor appear to be two eccentric circles—a lot of surface contact. The side of the conductor is pressed against the wall of the conduit by its weight or gravitational pull.

Figure 3-2 The percent-fill requirements of conduits and tubings is based on the cross-sectional area of the conduit or tubing.

For two smaller conductors installed in the same size conduit, gravitational pull creates three pressure points along the length of conductors and conduit with a minimum amount of surface area at each. The two conductors appear as the double barrel of a shotgun laid inside of a circular pipe. The percent-fill allowance is reduced to 31% to lower the pressure at the contact points.

For three still smaller conductors installed in the same size conduit, the angle of the pressure points change. One of the conductors lays at the center bottom of the pipe like the eccentric circles of the one conductor installation. The two shotgun barrel conductors (of the other installation) seem to straddle the single conductor, removing the high-pressure or high-stress contact points found in the two-conductor installation. Since the angle of pressure, due to the conductor weight, has changed, the more-than-two percentage is increased to 40% of the available conduit c.s.a.

The notes to Table 1 place both restrictions and exemptions on the intended conduit or tubing installation. Note 2 states that the maximum number of conductors indicated for a given size and type of conduit or tubing, as shown in Tables C1 through C12A (found in Appendix C of the NE Code), apply only to complete raceway systems. There is no percent-fill restrictions applied to short sections of conduit or tubing that are installed solely for the protection of conductors or cables from physical damage. For example, Sec.

336-6(b) requires Romex cable to be installed in conduit or tubing from where it passes through the floor up to six inches above the finished floorline.

Note 3 requires either equipment-grounding conductors or bonding jumpers to be counted in the percent-fill calculations for both conduit and tubing. Whether the equipment-grounding conductor is insulated or bare, and whether it is full size or reduced in wire size as compared to the other circuit conductors, will have a bearing as to which tables will be used in Part A of Chapter 9 and Appendix C to correctly size the conduit.

Note 4 allows conduit or tubing nipples that are two feet or less in length, and run between boxes, panels, or equipment enclosures, to be sized at a 60%-fill allowance instead of the 40%-fill required by Table 1 for "more than two conductors" in the conduit or tubing.

To meet the grouping requirements of Sec. 300-20(a), all circuit conductors, including the phase or ungrounded-circuit conductors, and where used the circuit neutral or other grounded-circuit conductor, and the equipment-grounding conductors must be installed in the same conduit. The conduit fill may be limited to a single circuit installation, or several circuits may be pulled in a single conduit or tubing.

There are two distinct divisions of sizing conduit- or tubing-type raceway. Both are determined by the required conductor c.s.a., versus a percent-fill of the available conduit or tubing c.s.a.:

1. The conductors have the *same* wire size, and the *same* (or physically equivalent) conductor insulation.
2. There will be a *mixture* of conductor sizes and/or insulations.

According to Note 1, if the circuit conductors are all the same wire size and insulation, the correct conduit or tubing size can be determined directly from Tables C1 through C12A without calculation. Which table to use is determined by the type of conduit or tubing, and whether the conductor stranding is concentric or compact.

According to Note 6, if the circuit conductors are of different wire sizes, or the same wire sizes but different types of insulations are to be run in a common (single) conduit or tubing, or if the cross-sectional area according to wire size is not found in the "same size" tables, two steps are required to arrive at the correct conduit or tubing size.

1. The individual conductor c.s.a., whether insulated, covered, or bare, must be determined by either Table 5, 5A, or 8.
2. The individual conductor c.s.a.s must be added together. The conduit or tubing must be sized according to the c.s.a.s listed in Table 4 for the different types of conduit.

3.2 SAME WIRE SIZE—SAME INSULATION

Tables C1 through C12A are constructed as a grid of columns and rows. The rows have two headings. The first heading lists the types of insulations that are physically equivalent (in c.s.a.). The second row heading (in the second column) identifies the wire sizes that correspond to the insulations listed in the first row heading.

For example, Table C1 is used to determine the number of same-size circuit conductors that can be installed in EMT (Electrical Metallic Tubing). In wire sizes No. 14 and No. 12 AWG (second row heading) insulations RHH, RHW, and RHW-2 are physically equivalent. In comparison RH is a thinner insulation for these two wire sizes (first row heading). For any wire size No. 10 AWG or larger, insulations RH, RHH, RHW, and RHW-2 are considered equivalent for a given wire size.

The conduit or tubing trade sizes are listed as individual column headings at the top of the table. The standard size of the conduit or tubing and the number of conductors for a given wire size increases from left to right across the table.

To use the tables, the table must be first identified as to the type of conduit or tubing. Then the type of insulation and wire size must be identified in the row headings. Once the row is established, the maximum number of conductors found in that row must be greater than or equal to the count or actual number of conductors to be installed in the conduit. The column containing this value (count or number) identifies the correct conduit size by column heading.

Example 3–1. Determine the NE Code-minimum size of EMT allowed for the installation of 36 No. 14 AWG THW copper conductors.

For electrical metallic tubing: Ref. App. C, T-C1
Table-listed number of conductors: 39
Tubing size: 1½ inch

EMT is the acronym for electrical metallic tubing, which identifies Table C1 in Appendix C to Chapter 9. THW insulation is found on page 2 of Table C1 in the second insulation row heading. Wire size No. 14 AWG is found within this insulation heading in the fifth row of the wire size column. The number of conductors in this row progresses from left to right as "6, 10, 16, 28, 39, 64, 112, 169, 221, and 282." The count of 39, which is the first value greater than or equal to 36 in the example, corresponds to (fifth column under conduit sizes) a 1½ inch conduit.

Code challenge 3–1. Determine the NE Code-minimum size of IMC allowed for the installation of 24 No. 10 AWG THHW copper conductors.

For intermediate metal conduit: Ref. (C)________________
Table-listed number of conductors: (B) __________
Conduit size: (C) __________ inch

The tables in Appendix C apply to either solid or concentric-stranded, insulated circuit conductors made of either copper, aluminum, or copper-clad aluminum.

Example 3–2. Determine the NE Code-minimum size of rigid schedule 40 PVC conduit allowed for the installation of four 350 kcmil THW aluminum conductors.

For rigid schedule 40 PVC nonmetallic conduit: Ref. App. C, T-C10
Table-listed number of conductors: 5
Conduit size: 3-inch

Code challenge 3–2. Determine the NE Code-minimum size of rigid metal conduit allowed for the installation of three 400 kcmil THHN copper-clad aluminum conductors.

For rigid metallic conduit: Ref. (A)____________________
Table-listed number of conductors: (B) ____________
Conduit size: (C) ____________ inch

As described in T310-13, rubber compound insulations RHH and RHW may be manufactured with an outerjacket or nonmetallic covering. As such, the c.s.a. of a given wire size with outer covering would be greater than the c.s.a. of the same conductor wire size without the outer covering.

Looking at the tables in Appendix C, the number of conductors with either RHH or RHW insulations not footnoted (asterisk) are RHH or RHW conductors with outer coverings. The number of conductors with RHH or RHW insulations that are footnoted are RHH or RHW conductors without outer coverings. The number or count of RHH or RHW insulated conductors without outer covering are normally higher for the same conduit size.

Example 3–3. Determine the NE Code-minimum size of rigid metal conduit allowed for the installation of six 250 kcmil RHW copper conductors *with* outer covering.

For rigid metallic conduit: Ref. App. C, T-C8
Table-listed number of conductors: 6
Conduit size: 3½ inch

Code challenge 3–3. Determine the NE Code-minimum size of rigid schedule 40 PVC conduit allowed for the installation of six 500 kcmil RHW aluminum conductors *with* outer covering.

For rigid schedule 40 PVC nonmetallic conduit: Ref. (A)____________________
Table-listed number of conductors: (B) ____________
Conduit size: (C) ____________ inch

Insulated conductors employing Type RHH or RHW insulation are more commonly produced (manufactured) without an outer covering. As such, Type RHH would not be moisture resistant.

Normally, these conductors will be manufactured with a triple-rated insulation such as RHH-RHW-USE. This allows the insulation to be rated as moisture resistant, but restricted to a 75°C temperature rating (for RHW or USE). The 90°C temperature rating can only be used in dry or damp locations (as RHH—T310-13). Insulation temperature ratings govern the table-rated ampacity of the conductors as determined by the Art. 310 tables.

Example 3–4. Determine the NE Code-minimum size of rigid metal conduit allowed for the installation of six 250 kcmil RHW copper conductors *without* outer covering.

For rigid metallic conduit: Ref. App. C, T-C8
Table-listed number of conductors: 6
Conduit size: 3-inch

Code challenge 3–4. Determine the NE Code-minimum size of rigid PVC (schedule 40—nonmetallic) conduit allowed for the installation of six 500 kcmil RHW aluminum conductors *without* outer covering.

For rigid schedule 40 PVC nonmetallic conduit: Ref. (A)____________________
Table-listed number of conductors: (B) ___________
Conduit size: (C) ___________ inch

Because many insulations have the same mil thickness for the same voltage rating and size conductor, the Appendix C tables can often be used when there will be a mixture of insulations for the same wire size. If the acronyms for the insulations are found in the same row heading, they are considered equal thickness or physically equivalent.

Example 3–5. Determine the NE Code-minimum size of flexible metal conduit allowed for the installation of four No. 4 AWG THHW conductors, six No. 4 AWG THW conductors, and six No. 4 AWG RHW conductors without outer covering.

For flexible metallic conduit: Ref. App. C, T-C3
For the same wire size and equivalent thickness insulation; total number of conductors determination:

$$4 + 6 + 6 = 16 \text{ conductors}$$

Table-listed number of conductors: 20
Conduit size: 2½ inch

Code challenge 3–5. Determine the NE Code-minimum size of liquid-tight flexible metal conduit allowed for the installation of eight No. 12 THW, and ten No. 12 THHW copper conductors.

For liquid-tight flexible metallic conduit: Ref. (A)____________________
For the same wire size and equivalent thickness insulation; total number of conductors determination:

___________ + ___________ = (B) ___________ conductors

Table-listed number of conductors: (C) ___________
Conduit size: (D) ___________ inch

If the conductors are compact stranded, with the same wire size and same type of insulation, the A tables of C1 through C12 must be used. The A tables are the last page of each respective table.

The definition of compact-stranded conductors is found as a footnote to Table C1A. A stranded conductor of several smaller concentric (round) conductors are drawn with a limited twist per inch to spiral bind these individual conductors into the required wire size (either AWG or kcmil). For compact stranding, this originally concentric-stranded conductor is drawn through a die process that flattens the peak of the curved surface on the outer strands. The compact-stranded wire is still round but the voids (interstices) between the strands have been filled. The outer surface of the conductor has the appearance of a commutator on the armature of a dc motor.

Compact stranding provides greater flexibility of the wire along its length: The melted and then extruded insulation is no longer trapped between two adjacent concentric-stranded conductors.

For ac circuits, especially at higher frequencies, the electrical energy is concentrated along the surface of the conductor. The transfer of this energy is referred to as the skin effect. For some applications, this flattened outer surface lowers the amount of resistance along the length of the wire. In other words the rated-load ampacity of a No. 4 AWG compact-stranded conductor would be greater than a No. 4 AWG concentric-stranded circuit conductor for the same type of insulation at higher frequencies.

Example 3–6. Determine the NE Code-minimum size of IMC allowed for the installation of ten No. 4 AWG compact-stranded XHHW conductors.

For intermediate metallic conduit: Ref. App. C, T-C4A

Table-listed number of conductors: 12

Conduit size: 1½ inch Ref. App. C, T-C4A

Code challenge 3–6. Determine the NE Code-minimum size of EMT conduit allowed for the installation six No. 1 AWG compacted-stranded THW conductors.

For electrical metallic tubing: Ref. (A)____________

Table-listed number of conductors: (B) ____________

Tubing size: (C) ____________ inch

3.3 MIXTURE OF WIRE SIZES AND/OR INSULATIONS

Any mixture of conductor sizes, and usually any mixture of insulations, prohibits the use of Tables C1 through C12A in Appendix C. Instead, either Table 5, 5A, and/or Table 8 of Chapter 9 must be used to determine the individual cross-sectional areas (c.s.a.s) of the conductors by wire size and type insulation. Once the c.s.a.s have been determined, they are simply added up for a total required conductor cross-sectional area. The total required area is taken into Table 4 to determine the NE Code-minimum conduit or tubing size.

Table 5 lists both the diameter and c.s.a of concentric-stranded insulated conductors. As with the Appendix C tables, the row is selected by both the wire size and the type of insulation. One of two columns must be selected according to either diameter or c.s.a.

Equivalent insulations, zoned by bold row lines, are listed in column fashion down the row headings. For example, on page 1 of Table 5, insulations RH, RHH, RHW, and RHW-2 (six-row heading) have the same diameter and cross-sectional area values for wire sizes No. 10 AWG through 2,000 kcmil. In wire sizes No. 14 and No. 12 AWG, conductors insulated with RH (second-row heading) are physically smaller than wire sizes No. 14 and No. 12 AWG conductors insulated (third-row heading) with either RHH, RHW, or RHW-2.

Example 3–7. Determine the cross-sectional area of a No. 14 AWG THWN copper conductor.

Concentric-stranded conductor cross-sectional areas: Ref. Ch. 9, T-5

No. 14 AWG THWN c.s.a. = 0.0097 square inches

Code challenge 3–7. Determine the cross-sectional area of a No. 10 AWG XHHW-2 aluminum conductor.

Concentric-stranded conductor cross-sectional areas: Ref. (A) ________________

No. 10 AWG XHHW-2 c.s.a. = (B) ____________ square inches

As discussed in sizing conduit or tubing for the same wire size, RHH and RHW insulations may be manufactured with an outerjacket or nonmetallic covering. As such, the c.s.a. or diameter of a given wire size with outer covering would be greater than the c.s.a. or diameter of the same wire size without the outer covering.

In Table 5, the c.s.a. or diameter for conductors with either RHH or RHW insulations not footnoted (asterisk) are the areas for circuit conductors with outer coverings. The c.s.a. for conductors with RHH or RHW insulations that are footnoted are the areas for circuit conductors without outer coverings.

Example 3–8. Determine the cross-sectional area of a No. 14 AWG RHH conductor with outer covering, and of a No. 14 AWG RHH conductor without outer covering.

Concentric-stranded conductor cross-sectional areas: Ref. Ch. 9, T-5

No. 14 AWG RHH *with* outer covering: c.s.a. = 0.0293 square inch

No. 14 AWG RHH *without* outer covering: c.s.a. = 0.0209 square inch

Code challenge 3–8. Determine the cross-sectional area of a No. 12 AWG RHW conductor with outer covering, and of a No. 12 AWG RHW conductor without outer covering.

Concentric-stranded conductor cross-sectional areas: Ref. (A) ________________

No. 12 AWG RHH *with* outer covering: c.s.a. = (B) ____________ square inch

No. 12 AWG RHH *without* outer covering: c.s.a. = (C) ________________square inch

When a variety of conductors with mixtures of either wire sizes and/or insulations are to be run in a single conduit or tubing, the two-step operation is best computed as follows:

First, determine the total conductor c.s.a. required.

Count	Wire size and insulation	c.s.a. for one (square inches)	Extended c.s.a. (square inches)
Given	Given	Ch. 9, T-5	Product of "Count"
—	—	or T-5A	times "c.s.a. for one"
—		—	—
			Total conductor c.s.a.

Second, find the cross-sectional area listed in Chapter 9, Table 4 according to the type of conduit or tubing and the number of conductors (usually "Over 2 Wires, 40%"). The available conductor c.s.a. value found in the table, according to type of conduit or tubing and nominal size must be greater than or equal to the computed conductor c.s.a. value in step 1.

Example 3–9. Determine the NE Code-minimum size of rigid metal conduit allowed for the installation of six No. 12 AWG THW, four No. 8 AWG TW, eight No. 4 AWG THHN, and two No. 1/0 AWG RH copper conductors.

Total conductor c.s.a. computation: Ref. Ch. 9, T-5

Count	Wire size and insulation	c.s.a. for one (square inches)	Extended c.s.a. (square inches)
6	No. 12 AWG THW	0.0260	0.1560
4	No. 8 AWG TW	0.0437	0.1748
8	No. 4 AWG THHN	0.0824	0.6592
2	No. 1/0 AWG RH	0.3039	0.6078
		Total conductor c.s.a.:	1.5978

For rigid metallic conduit: Ref. Ch. 9, T-4

Table-listed conductor c.s.a.: 1.946 square inches

Conduit size: 2½ inch

Code challenge 3–9. Determine the NE Code-minimum size of EMT allowed for the installation of four No. 10 AWG THW, four No. 6 AWG THWN, ten No. 4 AWG TW, and four No. 2/0 AWG RHW without outer covering copper conductors.

Total conductor c.s.a. computation: Ref. (A) ________________

Count	Wire size and insulation	c.s.a. for one (square inches)	Extended c.s.a. (square inches)
4	No. 10 AWG THW	(B) ____________	____________
4	No. 6 AWG THWN	(C) ____________	____________
10	No. 4 AWG TW	(D) ____________	____________
4	No. 2/0 AWG RHW (w/o)	(E) ____________	____________
		Total conductor c.s.a.: (F)	____________

For electrical metallic tubing: Ref. (G) ____________

Table-listed conductor c.s.a.: (H) ____________ square inches

Conduit size: (I) ____________ inch

If aluminum building wires or other compact-stranded conductors are to be installed in the conduit or tubing, the c.s.a. or diameters of these conductors will be found in Table 5A of Chapter 9. Once the total conductor cross-sectional area is determined for compact-stranded conductors, the conduit or tubing raceway is sized in the same manner as for concentric-stranded conductors using Table 4 according to the type of conduit or tubing to be installed.

Example 3–10. Determine the NE Code-minimum size of IMC allowed for the installation of six No. 1/0 AWG THW and four No. 4/0 XHHW compact-stranded aluminum building wires.

Total conductor c.s.a. computation: Ref. Ch. 9, T-5A

Count	Wire size and insulation	c.s.a. for one (square inches)	Extended c.s.a. (square inches)
6	No. 1/0 AWG THW	0.1963	1.1778
4	No. 4/0 AWG XHHW	0.2733	1.0932
		Total conductor c.s.a.:	2.2710

For imtermediate metal conduit: Ref. Ch. 9, T-4

Table-listed conductor c.s.a.: 3.169 square inches

Conduit size: 3-inch

Code challenge 3–10. Determine the NE Code-minimum size of rigid schedule 40 PVC conduit allowed for the installation of four 250 kcmil THHN and eight 500 THHW compact-stranded aluminum building wires.

Total conductor c.s.a. computation: Ref. (A) ____________

Count	Wire size and insulation	c.s.a. for one (square inches)	Extended c.s.a. (square inches)
4	250 kcmil THHN	(B) ____________	____________
8	500 kcmil THHW	(C) ____________	____________
		Total conductor c.s.a.: (D)	____________

For rigid schedule 40 PVC nonmetallic conduit: Ref. (E) ____________

Table-listed conductor c.s.a.: (F) ____________ square inches

Conduit size: (G) ____________ inch

If bare conductors are to be installed in conduit or tubing with insulated conductors, the cross-sectional area of concentric-stranded bare conductors must be determined by wire size as listed in Table 8 of Chapter 9.

Table 8 lists conductor properties and has two main column headings, "Conductors" and "DC Resistance at 75°C (167°F)." Under the main heading of "Conductors" are found two subheadings, "Stranding" and "Overall." Under the subheading of "Overall" are two subheadings, "Diam. In." (Diameter in Inches) and "Area In^2" (Area in Square Inches or c.s.a.). "Conductor Overall Area in Square Inches" identifies the sixth column of Table 8. The first column in Table 8 lists the row headings according to wire size (sizes No. 18 through No. 4/0 AWG, and 250 through 2,000 kcmil). The cross-sectional area of the bare conductor for a given wire size is found where the selected row and sixth column intersect.

Two cross-sectional area values are listed for wire sizes No. 18 AWG through No. 8 AWG bare conductors. The reason is found under the conductor subheading of "Stranding." The term stranding means the wire can be a single, concentric solid conductor, or the wire can be made up of several smaller-size concentric solid conductors that are spiral-wrapped around each other to yield the given wire size (in concentric stranding).

If the bare wire is a concentric solid conductor, the cross-sectional area found in the row identified by the given wire size and "1" in the stranding column is the correct area value. If the bare wire is a concentric-stranded conductor, the cross-sectional area found in the row identified by the given wire size and "7" in the stranding column is the correct area value.

Example 3–11. Determine the cross-sectional area of the following bare conductors:

Bare concentric-stranded conductor cross-sectional areas:

		Ref. Ch.9, T-8
No. 14 AWG solid:	c.s.a. = (row 5, column 6)	0.003 square inch
No. 14 AWG stranded:	c.s.a. = (row 6, column 6)	0.004 square inch
No. 8 AWG solid:	c.s.a. = (row 11, column 6)	0.013 square inch
No. 8 AWG stranded:	c.s.a. = (row 12, column 6)	0.017 square inch
No. 2 AWG:	c.s.a. = (row 16, column 6)	0.067 square inch
350 kcmil:	c.s.a. = (row 24, column 6)	0.364 square inch

The cross-sectional area for bare conductors in wire sizes No. 6 AWG and larger is listed in Table 8 as stranded conductor areas only. It should be noted that wire sizes No. 6 and No. 4 AWG solid are available (although not listed), and often used in the construction of the service grounding-electrode system.

Code challenge 3–11. Determine the cross-sectional area of the following bare conductors:

Bare concentric-stranded conductor cross-sectional areas:

Ref. (A) ____________________

No. 18 AWG, solid: c.s.a. = (B) ____________ square inch
No. 10 AWG, stranded: c.s.a. = (C) ____________ square inch
No. 1/0 AWG: c.s.a. = (D) ____________ square inch
250 kcmil: c.s.a. = (E) ____________ square inch
1,000 kcmil: c.s.a. = (F) ____________ square inch

The circuit conductors for a multiwire feeder circuit may be of the same wire size and type insulation but consists of different wire sizes. The feeder equipment-grounding conductor is normally smaller in wire size than any of the circuit conductors. The feeder grounded neutral conductor is normally smaller than any of the ungrounded phase conductors.

Example 3–12. Determine the NE Code-minimum size of rigid schedule 80 PVC conduit allowed for the installation of the following: Three 350 kcmil THW, one No. 4/0 AWG THW, and one bare No. 2 AWG copper conductor.

Total conductor c.s.a. computation:

Ref. (insulated) Ch. 9, T 5 Ref. (bare) Ch. 9, T-8

Count	Wire size and insulation	c.s.a. for one (square inches)	Extended c.s.a. (square inches)
3	350 kcmil THW	0.5958	1.7874
1	No. 4/0 THW	0.3718	0.3718
1	No. 2 bare	0.0670	0.0670
		Total conductor c.s.a.:	2.2262

For rigid schedule 80 PVC nonmetallic conduit: Ref. Ch. 9, T-4
Table-listed conductor c.s.a.: 2.577 square inches
Conduit size: 3-inch

Code challenge 3–12. Determine the NE Code-minimum size of rigid steel conduit allowed for the installation of the following: Three 500 kcmil THHN, one 250 kcmil THHN, and one No. 1/0 AWG bare copper conductor.

Total conductor c.s.a. computation:
Ref. (insulated) (A) ____________ Ref. (bare) (B) ____________

Count	Wire size and insulation	c.s.a. for one (square inches)	Extended c.s.a. (square inches)
3	500 kcmil THHN	(C) ____________	____________
1	250 kcmil THHN	(D) ____________	____________
1	No. 1/0 AWG bare	(E) ____________	____________
		Total conductor c.s.a.: (F)	____________

For rigid metallic conduit: Ref. (G)____________________

Table-listed conductor c.s.a.: (H) ____________ square inches

Conduit size: (I) ____________ inch

As a rule, the sizing of conduit or tubing is in conjunction with sizing the conductors for a given load. Sizing conductors according to ampacity of the load and the prevailing conditions of the conductor installation were discussed in Chapter 2 of this text.

Example 3–13. Determine the NE Code-minimum size of THHN copper conductors and EMT raceway allowed to supply a 125-amp, 2-wire, 240-volt, 1-phase continuous load. The EMT conduit will pass through an ambient air temperature of 94°F, and contain a total of six conductors. Three of the other circuit conductors will be No. 4 AWG THW copper, which supply a 3-phase load. The sixth wire will be a bare No. 6 AWG copper equipment-grounding conductor.

For load requirements:

Higher temperature derate: 0.96 Ref. T310-16, CF

Total number of circuit conductors: 5 Ref. T310-16, N-11

More-than-three derate: 80% Ref. T310-16, N-8(a) Tbl

Continuous-load overrate: 125% Ref. 220-3(a)

Temperature-compensated table-rated ampacity determination:

Overrate computation for higher ambient temperature:

$$125 \text{ amps} \div 0.96 = 130.2 \text{ amps}$$

Overrate computation for more-than-three circuit conductors:

$$130.2 \text{ amps} \div 0.80 = 162.8 \text{ amps}$$

Table-rated ampacity: 170 amps

Wire size: No. 1/0 AWG THHN copper Ref. T310-16

Wire terminal temperature limitations: 75°C Ref. 110-14(c)(2), Ex. 1

Overrate computation for continuous load: Ref. 220-3(a)

$$1.25 \times 125 \text{ amps} = 156.3 \text{ amps}$$

Table-rated ampacity: 175 amps

Wire size: No. 2/0 AWG THHN copper Ref. T310-16

NE Code-minimum: Compare for larger Ref. 310-15(c)

Wire size: No. 2/0 AWG THHN copper Ref. T310-16

For raceway requirements: Ref. 348-6

Total conductor c.s.a. computation:

Ref. (insulated) Ch. 9, T 5 Ref. (bare) Ch. 9, T-8

Count	Wire size and insulation	c.s.a. for one (square inches)	Extended c.s.a. (square inches)
2	No. 2/0 AWG THHN	0.2223	0.4446
3	No. 4 AWG THW	0.0973	0.2919
1	No. 6 AWG bare	0.0270	0.0270
		Total conductor c.s.a.:	0.7635

For electrical metallic tubing: Ref. Ch. 9, T-4
Table-listed conductor c.s.a.: 0.814 square inches
Conduit size: 1½-inch

Code challenge 3–13. Determine the NE Code-minimum size of XHHW-2 copper conductors and IMC raceway allowed to supply a 110-amp, 480-volt, 3-phase, 3-wire, continuous load. The conduit will pass through an ambient air temperature of 105°F, and contain a total of 7 conductors. Three of the other circuit conductors will be No. 1/0 AWG THW copper, which supply another 3-phase load. The seventh wire will be a No. 3 AWG bare copper equipment-grounding conductor.

For load requirements:
Higher temperature derate: (A) ____________ Ref. (B) ________________
Total number of circuit conductors: (C) ____________ Ref. (D) ________________
More-than-three derate: (E) ____________ % Ref. (F) ________________
Continuous-load overrate: (G) ____________ % Ref. (H) ________________
Temperature-compensated table-rated ampacity determination:
Overrate computation for higher ambient temperature:

____________ amps ÷ ____________ = (I) ____________ amps

Overrate computation for more-than-three circuit conductors:

____________ amps ÷ ____________ = (J) ____________ amps

Table-rated ampacity: (K) ____________ amps
Wire size: (L) ____________ XHHW-2 copper Ref. (M) ____________
Wire terminal temperature limitations: (N) ____________ °C Ref. (O) ____________
Overrate computation for continuous load:

____________ × ____________ amps = (P) ____________ amps

Table-rated ampacity: (Q) ____________ amps
Wire size: (R) ____________ XHHW-2 copper Ref. (S) ________________
NE Code-minimum: Compare for (T) ____________ Ref. (U) ________________
Wire size: (V) ____________ XHHW-2 copper
For raceway requirements: Ref. (W) ________________

Total conductor c.s.a. computation:

Ref. (insulated) (X) ____________ Ref. (bare) (Y) ________________

Count	Wire size and insulation	c.s.a. for one (square inches)	Extended c.s.a. (square inches)
3	(Z) ____________ XHHW-2	(1A) ____________	____________
3	No. 1/0 AWG THW	(1B) ____________	____________
1	No. 3 AWG bare	(1C) ____________	____________
		Total conductor c.s.a.: (1D)	____________

For intermediate metallic conduit: Ref. (1E) ________________

Table-listed conductor c.s.a.: (1F) ____________ square inches

Conduit size: (1G) ____________ inch

3.4 SUPPORT OF CABLE, CONDUIT, AND TUBING

Multiconductor cable wiring methods, such as BX cable (Art. 333), Romex cable (Art. 336), SE cable (Art. 338), or UF cable (Art. 339) are required by Secs. 333-7, 336-18, 338-4 and 339-3(a)(4) to be supported at intervals not to exceed 4½ feet, and within 12 inches of all terminations (Fig. 3-3).

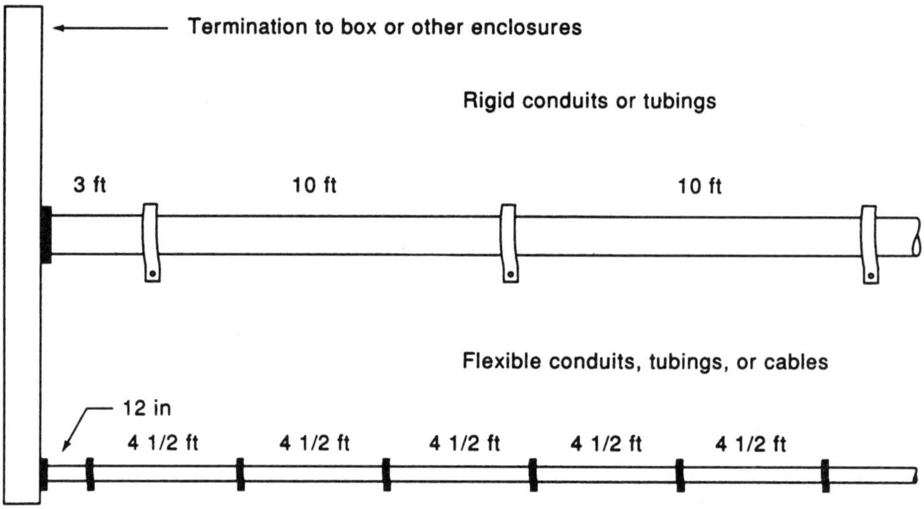

Figure 3-3 Cables, conduits, or tubings must be supported according to guidelines specified within the National Electrical Code.

Metal conduits, which include EMT (Sec. 348-12), IMC (Sec. 345-12), and Rigid (Sec. 346-12), are generally supported for each 10 feet of length, and within 3 feet of all terminations.

When the mounting for junction boxes or other terminations for metal conduits is not within 3 feet of another structural member (support other than what the box is mounted on), the 3-foot interval for support of the conduit or tubing can be increased to a maximum of 5 feet. Other exceptions fall under the specific sections or articles.

Nonmetallic conduit, which is usually PVC, must be supported within 3 feet of all terminations (Sec. 347-8). The linear support of rigid nonmetallic conduit is governed by the nominal size of the conduit, and the maximum intervals specified in T347-8.

Flexible metallic conduit (Sec. 350-18) and liquid-tight flexible metallic conduit (Sec. 351-8) must be supported at intervals not to exceed 4½ feet, and within 12 inches of all terminations. Liquid-tight flexible nonmetallic conduit (Sec. 351-27) must be supported at intervals not to exceed 3 feet, and within 12 inches of all terminations.

Flexible metallic tubing (Art. 349) has greater flexibility than flexible metallic conduit. Flexible metallic tubing is designed mainly for fixture whips [Sec. 349-10(a) Ex. No. 2, and Sec. 410-67(c)]. The construction of flexible metallic tubing (Sec. 349-1) makes it liquidtight without a nonmetallic outer jacket; whereas flexible metallic conduit [Sec. 350-5(1)] has limited approval for wet locations: The contained conductors must have a wet (W) designation in their insulation rating, and the flexible metallic conduit must be installed to prevent water entry (by angle or location) into either the tubing or connected enclosures.

The support intervals of flexible metallic tubing are not specified in Art. 349 since support along its length, as with flexible cords (Art. 400), would defeat their reason for being.

The bending radius of the flexible metallic tubing is the most critical concern of its installation. Section 349-20 determines the minimum bending radii according to the nominal trade size and whether the installation will be a "fixed bend," or if "infrequent flexing," due to the service requirements of the connected load equipment, is required. Flexible metallic tubing must always be terminated in approved fittings (Sec. 349-18), since this is usually its sole means of support.

3.5 CABLE TRAYS

Cable trays are not recognized as raceway by the NE Code. Instead, cable tray is a means of lateral or vertical support of either multiconductor or single cables. Many cable trays have the appearance of an aluminum extension ladder hung on threaded rod or other means of support run across an open ceiling of a structure. The conductors are not totally enclosed (this is what disqualifies the cable tray as a raceway). Protection from physical damage for the circuit conductors laid inside a cable tray is provided by the relative location of the assembly—high in the ceiling area.

Article 318 governs the installation and use of cable tray. Cable tray systems can be either metal or nonmetal, and can be made of ladders, troughs, channels, solid bottom trays, or other similar structures. Section 318-6(a) requires the cable tray to be installed as a complete system before any cables are installed. The cable tray layout must afford adequate access for installing and maintaining the cables.

Section 318-8 governs the installation of cables in cable trays. Cables can be spliced in the tray provided the splices are accessible and do not protrude above the railing of the tray [Sec. 318-8(a)]. The individual or multiconductor cables must be fastened to each "rung" or transverse member of the cable tray when vertical runs are installed [Sec. 318-8(b)].

When parallel conductors are run in cable trays as per Sec. 310-4, the single-conductor cables must be installed in groups consisting of no more than one conductor per phase and the neutral where used [Sec. 318-8(d)]. Single conductors must be bound in circuit groups to prevent excessive movement if and when fault overcurrent should occur. Section 300-20(a), which requires this conductor grouping to prevent inductive heating of metal enclosures that house the conductors, also requires the equipment-grounding conductor (where used) to be contained in the common group.

The type of installation for conductors in cable tray falls within two basic categories: Either multiconductor cables or single-conductor cables. The required size of cable tray for multiconductor cables is governed by Section and Table 318-9. The required size of cable tray for single-conductor cables is governed by Section and Table 318-10. A study of these two sections indicates that single-conductor cables and multiconductor cables are not allowed to be mixed in a given tray run.

To effectively size cable tray for multiconductor cables, a manufacturer's specification book for both the cable tray and the multiconductor cable are required. The No. 4/0 AWG hinge point described in Sec. 310-9(a), Conditions (1), (2), and (3) is the actual size of the conductors contained in the multiconductor cable.

The overall dimensions of either diameter or cross-sectional area of the cable must allow for the individual conductor insulations and for the outer insulation and/or covering of the multiconductor cable itself. The NE Code does not list these values in table form since the variety of market-available multiconductor cables is so vast.

In this text, sizing cable tray will be limited to individual, insulated, single-conductor cables. The tables in Chapter 9 list both the cross-sectional area and the diameter for these conductors. The 1,000 kcmil hinge point described in Sec. 318-10(a), Conditions (1), (2), and (3) is used in the same manner as the No. 4/0 AWG hinge point for multiconductor cable.

When all the single conductors to be installed in a cable tray are wire size 1,000 kcmil or larger, Sec. 318-10(a)(1) stipulates that all the diameters of the conductors laid side-by-side (Fig. 3-4) cannot exceed the manufactured tray width.

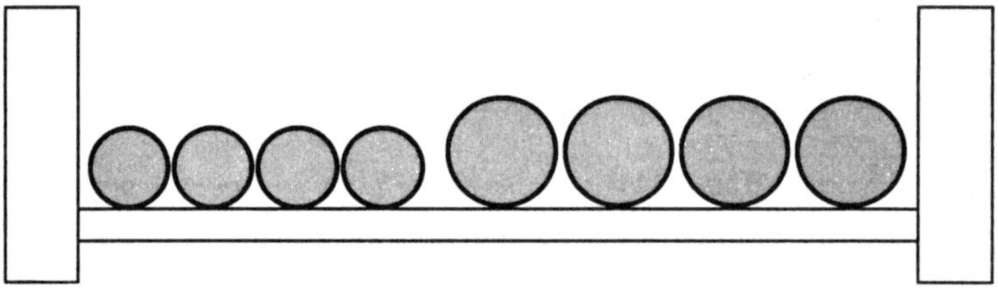

Figure 3-4 Section 318-10(a)(1) requires single conductors wire size 1,000 kcmil or larger to be laid side-by-side in the cable tray.

The trade diameter and cross-sectional area of conductors are listed in Tables 5, 5A, and 8 of Chapter 9, according to the type of insulation and wire size. In Table 5, the type of insulation (column 1) and the wire size (column 2) are both row headings. The diameter of the conductor is listed as its approximate diameter in inches in the third column. The c.s.a. of the same conductor is listed as square-inch area in the fourth column.

In Table 5A, the wire size identifies only the row. The diameters of compact-stranded insulated aluminum conductors (building wires) are listed by wire size as approximate diameters in inches on the left side of the insulation columns. The diameters of compact-stranded bare aluminum conductors are listed by wire size on the right side of the bare conductor column.

The cross-sectional areas of insulated compact-stranded conductors are listed in the right column under the types of insulation. The cross-sectional area of a bare compact-stranded conductor can be computed by the formula "πD^2 divided by 4" [$(3.1416 \times D^2) \div 4$ – area of a circle].

To determine the required cable tray width for the larger cables, first determine the total conductor diameter required:

Count	Wire size and insulation	Diameter for one (inches)	Extended diameter (inches)
Given	Given	Tables 5 through	Product of "Count"
—	—	8 of Chapter 9	Times "Diameter
—	—	—	for one"
			Total conductor diameter

Second, the inside tray width of the cable, as listed in Table 318-10 must be greater than or equal to the required (computed) tray width.

Example 3–14. Determine the NE Code-minimum cable tray width allowed for the installation of six 1,000 kcmil THHN, eight 1,500 kcmil THW, and four 1,750 kcmil RHW single-conductor copper cables.

When all the circuit conductors are 1,000 kcmil or larger, the cable tray width must at least equal the sum of the single-layer diameters of all the single-conductor cables.

Ref. 318-10(a)(1)

Total conductor diameter computation: Ref. Ch. 9, T-5

Count	Wire size and insulation	Diameter for one (inches)	Extended diameter (inches)
6	1,000 kcmil THHN	1.310	7.860
8	1,500 kcmil THW	1.662	13.296
4	1,750 kcmil RHW	1.776	7.104
		Total (single-layer) diameter in inches:	28.260

Table-listed cable tray width: 30 inches Ref. T318-10

Code challenge 3–14. Determine the NE Code-minimum cable tray width allowed for the installation of nine 1,250 kcmil THW and nine 1,500 kcmil TW single-conductor copper cables.

When all the circuit conductors are (A) ___________ kcmil or (B) ___________ , the cable tray width must at least equal the sum of the single-layer diameters of all the single-conductor cables.

Ref. (C) ___________

Total conductor diameter computation: Ref. (D) ___________

Count	Wire size and insulation	Diameter for one (inches)	Extended diameter (inches)
9	1,250 kcmil THW	(E) ___________	___________
9	1,500 kcmil RHW	(F) ___________	___________
		Total (single-layer) diameter in inches: (G)	___________

Table-listed cable tray width: (H) ___________ inches Ref. (I) ___________

When the wire size of all the single-conductor cables to be installed in a cable tray are within a range of 250 to 1,000 kcmil, the required cable tray width is based on the total cross-sectional area of the conductors. Section 318-10(a)(2) will allow the total conductor c.s.a. to be less than or equal to the area values specified in T318-10, Column 1 according to cable tray width.

To determine the required cable tray width for circuit conductors within this range of single-conductor cable wire sizes, first determine the total conductor cross-sectional area.

Count	Wire size and insulation	c.s.a. for one (square inches)	Extended c.s.a. (square inches)
Given	Given	Tables 5 through	Product of "Count"
—	—	8 of Chapter 9	times "c.s.a for
—	—	—	one"
			Total conductor c.s.a.

Second, the c.s.a. value listed in Column 1 of Table 318-10 corresponds to the required tray width. The table-listed value must be greater than or equal to the computed conductor c.s.a. value.

Example 3–15. Determine the NE Code-minimum cable tray width allowed for the installation of six 250 kcmil THW, six 300 kcmil THW, and twelve 350 kcmil THWN single-conductor copper cables.

When the wire size of all the circuit conductors fall within a range of 250 to 1,000 kcmil, the total computed conductor c.s.a. cannot exceed the available conductor c.s.a. listed in Column 1 of T318-10 by tray width.

Ref. 318-10(a)(2)

Total conductor c.s.a. computation: Ref. Ch. 9, T-5

Count	Wire size and insulation	c.s.a. for one (square inches)	Extended c.s.a. (square inches)
6	250 kcmil THW	0.4596	2.7576
6	300 kcmil THW	0.5281	3.1686
12	350 kcmil THWN	0.5242	6.2904
		Total conductor cross-sectional area:	12.2166

Table-listed conductor c.s.a.: 13.0 square inches

Required cable tray width: 12-inch Ref. T318-10, Col. 1

The maximum allowed cross-sectional area is not based on the conductors laid in single layers. Instead, the single conductor cables can be bundled or laid on top of each other (Fig. 3-5). Section 300-20(a) would require that all the phase conductors and neutral contained in a set of feeders be bundled or grouped together when installed in a cable tray. For single-conductor cables larger than 1,000 kcmil, the cables in a set of feeders would have to be laid side by side.

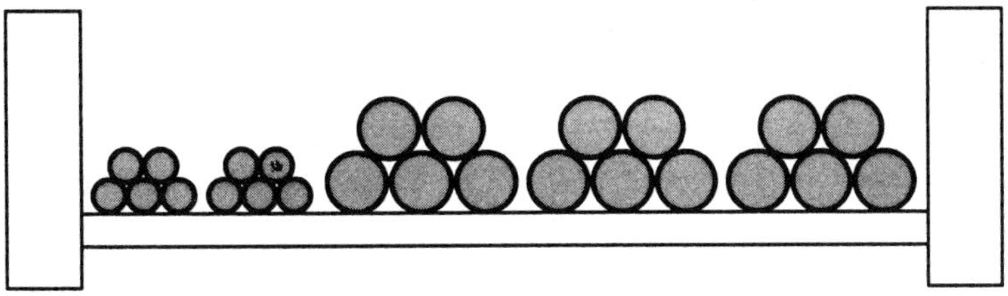

Figure 3-5 Section 318-10(a)(2) will allow single conductors smaller than 1,000 kcmil to be stacked or bundled in the cable tray.

Code challenge 3–15. Determine the NE Code-minimum cable tray width allowed for the installation of twelve 250 kcmil RHW, fifteen 300 kcmil RHW, nine 350 kcmil THW, and six 500 kcmil THWN single-conductor copper cables.

When the wire size of all the circuit conductors fall within a range of (A) ____________ to (B) ____________ kcmil, the total computed conductor c.s.a. cannot exceed the available conductor c.s.a listed in Column 1 of T318-10 by tray width.

Ref. (C)____________

Total conductor c.s.a. computation: Ref. (D)____________

Count	Wire size and insulation	c.s.a. for one (square inches)	Extended c.s.a. (square inches)
12	250 kcmil RHW	(E) ____________	____________
15	300 kcmil RHW	(F) ____________	____________
9	350 kcmil THW	(G) ____________	____________
6	500 kcmil THWN	(H) ____________	____________
		Total conductor cross-sectional area: (I)	____________

Table-listed conductor c.s.a.: (J) ____________ square inches
Required cable tray width: (K) ____________ -inch Ref. (L) ________________

Condition (3) of Sec. 318-10(a) creates a dual installation of single-conductor cables in a single-tray run (Fig. 3-6). Cables wire size 1,000 kcmil or larger must be laid side-by-side in a single layer with no other cables placed on top of them. Cables that are larger than or equal to 250 kcmil, but less than 1,000 kcmil can be bundled or stacked on top of each other beside the larger single-layered cables. The stack or bundle height of the smaller cables (in the rest of the cable tray) cannot exceed the rail height.

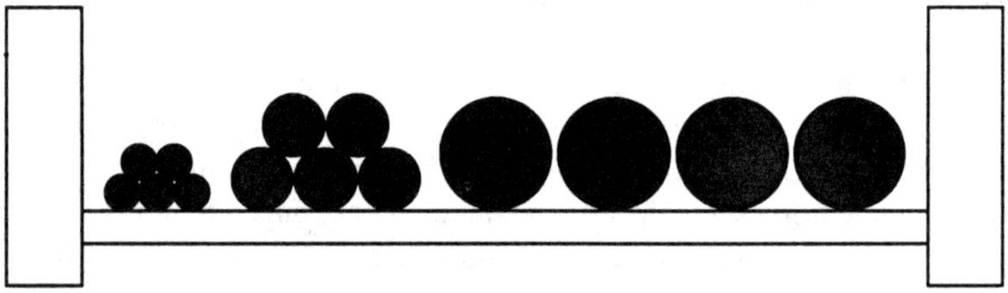

Figure 3-6 Section 318-10(a)(3) will allow the smaller conductors to be bundled separately from the larger conductors that must be laid side-by-side.

To size the cable tray correctly, two subtotals must be determined: One for the diameters of the conductors that are greater than or equal to 1,000 kcmil, and the other, for the cross-sectional areas of the conductors smaller than 1,000 kcmil.

The "Sd factor" found in Column 2 is used to modify the diameters for the larger cables into area values. This is accomplished by multiplying the Sd factor of 1.1 "inches" times the larger cable diameters. Inches times inches will yield a square-inch area value (actually an adjusted area), which can be added to the cross-sectional area of the smaller conductors. The correct cable tray width is determined by comparing the total computed conductor c.s.a. to the conductor c.s.a.s listed by tray width in Column 2 of T318-10.

The available conductor cross-sectional area listed in the table according to tray width must be greater than or equal to the calculated total cross-sectional area of the conductors.

Example 3–16. Determine the NE Code-minimum cable tray width allowed for the installation of four 250 kcmil TW, six 500 kcmil THHN, four 750 kcmil THHN, eight 1,000 kcmil RHW, and six 1,250 kcmil THW single-conductor copper cables.

For circuit conductors 1,000 kcmil or larger, that portion of the cable tray c.s.a. must be based on total conductor diameter adjusted to c.s.a.

Ref. 318-10(a)(3)

Total conductor diameter computation: Ref. Ch. 9, T-5

Count	Wire size and insulation	Diameter for one (inches)	Extended diameter (inches)
8	1,000 kcmil RHW	1.372	10.976
6	1,250 kcmil THW	1.539	9.234
		Total (single-layer) diameter:	20.210

Sd correction-factor computation: Ref. T318-10, Col. 2

1.1 inches × 20.21 inches = 22.231 square inches

For the circuit conductors greater than 250 kcmil, but less than 1,000 kcmil, the total conductor c.s.a. must be computed.

Ref. 318-10(a)(3)

Total conductor c.s.a. computation: Ref. Ch. 9, T-5

Count	Wire size and insulation	c.s.a. for one (square inches)	Extended c.s.a. (square inches)
4	250 kcmil TW	0.4596	1.8384
6	500 kcmil THHN	0.7073	4.2438
4	750 kcmil THHN	1.0496	4.1984
		Total conductor cross-sectional area:	10.2806

For the combined cross-sectional area of all the circuit conductors:

22.231 + 10.2806 = 32.5116 square inches

Table-listed conductor c.s.a.: 39.0 square inches
Required cable tray width: 36-inch Ref. T318-10, Col. 2

Code challenge 3–16. Determine the NE Code-minimum cable tray width allowed for the installation of eight 350 kcmil THWN, four 500 kcmil THW, eight 750 kcmil THHN, four 1,000 kcmil THHN, and four 1,250 kcmil RHW single-conductor copper cables.

For circuit conductors (A) ___________ kcmil or (B) ___________, that portion of the cable tray c.s.a. must be based on total conductor diameter adjusted to c.s.a.

Ref. (C)___________

Total conductor diameter computation: Ref. (D)___________

Count	Wire size and insulation	Diameter for one (inches)	Extended diameter (inches)
4	1,000 kcmil THHN	(E) ___________	___________
4	1,250 kcmil RHW	(F) ___________	___________
		Total (single-layer) diameter: (G)	___________

Sd correction-factor computation: Ref. (H)___________

(I) ___________ inches × ___________ inches = (J) ___________ square inches

For the circuit conductors greater than (K) ___________ kcmil, but less than (L) ___________ kcmil, the total conductor c.s.a. must be computed.

Ref. (M) ___________

Total conductor c.s.a. computation: Ref. (N)___________

Count	Wire size and insulation	c.s.a. for one (square inches)	Extended c.s.a. (square inches)
8	350 kcmil THWN	(O) ____________	____________
4	500 kcmil THW	(P) ____________	____________
8	750 kcmil THHN	(Q) ____________	____________
		Total conductor cross-sectional area: (R)	____________

For the combined cross-sectional area of all the circuit conductors:

____________ + ____________ = (S) ____________ square inches

Table-listed conductor c.s.a.: (T) ____________ square inches
Required cable tray width: (U) ____________ inch Ref. (V) ____________

If the cable tray will contain single-conductor cables within the range of wire size No. 1/0 AWG through No. 4/0 AWG, Sec. 318-10(a)(4) requires the cable tray to be sized in the same manner as cables wire size 1,000 kcmil or larger [Sec. 318-10(a)(1)]: total diameter of the cables laid side-by-side cannot exceed the tray width. By exception to Sec. 318-8(e), the smaller conductors (No. 1/0 through 4/0 AWG) can be bound and bundled in accordance with Secs. 318-11(b)(4) and 300-20(a).

Section 318-11 governs the ampacity of cables rated at 2,000 volts or less which are installed in cable trays. Section 318-11(b) puts restrictions or imposed derates on the ampacities of single conductor cables that are "triplexed" (group of three conductors spiral-wrapped around each other) or "quadraplexed" (group of four spiral-wrapped around each other) when installed in cable trays. Ampacity restrictions or derates are also required when the cable tray will have a solid bottom or solid top (covered).

REVIEW QUESTIONS

1. Raceways are defined as enclosed channels designed to hold or contain several types of conductors, but mainly single-conductor cables (wires), multiconductor cables, or busbars.

True or False

2. The types of conduit and tubing listed in the definition of raceway in the NE Code (Art. 100) are governed by individual articles contained in Chapter ____________.

3. The type of conduit or tubing that can be installed for a branch or feeder circuit is governed by the environment of the intended installation.

True or False

4. A hazardous location is defined as a location where fire or explosion hazards may exist due to the presence (concentration or accumulation) of flammable gases or vapors, flammable liquids, combustible dust, or ignitable fibers or flyings.

True or False

5. The required size of conduit or tubing is based on the type of raceway, the selected wire size and insulation, and count of individual conductors to be pulled.

   True or False

6. The total number of conductors that can be installed in a given trade size of rigid metal conduit is based on the percent-fill allowances listed in Table 1 of Chapter ____________.
7. The total number of conductors that can be installed in a given trade size of intermediate metal conduit (IMC) is based on the percent-fill allowances listed in Table 1 of Chapter ____________.
8. The total number of conductors that can be installed in a given trade size of electrical metallic tubing (EMT) is based on the percent-fill allowances listed in Table 1 of Chapter ____________.
9. The total number of conductors that can be installed in a given trade size of rigid nonmetallic conduit, such as PVC, is based on the percent-fill allowances listed in Table 1 of Chapter ____________.
10. When conduit or tubing type raceway is sized in accordance with the tables found in Chapter 9 and Appendix C, the total cross-section area of three circuit conductors contained in a single conduit or tubing cannot exceed 40% of the conduit or tubing's interior cross-sectional area.

    True or False

11. When wiring methods such as Romex or UF cable are installed in short sections of conduit or tubing to protect them from physical damage, the "percent-fill" requirements of Chapter 9 must be applied.

    True or False

12. Neither equipment-grounding conductors nor bonding jumpers are counted in the total number of conductors when sizing the conduit or tubing within the guidelines of Chapter 9.

    True or False

13. When conduit or tubing nipples, which are not more than two feet in length, are run between boxes, panels, or equipment enclosures, the percent-fill allowances of Chapter 9 are increased to 60% of the cross-sectional area of the raceway.

    True or False

14. When conductors carrying alternating current are installed in metal raceway, or pass through openings in metal enclosures, all the ungrounded phase conductors, the grounded neutral conductor, and where used the equipment-grounding conductors must be grouped together to avoid the effect of inductive heating.

    True or False

15. If concentric-stranded conductors, all of the same wire size and same or physically equivalent insulation (c.s.a.), will be installed in rigid metal conduit, the conduit can be sized by Table ____________ in Appendix ____________.
16. If compact-stranded conductors, all of the same wire and same or physically equivalent insulation (c.s.a.), will be installed in flexible metallic conduit (greenfield), the conduit can be sized by Table ____________ in Appendix ____________.
17. If several conductors, which have a mixture of either wire sizes and/or insulation types (different c.s.a.s), will be installed in a single run of EMT conduit, the individual conductor cross-sectional areas required must be determined by Tables ____________ through

__________ (including bare conductors) in Chapter ______________. The conductor cross-sectional areas must be totaled, and the conduit sized in accordance with Table ______________.

18. If a mixture of either wire sizes and/or insulations of compact-stranded aluminum building wire will be installed in a single run of rigid metal conduit, the individual conductor cross-sectional areas required must be determined by Table ______________ in Chapter ____________. The conductor cross-sectional areas must be totaled, and the conduit sized in accordance with Table ______________.

19. Cross-sectional areas of bare, concentric-stranded conductors are listed in Table ____________ of Chapter ______________ according to wire size.

20. Nominal diameters of bare, compact-stranded conductors are listed in Table ____________ of Chapter ______________ according to wire size.

21. In general, conduits and tubings must be supported at intervals not to exceed 10 feet of length, and within 3 feet of all terminations.

    True or False

22. In general, multiconductor cables such as Romex and BX, and flexible metallic tubing must be supported every 4½ feet, and within 12 inches of all terminations.

    True or False

23. A cable tray is a unit or assembly of units or sections, and associated fittings, that when installed form a rigid structural system used to support cables.

    True or False

24. When multiconductor cables are installed in ventilated cable tray, the width of the tray is governed by the size of the cables. When a mixture of cable sizes are installed, cables containing No. 4/0 AWG or larger cables must be installed in single layers. Cables containing conductors smaller than No.4/0 AWG can be bundled or stacked on top of each other.

    True or False

25. When single-conductor cables are installed in ventilated cable tray, the width of the tray is governed by the size of the cables. When a mixture of cable sizes are installed, cables sized 1,000 kcmil or larger must be laid in a single layer, side-by-side. Cables larger than or equal to 250 kcmil, but not larger than 1,000 kcmil, can be stacked or bundled within the tray width.

    True or False

REVIEW PROBLEMS

1. Determine the NE Code-minimum size rigid metal conduit allowed for the installation of 21 No. 12 AWG TW copper conductors.
2. Determine the NE Code-minimum tubing size of EMT allowed for the installation of 16 No. 6 AWG THHN aluminum conductors.
3. Determine the NE Code-minimum conduit size of IMC allowed for the installation of four 500 kcmil RHW copper conductors with outer covering.
4. Determine the NE Code-minimum size of flexible metallic conduit allowed for installation of three 350 kcmil RHH copper conductors without outer covering.

5. Determine the NE Code-minimum size rigid schedule 40 PVC nonmetallic conduit allowed for the protection of four No. 6 AWG TW and four No. 6 AWG THW conductors.
6. Determine the NE Code-minimum size of liquid-tight flexible metallic conduit allowed for the installation of eight No. 1/0 AWG compact-stranded THHN conductors.
7. Determine the cross-sectional area of a No. 10 AWG THWN copper conductor.
8. Determine the cross-sectional area of a No. 10 AWG RHH copper conductor with outer covering, and a No. 10 AWG RHH copper conductor without outer covering.
9. Determine the NE Code-minimum size of flexible metallic conduit allowed for the installation of four No. 14 AWG TW, four No. 14 THWN, six No. 10 AWG XHHW, and four No. 6 AWG THW copper conductors.
10. Determine the NE Code-minimum tubing size of ENT (electrical nonmetallic tubing) allowed for the installation of four No. 2/0 AWG THW and six 250 kcmil THHN compact-stranded aluminum building wires.
11. Determine the cross-sectional area of the following concentric-stranded bare conductors: No. 16 AWG solid, No. 16 AWG stranded, No. 12 AWG solid, No. 12 AWG stranded, No. 4 AWG, No. 3/0 AWG, 250 kcmil, and 600 kcmil.
12. Determine the NE Code-minimum conduit size of IMC allowed for the installation of the following conductors: Three 250 kcmil THW, one No. 2/0 AWG THW, and one No. 2 AWG bare copper conductor.
13. Determine the NE Code-minimum size of XHHW aluminum conductors and rigid schedule 40 PVC conduit allowed to supply a 70-amp, 2-wire, 208-volt, single-phase continuous load. The PVC conduit, which passes through an ambient air temperature of 100°F, will also contain six No. 6 AWG THHW concentric-stranded aluminum current-carrying conductors and a No. 6 AWG bare aluminum equipment-grounding conductor.
14. Determine the NE Code-minimum cable tray width allowed for the installation of four 1,250 kcmil THW and twelve 2,000 kcmil THW single-conductor copper cables.
15. Determine the NE Code-minimum cable tray width allowed for the installation of eight 300 kcmil XHHW, twelve 400 kcmil XHHW, eight 600 kcmil XHHW, and sixteen 750 kcmil XHHW single-conductor copper cables.
16. Determine the NE Code-minimum cable tray width allowed for the installation of twelve 500 kcmil XHHW, four 750 kcmil XHHW, four 1,250 kcmil XHHW, and four 1,750 kcmil XHHW single-conductor copper cables.

Chapter 4

Outlet and Device Boxes

The circuit conductors, as sized by Chapter 2 of this text, must be installed in protective conduit, tubing, or other raceway as sized by Chapter 3.

The circuit conductors normally originate from the load side of a fused disconnect or circuit breaker (overcurrent protection devices) located in a service panel, a subfed panel, on busduct, in a load center, or within some other enclosure of the electrical supply.

The circuit conductors are normally terminated in a box or some other type of enclosure for field splice or other makeup to the load conductors. The conduit, or other raceway, which protects the circuit conductors from physical abuse, and persons from contact with the circuit conductors, is normally installed continuous from point of supply to connected load.

The installation of electrical boxes, whether outlet, device, pull, or junction, are governed by Art. 370. Correctly sizing the box depends on the intended use of the box, and the wire size of the conductors that will either terminate in, or be pulled through the box.

The main concern of sizing any box is to ensure that the bending of the conductors inside the box will not damage the conductor insulation due either to too sharp of a bend or to cuts and scrapes against the interior wall of the box.

A bend that is too sharp can stretch or tear the insulation on the back of the bend and expose the bare conductor. Trying to put too much conductor (or too many conductors) into a box that is too little can cause cuts or scrapes that will also expose the bare conductor. A break in circuit-conductor insulation can cause high-magnitude short-circuit or ground-fault overcurrent conditions in the circuit.

Determining the required size of an electrical box has two major divisions: When the raceway entries will contain conductors wire size *No. 6 AWG or smaller,* the size of the box is governed by required conductor volume and available box volume within the guidelines of Sec. 370-16. When raceway entries will contain conductors wire size *No. 4 AWG or larger,* the size or dimensions of the box is (are) governed by conductor deflection requirements (bending of the conductor) and Sec. 370-28 (600 volts or less).

Chapter 4 will concentrate on the smaller-conductor boxes (Sec. 370-16). Chapter 5 will concentrate on the larger-conductor boxes (Sec. 370-28).

4.1 OUTLET, DEVICE, OR JUNCTION BOXES

When a branch circuit supplies lighting outlets, the continuous power is first routed to a switch box (Fig. 4-1). The contained switch controls the status of the lights (on or off). The switch box is referred to as a device box, since it is not a means or point of power supply (not an electrical outlet box, by definition). The body of the switch is normally recessed inside the box for make-up between the switch terminals and the circuit conductors.

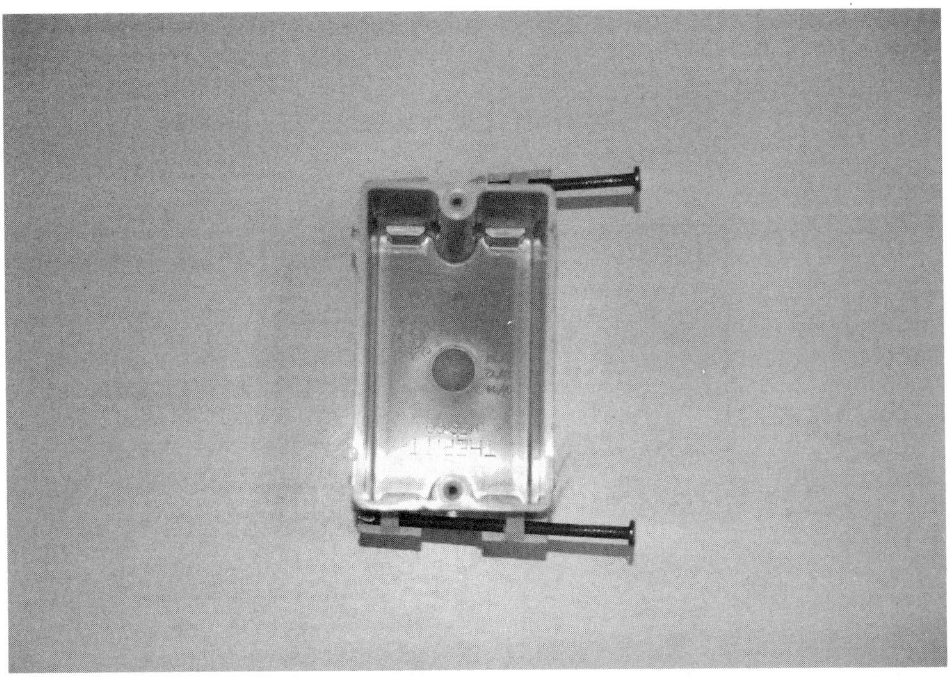

Fig. 4-1 A standard 3-by-2 device box is used to house either light switches or receptacles.

If several lighting outlets are supplied by a single branch circuit and controlled by the same switch (multioutlet branch circuit), then junction boxes must be set at various locations to tap or branch off to supply the individual lighting outlets. Often the lighting outlet box, recessed in the ceiling, is used for both make-up between the branch-circuit conductors and the lighting fixture leads (conductors), and for junction or splice to the downstream circuit conductors. The ceiling box is being used as both an outlet and a junction box.

Lighting of a room can also be provided by lamps, which are cord and plug connected to a receptacle. The receptacle must be installed in an outlet box for make-up of the receptacle terminals to the branch-circuit conductors. As with the lighting outlet box in the ceiling, receptacle outlet boxes are often used as junction boxes to supply downstream receptacle outlets.

For most power distribution systems within a building or other structure, the service equipment or other point of supply is located in one part of the building, and the connected loads are located throughout the building. Conduit or other raceway is run to connect the supply and load points together for installation of the circuit conductors.

Conduit or other piping-type raceways are normally installed to follow the surface of the building. This means several bends or turns in the piping to get from the point of supply to the connected load. The conduit must be installed as a complete system before the circuit conductors can be drawn through the pipe. Every bend or turn in the conduit means a deflection point for the conductors which will introduce friction with the conductors dragging against the wall in the bend when being pulled on the initial installation.

Fig. 4-2 Four-inch square boxes can be used as either a junction box that contains splices or as a pull point that does not contain splices.

Fig. 4-3 Condulet bodies, which are interposed in a piping raceway to establish a pull point, are available in a variety of piping figurations.

Too many bends or turns will add so much friction that the insulation may be scraped or cut when passing through the bend. In some instances, the tension introduced as more force is added to the pull can stretch or pull the conductors apart because the tensile strength of the conductors has been exceeded.

To protect the drawn conductors and their insulation from such abuse, any type of piping raceway is required to have a pull point (access opening) installed when the amount of conductor deflection (bends or turns) will exceed 360°. The pull point can be in the form of a box (Fig. 4-2) or a conduit body (Fig. 4-3).

As with sizing the protective conduit, the NE Code places restrictions on how many circuit conductors can be installed in a single device, outlet, junction, or pull box, or conduit body.

For wire size No. 6 AWG or smaller, Sec. 370-16 requires these type boxes or conduit bodies to be of sufficient size to provide free space for all enclosed conductors. The free space is based on the volume of the box.

The volume available from a box is listed as minimum cubic-inch capacity in the second column of T370-16(a). The types of manufactured boxes are listed as the row headings. The available volume of that specific box is found where the box row and cubic-inch column intersect.

The volume required per conductor is listed in T370-16(b) according to wire size. Wire size No. 18 AWG requires a minimum of 1.5 cubic inches per conductor. A 4 by 1¼ inch octagon box, first row in T370-16(a), has an available volume of 12.5 cubic inches:

12.5 ÷ 1.5 = 8. A maximum of eight No. 18 AWG conductors can be terminated or pulled through this 4 by 1¼ inch octagon box.

This same box can contain a maximum of five No. 12 AWG circuit conductors. A No. 12 AWG conductor, by T370-16(b), requires a minimum of 2.25 cubic inches of volume: 12.5 ÷ 2.25 = 5.55. Since the electrician cannot install part of a No. 12 conductor (0.55), the number is rounded down to five.

Instead of computing each installation in terms of volume, T370-16(a) lists in additional columns the number of conductors by wire size that can be terminated or pulled through a specific box. Technically any box listed in T370-16(a) can be used as a pull box.

The maximum number or count of conductors is restricted to a single wire size in any one box. For example, either five No. 18, four No. 16, three No. 14, three No. 12, three No. 10, two No. 8, or one No. 6 AWG conductor can either terminate or pass through a 3 by 2 by 1½ device box (tenth row).

The asterisk footnote to T370-16(a) stipulates that these maximum numbers of conductors, for any one wire size, do not take into account the volume allowances (conductor deductions) as required by Secs. 370-16(b)(2) through (b)(5).

4.2 VOLUME DEDUCTIONS

Section 370-16(b) lists the guidelines on how to size an outlet or device box. Restrictions are placed on the maximum number of conductors that can be contained when the box also contains devices, fittings, grounding wires, fixture studs, and etc. Allowances for these are represented as deductions of the number of conductors contained in the box.

To correctly size the box, it is easier to increase the total number of conductors to be installed in the box by the deductions (volume allowances) listed. The increase of conductors according to what else is to be installed in the device or outlet box should be determined as follows:

1. *Each* conductor that *passes* through the box without splice should be counted as *one* conductor [Sec. 370-16(b)(1)].
2. *Each* conductor that originates outside the box and terminates inside the box should also be counted as *one* conductor [Sec. 370-16(b)(1)].
   a. Termination can be to a device terminal, a terminal strip mounted in the box, or as a splice to another conductor or conductors.
   b. Splices would constitute two or more conductors that originate outside the box and terminate inside the box.
   c. Fixture wires (leads) smaller than wire size No. 14 AWG, unless their count is more than four, should not be counted [Sec. 370-16(b)(1), Ex.].
3. Conductors, such as pigtails and/or jumpers that originate and terminate within the box, should *not be counted* in the total number of conductors [Sec. 370-16(b)(1)].
4. *All* (one or more) regular equipment-grounding conductors should be counted as *one* additional conductor installed in the box. For a mixture of wire sizes, the single deduction for the regular equipment-grounding conductors should be based on the largest wire size [Sec. 370-16(b)(5)].

5. *All* (one or more) isolated equipment-grounding conductors, as allowed by Sec. 250-74, Ex. No. 4, should be counted as *one* additional conductor installed in the box. For a mixture of wire sizes, the single deduction for the isolated equipment-grounding conductors should be based on the largest wire size. If both grounding means are present (isolated and regular), two additional conductors, one for each type, should be counted. [Sec. 370-16(b)(5)]
6. *All* (one or more) internal cable clamps, where the tightening screws are mounted inside the box, should be counted as *one* additional conductor installed inside the box [Sec. 370-16(b)(2)].
7. Internal support fittings such as fixture hickeys, fixture studs, or fixture stems (assembly of the three that passes through the box to attach a hanging-type fixture to the back support of a box) should be counted as two additional conductors (one for the stud and one for the hickey) installed in the box. The stem does not take up enough box volume to be counted [Sec. 370-16(b)(3)].
8. Strap- or yoke-mounted devices, such as switches, receptacles, or pilot lights, mounted across the opening of the box, should be counted as *two* additional conductors for *each* strap [Sec. 370-16(b)(4)].

Although the threaded end of a conduit entry will normally protrude through a box knockout (hub or prepunched opening) at least far enough to mount a locknut and a bushing (to cover the exposed threads), Sec. 370-16(b) stipulates that no allowance (in terms of additional conductors) need be made for conduit entries. The same exemption is made by Sec. 370-16(b)(2) for two-screw cable squeeze connectors, which also have a threaded end that protrudes into the interior of the box.

No allowance is required for fasteners such as wire nuts or crimp-sleeve connectors with boots. Both are splicing means for cut wire. The volume of these fastening means are compensated for in determining the number of conductors that either pass through or terminate in the box.

When a single conductor passes through the box without splice, it is counted as one. When the same conductor is spliced, it becomes two conductors which originate outside, but terminate inside the box. Circuitwise, it's still a single conductor passing through the box. The fact that it was cut and spliced, made it two conductors: The wire nuts have been accounted for.

No allowance is required for fasteners such as grounding lugs (Fig. 4-4), grounding screws, box-mounting screws, or nails which may pass through the box (metal nail-in boxes). Like the fixture stem, which supports the hanging fixture from the fixture hickey, these types of support fasteners are exempt by Sec. 370-16(b)(3).

What happens when the total number of conductors for the given box installation exceeds the maximum count as listed in T370-16(a) for a selected wire size? Well, you need a bigger box. Three depth sizes are listed for the 4-inch round or octagon box, for the 4-inch square box, and for the $4\frac{11}{16}$ inch square box. Six depth sizes are listed for the 3 by 2 device box, and three are listed for the 4 by $2\frac{1}{8}$ handy box (Fig. 4-5). Increasing the depth increases the box volume, which increases the number of conductors that can be installed.

The minimum cubic-inch capacity of the 4-inch round, octagon, or square box, and

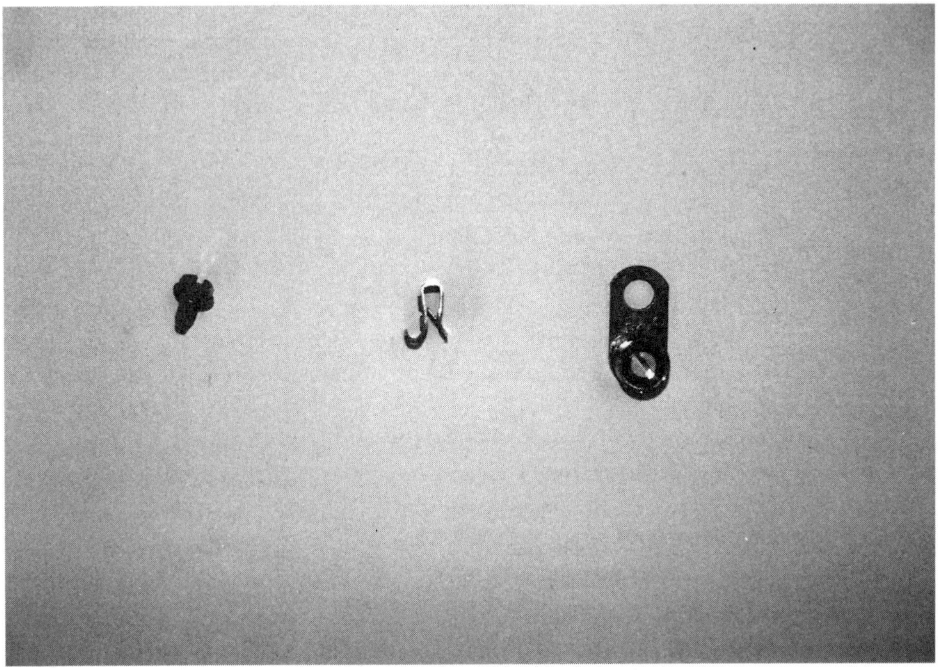

Fig. 4-4 No deduction is required for devices such as grounding screws, clips, or lug terminals that are used to terminate the grounding conductor to the metal box.

the 4$^{11}/_{16}$ square box (Fig. 4-6) can be increased with the use of box extensions, plaster rings or adapters, domed covers, and the like. These box bodies are required to have their cubic-inch volume etched or otherwise marked externally. Their dimensions must match the dimensions of the box their volume will increase. In other words a 4-inch square plaster ring will not fit a 4$^{11}/_{16}$-inch square box.

The 3 by 2 device boxes are often ganged together to accommodate several switches or receptacles or a combination of these at a common location. Ganging metal device boxes (construction boxes) requires removing the side plates and interlocking the shells to make the boxes appear as two, three, or more boxes installed side by side.

When these boxes are ganged, the cubic-inch capacity of the resulting single box is the sum of the individual volumes as listed in T370-16(a). Said another way, two-ganging a 3 by 2 by 2½ device box will effectively double the count of 5 No. 12 AWG conductors that can be installed in a single box, allowing a count of 10.

Octagon boxes are normally used for overhead lighting outlets. Round boxes are restricted in their use (Sec. 370-2). If the lighting outlet is also used as a junction point, the volume requirement for the extra conductors may exceed the volume available from even a deep (2⅛) octagon box. Square boxes may be used, and adapted with a cover or plate to mount and support the lighting fixture. The cover, or adapter ring as it is often described, adapts the square box, either 4-inch, or 4$^{11}/_{16}$-inch, to the standard 4-inch round opening of the lighting outlet box.

As a general rule, boxes that will support lighting fixtures, receptacles, or switches,

Fig. 4-5 Handy boxes, which conceal the strap screws, can be used when switches or receptacles must be surface-mounted on a wall.

should not be installed with box extensions. Box extensions should be limited to either pull or junction boxes, where many conductors are brought to a central point for splice and distribution.

The lighting fixture is normally supported by the box [Sec. 370-27(a)], although weight restrictions are placed on this type of support by Secs. 410-16(a) (hanging fixtures—50 pounds) and 422-18 (paddle fan/light assemblies—35 pounds). The lighting fixture may be supported by a stud–hickey–stem combination with the stud mounted to a box hanger in the back of the box, or by a strap that mounts across the face of the box (offset box hanger). The base of some lighting fixtures (the canopy) is designed to be screw-mounted directly to the face of the box.

The use of device boxes is normally limited to the installation of switches or receptacles. Switches may control overhead lighting or receptacle outlets for lamp lighting in a room or other area. Switches are also used to control small motor loads, appliances, and other utilization equipment, especially in a dwelling structure.

Receptacle outlets can be the convenience (duplex) receptacles installed in the home, business, or workplace, or receptacle outlets can supply appliances or other utilization equipment that is cord and plug connected. Single receptacle outlets are normally installed in device boxes.

If the receptacle outlet box will also be used as a junction box for splicing of the distribution wiring for other receptacle or lighting outlets downstream from the receptacle (multioutlet branch circuit), the device box may be too small. Square boxes, either 4-inch

Fig. 4-6 Four and 11/16 square boxes offer more box volume than a 4-inch square box when used as a pull, junction, outlet, or device box.

or $4^{11}/_{16}$-inch, can be used for the volume requirements of the extra conductors, and adapted with a cover or plate that can be used to mount the single receptacle. The adapter ring (Fig. 4-7), as with the lighting outlet, is designed to convert the square box opening over to a 3-by-2 device opening, which will accommodate the installation of single-strap receptacles or lighting switches.

Square boxes are normally used for pull or junction boxes. The boxes may contain splices from one to one or more conductors, or the conductors may pass through the box without splice (pull point). If the box does not support a device or is not used as an outlet, the box is effectively closed with a blank plate.

If the volume requirements for the number of conductors is exceeded by the planned junction-box installation, box extensions (Fig. 4-8) for octagon or square boxes can be mounted to the open face of the box. Box extensions appear to be a box with the back or bottom wall cut out. They are available in the same dimensions as the octagon or square boxes listed in Table 370-16(a). As mentioned before, box extensions should not be used for switch, receptacle, or lighting fixture installations.

4.3 BOX SIZING—SAME SIZE CONDUCTOR

For the same size of conductor, the easiest method to size an outlet or device box, according to the number of conductors that will be installed and allow for device or fitting deductions, is as follows:

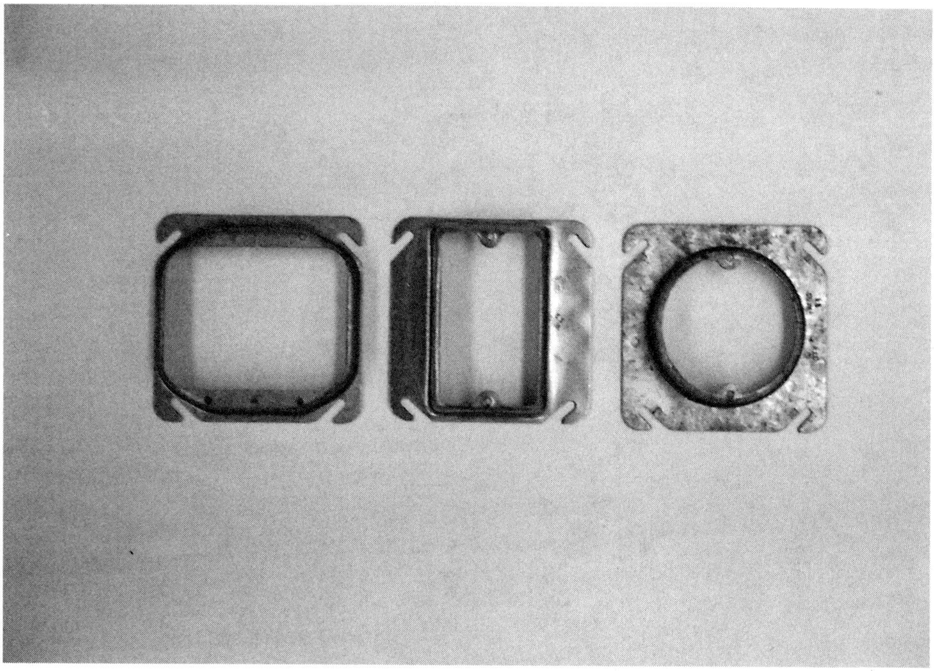

Fig. 4-7 Square boxes must be adapted to either a 3-by-4 opening, a 3-by-2 opening or a 3-inch or 4-inch round opening to accomodate the installation of switches, receptacles, or lighting fixtures.

First, determine the total number of conductors the box must contain by listing all circuit conductors as either pass-thru or termination, and then equating all volume deductions to additional conductors to be installed in the box.

Count	What	Total number of conductors
Given	Determined by example or	Determined by the requirements
—	planned installation.	of Sec. 370-16(b)
—	—	—
		Total number of conductors

Second, match the maximum number of conductors listed for the wire size and type box in T370-16(a).

1. The table-listed maximum number of conductors must be greater than or equal to the computed number of conductors.
2. The table-listed value corresponds by row to the NE Code-minimum size of the box required.

Example 4–1. Determine the NE Code-minimum size of octagon box required to support a lighting fixture supplied with a No. 12/2 w/GND Romex cable. The box will contain two internal cable clamps, and one grounding clip for a pigtail ground (equip-

Fig. 4-8 Box extensions can be used to add volume to a junction, pull, device, or outlet box.

ment-grounding of the box). The Romex cable will be spliced using wire nuts, and pigtailed to an externally mounted lighting fixture. The splice will be to another No. 12/2 w/GND Romex cable which supplies another lighting fixture downstream.

Total number of conductors computation: Ref. 370-16(b)

Count	What	Total No. 12s
2	No. 12/2 Romex cables for termination	4
2	No. 12 AWG equipment-grounding conductor terminations	1
2	Internal cable clamps	1
	Total count of No. 12s	6

NE Code-minimum box size determination: Ref. T370-16(a)

Table-listed maximum number of conductors: 6

Box size: 4 × 1½ inch octagon

Pigtails, the short jumpers used to supply the base of the lighting fixture from the splice of the branch circuit and to ground the metal box, were not counted in determining the total number of conductors (with deductions) that would be installed in the box. Neither were the wire nuts used to make up the circuit splice, nor the grounding clip used to ground the metal box.

Besides the two multiconductor cable terminations which yielded four circuit conductors, both the equipment-grounding conductors and the internal cable clamps were deductions that counted as two additional conductors (one for each type) toward the total count of conductors to be installed in the box.

Wire size No. 12 AWG held our selection to the fifth column in T370-16(a). An octagon fixture box limited our selection in this column to the first three rows. The count of 6 corresponded to the standard 4 by 1½-inch octagon box.

Had the cable clamps been specified as external cable clamps, meaning the threaded end of the connector was mounted through a ½-inch knockout in the box, and the two screws were held outside of the box, the conductor count would only have been five.

Often, the better design approach is to add at least one conductor for box volume lost to conduit entries or cable connectors that are mounted through the knockouts and secured with locknuts and bushings, with another conductor added for the wire nuts or other types of splicing means, and yet another for other fasteners or fittings such as grounding clips or grounding studs mounted within the box.

These additional allowances will require a larger box for most installations. Unfortunately, these allowances will prevent some types of boxes from being used for a particular application, and it's going over and beyond the requirements of Sec. 370-16(b). On an electrical exam, the student electrician is expected to meet or compute only NE Code-minimum in terms of sizing these device or outlet boxes.

In a real-world installation (back with good design), the student electrician should also remember that safe electrical practice demands that wires and/or devices not be jammed or forced into boxes. Jamming too many conductors into too little of a device or junction box creates the possibility of nicks or other damage to the conductor insulation which, in turn, poses the threat of short-circuit and ground-fault overcurrents in the circuit.

Code challenge 4–1. Determine the NE Code-minimum size of octagon box required to support a lighting fixture supplied with a No. 14/3 w/GND Romex cable. The box will contain two internal cable clamps, and one grounding clip for the pigtail ground. The Romex cable will be spliced using crimp-sleeve connectors and insulating caps (boots or diapers). Two of the No. 14 AWG conductors will be spliced and pigtailed to an externally mounted lighting fixture. The splice, including the third No. 14 AWG circuit conductor, will be to another No. 14/3 w/GND Romex cable which will supply another lighting fixture downstream.

Total number of conductors computation: Ref. (A) ____________

Count	What	Total No. 14s
(B) ________	No. 14/3 Romex cables for termination	(C) ________
(D) ________	No. 14 equipment-ground terminations	(E) ________
(F) ________	Internal cable clamps	(G) ________
	Total count of No. 14s:	________

NE Code-minimum box size determination: Ref. (H) ____________

Table-listed maximum number of conductors: (I) ________

Box size: 4× (J) ________ inch octagon

Octagon fixture boxes were required in both the example and the code challenge. Round boxes could have been used because internal cable clamps were specified, and multiconductor cable was the circuit wiring method.

When conduit or tubing is used as the grounding means for the outlet or device box, Sec. 370-2 will not allow round boxes to be used unless all conduit or cable entries are from the back of the box.

Threaded conduit entries use a double locknut system, one on the inside and one on the outside of the metal box to secure the conduit to the box (Fig. 4-9). Electrical metallic tubing, which cannot be threaded because of its thin wall, requires a box connector which consists of a threaded end and a barrel end. The threaded end has a shoulder at the back of the threads which rest against the sidewall of the box. A single locknut on the inside of the box secures the EMT connectors to the box in the same manner as threaded conduit.

The barrel end accepts the end of the conduit and either an indent, a set-screw, or a compression fitting is used to secure the conduit to the EMT connector.

A two-screw cable connector uses the same threaded-end and shoulder-stop assembly to secure the cable clamp to the box with a single locknut. A locknut requires a flat bearing surface in which the serrated cleats can penetrate the metal wall for electrical and mechanical continuity.

Octagon boxes (the open face is eight sided, similar to the common stop sign [Fig. 4-10]) have limited use with conduit or tubing raceway, especially when a lighting fixture outlet box will also be used as a junction box. Most octagon boxes are stamped or pre-

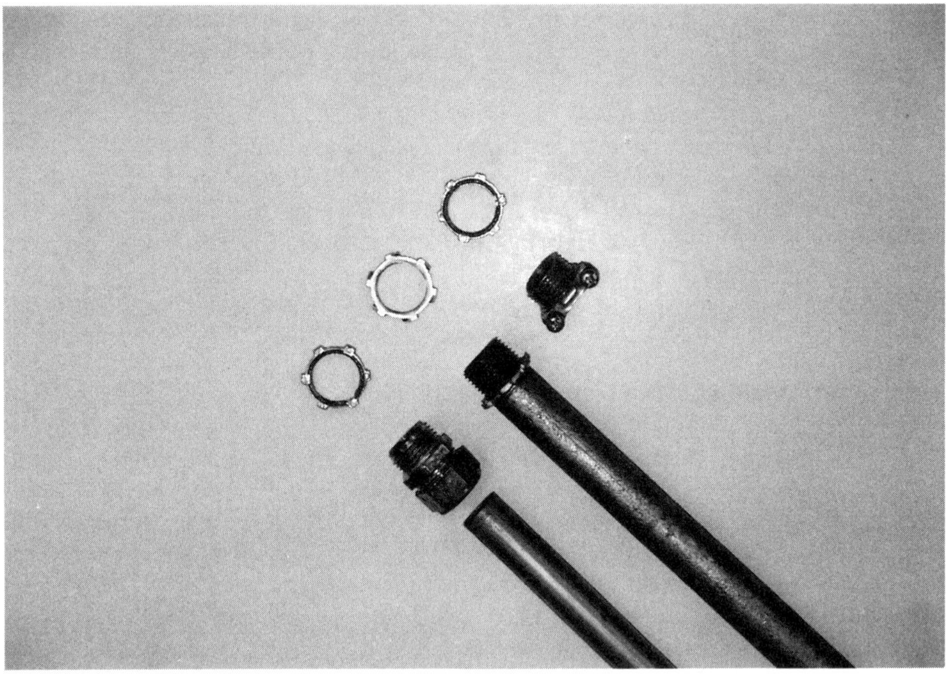

Fig. 4-9 Whether raceway or cable connectors, the short nipple that protrudes into the interior of the box is not counted as a conductor deduction.

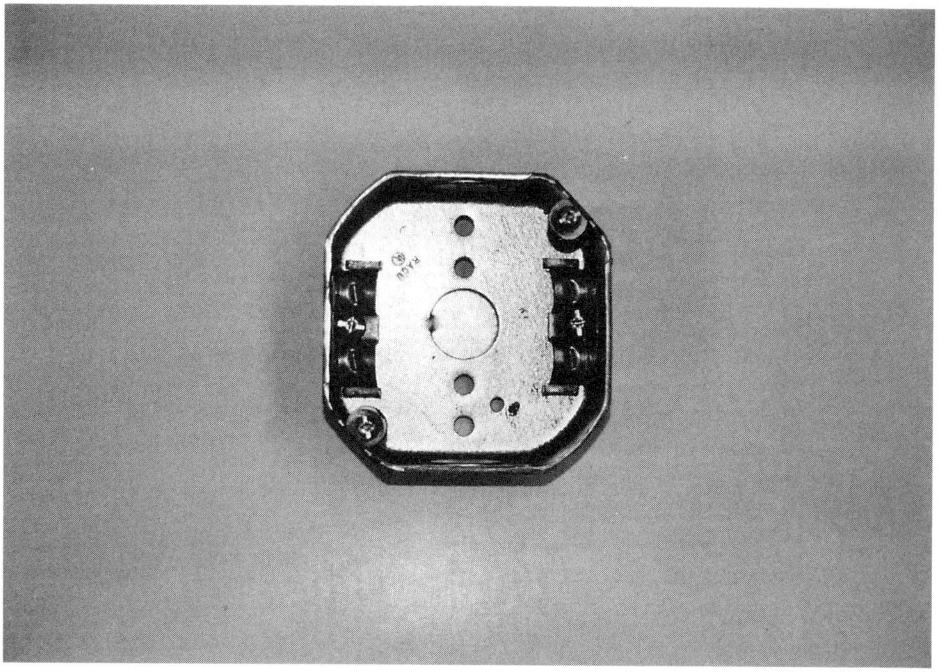

Fig. 4-10 Four-inch octagon boxes, which may contain both internal cable clamps and conduit knockouts in the walls of the box, are normally used as ceiling boxes for lighting fixtures.

punched with standard knockouts that will accept either ½ inch, ¾ inch, or in some instances 1-inch conduit entries.

Common 4-inch octagon boxes will have a minimum of a single knockout on each of the four side walls. This limits the total side wall conduit entries to a maximum of four. Some 4-inch octagon boxes have only two knockouts on opposite side walls. The other two walls contain smaller knockouts for cable entries (two on each side wall) with two internal cable clamps.

When the lighting fixture box will also be used as a junction box for multiple conduit entries, the 4-inch octagon box is often replaced with either a 4-inch, or a 4$^{11}/_{16}$-inch square box. Most square boxes have either two or three knockouts (according to the combination of sizes for the conduit entries) on each wall.

Besides removing the limitation of a single conduit entry being centered on one wall, as with the octagon box, these multiple knockouts allow two or more conduits to enter the same wall, or the single conduit entry can be offset to enter from either the left side or the right side of the wall of the box.

The face opening of either size square box has to be adapted back to a four-inch round opening to support the fixture.

Example 4–2. Determine the NE Code-minimum size of 4-inch square box required to support an externally mounted lighting fixture by adapter ring. The box will contain the splice of a 3-phase, 4-wire, multiwire branch circuit with a fifth-wire equipment-grounding conductor. Each No. 14 AWG THWN copper conductor will be spliced 1-to-1 between the two ½ inch EMT conduit entries. One phase conductor, the neutral, and the equipment ground (equipment-grounding conductor) will also be pigtailed for connection to the lighting fixture. The splices will be made with crimp-sleeve connectors and diapers (insulating caps).

Total number of conductors computation: Ref. 370-16(b)

Count	What	Total No. 14s
4	No. 14s spliced 1-to-1	8
1	No. 14 equipment ground spliced 1-to-1 and pigtailed	1
	Total count of No. 14s	9

NE Code-minimum box size determination: Ref. T370-16(a)
Table-listed maximum number of conductors: 9
Box size: 4 × 1¼ inch square

As used in this example, a 3-phase, 4-wire, multiwire branch circuit consists of three ungrounded phase conductors, phase A (Line 1), phase B (Line 2), and phase C (Line 3), which supply individual 2-wire branch circuit loads connected to a common fourth-wire neutral conductor.

As discussed with the ampacity of circuit conductors in Chapter 2, when the loads are balanced among the three circuits, the full-load current along the length of the fourth-wire common neutral conductor is near zero amps. (Ideally it is zero amps.)

Code challenge 4–2. Determine the NE Code-minimum size of 4$^{11}/_{16}$ inch square box required to support an externally mounted lighting fixture with adapter ring. The entry and exit ¾ inch conduits will contain two 3-phase, 4-wire, multiwire branch circuits with a ninth-wire common equipment-grounding conductor. Except for the phase conductor, the neutral, and the equipment ground that will be spliced 1-to-1 and pigtailed to supply the lighting fixture, the other six No. 12 AWG XHHW copper circuit conductors will pass through the box without splice. The splices will be made with wire nuts. The pigtails will connect directly to the fixture.

Total number of conductors computation: Ref. (A) ____________

Count	What	Total No. 12s
(B) ________	No. 12s spliced 1-to-1 and pigtailed	(C ________
(D) ________	No. 12s that pass through the box	(E) ________
(F) ________	No. 12 equipment grounds	(G) ________
	Total count of No. 12s:	________

NE Code-minmum box size determination: Ref. (H) ____________
Table-listed maximum number of conductors: (I) ____________
Box size: 4$^{11}/_{16}$ × (J) ________ inch square

Splices count as a minimum of two conductors that originate outside, and terminate inside the box. Splices can count more than two when two or more loads are served by a single branch circuit.

For example, to supply the receptacles in the living room of a dwelling, a single junction box could be mounted on the floor joists directly under the center of the room. From a single home run (branch-circuit conductors extending from the service panel to the junction box), all the required receptacles could be supplied with individual cable runs. The splice would constitute several conductors that originate outside and terminate inside the box.

The better method of determining the number of conductors in a splice, whether it be 1-to-1, 1-to-2, or so on, is to determine the total number of conductors that will terminate in the box to the splice.

When sizing an outlet or device box, the conductive material or the insulation of the conductor has no bearing on the cubic-inch capacity required by T370-16(b) for each conductor. As with all NE Code-related subjects, some judgment is needed to determine compliance with the requirements of Sec. 370-16.

The box size determined may not be large enough, since the free space available (box volume) will also depend on the length of the conductor (left in the box enclosure), the type of conductor insulation (some conductor insulations are thicker than others), whether the wire method is single-conductor or multiconductor cable (Romex, BX, or UF), the splicing method employed, and the conductor arrangement in the box.

Section 300-14 requires a minimum of 6 inches of conductor length at all outlet or device boxes when the conductors are to be spliced or terminated within the box. The 6 inches is measured from the open face of the box whose edge will be flush with the finished wall or ceiling surface. The conductor entry is either to the side, or to the back of the box. Including the box depth, at least eight inches of free conductor will have to be put back into the box.

"Type of conductor" not only identifies the conductor material and its insulation, but also whether the conductor is stranded or solid. All these considerations determine the conductor's flexibility. Thick insulations (such as rubber compounds) restrict the conductor's bending radius, aluminum conductive material is stiffer than copper, and stranded conductors are more flexible than solid conductors of the same wire size.

The 3-by-2 device boxes listed in the middle of the table (Fig. 4-11) are normally used to install a single light switch or a single receptacle. Often in wooden-framed structures these boxes are furnished as nail-in boxes. Some versions have external braces for the nails to pass through and secure the box to the frame. Others have pattern holes which allow the nails to pass through the interior of the box.

By design, one conductor should be added toward the total count of conductors installed in the box for nails that pass through the box. The nails take up volume that was available for the conductors. In reference to Sec. 370-16(b)(3), the nails are not support fittings that require a volume allowance (deduction).

If the nails pass through the box for mounting (Fig. 4-12), Sec. 370-23(b)(1) requires the nails to be located within ¼ inch of either the back of the box, or the front opening (leading edge) of the box. Installing the nails in the front makes the better installation. Nails passing through the back of the box may interfere with internal cable clamps, or pass too close to the wires that are secured by the cable clamps.

Fig. 4-11 The tyipcal 3-by-2 device box which contains internal cable clamps is often referred to as a "construction box."

Example 4–3. Determine the NE Code-minimum size of a 3-by-2 device box required for the installation of a single split (one switched, one not) duplex receptacle outlet. The box will contain a single No. 12/3 w/GND Romex cable entry, secured to the box by an internal cable clamp. The equipment-grounding conductor will be pigtailed twice from a single wire nut. One pigtail will be connected directly to the grounding terminal of the receptacle. The other pigtail will be connected to the metal box with a grounding clip.

Total number of conductors computation: Ref. 370-16(b)

Count	What	Total No. 12s
1	No. 12/3 Romex cable for termination	3
1	No. 12 AWG equipment-grounding conductor termination	1
1	Internal cable clamp	1
1	Duplex receptacle (strap)	2
	Total count of No. 12s	7

NE Code-minimum box size determination: Ref. T370-16(a)

Table-listed maximum number of conductors: 8

Box size: 3 × 2 × 3½ inch device

Fig. 4-12 The 3-by-2 construction boxes are also referred to as "nail-in" boxes. The nails must pass within 1/4-inch of the front or back of the box.

The term "split receptacle" in this example indicates that one half of the standard duplex receptacle will be supplied by a switched power source. The other half will be supplied by a constant power source.

Code-challenge 4–3. Determine the NE Code-minimum size of 3-by-2 device box required for the installation of a single-duplex receptacle outlet. The box will contain two No. 12/2 w/GND Romex cable entries that consists of a homerun (back to the supply panel) and a downstream feed (to another receptacle outlet). The two Romex cables, secured to the box with internal cable clamps, will be spliced with wire nuts and pigtailed to supply the receptacle. The equipment-grounding conductor will be pigtailed twice. One pigtail will be connected directly to the grounding terminal on the receptacle. The other pigtail will be connected to the metal box with a grounding clip.

Total number of conductors computation: Ref. (A) ______________

Count	What	Total No. 12s
(B) __________	No. 12/2 Romex cables for termination	(C) __________
(D) __________	No. 12 equipment-ground terminations	(E) __________
(F) __________	Internal cable clamps	(G) __________
(H) __________	Duplex receptacle (strap)	(I) __________
	Total count of No. 14s:	__________

NE Code-minimum box size determination: Ref. (J) ____________________

Table-listed maximum number of conductors: (K) ____________________

Box size: 3 × 2 × (L) ____________ inch device

When the number of conductors to be installed exceeds the maximum number of conductors allowed for any of the 3-by-2 device boxes, 4-inch or 4$^{11}/_{16}$-inch square boxes can be adapted to single-outlet dimensions. Since most device boxes are recessed, the square box must be adapted with a plaster ring which will allow the wall finish to cover the face of the box, except for the device opening in the plaster ring.

Example 4–4. Determine the NE Code-minimum size of a recess-mounted, 4-inch square box with single-gang plaster ring (adapter ring) required for the installation of a duplex switch. The box will contain three ½ inch conduit entries: one homerun and two downstream feeds. The homerun will be a 2-wire branch circuit with a separate equipment-grounding conductor consisting of No. 12 AWG THW copper conductors that are spliced to both downstream feeds (same size conductors). Each downstream feed will also contain a fourth-wire switch leg. The ungrounded side of the circuit will be spliced and pigtailed to supply the switch. All the splices will be made with wire nuts.

Total number of conductors computation: Ref. 370-16(b)

Count	What	Total No. 12s
8	No. 12s for termination	8
3	No. 12 equipment grounds for termination	1
1	Duplex switch (strap)	2
	Total count of No. 12s	11

NE Code-minimum box size determination: Ref. T370-16(a)

Table-listed maximum number of conductors: 13

Box size: 4 × 2⅛ inch square

Code-challenge 4–4. Determine the NE Code-minimum size of a recess-mounted, 4-inch square box with single-gang plaster ring required for the installation of a single split duplex, grounding-type receptacle. The box will contain two No. 12/3 w/GND Romex cables supported with two-screw squeeze connectors (external cable clamps). One Romex cable will be the supply from a switch box. The other cable will downstream feed other receptacle outlet boxes. The three circuit conductors will be spliced and pigtailed to the respective terminals on the receptacle. The No. 12 bare equipment-grounding conductor will be spliced and pigtailed to a grounding lug mounted in the 4-inch box. The box will be back-mounted on an adjustable hanger with a locking stud and nut. Splices will be made with wire nuts.

Total number of conductors computation: Ref. (A) ____________

Count	What	Total No. 12s
(B) ________	No. 12/3 Romex cables for termination	(C) ________
(D) ________	No. 12 equipment-ground terminations	(E) ________
(F) ________	Duplex receptacle (strap)	(G) ________
	Total count of No. 12s:	________

NE Code-minimum box size determination: Ref. (H) ____________

Table-listed maximum number of conductors: (I) ____________

Box size: 4 × (J) ________ inch square

When two-gang switches or receptacles will be installed (two single switches or two duplex receptacles installed side by side at a common location), either a 4-inch or $4\frac{1}{16}$-inch square box can be installed with a two-gang plaster ring, as in this example. Or, two 3-by-2 device boxes of the same depth can be ganged for a recessed installation.

As discussed in Sec. 4-2 of this chapter, ganging two device boxes together effectively doubles the number of conductors that can be terminated in the box.

Example 4–5. Determine the NE Code-minimum size of a recessed-mounted, $4\frac{11}{16}$-inch square box with two-gang plaster ring required for the installation of a single lighting switch ganged with (installed beside) an isolated ground-type duplex receptacle. The ½ inch conduit entries will consist of a homerun and a downstream feed. The homerun will contain a total of six No. 12 AWG THW copper conductors: two 2-wire branch circuits and two separate equipment-grounding conductors. The ungrounded side of one two-wire circuit will terminate to the switch. The grounded side of the two-wire lighting circuit and the solid green equipment-grounding conductor will be spliced and run with the switch-leg to downstream lighting outlets. The conductors of the other two-wire branch circuit and an isolated green with a yellow-stripe equipment-grounding conductor will be terminated to the receptacle. The splices will be made with wire nuts.

Total number of conductors computation: Ref. 370-16(b)

Count	What	Total No. 12s
6	No. 12s for termination	6
1	No. 12 isolated equipment ground for termination	1
2	No. 12 regular equipment ground for termination	1
1	Duplex switch (strap)	2
1	Duplex receptacle (strap)	2
	Total count of No. 12s	12

NE Code-minimum box size determination: Ref. T370-16(a)

Table-listed maximum number of conductors: 13

Box size: $4\frac{11}{16} \times 1\frac{1}{2}$ inch square

Code challenge 4–5. Determine the NE Code-minimum size of a recessed-mounted 4-inch square box with a two-gang plaster ring required for the installation of two isolated-ground receptacles. The box will contain a total of seven No. 12 AWG XHHW copper conductors that terminate from the homerun feed. A red phase conductor, a white neutral, and a green with one yellow stripe equipment ground will terminate to one receptacle. A blue phase conductor, a white with blue stripe neutral, and a green with two yellow stripes equipment ground will terminate to the other receptacle. A third solid green equipment ground will terminate to a grounding lug in the box. The box will contain only one ½ inch conduit entry.

Total number of conductors computation: Ref. (A) ____________

Count	What	Total No. 12s
(B) ________	No. 12s for termination	(C) ________
(D) ________	No. 12 isolated grounds for termination	(E) ________
(F) ________	No. 12 regular grounds for termination	(G) ________
(H) ________	Duplex receptacles (straps)	(I) ________
	Total count of No. 12s:	________

NE Code-minimum box size determination: Ref. (J) ____________

Table-listed maximum number of conductors: (K) ____________

Box size: 4 × (L) ________ inch square

The 4-by-2⅛ boxes listed below the 3-by-2 device boxes in the table are normally referred to as handy boxes. These boxes are intended for surface mounting of single switches or receptacles (single strap). In comparison, device boxes are normally recessed within the wall covering, and concealed from view when the structure is completed.

The standard strap for either a switch or receptacle has mounting slots on both ends for direct connection to the box. The slots are located 3½ inches on center. When mounted to a nail-in or other type of 3-by-2 recessed device box, threaded tabs located in the center of the top and bottom walls are turned out from the box wall. The threaded holes are located ¼ inch past the wall of the box, which yields the required 3½ inches (Fig. 4-13).

In the 4-inch handy box, the same threaded tabs are turned into the box, the ¼ inch offset of the threaded holes yield the required 3½ inches.

Handy boxes require a different faceplate as compared to device boxes (Fig. 4-14). The handy box faceplate is designed to overlap the face of the box approximately ⅟₁₆ inch. Device box faceplates are more concerned with the decor of the building. Their faceplates cover both the box and the wall cutout hole (⅛ inch air gap maximum; Sec. 370-21) for the box.

Handy boxes are normally manufactured with prepunched knockouts in the walls and back of the box that will accept either ½ or ¾ inch conduit, or two-screw cable squeeze connectors.

Example 4–6. Determine the NE Code-minimum size of a 4-inch handy box required for the surface mount of a single lighting switch. The single ½ inch conduit entry to the box will contain a total of three No. 10 AWG THW copper conductors. Two of the

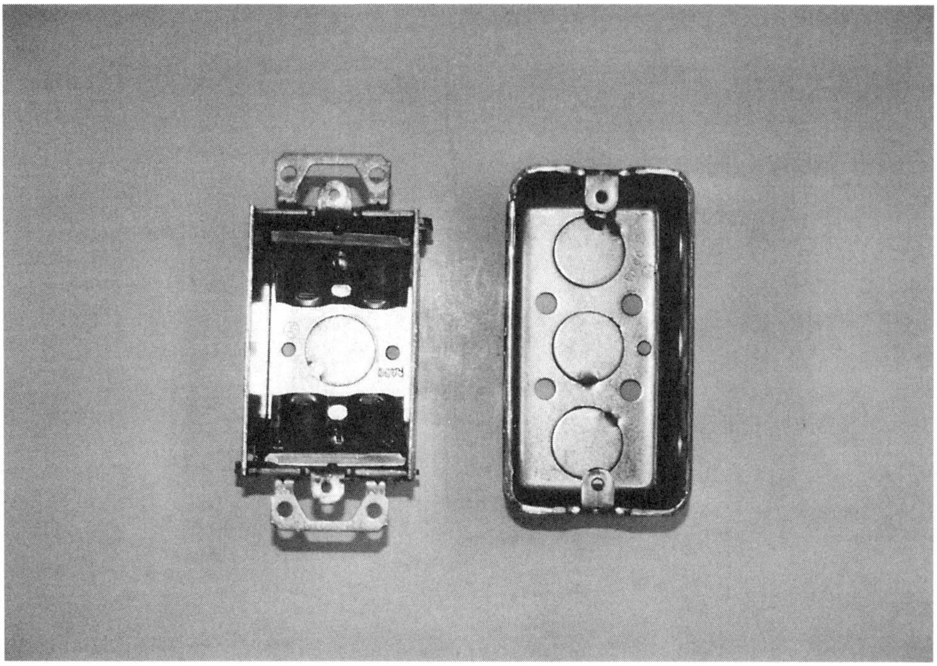

Fig. 4-13 Standard 3-by-2 device boxes and 4-by-2⅛ handy boxes are both designed to accept the 3½-inch strap of a switch or receptacle.

No. 10s are the switch drop. Both are ungrounded-circuit conductors. One will be the continuous power supply. The other will be the switch-leg or switch return. The third No. 10 is an equipment-grounding conductor that terminates to a grounding-screw terminal on the switch strap.

Total number of conductors computation: Ref. 370-16(b)

Count	What	Total No. 10s
2	No. 10s for termination	2
1	No. 10 equipment ground for termination	1
1	Lighting switch (strap)	2
	Total count of No. 10s	5

NE Code-minimum box size determination: Ref. T370-16(a)
Table-listed maximum number of conductors: 5
Box size: 4 × 2⅛ × 1⅞ inch handy

Code challenge 4–6. Determine the NE Code-minimum size of a 4-inch handy box required for the surface-mount of a split-wired single duplex receptacle. The single ½ inch conduit entry to the box will contain a total of four No. 12 AWG THW copper

Fig. 4-14 The device box faceplate (on the left) must close the recessed box wall opening. The handy box faceplate must only close the box.

conductors. Two of the No. 12s will terminate to the respective ungrounded terminals on each half of the split receptacle. A third No. 12 will connect to one of the common grounded terminals on the duplex receptacle. The fourth No. 12 will connect to the equipment-grounding terminal on the receptacle.

Total number of conductors computation: Ref. (A) ______________

Count	What	Total No. 12s
(B) __________	No. 12's for termination	(C) __________
(D) __________	No. 12 equipment ground for termination	(E) __________
(F) __________	Duplex receptacle (strap)	(G) __________
	Total count of No. 12s:	__________

NE Code-minimum box size determination: Ref. (H) ______________

Table-listed maximum number of conductors: (I) ______________

Box size: $4 \times 2\frac{1}{8} \times$ (J) __________ inch handy

When the number of conductors to be installed exceeds the maximum number of conductors allowed for a handy box, or when two-gang switches or receptacles must be surface-mounted, either a 4-inch, or a $4\frac{11}{16}$-inch square box can be adapted to either single, or two-gang outlet dimensions. Instead of a plaster ring used to adapt a recessed square box to either a single, or two-gang opening, the surface-mounted square box must be fitted with a raised cover mounted to the face of the box that will support the switch or receptacle (Fig. 4-15).

As discussed before, the primary purpose of either 4-inch or $4\frac{11}{16}$-inch square boxes is for pull or junction boxes. The term pull indicates no splices, as compared to junction, which indicates splices. As a general rule, the number of conductors passing through a pull box will not exceed the maximum number of conductors that can be installed in the square box.

When used as a junction box, the number of conductors installed or contained in the box can quickly exceed the table cutoff. When this happens, a box extension of the same depth can be used to effectively double the number of conductors allowed in the same manner that ganging device boxes doubled the number allowed.

Box extensions are designed primarily for either octagon or square boxes that are to be used as junction boxes. The size of the box and a box extension can be mixed or matched (shallow, $1\frac{1}{4}$; standard, $1\frac{1}{2}$; or deep, $2\frac{1}{8}$) to attain the required conductor count for a given installation.

Example 4–7. Determine the NE Code-minimum size of 4-inch box and single 4-inch box extension required for the splice of two 3-phase, 4-wire multiwire branch circuits. The two branch circuits will leave the box in separate conduits. A total of nine No.

Fig. 4-15 When a square box is used to surface mount a single receptacle, a raised cover is used to close the box opening and support the receptacle.

12 AWG XHHW copper conductors will be installed in the homerun, and five No. 12 AWG XHHW copper conductors will be installed in each of the two exit conduits. The No. 12 AWG equipment-grounding conductor will be spliced 1-to-2, and pigtailed to a grounding stud in the box. The splices will be made with crimp-sleeve connectors and boots.

Total number of conductors computation: Ref. 370-16(b)

Count	What	Total No. 12s
8	No. 12s for termination of homerun	8
8	No. 12s for termination of downstream feeds	8
3	No. 12 equipment ground terminations	1
	Total count of No. 12s	17

NE Code-minimum box size determination: Ref. T370-16(a)

Table-listed maximum number of conductors for box: 9
Table-listed maximum number of conductors for box extension: 9
Total: 18

Box size: 4 × 1½-inch square
Box extension size: 4 × 1½-inch square

Code challenge 4–7. Determine the NE Code-minimum size of 4¹¹⁄₁₆-inch square box and single 4¹¹⁄₁₆-inch box extension required for the splice of three 3-phase, 4-wire multiwire branch circuits, and a common equipment-grounding conductor. The three branch circuits will leave the box in three separate conduits. A total of thirteen No. 14 AWG conductors will be installed in the homerun, and five No. 14 AWG conductors will be installed in each of the three exit conduits. The No. 14 AWG equipment-grounding conductor will be spliced 1 to 3. The box will be grounded by the metal conduit. The splices will be made with wire nuts.

Total number of conductors computation: Ref. (A) ____________________

Count	What	Total No. 14s
(B) __________	No. 14s for termination of homerun	(C) __________
(D) __________	No. 14s for termination of downstream feeds	(E) __________
(F) __________	No. 14 equipment-ground terminations	(G) __________
	Total count of No. 14s:	__________

NE Code-minimum box size determination: Ref. (H) ____________________

Table-listed maximum number of conductors for box: ______________
Table-listed maximum number of conductors for box extension: ______________
Total: (I) ______________

Box size: $4^{11}/_{16} \times$ (J) ___________ inch square
Box extension size: $4^{11}/_{16} \times$ (K) ___________ inch square

4.4 BOX SIZING—MIXTURE OF WIRE SIZES

A mixture of conductor sizes will not allow a device, outlet, or small junction box to be sized directly from Table 370-16(a). Sections 370-16(b)(1) through 370-16(b)(5) require the "volume per conductor" to be determined by Table 370-16(b), with any volume deductions being taken for the largest conductor contained in the box.

The volume deduction for switches, receptacles, and other strap-mounted devices (two conductors for each strap) should be based on the largest conductor terminated to the device [Sec. 370-16(b)(4)].

To correctly size the box, the procedure used to size a box for a single wire size must be modified.

First, determine the total number of conductors the box must contain by listing all circuit conductors as either pass thru or termination according to wire size, and then equating all conductor deductions to additional conductors to be installed in the box. The deductions must be equated to the largest wire size that will be installed in the box.

Count	What	Wire size	Total number of conductors
Given	Determined by example or	Determined by the requirements	
—	planned installation.	of Sec. 370-16(b)	
—	—	—	—

Second, determine the total conductor volume.

Count	Wire size	Volume for one (cubic inches)	Extended volume (cubic inches)
Determined	Determined	T370-6(b) according	Product of count
by step one	by step one	to wire size	time volume for one
—	—	—	—
			Total conductor volume

Third, compare the total conductor volume to the available box volumes listed in Table 370-16(a), col. 2. The first box volume, according to type of box selected, that is greater than or equal to the total conductor volume, will correspond by row to the NE Code-minimum box size.

As when sizing the box for a single size of circuit conductor and the maximum number of conductors was exceeded, if all the box volumes listed for a particular type of box is exceeded by the required conductor volume, a single box extension, an extension ring, a plaster ring, or a dome cover can be used to increase the available box volume.

Box extensions are available with the same depths as the boxes they are to extend.

For example, a single 4-inch square by 1¼ inch box extension has a volume of 18 cubic inches. If mounted on the face of a 4-inch square by 2⅛-inch box, the box volume would be increased to 48.3 inches.

Only one box extension should be used to increase the volume of a junction box. If the 4-inch or 4¹¹⁄₁₆-inch square box will be adapted to mount devices like switches or receptacles, or to support a lighting fixture, then box extensions should not be used.

If the adapter ring has its cubic-inch volume etched on its surface, that volume can be used to increase the box volume to meet NE Code requirements. Domed covers and extension rings must also be etched or otherwise labeled with the available volume, if that volume is to be used in sizing the box.

Example 4–8. Determine the NE Code-minimum size of a 4¹¹⁄₁₆-inch square junction box required for the splice of five No. 8 AWG THHN copper conductors to five No. 12 AWG THHN copper conductors. Each circuit is the same 3-phase, 4-wire branch circuit with a fifth-wire equipment-grounding conductor. The No. 12s are the homerun to the panel. The No. 8s, which are used to compensate for voltage drop, are entended to the connected load. The box contains two conduit entries. The splices are made with crimp-sleeve connectors and insulating caps. The equipment-ground splice will also be pigtailed to a grounding lug installed in the box. The box will be mounted to a steel beam with beam clamps, and two 1/4-20 hexbolts.

Total number of conductors computation: Ref. 370-16(b)

Count	What	Wire size	Total number of conductors
4	No. 8s for termination	No. 8	4
4	No. 12s for termination	No. 12	4
2	Equipment grounds for termination	No. 8	1

Conductor to volume conversion: Ref. T370-16(b)

Count	Wire size	Volume for one (cubic inches)	Extended volume (cubic inches)
5	No. 8s	3.00	15.00
4	No. 12s	2.25	9.00
			Total conductor volume: 24.00

NE Code-minimum box size determination: Ref. T370-16(a)

Table-listed volume: 25.5 cubic inches

Box size: 4¹¹⁄₁₆ × 1¼-inch square

Code challenge 4–8. Determine the NE Code-minimum size of a 4¹¹⁄₁₆-inch square box required for the splice of a single 3-phase, 4-wire branch circuit with fifth-wire equipment ground to two downstream feeds. The box will contain one ¾ inch conduit entry and two ½ inch conduit entries. The ¾ inch homerun will contain five No. 10 AWG THWN copper conductors that splice 1-to-2 to ten No. 12 AWG THWN copper conductors that exit five each in the two ½ inch conduit runs. The conductors will be spliced with wire nuts. The equipment-ground splice will also be pigtailed to a grounding lug installed in the box. The box will be surface mounted with two ¼-by-¾ lag bolts.

Total number of conductors computation: **Ref. (A)** ____________

Count	What	Wire size	Total number of conductors
(B) ________	No. 10s for termination	No. 10	(C) ________
(D) ________	No. 12s for termination	No. 12	(E) ________
3	Equipment grounds for termination	(F) ________	(G) ________

Conductor to volume conversion: Ref. (H) ____________

Count	Wire size	Volume for one (cubic inches)	Extended volume (cubic inches)
(I) ________	No. 10s	(J) ________	________
(K) ________	No. 12s	(L) ________	________
		Total conductor volume: (M)	________

NE Code-minimum box size determination: Ref. (N) ____________

Table-listed volume: (O) ________ cubic inches

Box size: 4¹¹⁄₁₆ × (P) ________ inch square

Example 4–9. Determine the NE Code-minimum size of an octagon fan box required for the installation of a paddle fan/lighting fixture assembly. The fan/light will be supported with an external-mounted strap and fan-stem combination. The box will be supplied with a No. 12/3 w/GND Romex cable supported with an internal cable clamp. The fan/light supply will consist of three No. 14 AWG MTW circuit conductors and one No. AWG 14 MTW equipment-grounding conductor. All splices will be made with wire nuts.

Total number of conductors computation: Ref. 370-16(b)

Count	What	Wire size	Total number of conductors
1	No. 12/3 cable termination	No. 12	3
3	No. 14s for termination	No. 14	3
2	Equipment grounds for termination	No. 12	1
1	Internal cable clamp	No. 12	1

Conductor to volume conversion: Ref. T370-16(b)

Count	Wire size	Volume for one cubic inches	Extended volume cubic inches
5	No. 12s	2.25	11.25
3	No. 14s	2.00	6.00
			Total conductor volume: 17.25

NE Code-minimum box size determination: Ref. T370-16(a)
Table-listed volume: 21.5 cubic inches
Box size: 4 × 2⅛-inch octagon

If the supply leads to the fan/light assembly had been wire size No. 16 instead of No. 14 AWG, the octagon box would have been sized for a count of only five No. 12 AWG circuit conductors—4 by 1¼-inch octagon box could have been used.

Code challenge 4–9. Determine the NE Code-minimum size of an octagon fixture box required to support a chandelier. The chandelier will be supported by a fixture-stud, -hickey, and -stem combination inside the box. The light will be supplied with a single No. 12/2 w/GND Romex cable entry supported by a two-screw squeeze connector. Wiring for the chandelier will consist of two No. 14 AWG circuit conductors and one No. 14 equipment-grounding conductor. Splices will be made with wire nuts. The equipment-ground splice will contain a No. 12 pigtail connected to a grounding screw inside the box.

Total number of conductors computation: Ref. (A) ____________

Count	What	Wire size	Total number of conductors
(B) ____________	No. 12/2 cable termination	No. 12	(C) ____________
(D) ____________	No. 14s for termination	No. 14	(E) ____________
2	Equipment grounds for termination	(F) ____________	(G) ____________
1	Fixture stud	(H) ____________	(I) ____________
1	Fixture hickey	(J) ____________	(K) ____________

Conductor to volume conversion: Ref. (L) ____________

Count	Wire size	Volume for one cubic inches	Extended volume cubic inches
(M) ____________	No. 12s	(N) ____________	____________
(O) ____________	No. 14s	(P) ____________	____________
			Total conductor volume: (Q) ____________

NE Code-minimum box size: Ref. (R) ____________

Table-listed volume: (S) ________ cubic inches

Box size: 4 × (T) ________ inch octagon

Section 370-16(a)(2) requires manufactured boxes of 100-cubic-inch volume or less, and all nonmetallic boxes to be durably and legibly marked by the manufacturer with their cubic-inch volume. A 6 by 6 by 4 junction box (6 inches square by 4 inches deep) would have a cubic-inch volume of 144 cubic inches.

When manufactured boxes are used as junction boxes for conductors in wire sizes No. 6 AWG or smaller, the box volume must be greater than or equal to the required total conductor volume as determined by Sec. 370-16(b).

When conduit bodies are used for pull points in a conduit run (Fig. 4-16), Sec. 370-16(c) requires their cross-sectional area to be at least twice the cross-sectional area of the largest conduit to which they are attached. The maximum number of conductors that passes through the conduit body is based on the percent-fill allowance, determined by Table 1 of Chapter 9, for the largest conduit to which they are attached.

As a general rule, Sec. 370-16(c) will not allow conduit bodies to contain conductor splices unless the cubic volume is marked on the body, this volume can meet the volume requirements of a junction box (for the contained conductors) as determined by Sec. 370-16(b), and the conduit body is supported in a rigid and secure manner.

Short-radius conduit bodies, such as "jake-90's" or service-entrance elbows (Fig.

Fig. 4-16 As a general rule, conduit bodies must be the same size as the conduit they access and cannot contain any splices.

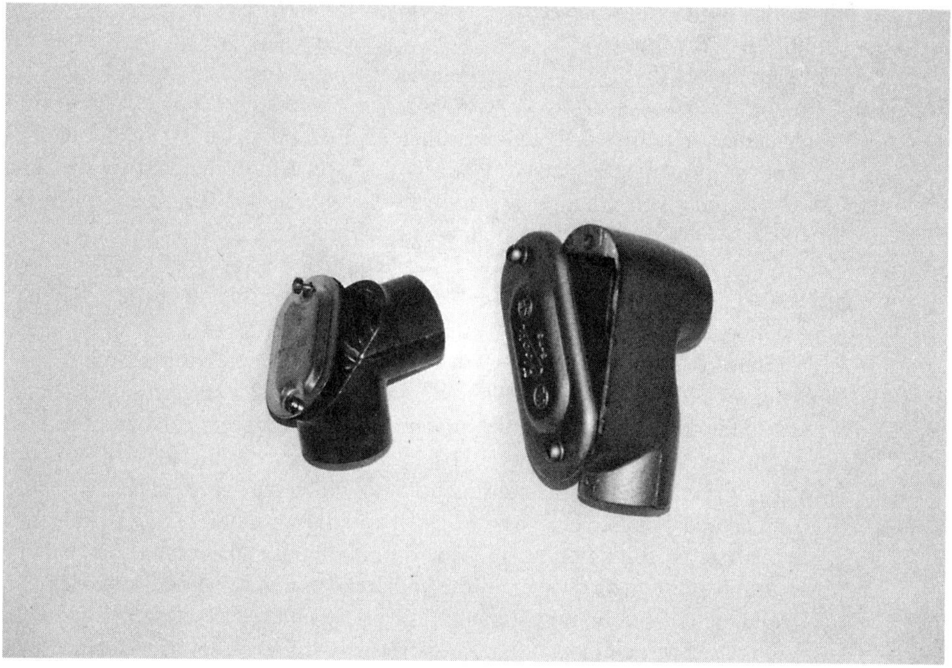

Fig. 4-17 Jake 90s (left) or service-entrance ells (right) cannot contain any splices due to the reduced body volume.

4-17), are restricted by Sec. 370-5 from containing any splices, taps, or other circuit devices.

REVIEW QUESTIONS

1. The installation of electrical boxes, whether outlet, device, pull, or junction, are governed by Art. ____________.
2. Correctly sizing an electrical box depends on the intended use of the box, and the wire size of the conductors that will either terminate in or be pulled through the box.

   True or False

3. The main concern of sizing any box is to ensure that the bending of the conductors inside the box will not damage the conductor insulation due either to a bend that is too sharp or to cuts and scrapes against the interior wall of the box.

   True or False

4. Trying to put too much conductor (or too many conductors) into a box that is too little can cause high-magnitude short-circuit or ground-fault overcurrent conditions in the circuit.

   True or False

Questions 5 and 6: Sizing of an outlet, device, pull, or junction box has two major divisions:

5. When the raceway entries will contain conductors wire size No. 6 AWG or smaller, the size of the box is governed by required conductor volume, available box volume, and Sec. ____________.
6. When the raceway entries will contain conductors wire size No. 4 AWG or larger, the dimensions of the box are governed by conductor deflection (bending of the conductor) and Sec. ____________ (600 volts or less).
7. The typical light switch box is referred to as a device box, since it is not a means or point of power supply (outlet location).

   True or False

8. Insulated circuit conductors can fall victim to short-circuit and ground-fault overcurrents if too many bends or turns are placed in a single conduit run.

   True or False

9. A pull box or conduit body must be placed in series with a run of rigid metal conduit pipe when the total deflection of the circuit conductors (turns, bends, or offsets in the pipe) will exceed ____________°.
10. A pull box or conduit body must be placed in series with a run of EMT (electrical metallic tubing) when the total deflection of the circuit conductors (turns, bends, or offsets in the pipe) will exceed ____________°.
11. The number of conductors that can be installed in a box intended to house devices or to serve as a junction or pull box for conductors wire size No. 6 AWG and smaller is governed by Sec. ____________ and Table ____________.
12. When sizing outlet or device boxes, conductors that pass through the box without splice are counted one for one (one conductor each). The box is being used only as a pull box.

    True or False

13. Table 370-16(a), which lists the number of conductors that can be installed in an outlet or device box according to a particular wire size and box dimensions, is based on the requirements of a pull box with no devices, fittings, conductor splices, or conductor terminations contained.

    True or False

14. The box volume or cubic-inch capacity available from an outlet or device box can be determined by Table ____________.
15. The volume or cubic-inch capacity required for a conductor to be installed in an outlet or device box can be determined according to wire size (No. 18 up to No. 6 AWG) in Table ____________.
16. When conductors enter and are terminated in an outlet or device box, they are counted one for one (one for each conductor), such that splices represent two or more conductors that terminate in the box.

    True or False

17. When sizing outlet or device boxes, fittings such as external cable clamps, wire nuts or other fasteners, box-grounding clips, studs, or connectors, and conduit entries (locknuts and bushings) require a reduction by one in the number of conductors that can be installed in the outlet or device box. This reduction is required for each type of fitting.

    True or False

18. When sizing outlet or device boxes, internal cable clamps are counted as a conductor reduction (volume allowance) one for one (one conductor for each clamp) as to the number of conductors that can be installed in the outlet or device box.

    True or False

19. When sizing outlet or device boxes, fixture-mounting equipment, such as fixture studs and hickeys, require a reduction by one in the number of conductors that can be installed in the outlet or device box. This reduction is required for each type of fitting.

    True or False

20. When devices, such as receptacles or lighting toggle switches, are to be installed in an outlet or device box, the number of conductors that can be installed in the box, according to wire size and box dimensions, must be reduced by a count of one for each strap that contains such devices.

    True or False

21. When determining the total number of conductors installed in an outlet or device box, one or more regular equipment-grounding conductors installed in the box counts as only one conductor.

    True or False

22. If an outlet or device box will contain both a regular equipment-grounding conductor and an isolated (insulated) equipment-grounding conductor, the total number of grounding conductors installed in the box is counted as only one conductor.

    True or False

23. When a mixture of wire sizes will be installed in an outlet or device box, the box dimensions must be based on the total conductor volume required. Volume for conductor deduction requirements must be based on the smallest conductor installed in the box.

    True or False

24. When conduit or tubing is used as the grounding means for the electrical box, round metal boxes can be used for either outlet, device, pull, or junction boxes, provided all the conduit entries are to the ____________ of the box.
25. A 3-phase, 4-wire, multiwire branch circuit consists of three ungrounded phase conductors, one from phase A, one from phase B, and one from phase C, which supply individual two-wire branch-circuit loads connected to a common fourth-wire neutral conductor. If a separate equipment-grounding conductor is installed, it becomes the fifth wire in the conduit.

    True or False

REVIEW PROBLEMS

1. Determine the NE Code-minimum size of an octagon box required for the installation of five No. 14 AWG THHN copper conductors. Two of the conductors will pass through the box without splice. The other three will be spliced and pigtailed to supply a face-mounted lighting

assembly. The splices will be made using crimp-sleeve connectors with insulating caps. The box will be grounded by the conduit.

2. Determine the NE Code-minimum size of a 4-inch square box required to support a face-mounted lighting fixture by adapter ring. The box will contain the splice of a single-phase, 3-wire, multiwire branch circuit of No. 12/3 w/GND Romex cable. One phase conductor, the neutral, and the equipment ground will also be pigtailed with No. 12 THW copper conductors for connection to the lighting fixture. The splices will be made with crimp-sleeve connectors and diapers.
3. Determine the NE Code-minimum size of a 3-by-2 device box required for the installation of a single triplex (triple) receptacle outlet. The box will contain two No. 14/3 w/GND Romex cable entries; the homerun and the downstream feed. The two Romex cables, secured to the box with external cable clamps, will be spliced with wire nuts and pigtailed to supply the receptacle. The equipment-grounding conductor will be pigtailed twice. One pigtail will be connected directly to the grounding terminal on the receptacle. The other pigtail will be connected to the metal box with a grounding clip.
4. Determine the NE Code-minimum size of a 4-inch square box with a plaster ring required for the installation of a two-gang split-receptacle outlet. The box will contain two ½ inch conduit entries, and the splice of a single-phase, 3-wire, multiphase branch circuit, with fourth-wire equipment ground. The four No. 12s in the homerun will be spliced 1-to-1 and pigtailed twice to feed the two receptacles. The splice to the downstream feed will be made using crimp-sleeve connectors and boots.
5. Determine the NE Code-minimum size of a 4 11/16-inch square box with a two-gang plaster ring required for the installation of two duplex receptacles—one isolated-ground type, and one regular grounding type. The homerun feeding the box will contain six No. 12 AWG THW copper conductors installed in ¾ inch conduit. The regular equipment ground will be solid green. The isolated ground will be green with one yellow stripe. One neutral will be white. The other will be white with a yellow stripe. One phase conductor will be yellow. The other will be red.
6. Determine the NE Code-minimum size of a 4-by-2⅛ handy box required for the surface mount of a single duplex switch. The single ½ inch conduit entry to the box will contain a total of four No. 14 AWG THW copper conductors. Three of the No. 14s will terminate to the respective terminals on the duplex switch as a hot feed (continuous power) and two switch-legs. The fourth No. 14 will connect to the equipment-grounding terminal on the switch strap.
7. Determine the NE Code-minimum size of a 4-inch square box and single box extension required for the splice of two 3-phase, 4 11/16-wire, multiwire branch circuits, and a common equipment-grounding conductor. The two branch circuits will leave the box in separate conduits. A total of nine No. 10 AWG THW copper conductors will be installed in the 1-inch conduit homerun, and five No. 10 AWG conductors in each of the two ¾ inch exit conduits. The No. 10 equipment-grounding conductor will be spliced 1-to-3 and pigtailed to a grounding lug mounted inside the box. The splices will be made with wire nuts.
8. Determine the NE Code-minimum size of a 4-inch square box required for the splice of a No. 10 AWG THW 3-phase, 4-wire homerun to two separate No. 12 AWG THW 3-phase, 4-wire, multiwire branch circuits. The two downstream branch circuits, which exit the box in separate ½ conduits, will each contain a fifth-wire No. 12 THW equipment-grounding conductor that is spliced common to a single No. 10 AWG THW equipment-grounding conductor installed the single ½ inch homerun conduit. The splices will be made with wire nuts. The box will be grounded by the ½ inch homerun conduit entry.
9. Determine the NE Code-minimum size of a 4-inch octagon box required for the installation of a 3-speed, reversing paddle fan. The box will be supplied with a No. 10/3 w/GND Romex ca-

ble, which is supported by an internal cable clamp. The fan will be externally box mounted and supplied with three No. 14 AWG MTW conductors. The bare equipment ground in the Romex cable will be spliced to a fourth No. 14 AWG equipment-grounding conductor from the fan assembly and pigtailed to a grounding clip mounted on the side wall of the box. Splices will be made with wire nuts.

Chapter 5

Pull and Junction Boxes

Outlet and device boxes, as well as junction and pull boxes for raceway entries containing wire sizes No. 6 AWG and smaller were discussed in Chapter 4. The NE Code-minimum box size was determined based on box volume available and required conductor volume.

Boxes containing conductors wire size No. 4 AWG and larger are not to be considered as device or outlet boxes. The dimensions of these pull or junction boxes must be determined according to the inner diameters (ID) of the conduit entries. Section 370-28 governs the installation and use of these larger pull and/or junction boxes. The minimum dimensions are based on conductor deflection, or bending of the wire, so as not to damage the conductor insulation during installation.

Auxiliary gutters and wireways, which can contain conductors of any wire size, will also be discussed in this chapter. For smaller circuit conductors, as with outlet or device boxes, the maximum number of conductors that can be contained or installed in either the gutter or wireway is based on the volume available.

Actually, for smaller wire sizes, the gutter or wireway is sized more on the order of piping-type raceway, using available cross-sectional area. The volume comparison could be based on the volume available at any one point or cross section.

For larger circuit conductor installations, as with pull and junction boxes, the size or dimensions of the auxiliary gutter or wireway is based on the requirements of conductor deflection.

5.1 PULL AND JUNCTION BOXES

As the name implies, a pull box is intended only as a *pull point* in the raceway run for the conductors, and as a general rule will not contain splices. In other words, the conductors are pulled by fish tape, rope, or cord from an entry point in the raceway to the first access point or pull box. The fish tape is unhooked from the conductors and extended from the next access point (pull box or termination enclosure) back to the first access point. Reconnecting the fish tape allows the conductors to be pulled to the next access point.

In a straight pull (Fig. 5-1), the pull box is used *only* as an access point to reconnect the fish tape and to relubricate the conductors. (Conductors are normally lubricated with a soap compound to reduce the friction between the interior wall of the conduit and the conductor insulation.) Sometimes conduit bodies are used for access points instead of pull boxes.

If the pull box is used for an angle pull (a change of direction for the raceway run), the pull box is used both as a pull point and as a guide point to ensure the conductor insulation is not scraped, cut, or somehow damaged in the change of direction. An angle pull

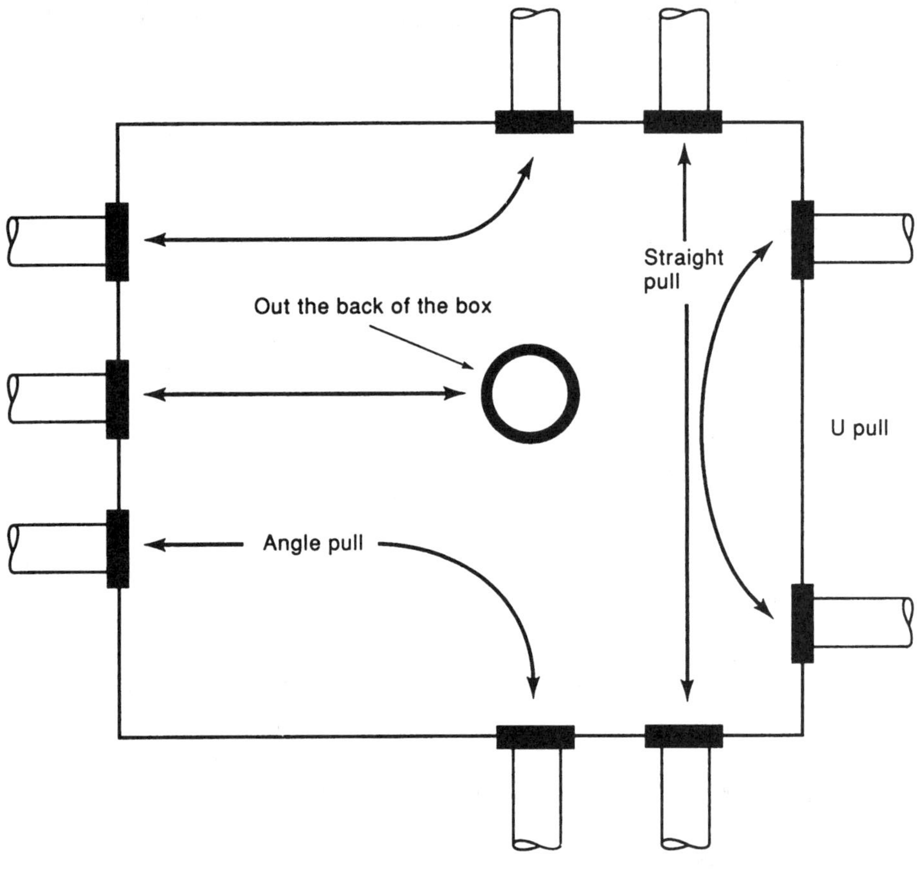

Fig. 5-1 Conductor pulls into a pull or junction box can be described as either a straight pull, an angle pull, or a U pull.

(Fig. 5-1) is described as a pull from a raceway-entry wall to another raceway-exit wall other than the wall opposite the entry wall. A straight pull is described as a pull from raceway-entry wall to raceway-exit in the opposite wall.

A pull from side wall entry to back wall exit, or vise-versa, would also be described as an angle pull. The back wall of a pull or junction box is the wall opposite the removable cover. In a U pull, the raceway containing the same conductors enters and exits the same wall of the box—either a side wall or the back wall.

Pull boxes are also required when more than 360° of conductor deflection (bends, turns, or offsets) will be installed in a single conduit or tubing run (piping raceway [Secs. 331-10, 345-11, 346-11, 347-14, and 348-10]). A junction box, in comparison, can be used for pulling and/or splicing of conductors.

Box dimensions as used in this text: For clarity, the dimensions of boxes used in the examples or problems shall be determined as looking into the page of the book, with the following criteria:

Length: as measured from the left wall to the right wall of the box

Height: as measured from the top wall to the bottom wall of the box

Depth or **width:** as measured from the surface into the paper (usually requires an isometric or other angle drawing)

Diagonal: for angle pulls, the distance measured between the centerlines of two raceways (conduits) that contain or enclose the same conductors

Box dimensions: will be specified as length × height × depth

5.2 STRAIGHT PULLS

For a straight pull, Sec. 370-28(a)(1) requires the overall length of the pull or junction box to be greater than or equal to eight times the trade diameter (8td) of the largest raceway being utilized. As a general rule, the conduit (or other pipe-type raceway) size will not change when only a single conduit run is installed.

For a straight pull in a single raceway run, Sec. 370-28(a)(1) does not specify the width or depth of the pull or junction box. These dimensions would be governed by the trade diameter of the raceway. Unless the raceway is connected to the wall of the junction or pull box by means of a threaded hub, Sec. 300-10 requires the raceway to both enter the box and be mechanically secured to it (locknuts). The accepted trade practice of determining these dimensions of a flat-sided box (pull or junction) is to add two inches to the trade diameter (inner diameter [ID]) of the entry raceway (td + 2 inches). Two inches, one inch on either side of the conduit or tubing, will normally allow access to the locknuts and bushings for installation and tightening.

Example 5–1. Determine the NE Code-minimum dimensions of a rectangular metal box used for a straight pull in 3-inch rigid conduit.

NE Code-minimum length computation: Ref. 370-28(a)(1)

$$8 \times 3 \text{ inches} = 24 \text{ inches}$$

Recommended minimum height and depth computations: Ref. 300-10

3 inches + 2 inches = 5 inches

In comparison to flat-sided metal boxes, conduit bodies (often referred to as condulets, in the form of an LB, LL, LR, or C body) are cast with hubs for pipe raceway entry and exit. The hubs are threaded to accept either the threaded pipe end, or the straight connector for conduit or tubings that cannot be threaded. The td + 2 requirement for raceway makeup does not apply to conduit bodies.

Code challenge 5–1. Determine the NE Code-minimum dimensions of a rectangular metal box used for a straight-pull in 4-inch IMC conduit.

NE Code-minimum length computation: Ref. (A) ____________

(B) __________ × __________ inches = (C) __________ inches

Recommended minimum height and depth computations: Ref. (D) ____________

__________ inches + (E) __________ inches = (F) __________ inches

A central junction box can be used as the pull point for several parallel runs of raceway. When several straight pulls are contained on the same wall of a junction box, eight times the trade diameter of the largest raceway will decide the minimum box length to the opposite wall of the box.

To meet the requirements of Sec. 300-10, the height dimension of the box must house all the raceways that enter the same wall, plus allow two inches for raceway makeup of each raceway (number of raceways × 2 inches). The box's width or depth must house the largest conduit entry: td + 2 inches.

Example 5–2. Determine the NE Code-minimum box dimensions of a common pull box if three straight pulls of 4-inch, 3-inch, and 2-inch conduit are stacked top to bottom on the left wall of the box (refer to Fig. 5-2).

NE Code-minimum length computation: Ref. 370-28(a)(1)

8 × 4 inches = 32 inches

Recommended minimum height computations: Ref. 300-10

Sum of trade diameters: 4 inches + 3 inches + 2 inches =	9 inches
Makeup allowance: 3 × 2 inches =	6 inches
Total height:	15 inches

Recommended minimum depth computations: Ref. 300-10

4 inches + 2 inches = 6 inches

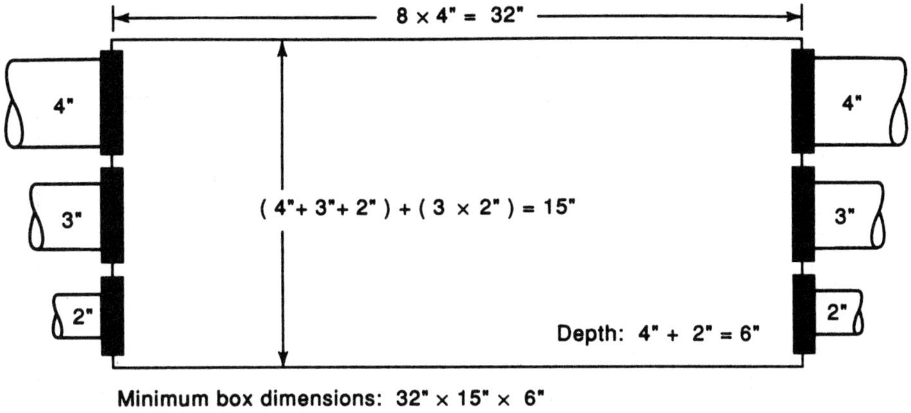

Fig. 5-2 For multiple straight pulls in a common box, the length to the wall opposite is based on the trade diameter of the largest raceway.

Code challenge 5–2. Determine the NE Code-minimum box dimensions of a common pull box if one straight pull of 6-inch conduit and two straight pulls of 4-inch conduit are stacked top to bottom on the left wall of the box (similar to Fig. 5-2).

NE Code-minimum length computation: Ref. (A) ______________

(B) __________ × __________ inches = (C) __________ inches

Recommended minimum height computations: Ref. (D) ______________
Sum of trade diameters: ________ + ________ + ________ = (E) ________ inches
Makeup allowance: ________ × (F) ________ inches = ________ inches
Total height: (G) ________ inches
Recommended minimum depth computations: Ref. (H) ______________

__________ inches + (I) __________ inches = (J) __________ inches

5.3 ANGLE PULLS

The minimum box dimensions of an angle pull (Fig. 5-3) are governed by Sec. 370-28(a)(2). The length of the box from the raceway-entry wall to the side wall opposite must be a minimum of six times the trade diameter (6td) of the raceway. This measurement will govern two dimensions of the box when the angle pull is from side to side: its length and height. The depth (width) of the box is governed by the trade diameter of the largest raceway plus two inches (td + 2, Sec. 300-10).

Example 5–3. Determine the NE Code-minimum dimensions of a metal box required for an angle pull of 6-inch rigid metal conduit from left to bottom wall (refer to Fig. 5-3).

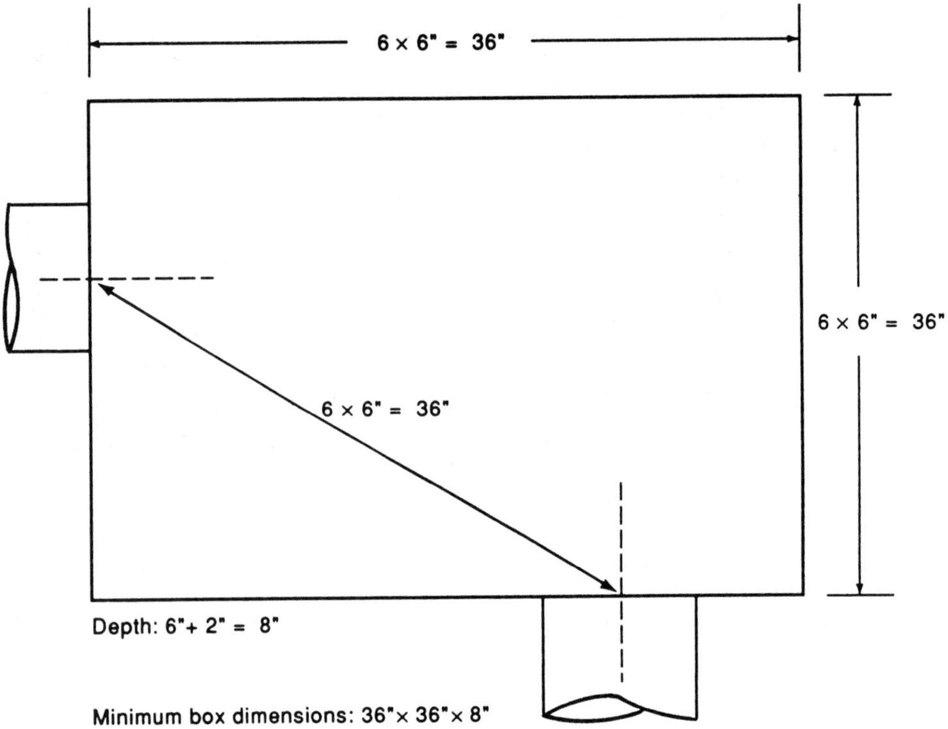

Fig. 5-3 For a single angle pull, the 6td requirement governs the length and height of the box and the relative location of the conduit entries.

NE Code-minimum length computation: Ref. 370-28(a)(2)

$$6 \times 6 \text{ inches} = 36 \text{ inches}$$

NE Code-minimum height computation: Ref. 370-28(a)(2)

$$6 \times 6 \text{ inches} = 36 \text{ inches}$$

Recommended minimum depth computation: Ref. 300-10

$$6 \text{ inches} + 2 \text{ inches} = 8 \text{ inches}$$

Code challenge 5–3. Determine the NE Code-minimum dimensions of a metal box required for an angle pull of 3½-inch EMT conduit from the left to bottom wall (similar to Fig. 5-3).

NE Code-minimum length computation: Ref. (A) ____________________

(B) ____________ × ____________ inches = (C) ____________ inches

NE Code-minimum length computation: Ref. (D) ____________________

(E) ____________ × ____________ inches = (F) ____________ inches

Recommended minimum depth computations: Ref. (G) ____________

____________ inches + (H) ____________ inches = (I) ____________ inches

In Figure 5-3, the diagonal measurement between the two centerlines, calculated at 36 inches minimum, is also an NE Code requirement that may affect the box dimensions as computed in Example and Code challenge 5-3. The second paragraph of Sec. 370-28(a)(2) (following the Exception) stipulates that the distance between raceway entries containing the same circuit conductors must be a minimum of 6td apart. This requirement applies to both angle pulls and U pulls. In the angle pull, this measurement is referred to as the diagonal.

The diagonal is measured from the raceway centerlines, simply because the center of the hole to be drilled, bored, or punched in the side wall of the box to accept the raceway is normally drawn on centers between the height of the wall and the length of the wall.

Although centering in this fashion may look well cosmetically for the electrical installation, the box dimensions computed for Figure 5-3 will not allow the raceways to be centered in the wall length of the box and meet the 6td diagonal requirement. Notice that both raceway entries are offset.

The two raceway entries of an angle pull from side to side in the walls of a box form a right angle with the common corner. The relative location of where the two raceways can enter the side walls is governed by the diagonal length of 6td and the Pythagorean theorem.

The measured length from the common corner to each centerline of the two raceway entries serves as the two bases or sides of the triangle (refer to Fig. 5-4). The Pythagorean theorem states that the square of the hypotenuse (diagonal) of a right triangle is equal to the sum of the squares of the two sides. If the two sides and diagonal of the right triangle are labeled as A, B, and C respectively, then by formula:

$$A^2 + B^2 = C^2$$

and

$$C^2 = A^2 + B^2$$

The box dimensions of Figure 5-3 were computed at 36 inches by 36 inches. To center the raceway entries, the 36 inches for each wall length would be divided by 2, equaling 18 inches measured from the common corner to the centerlines. Plugging these values into Pythagorean's theorem:

$$C^2 = (18 \text{ inches})^2 + (18 \text{ inches})^2$$

$$C^2 = 324 \text{ square inches} + 324 \text{ square inches}$$

$$C = (648 \text{ square inches})^{1/2}$$

$$C = 25.5 \text{ inches}$$

Eighteen inches squared equals 324 square inches. The side lengths would be equal, so the square of the diagonal is equal to 648 square inches. Taking the square root of both sides of the equation, the diagonal length, with the conduit entries centered in the 36-by-36 pull box, is only 25.5 inches. This does not meet the 6td diagonal requirement of Sec. 370-28(a)(2).

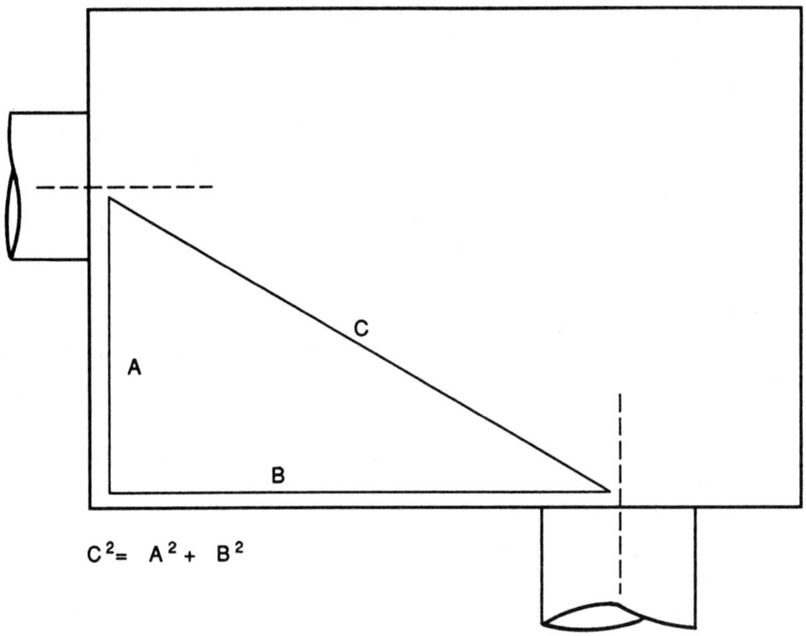

Fig. 5-4 Using the Pythagorean theorem, the relative location of the conduit entries can be found by the known length of the diagonal.

As drawn, the minimum lengths from the common corner to the centerlines could have been computed by assigning the 6td minimum computation for C in the formula. If the angle pull is to be offset, yet symmetrical, the two side lengths should be set equal, such that A = B.

$$A^2 + B^2 = C^2$$

$$A^2 + A^2 = C^2$$

$$2A^2 = C^2$$

$$A^2 = C^2 \div 2$$

$$A^2 = (36 \text{ inches})^2 \div 2$$

$$A^2 = 1{,}296 \text{ square inches} \div 2$$

$$A^2 = 648 \text{ square inches}$$

$$A = (648 \text{ square inches})^{1/2}$$

$$A = 25.4 \text{ inches}$$

The raceway-entry centerlines would have to be drilled at 25.4 inches from the common corner in the angle pull to meet NE Code-minimums for length, height, and diagonal measurements of the 36-by-36 box.

As shown in Figure 5-5, if symmetry is not a concern, or if a rectangular box will

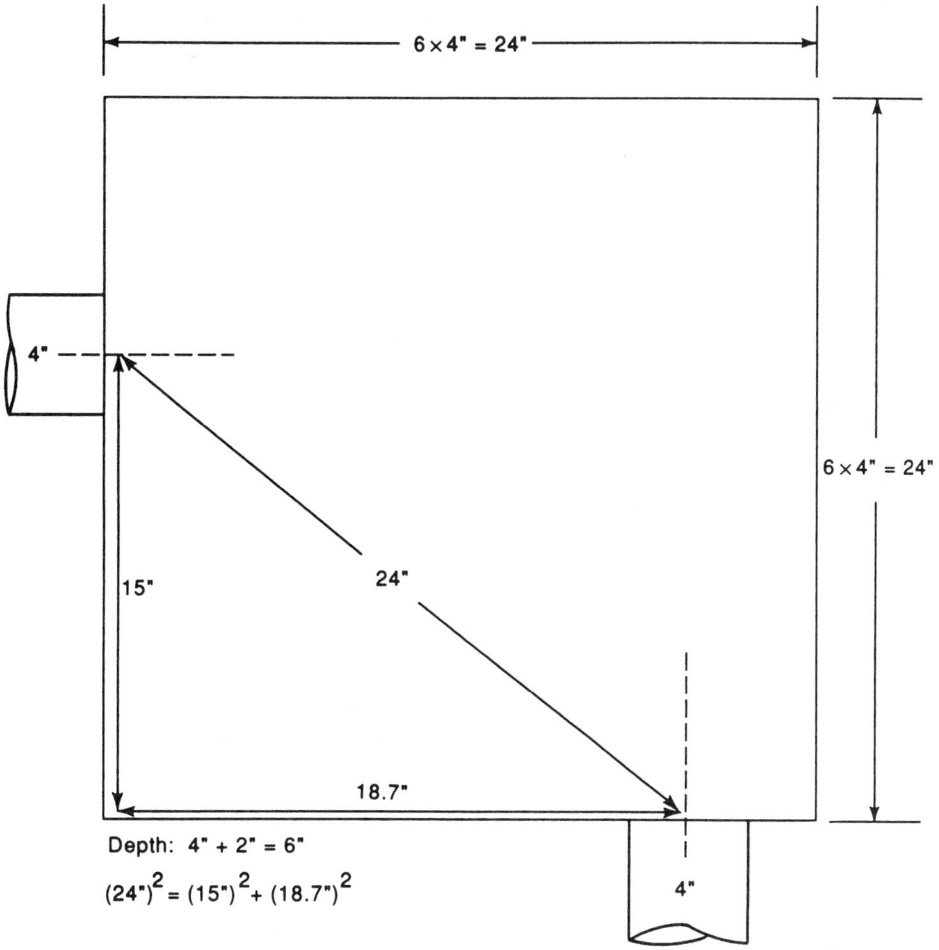

Fig. 5-5 The relative location of the conduit entries is governed by the common corner, the Pythagorean theorem, and the 6td diagonal requirement.

be installed, one side length can be assigned, and the other side measurement can be computed by solving the Pythagorean theorem for either side A or side B.

Example 5–4. Determine the NE Code-minimum box dimensions and the relative location of the bottom-exit conduit, if a 4-inch angle pull enters the left wall of the box with conduit centerline located at 15 inches from the bottom left corner of the box (refer to Fig. 5-5).

NE Code-minimum length computation: Ref. 370-28(a)(2)

$$6 \times 4 \text{ inches} = 24 \text{ inches}$$

NE Code-minimum height computation: Ref. 370-28(a)(2)

$$6 \times 4 \text{ inches} = 24 \text{ inches}$$

Recommended minimum depth computation: Ref. 300-10

$$4 \text{ inches} + 2 \text{ inches} = 6 \text{ inches}$$

NE Code-minimum diagonal computation: Ref. 370-28(a)(2)

$$6 \times 4 \text{ inches} = 24 \text{ inches}$$

Relative location of bottom-exit conduit: Ref. 370-18(a)(2)

If

A = left side wall measurement

B = bottom wall measurement

C = diagonal measurement

Then

$$B = (C^2 - A^2)^{1/2}$$
$$= [(24 \text{ inches})^2 - (15 \text{ inches})^2]^{1/2}$$
$$= (576 \text{ square inches} - 225 \text{ square inches})^{1/2}$$
$$= (351 \text{ square inches})^{1/2}$$
$$= 18.7 \text{ inches}$$

The diagonal measurement between two angle-pull raceways containing common conductors does not always affect the overall measurements of the pull or junction box, but it will always dictate the relative location of the two raceway entries into the box. The bases or legs of the triangle for an angle pull, from one wall to another wall other than the wall opposite, are relative to each other as well as to the length of the diagonal.

Code challenge 5–4. Determine the NE Code-minimum box dimensions and the relative location of the bottom-exit conduit, if a 3-inch angle pull enters the left wall of the box with the conduit centerline located at 12 inches from the bottom left corner of the box (similar to Fig. 5-5).

NE Code-minimum length computation: Ref. (A) ____________________

(B) ____________ × ____________ inches = (C) ____________ inches

NE Code-minimum length computation: Ref. (D) ____________________

(E) ____________ × ____________ inches = (F) ____________ inches

Recommended minimum depth computations: Ref. (G) ____________________

____________ inches + (H) ____________ inches = (I) ____________ inches

NE Code-minimum diagonal computation: Ref. (J) ____________________

(K) ____________ × ____________ inches = (L) ____________ inches

Relative location of bottom-exit conduit: Ref. (M) ____________________

If

A = left side wall measurement

B = bottom wall measurement

C = diagonal measurement

Then

$B = (C^2 - A^2)^{1/2}$

B = {[(N) ____________ inches]^2 – (O) ____________ inches]^2} $^{1/2}$

B = (____________ square inches – ____________ square inches) $^{1/2}$

B = [(P) ____________ square inches] $^{1/2}$

B = (Q) ____________ inches

As shown in Figure 5-6, if more than one angle pull enters the same wall of a box, Sec. 370-28(a)(2) requires the 6td of the largest raceway length to the wall opposite to be increased by the sum of the trade diameters of the other raceways mounted in the same entry wall.

Example 5–5. Determine the NE Code-minimum dimensions of a junction box with 4-inch, 3-inch, and 2-inch angle-pull conduits stacked in the left wall. The 4-inch angle pull exits the top wall of the box. The 2- and 3-inch angle pulls exit the bottom wall of the box (refer to Fig. 5-6).

NE Code-minimum length computation: Ref. 370-28(a)(2)

6td requirement:	6 × 4 inches =	24 inches
Sum of other trade diameters:	3 inches + 2 inches =	5 inches
Total length:		29 inches

NE Code-minimum height computation: Ref. 370-28(a)(2)

Three conditions must be satisfied with the one measurement.

1. The conduit stacked on the left wall:

Sum of trade diameters:	4 inches + 3 inches + 2 inches =	9 inches
Makeup allowance:	3 × 2 inches =	6 inches
Total height:		15 inches

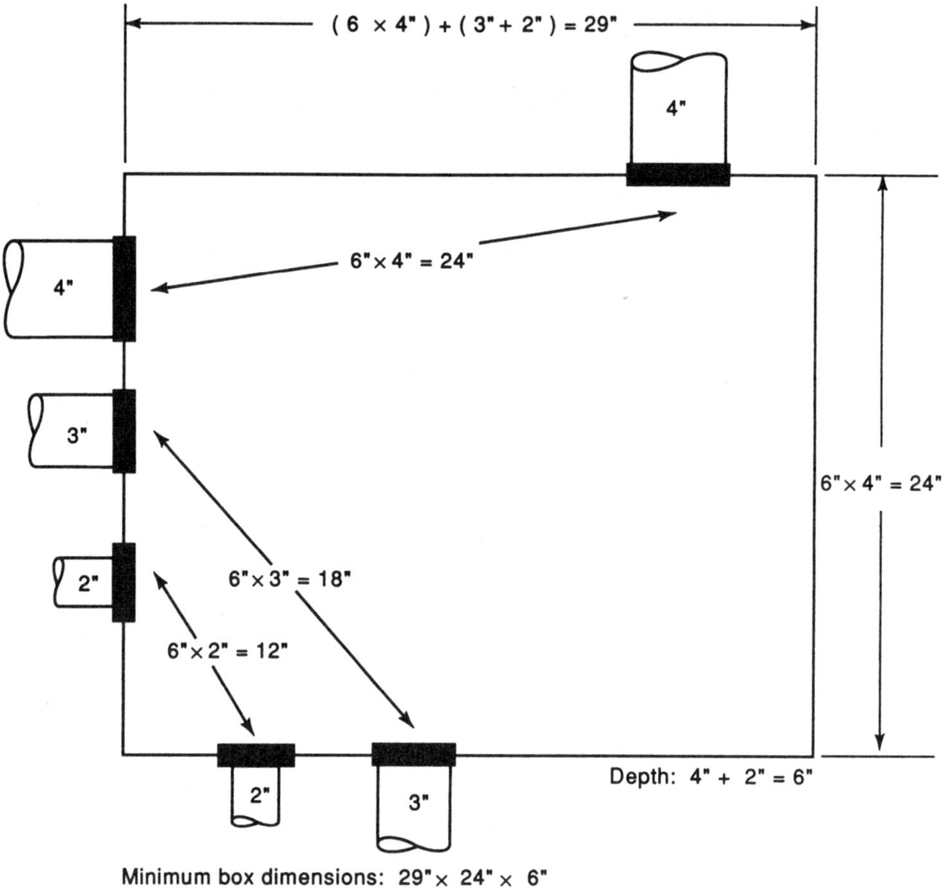

Fig. 5-6 Section 370-18(a)(2) requires the sum of the diameters of the other raceways to be added to the 6td of the largest diameter for the length to wall opposite.

2. The 4-inch angle pull from the top wall:
   6 × 4 inches = 24 inches

3. The 3-inch angle pull from the bottom wall:

6td requirement:	6 × 3 inches =	18 inches
Addition of the other trade diameter:		2 inches
Total height:		20 inches

NE Code-minimum height, compare for larger: 24 inches

Recommended minimum depth computation: Ref. 300-10

4 inches + 2 inches = 6 inches

Code challenge 5–5. Determine the NE Code-minimum dimensions of a junction box with 6-inch, 4-inch, and 3-inch angle-pull conduits stacked in the left wall. The 6-inch angle pull exits the top wall of the box. The 3- and 4-inch exit the bottom wall of the box (similar to Fig. 5-6).

NE Code-minimum length computation: Ref. (A) ______________

6td requirement: (B) __________ × __________ inches = __________ inches

Sum of other trade diameters: __________ + __________ = __________ inches

Total length: (C) __________ inches

NE Code-minimum height computation: Ref. (D) ______________

Three conditions must be satisfied with the one measurement.

1. The conduit stacked on the left wall:
   Sum of trade diameters:
   __________ + __________ + __________ = __________ inches
   Makeup allowance:
   (E) __________ × __________ inches = __________ inches
   Total height: (F) __________ inches
2. The 6-inch angle pull from the top wall:
   (G) __________ × __________ inches = __________ inches
3. The 4-inch angle pull from the bottom wall:
   6td requirement:
   (H) __________ × __________ inches = __________ inches
   Addition of the other trade diameter: __________ inches
   Total height: (I) __________ inches

NE Code-minimum height, compare for (J) __________: (K) __________ inches

Recommended minimum depth computation: Ref. (L) ______________

__________ inches + __________ inches = (M) __________ inches

5.4 ANGLE PULL OPPOSITE A REMOVABLE COVER

A problem exists when the raceway entry is from the back of the box (opposite the removable cover). A raceway entry into the back would constitute an angle pull to one of the side walls of the box. To meet the general requirements of Sec. 370-28(a)(2), the depth of the box would have to be at least six times the trade diameter of the raceway entry. For example, if the raceway is 3-inch rigid metal conduit, the depth of the box would have to be at least 6 × 3 inches, or 18 inches. This 18 inches would have to be increased still further if more angle-pull raceway runs enter the back of the box.

For most installations, this type of box is unrealistic. The Exception to Sec. 370-28(a)(2) allows the minimum depth of the box to be limited to the deflection requirements of the conductors in T373-6(a).

Deflection indicates the amount of bend required of the conductor in its routing from point of entry to point of termination, splice, or exit. The amount of bend is reflected in terms of radius of the bend. To protect the insulation from failure due to stretching on the outside of the bend, or from kinking on the inside of the bend, the radius of the bend is normally a multiple of the outer diameter (OD) of the conductor or cable.

Figure 5-7 shows a cut-away view of the long side of a box where the angle pull of the contained circuit conductors exits the back wall. The back wall is opposite a remov-

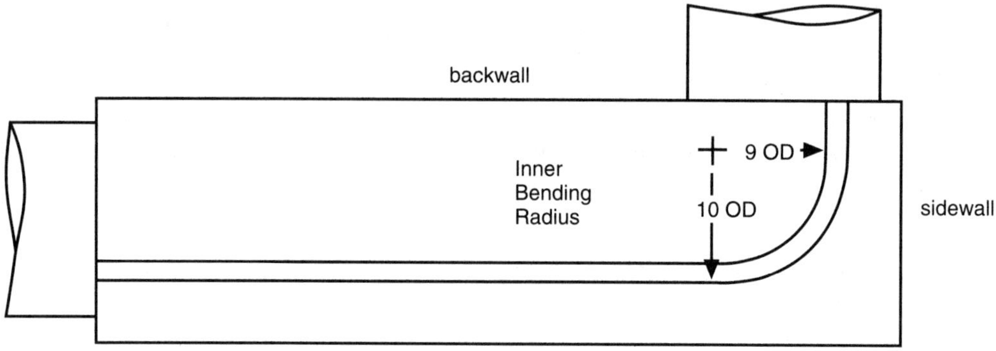

Fig. 5-7 The bending radius of a cable that enters or exits a wall opposite a removable cover is measured from a common point that is equidistant from every point around the bend.

able cover. If the bend in the wire is a full 90°, the bending radius is measured from a central point or origin, to where the arc is the same measured length from this point to any point along the inner curved surface of the conductor.

If the inner bending radius of the conductor is a multiple of 9 times the conductors OD, the box depth, as measured from the edge or lip of the open face of the box (removable cover) to the back wall, would have to be a minimum of 10 OD to totally enclose the circuit conductors.

For example, as listed in Table 5 of Chapter 9, a No. 1 AWG THHN conductor has an outer diameter of 0.446 inches. If the bending radius is set at a minimum of nine times the conductor OD, then the box depth would have to be a minimum of 4½ inches (10 OD) to meet conductor deflection (bending) requirements.

By the general requirement for an angle pull, if this No. 1 AWG THHN conductor was one of a four-wire feeder set in EMT conduit, Table C1 in Appendix C would require 1¼-inch conduit. Section 370-28(a)(2) would require a minimum box depth of 6td, or 7½ inches. By exception, the box depth is set at only 4½ inches—conductor deflection.

Instead of computing the required bending space for the conductors for every angle pull opposite a removable cover, Sec. 370-28(a)(2), Exception refers the electrician to the gutter space requirements of cabinets and cutout boxes in Art. 373. For all practical purposes, the required bending space for the circuit conductors, and hence the required depth of the box, is a look-up value as listed in T373-6(a).

Section 373-6 deals with the deflection of conductors either entering or leaving a cabinet or cutout box (both are load centers, which contain overcurrent protection devices or other electrical utilization equipment).

The open space between the terminal on a mounted circuit breaker or fuseholder and the wall opposite is referred to as the gutter space within the cabinet or cutout box.

Technically, Sec. 373-6(a) requires a wire gutter space, in which conductors are either bent for termination or routed through for termination at another location (within the same enclosure), on all sides of the cabinet or cutout box: on the sides, between the breakers or fused switches and side walls; at the bottom, between the neutral busbar (where used) or the last breaker slot and the bottom wall of the box; at the top, between the edge of the main breaker (or top of the lugs in a "main only" panel) and the top wall of the box.

Section 373-6(a) stipulates that the width of the wire gutter on all sides of the cabinet or cutout box shall not be less than the distance as listed in Table 373-6(a) according to wire size.

Section 373-6(b) governs the wire bending space about the connection to either the fuseholder, the breaker, or the busbar terminals. Subsection (1) stipulates that T373-6(a) shall be used to establish the minimum gutter space between the wall opposite the terminals and the terminals for an L or S bend in the conductors (Fig. 5-8). The number of "wires per terminal" to use for the minimum gutter space is selected according to the number of parallel conductors per phase in the feeder.

An L or S bend is indicated by the requirement that the conductors enter or leave any cabinet or cutout box wall other than the wall opposite their termination.

When the feeder conductors enter or leave the wall opposite the terminals on a fused disconnect, a circuit breaker or busbar, more gutter space is required in that section

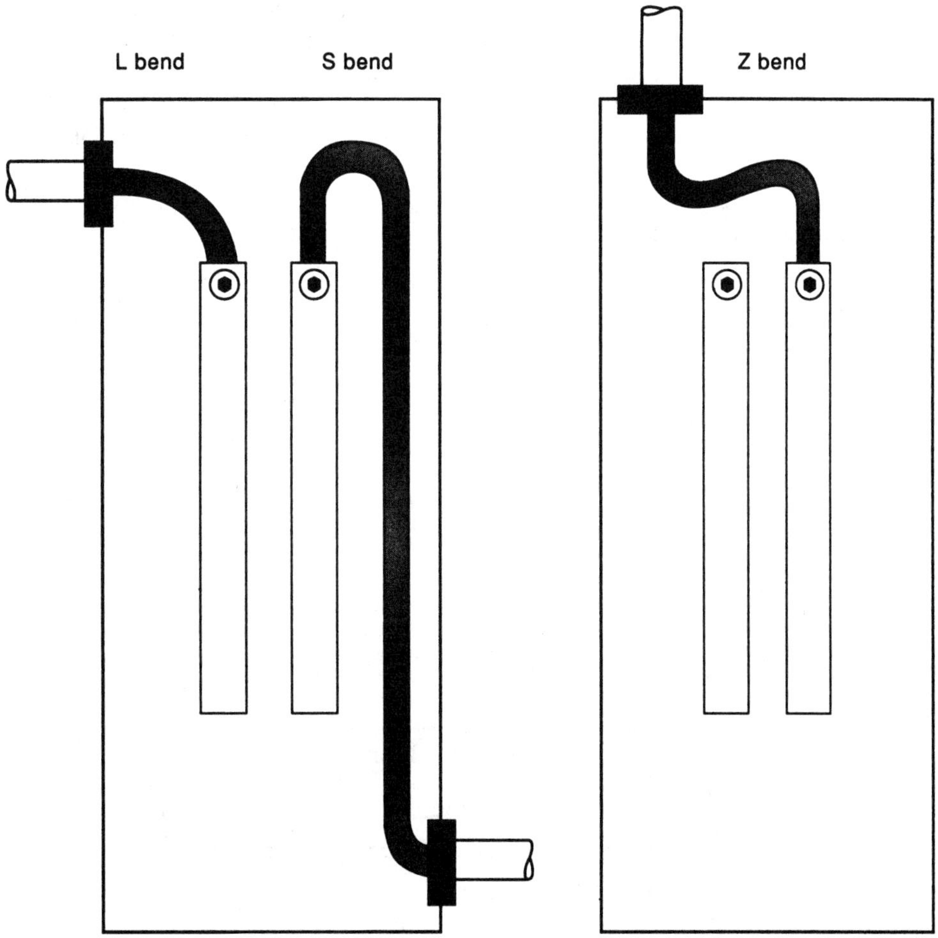

Fig. 5-8 Sections 373-6(a) & (b) govern the gutter space in a cabinet or cutout box according to the type of conductor bend.

of the cabinet or cutout box. The conductors are normally installed in a Z-bend configuration for proper connection to the terminal. The Z bend allows for thermal expansion without wedging the conductors against the bushing on the raceway entry.

The gutter width at the terminals for a Z-bend installation is governed by Sec. 373-6(b)(2) and T373-6(b). This gutter width is normally double the width required for an L or S bend as listed in T373-6(a).

To apply the requirements of S & T 373-6(a) to the proper sizing of a pull or junction box with raceway entry or exit opposite a removable cover (an L bend), the contained conductors in the conduit in question must be known. If only single conductors are contained in the feeder set (not run in parallel), the wire size establishes the row in T373-6(a). Section 370-28(a)(2), Exception specifies that the column should be located as "one wire per terminal." Where the row and column intersect or cross is the minimum depth value of the junction or pull box.

For example, if a 3-inch conduit, which contains four 350 kcmil THW copper conductors, exits as an angle pull out the back of a pull box, the minimum depth (measured from the open face to the back) of the box as found in column 1 of T373-6(a) is five inches.

The 5-inch depth, as determined by the table, will also allow for conduit makeup (locknuts and bushings: td + 2) of the conduit entry from the wall not opposite the removable cover. Five inches is also the minimum width of this pull box (again recommended minimum: td + 2).

The length of the pull box for the angle pull out the back is governed by the 6td minimum of an angle pull between two conduit entries containing the same conductors, and by conduit makeup requirements.

In Figure 5-9, the length of the diagonal between the two raceway centerlines must be a minimum of 6 × 3 inches, or 18 inches. If the sidewall entry of 3-inch conduit is cen-

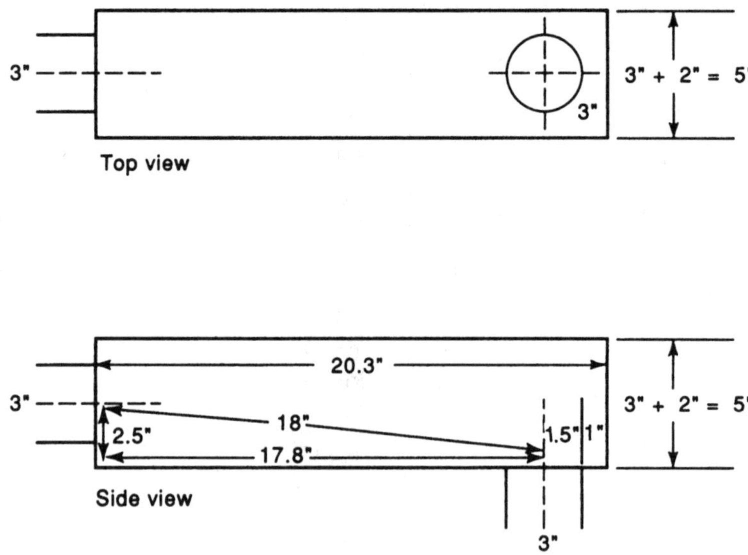

Fig. 5-9 The depth of the box as measured from the wall opposite a removable cover is greatly reduced by the exception to Sec. 370-18(a)(2).

tered in the 5-inch depth and width of the box, the distance between the corner of the box and the centerline of the 3-inch conduit entry in the back, using the Pythagorean theorem, will be 17.8 inches.

This measurement of 17.8 inches allows for only half of the conduit's internal diameter. For 3-inch conduit, 1.5 inches must be added to the 17.8 inches. The box length must also allow for conduit makeup. The 2-inch rule we adopted allows 1 inch on each side of the conduit. This installation needs only one side. The minimum length of the pull box would be computed as:

$$17.8 \text{ inches} + 1.5 \text{ inches} + 1 \text{ inch} = 20.3 \text{ inches}$$

The minimum box dimensions, as determined in accordance with Sec. 370-28(a)(2) would be:

$$20.3 \text{ inches} \times 5 \text{ inches} \times 5 \text{ inches}$$

As a rule, when the angle pull is opposite a removable cover, computing the measurement on the back wall from the common corner to the raceway centerline is not worth the effort because of the reduced box depth. In Figure 5-9, the required length of the diagonal was 18 inches. The length of the long side (of the triangle–back wall length) to the raceway centerline was 17.8 inches.

For this type of angle pull, the length of the diagonal, plus one-half the conduit trade diameter, plus one inch will always meet NE Code requirements.

Example 5–6. If a 2-inch rigid metal conduit will contain four No. 4/0 AWG THHN copper conductors, determine the NE Code-minimum box dimensions for an angle pull from the left side wall to the back wall of the box (similar to Fig. 5-9).

NE Code-minimum length computation: Ref. 370-28(a)(2)

Must equal a composite of three separate measurements.

1. Diagonal/back wall length to raceway centerline:
   $6 \times 2 \text{ inches} = 12 \text{ inches}$
2. One-half the width of the raceway td:
   $0.5 \times 2 \text{ inches} = 1 \text{ inch}$
3. One-half of the td + 2 raceway makeup: 1 inch

   Total length: 14 inches

Recommended minimum height computation: (top view) Ref. 300-10

Raceway makeup: 2 inches + 2 inches = 4 inches

Minimum depth computations: (side view)

Two conditions must be satisfied with the one measurement.

1. NE Code-allowed conductor deflection: Ref. 370-28(a)(2), Ex.

   T373-6(a): 4 inches

2. Recommended raceway makeup: Ref. 300-10

2 inches + 2 inches = 4 inches

Minimum depth, no comparison, equal: 4 inches

Code challenge 5–6. If a 4-inch rigid metal conduit will contain four 750 kcmil THHN copper conductors, determine the NE Code-minimum box dimensions for an angle pull from the left side wall to the back wall of the box (similar to Fig. 5-9).

NE Code-minimum length computation: Ref. (A) ____________
Must equal a composite of three separate measurements.

1. Diagonal/back wall length to raceway centerline:
(B) ________ × ________ inches = ________ inches

2. One-half the width of the raceway td:
________ × ________ inches = (C) ________ inches

3. One-half of the td + 2 raceway makeup: ________ inch

Total length: (D) ________ inches

Recommended minimum height computation: (top view) Ref. (E) ________

Raceway makeup: ________ inches + ________ inches = (F) ________ inches

Minimum depth computations: (side view)
Two conditions must be satisfied with the one measurement.

1. NE Code-allowed conductor deflection: Ref. (G) ____________

Table (H) ________ : ________ inches

2. Recommended raceway makeup: Ref. (I) ____________

________ inches + ________ inches = (J) ________ inches

Minimum depth, compare for (K) ________ : (L) ________ inches

5.5 U PULLS

For a U pull, two dimensions of the box are governed by the 6td requirement of Sec. 370-28(a)(2). As shown in Figure 5-10, the length (or height, according to the wall of entry) between entry wall and side wall opposite, must be at least six times the trade diameter of the largest raceway common to the U pull plus the diameter of the other raceway. The

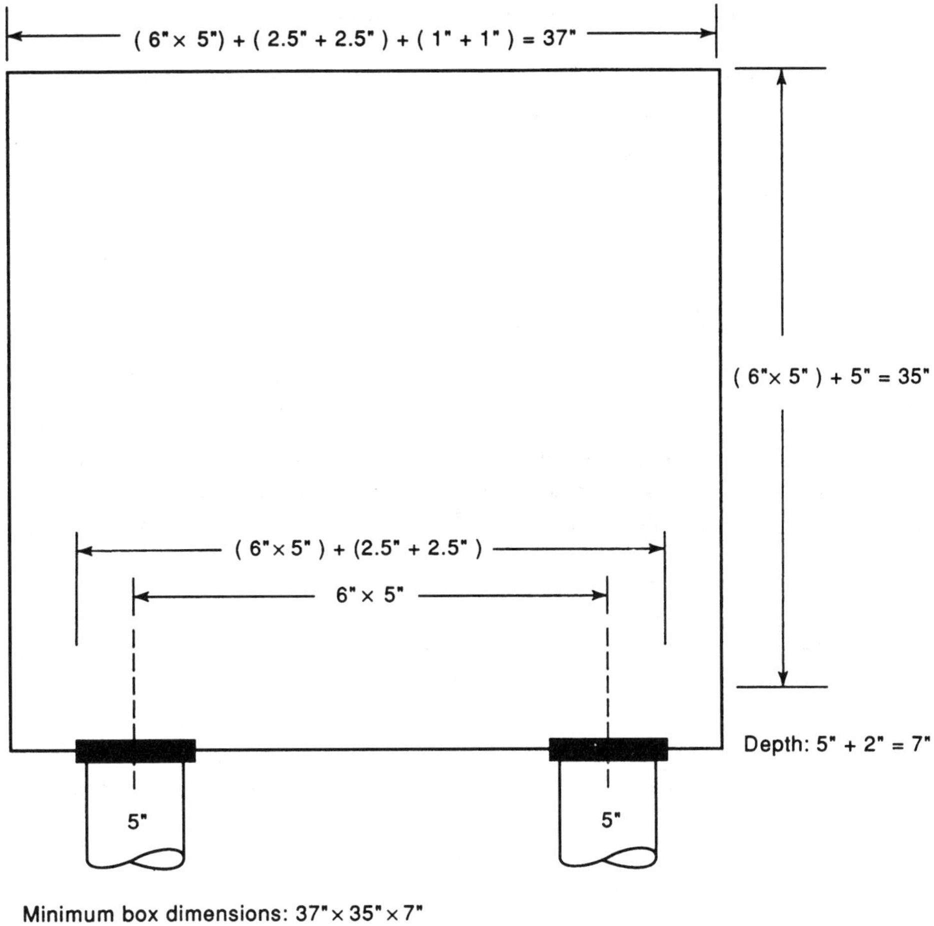

Fig. 5-10 For a U pull, the box dimensions must meet all the requirements of Sec. 370-18(a)(2) and house (support) the installed raceways.

length (or height) of the side wall containing the U pull is governed by the 6td minimum distance between the two raceway centerlines, the diameter of the conduits, and the space required to install and secure the locknuts and bushings (makeup).

Example 5–7. Determine the NE Code-minimum box dimensions for two 5-inch U pull conduit entries in the bottom of the box (refer to Fig. 5-10).

NE Code-minimum length computation: Ref. 370-28(a)(2)

Must equal a composite of three separate measurements.

1. The distance between the two raceway centerline: 6×5 inches = 30 inches

2. The other two halves of the raceway past the 6td: (1td) 5 inches

3. Raceway makeup: (each side in the td + 2) 2 inches

Total length: 37 inches

NE Code-minimum height computation: Ref. 370-28(a)(2)

6td requirement: 6 × 5 inches = 30 inches

Addition of the other trade diameter: 5 inches

Total height: 35 inches

Recommended minimum depth computation: Ref. 300-10

5 inches + 2 inches = 7 inches

When a U pull is the only raceway entry into a junction or pull box, the box dimensions can be computed rather quickly in the following manner: The length measured to the wall opposite will always be 7td (6td plus the other raceway diameter). The length of the box across the U pull will always be 7td + 2: 6td between the raceway centerlines, the other two halves of the raceway (1td) plus 1 inch between the other side of the two raceways and the corner on each side of the box (2 inches). The depth of the box (when the U pull enters/exits a side wall) is automatically td + 2 (trade diameter plus 2 inches).

Code challenge 5–7. Determine the NE Code-minimum box dimensions for two 4-inch U-pull conduit entries into the bottom of the box (similar to Fig. 5-10).

NE Code-minimum length computations: Ref. (A) ____________________

Must equal a composite of three separate measurements.

1. The distance between the two raceway centerline:
(B) ____________ × ____________ inches = ____________ inches

2. The other two halves of the raceway past the 6td: (1td) ____________ inches

3. Raceway makeup: (each side in the td + 2) (C) ____________ inches

Total length: (D) ____________ inches

NE Code-minimum height computation: Ref. (E) ____________________

6td requirement: (F) ____________ × ____________ inches = ____________ inches

Addition of the other trade diameter: ____________ inches

Total height: (G) ____________ inches

Recommended minimum depth computation: Ref. (H) ____________________

____________ inches + ____________ inches = (I) ____________ inches

5.6 COMBINATION PULLS

As shown in Figure 5-11, when a junction or pull box is used for a central dispersion point of several raceway, which might be in the form of a straight pull, an angle pull (to the side or out the back), or a U pull, all the conditions as listed in Sec. 370-28(a)(1) and (2) must be met.

Example 5–8. Determine the NE Code-minimum size of a junction box that will contain a 3-inch U pull in the left wall, a 4-inch straight pull centered from top to bottom wall, and a 3½-inch angle pull from the right bottom wall to the right side wall (refer to Fig. 5-11).

NE Code-minimum length computations: Ref. 370-28(a)(2)

Three conditions must be met with one measurement:

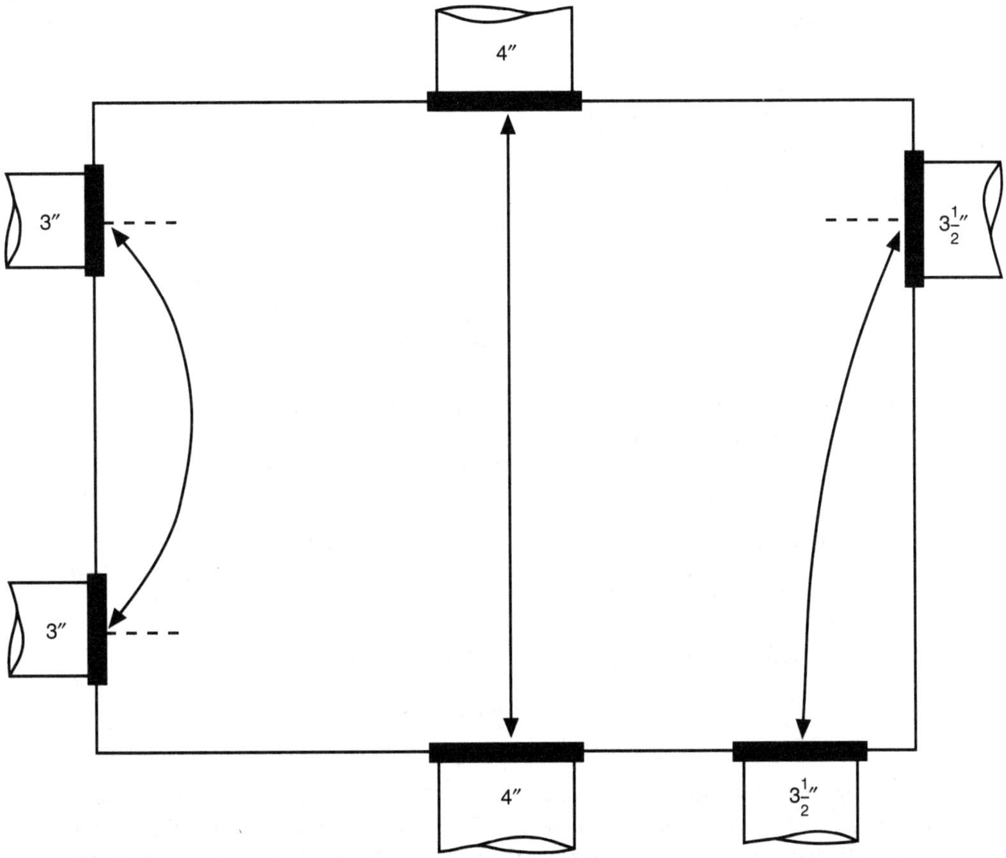

Fig. 5-11 To size a junction box correctly, all the conditions sited in Sec. 370-18 must be met.

1. U pull of the 3-inch conduit, left wall:

6td requirement: 6 × 3 inches =	18 inches
Addition of the other trade diameter:	3 inches
Total length:	21 inches

2. Angle pull of the 3½-inch conduit, right to bottom wall:

$$6 \times 3.5 \text{ inches} = 21 \text{ inches}$$

3. Conduit stack along the bottom wall:

Sum of trade diameters: 4 inches + 3.5 inches =	7.5 inches
Makeup allowance: 2 × 2 inches =	4.0 inches
Total length:	11.5 inches

NE Code-minimum length, compare for larger: 21 inches

NE Code-minimum height computations: Ref. 370-28(a)(1) & (2)

Three conditions must be met with one measurement:

1. U pull of the 3-inch conduit, left wall:

6td requirement: 6 × 3 inches =	18 inches
Addition of the other trade diameter:	3 inches
Conduit makeup:	2 inches
Total height:	23 inches

2. Straight pull of the 4-inch conduit, top to bottom wall:

$$8 \times 4 \text{ inches} = 32 \text{ inches}$$

3. Angle pull of the 3½-inch conduit, bottom to right wall:

6td requirement: 6 × 3.5 inches =	21.00 inches
Addition of the other half of trade diameter:	1.75 inches
Conduit makeup:	1.00 inch
Total height:	23.75 inches

NE Code-minimum height, compare for larger: 32 inches

Recommended minimum depth computation: Ref. 300-10

$$4 \text{ inches} + 2 \text{ inches} = 6 \text{ inches}$$

Code challenge 5–8. Determine the NE Code-minimum size of a junction box that has a 4-inch U pull in the left wall, a 3½-inch straight pull centered from top to bottom wall, and a 5-inch angle pull from the right bottom wall to the right side wall (similar to Fig. 5-11).

NE Code-minimum length computations: Ref. (A) ____________
Three conditions must be met with one measurement:

1. U pull of the (B) ________-inch conduit, left wall:
   6td requirement: ________ × ________ inches = ________ inches
   Addition of the other trade diameter: ________ inches
   Total length: (C) ________ inches

2. Angle pull of the (D) ________-inch conduit, right to bottom wall:

   ________ × ________ inches = (E) ________ inches

3. Conduit stack along bottom wall:
   Sum of tds: ________ inches + ________ inches = ________ inches
   Make-up allowance: ________ × ________ inches = ________ inches
   Total length: (F) ________ inches

NE Code-minimum length, compare for (G) ________: (H) ________ inches

NE Code-minimum height computations: Ref. (I) ____________
Three conditions must be met with one measurement:

1. U pull of the (J) ________-inch conduit, left wall:
   6td requirement: ________ × ________ inches = ________ inches
   Addition of the other trade diameter: ________ inches
   Conduit makeup: ________ inches
   Total height: (K) ________ inches

2. Straight pull of the (L) ________-inch conduit, top to bottom wall:

   ________ × ________ inches = (M) ________ inches

3. Angle pull of the (N) ________-inch conduit, bottom to right wall:
   6td requirement: ________ × ________ inches = ________ inches
   Addition of the other half of trade diameter: ________ inches
   Conduit makeup: ________ inch
   Total height: (O) ________ inches

NE Code minimum height, compare for (P) ________: (Q) ________ inches

Recommended minimum depth computation: Ref. (R) ____________

________ inches + ________ inches = (S) ________ inches

5.7 BOX SUPPORT

The correct size of either an outlet, device, pull, or junction box is often a very large box when all NE Code requirements are met. Installing the box in a rigid and secured manner is also an NE Code requirement. Acceptable means of support for these boxes are listed in the subsections of Sec. 370-23.

When the raceway is surface mounted, the box is normally surface mounted as well. For solid walls of wood, masonry, or concrete, the box is simply bolted or otherwise fastened to the wall surface. For hollow walls, such as sheetrock or paneling, Sec. 370-23(a) requires the box to be mounted on a metal or wooden brace, or else to be mounted over a wall stud, ceiling joist, or other structural member as described in Sec. 370-23(b).

As shown in Figure 5-12, recessed outlet or device boxes installed in hollow walls are allowed by Sec. 370-23(b) to be nailed or otherwise fastened to the side of the wood

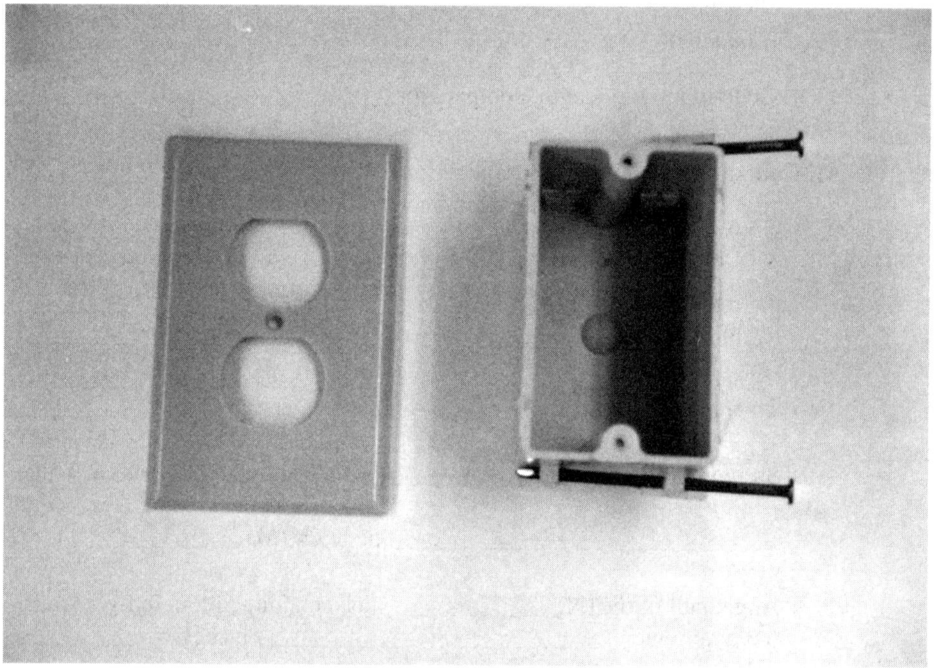

Fig. 5-12 Construction boxes are normally nailed to the side of a wooden wall stud for recessed mounting.

or other structural member. If the nails pass through the box (typical metal construction boxes are often referred to as "nail-in" boxes), the nail must pass within ¼ inch of the front or back of the box, and within ¼ inch of the top or bottom wall (holes are normally predrilled by the manufacturer). Other boxes have external bracing for nail or screw mounting to the stud (Fig. 5-13).

As shown in Figure 5-14, recessed outlet or device boxes, especially the octagon fixture boxes, are often suspended in the chase between two ceiling joists to center the

Fig. 5-13 Device or handy boxes are also available with external braces for nail or screw mounting to a wall stud.

opening in the ceiling of the room. Section 370-23(b)(2) requires the adjustable metal brace (box hanger) to be a listed device. If field constructed, a minimum mil thickness of the metal and suitable protection against corrosion are required.

Section 370-23(c) decribes two types of nonstructural mounting. In the first part, rework boxes are normally recessed into the hollow chases of sheetrock or panels walls (Figs. 5-15 and 5-16). The rework box is fitted with some type of clamp or anchor with screws. When set into the wall, tightening the screws tightens the clamp and draws the box to the surface of the wall.

The second part of the section deals with the framing members of a suspended ceiling (lay-in ceiling). If the ceiling frame is deemed adequately supported, the box or other enclosure can be bolted or otherwise fastened to the framing member for support.

A special type of lay-in lighting fixture, which contains a partitioned portion within the fixture for makeup to the branch-circuit supply conductors, is normally used with this type of suspended ceiling. The fixture enclosure is considered a box in terms of support. Section 370-23(c) requires the lay-in fixture to be clipped or otherwise fastened to the framing member to ensure that the fixture cannot be bumped, tilted, or otherwise moved to where it could fall out of the ceiling during relamping or other service to the fixture.

In open bay areas of industrial installations, and above suspended ceilings in either commercial or office installations, the conduit or other pipe-type raceway is supported with clamps to the underside of the bar joist (on the underside of the next floor or the roof). Section 370-23(d) will allow a junction or pull box, not over 100 cubic inches (usu-

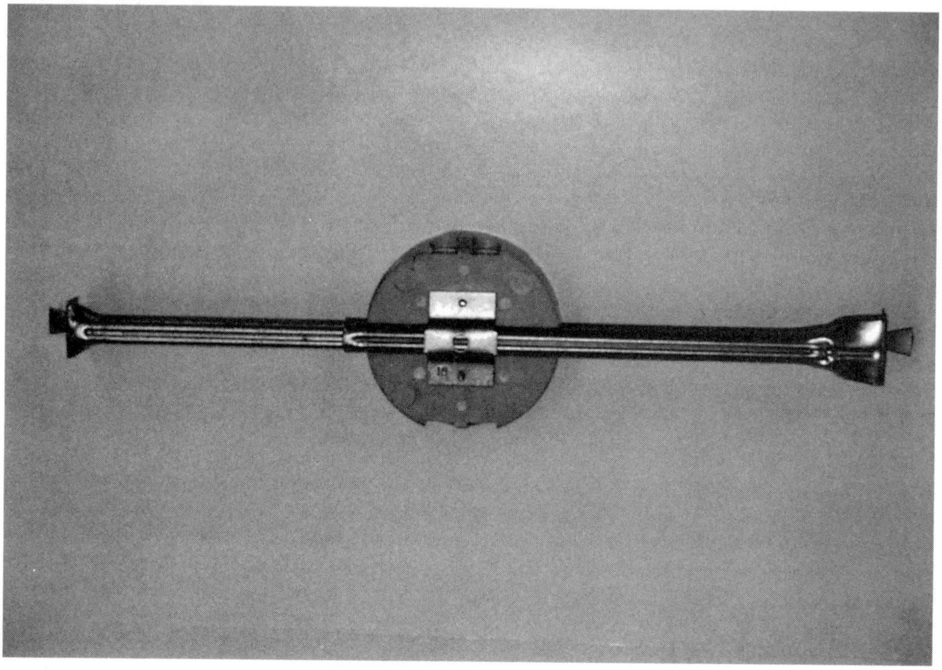

Fig. 5-14 To center a lighting fixture in a room, adjustable hanger bars allow the recessed octagon box to be mounted between two ceiling joists.

ally associated with the junction or pull boxes sized by Sec. 370-16), to be supported by the raceway entries when suspended between two adjacent bar joists or other structural members.

A minimum of two conduit entries, normally on opposite sides of the box, are required to be supported within 3 feet of the respective entry walls. If the box will contain devices or support lighting fixtures, Sec. 370-23(e) reduces the 3-foot raceway support of each entry to a maximum of 18 inches.

With or without devices, the raceway entries to raceway-supported boxes must be made up wrenchtight to prevent the box from turning. This is especially true when the support is limited to only the two entries as described. If a third entry forms a T, the likelihood of turning is greatly reduced, but Sec. 300-10 still requires all raceway entries to be made up wrenchtight for mechanical continuity, and for metal raceway and boxes, electrical continuity as well.

Section 370-23(f) will allow boxes listed for this use to be supported by being embedded in concrete or masonry, without further support. The connecting raceway is also embedded. The box and device or fixture support is mainly on the concrete slab or masonry wall.

In some instances, a flexible cord is dropped from a ceiling box to supply a device or outlet box on the other end. Section 370-23(g)(1) recognizes the device or outlet box as a pendant box. Support by the flexible cord is allowed. Some form of strain relief must

Fig. 5-15 Nometallic rework boxes have a collapsible brace that can be inserted into the wall box opening and catch on the backside of the wall to secure the box.

be installed at both ends of the cord. Section 370-23(g)(2) lists the installation and support requirements for boxes and lighting fixtures that will be pendant mounted on a drop of conduit or other piping raceway.

5.8 AUXILIARY GUTTERS AND WIREWAYS

In addition to device, outlet, pull, or junction boxes, conduit, tubing, and other types of piping raceway can also be terminated to either auxiliary gutters or wireways. In some instances, where a concentration of circuit conductors will occur in a limited space, such as process control on production equipment, wireway may be used in place of piping-type raceway.

Although similar in appearance, auxiliary gutters and wireways are quite different in their use or intended purpose. Wireway is a manufactured raceway (Art. 100) that is available in standard lengths of 1, 2, 3, 4, 5, and 10 feet. The different lengths afford flexibility in installing the wireway as a complete raceway system without field-cutting sections for correct fit. The sections are interlocking and leave a smooth surface inside and out when put together. Elbows, tee junctions, crosses, and similar-type fittings are available such that the raceway can be extended in several directions. Wireway installation and use are governed by Art. 362 in the NE Code.

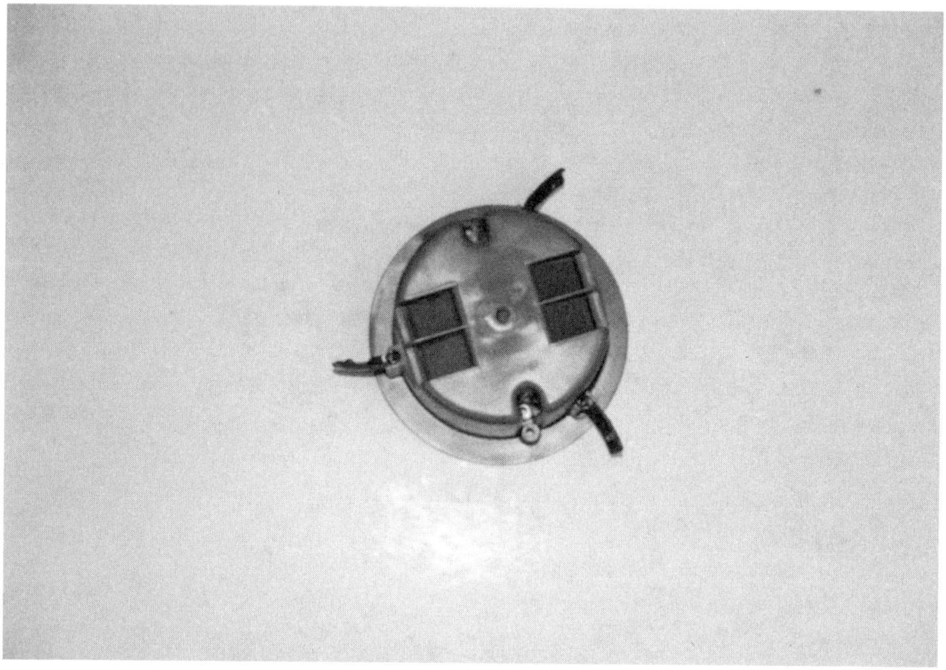

Fig. 5-16 Nonmetallic rework fixture boxes may have turn-out braces that overlap the back wall of the ceiling surface when the screws are tightened.

Auxiliary gutters, which are governed by Art. 374, are basically pull or junction boxes with exceptional lengths. As a rule, Sec. 374-2 restricts the length of an auxiliary gutter to a maximum of thirty feet beyond the equipment it serves. This restriction will normally keep load centers, transformers, disconnect enclosures, and/or controller enclosures that have common wiring within sight of each other (Fig. 5-17). There are no length restrictions on wireways.

Auxiliary gutters and wireways are available as both metallic and nonmetallic. The rules discussed in this section will be aimed primarily at the metallic type, although the same rules (different sectional addresses) apply to the nonmetallic type.

The number of conductors in auxiliary gutters [Secs. 374-5(a) and 374-6(a)] and wireways (Sec. 362-5) is restricted to a maximun count of thirty if the conductor's rated ampacity, as listed in Tables 310-16 through 310-19, is required. In other words, when more than three circuit conductors are installed, the derates required by Note 8(a) to these ampacity tables do not apply.

Any number of circuit conductors, exceeding thirty, can be installed in any auxiliary gutter (Sec. 374-5, Ex. 3), or wireway (Sec. 362-5, Ex. 1), if the more-than-three derate of the Note 8(a) Table is applied to the table-rated ampacity of each respective circuit conductor.

In terms of wire size and rated ampacity, a severe penalty is paid for adding that

Fig. 5-17 Auxiliary gutters are normally used when a concentration of electrical equipment will be located in a confined area.

thirty-first circuit conductor (current-carrying conductor) to the auxiliary gutter or wireway. If an installation will consist of No. 4 AWG THW copper conductors, Sec. 110-14(c)(1), Ex. 1 would restrict the table-rated ampacity (T310-16) of the first thirty circuit conductors to 70 amperes. If one more No. 4 AWG THW copper conductor is to be added, the 40% derate found in the Note 8(a) Table must be applied to the table-listed ampacity of 85 amps found in the 75°C column. The maximum load current of all the No. 4s is now restricted to only 34 amps.

As discussed in Chapter 2 of this text, if the original 70-amp rated ampacity is required by the load rating, then the conductors will have to be resized based on the 40% derate: 70 amps ÷ 0.40 = 175 amps. In the 75°C column, the wire size was increased from No. 4 AWG THW copper to No. 2/0 AWG THW copper. Another way of saying the maximum count of circuit conductors that can be installed at a given cross section of either auxiliary gutter or wireway is limited to thirty.

If the auxiliary gutter or wireway is installed in higher or lower ambient temperatures, the conductor ampacities must be derated or overrated as if they were installed or run in conduit. As with conduit, if the auxiliary gutter or wireway contains more than thirty current-carrying conductors and is installed in higher ambient temperatures, the contained conductors must be derated for each type of loss of effective heat dissipation (double derate).

5.9 20% FILL

Although the maximum number of conductors after ampacity derate is not restricted, the cross-sectional area for auxiliary gutters [Sec. 374-5(a)] and wireways (Sec. 362-5) is restricted to a maximum of 20% fill in terms of conductor cross-sectional area (c.s.a.). As with the maximum count of thirty circuit conductors, the 20%-fill restriction applies to any cross section or point along the auxiliary gutter or wireway.

The conductors or cables laying or racked evenly beside and on top of each other can occupy only one-fifth of the total interior c.s.a. of the gutter or wireway. Sections 374-8(a) and 362-7 increase the fill allowance to 75% where splices and/or taps are made (Fig. 5-18).

By Secs. 374-5(a), Ex. 2 and 362-5, conductors used for signaling circuits or motor-control circuits are not considered as current-carrying conductors and are not to be counted in the total number of contained conductors when derating for more than thirty current-carrying conductors; however, they are to be counted when determining the maximum conductor 20%-fill allowance.

When sizing auxiliary gutters or wireways, the cross-sectional area of the contained conductors is determined by or taken from either Table 5, 5A, or 8 in Chapter 9.

Since the total cross-sectional area occupied by the conductors is restricted to a maximum of 20% of the c.s.a. of the gutter or wireway, dividing the total conductor c.s.a. by 20% or 0.20 will yield the correct NE Code-minimum c.s.a. of the gutter or wireway. Some NE Code sources advise to multiply the total conductor c.s.a. by 5 to attain the NE Code-minimum gutter or wireway c.s.a. requirements. Either way is correct: 20% is ⅕ of the whole or 100%, such that dividing by 0.20 is equivalent to multiplying by 5.

The standard sizes of auxiliary gutters and wireways are normally given as the product of their width and height. To find the area of a particular size, simply multiply these two values. The more common manufactured sizes of wireways and gutters and their respective areas are:

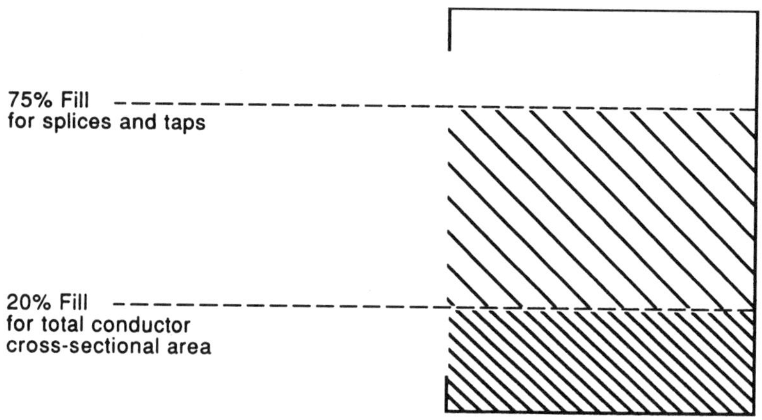

Fig. 5-18 The available cross-sectional area of both auxiliary gutters (Art. 374) and wireways (Art. 362) is restricted for total conductor cross-sectional area.

Size (in.)	Cross-sectional area (sq. in.)
2½ by 2½	6.25
4 by 4	16
6 by 6	36
8 by 8	64
10 by 10	100

To determine the total conductor cross-sectional area required when several conductors will be installed in a common auxiliary gutter or wireway, and the resulting NE Code-minimum cross-sectional area of the auxiliary gutter or wireway, the following formula is offered:

First, determine the total conductor c.s.a. by constructing a four-column table.

Count	Wire size and insulation	c.s.a. for one (sq. in.)	Extended c.s.a. (sq. in.)
Given — —	Given or determined from proposed installation	Tables 5, 5A, or 8 in Chapter 9	Product of "Count" times "c.s.a. for one"
			Total conductor c.s.a

Second, determine the NE Code-minimum auxiliary gutter or wireway cross-sectional area:

$$\text{total conductor c.s.a.} \div 0.20 \text{ (equivalent to 20\%)} = \text{minimum c.s.a.}$$

The available cross-sectional area of the selected gutter or wireway size should equal or exceed the computed total conductor cross-sectional area.

Example 5–9. Determine the NE Code-minimum size of auxiliary gutter required for the installation of thirty No. 12 AWG THHN, forty No. 10 AWG THHN, six No. 6 AWG TW, and three No. 2 AWG THW copper conductors.

Total conductor cross-sectional area computation: Ref. Ch. 9, T-5

Count	Wire size and insulation	c.s.a. for one (sq. in.)	Extended c.s.a. (sq. in.)
30	No. 12 AWG THHN	0.0133	0.3990
40	No. 10 AWG THHN	0.0211	0.8440
6	No. 6 AWG TW	0.0726	0.4356
3	No. 2 AWG THW	0.1333	0.3999
			Total conductor c.s.a.: 2.0785

NE Code-minimum auxiliary gutter c.s.a. computation: Ref. 374-5

$$2.0785 \text{ square inches} \div 0.20 = 10.3925 \text{ square inches}$$

The auxiliary gutter must have an interior cross-sectional area greater than or equal to 10.3925 square inches (next higher standard size as listed in text):

A 4-by-4 auxiliary gutter is required.

Code challenge 5–9. Determine the NE Code-minimum size of the wireway required for the installation of forty No. 14 AWG TW, thirty No. 12 AWG THW, fifteen No. 8 AWG THWN, and four No. 1/0 AWG THHN copper conductors.

Total conductor cross-sectional area computation: Ref. (A) ____________

Count	Wire size and insulation	c.s.a. for one (sq. in.)	Extended c.s.a. (sq. in.)
________	No. 14 AWG TW	(B) ________	________
________	No. 12 AWG THW	(C) ________	________
________	No. 8 AWG THWN	(D) ________	________
________	No. 1/0 AWG THHN	(E) ________	________
		Total conductor c.s.a.: (F)	________

NE Code-minimum wireway c.s.a. computation: Ref. (G) ____________

________ square inches ÷ ________ = (H) ________ square inches

A (I) ________ wireway is required.

5.10 CONDUCTOR DEFLECTION

As was stated before, auxiliary gutters and wireways differ mainly in their intended use. Section 374-9(d) requires auxiliary gutters and Sec. 362-6 requires wireways to also be sized according to the deflection of insulated conductors that enter the gutter or wireway. This requirement is aimed at the larger feeder or branch-circuit conductors (normally No. 4 AWG or larger). Conductor deflection was discussed for junction and pull boxes in Sec. 5.4 of this chapter.

"Deflection" indicates the amount of bend required of the conductor in its routing from point of entry to point of termination, splice, or exit. If the deflection of the conductor is over 30°, the auxiliary gutter or the wireway width, at the point of deflection (point of entry or exit), must meet the criteria of Sec. 373-6.

For auxiliary gutters or wireways, the conductors are deflected in either an L or S bend from side wall entry (along the length of the gutter or wireway [Fig. 5-19]). Section 373-6(a) requires the width at the point of deflection to be equal to or greater than the width specified by wire size and the number of wires per terminal in T373-6(a).

Example 5–10. Determine the NE Code-minimum size of wireway required for the splice of three 350 kcmil THW copper conductors, one each, to two No. 4/0 AWG THW copper conductors. The wireway length is five feet. The 350 kcmil THW copper conductors enter the top of the wireway in 3-inch conduit located 6 inches to center from the left-hand corner. The six No. 4/0 AWG THW copper conductors exit the bottom of

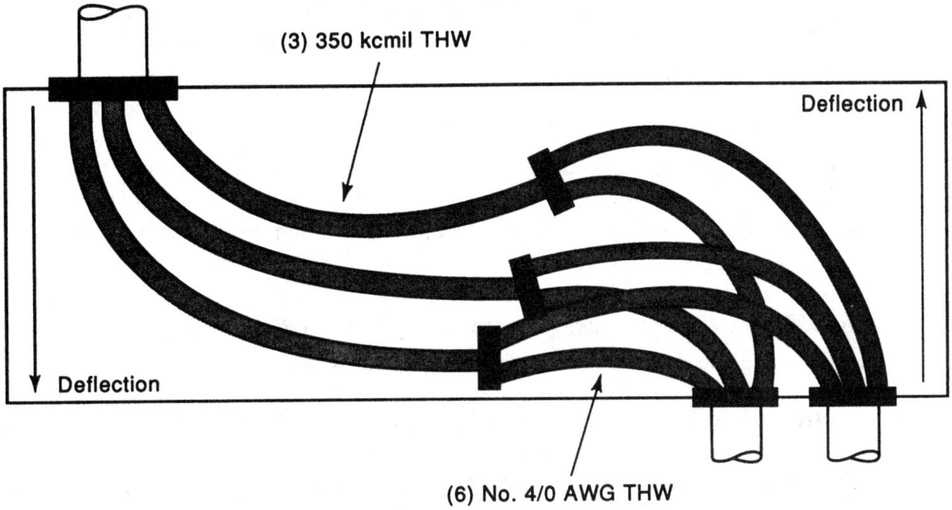

Fig. 5-19 Conductor deflection in an auxiliary gutter or wireway is also governed by Table 373-6(a).

the wireway, three each, in two 1-inch conduits, located 6 and 12 inches on center from the right-hand corner. The splices will be staggered so that the 75%-fill allowance of Sec. 362-7 will not be violated (refer to Fig. 5-19).

Wireway height for the conductor 20%-fill allowance: Ref. 362-5

Assuming a square wireway, the two points of concern for the 20%-fill allowance are the heights where the three 350 kcmil THW conductors lie before the splice, and where the six No. 4/0 AWG THW conductors lie after the splice.

Total conductor c.s.a. computation: Ref. Ch. 9, T-5

For the 350 kcmil THW conductors:

$$3 \times 0.5958 \text{ square inches} = 1.7874 \text{ square inches}$$

For the No. 4/0 AWG THW conductors:

$$6 \times 0.3718 \text{ square inches} = 2.2308 \text{ square inches}$$

Wireway c.s.a. computation: Ref. 362-5

Computation must be based on the largest conductor c.s.a.:

$$2.2308 \text{ square inches} \div 0.20 = 11.154 \text{ square inches}$$

A 4-by-4 wireway is required.

Wireway height for conductor deflection: Ref. 362-6 and T373-6(a)

For 350 kcmil and one wire per terminal: 5 inches

For No. 4/0 AWG and two wires per terminal: 6 inches

NE Code-minimum wireway size for both the 20%-fill allowance and conductor deflection (compare for larger; use next higher standard size as listed in text):

A 6-by-6 wireway is required.

Code challenge 5–10. Determine the NE Code-minimum size of auxiliary gutter required for the splice of three 600 kcmil THW copper conductors, one each, to two 250 kcmil THW copper conductors. The gutter length is six feet. The three 600 kcmil conductors enter the top of the gutter in a single 3-inch conduit located six inches on center from the upper left corner. The six 250 kcmil THW copper conductors exit the bottom of the gutter, three each, through two 2-inch conduits, located 6 and 12 inches on center from the lower right corner. The splices will be staggered so that the 75%-fill restriction of Sec. 374-8(a) will not be violated (similar to Fig. 5-19).

Gutter height for the conductor 20%-fill allowance: Ref. (A) ______________

Assuming a square gutter, the two points of concern for the 20%-fill allowance are the heights where the three 600 kcmil THW conductors lie (B) ____________ the splice, and where the six 250 kcmil THW conductors lie (C) ____________ the splice.

Total conductor c.s.a. computation: Ref. (D) ______________

For the 600 kcmil THW conductors:

____________ × (E) ____________ square inches = ____________ square inches

For the 250 kcmil THW conductors:

____________ × (F) ____________ square inches = ____________ square inches

Gutter c.s.a. computation: Ref. (G) ______________

Computation must be based on the (H) ____________ conductor c.s.a.:

(I) ____________ square inches ÷ ____________ = ____________ square inches

A (J) ____________ auxiliary gutter is required.

Gutter height for conductor deflection:

Ref. (K) ____________ and (L) ____________

For 600 kcmil and (M) ____________ wire per terminal: (N) ____________ inches
For 250 kcmil and (O) ____________ wires per terminal: (P) ____________ inches

NE Code-minimum auxiliary gutter size for both the 20%-fill allowance and conductor deflection (compare for (Q) ____________; use next higher standard size as listed in text):

A (R) ____________ auxiliary gutter is required.

Long auxiliary gutters must also meet the requirements of Sec. 370-28(b). The contained conductors must be cabled or racked when any dimension of the gutter (normally its length) is over six feet long.

Cabling of the conductors is more or less grouping the conductors of a commom raceway entry together with tie-wraps, tape, or spiral wrapping to ensure the support of the conductors will not depend solely on the side wall and/or removable front trim of the auxiliary gutter.

Racking is the insertion of wood or other nonconductive support strips in the outer lip of the auxiliary gutter so that support of the conductors will not depend solely on the attached removable trim. Racking can also be blocks of wood set under the conductors that lay in the bottom of the gutter to insulate the cables from the bottom of the gutter. Both racking and cabling will help to prevent high-pressure squeeze points against the grounded conductive surface of the gutter walls. The high-pressure squeeze points could weaken the conductor insulation, which in turn could cause high-magnitude ground-fault overcurrents.

Both auxiliary gutters [Sec. 374-3(a)] and wireways [Sec. 362-8(a)] are required to be supported every 5 feet. The length or distance between horizontal supports for wireway can be extended up to a maximum of ten feet apart, if allowed by the design of the wireway. [Section 110-3(b) requires electrical equipment to be installed according to any instructions furnished by the manufacturer with the equipment.] Section 362-8(b) will allow vertical runs of wireway to be supported at intervals not to exceed 15 feet. The 15-foot interval is restricted to a maximum of one joint per interval.

REVIEW QUESTIONS

1. The size or dimensions of pull or junction boxes that contain conductors No. 4 AWG and larger are governed by Sec. ____________.
2. When a pull or junction box is sized for a single straight pull of conductors No. 4 AWG or larger, the box length (as measured between the opposite walls containing the raceway entry) is based on ____________ times the trade diameter of the conduit, cable, or other raceway.
3. When a pull or junction box is sized for a single angle pull of conductors No. 4 AWG or larger, the box length from the raceway-entry wall to the wall opposite must be at least ____________ times the trade diameter of the raceway.
4. When a pull or junction box contains a U pull of conductors No. 4 AWG or larger, the length of the box to the wall opposite the raceway-entry wall must be a minimum of ____________ times the trade diameter of the raceway.
5. When a pull or junction box contains an angle pull or a U pull of conductors No. 4 AWG or larger, the distance measured between the two raceway centerlines can not be less than ____________ times the trade diameter of the raceway.
6. When a pull or junction box contains more than one angle pull of conductors No. 4 AWG or larger on a single wall, the length of the box from the raceway-entry wall to the wall opposite must be equal to ____________ times the trade diameter of the largest raceway in the angle pulls, plus the ____________ of the trade diameters of all the other raceways that enter the same wall.
7. When a pull or junction box contains an angle or U pull of conductors No. 4 AWG or larger that enters or exits the box in the wall opposite a removable cover, the length of the box from wall entry to the wall containing the removable cover is based only on conductor ____________.

8. When a pull or junction box contains combination pulls of conductors No. 4 AWG or larger, which can include straight pulls, angle pulls, or U pulls, the NE Code-minimum dimensions of the box are based on the largest measurements calculated for any pull into the box.

   True or False

9. If a pull or junction box has any dimension over six feet (length, height, or depth), all the contained conductors must be either cabled or racked.

   True or False

10. When an electrical box is surface mounted to a solid wall, the box is simply bolted or otherwise fastened to the wall surface.

    True or False

11. When an electrical box is surface mounted to a hollow wall, the box must be mounted on a metal or wooden brace, or else be mounted over a wall stud, ceiling joist, or other structural member.

    True or False

12. Recessed outlet or device boxes installed in hollow walls are allowed to be nailed or otherwise fastened to the side of the wood or other structural member. If the nails pass through the box, they must be located within ____________ inch of the front or back of the box, and within inch of the top or bottom wall of the box.

13. Rework boxes are normally recessed into the hollow chases of sheetrock or panels walls. The rework box is fitted with some type of clamp or anchor with screws. When set into the wall, tightening the screws tightens the clamp and draws the box to the surface of the wall.

    True or False

14. When a box will be supported by a framing member of a lay-in ceiling, the ceiling frame must be adequately supported, and the box must be bolted or otherwise fastened to the framing member for support.

    True or False

15. In open ceilings, or above suspended ceilings when the raceway is supported with clamps to the underside of the bar joist, the box can be supported by the raceway entries between two adjacent bar joists. If the box does not contain devices or fixture support, a minimum of ____________ (how many) raceway entries must be supported within ____________ feet of the box. If the box contains devices or fixture support, the raceway must be supported within ____________ inches of the box.

16. If the table-rated ampacities of the circuit conductors, as listed in Table 310-16, are required for the rating of the connected load, the number of lighting and power conductors that can be installed in an auxiliary gutter is restricted to a maximum count of forty.

    True or False

17. There is no restriction as to the number of current-carrying conductors that can be installed in an auxiliary gutter, provided the conductors' table-rated ampacities are derated as if the conductors were installed in a single conduit or other raceway.

    True or False

18. The total number of conductors installed in an auxiliary gutter for which the ampacity restrictions apply is based on current-carrying (load) conductors only. Signaling or control conductors are not to be counted.

    True or False

19. The total cross-sectional area of conductors installed in an auxiliary gutter is restricted to a maximum of 30% of the gutter's cross-sectional area.

    True or False

20. When feeder conductors enter from the long side of an auxiliary gutter, the gutter's width to the wall opposite the raceway entry must meet the minimum gutter space requirements for conductor deflection in Table ____________.
21. If the table-rated ampacities of the circuit conductors, as listed in Table 310-16, are required for the rating of the connected load, the number of lighting and power conductors that can be installed in a wireway is restricted to a maximum count of ____________.
22. There is no restriction as to the number of currrent-carrying conductors installed in a wireway, provided the conductors' rated ampacities are ____________ as if the conductors were installed in a single conduit or other raceway.
23. The total number of conductors installed in a wireway for which ampacity restrictions apply is not based on the number of current-carrying (load) conductors only. Signaling or control conductors must be counted.

    True or False

24. The total cross-sectional area of conductors installed in a wireway is restricted to ___________ % of the wireway's cross-sectional area.
25. When feeder conductors enter from the long side of a wireway, the wireway's width to the wall opposite the raceway entry is governed by conductor deflection, and the gutter width requirements of Table ____________.
26. When the length of an auxiliary gutter is over six feet, all the contained conductors must be either cabled or racked.

    True or False

REVIEW PROBLEMS

1. Determine the NE Code-minimum box dimensions of a pull box used for a straight pull in 3½-inch aluminum conduit.
2. Determine the NE Code-minimum box dimensions of a common pull box if four straight pulls of 3-inch, 3½-inch, 3-inch, and 2-inch are stacked top to bottom on the right wall.
3. Determine the NE Code-minimum dimensions of a pull box used for an angle pull of 4-inch conduit from right to top wall.
4. Determine the NE Code-minimum box dimensions and the relative location of the bottom-exit conduit, if a 5-inch conduit enters the right wall of the box with conduit centerline located 20 inches from the right bottom corner.

5. Determine the NE Code-minimum box dimensions of a pull box with three 3-inch conduit entries stacked on the right wall. The upper 3-inch angle pull exits the top wall of the box. The lower two 3-inch angle pulls exit the bottom wall of the box.
6. If a 3½-inch conduit will contain three 400 kcmil THW copper conductors, determine the NE Code-minimum box dimensions for an angle pull from the right side wall to the back wall of the box (opposite a removable cover).
7. Determine the NE Code-minimum box dimensions for two 2-inch U-pull conduit entries in the top wall of a box.
8. Determine the NE Code-minimum box dimensions of a junction box that has a 2½-inch U pull in the top wall, a 3-inch straight pull centered from left to right wall, and a 3½-inch angle pull from lower left wall to bottom wall.
9. Determine the NE Code-minimum size of wireway required for the installation of twenty-four No. 10 AWG THW, thirty-six No. 10 AWG THWN, twenty-two No. 8 AWG THHN, and four No. 4 AWG TW copper conductors.
10. Determine the NE Code-minimum size of an auxiliary gutter required for the splice of three 250 kcmil THW copper conductors, one each, to three No. 1/0 AWG THW copper conductors. The gutter length will be eight feet. The 250 kcmil conductors will enter the top of the gutter in a 2½-inch conduit located 8 inches to center from the right-hand corner. The nine No. 1/0 AWG conductors will exit the bottom of the gutter, three each, in three 1½-inch conduits located 8, 16, and 24 inches on center from the right-hand corner. The splices will be staggered so that the 75%-fill restriction will not be violated.

Chapter 6

Multiconductor Cable Wiring Methods

Any electrical circuit has three primary components: the power supply, the load, and circuit conductors to connect the two. Both the power supply and the load equipment are contained in some type of enclosure, either metallic or nonmetallic to protect persons from accidental contact with energized parts of the electrical system, and to protect the components of the electrical distribution system from physical damage or exposure to the elements (moisture, debris, contamination, etc.)

Insulated circuit conductors are used to convey the electrical power from the source to the load and back to the source. The information presented in Chapters 2 through 5 of this text have been concerned mainly with individual insulated conductors, installed as a group or set of circuit conductors in either piping-type raceway, cable tray, or wireway run between the power supply and the connected load.

As with the enclosure for the power supply and the connected load, the raceway has a dual role: to protect persons from accidental contact with energized parts of the electrical system, and to protect the individual conductors from exposure to physical damage, which could cause insulation failure.

When the circuit conductors will be installed in buildings or other structures constructed of masonry or concrete, raceway is required. The raceway can be run exposed along the surface of a building, or it can be installed, concealed within the hollow chases of block walls, or within the chases created on the surface of concrete or brick walls with furring strips and outer wall coverings, such as sheetrock or paneling.

When the building will be constructed of wood (stick frame), such as the average home or small shop, rigid piping- or tubing-type raceway run exposed along the finished wall surface just doesn't have the desired cosmetic effect. Technically, the raceway can be installed in the hollow chases of the walls, floors, and ceilings formed by the wood framing, but rigid raceway does not install easily within these type chases.

Flexible raceway, in the form of flexible tubing or conduit, is a possible alternative for concealed raceway within the chases. Any type of piping or tubing raceway affords the abilty to add circuit conductors to the existing electrical distribution system for future modification or addition of other circuits within the building. However, in the average home, flexible raceways are often cost prohibitive in terms of both time and materials.

An alterative to raceway installation in a wood frame structure is the use of multiconductor cables. Multiconductor cables have a set number of conductors installed at the point of manufacture in an outer sheath that binds the circuit conductors together. When installed concealed within the hollow chases and routed through holes or notches made in the individual wood members, the outer sheath protects the contained conductors from cuts and abrasions. The location, concealed within the chase, protects the conductors from physical damage in the same manner that piping or tubing raceway protects circuit conductors when run exposed.

When run exposed along the surface of wood members (attics, crawl spaces, or unfinished basements), the height or relative location of the cable provides the required physical protection (aids the outer sheath). In other instances, like run perpendicularly across the joists in an open attic, guard strips have to be installed beside the cable to protect it from physical damage, or else the multiconductor cable has to be installed in pipe- or tubing-type raceway to provide the required protection.

Multiconductor cables such as Romex, BX, SE, and UF are considered to be "wiring methods" by the NE Code. Section 110-8 stipulates that the National Electrical Code only recognizes wiring methods that can be installed in any occupancy. Restrictions placed on a particluar wiring method are found under the Article that governs its use and installation.

Multiconductor cable wiring methods are governed by their respective articles in Chapter 3. Romex is a trade name for nonmetallic-sheathed cable, which is governed by Art. 336. BX is a trade name for armored cable (Type AC), which is governed by Art. 333. SE is both a trade name and an NE Code name for service-entrance cable, which is governed by Art. 338. UF is both a trade name and an NE Code name for underground feeder and branch-circuit cable, which is governed by Art. 339.

When installed in a dwelling or other structure, these cables are supported, protected, and sized in basically the same way. All can serve as either branch-circuit or feeder wiring methods, although, with the exception of SE cable, they cannot be used as service-entrance or other service conductors. Service conductors usually carry higher-amperage loads than do either feeders or branch circuits, since they carry the total or full current load of an occupancy.

6.1 ROMEX CABLE

Section 336-2 describes Romex as a cable manufactured with two or more insulated conductors that are jacketed with a nonmetallic material that is both flame-retardant and moisture-resistant (Fig. 6-1). Section 336-30(a) describes this cable as Type NM as com-

Fig. 6-1 Type NM Romex cable can be used for both exposed and concealed work in normally dry locations. (Courtesy of Southwire Company)

pared to Type NMC, which is also fungus- and corrosion-resistant. In addition to the insulated conductors [Sec. 336-30(b)], either type of cable may contain an insulated or bare conductor that is intended for equipment-grounding purposes only.

To meet the need for either control or signaling conductors in a smart-house installation, Type NMS Romex cable is new in the 1996 NE Code. Section 336-30(a)(3) describes Type NMS as Type NM cable, by reference to Art. 780 (Sec. 780-6), that contains two or more No. 24 AWG or larger signaling conductors.

The student electrician should note that Romex cable in either aluminum or copper-clad aluminum, as recognized by Sec. 336-30(b), has not been market available in the last five to seven years. Copper conductors are better suited in terms of splicing and terminal makeup for the smaller wire sizes.

Section 336-4 lists the permitted uses of Romex cable, and Sec. 336-5 lists the occupancies or conditions in which the use of Romex cable is strictly prohibited. Of particular interest is the fact that Romex cable is not restricted to wood frame structures as implied in the introduction.

Romex multiconductor cable can be fished in the voids of block walls. Type NMC can be run in a shallow chase of masonry, concrete, or adobe [Sec. 336-4(b)(3)], and covered with a plaster or adobe wall finish (must be protected from nail penetration). Romex cable, regardless of the type [Sec. 336-5(a)(8)], cannot be embedded in either conrete, masonry, or similar aggregate, or be installed in hazardous locations [Sec. 336-5(a)(9)] or in places of public assembly [theatres, restaurants, churches, auditoriums, and the like (Sec. 336-5(a)(4)].

The use of Romex cable is primarily in wooden dwelling structures. Section 336-5(a)(1) limits the height of the structure (Fig. 6-2) to a maximum of three floors above finished grade, with the first floor being defined as designed as living quarters or workspace such as offices or process equipment, and which has at least half (50%) of its perimeter level with or above finished grade.

If a basement or lower level of a building is used as a parking garage, for storage, or for equipment such as heating or air-conditioning that only requires servicing, the first level in this definition is the next level above this lower level.

Romex cable can be run as either concealed (Sec. 336-9) or exposed (Sec. 336-6) in a building or other structure. Exposed work is restricted by Sec. 336-6(c) to beams and joists in unfinished basements (or crawlspaces under dwelling or similar structures), and by Sec. 336-6(d) to accessible attics.

When run exposed in unfinished basements (Fig. 6-3), Sec. 336-6(a) requires the cable to follow the surface of the building finish or to be installed on running boards. Following the guidelines of Sec. 110-12, which requires all electrical equipment to be installed in a neat and workmanlike manner, Romex cables are required to be run either

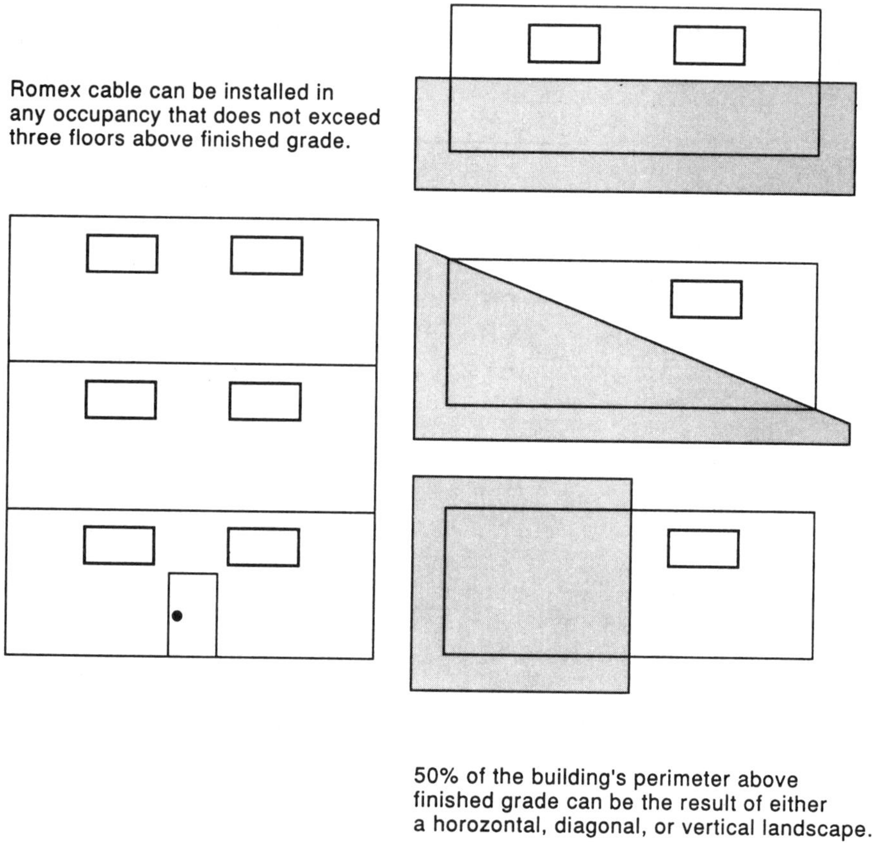

Fig. 6-2 The use of Romex cable is restricted by Sec. 336-5(1) according to "type occupancy."

parallel with or perpendicular to (across) joists, seals, and rafters. Diagonal runs of Romex cable across ceiling joists, between switching points and lighting fixtures, or between "looped" receptacles under floor joists, are normally not allowed by the local inspection authority (Sec. 90-4).

In conjunction with "following the surface," Sec. 336-6(b) requires Romex cable to be protected from physical damage, normally by installation in either conduit, tubing, or some type of piping raceway. If Romex cable is used to supply a water heater in an unfinished basement (Fig. 6-4), the cable has to be installed in raceway (can be installed in flexible conduit) from the point where the cable leaves the floor joist to where it enters the appliance enclosure.

When passing exposed through the floor (Fig. 6-5), Romex cable must be installed in raceway for a minimum of 6 inches above the floor line and to even greater heights (judgment call on the part of the electrician) if exposed to physical damage. This section and the six-inch raceway protection do not apply when the cable is run concealed through bored holes in the wall plate or floor seal to enter the chases created by wall studs, floor or ceiling joists, or roof rafters.

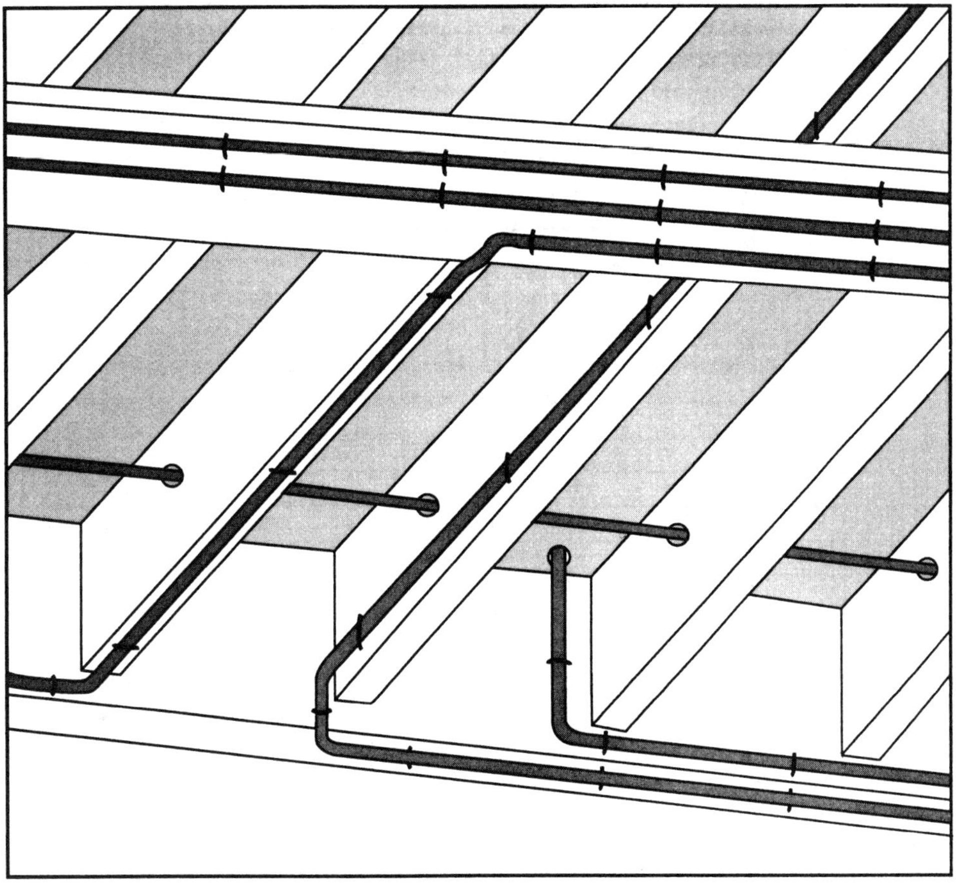

Fig. 6-3 In unfinished basements, the installation of multiconductor cables must "follow the surface" or be installed on running boards.

When run through joists, studs, or rafters (Fig. 6-6), Sec. 336-9 requires the cable installation to comply with Sec. 300-4. If the nonmetallic sheathed cable is run through bored holes in wooden frame members, Sec. 300-4(a)(1) requires the hole's edge to be a minimum of 1¼ inches from the nearest edge (face) of the wood member. A dressed 2-by-4 wooden wall stud has a measured width of 3½ inches. This section would restrict the hole size to a maximum of 1 inch, if it is bored in the center of the stud.

If the minimum distance of 1¼ inches cannot be established, or if a larger hole is required [Caution on larger holes, Sec. 300-4(a)(2), stipulates that the structure member cannot be weakened due to holes or notches cut for cable installation], the cable must be protected from the penetration of nails or screws by a steel plate or bushing (grommet) that is at least $\frac{1}{16}$-inch thick. The plate and/or bushing must be of appropriate length and width to cover the entire cable that passes through the wood member.

If the electrician elects to install the cable in notches on the face of the stud, joist, or rafter in such a manner that the wooden structural member will not be weakened, a

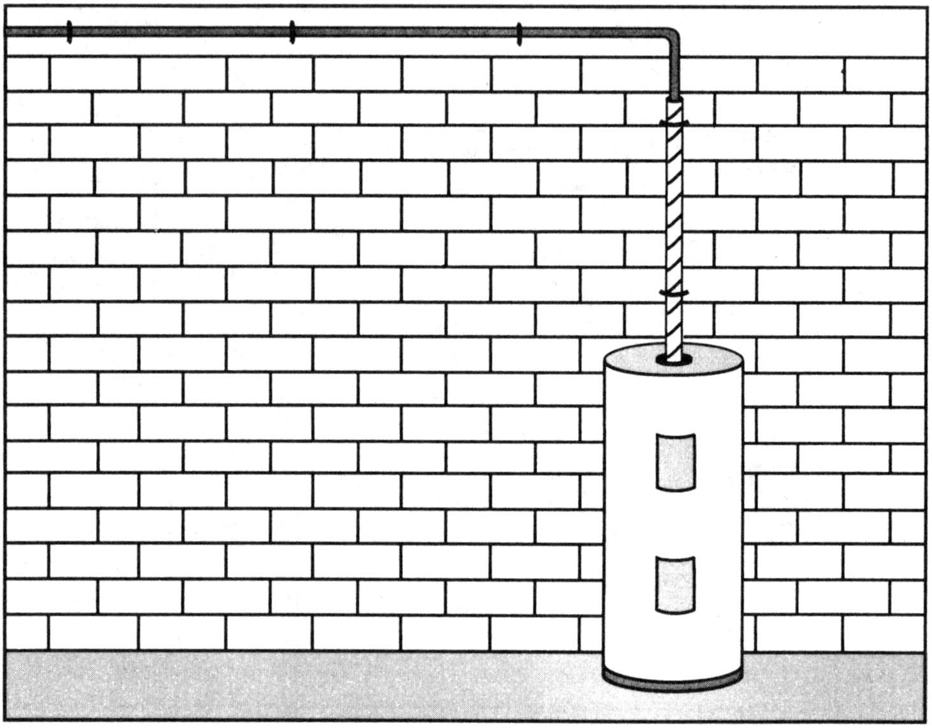

Fig. 6-4 In a basement, Romex cable used to supply a water heater must be protected from physical damage from the point where the cable leaves the joist.

steel plate with the same thickness and appropriate length must be installed over the cable before the building finish is applied [Sec. 300-4(a)(2)].

Bends in the cable are required when turning corners where the stud or rafter is butted to the joist, seal, or wall plate, or when leaving the joist or rafter to enter bored holes or notches in plates or seals. The bends for corners cannot be abrupt or sharp so as to "kink" the outer covering or crimp the contained conductors. Section 336-16 stipulates that any handing (installing) of nonmetallic sheathed cable shall be in such a manner that the outer covering (sheath) shall not be damaged.

The bending radius in a Romex cable is retricted to a minimum of five times the flat (elliptical, two or three conductor) or round (four conductor) diameter of the cable. Installed correctly, the cable will have a long "sweep" appearance as it turns corners. Installed incorrectly, the cable will appear to have its own corner (as if folded 90° like a piece of cardboard), with the outer sheath crinkled up in the inside of the bend.

As a general rule, Sec. 300-15(b) requires a box to be installed at any splice point when installing Romex cable. Exceptions to this are described in Sec. 336-21 as devices made of insulating material that are designed to splice the cable. These splice kits are commercially available and are used to a great extent in modular home construction where the sections of the rooms are prewired at the factory with Romex. The cables are spliced directly or "in line" as the structure is assembled on the site. Section 336-21 re-

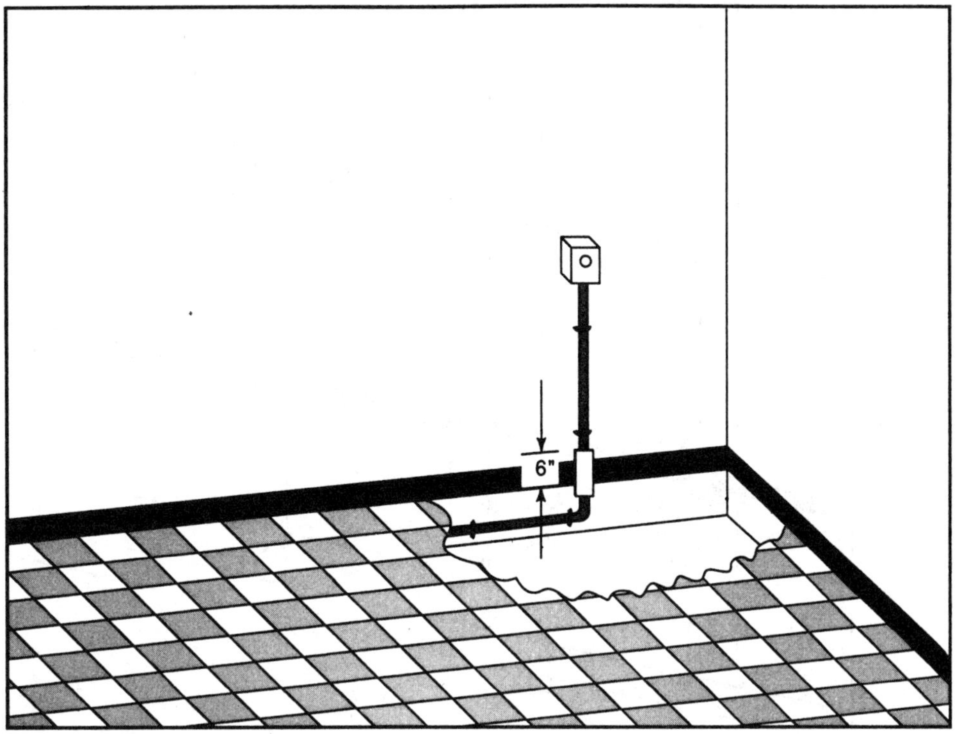

Fig. 6-5 When passing exposed through a floorline, Romex cable must be protected in pipe a minimum of six inches [Sec. 336-6(b)].

quires that all openings in these devices form a close fit around the outer sheath of the cable.

The box [Sec. 300-15(b)] can be metallic as described and sized by Sec. 370-16, with the cable being secured and supported by a cable clamp as it enters the box, or the box can be nonmetallic as allowed by Sec. 336-20.

When using nonmetallic boxes, the Exception to Sec. 370-17(c) will allow the Romex cable to be run to the box without being clamped (secured) to it. The box size is restricted to a maximum of 2¼ by 4 inches. The uncut outer sheath of the Romex cable must extend a minimum of ¼-inch inside the box. The cable must be stapled or otherwise supported within 8 inches of the box.

If the cable is clamped as it enters the box, the 8-inch interval is extended to a minimum of 12 inches. In all instances (whether clamped or unclamped, and whether metallic or nonmetallic boxes are used), the uncut sheath should extend inside the box at least ¼ inch.

Except where installed in rework where the cable is fished in finished walls, or in panels of prefabricated buildings where linear support is not practical, Sec. 336-18 requires Romex cable to be secured and supported by fasteners that will not damage the cable. This support must be installed at intervals not to exceed 4½ feet along the cable's length, and within either the 8 inches as described above, or in general within 1 foot (12

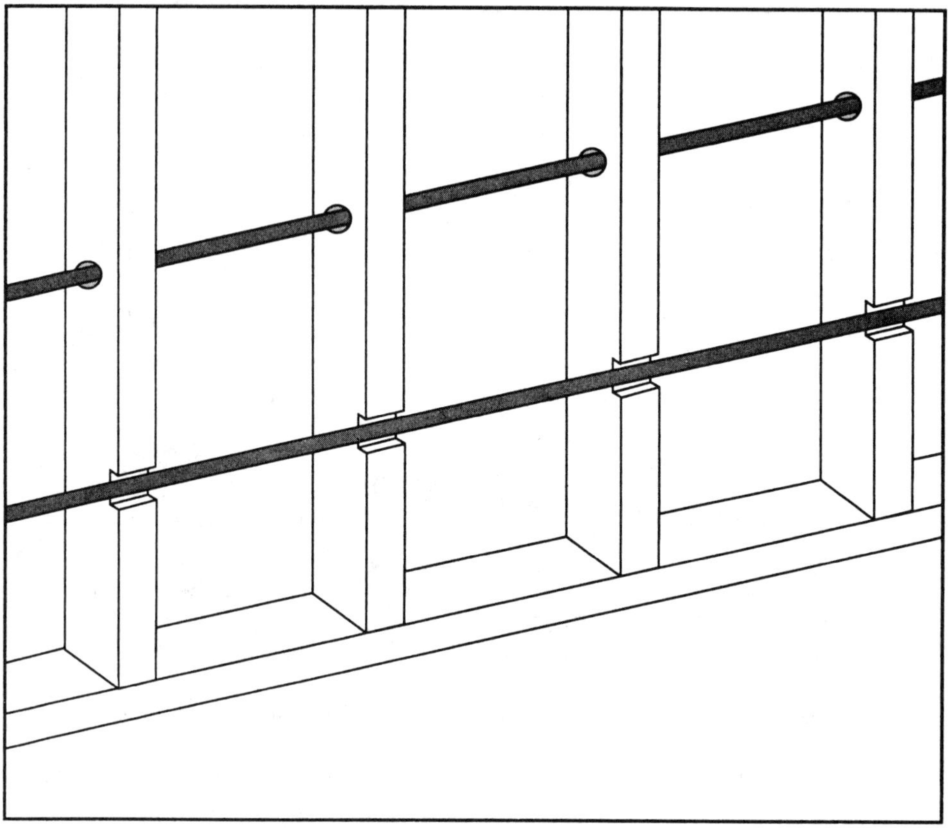

Fig. 6-6 When installed concealed within wooden studs, rafters, or joists, Romex cable can be installed in bored holes or notches in the wood member.

inches) of all cable termination points (cabinets, boxes, and/or conduit fittings or bodies).

When installed in unfinished basements, Romex cable must be installed in the manner described by Sec. 336-6(c). An unfinished basement would have an excavated floor with the height to the bottom of the floor joist large enough (usually seven feet or more) to accommodate the future installation of one or more habitable rooms. The floor could be fill dirt or finished concrete.

Where the cables are run parallel to the floor joists, the cables can be fastened to either the face or side of the joist (shown previously in Fig. 6-3). Where the cables are run perpendicular to the floor joists, they must be run through bored holes in the joists or installed on running boards (usually dress planks or boards with varying widths), and supported (stapled) at a minimum of 4½-foot intervals. Cables that are equivalent or larger than two No. 6 or three No. 8 AWG conductors (typical cable supply to an electric stove in the dwelling, or a feeder to a panel or other load center in any occupancy) can be run across and secured to the lower face (edge) of the floor joist without running boards.

An underfloor crawlspace would normally not be considered an unfinished basement. It is acceptable to most governing authorities (Sec. 90-4) to install Romex perpendicular to the face of the joist without running boards in underfloor crawlspaces. With this type of installation, it is imperative that the cables be protected against physical damage with either running boards or bored holes at all entrances or access points to the underfloor crawlspace.

When installed either concealed or exposed and run parallel to the sides of the joists, rafters, or studs, the lay of the cable cannot be closer than 1¼ inches to the nearest face of the wood member [Sec. 300-4(d)]. This limits the number of Romex cables that can be installed as parallel runs down the side of a wall stud to supply multiple gang lighting switching locations or receptacle outlets. If the average width of a Romex cable is ½ inch, then a maximum of two cables can be installed side by side in the center of the standard 2-by-4 wall stud.

Section 300-4(d) requires the same protection from nail penetration for the cables stapled to the sides of wood members that Sec. 300-4(a) places on bored holes and notches: When the 1¼-inch clearance cannot be maintained, a steel plate at least 1/16-inch thick must be placed over the face of the wood member, which will skirt and protect the cable stapled to the side of the wood member.

Stand-off cable hangers are available that suspend the cables side by side off from the center of the wood member when stacking is required. The hanger mounts over the face of the stud. The flat side of the cables are woven or laced into the fingers of the hanger.

When run in attics, Sec. 336-6(d) requires nonmetallic sheathed cable to be installed under the same guidelines as is required for BX cable in Sec. 333-12.

6.2 BX (ARMORED) CABLE

The installation and use of BX, or armored, cable is governed by Art. 333. Section 333-1 describes armored cable as a fabricated assembly of insulated conductors with an outer sheath of flexible metal. As indicated in Figure 6-7, BX is basically Romex cable installed in flexible metallic conduit (greenfield) instead of a nonmetallic outer sheath.

Section 333-19 requires armored cable to contain a bare copper or aluminum equipment-grounding conductor or bonding strip in contact with the armor the full length of the cable.

The permitted uses of BX multiconductor cable are listed in Sec. 333-3. The prohibited uses are listed in Sec. 333-4. Like Romex multiconductor cable, BX can be fished in the voids of aboveground block walls, provided it will not be exposed to excessive moisture or dampness.

BX multiconductor cable can be run in shallow chases of masonry, concrete, or adobe, and covered with a plaster or adobe wall finish (must be protected from nail penetration). However, it cannot be embedded in either concrete, masonry, or similar aggregate, or installed either in hazardous locations or in places of public assembly (theatres, restaurants, churches, auditoriums, etc.).

Like Romex cable, the use of BX cable is primarily in dwelling structures, but no height limit (stories of the structure) is imposed. BX multiconductor cable can be run as either concealed (Sec. 333-3) or exposed (Sec. 333-11) in a building or other structure. When exposed, Sec. 333-11 requires the cable to follow the surface of the building finish, or else be installed on running boards.

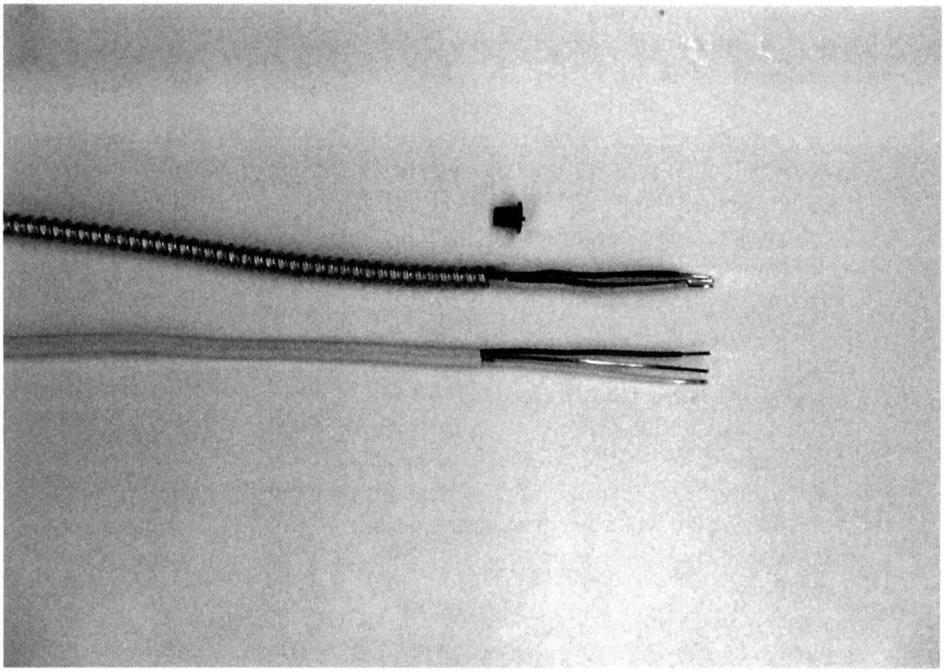

Fig. 6-7 BX cable is equivalent to Romex cable installed in flexible metal conduit for greater protection against physical damage.

Following the guidelines of Sec. 110-12, which requires all electrical equipment to be installed in a neat and workmanlike manner, BX cables are required to be run either parallel with or perpendicular to (across) joists, seals, and rafters. Diagonal runs of BX cable across ceiling joists, between switching points and lighting fixtures, or between "looped" receptacles under floor joists, are normally not allowed by the local inspection authority (Sec. 90-4).

In comparison to Romex cable and the requirement of running boards by Sec. 336-6(c), when installed in unfinished basements, Exception No. 2 to Sec. 333-11 will allow BX cable to be run perpendicular to the underside of floor joists without the use of running boards, provided the cable, regardless of wire size, is fastened (stapled) to every floor joist.

BX cable is not required (like Romex) to be installed in protective piping raceway when it leaves the the underside of floor joists to supply a water heater in an unfinished basement. The BX cable, from the point where it leaves the floor joist to where it enters the appliance enclosure, need only follow the surface of the wall to meet NE Code requirements.

When passing exposed through the floor, BX cable does not have to be sleeved with a piping raceway, as does Romex cable. The flexible metal enclosure is a protective raceway.

When run through joists, studs, or rafters, Sec. 333-10 requires the BX cable installation to comply with Sec. 300-4. The same criteria discussed for Romex cable in Sec.

6.1 applies to BX cable. The restriction of a minimum of 1¼ inches from the nearest edge of the wood member or else a sleeve or grommet to protect from nail penetration is required.

The bending radius in a BX cable is also retricted to a minimum of five times the diameter of the cable by Sec. 333-8. Installed correctly, the cable will have a long sweep appearance as it turns corners where the stud or rafter is butted to the joist, seal, or wall plate, or when leaving the joist or rafter to enter bored holes or notches in plates or seals. The bends for corners cannot be abrupt or sharp so as to rupture or otherwise open the flexible metallic enclosure, or crimp the contained conductors.

Sections 333-9 and 300-15(b) require a box to be installed at any splice point when installing BX cable. To protect the contained insulated conductors from nicks, cuts, or scrapes, Sec. 333-9 requires an antishort bushing to be installed between the cable armor and the insulated conductors at all points of termination.

The cable must also be fastened to the box in such an approved manner (by connector or clamp) that the insulating bushing is visible after installation. As compared to Romex cable, which in some instances does not require a box for splicing, outlets, or switching points (insulating devices), BX must be installed in a box at all points of connection or termination.

BX cable must be supported in the same manner (Sec. 333-7) as Romex cable: at a maximum of 4½-foot intervals, and within 12 inches of all terminations.

When installed in accessible attics, Romex and BX cables are installed in the same manner. When run perpendicular to (across) the top of floor joists (ceiling joists of the room below) Romex [Sec. 336-6(d)] or BX [Sec. 333-12(a)] multiconductor cable must be protected by suitable guard strips installed on both sides of the cable.

Guard strips (Fig. 6-8) are usually dress planks or boards whose thickness (height) must be greater than or equal to the height of the cable. NE Code sources recommend only one cable between the two guard strips (one on either side of the cable), since their purpose is to mechanically protect the cable from damage. If the guard strips are placed too far apart, or if two or more cables are run side by side, a person's foot could pass between the guard strips and put undue stress on the cable, which is suspended between (perpendicular to) the floor joist.

If it is evident by the location of the permanent stairs, ladders, or doors, and the size (area) of the attic, that the attic could be used for storage or could be floored at a later date, then guard strips should not be used. Romex or BX cables should be installed through bored holes below the face of the floor joist.

A scuttle hole in a clothes closet or utility room for access to the attic indicates that the attic is not permanently accessible. The owner or servicing (installing) electrician would have to use a ladder, stool, chair, or some other portable means to gain access to the attic. Sometimes the attic is only accessed from vents installed in gable ends of a dwelling.

When run perpendicular to (across) the face of rafters or studs in accessible attics, Romex or BX cables do not have to be protected by guard strips, unless their installation is within seven feet of the attic-access opening. The seven-foot perimeter applies to rafters or studs in a permanently accessible attic: disappearing staircases, stoop or hallway entrances, and the like. If the attic is not permanently accessible (scuttle hole), the "protected perimeter area" is reduced to only six feet. (The cables can pass across rafters or wall studs one foot closer without guard strips.)

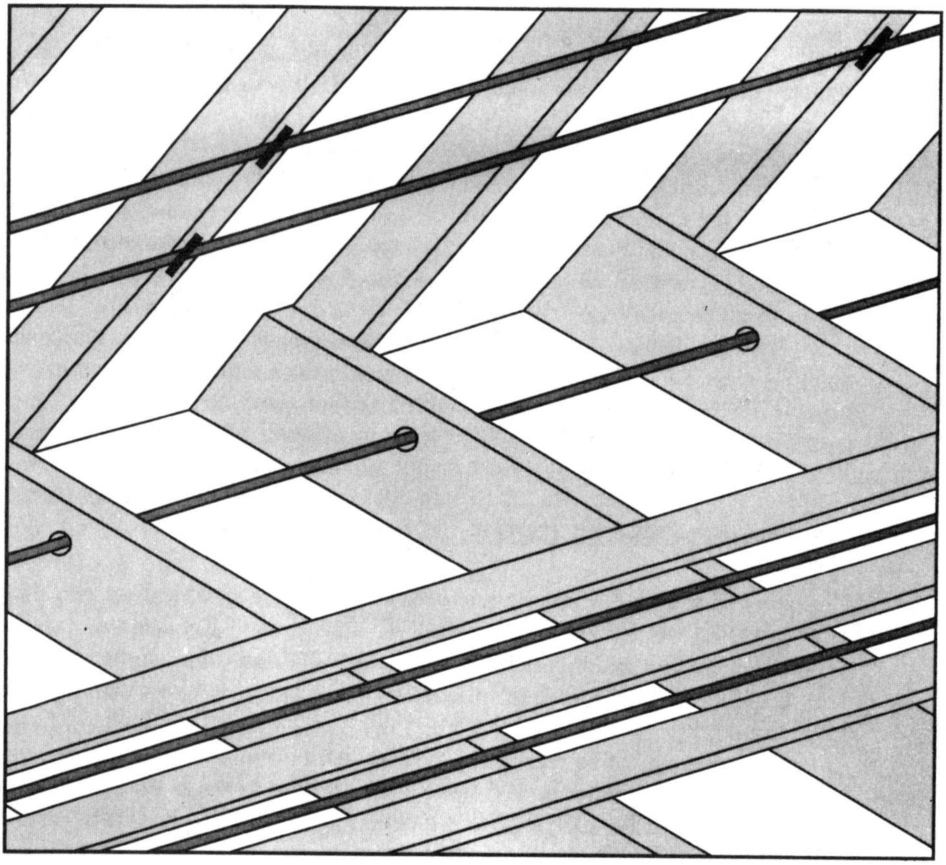

Fig. 6-8 When exposed to physical damage in attics, cables run perpendicular to the joists must be protected by location, bored holes, or guard strips.

Section 333-12(b) allows the cables to be installed or run along the sides of the rafters, studs, or joists without the added protection of running boards or guard strips. The cables cannot be installed within 1¼ inches of the face of the wood member without additional protection from nail penetration as required by Sec. 300-4(d). The cables still have to be supported at intervals not exceeding 4½ feet, and within 12 inches of box terminations. Neither BX nor Romex cable can just be laid or pulled through the insulation in the cavities (chases) between the floor joists.

6.3 SE CABLE

The installation and use of service-entrance cable, both Types SE (Figs. 6-9 and 6-10) and USE (Fig. 6-11, identified for underground use) are governed by Art. 338. Section 338-1 describes SE cable as a single conductor or multiconductor assembly that is used primarily for services.

When installed as service-entrance cable, Sec. 338-2 requires the cable installation to meet the requirements of Art. 230. The service entrance, as will be discussed in Chap-

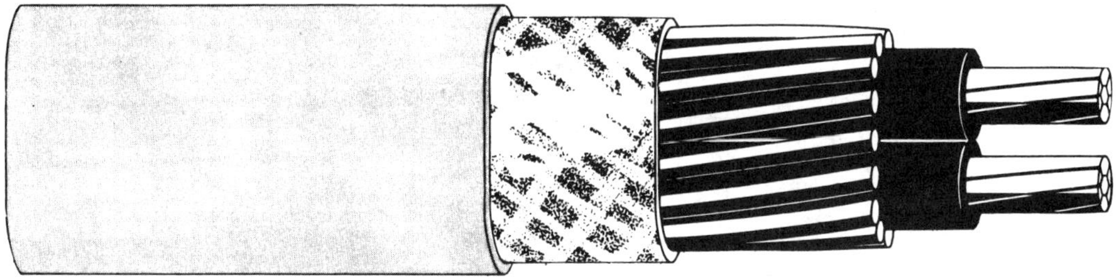

Fig. 6-9 This type of SE cable can be used between the service drop and the service equipment. (Courtesy of Southwire Company)

ter 7, extends from the point of attachment of an overhead service drop to the building to the metering base (enclosure), and on to the service disconnecting means.

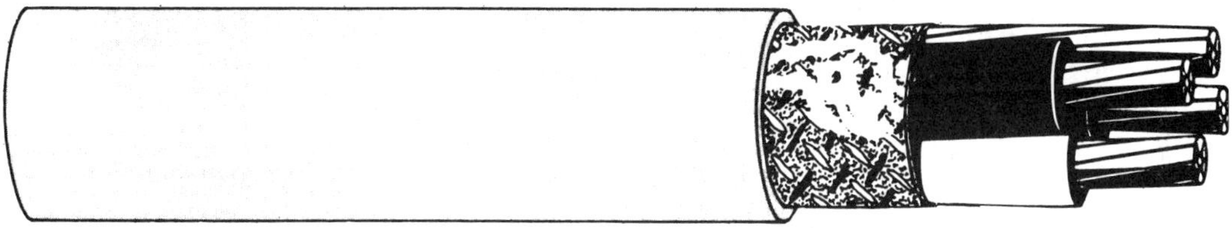

Fig. 6-10 This type of SE cable is used primarily as a panel feeder in a multiple-unit dwelling (fourth-wire equipment ground). (Courtesy of Southwire Company)

Due to the concentration of fault overcurrent available and the exposed location of the cable between the point of attachment of the service drop and the meter base, the installation of the SE cable must meet the requirements of Sec. 230-51(a). The cable must be supported at intervals not to exceed 2½ feet (30 inches) and within 12 inches (1 foot) of all terminations (Fig. 6-12).

When the service equipment, either a fused disconnecting means, a separate circuit breaker enclosure, or a service panel with main, is located inside the building, the SE cable from the metering base to the disconnecting means must be installed according to the guidelines of Romex cable.

Three-conductor SE cable w/GND is often used as a feeder from the main service

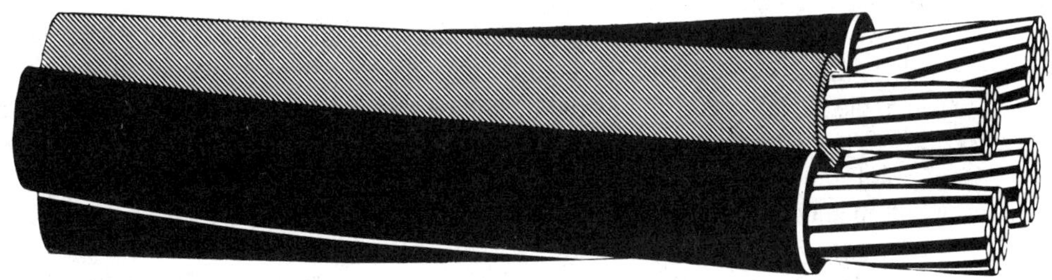

Fig. 6-11 Quadraplexed USE cable is service-entrance cable identified for direct burial without the use of raceway except where exposed when it emerges from the ground. (Courtesy of Southwire Company)

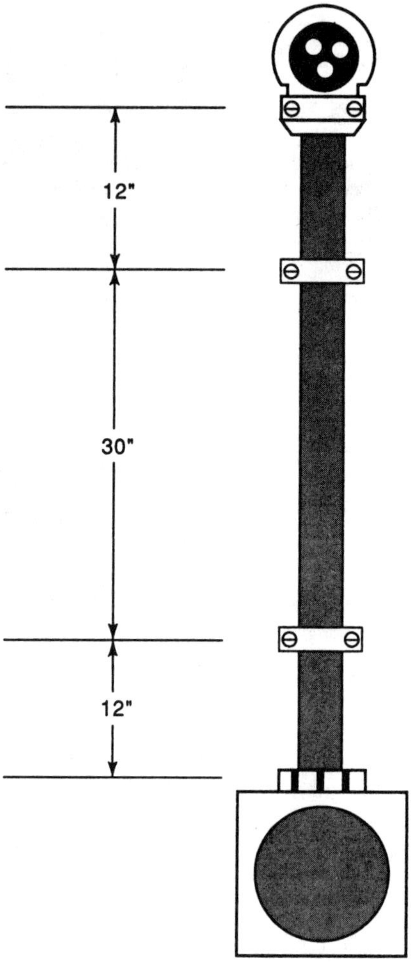

Fig. 6-12 SE cable must be supported (strapped) within 12 inches of all terminations, and at intervals not to exceed 30 inches [Sec. 230-51(a)].

panel to a lower-amperage panel located in another area, or on another floor of the structure (especially true for dwelling installations). Sometimes SE cable is used to supply an electric range load or other cooking equipment, or an electric clothes dryer in the dwelling structure.

When used in this manner, Sec. 338-4, as with the interior run of the service-entrance cable to the service equipment, requires these feeder or branch-circuit cables to be installed within the guidelines of Romex (nonmetallic-sheath) multiconductor cable, Art. 336.

When passing through bored holes in wood members (Sec. 336-9), Sec. 300-4(a)(1) will not allow the SE cable to pass within 1¼ inches of the nearest face (edge) of the wood member without additional nail penetration protection (1/16-inch steel plate, grommet, or equivalent).

When run along the sides of wall studs, plates, floor or ceiling seals or joists, or any wood framing member, the SE cable cannot be installed within 1¼ inches of the nearest face (edge) without additional skirting installed over the face that will sheild the cable from nail penetration.

When installed in notches, the steel plate is an automatic requirement for protection against driven nails. Caution should be taken when drilling holes or cutting notches for these larger SE cables [Sec. 300-4(a)(2)] not to weaken of the wood members.

When run perpendicularly on the underside of floor joists in an unfinished basement, running boards, as specified in Sec. 336-6(c), are normally not required, simply because of the size of the SE cable used. The equivalent of two No. 6 or three No. 8 AWG conductors in a single multiconductor cable can be nailed (stapled or clamped) directly to the underside of the joists.

When installed in an open attic, guard strips are required for perpendicular runs across the top of the floor joists, or across the face of rafters or studs if the SE cable passes within 7 feet of the attic opening (6 feet when not permanently accessible).

When run to follow the surface along the sides of wall studs, joists, or rafters, Type SE cable must be supported (stapled or otherwise fastened) at intervals not to exceed 4½ feet, and within 12 inches (1 foot) of all terminations (Sec. 336-18).

As will be discussed about grounding and bonding in Chapter 9, with the 1996 edition of the National Electrical Code, flat-type SE cable, as shown in Fig. 6-9, can no longer be used as a branch-circuit supply for dwelling electric cooking or clothes drying equipment. As allowed in previous editions of the NE Code (Sec. 250-60), the uninsulated, braided third conductor was used for both the grounded-neutral load-return conductor of the multiwire branch circuit and as the equipment ground-fault return path of the metal enclosure of either the electric cooking equipment or the clothes dryer.

The three insulated conductor with fourth-wire equipment ground-type SE cable, as shown in Fig. 6-10, is now required for the multiwire branch-circuit or feeder supply to any equipment or load center. The grounded-circuit conductor must be insulated from any type of equipment grounding. The bare fourth wire in the cable is used for equipment grounding only.

The second paragraph of Sec. 338-3(b) will allow the flat conductor type to be used for a two-wire circuit supply, such as a larger storage-type water heater that operates at the higher line-to-line voltage (240 volts in the typical dwelling), where the third braided wire is used solely for equipment grounding.

6.4 UF CABLE

The installation and use of underground-feeder (UF) and branch-circuit cable (Fig. 6-13) are governed by Art. 339. Section 339-3(a)(1) allows UF cable to be used for underground installations including direct earth burial without the protection of raceway.

Available in wire sizes from No. 14 AWG copper (the contained conductors) and No. 12 AWG aluminum up to No. 4/0 AWG [Sec. 339-1(a)], the use of this cable ranges

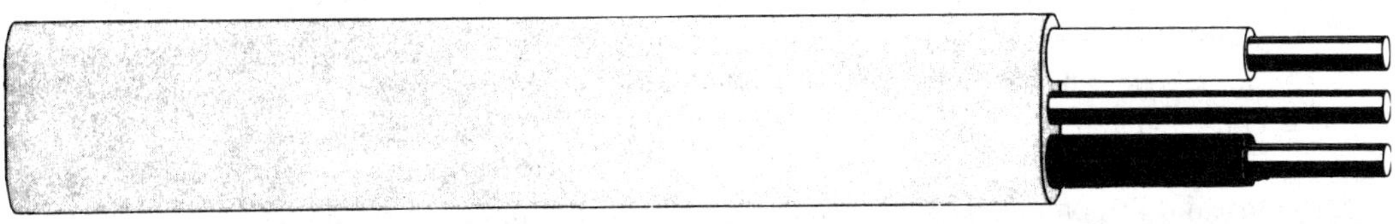

Fig. 6-13 Type UF cable is also approved for direct burial without the use of raceway except where exposed when it emerges from the ground. (Courtesy of Southwire Company)

from the switched supply of a driveway or walkway lamppost, to the branch-circuit supply of a small outbuilding, to the main feeder supply of a secondary building from a host building that contains the service equipment on a common premise (location or grounds).

Between the exit point of the building for the underground run and the entry point of the cable into the service equipment or other panel (load center) inside the building, Sec. 339-(a)(4) requires the installation of UF cable to follow the guidelines required for the installation of Romex cable.

UF cable is often used in place of Romex cable for interior wiring of branch or feeder circuits when the Romex cable (or other nonmetallic sheathed cable) is required to pass through or be exposed to wet or corrosive locations.

Like Romex cable, UF cable cannot be embedded in masonry, poured cement, concrete, or other aggregates. UF cable cannot be installed where exposed to direct sunlight unless the outer sheath is identified for such exposure. Sec. 339-3(b) lists other installations and uses where UF cable is prohibited.

Section 339-3(a)(3) requires the guidelines of Sec. 300-5 to be used for the underground installation of UF cable.

6.5 UNDERGROUND CABLE AND RACEWAY INSTALLATIONS

Article 300, Part A lists the general requirements of wiring methods recognized by the National Electrical Code. Section 300-5 deals specifically with underground installations.

"Underground installations" reflects direct burial of either cable or conduit containing conductors, or it can be other forms of raceway. Direct burial cable is normally limited to type UF or USE cables. Type UF cable normally has an overall outer sheath, similar to Romex cable. Type USE cable (underground service-entrance cable) is normally bound by spiral twisting of the individual insulated conductors (three-conductor cables are triplexed, four-conductor cables are quadraplexed) without an overall covering (outer sheath).

Direct burial ferrous-metal (steel) raceways include rigid and IMC. Section 300-6(a), as well as other sections within the article that governs the raceway, would require these raceways to be galvanized or otherwise coated for protection against corrosion. If the means of protection is solely enamel, then these raceways are not approved for direct burial. These raceways are also available in aluminum—Sec. 300-6(a) does not apply.

Nonmetallic conduits include cement (or concrete) and PVC. Most electricians don't think of cement when referring to nonmetallic underground raceways, but continuous form (poured as you go) cement or concrete ducts, as well as prestressed cement ducts (formed in sections similar to culvert pipes on roadways—assembled at the construction site) are often installed for larger underground feeders.

PVC and other types of plastic or vinyl piping are very common underground raceways. PVC is available in medium wall (schedule 40) and thick wall (schedule 80) types.

When the raceway is buried, Sec. 310-8(a) requires the installed or contained circuit conductors to have a "W" rating in their insulation acronym. Any underground piping will have some water accumulation due to either ground seepage around couplings that didn't completely seal or condensation of the moisture in the air when changes in temperature inside the building cause the air to migrate either into or out of the underground raceway. Section 300-18 requires the underground raceway, and where it rises for termination, to be installed as a complete system before the circuit conductors are installed.

Direct burial cables must be protected from physical damage by being installed in protective raceway (normally one of the piping or tubing conduits) when emerging from the ground. This is especially true when the service laterals (underground service conductors) are run up the wall outside the building or other structure for termination to the meter base or service-disconnecting means enclosure.

The minimum cover requirements for cables, conduits, and other raceways are listed in T300-5. The cover requirement is listed in inches according to the type of circuit, and according to the type of cable, conduit, or other raceway.

The cover requirement to the top of the cable for direct-burial cables (first row, first column) is two feet, or 24 inches. The cover requirement for either rigid or intermediate metal conduits (first row, second column), whether steel or aluminum, is only 6 inches. The cover for nonmetallic conduits (first row, third column), whether PVC or concrete, is 18 inches. The PVC must be recognized (listed) for direct burial without concrete encasement. Columns four and five are exceptions to these general rules with lesser depths or cover requirements for special-use cables or circuits.

The discussion will concentrate on the direct burial cables, since the cover requirements for complete raceway systems is straightforward. Triplexed or quadraplexed USE cable is often used for underground mobile home services or for services to other dwelling structures. The spiral-wrapped cables are approved for direct burial without raceway. The trench depth (Fig. 6-14), is 24 inches plus the cross-sectional diameter of the cable.

Note 3 to Table 300-5 allows lesser depth or cover requirements (less than 24 inches to the top of the cable) where the cable rises for termination, splice, or where other access may be required. In other words, the measured depth to the top of the triplexed USE is not required to be exactly two feet where the cable enters the ground coming off the service mast pole, or from the base of the meter can. The cable is allowed to enter the ground in a gentle sweep or bend to lay in the trench such that the bending radius will not damage the insulation, or the cable will not be held under pressure against the entry point to the protective conduit.

Section 300-5(d) requires direct-burial conductors or cables to be protected from physical damage when they emerge from the ground. The conductors or cables must be installed in conduit (or other types of piping raceway) from a maximum depth of 18 inches below the ground (in the trench) to a maximum height of 8 feet above the ground (finished grade).

If the conductors extend up a utility pole in the form of a service lateral, the conductors have to be protected the full eight feet. If the conductors or cable extends up to a meter can or service-disconnecting means enclosure, the conduit must extend from the lower side of the enclosure to a maximum of 18 inches below the finished grade.

Section 300-5(h) requires the raceway entry, both above and below ground, to be made up with a bushing that will protect the conductor insulation from cuts, scrapes, or abrasions during (conductor pull) and after (earth movement, due to expansion and contraction) installation. The inside edge of PVC or other plastic-type piping raceways can be reamed with a utility knife to create the same effect as the bushing. Metal pipes have to be fitted with a bushing.

If the raceway will terminate to an enclosure that contains live (energized) parts such as the meter base in Fig. 6-15, then Sec. 300-5(g) requires a raceway seal in addition

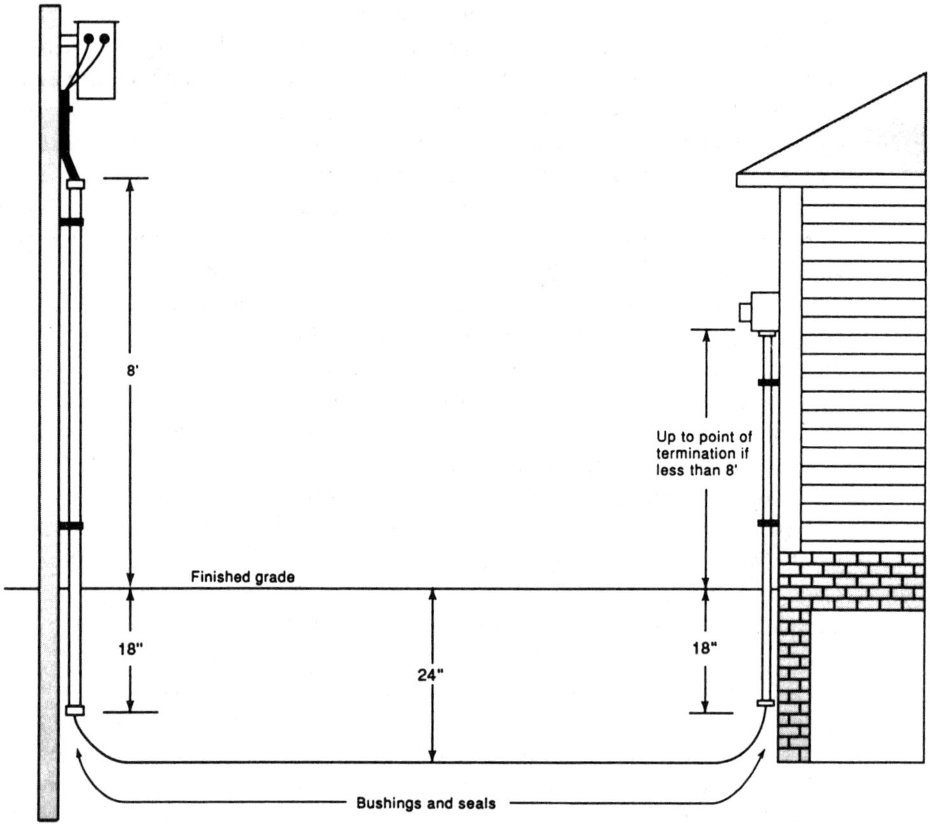

Fig. 6-14 When direct buried cable is installed in a trench, the requirements of Section and Table 300-5 must be met.

to the bushing, or a bushing-type fitting that incorporates the seal can be installed to terminate the pipe raceway.

The conduit extended up the side of the pole only requires a bushing at each end, unless the accumulation of rainfall in the conduit could have a standpipe effect, in which case water seeping into the ground at the base of the pole may undermine it. In this case, a bushing and seal should be installed on the top side.

A service feeder set for a mobile home (Fig. 6-15) is often a service drop to a service mast or utility pole. Although the service-disconnecting means is allowed by Sec. 380-8(a) to be mounted on the pole at a maximum height of 6 feet, 7 inches (2 meters) above the finished grade, the meter can is usually mounted at eye level (between 5 and 6 feet) and the service fused disconnect (usually a main circuit breaker) is mounted directly beneath it. The conduit for this type of installation must be extended from the bottom of the service-disconnect enclosure to a maximum of 18 inches below the finished grade.

If conduit is used for the service-entrance conductors above the meter enclosure, the eight-foot minimum height above finished grade (partially below and partially above the meter enclosure) would be an NE Code violation. If single insulated conductors are installed in raceway for the service entrance, the raceway must extend all the way up the

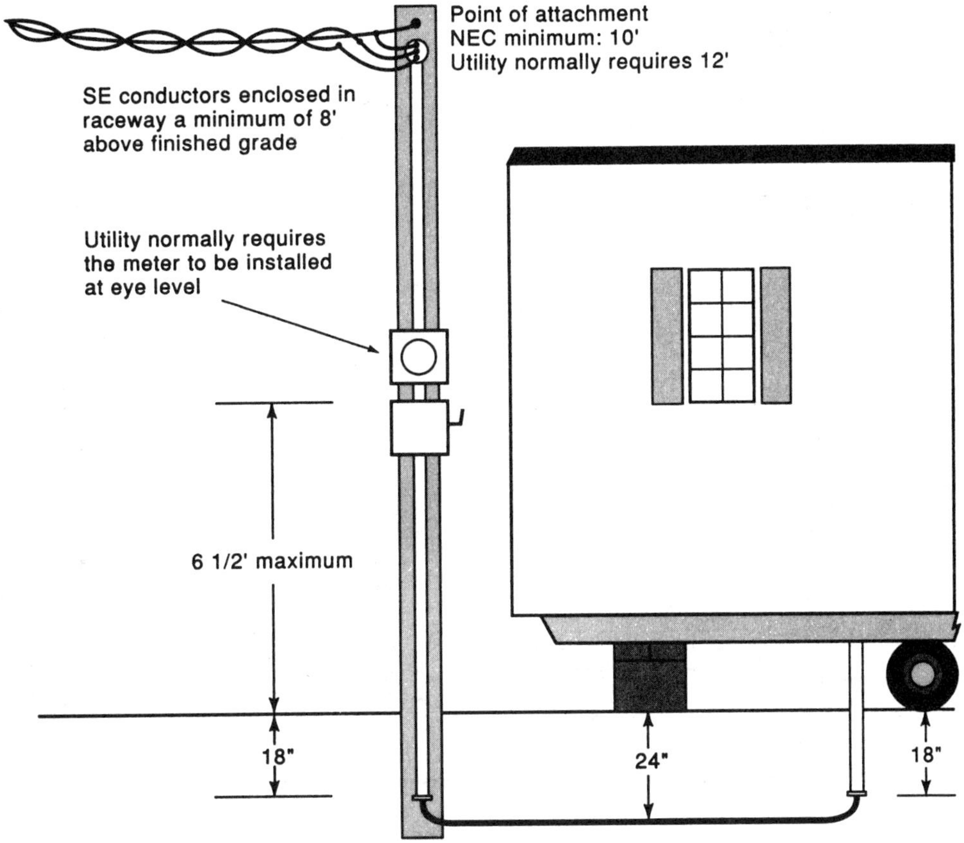

Fig. 6-15 The service installation of a mobile home must meet the requirements of Arts. 230, 240, 250, 300, 310, and 550.

pole to the weatherhead to meet the requirements of Sec. 230-54(a). Service drops will be discussed in detail in Chapter 7 of this text.

The 24-inch depth to the top of the cable is reduced to only 18 inches by the second row heading of T300-5 if a 2-inch concrete pad or equivalent protection is installed in the trench above the cable (Fig 6-16). This physical protection will prevent a person with a shovel or other hand tools from digging down and piercing the cable. The concrete pad or board (a 2-by-10 treaded board is often used for this type of "equivalent" protection) would be hit before the cable in this situation.

If the triplexed or quadraplexed USE cable supplying the dwelling passes under the driveway (sixth row, first column), the cover depth is reduced from 24 to 18 inches without the use of the concrete pad. The service lateral is considered as protected from digging by the driveway.

If a USE cable service lateral passes under a public road, whether to supply a dwelling or a nondwelling structure, the minimum cover depth is returned by T300-5 (fifth row, first column) to 24 inches. The public road or any other paved surface intended for truck traffic is subject to compaction due to the weight of these larger vehicles. The

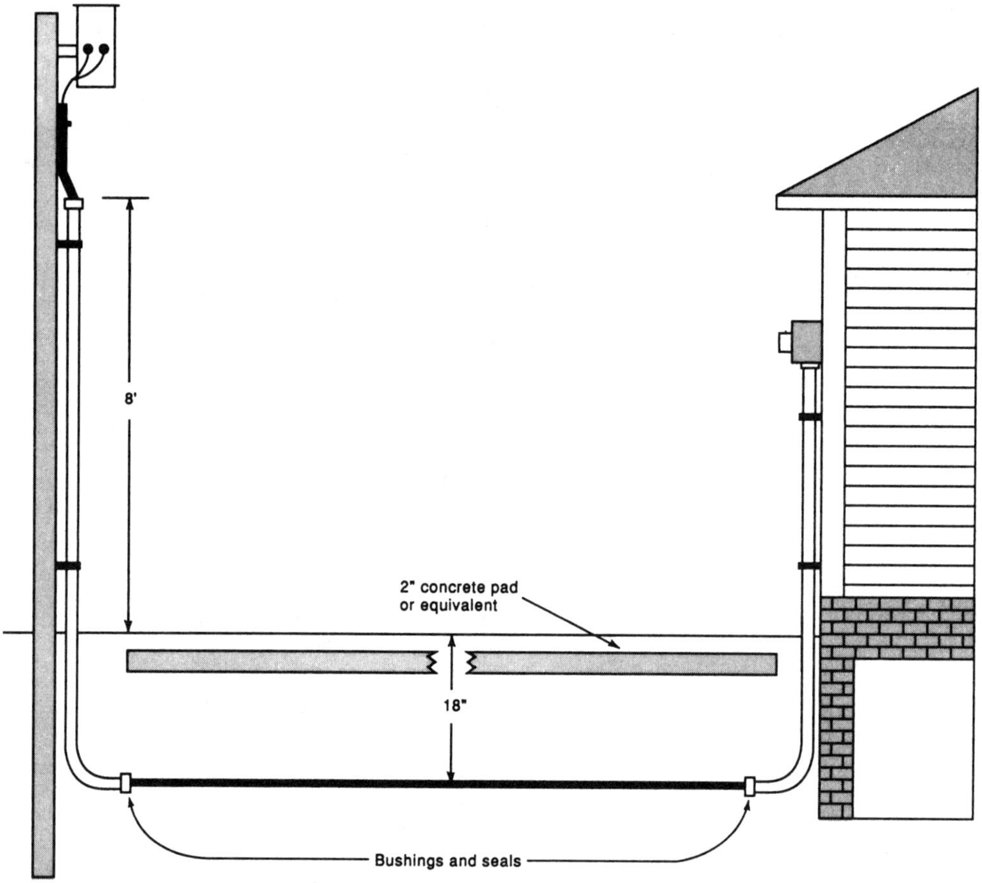

Fig. 6-16 The burial depth of cables can be reduced to 18 inches, if a 2-inch concrete pad or equivalent protection is installed above the cable.

minimum cover requirement for any column subject to truck traffic is 24 inches—say truck, say twenty-four.

If UF cable is used to supply an outbuilding or other structure, or if UF cable is used to supply outside lights, signs, or other pole installations, the general cover depth (to the top of the cable) is two feet, or 24 inches. Raceway protection is required in the same manner as the service laterals just discussed.

When the UF cable emerges from finished grade outside the building or other structure, it must be installed in protective raceway from a maximum of 18 inches below ground up to termination, to a maximum of 8 feet above ground. The same gentle sweep or bend in the cable described for the service lateral as it rises for termination from the trench is required for UF cable.

In a residential installation, if the UF cable is used to supply a branch circuit rated at 120 volts, 20 amps or less, and is GFCI-protected (such as a garage or a remote storage building), the cover depth without the concrete or equivalent protection (first row, fourth column) is only 12 inches. With protection (second row, fourth column), the cover is reduced to only six inches.

If the UF cable for this residential branch circuit passes under a driveway or parking area in the residential or dwelling yard (Fig. 6-17), the cover depth (sixth row, fourth column) would have to be maintained at 12 inches for that section of the burial regardless of whether or not the concrete pad or other protection is installed in the trench above the cable.

If UF cable is used to supply residential branch or feeder circuits that are rated greater than 120 volts, greater than 20 amps, or at any voltage or current that is not GFCI-protected (such as well pump houses, garage/workshops, or utility buildings that are often supplied with a 120/240-volt, three-wire circuit), the cover depth of column five is void. The cover depth as listed in the first column must be met.

Any remote underground feeder or branch circuit that extends from a nondwelling structure, regardless of voltage or current rating, and whether or not the circuit will be GFCI-protected, must have a minimum cover depth of 24 inches to the top of the cable (first row, first column). If the cable will not pass under a public road or similar area (fifth

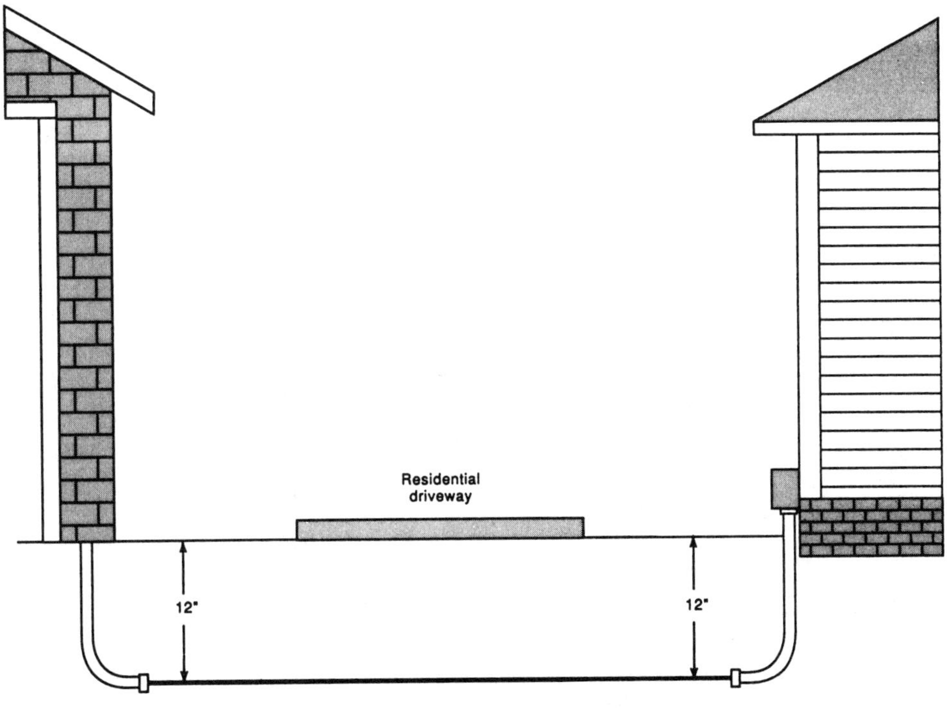

Fig. 6-17 Lesser depths are allowed for specific residential branch circuits wired with UF cable

row, first column) and will be suitably protected by a concrete pad or equivalent (second row, first column), the cover depth to the top of the cable is reduced to 18 inches.

If lampposts are installed at the end or along a walkway or driveway, they can be supplied with UF cable. When the UF cable enters the lamppost above ground, Sec. 300-5(d) requires the cable to be protected by (or installed in) conduit or other piping raceway from a maximum of 18 inches below finish grade up to the point where the cable enters the lamppost.

The cover depth to the top of the cable varies from 6 to 24 inches, depending on whether the installation is for a dwelling or nondwelling structure (or premise), and whether or not the cable in the trench is provided with extra protection from physical damage. Lesser depths for rise to termination are allowed by Note 3 to the table. A judgment call by the electrician is warranted here because the cable must be protected from physical damage (both bending radius and exposure to hand gardening tools, as well as lawnmowers and other lawn equipment), and no finite "lesser depths" are stipulated in the note.

If the lamppost is factory slotted with rounded or reamed metal edges (Fig. 6-18),

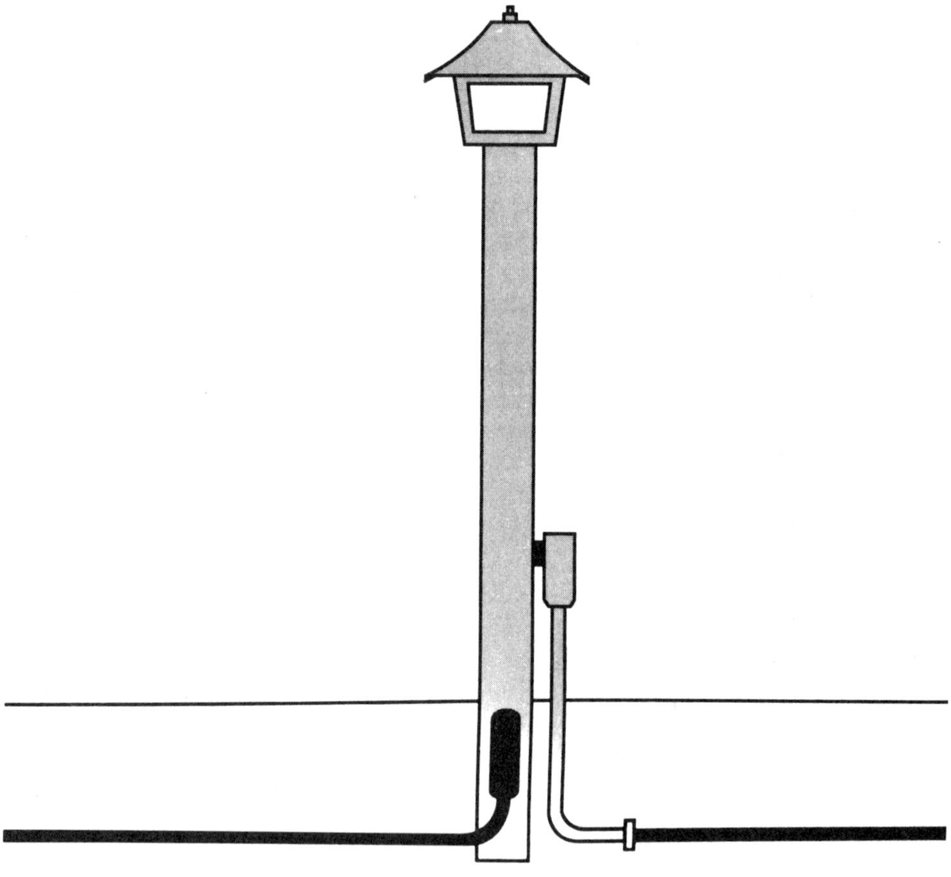

Fig. 6-18 Cable entry for a walkway-type lamppost can be above or below ground. Aboveground cable must be protected from physical damage.

the cable can enter the post without further protection. If a hole has to be bored to accept the cable, Sec. 300-4(b)(1) requires the bored hole to either be reamed and the edges rounded, or bushed to protect the cable from cuts and/or abrasions.

The lamppost may or may not be set in concrete, but regardless of concrete or fill dirt, the UF cable should not be extended down below the end of the pipe in the hole and then run up into the pipe. UF cable by Sec. 339-3(b)(8) cannot be installed (embedded) in concrete. If concrete is not used, the bend in the cable to enter the pipe in conjunction with the weight of the pipe sitting on the cable would damage the insulation and outer sheath of the cable.

The UF cable can be fished and left hanging loose in the lamppost [Secs. 339-3(a)(4) and 336-18, Ex. No. 1], but it must be secured and supported where it enters the fixture outlet box. The stress of the support by a squeeze connector (two-screw cable clamp) must be put on the outer sheath of the cable and not just on the individual conductor insulations emerging from the nonmetallic sheath.

The UF cable must contain an equipment-grounding conductor because the lamp fixture and (metal) post must be grounded as required by Sec. 250-42(b). Also, Sec. 250-91(c) stipulates that the earth cannot serve as the sole equipment-grounding means. (Grounding and bonding are discussed in Chapter 9 of this text.)

The lighting fixture must be approved for wet locations as required by Sec. 410-4(a). The better electrical installation would have the fixture GFCI (ground-fault circuit interruption) protected, but this is not a specific NE Code requirement. Refer to Sec. 210-8.

As a general rule, the buried cable should contain no splices, but if splices have to be installed in the trench, Sec. 300-5(e), by reference to Sec. 110-14(b), will allow the cable splice to be made without a box, provided an approved splicing kit or other materials and methods approved for the purpose are used. If a junction box is used for the underground trench splice, the box (or other conduit fitting) should be rated for a wet location or weatherproof installation.

The UF cable can either enter the outer wall of the structure at trench level (below ground), or it can be stubbed up in conduit or other piping raceway to enter at aboveground level. In either type of installation, Sections 300-5 (c) and (d) require the UF cable to be protected by approved raceway as it passes through the foundation wall or emerges from the ground to enter the structure.

If the UF cable passes through the foundation wall below ground level (Fig. 6-19), the conduit is required by Sec. 300-5(h) to be bushed at both ends to protect the cable's outer sheath from cuts and abrasions. At one end, usually in the trench, the conduit is required by Sec. 300-5(g) to be sealed (unless the bushing means is approved for sealing) to seal out moisture or seepage water that may enter the structure basement or underfloor crawlspace. The foundation wall should be sealed against the outer surface of the conduit to prevent seepage water from working its way into or under the structure as well.

If the protective conduit is stubbed up to enter the foundation or structural wall above ground (Fig. 6-20), the raceway is required to have a bushing on the trench end where the cable enters. The stub up is usually terminated in an LB conduit body.

From the LB, a straight-in run of conduit passes through the wall. The conduit passing through the wall can be terminated in a junction box [Sec. 300-15 (a) and (b)], and the UF cable then spliced to Romex cable for the homerun to the service or subfed panel.

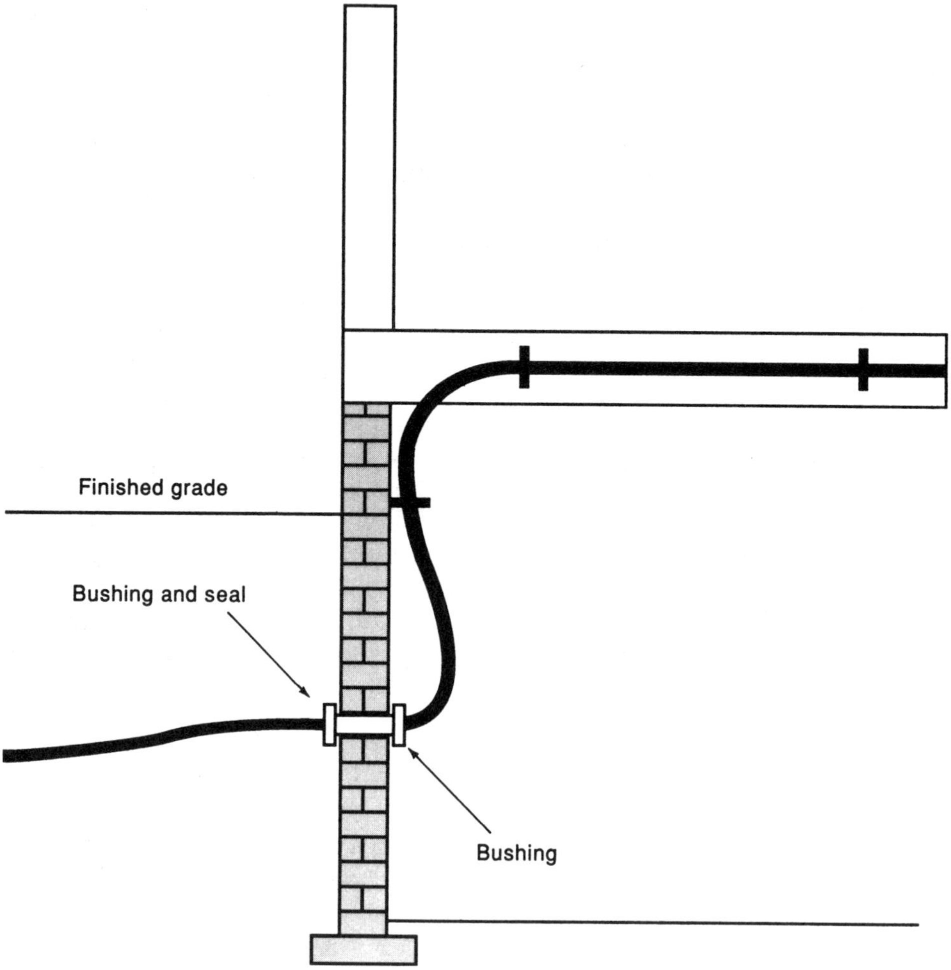

Fig. 6-19 Cable installations entering the foundation wall of a structure below finished grade must be protected from physical damage.

Grounding the box with the equipment-grounding conductor contained in the Romex cable will effectively ground the conduit stub extending into the trench.

If the conduit is just stubbed through the structure wall above ground and the UF cable run to the dwelling panel as the homerun without splice, the conduit should be terminated on the inside wall of the structure in a grounding-type bushing. The bushing should be grounded with a green insulated conductor [Sec. 310-12(b)] run to the nearest available grounding-electrode conductor or other equipment-grounding means, and sized according to the branch-circuit overcurrent protection as listed in T250-95.

Grounding of the conduit that enters at trench level (below ground) can be accomplished in the same manner, although the grounding of this raceway nipple is not as critical as with the aboveground conduit entry, since the entire nipple is below ground.

Installing the cable in a metal pipe nipple as it passes through the foundation wall,

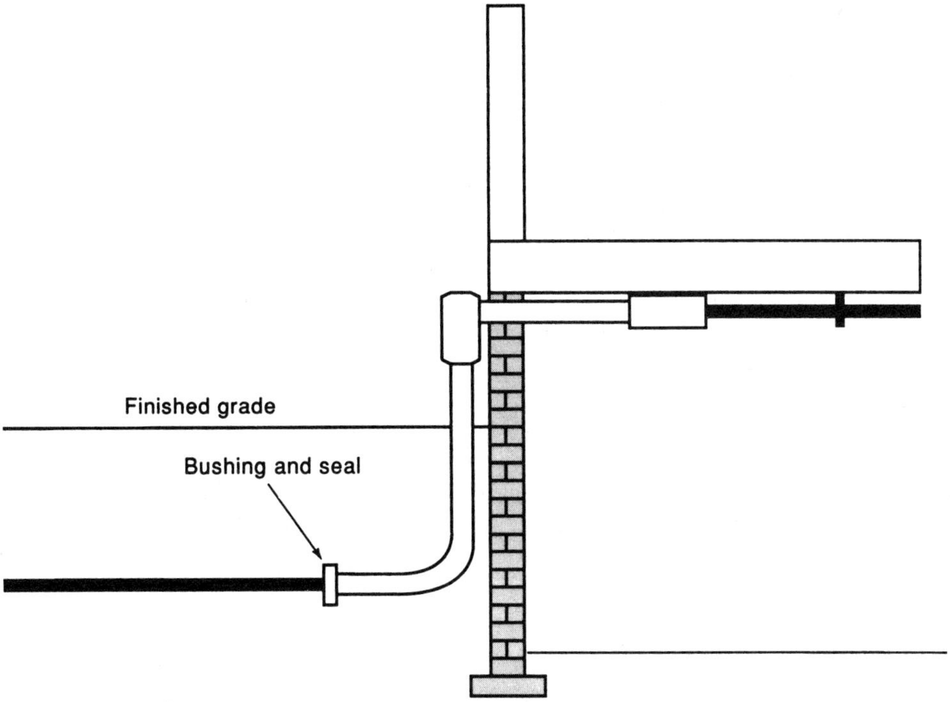

Fig. 6-20 Underground cable installations entering the foundation wall of a structure above finished grade must also be protected from physical damage.

or in conduit when stubbed up above ground to enter the foundation wall, protects the cable from physical damage. Grounding this type of raceway entry protects the individual from shock hazards should the cable develop a ground fault to the raceway. (Ground fault is an unintentional connection of a normally ungrounded current-carrying conductor to the grounded noncurrent-carrying metal parts of the power distribution.)

The metal conduit's being in physical contact with the earth does not ensure mechanical or electrical continuity (pressure connection) with earth ground, such that a potential difference (voltage) between the metal raceway and the earth could occur in terms of fault conditions. The grounding requirements can be avoided if the raceway protection is PVC or some other form of nonmetallic piping (conduit or tubing) raceway.

Where the stub-up conduit enters the foundation or other structure wall above ground, the bored or chiseled opening through the wall should be sealed to the raceway to prevent structure run-off water from entering the wall or underfloor crawlspace.

If the lamppost installed along the driveway or walkway is to be supplied with individual insulated conductors installed in an underground raceway, the contained conductors, as discussed for the service lateral earlier, will have to have a wet (W designation) insulation rating [Sec. 310-8(a)].

Metallic raceway can serve as the equipment-grounding means, but the better installation is to run an equipment-grounding conductor with the supply conductors. The separate equipment-grounding conductor is required with the use of nonmetallic raceway.

As with the installation of UF cable, Sec. 250-42(b) requires the (metal) lamp fixture and post to be grounded, and Sec. 250-91(c) will not allow the earth to serve as the sole equipment-grounding means.

The better electrical installation will not have any buried junction boxes or raceway fittings. If boxes or fittings must be used, they should be rated for wet or weatherproof installations (Sec. 300-6). If the boxes or conduit bodies are to be buried in the trench and contain splices, the electrician should go one step further and pack the junction box or conduit body with potting compound or electrical putty to isolate the made-up splices from moisture and seepage water accumulation in the box or body.

Section 370-16(c) restricts splicing in conduit bodies (such as LBs or condulet TEEs), unless the conduit body can meet the volume requirements for other boxes, as outlined in Secs. 370-16(a) and (b). With this ruling, a 90° ell conduit body or a straight pull-point conduit body (C body) cannot contain splices unless they can meet the volume per conductor requirements of Sec. 370-16(b) and the rigid and secure support requirements of Sec. 370-16(c). A conduit body or junction box used only for a pull point without conductor splicing would not require potting compound or any type of moisture-seal packing because the integrity of the conductor insulations is still maintained.

The buried conduit (or other underground raceway) should extend into the lamppost and be physically connected to the base of the lamp fixture. If the electrician elects to terminate the conduit to the side of the (metal) lamppost and fish the single conductors in the post up to the lighting fixture, the conduit must be terminated in a 90° ell conduit body to facilitate a pull point. The conduit body, using a short threaded nipple, must be mechanically secured to the lamppost [Secs. 300-10 and 300-12]. The nipple must be terminated on the inside of the lamppost with a bushing [Sec. 300-16(b)].

If the raceway enters the lamppost from underground and is terminated aboveground inside the post, the conduit must be terminated in a bushing [Sec. 300-16(b)]. Supplied in this manner, the lamppost must meet the requirements of Sec. 410-15(b) and have a covered access handhole at its base.

The interior of the lamppost cannot contain any conductor splices [Sec. 300-13(a)] unless the handhole is provided for either type of conduit entry.

The single conductors entering the base of the lighting fixture box must be protected from cuts and abrasions [Sec. 300-5(g)] by a bushing or grommet. A two-screw squeeze connector (cable connector) cannot be used to secure the fished single conductors to the base of the fixture outlet box, since Sec. 110-7 requires that the individual conductor's insulation integrity be free from short circuits and grounds when the installation is completed. A squeeze connector might impair the conductor insulation by making it thinner at the pressure point.

At the dwelling or other structure, the conduit or other underground raceway can enter the structural wall in one of two ways: The conduit can enter directly through the foundation wall at trench level (depth), or it can turn up and enter the foundation or structural wall above ground.

If the conduit enters the wall at trench level, the opening between the wall and the conduit should be adequately sealed with a compound to prevent the entry of moisture or seepage water into the basement or underfloor crawlspace.

If the conduit is rigid or intermediate metal conduit (IMC), the required cover depth to the top of the conduit (T300-5) is only 6 inches. If PVC or other rigid nonmetallic raceways are installed, the cover depth (first row, third column) is increased to 18 inches. If the rigid nonmetallic conduit is protected by a 2-inch concrete pad (or equivalent), the cover depth (second row, third column) is reduced to only 12 inches.

If the conduit or any other approved raceway or cable for an underground installation passes under driveways or other areas such as parking lots that are exposed to heavy vehicular traffic, the cover depth (fifth row, all columns) to the top of the raceway or cable must be increased to a minimum of 24 inches.

The conduit does not have to be run or extended all the way back to the panel. Instead, the conduit can terminate in a junction box in the basement or underfloor crawlspace, and the homerun made with Romex or UF cable containing conductors of the same size as dictated by the branch-circuit overcurrent device.

Underground raceway or cable installations for services or subfeeders to other buildings would be installed in basically the same manner as was described for this lamppost installation. Note that the third row of T300-5 does not specify a cover depth for raceway installed under a building.

Two other sections of interest for underground conduits or other raceway installations are Secs. 300-6 and 300-7. Section 300-6 requires ferrous metal raceway to be protected against corrosion. Section 300-7(a) requires the sealing of colder sections of raceway from warmer sections where moisture might be transmitted by condensation and/or convection.

REVIEW QUESTIONS

1. Multiconductor cables, such as Romex, BX, SE, and UF, are considered to be wiring methods by the NE Code, which, except for protection against physical damage, are not required to be installed in a raceway.

True or False

2. Nonmetallic sheathed cable can be used in any occupancy that does not exceed three floors above finished grade, with the first floor being defined as designed for human habitation, and which has 30% or more of its perimeter level with or above the finished grade of the exterior wall.

True or False

3. Romex can be embedded in concrete or other aggregate only if the outer sheath has not been damaged during the cable installation or the concrete pour.

True or False

4. Romex cable can be installed concealed within wall spaces, or it can be run exposed in a building or other structure.

True or False

5. If Romex cable can be installed exposed in a building or other structure, it must be installed so as to follow the ____________.

6. If Romex cable is installed along the joists, studs, or rafters in attics, or along the seals or joists in underfloor crawlspaces, the cable must be run parallel and perpendicular to the wood members.

   True or False

7. If Romex cable is installed exposed in unfinished basements, the cable, regardless of wire size, can be run perpendicular to the floor joists without the use of running boards, provided the cable is stapled to each floor joist.

   True or False

8. When Romex cable is run through bored holes in either wooden floor or ceiling joists or wall studs, there must be a minimum of a one-inch clearance between the nearest edge of the hole and either edge of the wood member.

   True or False

9. When run along the sides of either wooden ceiling or floor joists or wall studs, Romex cable must be installed on running boards that will support and protect the cable from physical damage.

   True or False

10. When exiting the "protective floorline" on the underside of floor joists in an unfinished basement to supply an oil-fired furnace, Romex cable must be installed in raceway to protect it from physical damage.

    True or False

11. When passing through floors, Romex cable must be protected by conduit or tubing up to a point at least 12 inches above the finished floorline.

    True or False

12. When Romex cable is run through bored holes in wood members of a structure, the hole's edge must be a minimum distance of 1¼ inches from the face of the wood member, unless additional protection such as steel plates are also installed.

    True or False

13. Romex cable can only be installed in notches cut in the wood members of a structure when the structural member will not be weakened by the cut.

    True or False

14. When installed perpendicular to the face of joists in an attic, Romex cable must be run through holes bored in the sides of the floor joists, or else protected by guard strips and installed across the face (perpendicular) of the floor joists.

    True or False

15. When installed perpendicular to the face of rafters or studs in an attic, Romex cable can be run through bored holes or across the face with staples. If the cables are run across the face and the attic is permanently accessible, cables passing within ____________ feet of the opening must be protected with guard strips. If the attic is not permanently accessible, the measurement is reduced to only ____________ feet.

16. When installed parallel to framing members (sides of joists, rafters, or studs) in an attic, BX multiconductor cable must be stapled (fastened or otherwise located) a minimum of 1¼ inches

from the nearest face (edge) of the framing member, or else be protected from nail penetration by a steel plate.

True or False

17. When installed in interior locations for branch or feeder circuits, Type SE cable (service entrance) must be installed under the same guidelines as Romex cable.

True or False

18. When installed for its intended purpose (as service-entrance cable), SE cable must be installed in accordance with Art. ____________, and supported at a minimum of ____________-inch intervals, and within ____________ inches of all terminations.

19. When installed in interior locations, type UF cable must be installed under the same guidelines as Romex cable.

True or False

Questions 20 through 25: For circuit voltages of 600 volts or less

20. Underground installations of cables rated for direct burial, and both metal and nonmetallic raceways, are governed by the requirements of Section and Table ____________.

21. A direct-burial cable, without additional physical protection in the trench, must have a trench depth that will allow a minimum of ____________ inches of cover to the top of the cable.

22. Rigid or intermediate metal conduit must have a trench depth that will allow a minimum of inches of cover to the top of the raceway.

23. When buried under public streets or areas subject to truck traffic, whether conduit, cable, or other raceway, the minimum trench depth to the top of the wiring means or enclosure must be a minimum of ____________ feet.

24. When direct burial cables are run up a pole or structure surface from the ground, the cable must be protected from a maximum depth of 18 inches below finished grade to a maximum height of 18 feet above finished grade.

True or False

25. When conduit containing direct burial cable enters a structure wall below the finished grade, the conduit must be effectively sealed.

True or False

Chapter 7

Overcurrent Protection

Chapter 6 introduced the composition of all electrical circuits as three basic components: the power supply, the connected load, and the circuit conductors that connect the two. Actually, all electrical circuits are comprised of five components. In addition to those already introduced, two means of electrical control are required: some means of switching or disconnect and some form of protection against the effects of overcurrent.

The switch, which will be discussed in Chapters 11 and 12 of this text, can be described as the directional control of the circuit. The knife switch can be used as a disconnecting means to assure the power is off when the supplied equipment or the electrical distribution has to be serviced, modified, or removed.

A simple control switch, such as a light switch, is used to decide when the lights will be on (energized) or off (de-energized). Either a manual or an electromagnetic starter (both are motor controllers) can be used to direct the operation of a motor in the same manner that the light switch controlled the status of the lights.

An overcurrent protection device, as the name implies, is a means to protect the circuit from the devastating effects of overcurrent on the overall electrical circuit.

Overcurrent protection can be described as limit control. Individual wire sizes are assigned a maximum rated-load ampacity. The load current these circuit conductors must convey cannot exceed this rated-load value.

The load current can be controlled in two ways: limit the number of connected parallel loads such that the summation of the full-load currents does not exceed the rated-load ampacity of the circuit conductors, or install either a fuse, circuit breaker, or some other type of overcurrent protection device (OCPD) in series with the circuit conductors at the point of power supply to limit the available load current to a maximum value that is less than or equal to the rated-load ampacity of the circuit conductors.

This second method is especially important when the branch circuit supplies convenience receptacles in the average home or business. Several receptacles are supplied from the single branch circuit. The values of load current are technically unknown at the time of circuit installation. Since several plug- and cord-connected devices or other utilization equipment can be supplied simultaneously, the rating of the overcurrent protection device is set at the same value as the rated ampacity of the circuit conductors. When this value is exceeded, the protection device opens to protect the circuit conductors.

Circuit and equipment overcurrent protection is governed by Art. 240 of the NE Code. The primary purpose of an overcurrent protection device, as described in the fine print note (FPN) to Sec. 240-1, is to limit the temperature of a conductor to a value that will not damage the conductor or its insulation by limiting the amount (value) of current the conductor is required to convey. The current magnitude is limited by the opening of the fuse or circuit breaker to stop the current flow. Protecting the conductors against overcurrent inherently protects the supplied equipment and/or connected load.

7.1 OVERCURRENT

The circuit conductors are the primary target for effective overcurrent protection because they are installed inside raceway or concealed in the voids of walls (multiconductor cables). Overheated conductors, especially in the concealment of the walls, can cause structural fires.

The electrical power supply to a building or other structure is normally a transformer that steps down the voltage and steps up the current from a high wattage-rated utility distribution system.

The power rating (normally in kVA, or thousands of volt-amps) is an indication of the full-load current the transformer can deliver at rated voltage. When overcurrent is allowed to flow, the transformer will overheat because it is operating in excess of its rating. The circuit conductors and the power supply can both fail from an overcurrent condition in the circuit.

As defined in Art. 100, there are three principal divisions or types of overcurrent: overload, short-circuit, or ground-fault. All three can damage the power supply, the circuit conductors, and the circuitry within the connected load if not detected and stopped by the opening of the overcurrent protection device.

7.2 OVERLOAD OVERCURRENT

Overload overcurrent occurs when the connected load on a branch circuit, feeder, or service exceeds the rated ampacity of the respective circuit conductors.

Most power distribution systems function with the service equipment as the central power supply. If feeder circuits are extended to other panels or load centers, each feeder circuit is wired in parallel in the service equipment. If branch circuits also originate in the service equipment panel, they too are wired in parallel with each other and with the feeder circuits.

Wired in parallel, all circuits operate at the same voltage. The current drawn from the service conductors will increase as each load is connected. Besides maintaining the same voltage rating, adding loads in parallel has the effect of lowering the circuit resistance. This is why the supply current increases.

Adding lighting fixtures to an existing multioutlet branch circuit or trying to supply too many fixtures on the initial installation can overload the branch circuit.

If a branch circuit is rated at 20 amps, 120 volts, and lighting fixture rated at 2 amps, 120 volts are to be supplied, the circuit can supply a maximum of 10 fixtures: 20 amps ÷ 2 amps = 10. If the branch circuit is rated at 20 amps, then the rated-load ampacity of the circuit conductors and the rating branch-circuit fuse or circuit breaker is also 20 amps.

If an electrician elects to add another lighting fixture simply because a junction box on this branch circuit is in the immediate area where the light is to be installed, the circuit will be overloaded. If the circuit is protected by a fuse or circuit breaker rated at 20 amps. The fuse should melt open or the breaker should trip open because of the 2-amp overload. This protects the circuit conductors from overheating due to the overload overcurrent.

If the rated ampacity of the circuit conductors on this 20-amp circuit (determined by the rating of the OCPD) is computed to be only 16 amps because of higher operating temperature in the attic or ceiling above the lighting fixtures, the circuit conductors were overloaded on the initial installation of the ten lighting fixtures. The overload may lead to failure of the circuit conductors or a possible structure fire.

The overload overcurrent will cause the circuit conductors to operate continually at an elevated temperature, which may soften the insulation to the point that a short-circuit may occur either between two adjacent circuit conductors or between one of the ungrounded circuit conductors and a grounded metal box or raceway. Undersized conductors should not be protected from overcurrent conditions in a circuit with oversized protection devices.

A panelboard (electrical panel) is normally supplied by a feeder circuit. Sizing the feeder conductors for the rated ampacity of the panelboard and protecting both with a fuse or circuit breaker of the same current rating prevents the panel or the feeder circuit from being overloaded.

A panel is purchased with a standard ampere rating at a given supply voltage and phase. The contained busbars inside the panel are rated to convey this current (full-load rating) without overheating or failing. Often the amperage of the panel is used up with far less branch circuits or other feeder circuits than the available slots in the panel. When additional circuits are required in the area of the panel, the spare slots, for breakers or switched fuseholders, gives the false impression that the power (amperage) is available. If additional circuits are installed, the panel busbars could be overloaded if it were not for the feeder fuse or circuit breaker opening to stop the overload overcurrent condition.

The same analysis can be applied to the service rating (main power supply) to a building or other structure. Many dwelling structures built in the 1960s were supplied with a single-phase, 3-wire, 120/240-volt, 60-amp service. With the boom of appliances and electrical gadgets that have been developed for the home in the last thirty years, this service is probably overloaded.

It seems the main service fuses should open if this 60-amp service is overloaded in the same manner the branch circuit or feeder just described was protected from overload overcurrent. But often circuits are added by side-stepping the main fuse panel on these older dwelling services. The service-entrance conductors are paralleled on the load side of the meter movement with supply conductors extended to additional load centers for added equipment such as window air conditioners or clothes dryers, independent of the main service panel inside the dwelling. The 60-amp fuse panel inside the dwelling is not overloaded, but the service-drop conductors to the dwelling are.

The service drop is purposely fused at its load end instead of its supply end to reduce the number of outages (opening of the service fuses or circuit breaker) caused by lightning strikes on the service drop to the structure. If the lightning strike hits the grounded side of the transformer supply, the discharge is directly into the earth with only minimal disturbance to the power distribution inside the building. If the strike hits the ungrounded side of the transformer, the affected winding of the transformer and the affected side of the meter movement shunt the majority of the lightning strike to the earth ground. If the service drop is fused at the connecting point to the transformer, the fuse would open due to the parallel path to the meter movement—an unwanted power outage.

Unfortunately, the load-end overload protection does not prevent the side-stepping just described. Naturally, this is an NE Code violation.

Overload overcurrent can also be the result of overloaded motor-driven equipment: More torque is required than can be delivered by the motor. Torque is a direct factor of the horsepower rating of the motor. The horsepower determines the full-load current.

A motor draws its nameplate full-load current when operated at rated voltage and rated torque. If the motor is underloaded, a 5-horsepower motor doing a 3-horsepower job, then the motor will draw far less current than the full-load current indicated on the nameplate. If the motor is overloaded, the same 5-horsepower motor doing a 6-horsepower job, then the motor will draw overload overcurrent (more current than indicated as full load on the nameplate). The amount or percentage of overload overcurrent is dependent on the amount of work required.

The motor will also draw overload overcurrent when the coupling between the motor and driven equipment (chains, belts, gears, or sleeves) are either too tight or misaligned. The added bind causes friction in the bearing assemblies, which the motor must supply or overcome by drawing additional load current. A bearing suffering from the lack of lubrication (grease or oil) will overheat. The added friction will also cause the motor to draw overload overcurrent.

When a motor draws overload overcurrent, both the supply conductors and the winding conductors inside the motor (like the supply transformer mentioned earlier) will overheat due to overload overcurrent. Fuses or circuit breakers must be installed at the point of power supply for the motor circuit, or additional protection devices must be installed at the controller (often starter heaters) to protect the motor from "running" overload overcurrent.

Overcurrent due to overload is normally a controlled overcurrent situation of low magnitude, due to either bad design, poor judgment, or a failure within the utilization (load) equipment. Overcurrent due to either short circuit or ground fault is normally an uncontrolled overcurrent situation of high magnitude.

7.3 SHORT-CIRCUIT OVERCURRENT

The magnitude of current flow in an electrical circuit is governed by Ohm's law: $I = V \div R$. When a fixed voltage is applied to the circuit, the magnitude of current is determined by the circuit resistance. Sized correctly, the circuit conductor resistance is of such low value that is considered negligible. The circuit resistance is normally assigned the value of the load resistance, which in turn restricts the current flow. High-magnitude, short-circuit overcurrent flows when the load resistance is either accidentally or intentionally bypassed.

Normally a circuit is assigned an operating voltage and current rating. The current rating establishes the rating of the branch-circuit fuse or the setting of the circuit breaker. The current rating also establishes the wire size of the circuit conductors. In reference to Sec. 240-3, the rated-load ampacity of the circuit conductors has to be greater than or equal to the rating or setting of the circuit overcurrent protection device.

The circuit in Figure 7-1 is rated at 120 volts, 20 amps. If the circuit is fully loaded, then by Ohm's law, the load resistance is equal to 120 volts ÷ 20 amps, or 6 ohms. The wire resistance to supply the connected load may be less than one-tenth of an ohm. This is why the wire resistance is considered negligible.

Now for a short-circuit condition: The 6-ohm load is supplied by a minimum of two circuit conductors—one supply conductor and one return conductor. These insulated circuit conductors are pressed tightly against each other in an overfilled raceway or an undersized junction box. Constant pressure is being applied at a given point between the

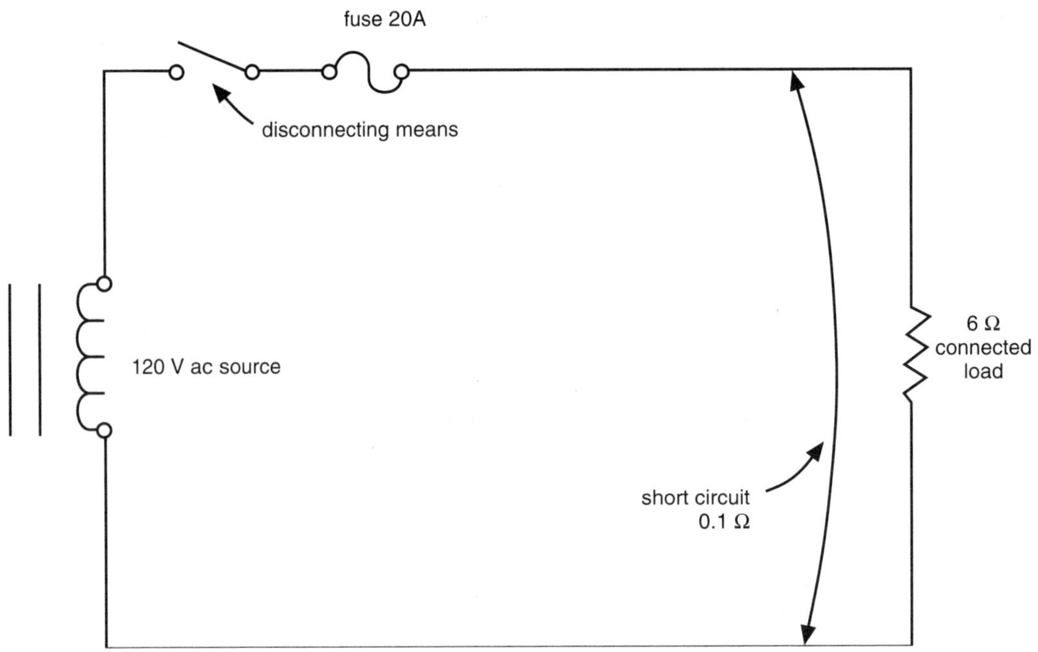

Fig. 7-1 In a short-circuit condition, the load resistance is paralleled with the wire resistance: High magnitude overcurrent is allowed to flow through the circuit.

two conductors and only their respective insulations are preventing metal-to-metal contact.

Due to either a low-value overload-overcurrent condition or a higher ambient temperature that wasn't compensated for (loss of heat dissipation, Chapter 2), both circuit conductors are operating at a higher ambient temperature. The heat is causing the conductor insulation to soften. As the heat continues to build, the pressure between the two circuit conductors causes the insulation to be squeezed out of the contact point and metal-to-metal contact between the two circuit conductors occurs. The branch circuit is forced into a short-circuit condition.

The 6-ohm resistance of the connected load is now forced in parallel with the wire resistance which we "guesstimated" to be about one-tenth of an ohm. The circuit resistance is reduced to less than the one-tenth ohm due to the short-circuit condition. If we again "guesstimate" that the circuit resistance goes down to one-twentieth of an ohm in the short-circuit condition (6-ohm load resistance in parallel with the one-tenth ohm wire resistance), then 120 volts/0.05 ohms = 2,400 amps: A short-circuit condition has forced the circuit into a high-magnitude, almost uncontrollable (without the fuse) overcurrent condition.

The fuse or circuit breaker has to respond to this fault condition before the possible 2,400 amps has time to destroy the circuit conductors or the supply transformer.

The low-value wire resistance is a must if the connected load is to receive its operating voltage (supply and load voltage ratings are the same in most applications) because the wire resistance is in series with the load resistance. In terms of short circuit, the low wire resistance is the culprit that causes high-magnitude overcurrent to flow. Both are controlled by the application of Ohm's law.

At the transformer supply (again Fig. 7-1), the two sides of the circuit, supply and return, are connected to the two respective ends of a transformer winding. Shorting across the two circuit conductors shorts the winding. Again the magnitude of the short-circuit current is controlled only by the wire resistance. The closer in wire length the short-circuit occurs in relation to the transformer, the higher the magnitude of short-circuit current will be. This is why a short-circuit condition on the service-drop conductors will appear like a celebration for the fourth of July (fireworks), as compared to a spark or flash on a long-run branch circuit.

When the power supply is a multiphase system (or polyphase, to use NE Code terms), more than one winding may be involved in the short circuit. More than one winding indicates more current available to the super-low-resistance path. In other words a phase-to-phase short-circuit, involving more than one transformer winding will be of higher magnitude than a single-phase, phase-to-return short-circuit. Either will generate high-magnitude fault overcurrents.

7.4 GROUND-FAULT OVERCURRENT

Ground-fault overcurrent is a special type of short-circuit overcurrent that involves one side of the supply circuit and the noncurrent-carrying metal parts of the power distribution system, which include metal load center enclosures, metal outlet, device, junction, or pull boxes, and metal raceways.

When a building or other structure is constructed of conductive materials such as concrete floors, block walls, and steel beams and joists, these conductive surfaces are bonded together to ensure that a stray voltage caused by friction of force air ventilation or different pressurized conveying systems within the building cannot exist.

The bonding does not stop the generation of static electricity (caused by the friction of two dissimilar objects or materials—exchange or imbalance of electrons), it only provides a low-impedance discharge (return) path.

Because of wind blowing against the outside of the building, and because the building (or other structure) normally projects out from the earth's surface, the bonding grid of the building is normally connected to a grounding-electrode system that will hold the different building surfaces at equal potential (no voltage difference) with the earth immediately adjacent to the building.

Working together, the bonding grid and grounding-electrode system ensure that static buildup along the building surface or a direct discharge from lightning cannot cause a difference of potential between two adjacent building surfaces, or between any building surface and the ground about the outside building surfaces. This is a safety factor to the people working or living within the building: Without a difference of potential (voltage), current cannot flow. The shock hazard is minimized.

Utility power (or commercial power) is normally generated at a central power station on a river or lake, and transmitted at high voltage across the land to the city or residential area for power distribution. High voltage warrants high clearances above the earth for public safety. The high clearance is normally provided by large steel towers, which are a natural attraction (like the high building) to lightning. To aid in lightning dissipation into the earth, and to provide a low-impedance return path between the different voltage systems (circuits) of the transmission and distribution power transformers, the multiphase power system is normally referenced to earth ground.

In other words, if the transmission lines are wired in a three-phase, four-wire wye configuration (more common), the fourth wire is the common connection to one end of all three of the transformer windings in both the primary and secondary circuits. The transmission line is referenced to earth ground by grounding these fourth-wire circuit neutral conductors at each steel tower and at each transformer location.

When the transmission voltage is stepped down to a working voltage and brought into the building or other structure as the service (Fig. 7-2), both the primary and secondary common (neutral) conductors are grounded to a grounding-electrode system at the transformer location to provide the final lightning discharge path along the transmission lines.

Since the common neutral conductor of both the primary high-voltage circuit of the transformer and the secondary lower working voltage circuit are connected to a single grounding-electrode system, the lower working voltage system is also protected from a high-voltage crossover should one of the ungrounded high-voltage conductors come in contact with one of the ungrounded low-voltage conductors. The low-voltage, high-current winding of the secondary circuit provides a low-impedance discharge path (short circuit) for the higher-voltage system. The primary fuse should open under the high-magnitude fault overcurrent condition.

Because the service supply is referenced to earth ground at the transformer location, and the building surfaces are referenced to earth ground by the structure grounding-

Fig. 7-2 The primary and secondary neutral conductors of the utility power entering a building must be grounded for protection from lightning strikes.

electrode system, a potential exists between the ungrounded supply conductors and the building surfaces. To provide a "fault-return path," should one of the ungrounded circuit conductor come in contact with the grounded surfaces of the building, the service neutral or other grounded service conductor is purposely regrounded at the building by connection to the structure grounding-electrode system.

Although mounting the noncurrent-carrying metal parts of the power distribution within the building (metal boxes, enclosures, or raceways) to the building surfaces appears to ground them, these metal enclosures require a positive grounding method. Since the ungrounded circuit conductors, some of which convey high values of load amperage, are more likely to come in contact with these metal surfaces, the noncurrent-carrying metal parts of the power distribution are also interconnected with the service neutral or other grounded circuit conductor at the service equipment to provide a low-impedance fault return path to the transformer. This set-up is referred to as the equipment grounding system. (Grounding and bonding will be discussed in Chapter 9.)

When a two-wire branch circuit operates at the lower phase-to-neutral voltage in a grounded distribution system, the two sides of the circuit, which we labeled earlier as supply and return, are actually the ungrounded side and the grounded side of the circuit.

The grounded circuit conductor, which is not a neutral on a two-wire circuit, has a common connection to the metal raceway, boxes, enclosures, or the grounded surfaces within the building: They are all held at "earth ground" (Fig. 7-3).

When the ungrounded circuit conductor accidentally comes in contact with the

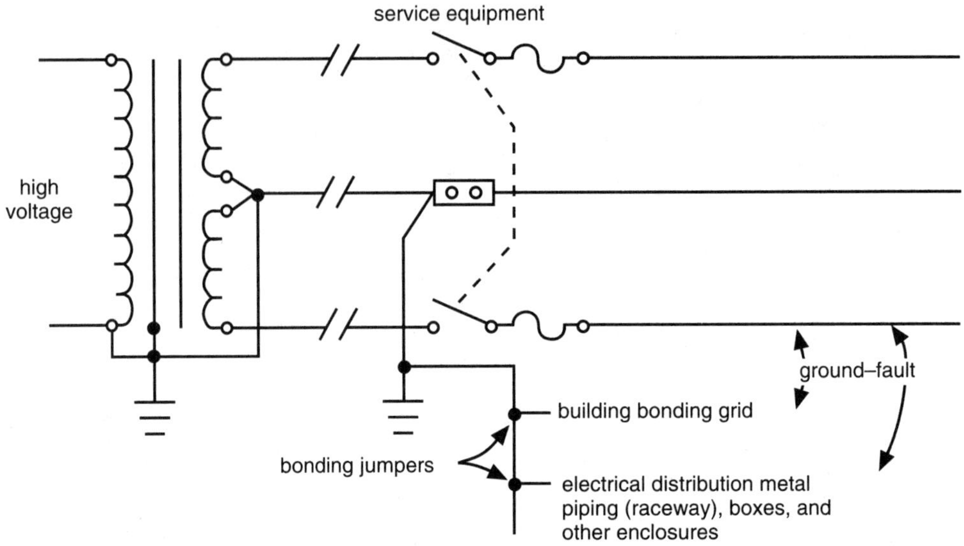

Fig. 7-3 Grounding the building grid and the noncurrent-carrying metal parts of the electrical distribution to the service neutral conductor provides a ground fault current return path back to the transformer supply.

grounded circuit conductor, this is a short-circuit conditon as described earlier: High-magnitude fault overcurrent is caused to flow, due to the low wire resitance.

When the ungrounded circuit conductor accidentally comes in contact with the grounded surface of the metal raceway, this too is a short-circuit condition, but it is described as a ground-fault condition because the normal noncurrent-carrying metal parts are involved in the return path. If the metal piping was joined or otherwise made up properly along the length of the circuit, high-magnitude ground-fault overcurrent is caused to flow, due to the low wire resistance and the low impedance of the fault-return path.

As shown in Figure 7-4, the equipment-grounding system (interbonding of all the noncurrent-carrying metal parts of the power distribution and the grounded service conductor) serves as the low-impedance fault-return path back to the service equipment. The service neutral or other grounded circuit conductor is the fault-return path back to the power supply.

In a polyphase system, a ground-fault overcurrent condition is normally of lower magnitude when compared to a true short-circuit overcurrent condition for two reasons: Normally the ground-fault condition involves only one transformer winding, whereas the short circuit may involve two or more windings—more power available.

Unless the equipment-grounding means is supplemented (paralleled) with a separate conductor, which offers low wire resistance comparable to the circuit conductors, the interconnection of metal pipe and boxes along the length of the circuit throughout the building normally has a higher value of resistance or impedance as compared to the ungrounded circuit conductor. More resistance in comparison to a wire-to-wire short-circuit means less ground-fault overcurrent. Both are high-magnitude fault overcurrents when compared to low-level overload overcurrents, or even lower level full-load currents of the circuit.

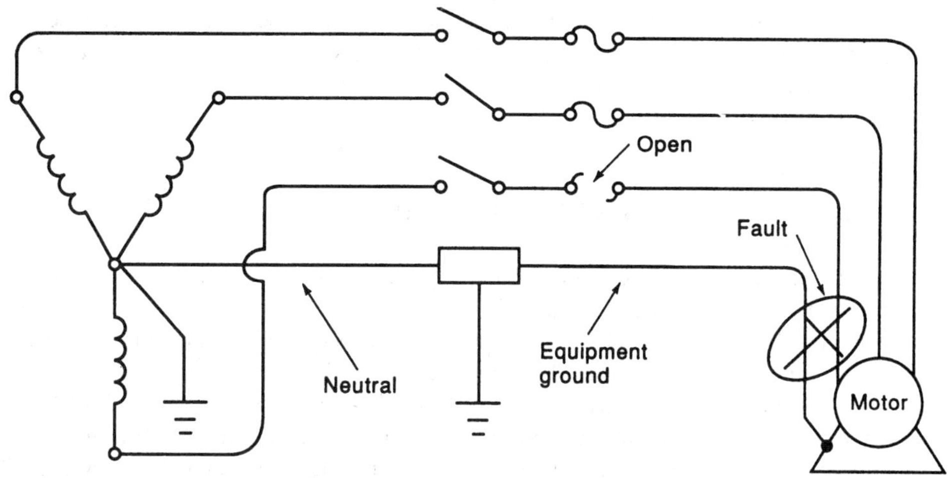

Fig. 7-4 In a grounded ac distribution system, a fault-to-ground by any phase conductor will cause high-magnitude fault current to flow.

Overcurrent due to ground fault will not have high-magnitude current flow unless the distribution system is referenced to earth ground. A 3-phase, 3-wire "floating delta" distribution, as shown in Figure 7-5, is not referenced to earth ground.

Although the building must have an established grounding-electrode system to discharge static electricity, and all the noncurrent-carrying metal parts of the power distribution system have to be connected to it, there is no interconnection between the transformer windings in a secondary floating delta configuration (3-wire) and earth ground. The ground-fault of only one phase conductor will not cause current to flow (no return

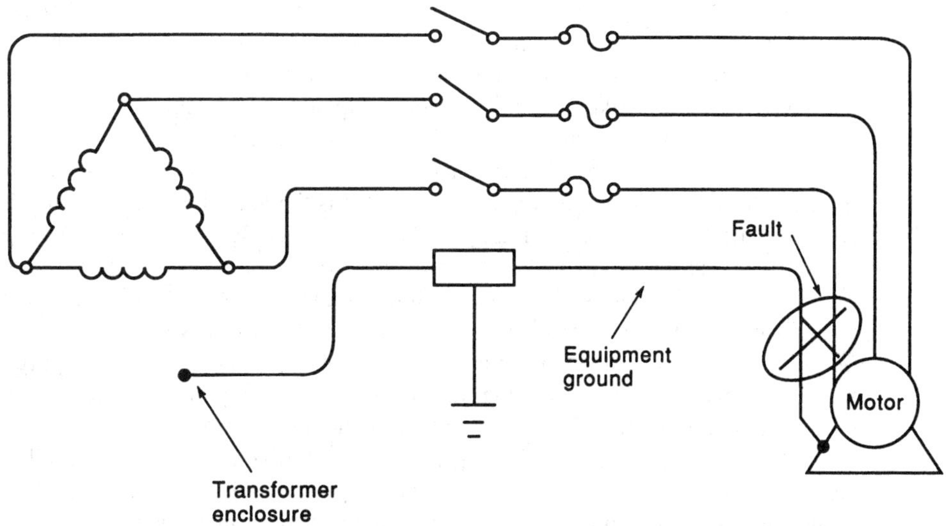

Fig. 7-5 In an ungrounded ac distribution system, a fault-to-ground by any one phase conductor will not cause high-magnitude fault current to flow.

path to supply). Short-circuit current will flow if a second phase conductor becomes grounded, because a low-impedance path is now established parallel to the connected phase-to-phase loads across the interconnected noncurrent-carrying metal parts of the distribution system.

7.5 OVERCURRENT PROTECTION DEVICES

The primary purpose of an overcurrent protection device is to protect the circuit conductors from overheating. Section 240-3 requires conductors to be protected against overcurrent at or below their table-rated ampacities as specified in Tables 310-16 through 310-19 (for working voltages of 2,000 volts or less).

The ampacity tables are based on an ambient temperature of 86°F (30°C). The correction factors listed at the bottom of these tables allows higher environmental temperatures for current-carrying conductors by restricting their rated-load ampacity. The conductor's rated-load ampacity is also restricted by the temperature limitations of Sec. 310-10. Conductor ampacity derates due to higher temperatures or other loss of effective heat dissipation were discussed in Chapter 2 of this text as "prevailing conditions."

Conductor protection against overcurrent is normally in the form of fuses or circuit breakers. Thermal devices, such as thermal cutouts or thermal relays (Sec. 240-9), can be used to protect conductors against overload, but normally these overcurrent protection devices are used in conjunction with (in series with) either fuses or circuit breakers because of their inability to clear or open the circuit against high-magnitude short circuit or ground-fault overcurrents.

The standard current or ampere ratings or settings of overcurrent protection devices are listed in Sec. 240-6(a) of the National Electrical Code. Notice in this section that both fuses and circuit breakers are recognized as standard overcurrent protection devices. This indicates that both fuses and/or circuit breakers offer full overcurrent protection against the effects of either overload, short-circuit, or ground-fault overcurrents. The two principal types of fuses are instant nontime delay and dual-element time delay. The two principal types of circuit breakers are instantaneous trip and inverse time.

Because of the varying levels of high-magnitude short-circuit or ground-fault overcurrents between the point of supply and connected load, fuses or circuit breakers are required to have two current or ampere ratings. One is the load-current rating. This is an indication of the full-load current rating of the fuse or setting of the circuit breaker. When overload overcurrent is drawn, the load rating or setting is the threshold of the response level.

The second rating, which indicates how well the fuse or circuit breaker will respond to high-magnitude short-circuit or ground-fault overcurrents is the interrupt rating. The interrupt rating is a high value setting of at least 5,000 amperes. The interrupt rating is the maximum level of fault current that the overcurrent protection device can respond to by clearing or opening the circuit and maintaining the open.

A fault overcurrent of 5,000 amperes, even at 120 volts, would have a full-bodied arc across the melted link of the fuse or the contact opening in the circuit breaker. The interrupt rating guarantees the extinguishing of such an intensified arc up to the current level specified.

There is no such thing as an instant fuse or circuit breaker. Any study of electricity will reveal that current flow is established in a circuit only after a period of time. The terminology of an instant-opening fuse or an instantaneous-trip circuit breaker indicates that additional devices have not been installed that would purposely hamper the response time of the overcurrent protection device to any current in excess of its load-current rating.

A fuse rated at 20 amps should begin to heat when the load current increases to 21 amps. If the 21-amp overload is sustained, the heat buildup would eventually melt the fuse link, opening the fuse. An overload of 22 amps would generate more heat; the response time of opening the circuit should be less. The response time is inversely proportional to the magnitude of the overload overcurrent. If the overload overcurrent is 25 amps instead of 22 amps, the response time will be even less. If the overcurrent increases to a 500-amp short-circuit or ground-fault overcurrent, the fuse will appear to open "instantly."

The same analysis can be applied to the instantaneous-trip circuit breaker. The response to the overload overcurrent is magnetic instead of thermal. A solenoid coil, wired in series with the connected load, is set to trip a latch switch, which is wired in series with both the solenoid and the load, when the load current exceeds the trip setting.

If the trip setting is 20 amps, a 21-amp overload may not trip the latch "instantly." A 22-amp overload would generate a stronger magnetic field, causing a quicker response as would a 25-amp overload. At the 500-amp fault level, the concentrated magnetic field of the solenoid would trip the circuit breaker latch "instantly."

Time-delay fuses and inverse-time circuit breakers have internal circuitry that purposely delays the response to low-level overload overcurrents. This delay is necessary to allow an inductive load to attain its operating status.

As mentioned earlier, there is no such thing as "instant" in an electrical circuit. As time is required for current flow to become established in an electrical circuit, additional time is required to reduce the current flow into an inductive circuit. An inductive circuit is technically any coil of wire. The magnetic field established along the length of wire by the flow of current creates a concentrated magnetic field along the inner and outer surfaces of the coil. Any change in the flow of current has to change the strength of the magnetic field.

The reluctance to any change in the magnetic field is referred to as inductive reactance. The reactance causes a phase shift between the supply voltage and current. Electrically the voltage is applied instantly, but the corresponding change in magnitude of the circuit current is delayed by the build up (or collapse when ac current reverses) of the magnetic field, hence a "phase shift."

In a dc circuit, the reluctance is only momentary. The length of time is referred to as the charge rate. The charge rate is five time constants. A time constant is the division of the value of inductance by the circuit resistance.

If only a coil of wire is placed across a dc supply, the momentary reluctance will first appear as an open circuit (no magnetic field) and transcend into a short circuit (concentrated magnetic field). The inductive reactance and the wire resistance within the coil of wire both resist the flow of circuit current during the charge period. Initially, the inductive reactance is almost infinite and the resistance is negligible. But once the charge is complete, the inductive reactance goes away with the saturation of the concentrated magnetic field. The resistance of the coil of wire (normally referred to as the winding resistance) is the only opposition presented to the circuit current.

As discussed earlier, wire resistance is normally held at an extremely low value. A coil of wire on a dc source will normally draw high-magnitude short-circuit overcurrent and open either its internal windings or the protective fuse or circuit breaker almost instantly.

If the magnetic field is caused to do useful work (such as holding relay contacts open or a piping valve open against spring tension by magnetizing a ferrous metal frame and movable armature), the magnetic reluctance to the flow of the magnetic field in the iron is reflected into the dc electrical supply as reactance, and the circuit current is held at a lower working-load (rated-load) current value.

In an ac circuit, the coil or winding of wire always offers some reluctance to the flow of ac current because the alternating current magnitude is sinusoidal, meaning it is always changing magnitude and periodically reversing direction of flow. The magnetic field can never reach a steady state.

The ac coil still has the problem of high-value circuit current if the concentrated magnetic field is not caused to do useful work. The useful work is reflected as impedance back into the ac supply to lower the load current to a value less than or equal to the load-current rating of the circuit breaker or fuse rating, which is also the ampacity rating of the circuit conductors.

In an ac circuit, the coil of wire may be the winding of a solenoid that operates either a relay or piping valve, as in the dc circuit. More commonly, the ac winding will be the primary circuit of a transformer, a lighting ballast, or an induction motor. Electrically, all three function as a transformer.

The basic transformer consists of two coils of wire held in close proximity to each other (side by side). When ac power is applied to one winding, the generated magnetic field, which is also constantly changing in magnitude (strength) and periodically reversing (in polarity), induces an alternating voltage in the other winding. If a load is connected to the other winding (complete electrical path), a secondary current is caused to flow. The impedance in the secondary circuit of the transformer is reflected into the primary circuit to limit the primary-circuit current flow.

Like the charging of the dc coil, until the secondary circuit impedance is established and reflected into the primary circuit, high-level in-rush currents flow into the primary circuit. Once the secondary circuits are established, the motor attains base-line speed, the lights are on, or the secondary load of a transformer is established, the high-value in-rush currents quickly diminish to the rated-load or full-load current condition of these inductive loads.

If these momentary high-value overload overcurrents are not allowed to pass, the solenoid could not move the armature to do useful work, the motor would never attain running speed, the lights would never get past a couple of flashes or flickers, and the transfomer could never establish supply power for the secondary panel, load center, or individual load before the supply fuse or circuit breaker opened.

The time delay for these inductive loads is only in the overload range of the fuse or circuit breaker. Inverse time, as used to described the circuit breaker, means the time delay is inversely proportional to the magnitude of the in-rush overcurrent. If the in-rush overcurrent is of low value (in the overload range of the device as compared to its interrupt rating or fault range), a long time delay occurs before the fuse or breaker opens the circuit. If the value of in-rush overcurrent is high, the time delay is short. If the magnitude of the in-rush overcurrent reaches short-circuit or ground-fault levels, the circuit breaker or fuse, according to the interrupt rating, will respond almost instantly.

Overcurrent protection devices must also have the correct voltage rating. Circuit breakers have voltage ratings of 600, 480/277, and 240-208/120 volts. Fuse voltage ratings are broken down into three basic ranges: high voltage, over 34,500 volts; medium voltage, between 601 and 34,500 volts; and low voltage, 600 volts or less. The low-voltage ratings are further subdivided into voltage ratings of 600, 300, 250, and 125 volts.

The molded case of circuit breakers will have different rejection slots on their bases (where they mount to the panel busbars), as well as varying overall widths and lengths, to prevent the interchange of different voltage ratings. Manufacturers of panels and circuit breakers will also use these rejection features to prevent breakers from other manfacturers being used in their panels, and vice versa. Fuses use physical length and barrel diameter to discriminate between the different voltage and current ratings.

Both fuses and circuit breakers are manufactured in current ranges assigned to a voltage rating. For example, fuses rated at 250 volts are the same physical size up to 30 amps. Fuses rated from 31 to 60 amps will be larger for the same 250-volt rating. Circuit breakers rated at 250 volts are the same physical size up to 100 amps.

Overcurrent protection devices must also be marked with their maximum current-interrupt rating. Section 110-9 requires overcurrent protection devices intended to clear fault-level currents to have an interrupt rating sufficient for the system voltage and the current available at the distribution equipment (panels, loadcenters, switchgear, etc.). The interrupt rating (sometimes labeled as I_{rms} symmetrical) is the highest level of current, at rated voltage, that an overcurrent protection device can clear or break the conductive path and maintain the open circuit.

Circuit breakers (Fig. 7-6) have interrupt ratings of 5,000, 10,000, 22,000, 40,000,

Fig. 7-6 The interrupt rating of a circuit breaker, which has to be a minimum of 5,000 amperes, must be marked or otherwise labeled on the circuit breaker assembly (body).

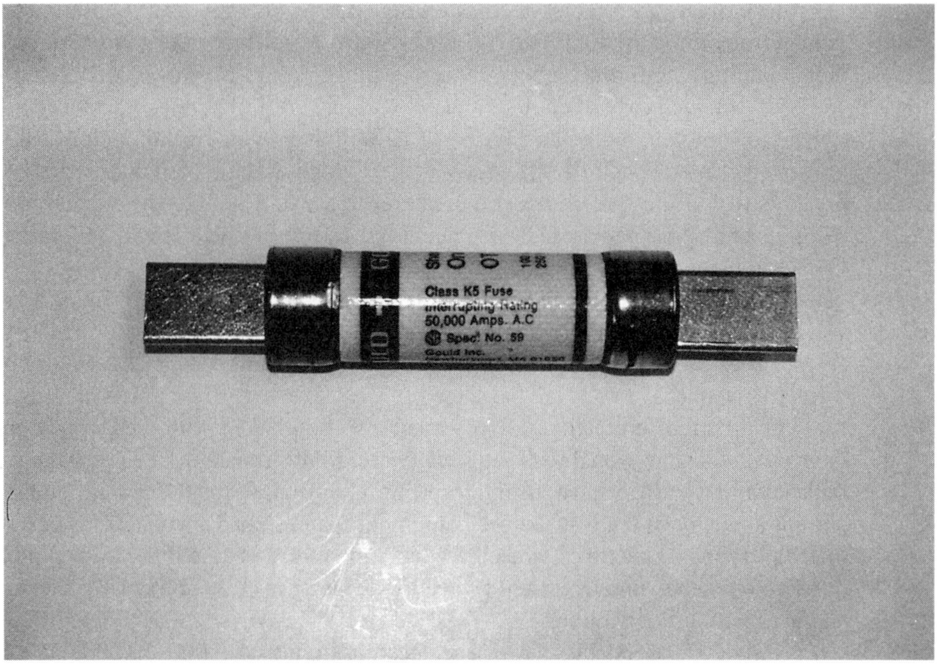

Fig. 7-7 The interrupt rating of a fuse, which is a minimum of 10,000 amperes, must also be marked or otherwise labeled on the fuse assembly (body).

and 100,000 amps. Breakers of up to 40,000-amp interrupt ratings are usually intended for branch circuits or small feeders for subpanels. Circuit breakers intended for use as mains for panels (panelboards) or other service enclosures will normally have 100,000-amp interrupt ratings.

Fuses (Fig. 7-7) have interrupt ratings of 10,000, 50,000, 100,000, and 200,000 amps. The 10,000-amp interrupt rating is normally restricted to the 125-volt plug-type (Edison-base or S-type of screw shell) fuse. The higher interrupt ratings (50,000 to 200,000) are available for different current and voltage ratings from many manufacturers.

7.6 SELECTIVE COORDINATION OF OVERCURRENT PROTECTION DEVICES

The basic purpose or function of an overcurrent protection device is to clear (open) the electrical circuit before an overcurrent condition can cause extensive damage to the electrical equipment within the circuit. In other words, it is normally up to the installing electrician to ensure that fuses or circuit breakers selected for the installation are capable of interrupting the available current, at the rated voltage, that may flow under any overcurrent condition (overload, ground fault, or short circuit) in the circuit. The interrupt should occur with electrical and mechanical safety to personnel, and without damage to switchgear, conductors, load centers, or the overcurrent protection devices themselves.

Electrical safety means prevention of shock hazard by proper installation procedures and techniques. Mechanical safety stems from the fact that overcurrent protection devices, especially breakers with inadequate interrupt ratings, have literally exploded when subjected to high-magnitude short-circuit and/or ground-fault overcurrents.

Sections 110-10 and 240-12 require the coordination of overcurrent protection devices in an electrical distribution system to be selective. In terms of building power distribution, power is initially applied to the service equipment. Feeder circuits convey the power to secondary load centers, and branch circuits convey power to the connected load. In this analysis, a minimum of three fuses or circuit breakers, or a combination of the two are in series with each ungrounded circuit conductor between the transformer winding supply and the connected load.

As outlined in the discussion of short-circuit overcurrent, the greatest concentration of available fault overcurrent is at the secondary terminals of the supply transformer, where the least amount of circuit conductor length and corresponding wire resistance will occur. As the wire resistance increases moving away from the transformer terminals (increasing the circuit conductor length), the available fault overcurrent decreases.

Since overcurrent protection devices are available with different interrupt ratings for the same load-ampere rating, and the available fault current diminishes as the length of conductor increases, the lowest interrupt rating should be at the branch-circuit level. The feeder protection should have a higher interrupt rating than the branch-circuit protection. And the service overload overcurrent protection device should have the highest interrupt rating of all three.

As defined in the fine print note to Sec. 240-12, the opening of the overcurrent protection device should be localized to only the area or section where the fault condition (overcurrent condition) exists (Fig. 7-8). The restrictive outages (loss of power), in turn, will affect only the equipment supplied by that section (circuit or circuits if the fault is at the feeder level). This is referred to as a selectively coordinated system, in which only the overcurrent protection device immediately upstream from the fault opens.

It is not an NE Code requirement for the interrupt ratings of fuses or circuit breakers to be increasing from the connected load, back up to the central service-equipment rating, as do the load-ampere ratings. Interrupt ratings are a guaranteed clearing value. The fuse or circuit breaker is going to respond to any current in excess of its load-ampere

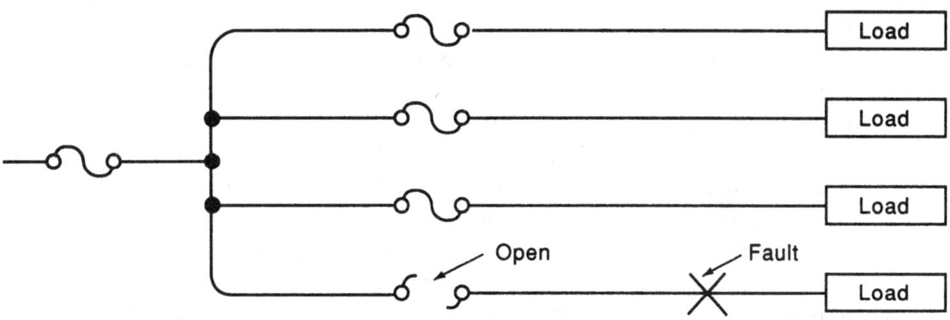

Fig. 7-8 If a distribution has been selective coordinated, only the overcurrent protection immediately upstream of a fault should open.

rating to open the circuit. The interrupt rating is the maximum value of fault current it can clear, not the minimum.

Selective coordination of the series overcurrent protection devices ensures that the interrupt rating is adequate for the fault overcurrent available at the point of insertion in the circuit. If 5,000 amps of fault overcurrent is available at the branch-circuit level, then the fuse protecting a 20-amp branch circuit must have a 20-amp load rating and a minimum of a 5,000-amp interrupt rating. If 12,000 amps of fault overcurrent is available, the 5,000-amp interrupt rating would cause the selective system to go nonselective, because the fuse, even though it melts open, may not clear the circuit when the fault occurs.

In this situation, the feeder overcurrent protection device, which may have an interrupt rating of 50,000 amps, opens the supply to the panel or other load center to clear the branch-circuit fault condition (Fig. 7-9). The whole panel suffers the power outage due to a fault or overload condition on a single branch circuit.

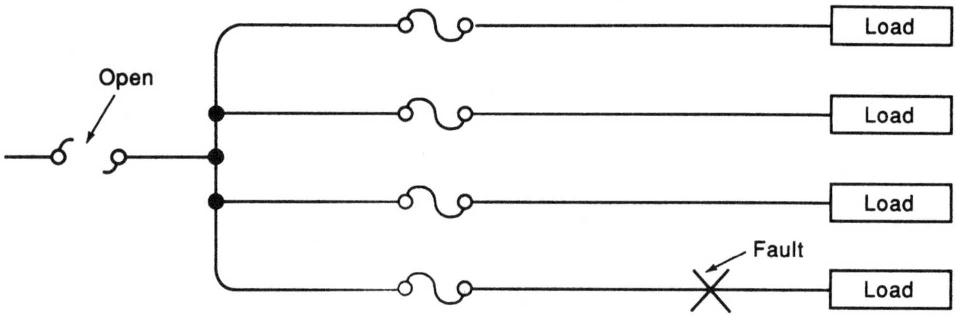

Fig. 7-9 If the distribution is nonselective, the power outage is not restricted or localized as required by Sec. 110-10.

Nonselective coordination really takes its toll if the feeder overcurrent protection device ahead of the branch-circuit fault that was nonselective also has an inadequate (too low a value) interrupt rating and the service equipment overcurrent protection device responds to the fault condition.

Information on how to determine the amount of fault overcurrent available at a certain point along the power distribution system within a building or other structure is not within the scope of the National Electrical Code, nor is information on how to determine the interrupt ratings of the fuses or circuit breakers.

It is normally enough for the electrician to know that the greatest concentration of fault overcurrent available is at the service equipment, and the least amount of fault overcurrent available is at the load end of the longest branch circuit in the system. Installing fuses with the maximum interrupt rating of 200,000 amps for all three distribution points with descending load-ampere ratings according to respective full-load currents is a sure cure for the lack of knowledge on interrupt ratings.

Normally the major concern for interrupt ratings occurs on the initial design of the building (or other structure) power distribution. Selective coordination and adequate interrupt ratings are of utmost importance at this stage of the power distribution. The concern for the practicing electrician comes after the fact—after the installation is in operation and one of the circuit overcurrent protection devices opens.

A selectively coordinated system can be made nonselective when a failed circuit breaker or blown fuse is replaced with a breaker or fuse that has a lower interrupt rating. It is imperative that a circuit breaker or fuse with not only the same interrupt rating, but the same voltage rating and same load-current rating be used for the replacement.

Identification numbers by the manufacturer are also of utmost importance. Circuit breakers and fuses are often designed for a particular task. The removed circuit breaker and the intended replacement circuit breaker may have the same dimensions as for their molded case, but their ID numbers may not be the same.

For example, the circuit breaker removed from a solid-state motor controller may be of the instantaneous type. If an inverse-time circuit breaker with the same current-, voltage-, and interrupt-ratings were to be installed, the critical short-circuit and ground-fault protection of the solid-state controller would not be correct. The two different circuit breakers are designed for two different types of response—the selective system has now become nonselective.

Fuses are more susceptible to nonselectivity when compared to circuit breakers. Manufacturer identification numbers must be followed to the letter. For example, a current-limiting fuse with the suffix R is designated as a motor fuse (Gould Shawmut Inc.). With any system involving R-rated fuses, a contactor and overload relay must be employed to open on low-value overload overcurrents.

Time delay is purposely introduced in the manufacture of R-rated fuses to allow for the harmless momentary overloads (high in-rush overcurrents discussed earlier) drawn on motor starting. If R-rated fuses are replaced with instant (nontime) primary fuses (Gould Shawmut identifies their primary fuses as type E), the overload relay protecting the motor would not be allowed to do its intended job. For starting under an overload condition, the primary fuse or fuses would probably open before the running overload protection (starter heaters) could clear the circuit. In other words, the branch-circuit protection supplying the motor has become nonselective.

The situation with motors and nonselective fuses becomes a critical situation at the feeder level. Instant-type fuses used to replace R-rated fuses protecting the motor feeder will often open to protect the feeder circuit when one or more motors is starting under normal loads conditions. The response is nonselective, because the normal starting of the motors may have caused extensive downtime.

A selectively coordinated system is most critical in health care facilities. Article 517 continually stresses this point, since the loss of power in critical-care areas of hospitals or nursing homes could mean the loss of life. Sections 230-95, Exception No. 1, and 240-12 both stress the importance of selectivity in industrial installations, where a nonorderly shutdown could introduce additional safety hazards.

Selectivity is based on both the peak current let-through charts of the overcurrent protection device (the fuse or the circuit breaker), and the short-circuit overcurrent available from the supply line (power distribution) at the point where the fuse or circuit breaker is installed. The peak let-through charts are published by the fuse or circuit breaker manufacturer on their type of overcurrent protection device. The charts are based on the time in seconds (or cycles, which would be a fractional multiplier of seconds on a 60-Hz system) as compared to the magnitude of the available short-circuit overcurrent. In other words, the overcurrent protection device has a definite timed response to the magnitude of the overcurrent condition.

The available short-circuit overcurrent in an electrical distribution system can only be calculated at the secondary terminals of the supply transformer, or at the load terminals of a generator for a separately derived source. As the distance or length of the conductors increases from the tie point of the secondary or load terminals, the available short-circuit overcurrent decreases. The decrease is attributed to the choking effect of the conductors.

From the design standpoint of an electrical distribution system, the available short-circuit overcurrent downstream is controlled not only by the distance or length of the circuit conductors, but also by the wire or conductor size (cross-sectional area), the type of conductor material (whether copper, aluminum, or copper-clad aluminum), and by the type of material used to house the conductors (nonmetallic versus metallic raceways or other enclosures, and when metallic, whether the enclosure is ferrous metal that can be magnetized, or nonferrous metal, such as aluminum, that cannot be magnetized).

The kVA rating of a generator and its rated impedance will determine its available short-circuit overcurrent. The kVA rating of a transformer, its rated impedance, and whether it is fed from the infinite bus of the utility company, or subfed from the service switchgear will determine its available short-circuit overcurrent. The infinite bus of a utility company means the power supplied from the utility primary lines are considered to have an unlimited amount of current available.

The adequacy of overcurrent protection device interrupt ratings and the degree of short-circuit current available to the overcurrent protection device are beyond the scope of this text.

7.7 THERMAL OVERCURRENT PROTECTION DEVICES

Electrical circuits or equipment can be protected against overcurrent due to overload by thermal devices, such as thermal cutouts or thermal relays, which are temperature sensitive. These devices are designed to open the supply or control circuit when the temperature of the equipment they are either mounted in or fastened to reaches a predetermined high-degree value.

A thermal relay (Fig. 7-10) uses a bimetallic strip to sense temperature in the same manner as the room thermostat for a furnace or air-conditioner in a building. The two dissimilar metals bonded together in the form of a strip or a disc (button) have different expansion or contraction rates due to a rise or fall in temperature. The bonded strip will bend or flex in one direction or the other as the surrounding temperature changes.

The flexing motion is used to make or break a set of contacts mounted on one end of the strip. The other end of the strip is normally held in a fixed position. The supplied load or its electrical control is placed in series with the switching action of the bimetallic strip. As the temperature of the equipment reaches the operating temperature of the strip, the expansion of one of the metals will open the supply circuit.

A thermal cutout (Fig. 7-11) is normally a solid-state semiconductor device (no moving parts). A resistor is a solid-state semiconductor device that offers resistance to the flow of current in a circuit. It is called a semiconductor because it does allow some current to flow in the circuit. A conductor is a semiconductor with a very low resistance value. An insulator is a semiconductor with a very high resistance value. A thermal resis-

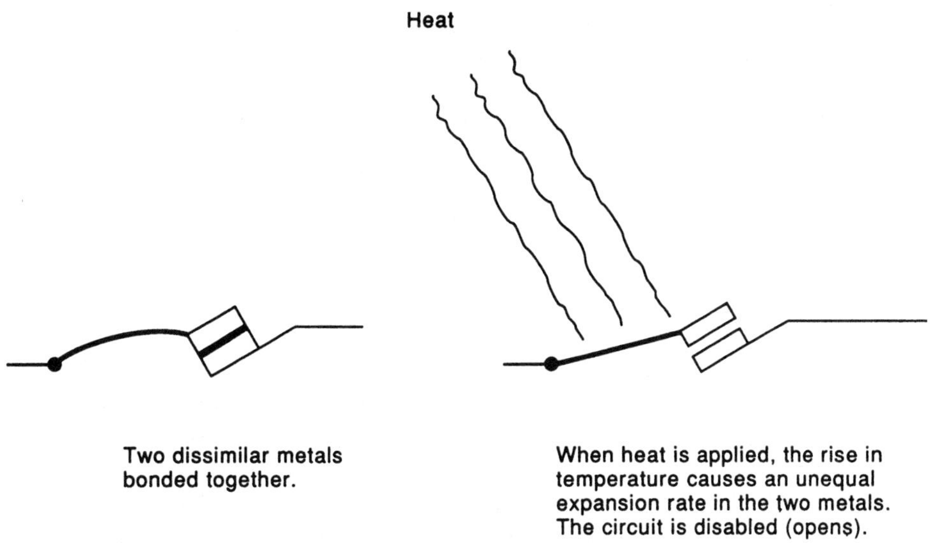

Fig. 7-10 The opening of a thermal relay may be the result of a rising ambient temperature, or electrical equipment forced into an overload condition.

tor is a resistor whose ohmic value of resistance changes with the temperature of the resistor.

If a thermal resistor has a positive temperature coefficient, the resistance of the semiconductor device will increase as its temperature increases. If the thermal resistor has a negative temperature coefficient, the resistance of the semiconductor will decrease as its temperature increases.

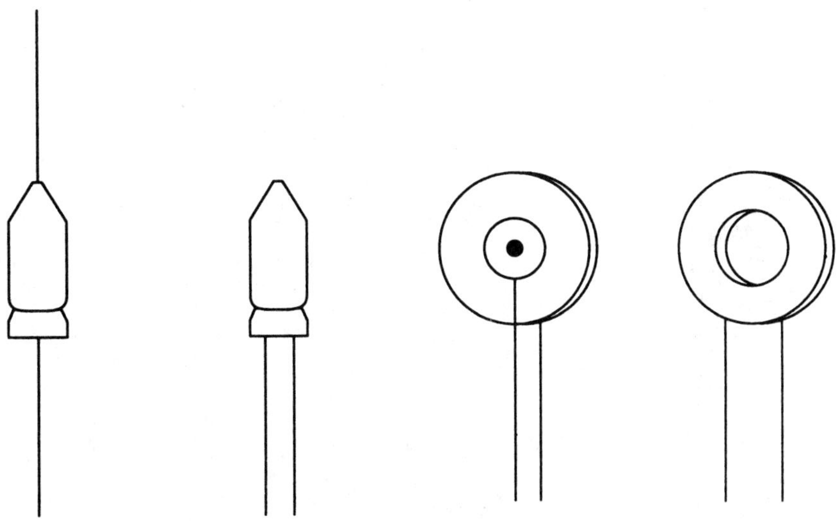

Fig. 7-11 A thermal cutout is normally an embedded solid-state device used to sense an over-temperature situation in a motor, ballast, or transformer.

A thermal cutout will normally have a positive temperature coefficient. When the temperature of the equipment it is designed to protect reaches a predetermined high-degree value, the resistance will be of such high value it will appear as an insulator and open the supply circuit to the equipment.

Equipment such as motors, transformers, or lighting ballasts often use thermal devices to protect the equipment from overheating. Overheating can be caused by the lack of adequate ventilation for the equipment, but normally the excess heat is generated internally by an overload-overcurrent condition of the equipment.

Overload protection for a motor can be provided by thermal cutouts or overload relays (Sec. 430-40). Overload relays (Fig. 7-12) do not sense heat buildup in a motor. Instead the overload relay uses heat generated by excess current drawn by the motor to detect the overload situation.

The overload overcurrent is drawn because the mechanically overloaded motor cannot deliver its rated impedance to the power supply lines. Lower impedance (Ohm's law) means more current drawn for an applied voltage. The value of the motor's impedance increases as the speed of the rotor increases, and decreases as the rotor is forced to slow down (mechanical overload). The magnitude of the overload overcurrent is inversely proportional to the motor's impedance. Rated impedance is attained at the motor's base or line speed.

Failure to start (especially in a single-phase situation) may increase the magnitude of the motor's overload overcurrent to that of short-circuit fault overcurrent. In this situation, the branch-circuit short-circuit, ground-fault overcurrent protection (fuses or circuit

Fig. 7-12 A separate overload relay for a motor, which provides running overload protection for a motor, is normally located on the load side of the starter.

breaker) will normally clear the circuit before the overload overcurrent protection device has time to respond.

Thermal devices are used to protect both equipment and people from the hazards of overheating due to overload. Section 240-9 stipulates that unless a thermal device is specifically designed to open (clear) the high-magnitude short-circuit or ground-fault overcurrents of an electrical circuit, it shall not be used as overcurrent protection for branch-circuit conductors.

7.8 CIRCUIT BREAKERS

Circuit breakers have two main categories: instantaneous trip and inverse time. The instantaneous-trip type, as the name implies, will open the circuit before its interrupt rating is reached. It will begin to react to overload when the overcurrent level reaches and passes its load-current rating. The instantaneous-trip circuit breaker does not have any built-in tolerance for overload.

The inverse-time circuit breaker has a dual mechanism: one to allow for a percentage of overload overcurrent to pass without tripping, and one to open or interrupt the circuit when overcurrent levels reach fault conditions.

The overload mechanism (Fig. 7-13) is a latch switch with contact pads that func-

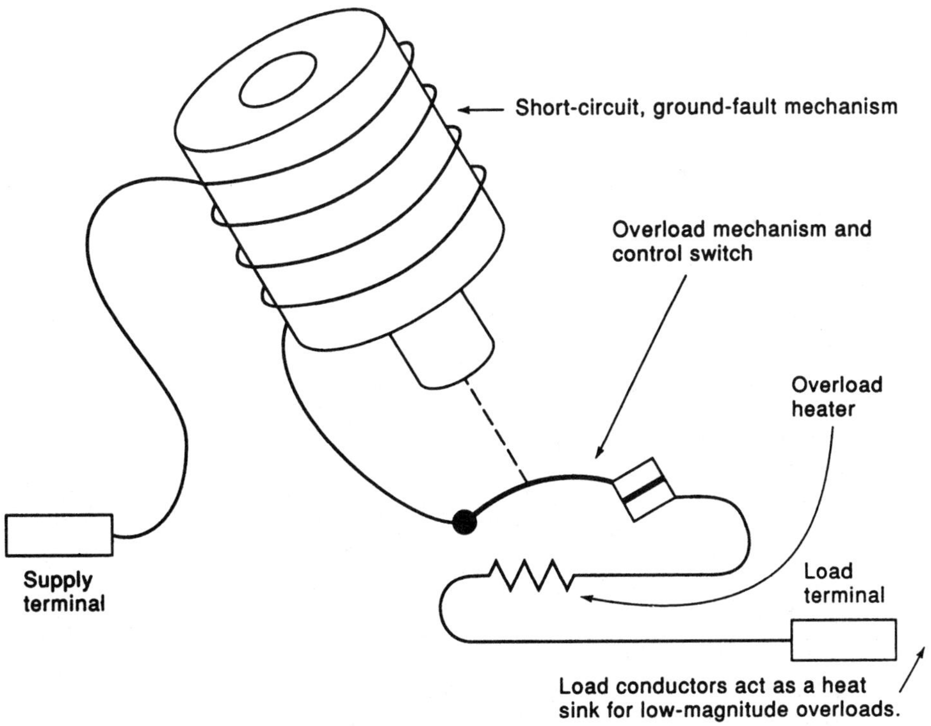

Fig. 7-13 The operating principle of an inverse-time circuit breaker requires two control elements.

tions as a bimetallic thermostat mounted over a heating resistor. This resistor, in series with the connected load, will not heat up at the load-current rating of the circuit breaker, but when an overload occurs, the resistor will begin to heat and continue to get hotter until the bimetallic strip opens the latch switch and clears the circuit.

If the overload is removed and normal current (within the breaker's load rating) resumes, the circuit conductor connected to the breaker will serve as a heat sink and extract or remove the heat from the resistor assembly. The magnitude of the overcurrent due to overload determines the length or span of time delay sustained before the breaker trips. Under heavy or high-magnitude overload overcurrents, the overload element will respond quickly and clear the circuit.

The fault mechanism of the inverse-time circuit breaker is a solenoid which contains a metal plunger. The coil of the solenoid is wired in series between the latch-switch, the heating resistor, and the connected load. When a fault condition occurs, the high-magnitude overcurrent flowing through the coil retracts the plunger in a quick ramming motion that unlatches the bimetalllic switch.

The instantaneous-trip circuit breaker contains only the latch switch and solenoid or fault mechanism, which is fine tuned or calibrated to initiate plunger movement as the circuit current reaches and passes the breaker's load-current rating. It will respond almost instantly to even low-level overload overcurrents. In comparison, the fault mechanism of the inverse-time circuit breaker is set to trip (move the plunger to open the latch switch) at a fault or overload-overcurrent level that is ten times the load-current rating of the breaker.

Fig. 7-14 Any type of fuse must have a means of disconnect on its line or supply side.

For example, an instantaneous-trip circuit breaker with a load-current rating of 15 amps would trip when the load current passes a magnitude of 15 amperes. The fault mechanism of an inverse-time circuit breaker, with the same 15-amp load rating, would allow the fault or overload overcurrent to reach a magnitude of 150 amps before it would clear the circuit. According to the time duration of the 150-amp fault or overload overcurrent, the overload mechanism of the inverse-time circuit breaker would probably clear the circuit before the fault mechanism has time to respond.

Both instantaneous-trip and inverse-time circuit breakers are required by Sec. 240-80 to be trip free. Trip free means that when a circuit breaker is switched on or into an overcurrent condition, it will trip out (open the circuit) automatically and cannot be "held on." Reset of a circuit breaker is accomplished by first switching it off and then back on. A tripped condition is normally indicated by the handle staying in the center position.

Any type of fuse is required by Sec. 240-40 to have a disconnecting means on its supply side (Fig. 7-14). A circuit breaker (Sec. 240-80) is allowed to serve as both the overcurrent protection and the disconnecting means of the circuit (Fig. 7-15). Section 380-11 requires the circuit breaker, used as a switch (disconnecting means), to be either hand operable and equipped with a lever or handle, or to be power operated and capable of being locked open in the event of a power failure. Section 240-81 requires the circuit breaker to be indicating. The position of the handle or power operator must clearly indicate that the circuit breaker is either in the open (OFF) or closed (ON) position.

Because the load-circuit conductors connected to a circuit breaker serve as a

Fig. 7-15 Circuit breakers have a dual role of both an overcurrent protection device and a means of disconnect for the supply power.

heatsink for the overload-sensing element of the breaker, the temperature rating of the circuit breaker load terminals is used to restrict the rated ampacity of the circuit conductors.

The conductors are normally sized according to both the full-load current rating of the connected load and the load ampere rating of the circuit breaker, using ampacity Tables 310-16 through 310-19 (for circuit voltages of 2,000 volts or less).

Circuit conductors up to wire size No. 1 AWG, or circuit conductors protected by circuit breakers rated up to 100 amperes, are restricted by Sec. 110-14(c)(1) to the ampacities listed under the 60°C insulation temperature column according to conductive material and the respective wire sizes.

Circuit conductors larger than wire size No. 1 AWG, or circuit conductors protected by circuit breakers rated in excess of 100 amperes, are allowed by Sec. 110-14(c)(2) to be sized by the 75°C temperature column.

7.9 FUSES

Fuses come in many shapes and sizes. Besides voltage, current, and interrupt ratings, fuses are also classified as either instant (nontime-delay) or time-delay, regular (ordinary or general-purpose) or current-limiting, and renewable or one-time. They are also classified as either plug fuses or cartridge fuses.

7.10 INSTANT FUSES

A one-time, nontime-delay, general-purpose fuse (usually shortened to just instant or regular fuse) consists of a thin blade or sheet of conductive metal or fusible link suspended between the two end caps (Fig. 7-16). The fusible link is contained in an insulat-

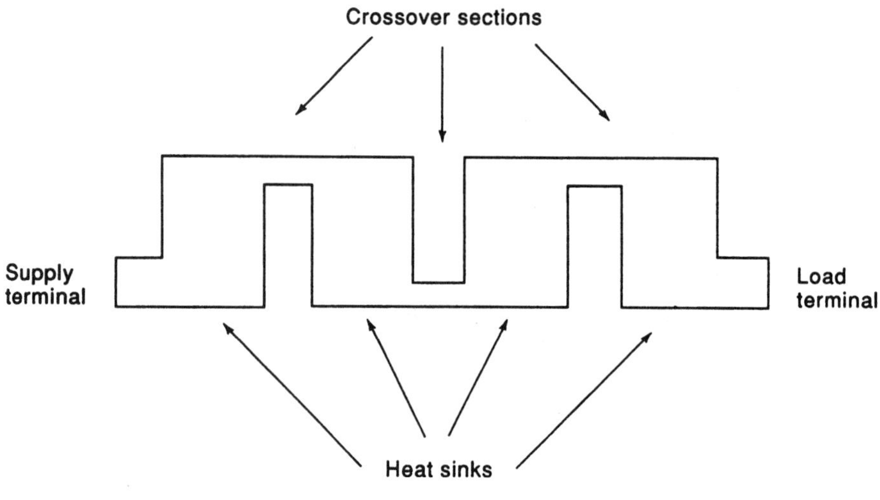

Fig. 7-16 The heat-sink blades of a fuse link (fuse element) remove some of the heat generated by current overload of an ordinary one-time fuse.

ing (nonconductive) barrel that also supports the end caps. The metal strip (fusible link) has one or more narrow crossover sections, at which points the majority of its width has been cut away at spaced intervals along its length.

The narrow strips are less conductive (due to less overall cross-sectional area) than the wider sections of the metallic strip. The current rating of the crossover sections correspond to the load-current rating of the fuse.

The fusible link is also referred to as the fuse element. The element (consisting of the wide heatsinks and narrow crossovers) is a calibrated conductor, for which the configuration, mass, and materials employed in its makeup are varied to achieve the desired electrical and thermal characteristics.

The element provides the only current path through the fuse. It generates heat at a rate that is dependent upon both its resistance and the magnitude of the load current that passes through it.

Heat dissipation is also a critical factor in the design of the fuse. When the heatsink portions (blades) of the fuse element absorb heat from the crossover sections, the heat must be transferred (dissipated) away from the fuse element. A filler, such as quartz crystals, may be added to the barrel of the cartridge fuse. The filler absorbs the heat from the fuse element and transfers it to the barrel and end caps of the cartridge. The heat is then transferred to the ambient air that surrounds the fuse. The quartz crystals also function as an "arc extinguisher" when the fuse opens due to high-magnitude fault overcurrents.

Some fuses may not contain crystal fillers. The heat from overcurrent is also dissipated from the fuse element through the end caps of the fuse, the busbar or clip connection on the supply side of the fuse, and the circuit conductors on the load side of the fuse.

When the fuse sustains an overload-overcurrent condition, the conductive rating of the crossovers is exceeded, and the crossovers begin to heat. If the overload condition is removed, the crossovers will cool down as they continue to conduct. If the overload persists or begins to climb in magnitude, one or more of the crossovers will melt, opening the circuit. The fuse is said to be blown.

When the fuse sustains an overcurrent condition from short circuit or ground fault, the crossovers will melt rapidly before the heatsinks have a chance to counteract the melting. The fuse offers inverse-time characteristics in that the greater the magnitude of current, the lesser the required time for the crossovers to melt.

Should only one crossover melt, the overcurrent could be of such magnitude that it will arc and form its own bridge between the two heatsinks. The current will continue to flow. When this happens, another crossover will quickly melt, and the heatsink, originally suspended between the two crossovers, will fall out of the current path. This action will continue to recur until the gap is widened to the point that the arc is extinguished and the circuit is opened (cleared).

With this type of fuse, several cycles (alternating current is cyclic in nature, based on either 50 or 60 Hertz or "cycles per second") of high-magnitude fault overcurrent can be passed before the circuit is cleared. Filling the fuse with quartz or other types of nonconductive crystals will help surpress the arc at higher voltage ratings. The higher the potential or voltage rating of the circuit, the longer the length of the arc in a fault condition.

7.11 TIME-DELAY FUSES

Motors and transformers draw high in-rush overcurrents, sometimes of the magnitude of fault-level current, when they are initially energized. As you could expect, this high in-rush overcurrent (starting currents) will normally blow instant fuses. To avoid nuisance blowing of instant fuses, the dual-element, time-delay fuse was developed.

A dual-element, time-delay fuse has a thermal-cutout element suspended between two fuse-link elements (Fig. 7-17). The thermal-cutout element, which will allow the harmless in-rush overcurrents to flow for a short period of time before it opens, serves as the circuit's overload-overcurrent protection. The fuse links at both ends serve as the circuit's short-circuit and ground-fault overcurrent protection.

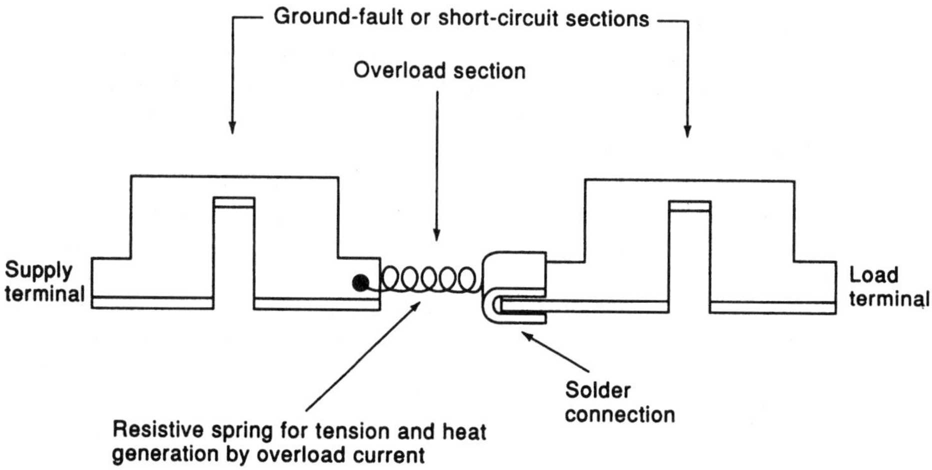

Fig. 7-17 Dual-element, time-delay fuses are required for the starting of a motor.

The thermal-cutout element is basically two small conductive plates that are soldered together. One plate is the end of one of the fuse links. The other plate is spring-connected to the end of the other fuse link. Besides putting some tension on the soldered joint of the two plates, the spring serves as a heating element when overload overcurrents are drawn by the connected load.

If overcurrent is drawn due to overload of the circuit, the spring will heat up, melting the solder. When the solder turns to a liquid, the spring tension will pull the two plates apart and the fuse is blown (circuit opens). The load-current rating of the fuse is based on the melting point of the solder.

If the thermal-cutout element is not capable of interrupting the high-magnitude overcurrents of either short circuits or ground faults, the fusible links at both ends of the thermal element will function as an ordinary fuse and clear the circuit. The interrupt rating of the fuse links in a dual-element, time-delay fuse is set much higher than the overload rating of the thermal element.

Not all time-delay fuses have dual elements. "Dual-element" is actually a manufacturer's term describing a fuse element construction. The two terms have became associated together through NE Code and trade usage. The main requirement of a fuse identified as time-delay is that it offer inverse-time response to overcurrent with extended time delay for momentary overloads.

7.12 RENEWABLE FUSES

Renewable fuses are simply cartridge fuses with removable ferrules or end caps. When the fuse blows, the fuse link (fuse element) can be replaced and the fused is "renewed." Renewable fuses are available either as instant-acting or time-delay for motor loads.

The main drawback to the use of renewable fuses is the lack of discrimination between the different load-current ratings. In other words, the same fuse cartridge will accept fuse links of greater or lesser load-current ratings. This presents a serious problem because the circuit cannot be properly protected. For example, No. 14 AWG copper circuit conductors are protected by a renewable fuse containing a 15-amp fuse link. There's nothing to prevent the inexperienced or neglectful electrician from installing a 30-amp fuse link "to get the circuit going again." The overload overcurrent, if less than 30 amps, will quickly overheat the No. 14 AWG copper circuit conductors and eventually destroy their insulation. A 30-amp, 250-volt renewable fuse cartridge will accept fuse links with ratings from 1 to 30 amps.

7.13 CURRENT-LIMITING FUSES

With the introduction of solid-state motor controllers into the electrical world came a new problem in overcurrent protection. The fuses discussed up to this point will pass serveral cycles of fault or overload overcurrent (with inverse-time characteristics) before they will clear the circuit. Solid-state circuitry has a maximum current rating that cannot be exceeded without total destruction to the individual components it is made of (such as the peak inverse rating of a power diode). This fact brought about the development of current-limiting fuses.

Current-limiting fuses are designed with crossover sections that are fine-tuned or calibrated to melt open at the threshold of their interrupt current rating. Section 240-11 gives a simple, yet vague, definition of current-limiting overcurrent protection devices. For a fuse to be current-limiting, it must begin to melt within 90° (¼ cycle) and completely clear the circuit within 180° (½ cycle, within the same alternation), when subjected to fault overcurrents approaching its interrupt rating at its rated voltage.

As required by Sec. 240-60(b), current-limiting fuseholders will normally have some type of rejection capability in their construction. The rejection-type fuseholder (Fig. 7-18) will not accept a regular (ordinary or general-purpose) fuse that is not current-limiting, thus the current-limiting ability of a replacement fuse is assured.

Section 240-60(b) will allow a regular fuseholder (general-purpose, nonrejecting type) to accept either noncurrent-limiting or current-limiting fuses. In this manner, the benefits of current-limiting fuses can be provided but not assured (in replacement).

For 600-volt blade-type cartridge fuses (Fig. 7-19), the rejection feature of the current-limiting fuse and fuseholder may be a pressure clip that has a dowel or pin run

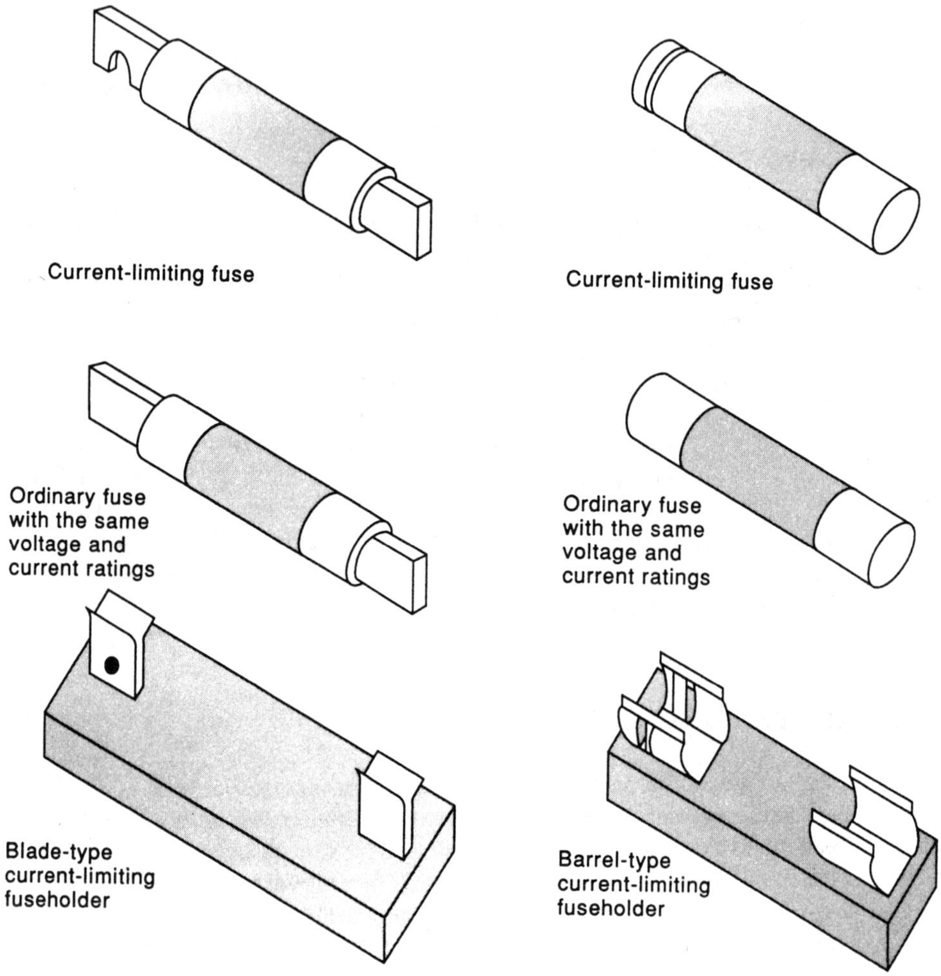

Fig. 7-18 Current-limiting fuses and fuseholders have rejection features that will exclude noncurrent-limiting fuses of the same ratings from being installed.

perpendicularly through the wall of both clips (on one end of the fuseholder). One blade on a current-limiting fuse has a slot cut in it that will straddle the rejecting pin. A regular (oridinary, noncurrent-limiting) fuse has a solid blade on both ends.

Barrel- or ferrule-type cartridge fuses (250-volt fuses) normally employ a groove or recess next to the end on one of the ferrule caps (Fig. 7-20). The rejection feature on the ferrule-type fuseholder is a straight bar that is placed across the curvature of the clips. Unless the fuse has the recess in the end cap that matches the straight bar in the clip, the fuse cannot be installed in the fuseholder. Once again, continuation of the current-limiting ability of a replacement fuse is assured—noncurrent-limiting fuses have straight ferrules on both ends.

Current-limiting fuses with higher voltage and/or current ratings may be of the bolt-in type. The rejection feature of a bolt-in fuse is determined by the pattern of the

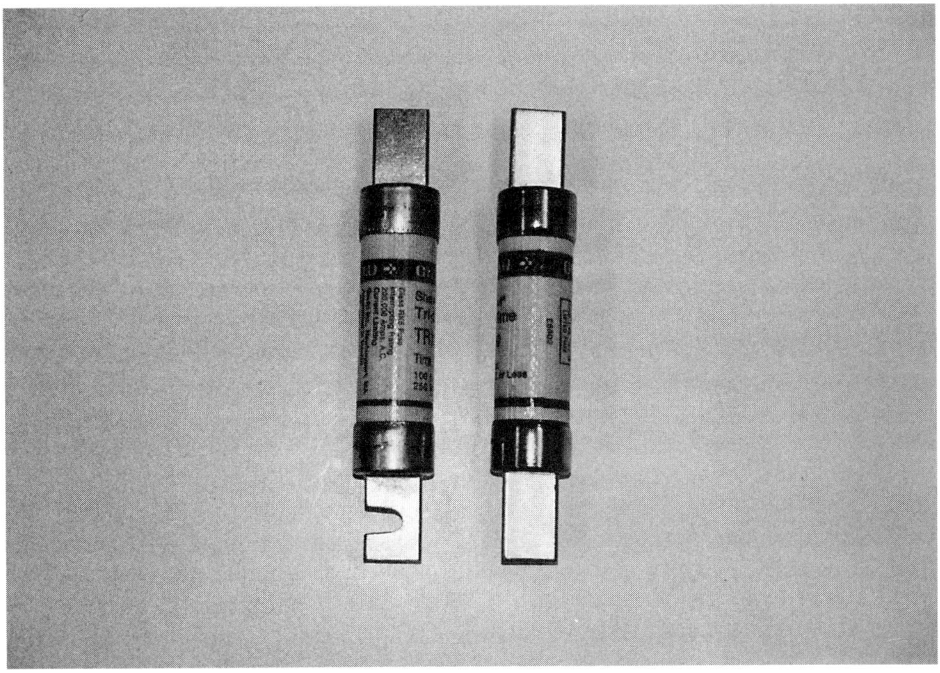

Fig. 7-19 One blade on a 600-volt current-limiting cartridge fuse is purposely slotted to straddle the rejection pin on the fuseholder.

holes in the flat bars (wide, thick blades) of the fuse ends. The same discriminating or rejecting pattern of the holes on the fuse will be found in the mounting bolts on the fuseholder. The ability of current-limiting with a replacement fuse is assured.

Current-limiting fuses are specially designed for their interrupt rating. They are available in both instant and nontime-delay types, and in the dual-element or single-element time-delay types. They are normally not available as renewable fuses.

It should be noted that a current-limiting fuse has inverse-time characteristics between its load-current rating (overload rating) and its interrupt rating. In other words, the currrent-limiting fuse may pass several cycles of low-level overload overcurrent to the equipment it is intended to protect before the overload element clears the circuit.

7.14 PLUG FUSES

The installation and use of plug-type fuses (also referred to as either Edison-base or screwshell fuses and type S fuses) are governed by Part E of Art. 240.

Plug fuses are restricted by Sec. 240-50(a) to a maximum operating voltage of 125 volts between current-carrying (circuit) conductors. The Exception removes the 125-volt maximum for distribution systems that contain a grounded neutral conductor, and the phase voltage is limited to a maximum of 150 volts to ground (a single-phase, 3-wire 120/240-volt system, or a 3-phase, 4-wire, 208Y/120-volt system).

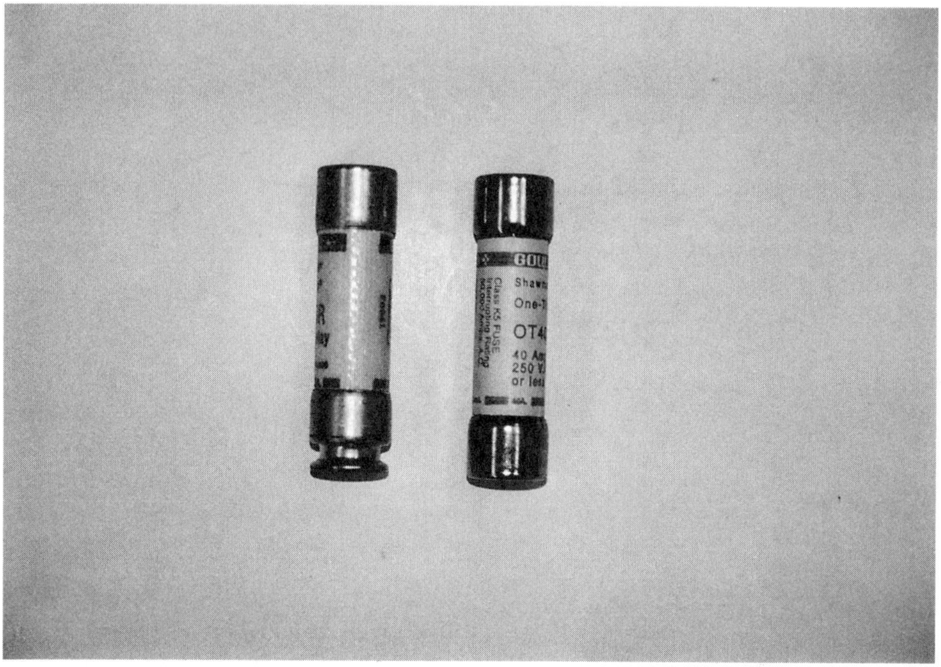

Fig. 7-20 One ferrule on a 250-volt current-limiting fuse is indented to fit within the rejection bars in the clip of the fuseholder.

Edison-base fuses are restricted by Sec. 240-51(a) to a maximum load-current rating of 30 amperes. Edison-base fuseholders cannot be installed in either existing rework or new installations (Sec. 240-52), unless adapted to accept only Type S fuses (Fig. 7-21).

Ordinary plug fuses suffer the same problem as renewable cartridge fuses in that fuses of greater load-current rating can be installed to protect branch circuits of lesser load-current ratings. As a result, the branch-circuit conductors and the connected load they serve are not properly protected against overload overcurrent.

Section 240-51(b) will allow Edison-base fuses to be installed as replacement fuses in existing installations. If there is evidence of tampering or overfusing, the fuseholder must be either replaced and modified, or at least modified to accept only type S fuses.

Type S fuses are discriminating [Sec. 240-53(b)] fuses that are classified [Sec. 240-53(a)] in ranges of 0 to 15, 16 to 20, and 21 to 30 amps. As outlined in Sec. 240-54, type S fuse adapters must be designed and manufactured to fit Edison-base fuseholders. The adapter, a threaded barrel whose outside threads are equivalent to a screwshell fuse, and whose inside threads (curvature of the end lip of the barrel insert) match the threads of the type S fuse classification for which it was designed, must have a locking means such that when installed in an Edison-base fuseholder it cannot be easily removed.

This locking means is normally a pointed pin that protrudes through the outer threads of the screwshell adapter. The locking pin will drag against the inner threads of an Edison-base fuseholder as it is being inserted, similar to the insertion of the needle point of a safety pin when closed into the holder. When an attempt is made to remove the

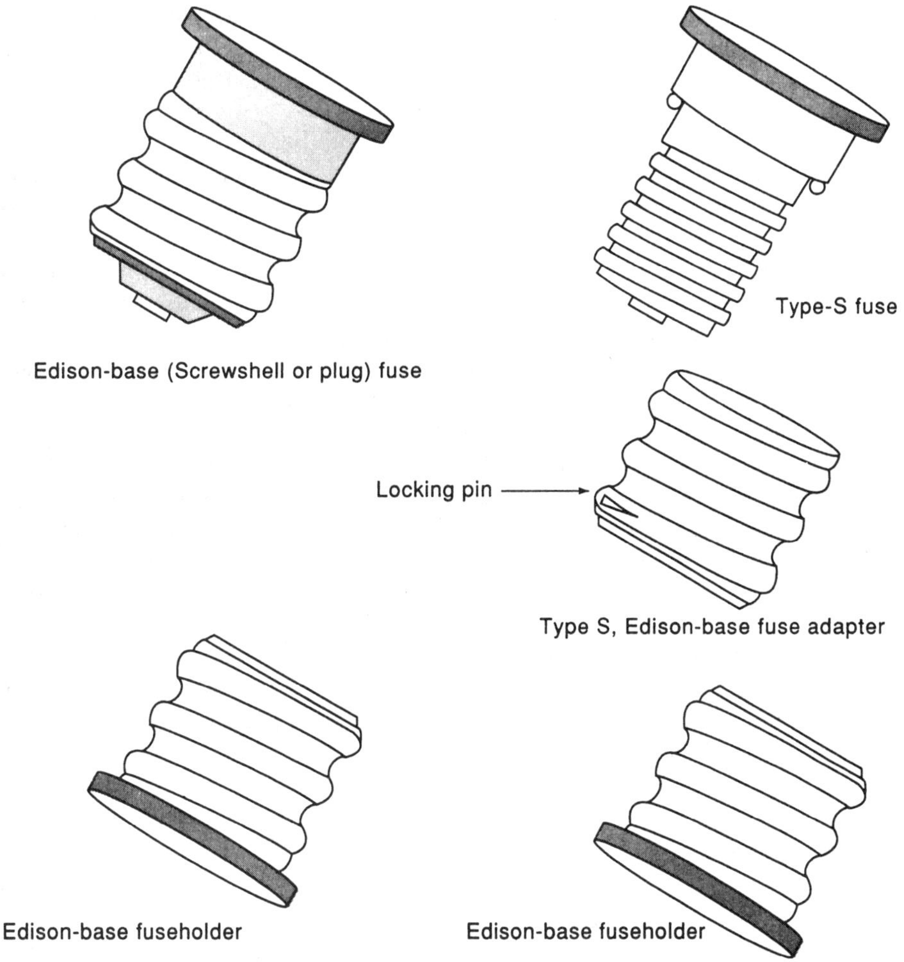

Fig. 7-21 Edison-base (plug) fuseholders cannot be installed on new construction unless they are adapted to the discriminating Type S fuse.

type S fuse adapter, the locking pin will dig in or pierce the screwshell threads of the Edison-base fuseholder. In most instances, attempted removal of the type S fuse adapter results in replacement of the Edison-base fuseholder.

7.15 CIRCUIT LOCATION

Sections 240-20, 240-21, and 240-24 require all ungrounded conductors to be protected by a fuse or circuit breaker at the point where the conductor receives its supply. The overcurrent protection device must be installed in a readily accessible location that is free from physical damage. Overcurrent protection devices cannot be located in the vicinity of flammable or easily ignitible materials.

Overcurrent protection must be either installed or contained in enclosures, or be suitably guarded and protected as allowed by Sec. 240-41(b), such that the sudden movement of parts (the handles on circuit breakers or power-operated disconnects) cannot injure a person in the vicinity when they are activated by overcurrent.

Beginning with subsection (b), the subsections of Sec. 240-21 are known as the general tap rules. The different installations described allow circuit conductors to be connected or tapped to higher-amperage feeder circuits without overcurrent protection devices installed at the point of tap.

Section 240-21(b) is known as the "ten-foot tap rule." Feeder conductors of any smaller ampacity can be tapped without overcurrent protection at the point of tap to a feeder or connection to the secondary of a transformer, if the conductors are not over ten feet long and all the other listed conditions are met. Of main concern, the rated-load ampacity of the tap conductors has to be greater than or equal to 10% (1/10) of the feeder OCPD setting or rating [Sec. 240-21(b)(5)].

Section 240-21(c) is known as the "25-foot tap rule." Feeder conductors can be tapped to larger (higher-ampacity) feeder conductors without overcurrent protection at the point of tap if the conductors are not over 25 feet long and all the other listed conditions are met. As with the 10-foot tap rule, the main concern is the rated-load ampacity of the tap conductors. Section 240-21(c)(2) requires the rated-load ampacity of the tap conductors to be greater than or equal to 33% (⅓) of the feeder OCPD setting or rating.

The more common installations follow the general rule of Sec. 240-21(a), circuit conductors are series wired with or from an overcurrent protection device at the point of power supply. For branch circuits, the overcurrent protection device will be installed at the point of connection to the feeder conductors. Branch circuits normally originate in a panel that has many slots (openings or positions) for individual switched fuseholders or circuit breakers. Feeders supplying panels normally originate from a fused disconnect that is contained in a power panel (subfeeders) or mounted on a busway (Art. 364) distribution system.

The fused disconnect can be the service disconnecting means for a dwelling or other structure. These fuses provide short-circuit and ground-fault overcurrent protection for any circuits on their load side, but only overload overcurrent protection for the service conductors on their supply side. Short-circuit and ground-fault overcurrent protection for the service conductors is provided by the fuses installed ahead of the service drop or service lateral. For a service supplied by a utility, the fuses may be on the primary side of the power transformer.

7.16 PREMISE LOCATION

The overcurrent protection devices and the enclosures that house them are required by Sec. 240-24(a) to be readily accessible. By definition (Art. 100), readily accessible means a person requiring access to the overcurrent protection device does not have to climb over obstacles such as cabinets or utilization equipment (clothes washers, dryers, etc.) to reach the panel and contained overcurrent protection devices. Readily accessible also means the person requiring access does not have to resort to portable means, such as chairs, boxes, or ladders, to reach the panel or contained overcurrent protection devices.

To be readily accessible [Sec. 380-8(a)], the handle of the switch or circuit breaker can be located at a maximum of 6 feet, 7 inches (2 meters) above the ground (finished grade), floor (finished floorline), working platform (mezzanine), or other surface that the panel or other enclosure is mounted above. The area under the switch or circuit breaker enclosure should be clear and unobstructed and extend from the handle to the floor where the person stands [Secs. 110-16 (a) and (b)].

Section 240-24(b) requires each occupant in a structure to have ready access to the overcurrent protection devices that supply circuits in their occupancy. This means a panel in a single or one-family dwelling must be located in a readily accessible place that is not exposed to physical damage [subsection (c)], and is not in the vicinity of readily ignitible materials such as clothes closets [subsection (d)]. In multifamily dwellings, such as apartments or condominiums, each occupancy is to have the panel or other overcurrent protection device enclosure located in the same manner as for a single dwelling [readily accessible in each apartment—note Excp. No. 1 to Sec. 240-21(b)].

Section 240-24(e) is new in the 1996 edition of the NE Code. Overcurrent protection devices such as contained in service equipment or subfed panels cannot be installed in bathrooms of dwelling structures or similar occupancies.

7.17 OCPD RATINGS

Section 240-6(a) lists the accepted or standard ratings or settings of overcurrent protection devices (OCPDs). In conjunction with the Exception, the ampere ratings of fuses and circuit breakers range from 1 ampere to a maximum of 6,000 amperes. As a general rule, Sec. 240-8 will not allow fuses or circuit breakers to be connected in parallel to provide protection against overcurrent on their load side. A single fuse or circuit breaker is required by Sec. 240-20(a) to be installed one for one in series with each ungrounded circuit conductor in an electrical power distribution.

The ampere rating of the overcurrent protection device is determined by the rated ampacity of the load side branch-circuit or feeder conductors and the full-load current rating of their connected loads. Section 240-3 requires conductors to be protected in accordance with their listed ampacities as determined by the ampacity tables in Art. 310 of the NE Code.

The rated ampacity of a conductor is determined mainly by the integrity of its insulation. The insulation must be able to dissipate or at least withstand the heat generated by the current flow in the conductor or adjacent conductors, or the heat of the surrounding air or earth about the conductor [temperature parameters (determinants) are listed in Sec. 310-10].

Often the rated ampacity of a branch circuit or feeder conductor will not correspond to a standard size (setting or rating) of overcurrent protection device as listed in Sec. 240-6(a). The general rule of Sec. 240-3(c) requires the rating of the OCPD to be sized greater than or equal to the full-load current rating of the circuit and the circuit conductors to be sized with a rated-load ampacity greater than or equal to the rating of the OCPD.

For some applications, if the rated-load ampacity of the circuit conductors is greater than or equal to the full-load current rating of the circuit [Sec. 240-3(b)(2)], and neither are equal to a standard rating of a fuse or setting of a circuit breaker, the next higher standard size of OCPD can be used without a further increase in wire size (conductor rated-load ampacity).

Section 240-3(b) list the conditions that must be met. The full-load current must be known and controlled to protect the circuit conductors from overload: Section 240-3(b)(1) forbids the circuit to be a multioutlet branch circuit that supplies convenience or other receptacle loads. The receptacles supply plug- and cord-connected equipment of unknown full-load current rating. This exception to the general rule is restricted by Sec. 240-3(b)(3) to circuits that are rated at 800 amps or less.

In general, conductors must be protected by fuses or circuit breakers at the point where they receive their electrical supply. The overcurrent protection device must be installed in accordance with the requirements of Art. 240, and both the conductors and the connected load they serve must be protected at the correct amperage. The size of the circuit conductors, the connected load, and the circuit OCPD is actually a two-way street: The full-load current of the connected load dictates the size (by rated ampacity) of the supply conductors and the rating or setting of the overcurrent protection device. By the same token, the rating or setting of the overcurrent protection device and the rated ampacity of the load side circuit conductors dictates the maximum full-load current that can be supplied through a branch or feeder circuit.

REVIEW QUESTIONS

1. Overcurrent in an electrical circuit can be the result of ____________, ____________-____________, or ____________-____________.
2. Overcurrent due to ____________ is normally a controlled overcurrent of low magnitude that is the result of either driven-machine malfunction or poor design.
3. Overcurrent due to ____________-____________ occurs when a shunt path around a connected load is established between two or more circuit conductors.
4. Overcurrent due to ____________-____________ occurs when a shunt path around a connected load is established between an ungrounded circuit conductor and earth or some conductive surface that serves in place of the earth.
5. The ungrounded circuit conductors of an electrical distribution system can not be protected from the effects of high-magnitude ____________-____________ currents unless one of the circuit conductors in the power distribution is effectively grounded.
6. The primary purpose of an overcurrent protection device is to protect circuit conductors from overheating by allowing current in excess of their rated-load ampacity to be conveyed.

   True or False

7. The standard ampere ratings of overcurrent protection devices are listed in Sec. ____________.
8. Full overcurrent protection (protection against either overload, short-circuit, or ground-fault overcurrent) can be provided for circuit conductors and their connected load by either ____________ or ____________ ____________.
9. When supplementary overload-overcurrent protection (such as a thermal cut-out or a thermal relay) is installed for equipment such as lighting fixtures, appliances, or the internal circuits of

utilization equipment, this supplementary overcurrent protection can serve as the required means of branch-circuit overcurrent protection even if it does not have an interrupt rating.

True or False

10. An electrical distribution system is considered to be selectively coordinated when the overcurrent protection device immediately upstream from the fault condition is the only device that opens to clear the circuit or faulted condition.

True or False

11. Electrical circuits or equipment, such as motors or lighting ballasts, can be protected against overcurrent owing to overload or overheating by devices such as thermal ____________ or thermal ____________.

12. The types of circuit breakers include ____________-____________ and ____________-____________.

13. All circuit breakers, regardless of type, must be rated as trip free in order to be used as the circuit overcurrent protection device.

True or False

14. Fuses can be classified as instant or ____________-delay, regular or current-____________, and one-time or ____________.

15. A fuseholder that is designed for current-limiting fuses will accept (hold) regular fuses that are not current limiting.

True or False

16. A regular fuseholder that is not designed as current limiting will accept current-limiting fuses. The effect of current limiting can be implemented but not assured.

True or False

17. In existing installations, where there is no evidence of tampering or overfusing of the load center, replacement Edison-based fuses can be installed without conversion to type S fuses.

True or False

18. Edison-base, plug-type fuseholders cannot be installed for new structure installations, or for rework of existing structure installations unless the fuseholder is adapted to accept only type S fuses.

True or False

19. As a general rule, all ungrounded, current-carrying conductors in an electrical circuit must have overcurrent protection devices installed in series with each conductor at the point where the conductors receive their supply.

True or False

20. The "general tap rules" for the location of overcurrent protection devices in a circuit are found in Sec. ____________.

21. All fuses or circuit breakers intended as the branch-circuit, feeder, or service overcurrent protection must be installed and located so as not to be readily accessible in or on a premise.

True or False

22. Where cartridge fuses are accessible to other than qualified persons, each fuse must have a disconnecting means installed on its load side.

    True or False

23. For a circuit breaker to serve as the disconnecting means, the breaker must clearly indicate whether it is in the ____________ (OFF) or ____________ (ON) position.
24. When a circuit breaker is used as the disconnecting means (switch) for circuit conductors, it must be operated from a readily accessible location. For the breaker to be readily accessible, the center of the grip of the handle can be a maximum of ____________ feet, inches or ____________ meters above the floor, platform, or other surface it is located above.
25. When the full-load rating of the connected load for a feeder circuit is 870 amperes, the overcurrent protection device must be sized for the next higher standard rating or setting of 1,000 amperes. The feeder conductors must then be sized for a rated-load ampacity greater than or equal to the ____________ (870 or 1,000) amperes.

Chapter 8

Service Requirements

Regardless of the size of the structure or its required electrical needs, electrical power distribution systems are installed in the same basic manner. Power brought into the structure from the utility source (could be a single-phase transformer on a pole or a switchyard containing several polyphase transformers) to the initial distribution equipment is referred to as the service.

8.1 SERVICE EQUIPMENT

The service equipment is intended to be the single electrical source and cut-off point of the electrical power distribution in a building, structure, or occupancy.

The service equipment may be a single-fused disconnect, a service panel, a switchboard, or a unit substation. A single-fused disconnect, rated as service equipment, would protect feeder conductors that in turn supply a panel or panels that do not have to be rated as service equipment. The service panel will normally contain a main fused disconnect or circuit breaker protecting several feeder and/or branch-circuit overcurrent protection devices, which in turn supply feeder and/or branch circuits at a lower load-current rating. This type of power or lighting and appliance panel must be rated as service equipment.

Residential installations are normally supplied with either a single-phase, 3-wire, 120/240-volt distribution system, or in some hotter regions of the country where air conditioning loads are heavier, a 3-phase, 4-wire 240/120-volt distribution system (high-leg delta). The higher transmission voltages and associated transformers (which must be used to step down ac voltages) are normally installed and maintained by the utility company.

Small commercial buildings such as individual barber shops, quick-stop shops, and gas stations may be supplied in the same manner as a residence because the power consumption is not large scale.

Larger commercial and industrial complexes have such an enormous consumption of electrical power that high-voltage lines are normally brought into the building. The higher voltages, which help to curb or decrease losses due to voltage drop and high-value load currents, are usually part of a polyphase (either 2- or 3-phase) power distribution system because of the higher efficiency as compared to a single-phase system. The high-voltage lines are normally terminated in a unit substation (Fig. 8-1) that contains the high-voltage side, the transformer bank, and the low-voltage side (Fig. 8-2).

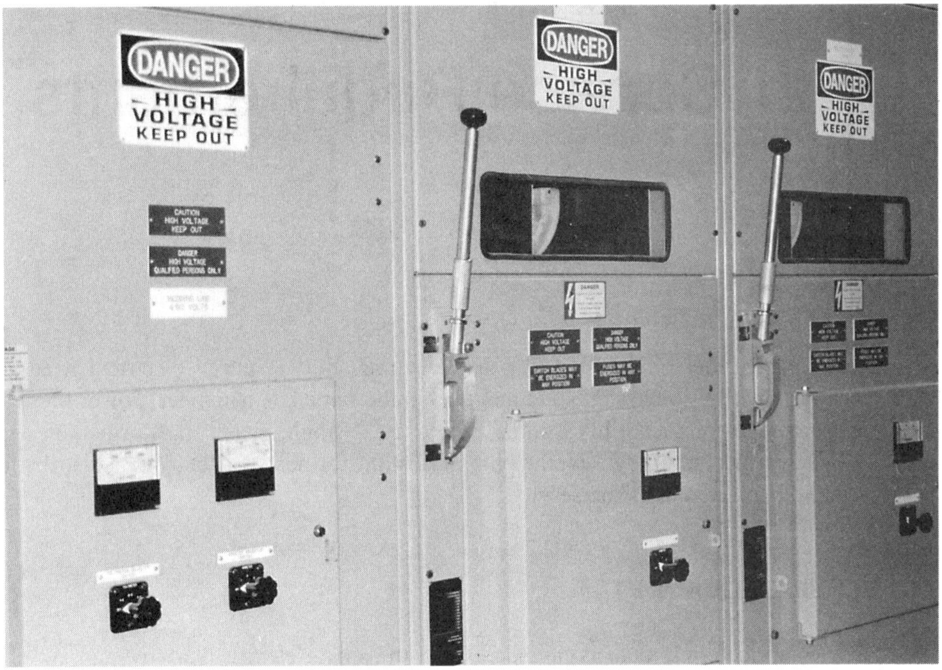

Fig. 8-1 Due to the enormous amount of power used in commercial and industrial concerns, high voltage is normally terminated to switchgear inside the building or other structure.

The high-voltage side consists of a 3-phase fusible disconnect that must have both the voltage rating and the interrupt rating of the service. Voltage divisions in the NE Code are normally "600 volts or less" and "over 600 volts nominal." The interrupt rating, as defined by Sec. 110-9, means the fusible member must have the capability of clearing or opening the large concentration of power being delivered to the transformer bank and, in turn, to the structural distribution system.

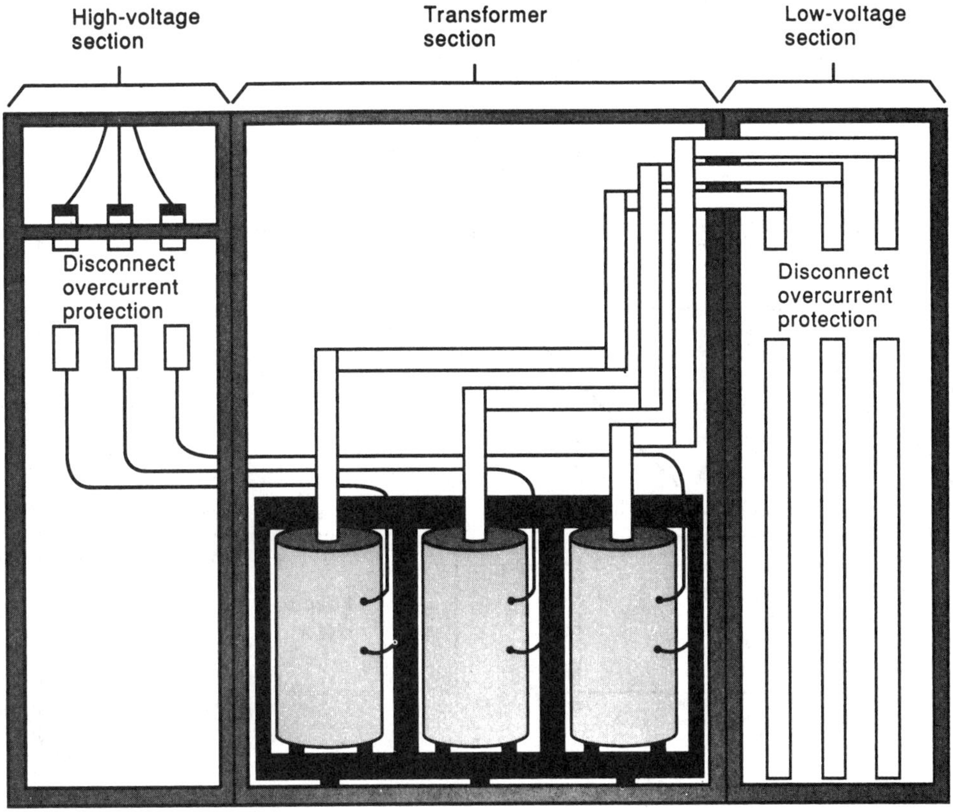

Fig. 8-2 The secondary distribution constitutes the service equipment of a unit substation (switchgear). It must contain the service-disconnecting means.

The transformer may be a single 3-phase, oil-filled or air-cooled transformer sized in accordance with the service loads, or the transformer bank may consist of three individual single-phase transformers that are wired in a standard configuration to transform the higher distribution voltage into the lower usable structural distribution voltage (working voltage normally 600 volts or less). The low-voltage windings of the transformers may be bolted to busbars for power delivery to the low-voltage side.

The low-voltage side is normally referred to as the switchgear, which may be protected by one main fused disconnect bolted to both the transformer busbars on its supply side and to the feeder busbars on its load side. The fusible disconnect must have the voltage rating and interrupt rating of the secondary service distribution. The secondary fused disconnect can protect the busbars of a secondary service power panel that contains smaller feeder overcurrent protection for the feeder circuits.

By definition (Art. 100 and Fig. 8-3), a service consists of the conductors and equipment that extend from the point of power supply (can be a public utility or a separately derived system such as generators) to the service-disconnecting means. Feeder circuits extend from the service equipment to the final overcurrent device. Branch circuits extend from the final overcurrent device to the connected load. A service equipment

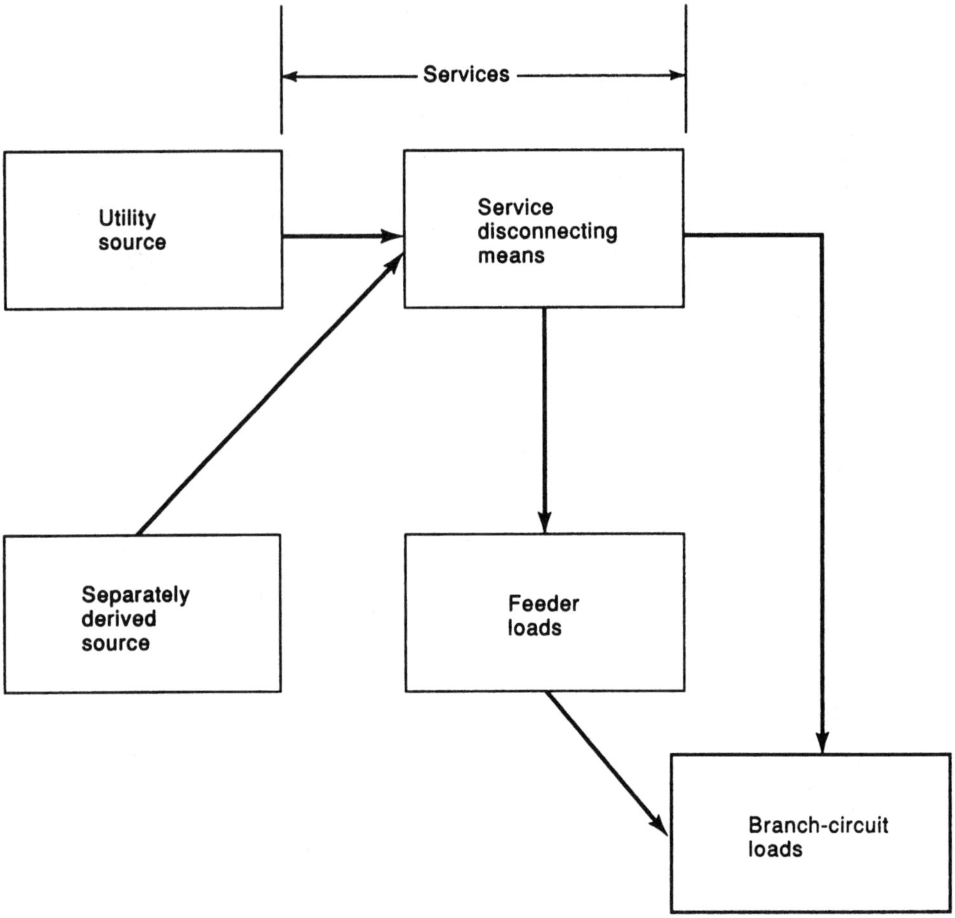

Fig. 8-3 The type of load conductor, whether service, feeder, or branch circuit, is defined in Art. 100.

lighting and appliance or power panel may contain only branch circuits and no feeder circuits, or it may contain both feeder and branch circuits.

Regardless of their size, most service distribution systems have several things in common:

1. They are all supplied by a service as recognized in Art. 230.
2. All conductors, whether service, feeder, or branch circuit must have overcurrent protection sized in accordance with Art. 240.
3. All conductors must be sized according to their rated ampacity (Art. 310), and the full-load current of the load served (Art. 220 for branch-circuit and feeder conductors; Arts. 220 and 230 for service conductors).
4. All circuits, whether on the supply side or load side of the service, must have an effective grounding system to protect personnel and equipment from the hazards of high-magnitude ground-fault overcurrent (Art. 250).

Not all power distribution systems are required to be referenced to earth ground and thus contain a grounded circuit conductor, but all distributions must have an effective equipment-grounding system to ensure that all noncurrent-carrying metal parts of equipment are maintained at a common potential in reference to earth or ground. Grounding and bonding are governed by Art. 250 of the NE Code and discussed in Chapter 9 of this text.

8.2 SERVICES

The guidelines for services are laid out in Art. 230. The service equipment usually consists of circuit breakers or fused switches located nearest the point of entrance of the service conductors into the building. The service-disconnecting means must disconnect all power from the building.

The service conductors (Art. 100), which extend from the power supply (service point) to the service-disconnecting means, consists of either a service drop or service lateral and the service-entrance conductors.

If the service conductors are run open and overhead, they are referred to as a *service drop*. If the service conductors are run underground, they are referred to as a *service lateral*. The *service-entrance* conductors extend from the point of connection to either the service drop or service lateral to the supply terminals of the service equipment. The service drop normally stops at the point of attachment to the building. The service lateral normally stops at the point of entry into the building.

For some residential or small commercial/industrial installations, the utility company may extend their service drop without splice into the supply side of the utility-owned meter can. With this type of installation, the service-entrance conductors (installed by the electrician) run from the load side of the meter can to the supply side of the service equipment.

When the service-entrance conductors are connected to a service lateral, the service-entrance conductors may be spliced to the service-lateral conductors at a pull point either inside or outside the building, or the service-lateral conductors may extend, without splice, on into the supply side of the meter can (same as the service drop installed by the utility). With this type of installation, the service-entrance conductors (installed by the electrician) run from the load side of the meter can into the supply side of the service equipment.

8.3 SERVICE DROPS

Overhead services are governed by Part B of Art. 230. Section 230-24 requires an overhead service drop to be installed accessible, but not readily accessible. Subsections (a) through (d) specify minimum vertical clearances that are allowed for service drop conductors rated 600 volts or less.

When a service mast with weather head is extended through the roof, the height of the mast is governed by both the phase-to-phase voltage (voltage between two adjacent phase conductors) and the slope or pitch of the roof. Generally (Fig. 8-4), Sec. 230-24(a)

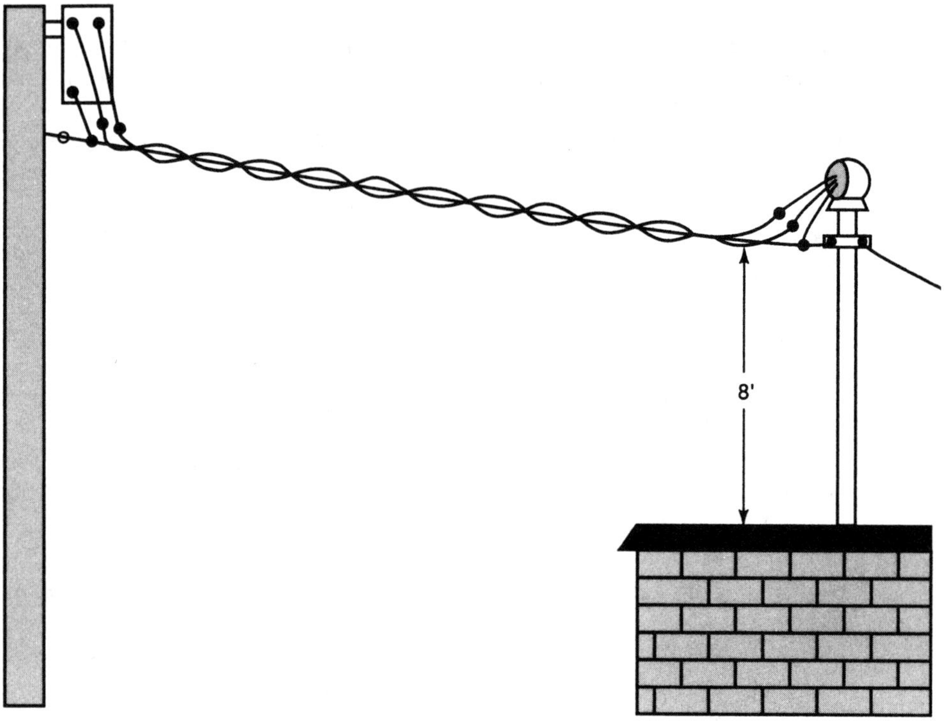

Fig. 8-4 Service drops are not allowed to be readily accessible.

requires the service-drop conductors to have a minimum of 8 feet clearance above all points on the roof and within a 3-foot perimeter past the edge of the roof.

Usually the lowest point of the service drop would be the lowest point of sag on a flat roof, and at the point of passing over the top of a gabled roof. If the connection to the service-entrance conductors is in the form of a drip loop as required by Sec. 230-54(f), the lowest point that must meet the 8-foot minimum clearance may be the bottom of the drip loop.

If the flat roof is subject to either pedestrian or vehicular traffic, the 8-foot minimum clearance is void by Exception No. 1 to Sec. 230-24(a). This type of roof is treated as an extension of the earth, and must meet the vertical clearance from ground requirements of Sec. 230-24(b). The clearance from the roof surface to the lowest point of the service drop (or drip loop) will range between ten and eighteen feet, according to what type of vehicular traffic is allowed on the roof, whether it is of private (dwelling) or commercial (public) use, and the maximum voltage to ground limitations.

For gabled roofs (Fig. 8-5), if the phase-to-phase voltage of the conductors does not exceed 300 volts, and the pitch (slope) of the roof is at least 4-in-12 [4 inches in vertical rise for every 12-inches of horizontal span or measurement (run)], Sec. 230-24, Excp. No. 2 reduces the 8-foot vertical clearance to a minimum clearance of only 3 feet between the service drop and the roof surface.

This type of installation is not restricted to aerial drops that extend between the pole and roof on the same side of the structure. A slope of 4-in-12 is considered steep enough

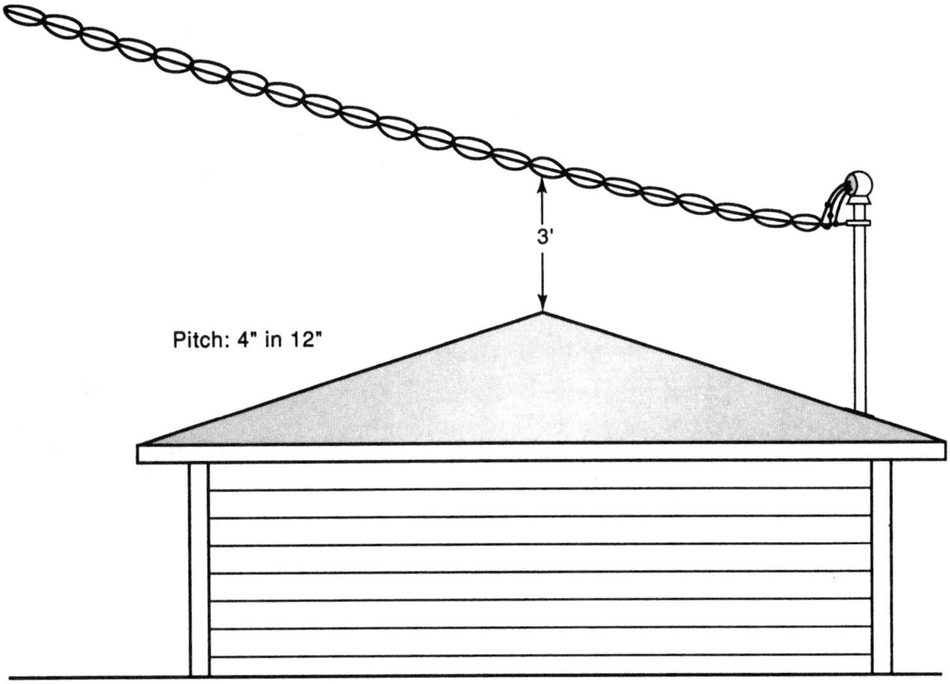

Fig. 8-5 The exceptions to Sec. 230-24(a) allow lower clearance heights for service drops over pitched roofs.

that a person or persons cannot normally stand up on the roof without difficulty. If the drop crosses the pitch of the roof, or if the drop must enter from an angle that crosses the gable end, which is located at a higher point than the entry point of the service mast, the minimum vertical clearance "from all points on the roof" would be 3 feet. As with any electrical installation, some judgment is required by the electrician as to the safety of personnel who require access to or around the electrical equipment.

If the service mast passes through the roof within 4 feet of the edge of the roof overhang (Fig. 8-6), the minimum vertical clearance "from all points on the roof" is reduced to 18 inches regardless of the pitch of the roof. The restriction of 300 volts or less between phase conductors (phase-to-phase) is still required; the service drop must cross or enter the roof from the overhang side, and Sec. 230-24(a) Excp. No. 3 stipulates that the service drop must be terminated to a through-the-roof raceway or other approved support. The length or span of the service-drop conductors above the 4-foot portion of the roof cannot exceed 6 feet.

If the point of attachment of the service drop is to the side wall of a building under the overhang, to the faceboard of the overhang (Fig. 8-7), or to the side wall in a gable end of the structure, Sec. 230-26 requires the point of attachment to be a minimum of 10 feet above the finished grade. Caution at this point: Some utility companies impose greater heights for the point of connection than what is described as minimum in the NE Code to meet the minimum vertical clearance-from-ground requirements of Sec. 230-24(b).

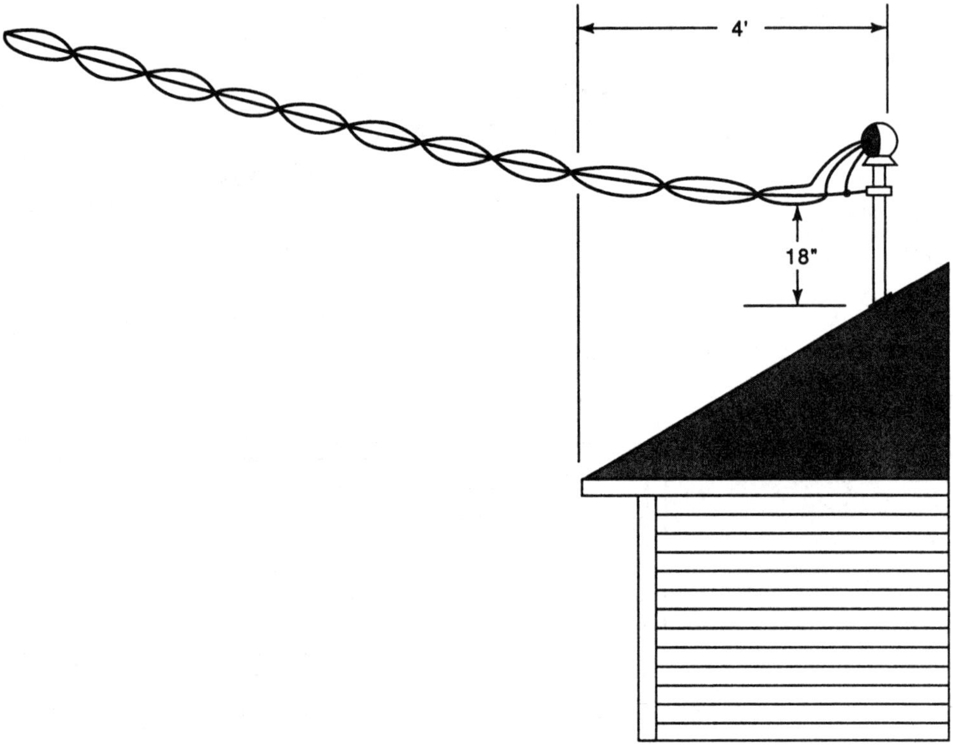

Fig. 8-6 If the service mast is within four feet of the edge of the roof, the minimum height at the point of attachment is reduced to 18 inches.

If the voltage to ground does not exceed 150 volts, Sec. 230-24(b) requires the lowest point of sag or drip loop measured from finished grade or other accessible points, such as covers on building entrance walkways and canopies (accessible surfaces), to have a minimum vertical clearance of 10 feet above finished grade (ground) (Fig. 8-8). This 10-foot minimum clearance is restricted to the triplexed or quadraplexed type SE cable that is supported by a messenger wire. The bare messenger wire is required to be the neutral (or other grounded) service conductor.

If the voltage is limited to 300 volts to ground, the service drop may pass over dwelling property or driveways or commercial areas that are not subject to truck traffic at a minimum vertical clearance of 12 feet. If the voltage to ground does exceed 300 volts, the 12-foot clearance is raised to a minimum vertical clearance of 15 feet.

If the service drop passes over public roads or other areas such as commercial farming, ranching, or logging concerns that are subject to truck traffic, the minimum vertical clearance above finished grade, regardless of the voltage-to-ground rating, must be a minimum of 18 feet—say truck, say 18 feet.

In addition to the minimum vertical clearances above finished grade, Sec. 230-24(c) refers the electrician to Sec. 230-9 for the relative location of the service-drop conductors or their point of attachment in reference to the outside of the building or other structure. Openings in the building, such as windows or doors, or appurtenances on the building,

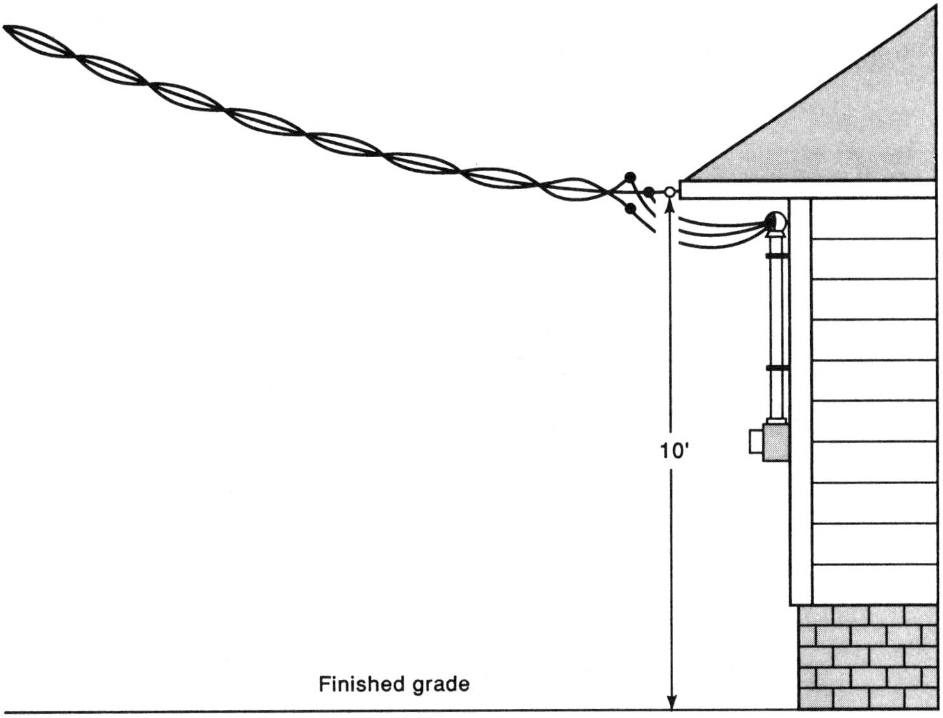

Fig. 8-7 The point of attachment of a service drop can never be less than 10 feet above the finished grade (Sec. 230-26).

such as balconies, porches, or stairs, could be used as a means of egress in times of emergency, such as fire. As shown in Figure 8-9, a minimum clearance of 3 feet, both horizontally and vertically, must be maintained for either the service-drop conductors or their point of attachment from such openings or appurtenances.

The second paragraph of Sec. 230-9 will not allow the service drop to pass under, or its point of attachment to be located under loft openings (second story openings in buildings such as commercial barns or warehouses) where materials or equipment can be unloaded for storage or loaded for use (Fig. 8-10).

This section points out openings in farm or commercial buildings, but the restriction also applies to two-story structures such as two-story dwellings or multifamily dwellings that have second-story windows. The second-floor windows could be used as a means of egress in times of emergency. Service drops in the vicinity of the windows would prove to be "detrimental to one's health" if used as an aid of escape. The last sentence of Sec. 230-9 stipulates that the installation of service drops cannot obstruct entrance to (or exit from) these type openings.

By Exception to Sec. 230-9, service drops above a window opening are not considered readily accessible, and the 3-foot clearance does not apply. This 3-foot clearance applies specifically to the service-drop conductors that are either open stranded or triplexed or quadraplexed cables that do not have an overall outer covering. If the service-drop conductors are spliced to SE cable (which has an overall covering), or enter a raceway at

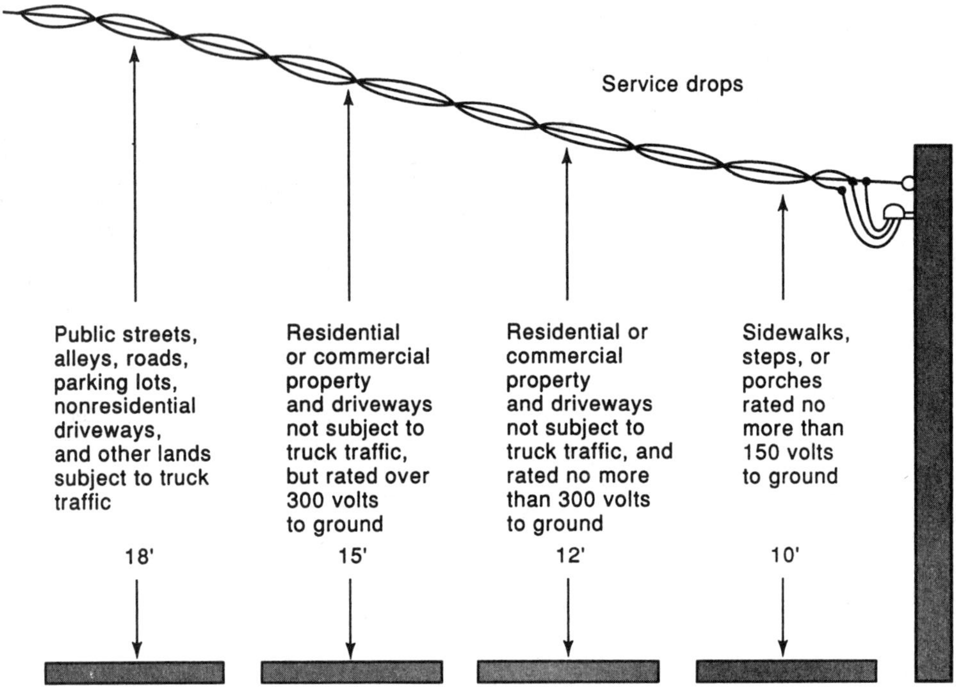

Fig. 8-8 The minimum height between final sag of a service drop and the finished grade is determined by area location and the type of vehicular traffic involved.

the point of attachment, the SE cable or service-entrance raceway can be installed or run within 3 feet of the structure opening.

Section 230-24(d) refers the electrician to Sec. 680-8 for clearances of service drops around swimming pools. As a general rule, service drops are never allowed to pass over a swimming pool, either permanent or portable, or within a 10-foot clearance measured horizontally from the pool's inside edge (Fig. 8-11).

Vertical distances outside this 10-foot outer perimeter are governed by the voltage-to-ground rating of the service-drop conductors and by the maximum height of pool equipment like diving boards, platforms, and observation towers. The vertical distances listed in the table (Figure 680-8, Exception No. 1), which range between 14 and 27 feet, would actually be the diagonal or hypotenuse measurement of the right triangle (Fig. 8-12) formed by the 10-foot minimum horizontal clearance of the table (row C) and the 12- to 18-foot minimum vertical clearance required by Sec. 230-24(b).

Section 230-27 requires the point of attachment for the messenger wire (normally doubles as the bare neutral conductor in a grounded distribution system) to be of adequate strength to support the service drop. The fittings, which are normally an eyebolt or closed-hook bolt for wall or overhang installations, must be identified for use with service conductors. The bolt or other approved fitting must be screwed (lagged) into a structural member (stud or joist) to provide the strength required to support the service drop. Even a short-span service drop can be extremely heavy, and the support fitting merely mounted into a face board or wall board would tend to give away under the weighted load.

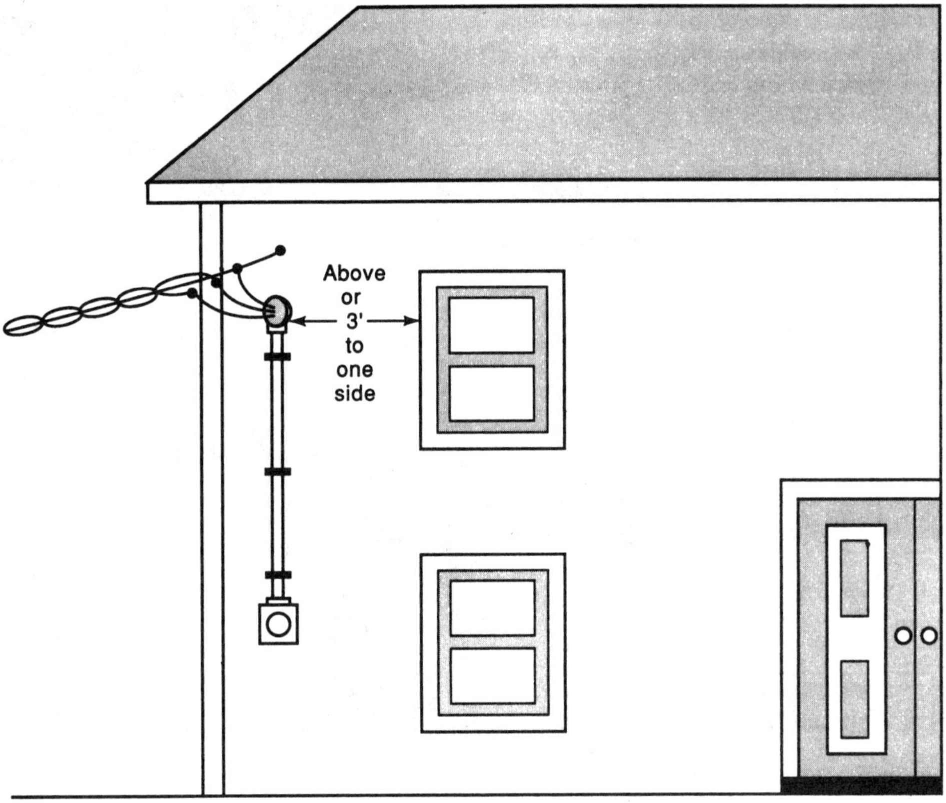

Fig. 8-9 Due to "means of egress" during an emergency, service conductors cannot be located near windows, doors, or similar openings.

When the service mast (through the roof) is used as the means of support for the service drop (tie point for the messenger wire), Sec. 230-28 requires the mast to be of adequate strength to support the service drop. If the strength of the service mast is questionable due to the weight of the service drop (long spans), Sec. 230-28 requires the mast to be supplemented with guy lines or braces. For raceway-type service masts, the fittings required for the installation must be approved for use with service masts.

PVC or other rigid nonmetallic conduit should never be used for the service mast (if it is to support the messenger wire). When rigid metal conduit is used, ideally it will be a single run (not two pieces coupled) that extends from the supply side of the meter can or service equipment through the roof up to the required height (Sec. 230-24) of the service mast. For most electrical installations, the installing electrician should check with both the local inspection authority (AHJ) and the serving public utility company before a service mast is permanently installed.

Individual open conductors on insulators are rare nowadays for service drops (Fig. 8-13). However, the student electrician may encounter this type of service drop on rework or repair projects for older dwellings. Section 230-51(c) gives the mounting, support, and clearance requirements for individual, open (not insulated) service conductors.

Table 230-51(c) lists the maximum distance in feet between supports for open-

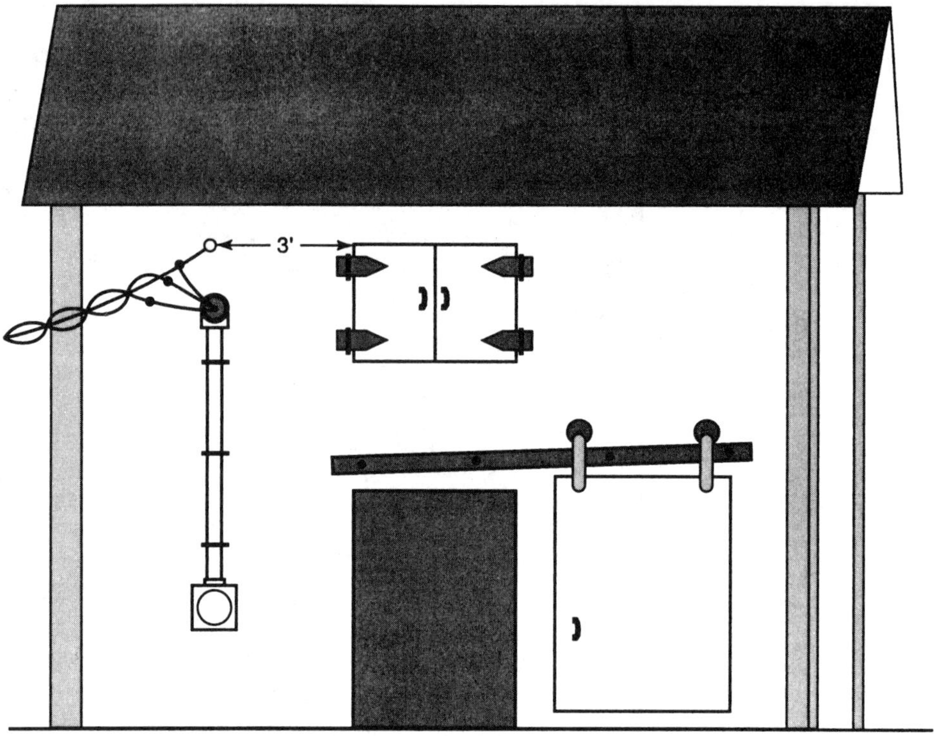

Fig. 8-10 The three-foot clearance restriction on service conductors also applies to loft openings in farm and commercial buildings.

conductor service drops. This maximum distance is governed by the voltage between the conductors, the clearance between the conductors, and the clearance between the conductors and other surfaces.

The service drop normally ends at the point of attachment to the building or other structure.

8.4 SERVICE LATERALS

The guidelines in Part C of Art. 230 for underground services are quite skimpy in terms of burial requirements and protection of the service conductors. As specified in Sec. 230-32 for protection against physical damage, a service lateral is installed as an underground installation and is governed by Sec. 300-5.

Table 300-5 lists the general cover depth or trench requirements for underground cable or raceway installations. Direct-burial cable such as USE (Art. 338) must be buried with a minimum cover depth of 2 feet to the top of the cable, PVC conduit (Art. 347) must be buried with a minimum cover depth of 18 inches to the top of the pipe, and rigid metal (Art. 346) or IMC (Art. 345) conduit must be buried with a minimum cover depth of 6 inches to the top of the pipe. Underground cable or raceway installations were discussed in Chapter 6 of this text.

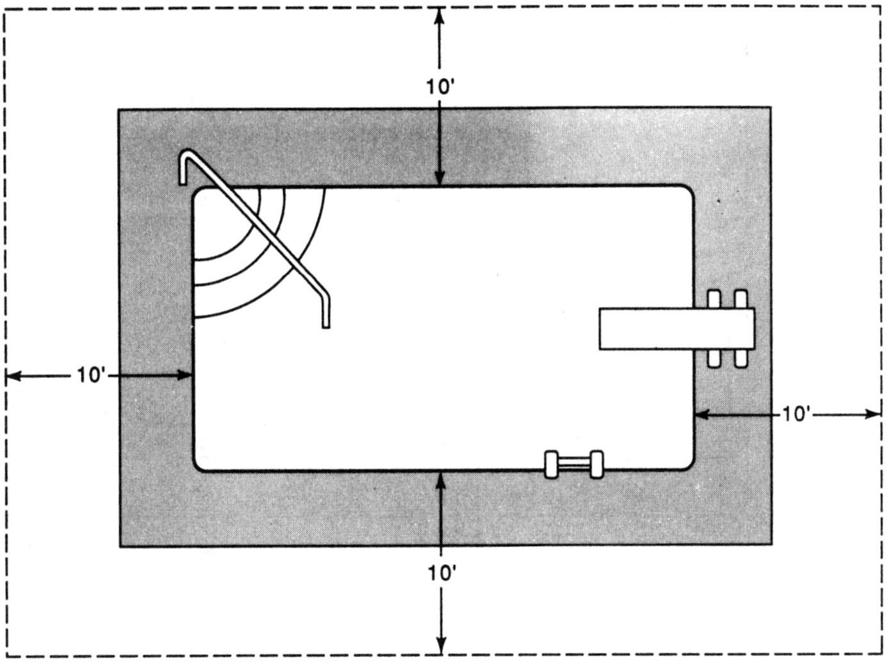

Fig. 8-11 Service-drop conductors are not allowed to pass over swimming pools [Secs. 230-24(d) and 680-8].

If the service lateral is composed of individual insulated conductors installed in conduit (or other piping raceway), the conduit must extend from the point of attachment to the pole [Sec. 300-5(d): at least 8 feet above finished grade] or at the street main, to the meter can or terminal box (auxiliary gutter) in or on the building or other structure.

If PVC is to be used for the service-lateral conduit, it must meet the criteria of Sec. 347-1. The two main concerns for rigid nonmetallic conduit (PVC) are approval for direct exposure to sunlight (where installation warrants) and approval for direct burial without concrete encasement. The concerns for PVC and direct burial installations are to ensure that the conduit material has the structural strength to withstand the continued loading of the earth. The earth is continually moving and settling due to expansion [heat and cold—Sec. 300-5(j)]. The conduit might also be exposed to compaction due to vehicular traffic across the trench area.

Either service drops or service laterals, when installed by the utility company, will usually be of aluminum conductor material. If these conductors are to be spliced to copper or copper-clad aluminum conductors, Sec. 110-14 will not allow conductors of dissimilar metals to be spliced or terminated in a common connector, unless the device is approved for the purpose.

Whether a service drop (Sec. 230-23) or a service lateral (Sec. 230-31), the service conductors are required to have an ampacity sufficient for the intended current load. The phase or ungrounded conductors are not allowed to be smaller than No. 8 AWG copper or No. 6 AWG aluminum or copper-clad aluminum conductors (note exceptions).

The neutral, which is required by Sec. 250-23(b) to be run to all service equipment

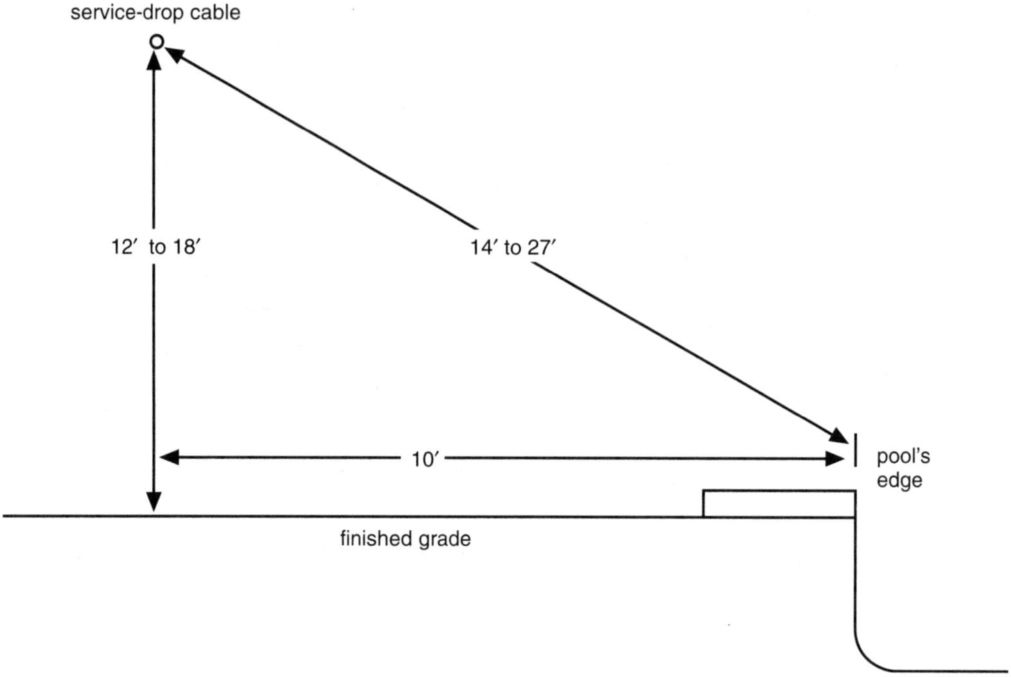

Fig. 8-12 When service drops pass near a swimming pool, the vertical, horizontal, or diagonal measurement that puts the service drop farthest away from the pool's edge is the NE Code-minimum clearance.

(of ac systems that contain a grounded circuit conductor), is not allowed to be smaller than the required grounding-electrode conductor as sized by S & T 250-94. The size of the feeder or service neutral conductor can also be reduced in size according to the guidelines of Sec. 220-22.

8.5 SERVICE-ENTRANCE CONDUCTORS

From the point of attachment to or point of entry into the building, a service drop or service lateral becomes the service-entrance conductors. If the service drop is spliced to SE cable (service-entrance cable, Art. 338) and run down to the meter can on the exterior of the building (Fig. 8-14), the cable must be supported within 12 inches of the service weather head, within 12 inches of the top of the meter can, and within intervals not to exceed 30 inches along its length [Sec. 230-51(a)]. If the SE cable is or will be exposed to any physical damage, Sec. 230-50(a) requires the multiconductor cable to be installed in conduit, raceway, or other approved means of protection.

The SE cable must be equipped with a raintight service head or weather head that is located above the point of attachment (messenger wire) of the service drop [Fig. 8-15 and Sec. 230-54(b)]. If it is impractical to install the weather head above the point of attachment, Sec. 230-54(c) Excp. will allow a lesser height that must be within two feet (verti-

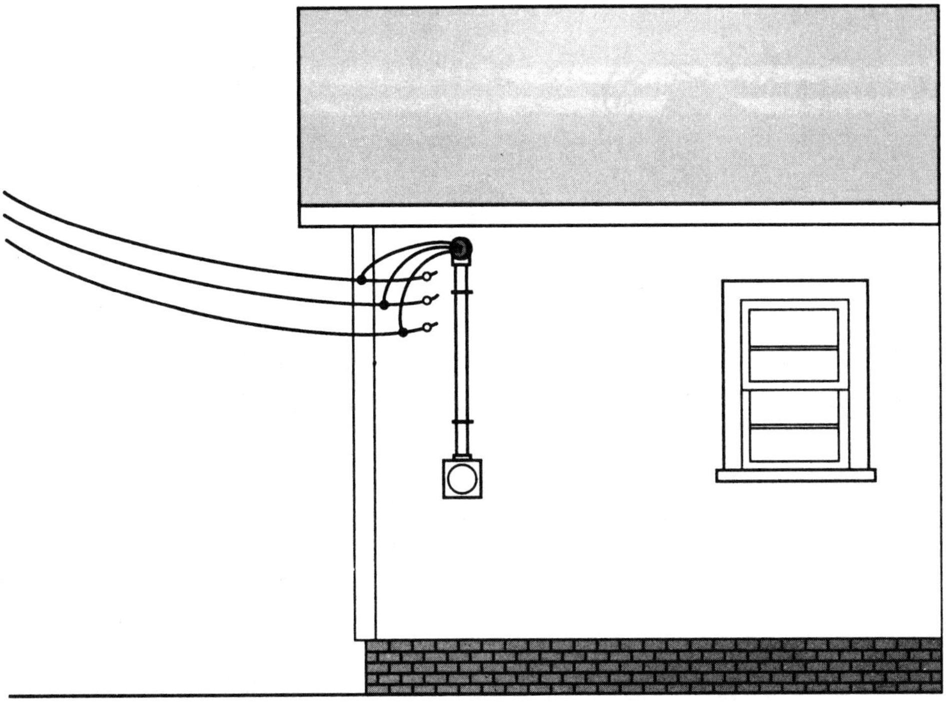

Fig. 8-13 Older services may have open conductors for the service drop. Section 230-51(c) lists the minimum clearances required for the service conductors.

cally and/or horizontally) of the point of attachment. The conductors of the SE cable must exit the weather head or service head in separately bushed holes as required by Sec. 230-54(e). The individual service conductors must be long enough to form a drip loop [Sec. 230-54(f)], which sags below the point of entry into the weather head, when connected to the service drop.

Section 230-54(b) will allow the SE cable to terminate in a gooseneck instead of a raintight service head (weather head). To form a gooseneck, the outer covering of the SE cable is removed from the end to between 18 and 24 inches of the end of the cable. The outer braided bare conductor [intended neutral, Sec. 338-3(b)] is unbraided to free the insulated conductors and then retwisted (spiral-wrapped on itself) for proper splice connection. The outer sheath should be suitably sealed with electrical putty and/or electrical tape that is rated as a self-sealing, weather-resistant thermoplastic.

The cable, with the outer sheath intact, should extend above the point of attachment for the service drop and be bent over to form a gooseneck appearance. The uncovered conductors should exit the outer sheath or covering upside down in reference to rainfall (or to the ground) for attachment to the service drop.

If individual insulated service-entrance conductors contained in raceway are used for the service-entrance run from the point of attachment to the meter can (in lieu of SE cable), the raceway must be terminated at the point of attachment in a raintight service head also [Sec. 230-54(a)]. Sections 230-54(f) and (g) require a drip loop to be formed

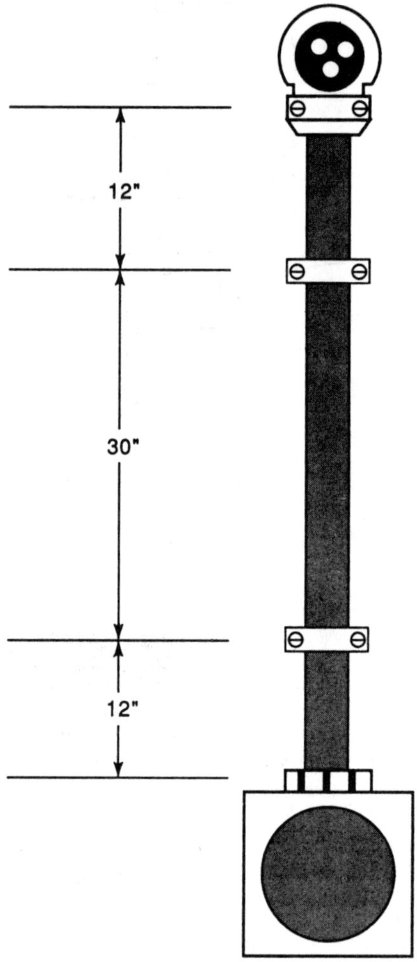

Fig. 8-14 SE cable must be supported (strapped) within 12 inches of all terminations, and at intervals not to exceed 30 inches [Sec. 230-51(a)].

for the individual service-entrance conductors whether they enter a raceway or are spliced to SE cable. A drip loop usually consists of extending the conductors in a loop fashion to where the curve or bend in the conductors extends 8 to 12 inches below the raintight service head (weather head).

The general rule of Sec. 230-40 is that a single service drop or service lateral shall supply only one set of service-entrance conductors. Multifamily dwellings or other multiple occupancies in a single structure supplied by a single service drop or lateral, are allowed by Sec. 230-40, Excp. No. 1 to have individual sets of service-entrance conductors run to each occupancy.

Where the two to six separate service disconnects, contained in two to six separate service enclosures, as allowed by Sections 230-71(a) and 230-72(a), are installed, Sec.

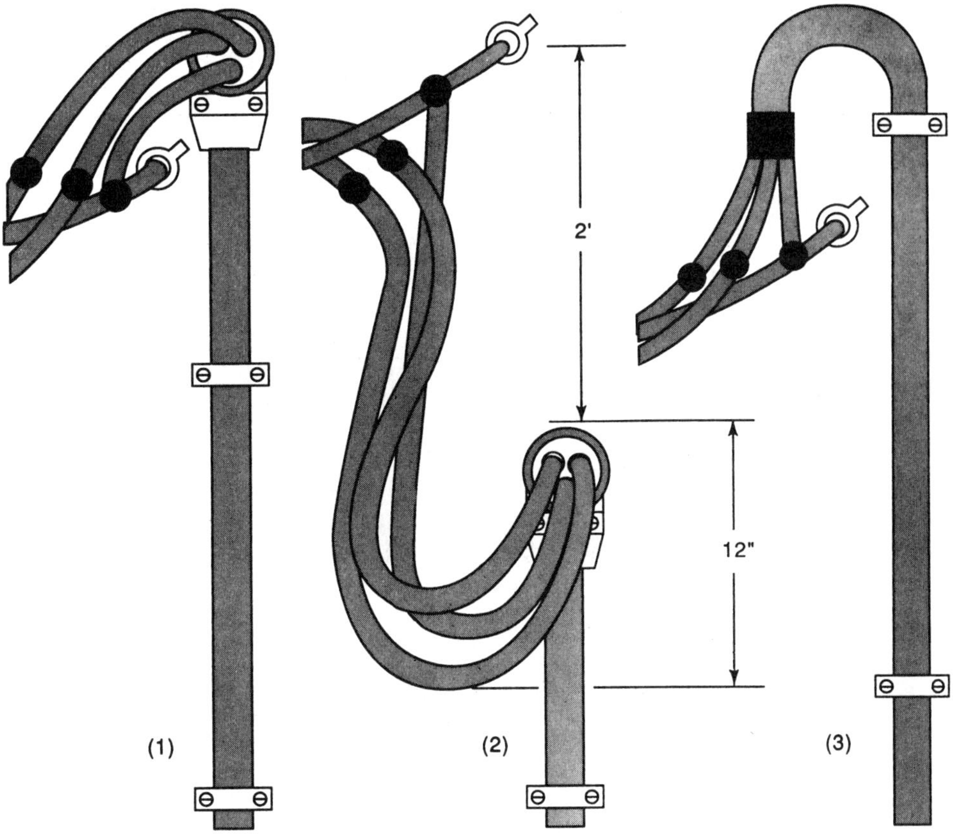

Fig. 8-15 At the point of attachment for the service drop, Sec. 230-54 will allow SE cable to be terminated in one of three ways.

230-40, Excp. No. 2 will allow all six disconnects to be supplied by a single set of service-entrance conductors. The individual disconnects are normally mounted to a common auxiliary gutter, and each disconnect is fed by an individual feeder tap from the single set of service-entrance conductors.

Sec. 230-40, Excp. No. 3 is new in the 1996 edition of the NE Code. A second set of service-entrance conductors can be run from a single service drop or service lateral to a single-family dwelling to supply a separate garage, shop, or utility building, independent of the service equipment on or within the dwelling structure.

The size of the service-entrance conductors, as well as the size of the service conductors contained in the service drop or service lateral, is dictated by the required ampacity of the service installation.

On many installations, the service drop or service lateral is furnished and installed by the public utility serving the area. The utility-installed conductors may not be of the same size or conductor ampacity as required by the NE Code for the service-entrance conductors.

In discussing the scope of the National Electrical Code, Sec. 90-2 discusses the areas of conductor installations covered by the NE Code. Section 90-2(b)(5) explains that the NE Code does not govern the installation of utility-owned and -maintained service conductors. These conductors, installed on the supply side of the utility company's metering equipment, normally extend up to the point of attachment to the building or other structure. In other words, the utility company is the governing authority for the wire size of the service drop or lateral when it is utility furnished up to the point of attachment.

The installing electrician should also realize that the rated ampacity of the overhead service-drop conductors would be determined by T310-17 because they are installed in free air. In comparison, the service-entrance conductors are installed in either raceway or a multiconductor cable with outer sheath (SE cable) from the point of attachment to the line side of the service-disconnecting means. Their rated-load ampacity must be determined by T310-16. The rate of heat dissipation is greater for free-air conductors; therefore, their rated-load ampacity is also greater.

Section 230-42(a) requires the size of the service-entrance conductors to be of sufficient ampacity to safely carry the current load as determined by Art. 220. The required conductor size is governed by the computed service current (amperage), the type of conductor material selected, the type of insulation material selected, the method of installation, and the type of environment where the service-entrance conductors are to be installed. All these factors govern the proper ampacity table to be selected from Tables 310-16 through 310-19.

Note 3 to Tables 310-16 through 310-19 (following T310-19) gives special consideration to the service-entrance and main feeder conductors for single-phase, 3-wire dwelling services. The conductors, as listed in the table of Note 3, have a higher current rating as compared to their listed ampacities in Table 310-16.

For instance, an SE cable containing No. 2/0 AWG THW copper has a listed ampacity in T310-16 of only 175 amps. The Note 3 table allows the No. 2/0 AWG THW copper conductors contained in SE cable to supply a dwelling service rated at 200 amps.

The grounded (neutral) service-entrance conductor is allowed to be smaller than the ungrounded service-entrance conductors, provided the requirements for minimum feeder-conductor ampacity rating (Sec. 215-2), for minimum neutral-conductor ampacity rating (Sec. 220-22), and minimum service-conductor ampacity rating (Sec. 230-42) can be met.

Service-entrance cable is commonly manufactured with a neutral reduced in size to 70% area as compared to the insulated phase (ungrounded) conductors. Section 338-5 requires SE cables containing a reduced neutral to be so marked on its outer sheath.

The neutral conductor carries only the unbalanced neutral current of a service (either 120-, 260-, or 277-volt loads). Section 220-22 will allow the neutral conductor to be sized according to the maximum unbalanced load. As a rule, this calculation will yield less current than is conveyed by the phase or ungrounded conductors of the same service.

Sections 230-42(c) and 250-23(b) require the reduced, grounded service-neutral conductor to be at least equivalent in wire size to the grounding-electrode conductor required for the service. The service grounding-electrode conductor is sized in accordance with Sec. and Table 250-94. This minimum-size grounded neutral conductor must be run to all service equipment, even if neutral current loads are not supplied by the service.

The grounded service-neutral conductor has a dual role: Load return, for the maximum unbalanced load current computed by Sec. 220-22, and fault return, the single return path for ground-fault overcurrent from the equipment-grounding system of the power distribution within the building or other structure.

Note 3 (to Tables 310-16 through 310-19) is restricted to dwelling structures only. Ampacity ratings listed in the Note 3 table can be used to size the feeder conductors that supply the service equipment interior panel from the outside meter can. Or, the Note 3 table ampacities can be used for sizing the feeder conductors supplying an interior panel if the service disconnect is located outside the dwelling structure.

Caution at this point: The panel on the inside of the structure would be a subfed panel with the fused service disconnect outside. This panel would be located on the load side of the service-disconnecting means. Sections 250-23(a) and 250-61(b) will not allow any grounding connections (equipment grounding) to a feeder or branch-circuit neutral or other grounded-circuit conductor on the load side of the service disconnect.

A fourth-wire equipment-grounding conductor, sized in accordance with S & T 250-95 would have to be added to this feeder run. A second equipment-grounding terminal strip (busbar) would have to be added to the panel enclosure. The grounded neutral terminal strip (busbar) would have to be isolated from the panel enclosure. (Refer to the discussion of the service grounding-electrode system in Chapter 9.)

Service conductors, especially in a service drop or service lateral, or service-entrance conductors or cables where exposed to weather or other high-moisture conditions, should have an insulation rated for wet locations. This type of insulation (T310-13) has the letter W in its insulation designation. Examples, as listed in Sec. 310-8(a), include RHW, TW, THW, THHW, THWN, and XHHW.

Section 230-42(b) lists the minimum service rating and corresponding wire size (by rated-load ampacity) of the service conductors for a building or other structure. Subsections (1) and (2) deal specifically with dwelling services and set the minimum service rating for a single-family dwelling at 100 amps.

If the computed loads of the dwelling service rating do not at least equal the requirements of either subsection (1) (service equipment supplies six or more two-wire branch circuits) or (2) (the computed service load is 10 kVA or more), then the dwelling service rating is set by subsection (3) at a minimum of 60 amperes.

Service ratings less than 60 amps are allowed by exception. When the service load consists of no more than two branch circuits, the minimum service rating can be lowered to the minimum feeder-circuit rating set by Sec. 215-2(a)—30 amps. Section 230-42(b), Excp. No. 1 will allow the service-entrance conductors to be downsized to either No. 8 AWG copper, or No. 6 AWG aluminum or c-c (copper-clad) aluminum.

When the service load consists of a single branch circuit, the minimum service rating can be lowered to the minimum branch-circuit rating set by Sec. 210-19(c) and the obelisk footnote to T310-16—15 amps. Section 230-42(b), Excp. No. 3 will allow the service-entrance conductors to be downsized to either No. 12 AWG copper, or No. 10 AWG aluminum or c-c aluminum. Of course, if a higher-amperage branch circuit is required, the exception will not allow the service conductors to be smaller than the required branch-circuit conductors. The same would hold true for the lower feeder-rated service of Exception No. 1.

As a general rule, Sec. 230-46 will not allow service-entrance conductors to be spliced. Note exceptions.

8.6 DISCONNECTING MEANS

Section 230-70 requires any building or structure to be provided with a means to disconnect all the building feeder or branch-circuit conductors from the service-entrance conductors.

Sections 240-20, -21, and -24(a) require that all ungrounded conductors have overcurrent protection, either fuses or circuit breakers, located readily accessible at the point where the building or other structure receives its supply. Section 230-91(a) correlates with these sections by requiring the service overload-overcurrent protection device to be located in a common enclosure with the service-disconnecting means. The service overcurrent protection device is required by Sec. 230-94 to protect all electrical equipment on the load side of the service-disconnecting means. Note Exceptions.

Section 230-74, in correlation with Sec. 380-11, will allow a circuit breaker to serve as the service-disconnecting means, provided it has the correct number of poles to protect all the ungrounded conductors and complies with Sec. 230-77. The circuit breaker handle must clearly indicate whether it is in the open (OFF) or closed (ON) position. Section 230-77 correlates with Sec. 240-81.

The service-disconnecting means can be a fused switch or circuit breaker in an individual service enclosure, or it can be the main circuit breaker or fused switch contained in a panel (panelboard) enclosure. If a panel is to be used for service equipment (either inside or outside the structure), Sec. 230-70(b) requires it to be permanently marked as a service disconnect, and Sec. 230-70(c) requires it to be identified (labeled) as a type suitable for use as service equipment. If service equipment (panels and/or disconnects) are located outside the building, their enclosures are required by Secs. 230-70(c), 373-2(a), and 384-17 to be approved for the pending environmental installation (wet, damp, raintight, and so on).

Section 230-70(a) will allow the fused service-disconnecting means to be installed either inside or outside the building structure at a readily accessible location nearest the entry point of the service-entrance conductors. The maximum footage allowed between the entry point of the service-entrance conductors and the point of service disconnect inside the building or other structure is not specified in this section. The student electrician should realize that the power utility companies provide only limited overcurrent protection on the primary side of the power transformer, such that the service-entrance conductors may be virtually unprotected against high-magnitude fault overcurrents up to the point of the fused service-disconnecting means within the structure.

A good word of advice is to check with the local authority having jurisdiction (AHJ, Sec. 90-4) before permanently installing the fused service-disconnecting means. Many local inspectors are very restrictive as to where the disconnecting means must be located (especially for power distribution within a wooden structure).

The general rule for service equipment located on the inside of a structure is back to back on the same wall for a meter can and an interior panel. Usually the service-entrance conductors are required to be in metal pipe (conduit) connecting the two. Some inspectors allow the pipe length between the outside meter can and the inside panel to be extended up to 3 feet when installing the service-entrance conductors. This allows an offset so the proper height for the meter can and the panel can be maintained. Other inspectors allow the 3-foot offset with SE cable, instead of individually insulated service conductors installed in raceway.

When these conditions cannot be met, the fused or overcurrent-protected (circuit breaker) service-disconnecting means must be installed outside the structure on the load side of the meter base (where used). When located outside (Fig. 8-16), the panel or other service enclosure must be located in a readily accessible location. Section 380-8(a) will allow the handle of the switch or circuit breaker in its highest position to be a maximum of 6 feet, 7 inches (2 meters) above finished grade.

The height of the meter base is normally required by the local utility company to be at eye level, which is nominally assigned a value of 5½ feet. A judgment call on the part of the electrician or a check with the local authority (having jurisdiction) is advisable for the correct installation heights. Some local authorities require all service equipment, including the panels containing branch-circuit protection, to be installed outside the structure.

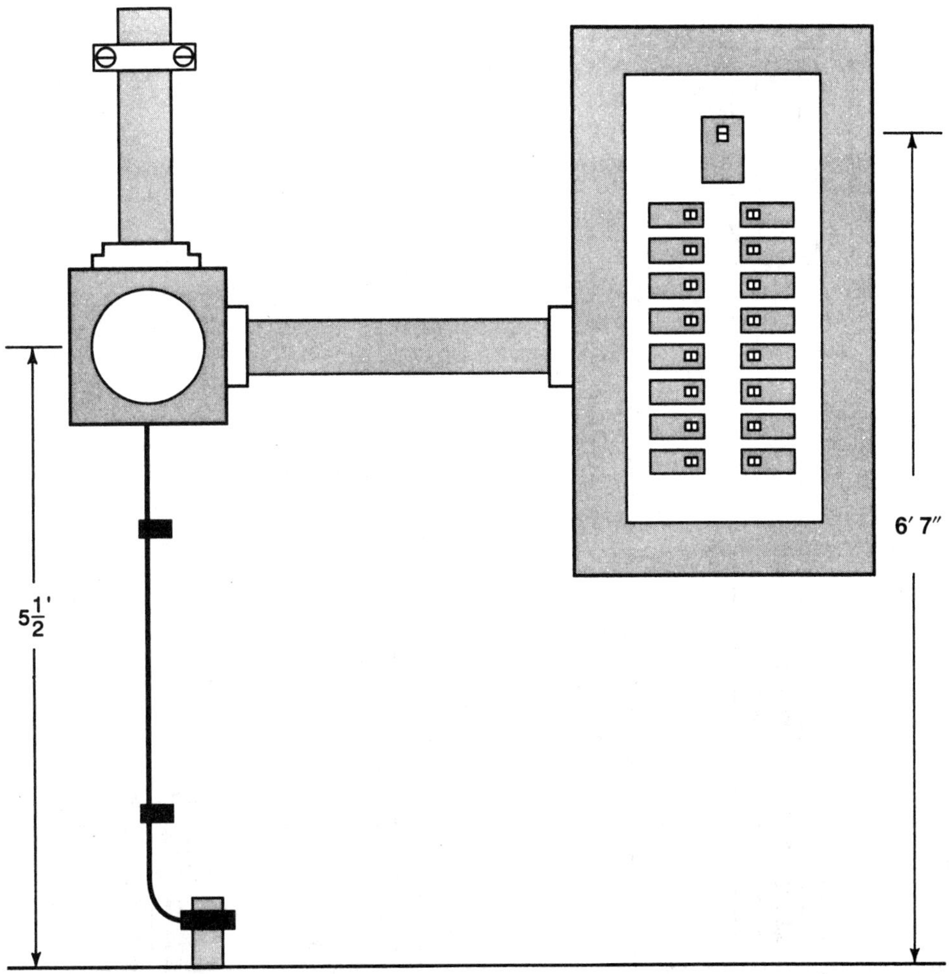

Fig. 8-16 When a panel is located outside, the meter height is normally governed by the supply utility. The disconnect height is governed by Sec. 380-8(a).

The service for a mobile home is required by Sections 550-5(a) and 550-23(a) to be installed independent of the mobile home. The mobile home service equipment and/or disconnecting means (Fig. 8-17) must be readily accessible to, within sight of, and located not more than 30 feet from the mobile home it serves.

The mobile home can be supplied by a service drop to a service pole (same type of installation as a temporary service although a utility or power pole is used instead of the normal temporary 4-inch-square pole). The meter can and service-disconnecting means are both located on the pole.

The mobile home can also be supplied by a service lateral. The better installation for a service lateral is to install a 4-foot by 4-foot sheet of plywood on two 4-by-4 treated posts. The meter can and service-disconnecting means are located side by side. The service lateral is extended up one of the 4-inch posts to the meter can. The mobile home un-

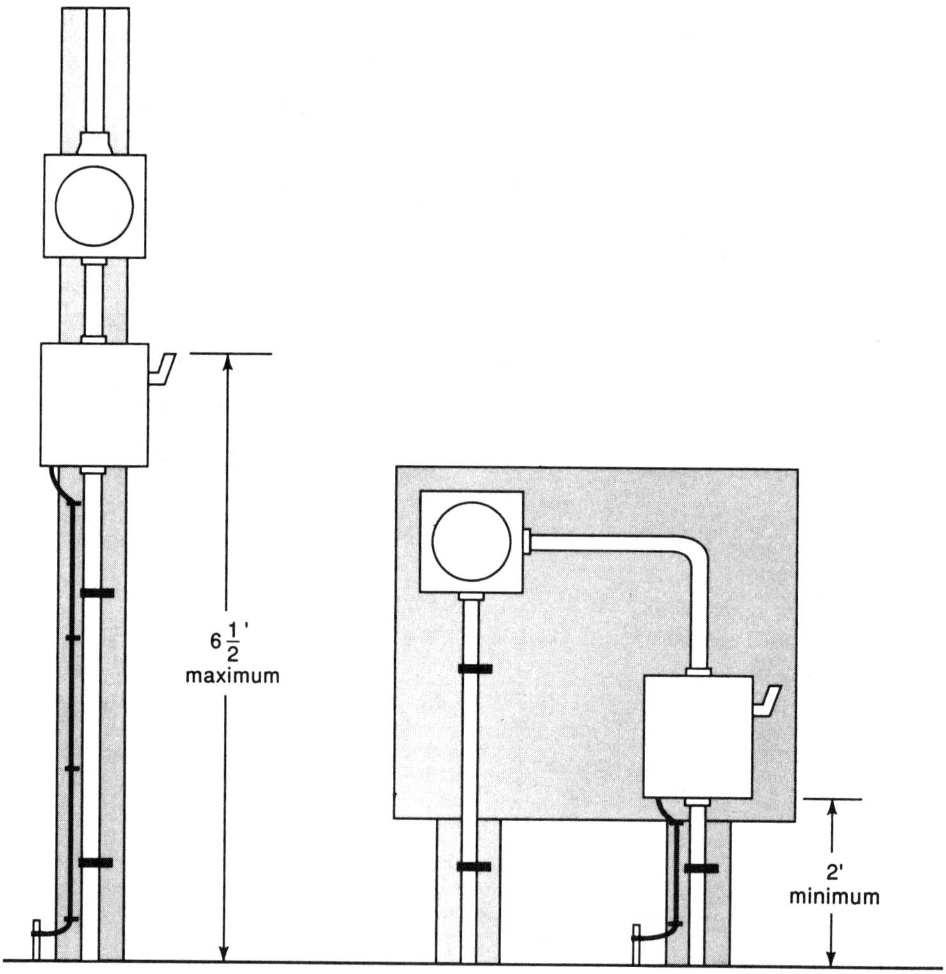

Fig. 8-17 The maximum and minimum disconnect heights for a mobile home service installation are governed by Secs. 380-8(a) and 550-23(e).

derground feeder (4-wire) is extended up the other 4-inch post to the service-disconnecting means.

Whether the mobile home service equipment is supplied by a service drop or a service lateral, the bottom of the service-disconnecting means enclosure is required by Sec. 550-23(e) to be located at least 2 feet above finished grade. The service-disconnecting means, when in its highest position [in correlation with Sec. 380-8(a)], cannot be located higher than 6½ feet above the finished grade. (The trend in the 1996 NE Code is to change the switch height maximum from 6 feet, 6 inches to 6 feet, 7 inches to equal 2 meters, which is the European standard for switch height.)

If the service-disconnecting means (often the panel main) is located inside the building, the maximum height of 6 feet, 7 inches and installation in a readily accessible location is still required by Sec. 380-8(a). The panel's location must also meet the requirements of Sec. 110-16. If the panel or other service enclosure is mounted at heights

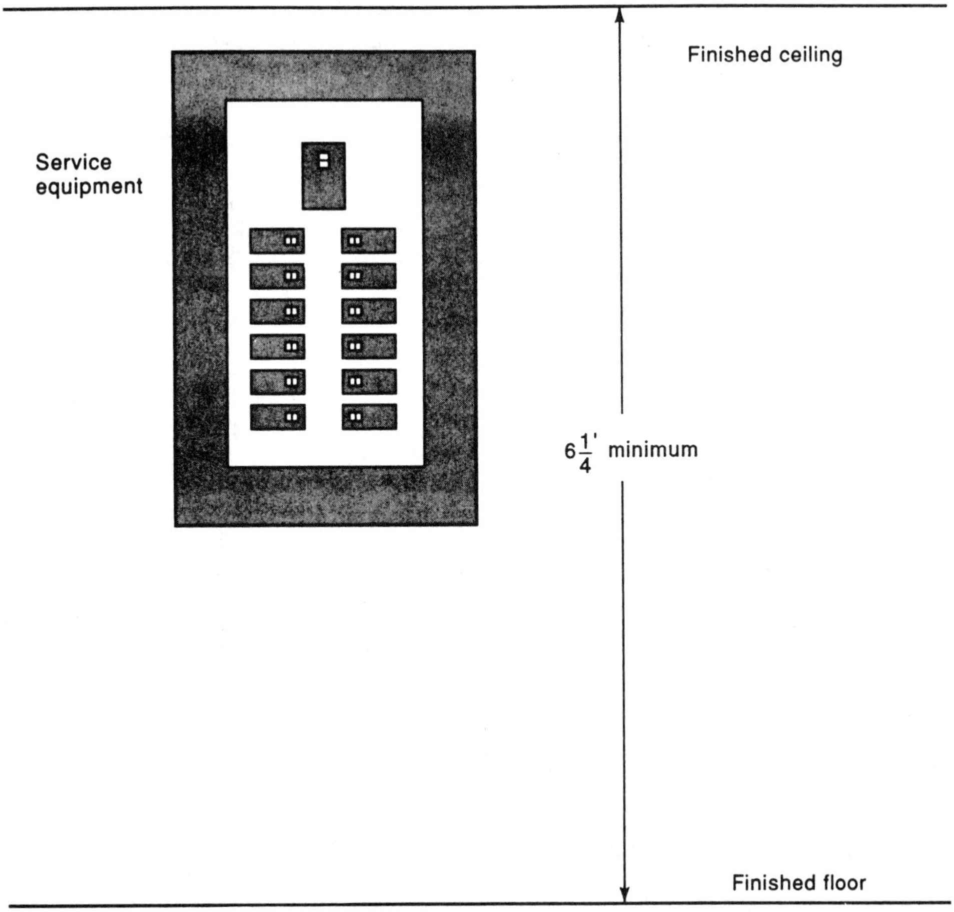

Fig. 8-18 The height of a finished ceiling located above service equipment is restricted by Sec. 110-16(e).

less than 6½ feet (Fig. 8-18), Sec. 110-16(e) requires the ceiling over the service equipment to be a minimum of 6½ feet high (minimum head room).

8.7 WORKING CLEARANCES

Section 110-16 lists the working clearances required for electrical equipment that is rated up to 600 volts. For lower-amperage panels or fused disconnects (usually 200 amps or less), subsection (a) requires the service equipment to be installed within a 30-inch wide workspace. The panel or disconnect enclosure can be mounted anywhere within this 30-inch clearance (Fig. 8-19) and does not have to be mounted in the center. The 30-inch wide wall space must be clear [cannot be used for storage, subsection (b)] or free from obstruction from the ceiling to the floor.

Many panels with lower-amperage ratings are manufactured with enclosures 14½ inches wide so they can be recessed within the wall studs set 16 inches on center. According to the type of surface or other equipment that is mounted in front of the panel or disconnect enclosure, and according to the maximum voltage to ground of the phase conductors contained in the panel or disconnect enclosure (service equipment), the 30-inch width must form a vertical column that extends unobstructed from the ceiling to the floor with a depth (measured facing the open panel) of 3 to 4 feet (Fig. 8-20). The conditions that require a clear space in front of the service equipment deeper than three feet, as outlined in T110-16(a), are listed below the table in the NE Code.

For larger industrial or commercial services that require switchgear and unit substations (which are normally located in a separate equipment room), Sec. 110-16(c) requires

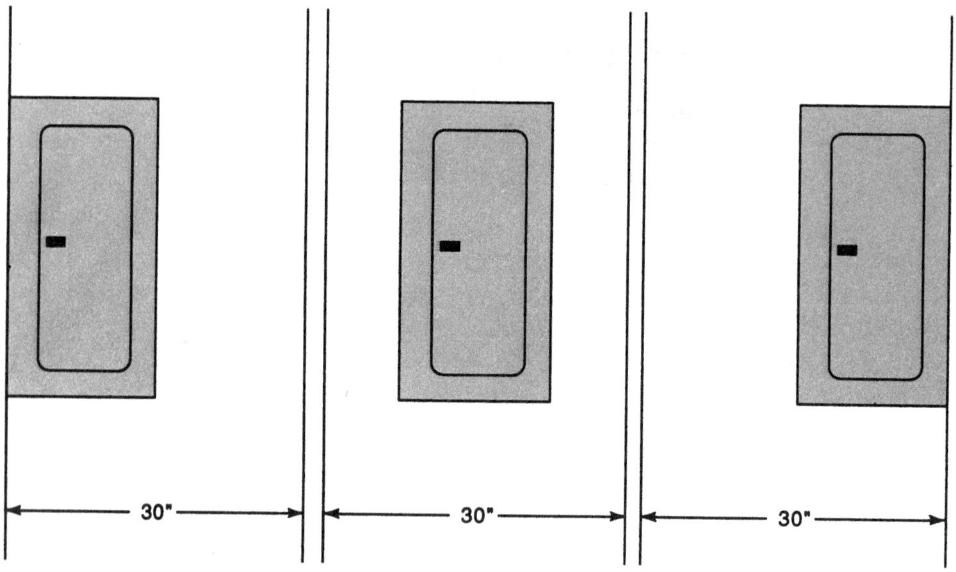

Fig. 8-19 The working clearance in front of panels, service equipment, or other electrical equipment rated 600 volts or less, must be at least 30 inches.

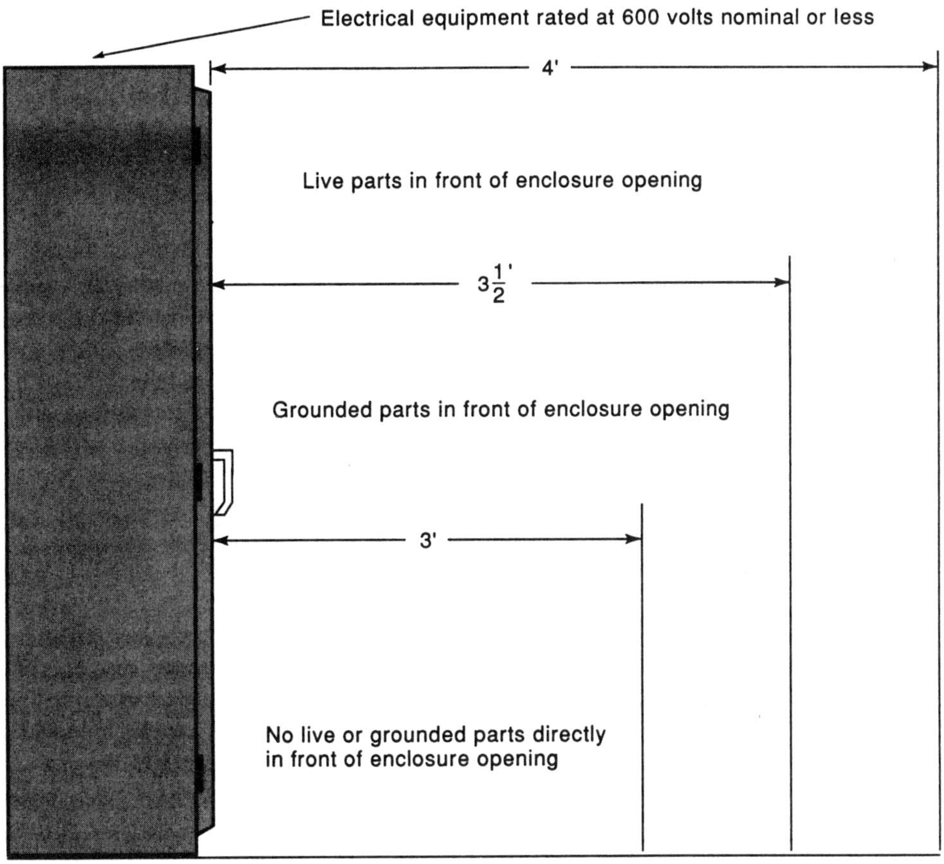

Fig. 8-20 A working clearance of at least three feet is required in front of electrical equipment that can be worked on (serviced) while energized.

at least one door entrance into the room. In some instances, two door entrances are required or the clear workspaces listed in T110-16 may have to be increased.

Switchgear or service equipment rooms containing equipment operating at over 600 volts nominal must satisfy the requirements of Sec. 110-34 for working clearances. The depth of the clear workspace is governed by the operating or nominal voltage of the switchgear as listed in T110-34(a).

REVIEW QUESTIONS

1. A service consists of the conductors and equipment that deliver energy from an electrical supply system to the wiring system of the structure or premise served.

True or False

2. A feeder or feeders consists of all the circuit conductors that extend between the service equipment or the source of a separately derived system and the final branch-circuit overcurrent protection device.

True or False

3. A branch circuit consists of the circuit conductors that extend between the final overcurrent protection device and the outlet or other load served.

True or False

4. Service equipment usually consists of circuit breakers or fused switches and their accessories that are located readily accessible near the point of entrance of the supply conductors to a building or other structure. This service equipment constitutes the main control or cutoff of the power distribution.

True or False

5. Service-drop conductors are the overhead service conductors that extend from the last pole or other aerial support to the point of attachment to the building and connection to the service-entrance conductors.

True or False

6. Service-lateral conductors are the underground service conductors that extend from the street main to the first point of connection to the service-entrance conductors.

True or False

7. Service-entrance conductors extend from the point of connection of either the service drop or service lateral to the supply terminals of the service equipment.

True or False

8. The service conductors are the supply conductors that extend from the source of power to the service equipment, and consist of either the service-drop or service-lateral and the service-entrance conductors.

True or False

9. The point of attachment to a building for a service drop must be at such a height that the service conductors are readily accessible at any point.

True or False

10. As a general rule, service-drop conductors must have a minimum clearance of 10 feet when passing over the (flat) roof of a structure.

True or False

11. Service-drop conductors passing over residential properties and driveways must be at a height not lower than 10 feet above finished grade.

True or False

12. Service-drop conductors passing over commercial properties, public streets, or other areas subject to truck traffic must have a minimum vertical clearance of ____________ feet above finished grade at the lowest point of sag.

13. Unless contained in conduit, outersheath, or some other form of raceway, service conductors must have a minimum clearance of not less than 3 feet from windows, doors, fire escapes, porches, or upper-level building entries.

    True or False

14. Service drops can pass within a 10-foot horizontal measurement from the water's edge of a swimming pool if the pool is permanently installed as an in-ground pool.

    True or False

15. The vertical clearance of service-drop conductors in the vicinity of a swimming pool is governed by the voltage-to-ground of the conductors, and other requirements listed or described in Figure ____________.
16. If SE cable is used as the service-entrance conductors that extend from a service drop, the SE cable must be supported within 12 inches of the weather head, within 12 inches of the meter can, and at intervals not to exceed 4½ feet in length.

    True or False

17. Any building or structure must be provided with a means to disconnect all building feeder or branch-circuit conductors from the service-entrance conductors.

    True or False

18. When a circuit breaker is used as the service-disconnecting means, it must have the correct ____________ of poles to protect all the ungrounded conductors, and must be marked to clearly indicate whether is in the ____________ (OFF) or ____________ (ON) position.
19. The service-disconnecting means can consist of no more than ____________ switches or circuit breakers installed in a single enclosure, or in a group of separate enclosures.
20. When installed either inside or outside a building or other structure, the service-disconnecting means must be located readily accessible at a maximum of ____________ feet, ____________ inches above the finished floorline, finished grade, or working platform surface.
21. Whether installed inside or outside the structure, sufficient working space must be provided in the vicinity of the service equipment to permit safe operation, inspection, and repairs. The working clearance around service equipment must meet the requirements of Sec. ____________.
22. A single-phase, 3-wire, 120/240-volt service panel rated at 200 amps will be installed in a two-story, single-family dwelling. The clear work space as measured across the front of the panel must be a minimum of ____________ inches wide.
23. A single-phase, 3-wire, 120/240-volt service panel rated at 200 amps will be installed in a two-story, single-family dwelling. The clear work space in front of the panel must have a depth (facing the open panel) that is a minimum of ____________ feet, regardless of whether grounded surfaces, energized live parts, or nothing but open space is in front of the panel.
24. A 3-phase, 4-wire, 480Y/277-volt service lighting panel will be installed facing a subfed 3-phase, 4-wire, 208Y/120-volt receptacle panel in the equipment room of a small commercial building. The NE Code-minimum clear workspace distance measured between the two panel fronts cannot be less than ____________ feet.
25. Where the workspace outlined by T110-16(a) is effectively doubled, only one entrance is required into an equipment room which contains a switchgear that is over 6 feet wide and rated at 2,000 amperes. The closest edge of the doorframe to the edge of the switchgear enclosure cannot be less than the single depth measurement (clear distance) determined by T110-16(a).

    True or False

Chapter 9

Electrical Grounding and Bonding

The generation of electricity, which is technically a forced flow of electrons, can be the result of the chemical reaction inside a storage-cell battery, the interaction (relative motion) of a magnetic field and a length of conductor, or the friction of two surfaces being rubbed against each other.

Electricity must either be controlled and commanded to do useful work or, as is the case with static electricity (friction between two surfaces), it must be discharged or dissipated through an effective (low-impedance) grounding or bonding (return) path.

The power available from either batteries or alternators is studied at length in electrical and electronic courses. Electricity produced by friction is normally mentioned only in passing, because static electricity cannot be effectively harnessed to do useful work.

9.1 STATIC ELECTRICITY

The storage of an electrical charge on a surface (static electricity) is normally owing to the *capacitance* that surface has with another surface. When a bolt of lightning is witnessed streaking down (or up) through the sky, this is actually a discharge of the difference of potential the cloud coverage had with the surface of the earth.

Wind blowing along the ground causes friction between the air and the surface of the earth or objects that protrude from it. The friction strips available (free) electrons from the materials on the earth's surface and transfers them to cloud accumulations suspended above the earth.

As the transfer of electrons from one surface to the other continues, a voltage buildup, or a difference of potential, occurs between the two surfaces. When the wind either slows down or changes direction, the static charge is maintained between the two surfaces because the slow or still air remaining between the two surfaces becomes less conductive and acts as a dielectric. A charged capacitor has been formed.

When the air is caused to move, either by another change in the direction of the wind or by a change in the barometric pressure (change in the air density), its dielectric strength also changes. A bolt of lightning occurs when the two charged surfaces discharge between each other. The two surfaces are at equilibrium (no charge between them) after the energy of the lightning is dissipated.

A static buildup can also occur between two clouds. In fact, the majority of lightning discharges are cloud to cloud versus cloud to ground.

When the wind blows across hills and valleys along the earth's surface, the air is compressed traveling over the hill and depressed (lowering of the pressure) as it crosses the valley. The friction strips free electrons from the hillside and deposits them along the length of the valley. The charge between the two sections of earth is referred to as *stray capacitance.* The voltage buildup, as it continues to increase, will eventually find a discharge path through the earth. This is referred to as *stray current* in the earth.

Because of the magnetic forces of the earth's gravity and the flow of underground rivers and streams, stray currents also occur as eddy currents between different soil compositions below the earth's surface.

Between the static buildup on the earth's surface and stray currents beneath it, current flow is an almost continual happening below and along the earth's surface.

Buildings project above the earth's surface. When the blowing wind is compressed against a building surface, free electrons are removed from that face (side) of the building (or other structure) and transferred to the low-pressure side (referenced by the blowing wind) of the building (Fig. 9-1). In other words, a static buildup (difference of potential or voltage gradient) now exists between the two sides of the building.

The amount of charge that can accumulate on the building surface, and the rate at which it will discharge, is determined by the building materials used in its construction. If the building is of wood construction, the number of free electrons that can be moved along the surface of the building is limited because dry wood will function as a good insulator (has few free electrons).

If the building is of brick, block, or concrete construction, the number of electrons that can be moved by the wind is increased, as compared to dry wood, because masonry or other aggregate (sand, stone, and so forth) holds moisture, and water is easily ionized (imbalance of electrons).

It seems the metal reinforcement in these types of building materials would tend to reduce the problem of static electricity by providing effective discharge paths. But if the individual metal members are not interconnected or if, when interconnected, the contact surfaces are not properly prepared for a low-impedance connection, the metal reinforcement can actually aid the different sections of the building to hold the charge.

Besides offering resistance to the blowing wind above the earth's surface, most buildings have lower levels that protrude below the earth's surface. Penetration of the

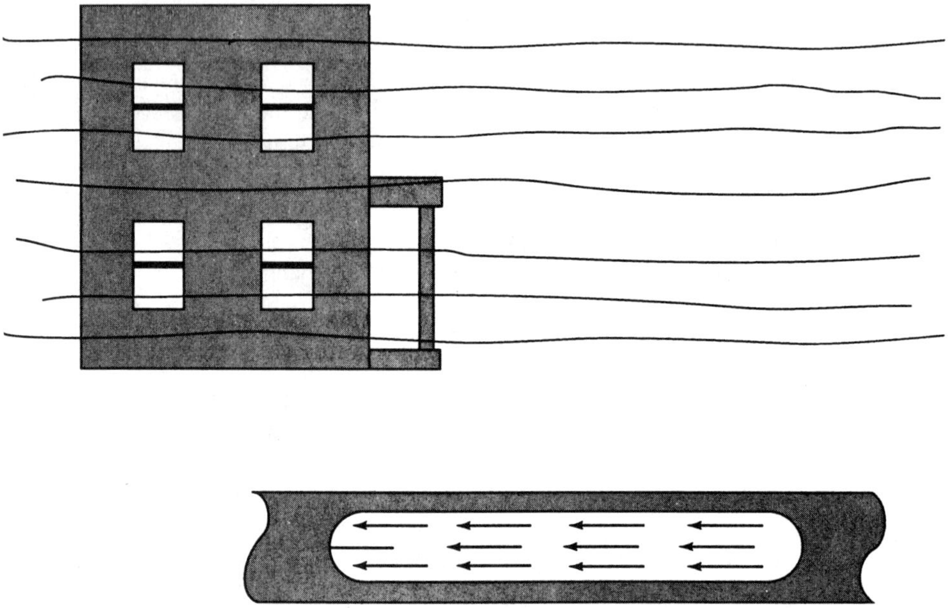

Fig. 9-1 Static electricity is created when two dissimilar objects are rubbed against each other.

earth's surface interrupts the natural discharge paths of the stray currents. Again, a charge could be maintained between two different sides or surfaces of the building.

Both metal and nonmetal piping systems are used to convey gas, steam, or water to different areas within a building or other structure. The friction between the flow of the liquid or gas within the confines of the pipe or tubing and its inner surface causes free electrons to be stripped from the pipe surface. A static buildup (potential difference) now exists between the contained fluid and the piping.

The static buildup on nonmetallic pipe is very noticeable, simply because the surface of the pipe does not offer a low-impedance discharge path. Static buildup can occur on metal piping systems as well.

The lengths of pipe in the metal piping systems are extended and routed through the structure with couplings, elbows, and other pipe fittings. If the exposed pipe threads on the end of the pipe and on the inside of the coupling are allowed to rust (oxidize) before installation, if too much pipe doping or other lubricant compounds are used in the assembly of the joint, or if the joint is not made up wrenchtight, the joints can produce a resistance connection between the successive sections of pipe. The charge placed on the fluid can be transferred to different sections of the piping, such that a difference of potential could exist between two sections of pipe.

The static buildup along sections of interior metal piping, especially nonmetallic piping, is often transferred to the interior surfaces of the building, leaving a difference of potential between the different surfaces.

There are many sources of static electricity. Since the energy produced cannot be effectively harnessed to do work, means (such as bonding and grounding) must be provided to dissipate the stored energy.

As defined in the NE Code (Art. 100), bonding is the permanent joining of two or more metal parts to form a low-impedance electrical path that can safely conduct any current that is likely to be imposed. Grounding (or as defined, "grounded") is a low-impedance path into the earth's surface, either by direct penetration or by direct connection to a bonding grid that has been effectively grounded through a grounding-electrode system.

9.2 BONDING

The seriousness of effective bonding to discharge the static sources of electricity is demonstrated in the NE Code by the bonding requirements of an in-ground swimming pool. Grounding and bonding are governed by Art. 250. Swimming pools are considered to be special equipment. The rules governing swimming pool installations are found in Art. 680.

Section 680-22(a) lists all the metal parts that go into the construction of the pool that must be effectively bonded together to form a common bonding grid. The list includes any type of metal that goes into the pool construction, whether it be rebar for reinforcement; metal rails for ladders, hand rails, or diving boards; or metal enclosures for electrical support equipment, such as convenience receptacles, lighting fixtures, circulating/filtration pumps, or driven rods that might be used for wall alignment or other structural support.

Section 680-22(a)(6) describes a perimeter of 5 feet horizontally and 12 feet vertically from the water's edge in the pool as the minimum limits or boundary of the bonding grid.

It is not enough just to ensure metal-to-metal contact between these different metal parts within the pool area. Section 680-22(b) requires all metal parts to be mechanically interconnected with a minimum wire size of No. 8 AWG bare copper conductor to form the overall bonding grid.

The bonding grid does not have to be connected to a grounding-electrode system, simply because the body of the pool is below the earth's surface. In essence, the concrete pour, especially when rebar reinforced, serves as the grounding-electrode system.

As explained in the fine print note to Sec. 680-22, the common grid connection with the No. 8 AWG copper conductor eliminates any voltage gradients (difference of potential) between surfaces within the pool area that can occur due to static electricity.

The common commercial building is built both below and above ground out of conductive materials such as steel, cement, concrete, brick, or stone. Bonding of the different sections of cement, concrete, brick, or stone is normally limited to the pressure provided at the different points of contact (brick walls constructed on top of the concrete footing, or poured concrete pressing against the base of a block wall).

The bonding grid, as shown in Figure 9-2, is normally formed by the interconnection of all the reinforcement metal (rebar) that goes into the construction of the building. The bonding grid is further enhanced when all the metal piping systems within the building (compressed air, air handling, steam, or gas) are also interconnected with the bonding grid.

Bonding of the building steel consists of cleaning the metal contact surfaces by re-

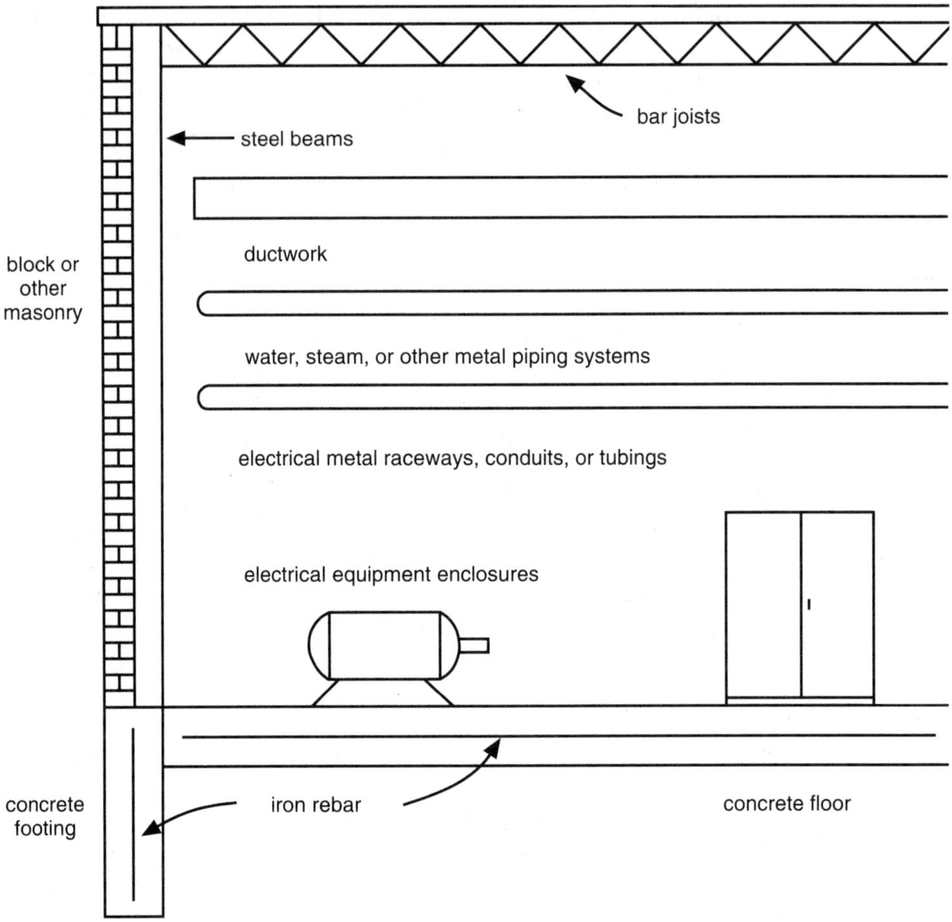

Fig. 9-2 All the metal parts used in the construction of a building and the metal it may contain should be electrically bonded together.

moving paint, rust, dirt, or other contaminants, such that when two structural members are bolted together, a low-impedance path is provided across the junction. Welding the two sections together normally forms a lower-impedance path as compared to a bolted pressure connection.

The sections of steel rod or wire embedded in concrete for reinforcement (structural rebar) should also be welded together. If welding is not possible, then some form of pressure connection (clamps, split bolts, the usual steel tie wires, or other fasteners) should be installed at each intersection of the individual bars.

If the rebar is in the form of short, independent sections of iron rod, these sections should first be bonded together with wire jumpers and pressure connectors, and then bonded to the building steel with additional wire jumpers and pressure connectors.

The joints of all metal piping systems within the structure should be made up wrenchtight to ensure a pressure connection between the individual pipe section threads and the coupling. The different metal piping systems (water, steam, air handling, and so on) should be bonded with additional wire jumpers and pressure connectors to the building steel. The straps or hangers mounted on or connected to the building steel, which are used to support the piping systems throughout the structure, cannot be used as an effective bonding means because they do not provide a pressure connection between the metal parts.

This interconnection or bonding of all the metal parts contained within a building or other structure (Fig. 9-3) ensures that a difference of potential (voltage gradients) cannot be maintained between adjacent parts or surfaces. The static sources of electricity are still there, and the transfer of electrons still occurs, but the bonding grid provides a low-impedance return path, such that the static charge is quickly discharged on a continual basis.

Fig. 9-3 All the exposed surfaces within a building must be electrically bonded together to ensure a voltage difference between any adjacent surfaces.

9.3 GROUNDING

The common bonding grid of a building or other structure effectively dissipates the energy of all the static sources except lightning and the buildup of voltage between the building's outer surface and the earth's surface. To discharge the lightning or surface voltage buildup, a grounding-electrode system, as shown in Figure 9-4, has to be established and interconnected with the building bonding grid.

If the materials used in the construction of the building are conductive (masonry, block, concrete, or metal siding), then the building, protruding above the earth's surface,

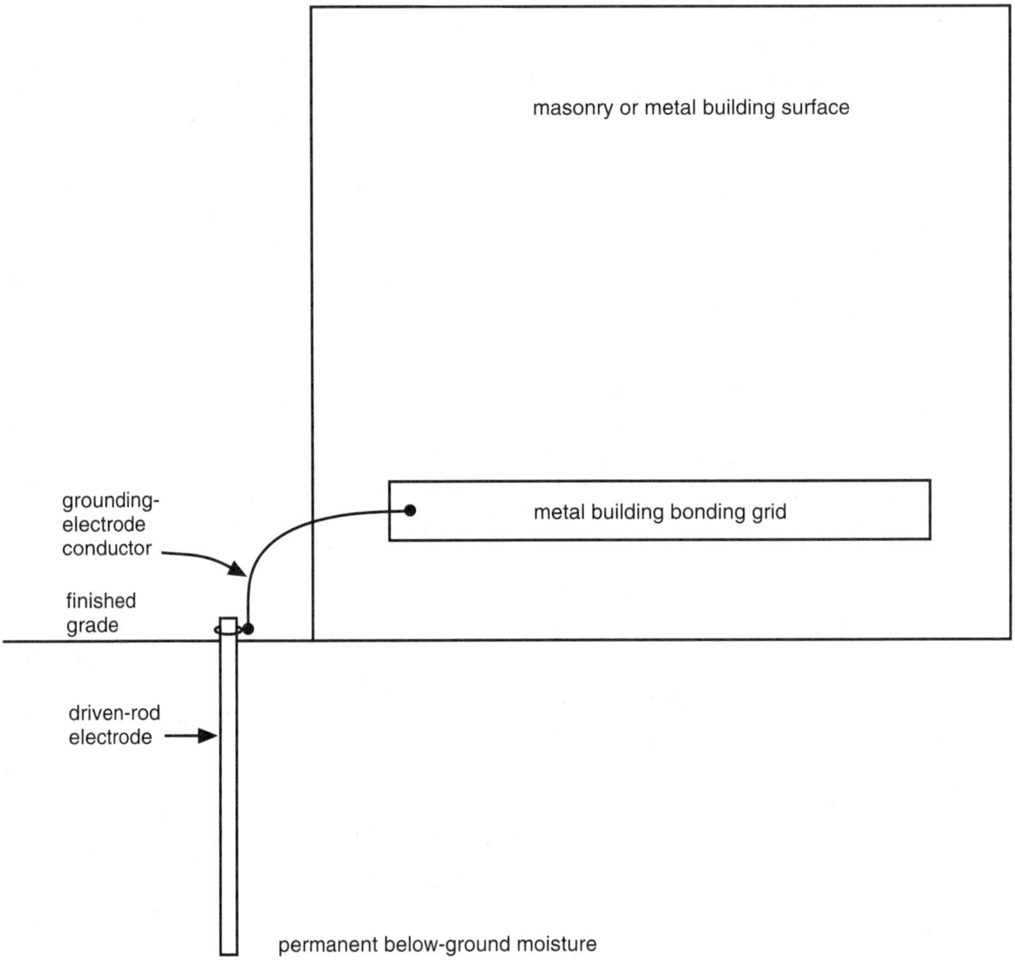

Fig. 9-4 The metal bonding grid of a building must be connected to a grounding-electrode system to reference the building surfaces at ground potential.

serves as a natural attraction to lightning because of the static buildup by the wind blowing against the building surface. The building piercing the air changes the air's dielectric strength (ability to insulate or block voltage). The lightning effectively discharges from the cloud to the building first, and then into the earth.

The bonding grid within the structure, connected to the grounding-electrode system, forms a low impedance path into the earth. As a result, lightning energy is rapidly dissipated into the earth and not between the different parts or surfaces of the building. The low-impedance path also forms a continual discharge part for any voltage buildup between the outer building surfaces and the ground immediately adjacent to the building, which in turn lessens the attraction of lightning.

The sheer weight of a building ensures continuous pressure contact with the earth in the concrete footings, which are normally below the earth's surface. In most instances, the steel rebars in the concrete footing are pressure connected to each other and to bonding jumpers that protrude above the footing. The bonding jumpers provide a means of direct (low-impedance) connection of the main building bonding grid to the grounding-electrode system. The reinforced concrete footings provide an excellent discharge path or grounding electrode for lightning and the outer surface static buildup.

In some types of building construction, the concrete footings are poured within a moisture barrier (such as plastic film or asphalt base compounds) to prevent ground water or moisture from migrating into the interior of the building. Moisture is a key ingredient in the conductivity level of concrete. In other words, dry concrete is not very conductive.

When the footings are sealed against moisture, the building bonding grid must be provided with an alternate, direct connection to permanent moisture beneath the earth's surface if lightning and outer building surface buildup are to be effectively dissipated. The bonding grid within this type of building would have to be grounded by a supplemental grounding-electrode system such as an underground metal potable water piping system (drinking water), rods or pipes driven into the ground, or metal plates buried in the ground. In some instances, a bare copper conductor will be buried around the outer perimeter of the building to provide the required grounding-electrode system.

The overall surface of the building, both inside and out, is electrically continuous and grounded (referenced to the earth), such that a potential difference (voltage) cannot exist between adjacent parts or surfaces within the building or between the building's outer surface and the earth next to the building.

9.4 UTILITY GROUNDING

Due to the enormous amount of electrical power consumed by the public, commerical power is distributed cross-country from central generating stations to various cities and industrial complexes on steel towers or various types of poles. The power is transmitted at high voltage to decrease line losses. High voltage (greater arcing distance) means the clearance above ground is also high, and, as a result, the high-tension (high-voltage) lines are subject to lightning.

Most high-voltage power distribution systems use a 3-phase, 4-wire, wye configuration. One end of the three windings, which are electrically and mechanically displaced by 120° in the alternators and transformers used in the generation and distribution of the electrical power, is taken as the three ungrounded phase or line conductors for transmission of the power. The other end of the three windings is tied common to each other and to a fourth distribution or transmission conductor.

The fourth wire in the power distribution is referred to as the common conductor because it is common to all three of the phases. In other words, the same value of voltage (coil or winding voltage) can be read from any one of the phase (line) conductors to this common conductor.

To aid in the discharge of lightning (Fig. 9-5), this common conductor is purposely connected to earth ground through successive grounding-electrode systems (at each tower) along the transmission lines. When grounded, the fourth conductor is also referred to as the neutral conductor, because it is neutral or has no voltage in reference to earth ground.

Because connected loads at the other end of the transmission line can be either three-phase or single-phase in configuration, the current magnitudes between the three

Fig. 9-5 The neutral conductor in a 3-phase, 4-wire power distribution is connected directly to the top of the metal tower. Each tower is connected to a grounding electrode.

line conductors will normally be unbalanced. If the transmission of the power is only three wire, the current imbalance, owing to the single-phase loads connected line to line (three-phase loads are automatically balanced), will cause a corresponding voltage imbalance between the phases (line conductors).

To correct the problem of voltage imbalance, in other words, to regulate the voltage supplied to the load when the respective line currents are not balanced, the single-phase loads, on the utility side of the distribution, are connected between one of the line conductors and the common neutral conductor instead of phase to phase (line to line). In this configuration, the fourth wire is used to maintain the same magnitude of voltage on each of the three ungrounded line conductors (voltage regulation) by presenting the respective generator or transformer winding voltage to the single-phase loads.

The fourth-wire neutral conductor provides a lower single-phase, line-to-neutral (line-to-ground) voltage as compared to the already available single-phase, line-to-line voltage. In a wye transformer configuration, the magnitude of the line-to-neutral voltage is equal to the line-to-line voltage divided by the square root of three.

The fourth-wire neutral conductor is normally run along the top of successive steel towers and mechanically connected directly to them. The base of each tower is then connected to a driven rod or other grounding-electrode system. This multiple grounding, at each tower, provides multiple paths of discharge for a lightning strike. A large lightning bolt can cause a severe voltage disturbance at both the generating and load ends of the transmission line if not quickly (low-impedance ground path) discharged.

Both the metal tower and the neutral conductor are connected to the driven rod through a grounding-electrode conductor. If the lightning should strike either the neutral conductor or the metal tower, the grounding electrode provides a direct low-impedance path into the earth.

The voltage rating of the individual phase windings within the alternator or transformer sets the maximum operating voltage at which the ungrounded line conductors can operate above the ground potential (zero volts). When lightning strikes an ungrounded line conductor (Fig. 9-6), the affected winding in the alternator or transformer offers a low-impedance path to the fourth-wire neutral conductor and finally into the earth through the grounding electrode.

The earth is common to both the primary and secondary circuits of the transformers used to step up the voltage (and step down the current) from the generating station. Should an accidental cross connection occur between the primary and secondary phase conductors (ungrounded line wires) of the transformer, the lower-voltage winding (larger wire size) will offer a low-impedance path to the ungrounded end of the higher-voltage winding through the grounding-electrode system. In other words, the affected higher-voltage winding of the transformer will see a short-circuit through the lower-voltage winding because of the common connection of the two windings in the earth through the grounding-electrode systems.

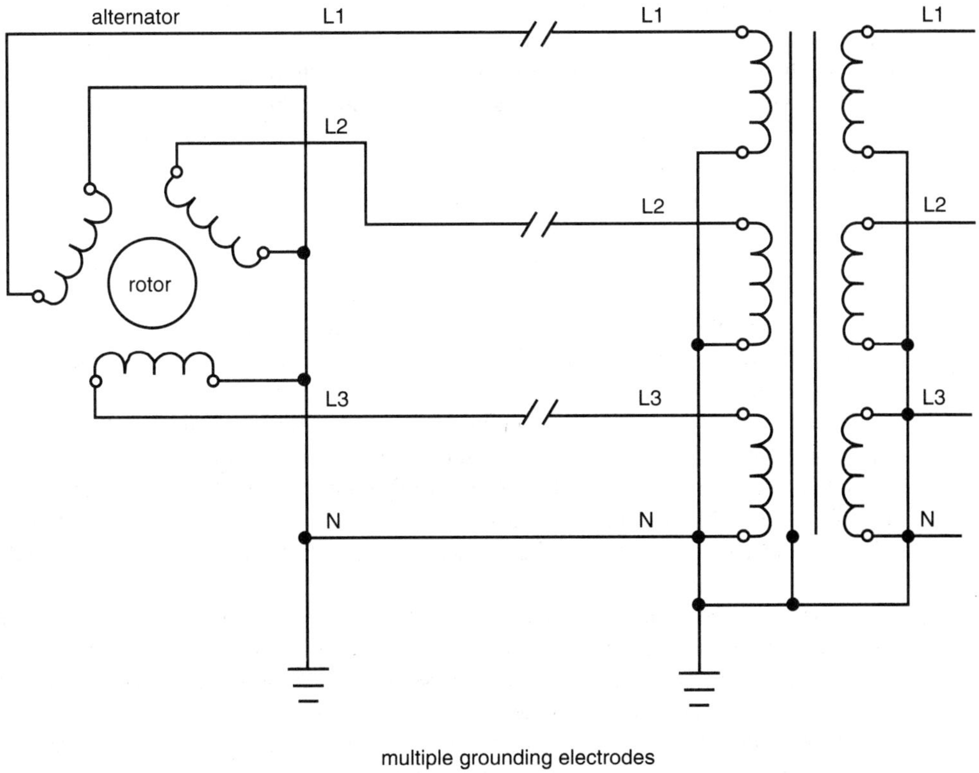

Fig. 9-6 In a 3-phase, 4-wire power distribution, the fourth-wire grounded neutral conductor provides voltage regulation for load imbalance and protects the system from damage due to lightning.

Along the transmission line, several transformers may be installed to step up the voltage even further to compensate for line losses. When the transmission line finally reaches the city or industrial complex, several more transformers are installed to step down the voltage (and step up the current).

If all the primary and secondary circuits of these successive transformers are connected in a wye configuration with the fourth-wire common grounded as a neutral conductor, then both systems are provided with a rapid discharge path for lightning, and the lower-voltage systems are protected from the accidental cross connection of the higher-voltage systems.

9.5 DUAL ROLE OF THE GROUNDED SERVICE CONDUCTOR

Within the city or industrial complex, the secondary of the transformer at the load end of the transmission line supplies the service conductors, which convey the electrical power into the building or other structure. The service equipment within the building, which normally consists of a disconnecting means, an overload-overcurrent protection device, and some type of load center that provides for further dissemination of the electrical power, is the final load for the power distribution system of the utility.

As discussed in the previous section, the fourth-wire neutral conductor on the secondary, 3-phase, 4-wire service (either wye or high-leg delta), the third-wire neutral conductor on the secondary single-phase, 3-wire service, or the B-phase conductor on a 3-phase, 3-wire delta, corner-ground service is required by the serving utility to be grounded with an electrode system (usually a driven rod) at the transformer location.

Establishing a grounding-electrode system at the transformer location has a twofold purpose: It protects the lower-voltage secondary circuit from the damaging effects of a high-tension crossover (high-voltage primary line accidentally comes in contact with the low-voltage secondary line), as it simultaneously provides a low-impedance discharge path for lightning.

The service of the building starts at this last transformer secondary. Article 250 lists the grounding requirements of the power distribution within the building or other structure. Part C deals specifically with the location of system grounding connections.

Section 250-23(a) is in agreement with the serving utility by requiring the grounded service conductor to be connected to a grounding-electrode system. If high voltage (actually medium voltage—less than 35 kV) is brought into the building to the switchgear (unit substation), the secondary intended neutral or other grounded-circuit conductor must be connected to the existing building grounding-electrode system as discussed in Sec. 9.3.

If the transformer is located outside the building or other structure (Fig. 9-7), Sec. 250-23(a) requires the intended grounded service conductor to have a direct connection to earth ground at both ends: at the transformer secondary and at the service-disconnecting means location (in the service equipment panel).

The regrounding of the service neutral or other grounded-circuit conductor places the earth between the two electrodes in parallel with the grounded service conductor. This seems to be a safety hazard since this is a current-carrying conductor, but technically it is not.

If the service neutral or other grounded service conductor is connected to a ground-

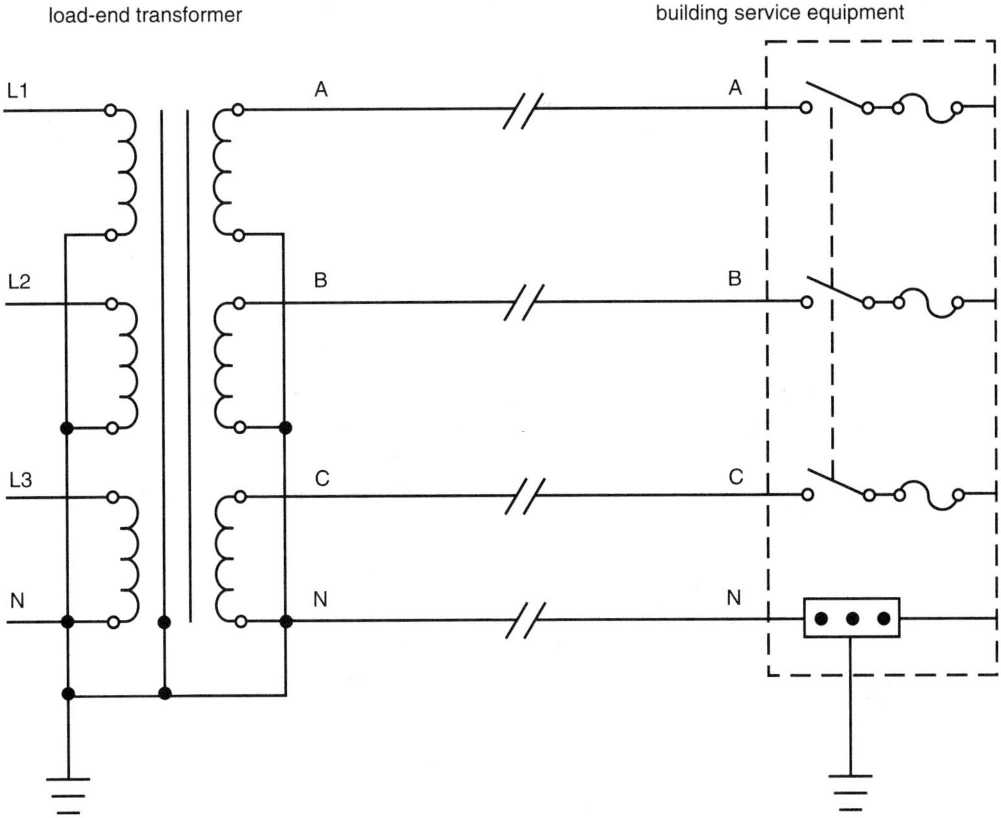

Fig. 9-7 The fourth-wire neutral conductor provides voltage regulation at the load end. The multiple grounding provides a low impedance path for lightning discharge.

ing electrode only at the transformer location (its supply end), it could have a voltage potential with the grounded metal parts of the service enclosure or load-side electrical equipment, or the surfaces of the building that are considered to be grounded when it conducts either load or fault current.

The voltage potential is actually the measured voltage drop along the length of the grounded service conductor in reference to earth ground at the transformer location. The magnitude of the neutral voltage drop is determined by the conductor's wire size (circular mil area), its overall length, and the magnitude of both imbalanced-load and fault currents. The actual difference of potential that can exist between the neutral conductor and the building surfaces depends on how conductive the earth is between the building location and the transformer location. The risk could be only minimal (low-voltage differential), or it could be lethal (high-voltage differential).

Establishing a second grounding-electrode system with the neutral at the building or other structure does not alleviate this neutral conductor voltage drop. Instead, common bonding of the grounding-electrode system and the equipment-grounding system of the building with the service neutral or other grounded service conductor will bring the earth next to the building and any grounded surfaces or parts within the building up to the potential of the service neutral conductor at that point (point of connection in the service equipment enclosure).

Section 250-23(a) will allow the connection of the grounding-electrode conductors or other bonding or grounding conductors or jumpers to the neutral or other grounded service conductor at any accessible or convenient point between the point of attachment (service drop) or entry (service lateral) into the building, up to the termination (disconnect) of the service-entrance conductors in the service equipment panel or other load center (switchgear).

Within the building or other structure (Fig. 9-8), the service neutral or other grounded service conductor (B phase in the corner-ground delta) has a two-fold purpose: to reference the supply power to earth ground for lightning discharge and voltage-drop compensation, and to provide a low-impedance return path for ground-fault overcurrents that can occur on grounded equipment enclosures or structural surfaces (Fig. 9-9).

The service neutral conductor in the building power distribution system is an intentionally grounded, current-carrying circuit conductor that is bonded to the established bonding grid and grounding-electrode system of the building or other structure. Because

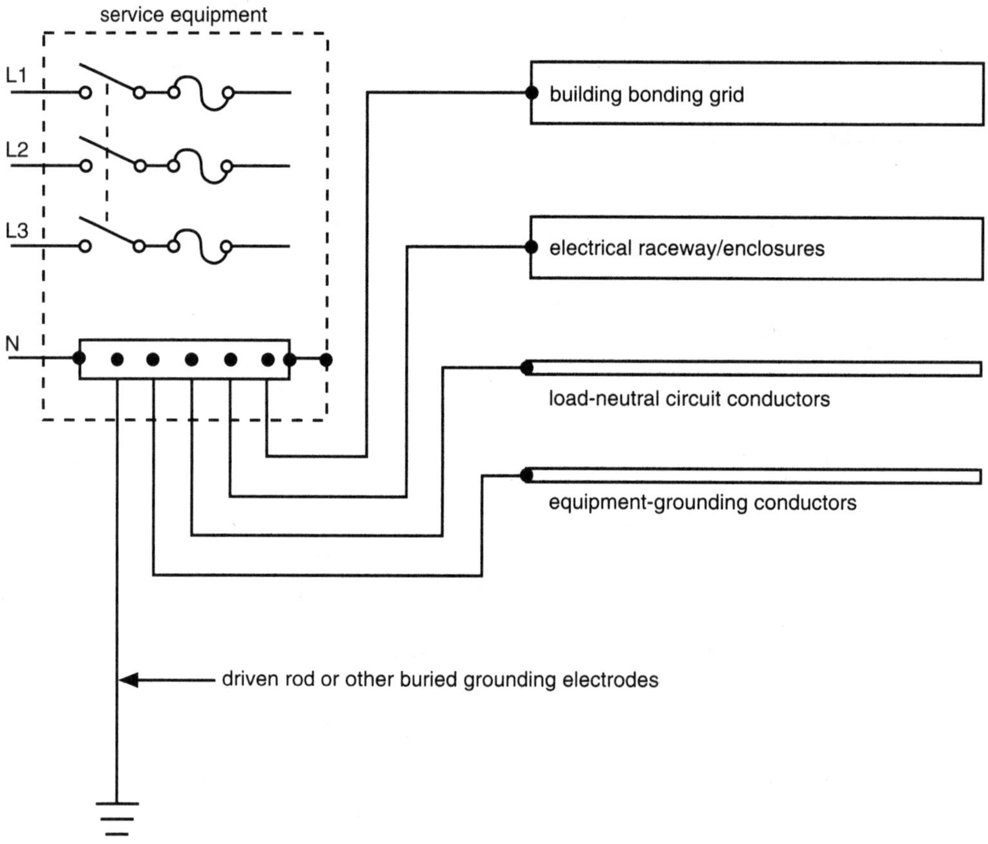

Fig. 9-8 The service neutral busbar is the common tie point for the bonding grid, the equipment-grounding grid, the grounding electrodes, and all circuit neutral conductors.

Fig. 9-9 Due to the close proximity of each other, both the equipment-grounding grid and the building-grounding grid are referenced to the grounded service neutral conductor.

of its common connection to the transformer windings, the neutral conductor also serves to regulate the load voltage for unbalanced currents by hold the incoming line voltage (service voltage) at the voltage rating of the transformer secondary winding(s).

Because it is grounded, the service neutral conductor serves to protect the ungrounded lower-voltage service conductors and connected loads from lightning and from the accidental cross-connection of the higher-voltage primary power lines.

If the building is constructed of nonconductive materials, such as a wood, which does not require a bonding grid or an established grounding-electrode system, then a service grounding-electrode system, such as a driven rod or pipe, must be installed and connected to the already grounded service neutral conductor to provide an effective equipment-grounding means.

9.6 GROUNDING ELECTRODES

Part H of Art. 250 lists the NE Code requirements for the grounding-electrode system. Electrodes that may be available in the building or other structure are listed in Sec. 250-81. If these electrodes are not available, Sec. 250-83 lists some made electrodes that can be furnished by the installing electrician to ground the noncurrent-carrying metal parts of the electrical power distribution system within the building or other structure (Fig. 9-10).

Section 250-81(a) recognizes metal underground water pipe, whether extended

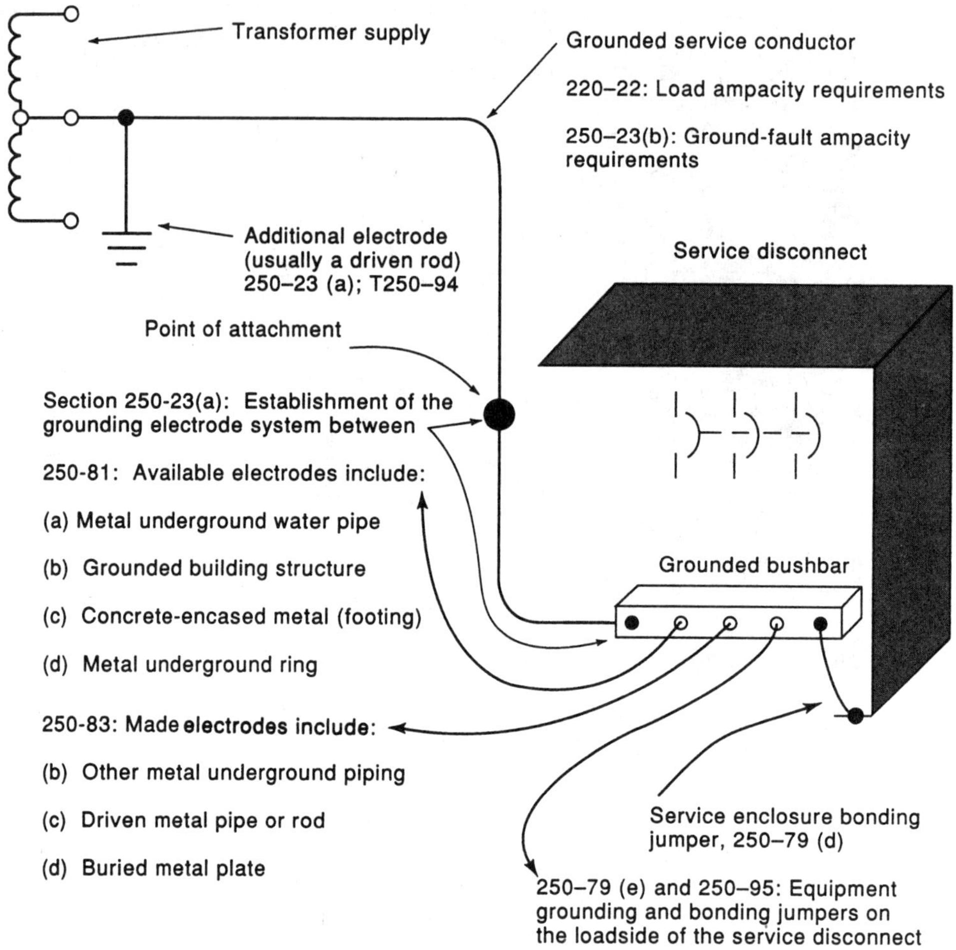

Fig. 9-10 The grounded service conductor is the "common tie point" for the grounding means, the bonding means, and the grounding electrode system.

from an individual well and pump system or from a municipal water system, as an available grounding electrode for the service within the building (Fig. 9-11). A minimum of 10 feet of metal pipe must be in direct contact with the earth, and any type of pipe connection, such as a water meter, a cut-off valve, or a filtering cannister (anything that could be removed), that is in-line between the grounding-electrode conductor connection to the water pipe and the 10 feet of earth must be jumpered (bonded) around to ensure conductor continuity.

Interior metal water pipe is required by Sec. 250-80(a) to be bonded to the service grounding-electrode system as if it were an electrode because the interior metal water pipe is often used as a common connection (bus) when multiple electrodes are to be connected to a single service.

Due to technological advancements in PVC and other types of nonmetallic tubing or piping, many of the metal water supply piping systems to the building and distribution

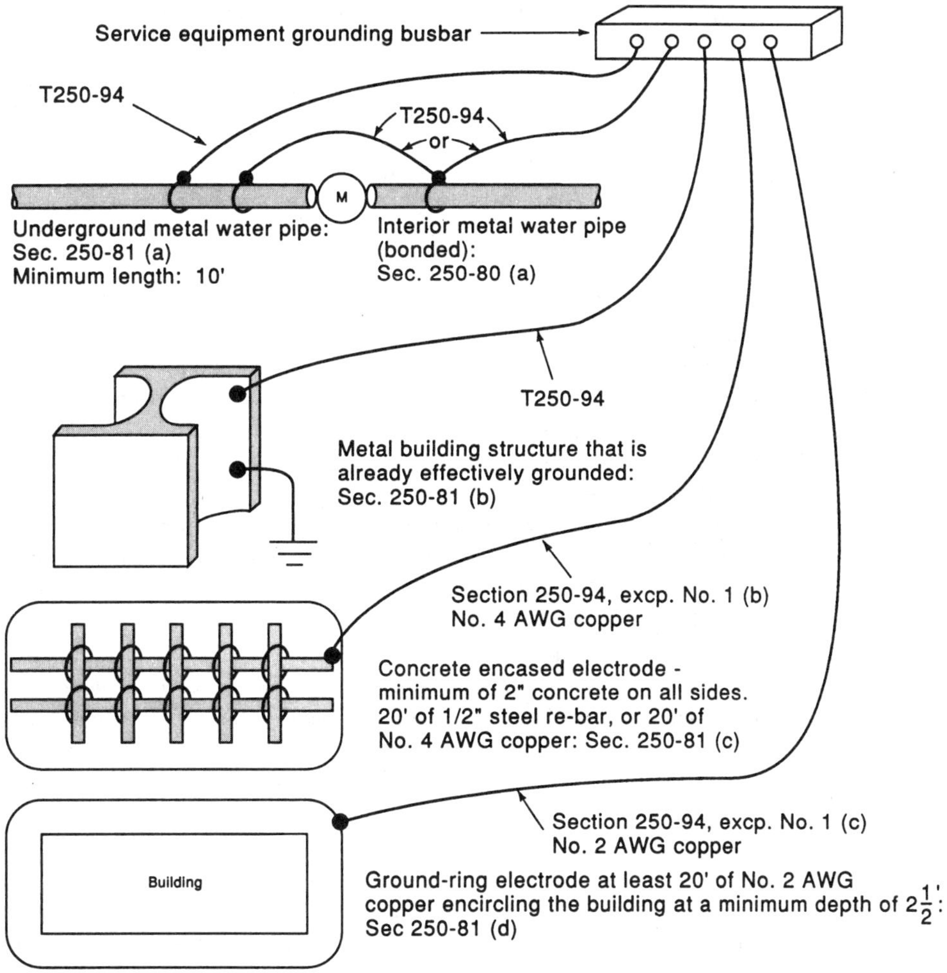

Fig. 9-11 The service grounding electrode system can be made from any of the available electrodes listed in Sec. 250-81.

systems within the building are being replaced with nonmetallic tubing. This advancement has brought about several restrictions on metal water pipe being used as a grounding-electrode system.

Out of all the available electrodes that will be discussed, the metal underground water pipe is the only one that must be supplemented with at least one additional electrode. The supplement stems from "here today and gone tommorow," when nonmetallic pipe is used for replacement or repair of deteriorated exterior metal water pipe.

In prior NE Code editions, the interior metal water pipe was used as the main grounding electrode throughout the building. If more than one switchgear or unit substation was installed in a building or other structure, the grounded service conductor of each switchgear was bonded to the nearest available interior metal water pipe.

In the 1996 NE Code edition, use of the interior metal water pipe as the common electrode bus is restricted by Sec. 250-81 to the first 5 feet of pipe entry into the building. In other words, the run of metal water pipe throughout the interior of the building is no longer recognized as a common grounding-electrode bus. The grounding-electrode conductor must be extended from the grounded service conductor or busbar to the grounding electrode.

The restriction applies to dwelling and light commercial structures. Section 250-81, Exception No. 2 allows the interior metal water pipe to be used as a common grounding electrode for industrial and commercial structures that have supervised maintenance departments—the metal pipe has to be visible over its entire length within the building.

Section 250-81(b) recognizes the structural steel bonding grid discussed previously as an available grounding electrode for the power distribution. It is up to the installing electrician to ensure that the building steel is effectively bonded together and interconnected to a grounding-electrode system. If it cannot be determined that the building steel has been effectively grounded, Sec. 250-80(c) requires the building steel to be bonded to the grounded service conductor as if it were a grounding electrode. The bonding jumper must be sized by T250-94.

Section 250-81(c) recognizes the rebar in the concrete footings of the building discussed earlier as a suitable service grounding electrode. This is normally used as the grounding electrode for the building steel bonding grid.

Restrictions on the rebar include a minimum of 20 feet of at least ½-inch rebar encased in a minimum of 2 inches of concrete. The rebar must be coated to ensure its conductivity (ferrous metals rust, the oxidation offers resistance to the flow of electricity) and located near the bottom of the concrete footing to ensure low resistance to the earth.

If the rebar is suspended in the footing (not at the bottom), or if the footing will not contain any rebar, then the electrician can install a minimum of 20 feet of No. 4 AWG or larger copper conductor in the form before the footing is poured. For wooden single-family dwelling structures this is an excellent method of providing the necessary grounding electrode for the service.

Section 250-81(d) describes a ground ring, which encircles the building, as a service grounding electrode. This type of electrode is normally installed in combination with driven rods around the perimeter of the building as the grounding electrode system for the building-steel bonding grid.

The ground-ring electrode must be a minimum of 20 feet long, buried not less than 2½ feet deep, and at least wire size No. 2 AWG bare copper.

Section 250-83 lists alternate grounding electrodes (Fig. 9-12) that can be used when those listed in Sec. 250-81 are not available at the building or other structure. Subsection (a) will not allow metal underground gas pipe to serve as a grounding electrode for the electrical service of a building.

Section 250-83(b) recognizes other types of metal underground pipe or tank systems as suitable grounding electrodes. Normally this is a judgment call on the part of the electrical design engineer or the installing electrician. If the tank farm is used to store volatile or flammable chemicals or solvents, then, as with metal underground gas piping, it should not be used as the required service grounding-electrode system. Another caution here: to meet government EPA requirements concerning ground seepage and tank leakage, many underground metal tank farms are being replaced with nonmetal storage systems.

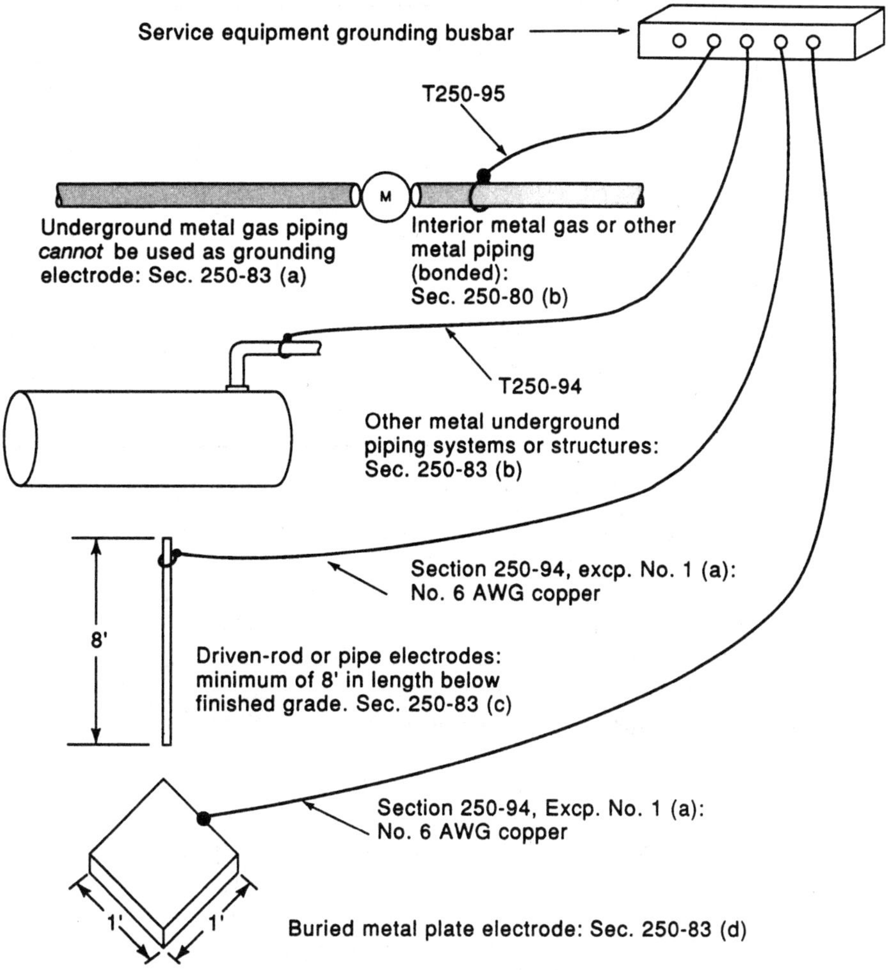

Fig. 9-12 Any of the made electrodes listed in Sec. 250-83 can be used in conjunction with or instead of the available electrodes listed in Sec. 250-81.

Section 250-83(c) recognizes the more common made electrode for the building service, especially a dwelling or small business structure that has a wooden frame. An 8-foot long rod or pipe can serve as the required grounding electrode. The preferred method of installation is to drive the rod or pipe vertically the full 8 feet into the earth's surface.

Section 250-83(c)(1) requires the pipe or electrical conduit to be at least ¾-inch trade size. The outer surface must be galvanized or otherwise metal coated to prevent corrosion.

The driven rod [Sec. 250-83(c)(2)] must be made of either iron or steel and be a minimum of ⅝-inch in diameter. The more common driven rod electrode is made of ½-inch round-stock steel that has been sleeved with a 1⁄16-inch copper sheath to attain both the corrosion protection and the minimum ⅝ requirement. Listed stainless steel rods are only required to be ½-inch in diameter.

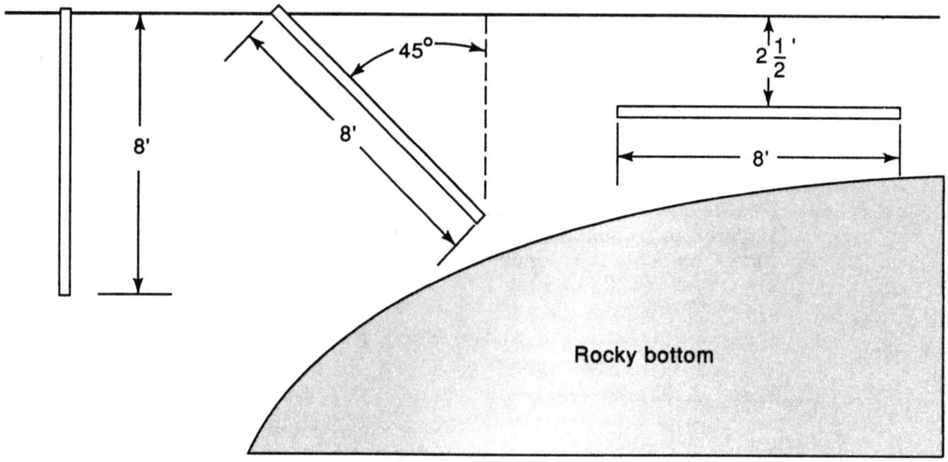

Fig. 9-13 For driven rod or pipe electrodes, a minimum of eight feet in length must be in direct contact with the earth.

When rock bottom under the earth's surface will not allow the rod or pipe to be driven vertically the full 8 feet (Fig. 9-13), Sec. 250-83(c)(3) will allow the rod or pipe to be driven at a 45° angle (from the vertical) the full eight feet into the ground.

If the encountered rock bottom will not allow the rod or pipe to be driven the full 8 feet at the 45° angle, Sec. 250-83(c)(3) will allow the rod to be buried horizontal in a 2½-foot trench.

The last sentence of Sec. 250-83(c)(3) stipulates that the upper end of the electrode shall be flush with or below ground to protect it from physical damage. This is an indication that the 8-foot rod can be bent when laid horizontally in the trench, such that one end is flush with the ground (finished grade). In other words, instead of the 8-foot horizontal, the buried rod or pipe would actually be 5½ feet horizontal and 2½ feet vertical.

Section 250-83(d) recognizes a buried plate as a suitable service grounding electrode. Although an NE Code-minimum depth is not specified in this section, the plate electrode must be buried in contact with permanent moisture (Sec. 250-83) within the earth (normally a minimum of 2½ feet).

The dimensions of the metal plate must generate a minimum of one square foot area on each side. The minimum thickness depends on the type of metal used. Iron or steel plates must be at least ¼-inch thick. Nonferrous metal plates, such as brass or copper, only need be 60-thousandths of an inch thick. As specified in Sec. 250-83(e), aluminum electrodes, whether pipe, rod, or plate, are not permitted.

Like the metal underground water pipe, unless the driven rod or pipe or buried plate is ohmed or megged (Fig. 9-14) to determine that the electrode-to-ground resistance is no more than 25 ohms, either type (driven or buried) must be supplemented with at least one additional electrode. Section 250-84 requires the resistance check and the additional electrode when the reading exceeds 25 ohms. This section can be read as a minimum of two made electrodes without the resistance check.

If the driven rod is supplemented with a second driven rod, Sec. 250-84 requires the two rods to be a minimum of 6 feet apart. The second rod should be bonded back to the first.

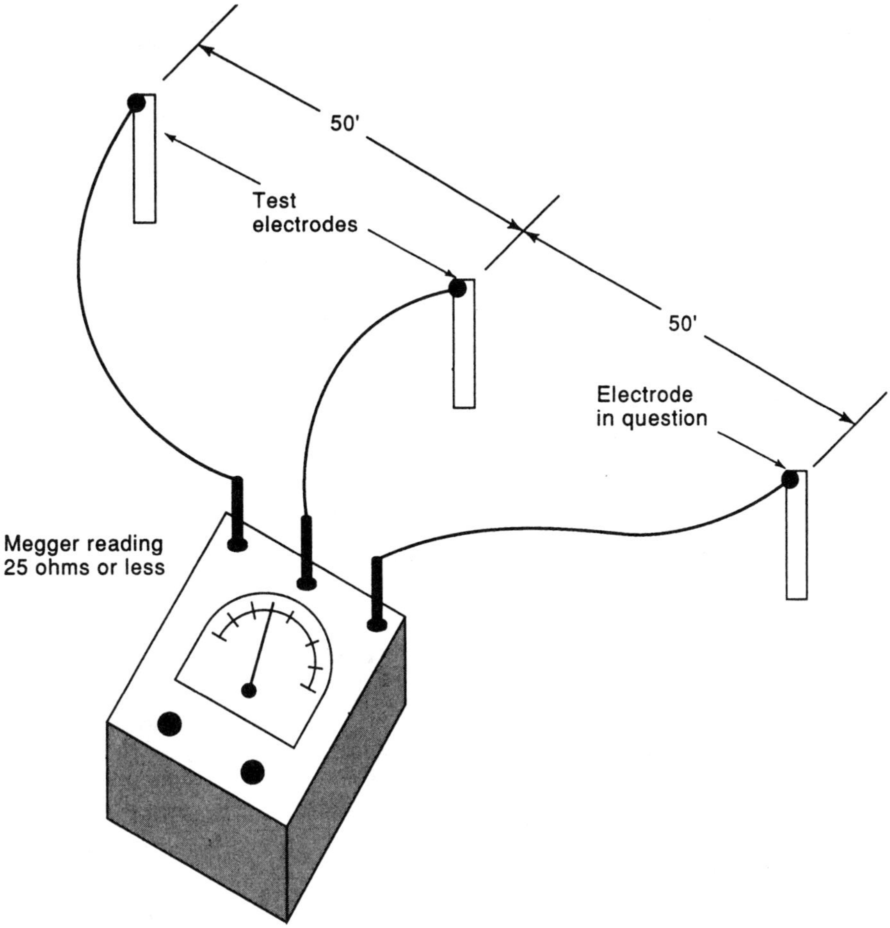

Fig. 9-14 If a made electrode has more than 25 ohms ground resistance, a second electrode must be installed a minimum of six feet away from the tested electrode.

For wooden structures, especially dwellings, 20 feet of No. 4 AWG copper can be laid along the bottom of the length of a poured wall footing where the meter base will be installed. If a length of the No. 4 is left above ground on the inital installation of the cement or concrete footing, when the meter base is installed the driven rod can be added and loop-connected to the concrete-footing (concrete-encased) electrode conductor. From the driven-rod connection, the No. 4 AWG copper conductor can be terminated to the neutral bus in the meter can to meet the requirements of Sec. 250-84.

Section 250-86 will not allow lightning-rod grounding-electrode systems to serve as the made electrodes of the electrical service. However, the second fine print note recommends bonding the two electrode systems together (normally at ground level). If the two systems are not effectively bonded together (Fig. 9-15), then Sec. 250-46 requires a minimum of 6-foot spacing between any of the conductive parts (grounding conductors, metal raceway, enclosures, and so on) of the two separate systems throughout the structure.

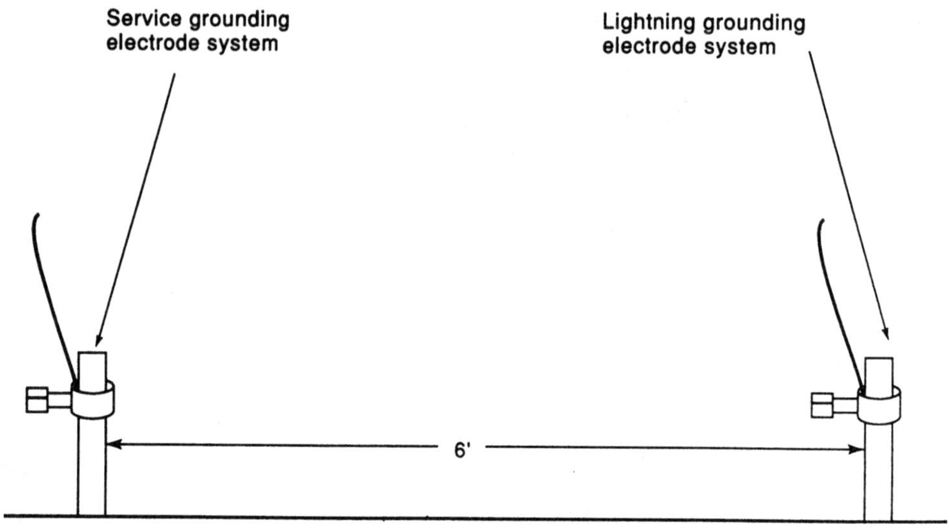

Fig. 9-15 Section 250-46 will not allow service electrodes (such as driven rod or pipe) to be closer than six feet to lightning electrodes.

9.7 GROUNDING-ELECTRODE CONDUCTORS

Section 250-81 requires the grounding-electrode conductors to be sized in accordance with S & T (Sec. and Table) 250-94, and connected by a method recognized by Sec. 250-115.

Because of the high-magnitude ground-fault overcurrent available to service conductors which are terminated with overload-overcurrent protection in the service equipment, the NE Code-minimum wire size of the grounding-electrode conductor(s) found in T250-94 is based on the wire size of the service-entrance conductors. The cross-sectional area of the ungrounded service conductors is an indication of the magnitude of ground-fault overcurrent available.

Example 9–1. Determine the NE Code-minimum wire size of a copper grounding-electrode conductor run to the building-steel grounding electrode, if a small retail store will be supplied with a 200-amp service. Use THW copper for the ungrounded service-entrance conductors.

Ungrounded service-entrance conductor wire size determination:
Wire terminal temperature limitations: 75°C — Ref. 110-14(c)(2)
Table-rated ampacity: 200 amps
Wire size: No. 3/0 AWG THW copper — Ref. T310-16
Grounding-electrode conductor wire size determination: Ref. 250-81(b)
Wire size: No. 4 AWG copper — Ref. T250-94

The rated-load ampacity of the ungrounded service-entrance conductors was technically not restricted by the wire terminal temperature limitations of Sec. 110-14(c)(2), since THW insulation already has a 75°C rating. The 200-amp service rating corresponded to the table-rated and rated-load ampacity of the No. 3/0 AWG THW copper conductors.

The reference for the grounding-electrode conductor wire size determination recognized the grounding electrode. Section 250-81 required the grounding-electrode conductor to be sized by T250-94.

Code challenge 9–1. Determine the NE Code-minimum wire size of a copper grounding-electrode conductor run to the building steel-grounding electrode, if an auto repair shop will be supplied with a 400-amp service. Use THWN copper for the service-entrance conductors.

Ungrounded service-entrance conductor wire size determination:
Wire terminal temperature limitations: (A) ____________ °C
Ref. (B) ______________

Table-rated ampacity: (C) ____________ amps
Wire size: (D) ____________ THW copper Ref. (E) ______________

Grounding-electrode conductor wire size determination: Ref. (F) ______________
Wire size: (G) ____________ copper Ref. (H) ______________

The use of copper service-entrance conductors does not warrant the use of copper grounding-electrode conductors. Nor does the use of aluminum service-entrance conductors warrant the use of aluminum grounding-electrode conductors. Table 250-94 is a composite table. If copper grounding-electrode conductors will be used for aluminum service conductors, the required size of copper grounding-electrode conductor is simply read directly from the row identified by the size of the aluminum service conductors.

From Chapter 2 of this text, the required wire size of the service-entrance conductors is based on the load-current rating of the service and any of the derate requirements due to temperature compensation or voltage drop.

Example 9–2. A small manufacturing concern will be supplied by a 3-phase, 4-wire, 208Y/120-volt service rated at 200 amps. If the THHN copper ungrounded service-entrance conductors will be installed in conduit run down the side of the building where the ambient temperature will reach 120°F, determine the NE Code-minimum wire size of the copper-clad (c-c) aluminum grounding-electrode conductor run to underground metal water pipe, that is connected in the interior of the building.

Ungrounded service-entrance conductor wire size determination:
Table-rated ampacity determination:
Higher temperature derate: 0.82 Ref. T310-16, CF
Minimum table-rated ampacity computation:

$$200 \text{ amps} \div 0.82 = 243.9 \text{ amps}$$

Table-rated ampacity: 260 amps
Wire size: No. 4/0 AWG THHN copper Ref. T310-16

Wire terminal temperature limitations: 75°C — Ref. 110-14(c)(2), Ex. 1
  Table-rated ampacity: 200 amps
  Wire size: No. 3/0 AWG THHN copper — Ref. T310-16
NE Code-minimum, compare for larger: — Ref. 310-15(c)
  Wire size: No. 4/0 AWG THHN copper
Grounding-electrode conductor wire size determination: Ref. 250-81(a)
  Wire size: No. 1/0 AWG c-c aluminum — Ref. T250-94

Code challenge 9–2. A small manufacturing concern will be supplied by a 3-phase, 4-wire, 480Y/270-volt service rated at 400 amps. If the XHHW-2 aluminum ungrounded service-entrance conductors will be installed in conduit run down the side of the building where the ambient temperature will reach 110°F, determine the NE Code-minimum wire size of the copper grounding-electrode conductor run to underground metal water pipe.

Ungrounded service-entrance conductor wire size determination:
Table-rated ampacity determination:
  Higher temperature derate: (A) ____________ Ref. (B) ____________
  Minimum table-rated ampacity computation:

____________ amps ÷ ____________ = (C) ____________ amps

  Table-rated ampacity: (D) ____________ amps
  Wire size: (E) ____________XHHW-2 aluminum Ref. (F) ____________
Wire terminal temperature limitations: (G) ____________ °C Ref. (H) ____________
  Table-rated ampacity: (I) ____________ amps
  Wire size: (J) ____________ XHHW-2 aluminum Ref. (K) ____________
NE Code-minimum, compare for larger: Ref. (L) ____________
  Wire size: (M) ____________ XHHW-2 aluminum
Grounding-electrode conductor wire size determination: Ref. (N) ____________
  Wire size: (O) ____________ copper Ref. (P) ____________

When the load-current rating of a single service exceeds the table-rated ampacity of the largest insulated conductor listed in T310-16 (normally larger than 600 amps), the service-entrance conductors must be paralleled to match the combined rated-load ampacity of the insulated conductors to the service rating.

Section 310-4 lists the requirements that must be met in the paralleling of circuit conductors. All the conductors paralleled to any one phase in the service must be of the same wire size, which is at least No. 1/0 AWG, have the same physical length, be of the same conductor material, have the same type of insulation, and be terminated in the same manner.

When the service conductors are paralleled, the size of the grounding-electrode conductor must be determined from the largest computed (summed) overall area of all the conductors paralleled to a given phase.

Example 9–3. A large manufacturing concern will be supplied by a 3-phase, 4-wire, 480Y/277-volt service rated at 800 amps. If the ungrounded service-entrance will consists of 250 kcmil THHN aluminum conductors paralleled four times per phase, determine the NE Code-minimum wire size of the aluminum grounding-electrode conductor run to the building steel.

Total circular-mil area (c.m.a.) per phase computation: Ref. 310-4

$$4 \times 250 \text{ kcmil} = 1{,}000 \text{ kcmil aluminum}$$

Grounding-electrode conductor wire size determination: Ref. 250-81(b)
Wire size: No. 4/0 AWG aluminum Ref. T250-94

Code challenge 9–3. A large manufacturing concern will be supplied by a 3-phase, 4-wire, 460Y/265-volt service rated at 600 amps. If the ungrounded service-entrance will consists of 250 kcmil THWN copper conductors paralleled three times per phase, determine the NE Code-minimum wire size of the copper grounding-electrode conductor run to the building steel.

Total circular-mil area (c.m.a.) per phase computation: Ref. (A) ____________

____________ × ____________ kcmil = (B) ____________ kcmil copper

Grounding-electrode conductor wire size determination: Ref. (C) ____________
Wire size: (D) ____________ copper Ref. (E) ____________

When wire sizes in the American Wire Gage (AWG) are to be paralleled to supply a service, the wire size must first be converted to circular-mil area and then multiplied by the number of times paralleled to correctly size the grounding-electrode conductor. The American wire gages are equated to circular-mil area values in Table 8 of Chapter 9 in the NE Code.

Example 9–4. A service rated at 600 amps, will be supplied with No. 1/0 AWG THW copper ungrounded service-entrance conductors paralleled four times per phase. Determine the NE Code-minimum wire size of the copper grounding-electrode conductor run to an underground metal tank farm.

Conversion from AWG to kcmil wire size: Ref. Ch. 9, T-8

$$\text{No. } 1/0 \text{ AWG} = 105{,}600 \text{ circular mils}$$

$$105{,}600 \text{ circular mils} \div 1{,}000 = 105.6 \text{ kcmil}$$

Total c.m.a. per phase computation: Ref. 310-4

$$4 \times 105.6 \text{ kcmil} = 422.4 \text{ kcmil copper}$$

Grounding-electrode conductor wire size determination: Ref. 250-83(b)
Wire size: No. 1/0 AWG copper Ref. T250-94

Code challenge 9–4. A service rated at 500 amps will be supplied with No. 2/0 AWG XHHW-2 c-c aluminum ungrounded service-entrance conductors paralleled four times per phase. Determine the NE Code-minimum wire size of the copper grounding-electrode conductor run to an underground metal tank farm.

Conversion from AWG to kcmil wire size: Ref. (A) ____________

No. 2/0 AWG = (B) ____________ circular mils

____________ circular mils ÷ ____________ = (C) ____________ kcmil

Total c.m.a. per phase computation: Ref. (D) ____________

____________ × ____________ kcmil = (E) ____________ kcmil copper-clad aluminum

Grounding-electrode conductor wire size determination: Ref. (F) ____________
Wire size: (F) ____________ copper Ref. (G) ____________

When the paralleling of service conductors yields an overall circular-mil area that is larger than 1,100 kcmil copper or 1,750 kcmil aluminum or c-c aluminum (the table cut-off), the grounding-electrode conductor is never required to be larger than No. 3/0 AWG copper or 250 kcmil aluminum or c-c aluminum.

Grounding-electrode conductors extended to either metal underground water pipe, other underground metal piping systems, or building steel must be full size as determined by S & T 250-94. Grounding-electrode conductors extended to the other either available or made electrodes are sized by the Exceptions to Sec. 250-94.

The concrete-encased electrode [concrete footing, Sec. 250-81(c)] must be a minimum of 20 feet of ½-inch rebar or 20 feet of No. 4 AWG copper. Section 250-94, Excp. (b) does not require the grounding-electrode conductor to be larger than No. 4 AWG copper.

The ground-ring electrode [encircles the exterior of the building, Sec. 250-81(d)] is not required to be larger than No. 2 AWG copper. Section 250-94, Excp. (c) does not require the connecting grounding-electrode conductor to be larger than the conductor used to make the electrode.

If the concrete-footing or ground-ring electrode is constructed of a larger copper conductor, the size of the grounding-electrode conductor must at least equal the size used, unless the larger electrode conductor exceeds the wire size specified in T250-94 according to the wire size of the service-entrance conductors. Table 250-94 lists the maximum wire size required for the grounding-electrode conductor. The size of the grounding-electrode conductor extended to the concrete footing or the ground ring does not have to be larger than the table listing, even if the actual electrode wire size is.

The grounding-electrode conductor for the driven rod or pipe, or for the buried-plate electrode [Secs. 250-83(c) and (d)], is not required to be larger than No. 6 AWG copper, regardless of the actual wire size of the ungrounded service conductors.

Section 250-94, Excp. (a) also allows No. 4 AWG aluminum, but the asterisk footnote to T250-94 refers the student electrician to Sec. 250-92(a) (installation of the grounding-electrode conductor). The second paragraph will not allow aluminum conductors, whether insulated or bare, to be installed in direct contact with masonry or the earth, or when installed outside, as most driven-rod electrodes are, not less than 18 inches above the earth.

These exceptions of reduced wire size apply to that part of the grounding-electrode conductor that is the sole connection between the grounding electrode and connection to the grounded service conductor. If a single grounding-electrode conductor is used to connect two or more electrodes to the grounded service bus or conductor, the exceptions are void; the grounding-electrode conductor wire size must be determined by T250-94.

If the ac power distribution is a 575-volt delta, or other voltage delta power distribution system that is not referenced to ground (floating—none of the transformer windings are connected to earth ground), Sec. 250-81 requires a grounding-electrode system to be established for the noncurrent-carrying metal parts of the building and the power equipment within the building. The NE Code-minimum wire size of grounding-electrode conductors required by Sec. 250-94 applies to both grounded and ungrounded power distribution systems.

Smaller grounding-electrode conductors, as listed in T250-94 or allowed by the Exception to Sec. 250-94, must be protected from physical damage. In some instances, installing to follow the surface of the building is protection enough. In other instances, they must be installed in conduit or some other type of piping raceway.

Section 250-92(a) lists the general protection requirements of grounding-electrode conductors. No. 8 AWG grounding-electrode conductors must always be installed in raceway. No. 6 AWG must be installed in conduit when exposed to any physical damage. No. 4 AWG must be installed in conduit when exposed to severe physical damage.

An example of any physical damage to a No. 6 AWG conductor can be found at the grounding of the service for the typical single-family dwelling. Normally, a No. 6 AWG copper grounding-electrode conductor is run exposed down the wall of the dwelling structure from the meter can to the driven-rod electrode. The connection of the conductor to the driven rod is required to be below ground level (finished grade) to protect it from physical damage. The conductor is still exposed to physical damage, due to the shield of lawn mowers or other lawn and gardening equipment, where it enters the ground off the surface of the foundation wall. The No. 6 grounding-electrode conductor must be installed in conduit.

If the No. 6 AWG is installed in metal pipe, Sec. 250-92(b) requires the metal pipe to be made part of the grounding-electrode conductor by bonding at both ends. The threaded pipe can be terminated with grounding-type bushings made up wrenchtight and jumpered to the terminations of the grounding-electrode conductor, or the bonding jumpers can be split-bolt connected to the grounding-electrode conductor when the metal pipe does not extend the full length of the grounding-electrode conductor.

If the metal pipe is not bonded to the conductor it protects, circulating currents can be induced in the pipe when current from a lightning discharge or other static source is conveyed along the length of the conductor. The pipe serves as an inductive "choke" which impedes the discharge path. The choke introduces an unwanted time delay in the response time of dissipation.

When the service neutral is conducting ground-fault overcurrent back to the transformer supply, some of the fault overcurrent is conducted along the parallel loop of the grounding-electrode system. The service-neutral conductor is purposely grounded to the earth at each end because of the wire resistance along the length of the neutral conductor. As discussed earlier, if the service neutral is not grounded at both ends, the voltage drop due to its wire resistance could appear as a voltage potential between the equipment grounding system (noncurrent-carrying metal parts of the distribution) and the grounded surfaces of the building.

If the length of protective metal pipe is not bonded to the grounding-electrode conductor, the choking effect of the ground-fault overcurrent by the metal pipe would cause the voltage potential of the neutral conductor to return in reference to the building surfaces until the fault condition is relieved.

If the grounding-electrode conductor is installed in PVC conduit, the choking affect is eliminated.

The No. 4 AWG copper grounding-electrode conductor can stand a little more abuse than the No. 6 grounding-electrode conductor. An example of its exposure to severe physical damage is also found in the single-family dwelling.

The interior metal water pipe in the dwelling structure must be bonded to the service grounding-electrode system as if it were a grounding electrode. Section 250-80(a) requires the bonding jumper to be sized according to the size of the service-entrance conductors and T250-94.

If No. 4/0 AWG aluminum SE cable is used for the service drop of a single-phase, 3-wire, 120/240-volt, 200-amp, dwelling service, T250-94 would require a No. 4 AWG copper bonding jumper to be extended from the grounded bus in the service-equipment panel to the interior metal water pipe.

If the No. 4 AWG conductor is run perpendicularly across the bottom of the floor joist in an unfinished basement, the wire would be exposed to possible severe physical damage in that hand garden tools or stored winter clothes could be physically hung from the wire. The wire would have to be installed in conduit or some other type of piping raceway.

On the other hand, if the same No. 4 AWG bonding jumper is run along the side of the same floor joist, or along the side of the floor seal or plate, it would not be exposed to severe physical damage and as a result, it would not have to be installed in protective raceway.

When the service panel is installed in an industrial plant, even wire size 250 kcmil copper-clad aluminum, the largest size required by T250-94 for the grounding-electrode conductor, can be exposed to severe physical damage. As shown in Figure 9-16, the protective raceway must be made a part of the grounding-electrode conductor by being connected common (parallel) to it at both ends.

9.8 BONDING OF INTERIOR METAL WATER PIPING

The advances in the use of PVC or other types of nonmetallic tubing or piping for potable water in a building or other structure has introduced additional problems or hazards besides impairing the use of metal underground water pipe as the main service grounding electrode.

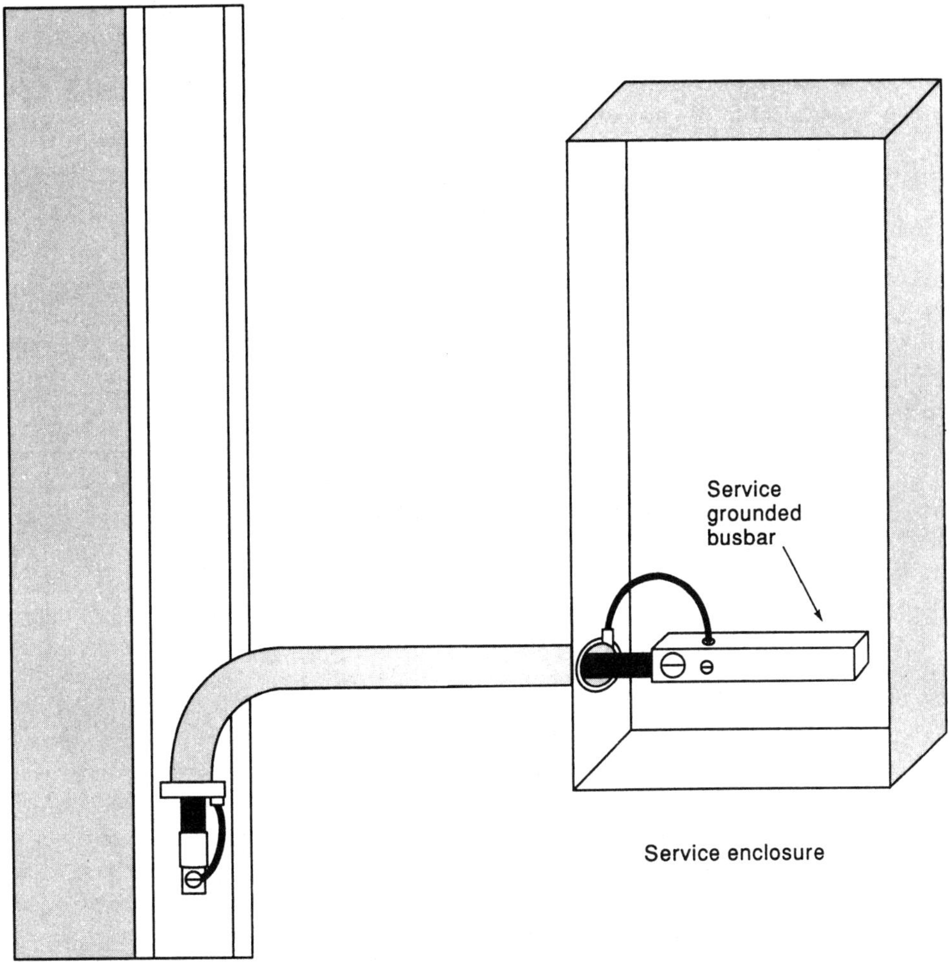

Fig. 9-16 Bonding jumpers are required at both ends of protective conduit, which encloses a grounding electrode conductor.

Section 250-80(a) requires interior metal water pipe to be bonded to the service grounding-electrode system because Sec. 250-81 recognizes it as the common bus for interconnection of multiple grounding electrodes (especially in the industrial/commercial sector). No mention is made of grounding PVC or other types of interior nonmetallic water piping because everyone knows that you can't ground PVC.

A serious shock hazard arises with the use of immersible heating elements in an electric water heater and nonmetallic potable water pipe—the water is not grounded.

Sealed resistive heating elements are immersed through the tank wall (either base bolted or screwed into a flange) into the inside of a storage-type water heater for direct element contact of the heating surface to the pressurized water. Direct contact yields maximum heat transfer.

The hazard arises when a lightning strike ruptures the heating element and the two ends of the now-open heating element are in contact with the water. The water is being held at line potential with any grounded surface in the building, or for outside spigots, with earth ground. The ruptured element will hold the water contained in both piping systems, the hot water distribution as well as the cold water inlet/distribution, at line potential because with the use of nonmetallic piping the water is no longer bonded to the service grounding-electrode system.

Good electrical design would purposely insert a short section of copper or other metal water piping into both the hot and cold nonmetallic water piping distribution systems to reference (bond) the conductive water back to the building service grounding-electrode system.

The ideal location would be as close to the potential ruptured element as possible, which would be a short stubup from the inlet and outlet ports of the electric water heater. This is not technically an NE Code requirement, although the last sentence of Sec. 250-80(a) requires the bonding connection when the sections of metal pipe are installed to be accessible.

The hazard of interior pressurized water being held at line potential can be realized at the kitchen sink of the single-family dwelling, which is more often than not a stainless steel sink that supports a garbage disposal. The sink is grounded through the separate equipment-grounding conductor in the branch-circuit supply of the disposal. A person merely washing his or her hands is subject to a severe electrical shock if the water is not purposely referenced with the service grounding-electrode system. Other scenarios can be created about the clothes washing machine in the laundry area or the wash basin/shower stall in the common bathroom.

9.9 WIRE SIZE—SERVICE NEUTRAL CONDUCTOR

As discussed in Sec. 9.5, the grounded service neutral conductor has a dual role in the power distribution: It is a current-carrying conductor under unbalanced load conditions of the building power distribution, and it is used as the ground reference or ground-return path for static source discharge (lightning) along the length of the service conductors.

Because it is referenced to earth ground, the service neutral conductor is also used as the ground-fault return path for the building service equipment and for the power distribution within the building when any ungrounded circuit conductor is accidentally or purposely placed in a faulted overcurrent condition with any grounded surface of the building or other structure.

In reference to Sec. 220-22, the grounded service neutral conductor must be sized according to the maximum unbalanced load current that can occur due to varying load conditions. In reference to Sec. 250-23(b), the same grounded service neutral conductor must also be sized according to the maximum ground-fault overcurrent available from the transformer at the load-end of the service conductors.

From the aspect of maximum unbalanced load current, the grounded neutral conductor is connected common to both windings in the single-phase, 3-wire service, common to all three windings in the 3-phase, 4-wire, wye services, and common to A and C windings in the 3-phase, 4-wire, high-leg delta service. The purpose of this common neutral connection is to conduct the unbalanced load current when the single-phase, line-to-neutral current loads are not of equal value across the phases (ungrounded circuit conductors—other ends of the transformer windings).

The maximum-load current of the service neutral conductor occurs when the phase with the highest value of neutral-load current is operating at full-load condition, and the neutral current loads on the other phase or phases are opened or turned off (no-load condition).

The minimum-load current occurs when the phase-to-neutral loads are not balanced (not equal in current value), but are all at their full-load condition. When the neutral loads are balanced (equal value) across the phases, the service neutral load current is equal to zero.

Section 220-22 lists the guidelines for determining the maximum unbalanced load current. Unbalanced loading will be discussed with the different types of occupancy services later in the text.

A corner-ground delta operates without a common neutral conductor or unbalanced load current. Instead, the B-phase service conductor is referenced to earth ground. For load requirements, the B-phase conductor normally carries the same load current as the ungrounded A and C conductors, and must be sized accordingly.

For ground-fault ampacity requirements, the service neutral or the B-phase grounded circuit conductor is sized by Sec. 250-23(b). Ground fault actually indicates the ground-path return of high-magnitude overcurrent should an ungrounded phase conductor come in contact with a grounded surface within the building or other structure, and the load impedance be effectively shunted (bypassed).

As discussed in Secs. 9.2 and 9.3, if the building is constructed of conductive materials, the building bonding grid and grounding-electrode system must be interconnected with (bonded to) the grounded service neutral conductor.

As the noncurrent-carrying metal parts of the electrical power distribution, which includes all the metal raceways, boxes, load centers (switchgear, panels, motor control centers, and so on), and other enclosures that either support wiring, or contain overcurrent-protection devices, outlet devices, or utilization equipment (motors, lighting fixtures, and so on), are installed within the building or other structure, they must be interconnected to each other to form a bonding grid.

This bonding grid (Fig. 9-17) must be purposely bonded with wire jumpers to the grounded terminal strip or busbar (service neutral) in the service equipment enclosure to ensure that all the noncurrent-carrying metal parts of the power distribution have a low-impedance, ground-fault overcurrent return path back to the transformer windings along the service neutral conductor. This is the equipment-grounding system (means).

The service neutral conductor is an intentionally grounded, current-carrying circuit conductor that is bonded to the established bonding grid of both the building surfaces and the noncurrent-carrying metal parts of the electrical equipment. The interconnection of the bonding grids and grounding-electrode systems with the service neutral conductor ensures that a voltage potential, between the earth and any metal enclosure, or between a grounded surface within the building or other structure and any metal enclosure cannot exist.

The equipment-grounding means for all the noncurrent-carrying metal parts of the distribution has to be established, regardless of whether or not the building or other struc-

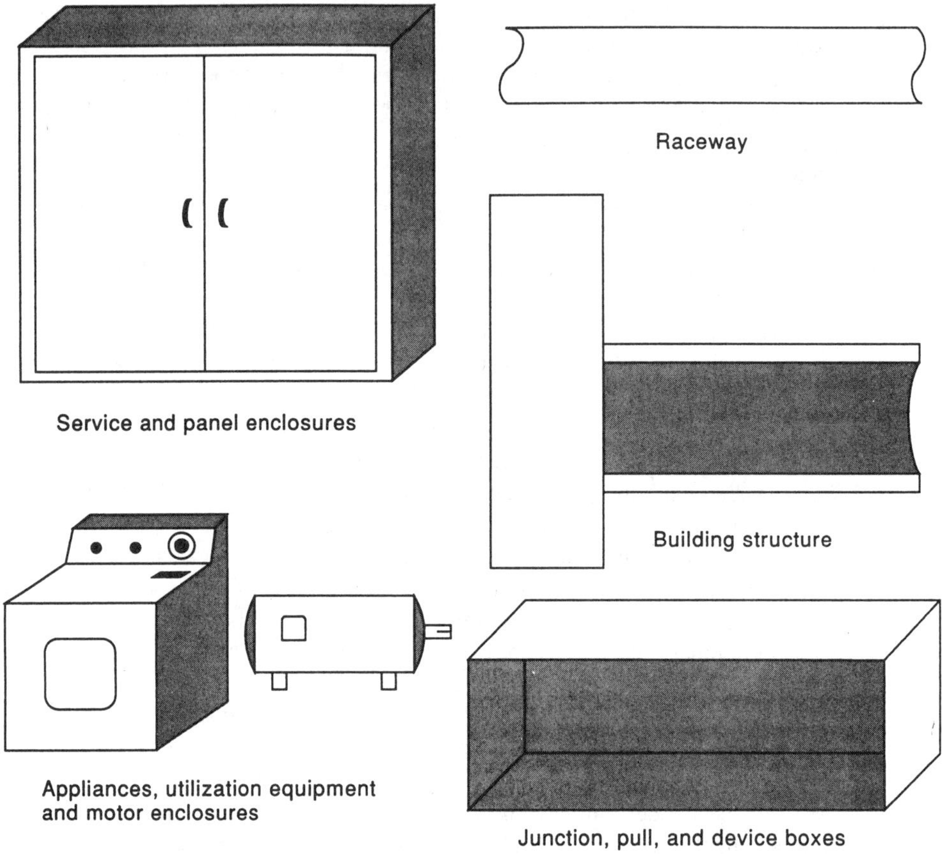

Fig. 9-17 All the noncurrent-carrying metal parts within a building must be grounded by means of the service grounding electrode system.

ture has an established bonding grid and grounding-electrode system. Wooden structures, such as the typical dwelling, are not constructed of materials that require a bonding grid and grounding-electrode system for lightning or other static discharge (other than older farmhouses with tin roofs). All the surfaces within a given building, such as the surface of the floor or walls, where a person could be standing or touching, are considered to be grounded surfaces in reference to the ungrounded power distribution conductors within the structure.

If the metal enclosure housing the circuit conductors should come in contact with one of the ungrounded circuit conductors (or vice versa) and the metal enclosure was not bonded (grounded) to the service neutral conductor, the enclosure would be held at the same voltage potential as the faulted circuit conductor. A person accidentally or unknowingly coming in contact with the metal enclosure is subject to severe shock and possible electrocution.

When the metal enclosure housing the circuit conductors is bonded to the service neutral conductor, the metal is provided with a low-impedance fault-return path back to the secondary of the transformer. In other words, the accidental contact of the ungrounded circuit conductor to the grounded metal enclosure places a short circuit across the affected winding of the transformer, which will cause high-magnitude, ground-fault overcurrent to flow. The fuse or circuit breaker installed at the supply end of the affected ungrounded circuit conductor will clear (open) the circuit because of the overcurrent condition.

Current must take every path when parallel paths are available to (or from) a power source. If a person should be touching the metal enclosure while the ground fault was in progress, it seems reasonable that he or she standing on the grounded floor of the building is providing a parallel path for current flow and as such would be subject to electrical shock. But the majority of electrical current is transferred by the path of least resistance. If the metal enclosure is effectively bonded to the service neutral conductor, it offers a very low-resistance path when compared to the body's exterior resistance. The shock would be minimal, if even felt.

In some instances a service will not have the lower voltage phase-to-neutral loads. It seems reasonable that the neutral is not required. Consider the service for a remote pumping station that consists of only three-phase motors operating at 480 volts, with lighting and required receptacle outlets supplied through a separately derived 208Y/120-volt system. The parent 480-volt system does not have a neutral load.

If this were a motor-circuit feeder, only the three ungrounded phase conductors would have to be extended to the power center. Because this is service equipment, Sec. 250-23(b) requires the neutral conductor for the ground-fault return of the equipment grounding means.

The service-entrance neutral conductor, from the point of entry or attachment to the building to termination to a bus in the service equipment is the common tie point for all ground-fault return on both the supply and load sides of the service equipment. For ground-fault ampacity requirements, Sec. 250-23(b) requires the wire size of the grounded service neutral conductor to at least equal the required wire size of the grounding-electrode conductor as determined by T250-94.

Example 9–5. A single-phase, 3-wire, 120/240-volt service rated at 200 amps, will be supplied with 250 kcmil THW aluminum ungrounded service-entrance conductors. If the service has only minimum phase-to-neutral loading, determine the NE Code-minimum wire size of the service-entrance neutral conductor for ground-fault ampacity requirements.

Service-neutral ground-fault ampacity requirements: Ref. 250-23(b)
Wire size: No. 2 AWG THW aluminum Ref. T250-94

Unlike the grounding-electrode conductor, the material for the service conductors should be the same: copper for copper, aluminum for aluminum, or copper-clad (c-c) aluminum for c-c aluminum.

Code challenge 9–5. A 3-phase, 4-wire, 240/120-volt service rated at 225 amps, will be supplied with No. 4/0 AWG RHW copper ungrounded service-entrance conductors. If the service has only minimum phase-to-neutral loading, determine the NE Code-minimum wire size of the service-entrance neutral conductor for ground-fault ampacity requirements.

Service-neutral ground-fault ampacity requirements: Ref. (A) ________________
Wire size: (B) ____________ RHW copper Ref. (C) ________________

Services that require ungrounded circuit conductors (phase conductors) larger than 750 kcmil are normally paralleled with smaller conductors to meet the required circuit ampacity. When conductors are paralleled as allowed by Sec. 310-4, the rated ampacity of the chosen wire size is multiplied by the number of times paralleled to determine the total conductor ampacity.

To determine the NE Code-minimum wire size of the grounded service neutral conductor for ground-fault ampacity requirements, the individual wire size or rather the equivalent circular-mil area, is multiplied by the number of times paralleled to determine the total single-conductor service-conductor circular-mil area per phase.

As outlined in Example 9-4, when AWG wire sizes are paralleled, Table 8 in Chapter 9 is used to convert the wire size into circular-mil area.

If all the paralleled phase conductors are installed in a common enclosure, such as an auxiliary gutter or wireway to meet the conductor grouping requirements of Sec. 300-20(a), a single service neutral conductor, size according to T250-94, can be installed to meet the ground-fault ampacity requirements of Sec. 250-23(b).

Example 9–6. A service rated at 1,000 amps, will be supplied with 250 kcmil THW copper ungrounded service-entrance conductors paralleled four times per phase. If the service has only minimum phase-to-neutral loading and all the parallel service-entrance conductors will be installed in a common gutter, determine the NE Code-minimum wire size of a single service-entrance neutral conductor for ground-fault ampacity requirements.

Total c.m.a. per phase computation: Ref. 310-4

$$4 \times 250 \text{ kcmil} = 1{,}000 \text{ kcmil copper}$$

Service-neutral ground-fault ampacity requirements: Ref. 250-23(b)
Wire size: No. 2/0 AWG THW copper Ref. T250-94

A single service-entrance neutral conductor can also be used for either aluminum or c-c aluminum service conductors, the window of T250-94, 1,750 kcmil, versus 1,100 kcmil is a bit larger for aluminum conductors.

Code challenge 9–6. A service rated at 1,000 amps, will be supplied with 350 kcmil THW aluminum ungrounded service-entrance conductors paralleled four times per phase. If the service has only minimum phase-to-neutral loading and all the parallel service-entrance conductors will be installed in a common gutter, determine the NE Code-minimum wire size of a single service-entrance neutral conductor for ground-fault ampacity requirements.

Total c.m.a. per phase computation: Ref. (A) ________________

____________ × ____________ kcmil = (B) ____________ kcmil aluminum

Service-neutral ground-fault ampacity requirements: Ref. (C) ________________
Wire size: (D) ____________ THW aluminum Ref. (E) ________________

When the largest wire size of a single ungrounded service conductor, or the largest total phase circular-mil area when the service conductors are paralleled by phase, exceeds 1,100 kcmil copper, or 1,750 aluminum or c-c aluminum, the grounding-electrode conductor, in reference to T250-94, is not required to be larger than either No. 3/0 AWG copper, or 250 kcmil aluminum or c-c aluminum.

In comparison, Sec. 250-23(b) requires the wire size of the grounded service neutral conductor for these higher-amperage services to be a minimum of ⅛ the circular-mil area of either the largest ungrounded phase conductor, or largest total c.m.a when the ungrounded circuit conductors are paralleled by phase.

When the one-eighth computation (multiplied by 0.125), which is equal to the 12½% requirement of Sec. 250-23(b), does not correspond to a standard wire size, Table 8 of Chapter 9 is used to select the next higher (larger) standard wire size for the service neutral.

Example 9–7. A service rated at 600 amps will be supplied with 1,500 kcmil THW copper ungrounded service-entrance conductors. If the service has only minimum phase-to-neutral loading, determine the NE Code-minimum wire size of the grounded service-entrance neutral conductor for ground-fault ampacity requirements.

Service-neutral ground-fault ampacity requirements: Ref. 250-23(b)
Minimum c.m.a. computation:

$$0.125 \times 1{,}500 \text{ kcmil} = 187.5 \text{ kcmil copper}$$

Wire size: (NHS) No. 4/0 AWG THW copper Ref. Ch. 9, T-8

Code challenge 9–7. A service rated at 650 amps will be supplied with 1,750 kcmil THW copper ungrounded service-entrance conductors. If the service has only minimum phase-to-neutral loading, determine the NE Code-minimum wire size of the grounded service-entrance neutral conductor for ground-fault ampacity requirements.

Service-neutral ground-fault ampacity requirements: Ref. (A) ________________
Minimum c.m.a. computation:

____________ × ____________ kcmil = (B) ____________ kcmil copper

Wire size: (NHS) (C) ____________ THW copper Ref. (D) ________________

If the ungrounded service conductors are paralleled in a common gutter or other large raceway enclosure, and the largest total c.m.a. per phase exceeds the cut-offs listed in T250-94, the single grounded service neutral conductor c.m.a. must be equal to at least ⅛ the total ungrounded conductor c.m.a. (same manner as with the larger single ungrounded service conductors).

Example 9–8. A service rated at 1,600 amps will be supplied with 300 kcmil THWN copper ungrounded service-entrance conductors paralleled six times per phase. If the service has only minimum phase-to-neutral loading and all the parallel service-entrance conductors will be installed in a common gutter, determine the NE Code-minimum wire size of the single grounded service-entrance neutral conductor for ground-fault ampacity requirements.

Total c.m.a. per phase computation: Ref. 310-4

$$6 \times 300 \text{ kcmil} = 1{,}800 \text{ kcmil copper}$$

Service-neutral ground-fault ampacity requirements: Ref. 250-23(b)
Single-neutral conductor c.m.a. computation:

$$0.125 \times 1{,}800 \text{ kcmil} = 225 \text{ kcmil copper}$$

Wire size: (NHS) 250 kcmil THW copper Ref. Ch. 9, T-8

If the phase conductors are paralleled more than eight times, the service neutral, because it is scaled at ⅛ of the largest phase area, would compute to a larger wire size than what is installed for the phase conductors. Exception No. 1 to Sec. 250-23(b) corrects this situation by not requiring the grounded service neutral conductor to be larger than the largest ungrounded service conductor.

Code challenge 9–8. A service rated at 2,400 amps will be supplied with 900 kcmil THW aluminum ungrounded service-entrance conductors paralleled six times per phase. If the service has only minimum phase-to-neutral loading and all the parallel service-entrance conductors will be installed in a common gutter, determine the NE Code-minimum wire size of the single grounded service-entrance neutral conductor for ground-fault ampacity requirements.

Total c.m.a. per phase computation: Ref. (A) ________________

___________ × ___________ kcmil = (B) ___________ kcmil aluminum

Service-neutral ground-fault ampacity requirements: Ref. (C) ________________
Single-neutral conductor c.m.a. computation:

___________ × ___________ kcmil = (D) ___________ kcmil aluminum

Wire size: (NHS) (E) ___________ THW aluminum Ref. (F) ________________

Although there's nothing in the NE Code that prevents all the parallel service conductors from being installed in a common gutter or wireway as described in these previous examples, the more common method of parallel conductor installation is a set of service conductors installed one (set) each in the same number of parallel conduits or other piping raceway. One each of the ungrounded service phase conductors and the grounded service neutral conductor are installed in each parallel raceway. All are terminated in the common gutter area at the transformer supply and at the service-equipment panel.

As with the single grounded service neutral conductor, Sec. 250-23(b) requires the total c.m.a. of all the paralleled service neutral conductors to be at least equal ⅛ of the largest total conductor c.m.a. connected to any one phase. The c.m.a. and resulting wire size of each individual parallel neutral conductor is determined by dividing the ⅛ computation by the number of times the service neutral is paralleled.

Since the total c.m.a. per phase is computed by multiplying by the total number of ungrounded service conductors in parallel per phase, and the individual, parallel service neutral conductor c.m.a. is computed by dividing the ⅛ computation by the same number, it is easier just to take the ⅛ of the wire size or c.m.a. of the largest ungrounded service conductor connected to any one phase in the parallel to compute the required individual grounded service neutral conductor circular-mil area.

The purpose of computing the total c.m.a. per phase then is only to determine if the table cut-off of 1,100 kcmil copper, or 1,750 aluminum or c-c aluminum has been exceeded and the ⅛ computation is warranted.

If the ⅛ computation of the individual parallel neutral conductor c.m.a. yields a wire size less than No. 1/0 AWG, the requirements of Sec. 310-4, by way of 310-15(c), override Sec. 250-23(b): Each individual service neutral conductor in the parallel cannot be smaller than wire size No. 1/0 AWG.

Example 9–9. A service rated at 1,600 amps will be supplied with 600 kcmil THHN aluminum ungrounded service conductors paralleled five times per phase. If the service has only minimum phase-to-neutral loading and all the parallel service-entrance conductors will be installed as sets in five parallel rigid conduits, determine the NE Code-minimum wire size of the individual parallel grounded service-entrance neutral conductors for ground-fault ampacity requirements.

Total c.m.a. per phase computation: Ref. 310-4

$$5 \times 600 \text{ kcmil} = 3{,}000 \text{ kcmil aluminum}$$

Service-neutral ground-fault ampacity requirements: Ref. 250-23(b)
Individual parallel-neutral conductor c.m.a. computation:

$$0.125 \times 600 \text{ kcmil} = 75 \text{ kcmil aluminum}$$

Wire size: (NHS) No. 1 AWG THHN aluminum Ref. Ch. 9, T-8
Wire size for paralleling: No. 1/0 AWG Ref. 310-4
NE Code minimum, compare for larger: Ref. 310-15(c)
Wire size: No. 1/0 AWG THHN aluminum

Code challenge 9–9. A service rated at 1,200 amps will be supplied with 250 kcmil THHN copper ungrounded service conductors paralleled five times per phase. If the service has only minimum phase-to-neutral loading and all the parallel service-entrance conductors will be installed as sets in five parallel rigid conduits, determine the NE Code-minimum wire size of the individual parallel grounded service-entrance neutral conductors for ground-fault ampacity requirements.

Total c.m.a. per phase computation: Ref. (A) ______________

__________ × __________ kcmil = (B) __________ kcmil copper

Service-neutral ground-fault ampacity requirements: Ref. (C) ______________

Individual parallel-neutral conductor c.m.a. computation:

__________ × __________ kcmil = (D) __________ kcmil copper

Wire size: (NHS) (E) __________ THHN copper Ref. (F) ______________

Wire size for paralleling: (G) __________ Ref. (H) ______________

NE Code minimum, compare for (I) __________: Ref. (J) ______________

Wire size: (K) __________ THHN copper

When run in parallel conduits, the individual service neutral conductors, again for ground-fault ampacity requirements, are never allowed to be smaller than No. 1/0 AWG to satisfy the requirements of Sec. 310-4. By the same token, the individual service neutral conductors would never be required to be larger than No. 1/0 AWG, unless the total c.m.a. of the largest individual paralleled phase conductor exceeds 800 kcmil: According to Table 8 of Chapter 9, the circular-mil area of a No. 1/0 AWG conductor is 105.6 kcmil. The one-eighth computation of anything larger than 800 kcmil will exceed this area value.

For the typical service installation, the maximum unbalanced load computations, outlined in Sec. 220-22, will normally require a wire size for the service neutral conductor that is larger than the minimum required for ground-fault return, but both should be checked. As indicated in Sec. 310-15(c), when two different load computations dictate the minimum wire size of a circuit conductor, the largest wire size by comparison becomes NE Code minimum.

9.10 THE EQUIPMENT GROUNDING SYSTEM

As discussed for the ground-fault ampacity requirements of the service neutral conductor, all the noncurrent-carrying metal parts of the electrical distribution system to or within a building or other structure must be effectively bonded together to form a grid. The grid in turn must be bonded to the grounded service conductor either in the service equipment panel or on the supply side of the service disconnecting means.

Caution must be taken when multiple connections to the service neutral conductor are used for equipment grounding. All the metal pipe and enclosures that contain circuit conductors are interconnected throughout the distribution. All grounding and bonding connections must be "single-ended" to the service neutral conductor to ensure a section of pipe raceway or metal enclosure is not haphazardly connected in parallel with a section of the service-entrance neutral conductor.

In the typical dwelling or small-business service, the service-entrance conductors are series fed through a meter can for measure of the amount of electrical power consumed. If the meter can is interconnected to the service panel by way of metal pipe, the metal pipe could become a parallel neutral conductor by grounding to the service neutral conductor at any convenient point.

The meter can is normally bonded to the neutral terminal in the meter can. The panel enclosure is normally bonded to the neutral grounded busbar in the service panel. Conveniently bonded in this manner, the metal pipe is technically wired as a parallel neutral by the individual grounding of the two separate metal enclosures. Since current must take every available path, the metal raceway is placed in parallel with the service neutral conductor—it becomes a current-carrying conductor.

The grounded bus in the service equipment panel should be used as the common tie point for all grounding and bonding of the equipment. All grounding connections should be single-ended to avoid the possible parallel neutral condition just described.

If the building or other structure already has an established bonding grid and grounding-electrode system (buildings constructed of brick, block, or metal normally already contain a bonding grid connected to a grounding-electrode system) and the service neutral conductor has been regrounded to it, then the noncurrent-carrying metal parts of the electrical distribution are technically grounded by their mechanical supports, which are either connected or mounted to the grounded surfaces of the building.

But mounting supports, such as metal straps, channel, or pipe hangers, cannot be depended upon as mechanical pressure connections. The noncurrent-carrying metal enclosures (parts) of the power distribution must be bonded common to each other (by jumpers and pressure-type wire terminals) to form a bonding grid. This second bonding grid is tied common (connected) to the building bonding grid and the building grounding-electrode system through the grounded busbar or terminal strip (neutral) in the service equipment enclosure.

The interconnection of metal raceway and other metal enclosures used in the power distribution, which house (contain) individual insulated circuit conductors, is allowed by Sec. 250-91(b) to serve as the required equipment grounding system for the metal enclosures of lighting fixtures, motors, and other equipment, provided the raceway is bonded (connected) to the grounded service neutral conductor or busbar at the circuit's supply end (in the service equipment enclosure).

If the electrical distribution is allowed by the NE Code to be a wiring method that does require raceway (cabling), or if the raceway is made of nonmetallic materials (such as PVC), the cabling, or the nometallic raceway must contain a separate equipment-grounding conductor to connect the remote metal enclosures of the various loads or other outlet points to the grounded service neutral busbar. This separate grounding means provides a low-impedance return path for ground-fault currents when an ungrounded circuit conductor comes in contact with the isolated metal surfaces.

Because the metal raceway interconnecting the remote metal enclosures is subject to both physical damage (when run exposed along the length of a wall) and high-resistance connections (rust or loose makeup connections), the better electrical installation will contain a separate equipment-grounding conductor that is installed in the raceway, as in PVC or cabling wiring methods, and run with the circuit conductors. Wired (connected) in parallel with the metal raceway grounding means by terminal connection to both the metal enclosure at the load end and to the service neutral conductor or bus at the supply end, this equipment-grounding conductor provides a positive (separate) grounding means (low-impedance return path) for the metal enclosures of the load equipment.

If the building contains metal piping systems such as interior metal water pipe [Sec. 250-80(a)] or metal ductwork [Sec. 250-80(b)], these metal piping systems, as well as exposed building steel [Sec. 250-80(c)], because they could become electrically energized, must also be bonded to the grounded service neutral conductor.

Because the service neutral conductor is interconnected with the bonding grid and the grounding-electrode system of the building or other structure, any ungrounded fused conductor of the distribution that has a potential difference with the neutral conductor, has the same potential difference with the grounded surfaces of the floor, walls, support beams of the building (building steel), and the metal raceway, metal boxes, or other metal enclosures that support the wiring distribution or the connected loads.

This interconnection with the service neutral conductor assures that neither stray voltages nor stray currents (referenced to ground), which are a shock hazard to personnel, can remain on the normally noncurrent-carrying metal parts of the power distribution or other grounded surfaces within the building, as it simultaneously provides a low-impedance, ground-return path for fault overcurrents of the distribution.

The wire size of the equipment-grounding conductor is determined by both the magnitude of the available ground-fault overcurrent and its duration. Equipment-grounding conductors or bonding jumpers used on the supply side of the service-disconnecting means have a higher magnitude of ground-fault overcurrent available as compared to subfeeder or branch-circuit equipment-grounding conductors (on the load side of the service equipment).

The greatest concentration of fault overcurrent available is at the secondary terminals of the transformer. Every conductor termination or foot of conductor length in the service conductors adds some minute resistance to the circuit. At fault levels, this minute resistance drastically reduces the magnitude of the overcurrent; however, a high value of ground-fault overcurrent is normally still available at the terminals of the service-disconnecting means.

As the power is further disseminated along the length of feeder or branch-circuit conductors, the available ground-fault overcurrent dwindles due to the increasing wire resistance. The resistance involved at the point of contact also serves to limit the ground-fault overcurrent.

Another governing factor that requires a distinction between line-side (supply side) and load-side equipment-grounding conductors is the location of the overcurrent-protection device that must open to relieve the fault. The circuit breaker or fuses that are integral with the service-disconnecting means serve only as overload-overcurrent protection for the service conductors (load end). The short-circuit, ground-fault overcurrent protection of the service conductors must be located ahead of the service conductors.

Due to lightning strikes, the secondary of the service-supply transformer, especially those that are pole-mounted, is normally not fused at the transformer location. Short-circuit or ground-fault overcurrent protection of the secondary service conductors is provided at the supply end of the primary trunk line conductors.

The service conductors are also subject to a longer duration of the high-magnitude fault overcurrent. In addition to the passing of lightning discharge, the primary overcurrent-protection device (OCPD) is designed with the thought of "self clearing." In other words, to reduce the occurrence of service interrupt or outage, utility companies normally require the primary overcurrent-protection device to pass fault overcurrent through a repeater set-up to allow the fault, which could be a limb of a tree or other vegetation, to literally burn off the line before the fuse or other protection device (current-sensing relays) opens to electrically disable the line.

The service overload OCPD provides full overcurrent protection for any fault condition on its load side (panel busbars). In reference to Sec. 240-3, subfeeder or branch circuits can only be tapped to the ungrounded service busbars through another OCPD. Any load-side faults on feeder or branch circuits are actually series-fed through a minimum of two OCPDs: the service overload-overcurrent protection device and the feeder or branch-circuit full overcurrent protection device.

As discussed in Chapter 7, the ratings of the series-fed OCPDs must be downscaled in terms of load current and interrupt ratings to maintain selective coordination of the devices (outage by the fault must be localized by response of only the OCPD immediately upstream).

Section 250-79 divides the computation or determination of equipment-grounding conductors into two major categories: either line side or load side of the service overcurrent-protection device.

On the line or supply side of the service equipment [Sec. 250-79(d)], equipment-grounding conductors or bonding jumpers are sized according to the largest wire size (total circular-mil area when paralleled) of the ungrounded service-entrance conductors.

On the load side of the service equipment [Sec. 250-79(e)], equipment-grounding conductors or bonding jumpers are sized according to the rating of the feeder or branch-circuit OCPD, which must open in the event of a fault.

9.11 LINE-SIDE EQUIPMENT GROUNDING

The wire size of bonding jumpers or equipment-grounding conductors on the line side of the service-disconnecting means is determined in the same manner as for the grounded service neutral conductor in terms of ground-fault ampacity requirements. Section 250-79(d) requires the bonding jumper or other grounding conductor to be at least equal in wire size to the required grounding-electrode conductor as determined by T250-94 up to 1,100 kcmil copper or 1,750 aluminum or c-c aluminum.

Example 9–10. A 3-phase, 4-wire, 240/120-volt, high-leg delta service rated at 400 amps will be supplied with 600 kcmil XHHW-2 copper ungrounded service-entrance conductors. Determine the NE Code-minimum wire size of the copper bonding jumper used to ground the service-equipment enclosure.

Line-side equipment-grounding requirements:	Ref. 250-79(d)
Wire size: No. 1/0 AWG copper	Ref. T250-94

Code challenge 9–10. A 3-phase, 4-wire, 208/120-volt, wye service rated at 225 amps will be supplied with 300 kcmil RHW c-c aluminum ungrounded service-entrance conductors. Determine the NE Code-minimum wire size of the c-c aluminum bonding jumper used to ground the service-equipment enclosure.

Line-side equipment-grounding requirements: Ref. (A) ____________
Wire size: (B) ____________ c-c aluminum Ref. (C) ____________

As with grounding-electrode conductors, the bonding jumper does not have to be of the same material as the service-entrance conductors. Instead of using the table directly as before with the wire size of the ungrounded service conductors establishing the row and the grounding-electrode conductor material establishing the column, Sec. 250-79(d) stipulates that when the phase conductors and supply-side bonding jumpers are of different materials, the wire size for the phase conductors has to be recomputed using the same conductive material as the bonding jumper.

Example 9–11. As allowed by the Note 3 table to T310-16, a 200-amp dwelling service can be supplied with No. 4/0 AWG aluminum SE cable. Determine the NE Code-minimum wire size of a copper bonding jumper run to ground the meter can.

Line-side equipment-grounding requirements: Ref. 250-79(d)
Ungrounded service-entrance conductor wire size based on copper:
Wire size: No. 2/0 AWG copper Ref. T310-16, N-3 Tbl
Bonding jumper wire size: No. 4 AWG copper Ref. T250-94

Code challenge 9–11. As allowed by the Note 3 table to T310-16, a 100-amp dwelling service can be supplied with No. 2 AWG aluminum SE cable. Determine the NE Code-minimum wire size of a copper bonding jumper run to ground the meter can.

Line-side equipment-grounding requirements: Ref. (A) ____________
Ungrounded service-entrance conductor wire size based on (B) ____________:
Wire size: (C) ____________ copper Ref. (D) ____________
Bonding jumper wire size: (E) ____________ copper Ref. (F) ____________

When the service-entrance conductors are paralleled in parallel metal raceways, locknuts, used to interconnect the pipe raceway to the metal service enclosure(s), cannot be used as the sole bonding/grounding means of the service raceway. Again, due to the magnitude of the ground-fault overcurrent available and the limited overcurrent protection of service conductors, Sec. 250-72 requires a pressurized (positive) connection of the bonding/grounding conductor to the metal pipe. Normally, a bonding-type locknut or a standard locknut used in conjunction with a bonding-type bushing or locknut wedge is used to bond service raceway. (See also Sec. 250-113.)

If individual bonding jumpers are extended to the parallel conduits, the last sentence in the first paragraph of Sec. 250-79(d) will allow the wire size of the bonding jumpers to be based on the wire size of the service-entrance conductors contained in the pipe to be bonded.

Example 9–12. A service rated at 1,000 amps will be supplied with 350 kcmil THW c-c aluminum ungrounded service-entrance conductors paralleled four times per phase. Determine the NE Code-minimum wire size of the c-c aluminum bonding jumpers extended one each to the four parallel conduits.

Line-side equipment-grounding requirements: Ref. 250-79(d)
Bonding jumper wire size: No. 1/0 AWG c-c aluminum Ref. T250-94

Unlike the grounded service neutral conductors when paralleled in individual piping raceway and each individual neutral conductor has to be at least wire size No. 1/0 AWG, individual bonding jumpers run to the individual conduits are not considered to be in parallel. (Technically they are, since the metal raceways are mechanically made up to the common service enclosure by the cleats of the double locknuts.) The required circular-mil area or wire size of the individual bonding jumpers is not based on the overall equivalent area of the service conductors connected to any one phase, but on the wire size or circular-mil area of the largest ungrounded service-entrance conductor contained in any one of the parallel piping raceways.

Code challenge 9–12. A service rated at 1,000 amps will be supplied with 250 kcmil THW copper ungrounded service-entrance conductors paralleled four times per phase. Determine the NE Code-minimum wire size of the copper bonding jumpers extended one each to the four parallel conduits.

Line-side equipment-grounding requirements: Ref. (A) ________________
Bonding jumper wire size: (B) ____________ copper Ref. (C) ________________

A single bonding jumper loop-connected to ground all four conduits for the same service would be sized in the same manner as the equipment-grounding conductor run to the metal service enclosure (Fig. 9-18).

Example 9–13. A service rated at 1,000 amps will be supplied with 350 kcmil THW c-c aluminum ungrounded service-entrance conductors paralleled four times per phase. Determine the NE Code-minimum wire size of a single c-c aluminum bonding jumper loop-connected to ground all four parallel conduits.

Total c.m.a. per phase computation: Ref. 310-4

$$4 \times 350 \text{ kcmil} = 1{,}400 \text{ kcmil c-c aluminum}$$

Line-side equipment-grounding requirements: Ref. 250-79(d)
Bonding jumper wire size: No. 4/0 AWG c-c aluminum Ref. T250-94

Code challenge 9–13. A service rated at 1,000 amps will be supplied with 250 kcmil THW copper ungrounded service-entrance conductors paralleled four times per phase. Determine the NE Code-minimum wire size of a single copper bonding jumper loop-connected to ground the four parallel conduits.

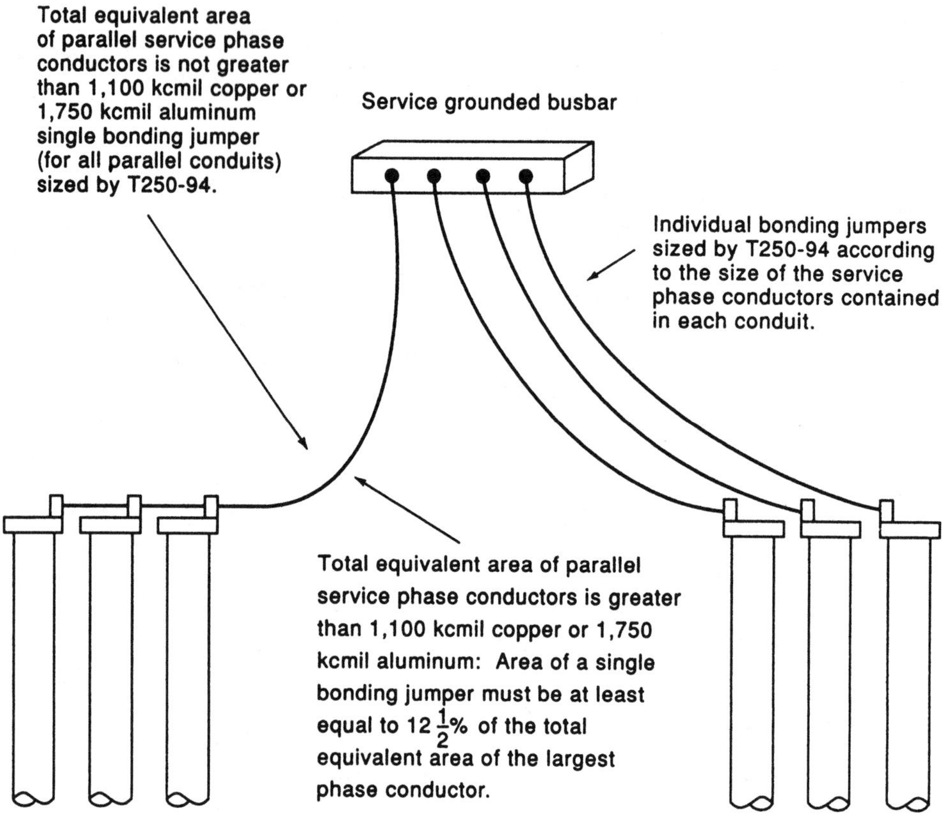

Fig. 9-18 Supply-side equipment-bonding or grounding jumpers (conductors) must be sized in accordance with Sec. 250-79(d) and Sec. and Table 250-94.

Total c.m.a. per phase computation: Ref. (A) ____________

____________ × ____________ kcmil = (B) ____________ kcmil copper

Line-side equipment-grounding requirements: Ref. (C) ____________

Bonding jumper wire size: (D) ____________ copper Ref. (E) ____________

Section 250-79(d) requires the wire size of a single bonding jumper extended to any metal service enclosure to be based on the wire size of the service-entrance conductors. When the wire size of the largest ungrounded service-entrance conductor exceeds 1,100 kcmil copper or 1,750 aluminum or c-c aluminum, T250-94 is void, and the wire size of the bonding jumper must at least equal ⅛ the wire size of the largest ungrounded service-entrance conductor. The same would hold true for a single loop-connected bonding jumper for the parallel service conduits just discussed.

As with the sizing of the grounded service neutral conductor, when the one-eighth computation yields a nonstandard wire size, Table 8 in Chapter 9 is used to determined the next higher standard wire size.

Example 9–14. A service rated at 500 amps will be supplied with 2,000 kcmil RH aluminum ungrounded service-entrance conductors. Determine the NE Code-minimum wire size of the aluminum bonding jumper run to the service conduit.

Line-side equipment-grounding requirements: Ref. 250-79(d)
Bonding jumper c.m.a. computation:

$$0.125 \times 2{,}000 \text{ kcmil} = 250 \text{ kcmil aluminum}$$

Wire size: 250 kcmil aluminum Ref. Ch. 9, T8

Code challenge 9–14. A service rated at 600 amps will be supplied with 1,500 kcmil THW copper ungrounded service-entrance conductors. Determine the NE Code-minimum wire size of the copper bonding jumper run to the service conduit.

Line-side equipment-grounding requirements: Ref. (A) ____________________
Bonding jumper c.m.a. computation:

____________ × ____________ kcmil = (B) ____________ kcmil copper

Wire size: (NHS) (C) ____________ copper Ref. (D) ____________________

When the service-entrance conductors are paralleled and the total conductor circular-mil area connected to any one phase exceeds the table cutoff of 1,100 kcmil copper or 1,750 aluminum or c-c aluminum, the wire size of the single bonding jumper or other equipment-grounding conductor must be at least equal to ⅛ the wire size of the largest equivalent area ungrounded service-entrance conductor. The same method is required as is used for a single grounded service-neutral conductor area computation when the ungrounded service-entrance conductors are paralleled in a common gutter or other enclosure.

Example 9–15. A service rated at 2,400 amps will be supplied with 700 kcmil THW c-c aluminum ungrounded service-entrance conductors paralleled seven times per phase. Determine the NE Code-minimum wire size of the copper equipment-grounding conductor used to bond the metal gutter of the service.

Service conductor wire size based on copper: Ref. 250-79(d)
Individual conductor amperage computation: Ref. 310-4

$$2{,}400 \text{ amps} \div 7 \text{ conductors} = 342.8 \text{ amps per conductor}$$

Wire size: 500 kcmil THW copper Ref. T310-16
Total c.m.a. per phase computation: Ref. 310-4

$$6 \times 500 \text{ kcmil} = 3{,}000 \text{ kcmil copper}$$

Line-side equipment-grounding requirements: Ref. 250-79(d)
Bonding jumper c.m.a. computation:

$$0.125 \times 3{,}000 \text{ kcmil} = 375 \text{ kcmil copper}$$

Wire size: (NHS) 400 kcmil copper Ref. Ch. 9, T-8

Code challenge 9–15. A service rated at 2,000 amps will be supplied with 600 kcmil THWN copper ungrounded service-entrance conductors paralleled five times per phase. Determine the NE Code-minimum wire size of the aluminum equipment-grounding conductor used to bond the metal gutter of the service.

Service conductor wire size based on (A) ___________: Ref. (B) ___________
Individual conductor amperage computation: Ref. (C) ___________

___________ amps ÷ ___________ conductors = (D) ___________ amps/conductor

Wire size: (E) ___________ kcmil THW aluminum Ref. (F) ___________
Total c.m.a. per phase computation: Ref. (G) ___________

___________ × ___________ kcmil = (H) ___________ kcmil aluminum

Line-side equipment-grounding requirements: Ref. (I) ___________
Bonding jumper c.m.a. computation:

___________ × ___________ kcmil = (J) ___________ kcmil aluminum

Wire size:(NHS) (K) ___________ kcmil aluminum Ref. (L) ___________

9.12 LOAD-SIDE EQUIPMENT GROUNDING

Line-side equipment grounding must be by means of separate equipment-grounding conductors or bonding jumpers that provide a positive connection to the service neutral or other service grounded-circuit conductor. In comparison, load-side equipment grounding can be by means of the interconnection of the metal piping raceway with the metal enclosure of gutters, junction, pull, or device boxes, or the terminal housings on motors or other types of utilization equipment.

Section 250-91(b) lists the different types of load-side equipment-grounding conductors recognized by the NE Code. With the exception of liquidtight flexible metallic conduit, the interconnection of the metal enclosures is recognized as a sufficient equipment-grounding means.

Greenfield, which is the trade name for flexible metallic conduit, can be run throughout the structure in terms of a raceway wiring method and equipment-grounding means, if both the conduit and the fittings are listed for grounding [Sec. 250-91(b)(5)]. If not listed, it falls in the ranks with flexible metallic tubing and liquidtight flexible metal conduit. When used as the sole grounding means of the remote metal enclosures of the equipment served, the exceptions to Sec. 250- 91(b) limit the length of these conduits or tubings to a maximum of 6 feet. If this length is exceeded, a separate equipment-grounding conductor, sized as indicated in the following paragraphs, must be pulled with the contained circuit conductors.

Interconnection of the metal raceway and enclosures as the sole equipment-grounding means can pose problems. The purpose of the load-side equipment-grounding means is to provide a low-impedance return path for ground-fault overcurrent to the grounded service neutral conductor for further conveyance back to the transformer supply, as it simultaneously ensures that any stray voltages from static electrical sources throughout the building cannot remain on the noncurrent-carrying metal parts of the power distribtuion system.

Every joint or other makeup of these metal parts is subject to oxidation prior to installation. A rusty joint (in ferrous metal raceways) increases the resistance of the length of the pipe. The pipe is made up to metal boxes or other enclosures by means of double locknuts, or a shouldered connector and a single locknut. Both the joints (couplings) and the locknuts have to be wrenchtight to ensure a low-impedance connection. Any loose connection along the length of the pipe or other raceway will further impede the fault-return path.

Another concern is physical damage. When rigid metal pipe is run exposed along the length of a wall in an industrial concern, the pipe may be exposed to physical damage by forklifts (tow motors) or other moving machinery within the building. Any break in the pipe due to physical damage will create a severe safety hazard should an ungrounded-circuit conductor fault downstream of the break.

The fault will bring the sections of pipe downstream of the break to line potential, but fault overcurrent will not flow simply beause of the break in the fault-return path. The personnel hazard is at the break: The downstream pipe section is being held at line potential in reference to the upstream section. Anyone who purposely or accidentally touches both sections (such as trying to repair the break) will be exposed to a possible severe electrical shock.

The safer electrical distribution contains a separate equipment-grounding conductor routed with the circuit conductors in the metal pipe. If the fault downstream of the break just described should occur, the separate equipment-grounding conductor, grounded (connected) in parallel to the pipe, maintains a low-impedance fault-return path, such that the fuse will quickly blow or the affected conductor circuit breaker will open, alleviating the safety hazard.

Providing the separate equipment-grounding conductor does not eliminate the need of effectively bonding the piping and the metal enclosures together by making them up wrenchtight. Sections 250-75 and 250-76 require all the noncurrent-carrying metal parts that go into the makeup of a power distribution system to be effectively bonded together.

Metal enclosures on the load side of the service equipment that are not interconnected with metal piping or other types of metal raceway, such as the use of PVC or other nonmetallic conduit, or wiring methods such as BX or Romex cable, must be provided with a separate equipment-grounding conductor that is run (contained in the cable) with the circuit conductors.

The wire size of equipment-grounding conductors or bonding jumpers on the load side of the service equipment is governed by Secs. 250-79(e) and 250-95. The actual wire size is determined by the rating of the OCPD protecting the feeder or branch circuit supported by the equipment-grounding conductor or bonding jumper. As listed in T250-95, the circuit conductors and equipment grounding or bonding conductors can be either copper, aluminum, or c-c aluminum. Copper circuit conductors do not technically warrant a copper bonding jumper or equipment-grounding conductor.

Example 9–16. Determine the NE Code-minimum wire size of a copper equipment-grounding conductor run to the metal enclosure of a subfed 200-amp panel in an industrial concern.

Load-side equipment-grounding requirements:	Ref. 250-79(e)
Wire size: No. 6 AWG copper	Ref. T250-95

Code challenge 9–16. Determine the NE Code-minimum wire size of an aluminum equipment-grounding conductor run to the metal enclosure of a subfed 400-amp panel in an industrial concern.

Load-side equipment-grounding requirements:	Ref. (A) ______________
Wire size: (B) __________ aluminum	Ref. (C) ______________

When several branch-circuit and/or feeder conductors are pulled in a single conduit, or terminate in a single wireway or gutter, individual equipment-grounding conductors are not required. Instead, Sec. 250-95 will allow a single equipment-grounding conductor to support (ground or bond) all the contained circuits.

If the multiple branch or feeder circuits are pulled in individual metal conduits, Sec. 250-79(e) will allow the use of a single, continuous bonding jumper to loop-connect and ground all the load-side metal conduits (Fig. 9-19), because the equipment ground is a ground-fault return conductor. The odds of two branch or feeder circuits faulting simultaneously in a single enclosure are high enough to make the occurence not likely. Sections 250-79(e) and 250-95 require the wire size of the single equipment-grounding conductor to be based on the largest branch circuit or feeder OCPD served.

Example 9–17. Determine the NE Code-minimum wire size of a c-c aluminum equipment-grounding conductor used to ground a single wireway that contains circuits rated 15, 20, 50, and 70 amps.

Load-side equipment-grounding requirements:	Ref. 250-79(e)
Wire size must be based on the 70-amp OCPD rating.	Ref. 250-95
Wire size: No. 6 AWG aluminum	Ref. T250-95

Seventy amperes is not listed in T250-95 as a circuit OCPD rating. The next higher standard size listed is 100 amps. To read the table correctly, a No. 6 AWG aluminum or c-c aluminum conductor is good for circuits rated from 61 to 100 amperes. The same comparison can be made for No. 8 AWG copper.

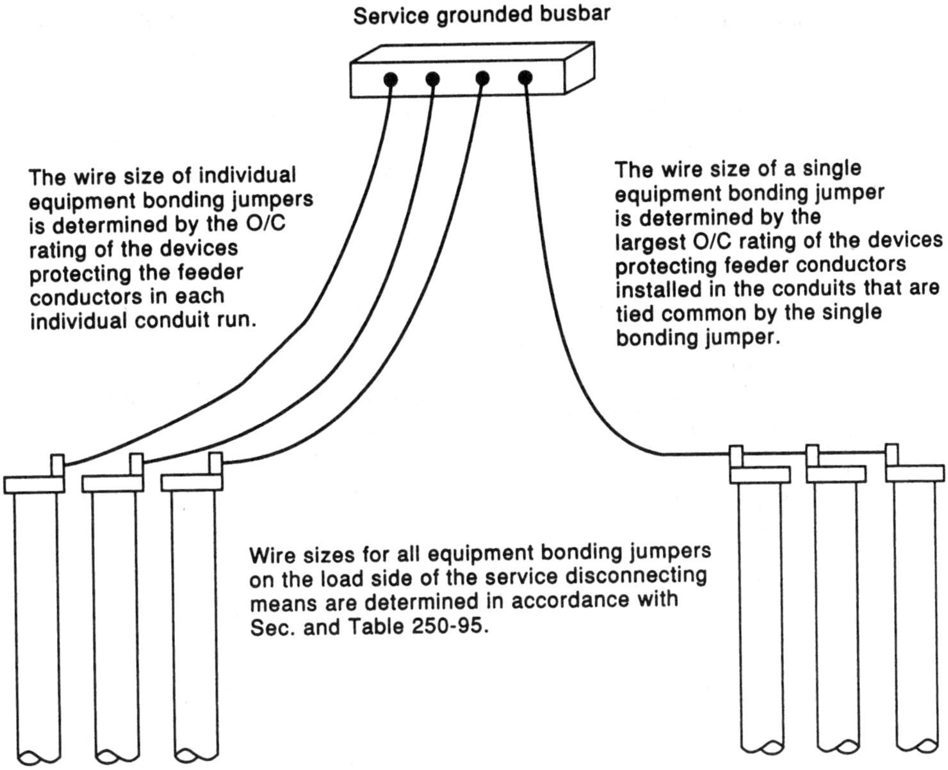

Fig. 9-19 Load-side equipment-grounding jumpers (conductors) must be sized in accordance with Sec. 250-79(e) and Sec. and Table 250-95.

Code challenge 9–17. Determine the NE Code-minimum wire size of a copper equipment-grounding conductor used to ground a single auxiliary gutter that contains circuits rated 30, 50, 60, and 90 amps.

Load-side equipment-grounding requirements: Ref. (A) ________________

Wire size must be based on the (B) __________-amp OCPD rating.

Ref. (C) ________________

Wire size: (D) __________ copper Ref. (E) ________________

When higher-amperage feeder circuits are constructed of smaller wire size conductors paralleled to attain the required conductor amperage, the second paragraph of Sec. 250-95 requires the wire size of either the single equipment-grounding conductor installed when the parallel conductors are contained in a common gutter or metal raceway enclosure, or the individual equipment-grounding conductors, when the parallel feeder conductors are installed as multiple sets of the ungrounded phase conductors and the feeder grounded neutral conductor in each parallel conduit or multiconductor cable, to be based on the rating of the feeder-circuit OCPD.

In other words, the multiple equipment-grounding conductors, installed to meet the grouping requirements of Sec. 300-20(a) (reduction of induced heating), cannot be reduced in wire size for "their share" of the available ground-fault overcurrent.

Example 9–18. The feeder circuit for a 3-phase, 4-wire, 480Y/277-volt lighting panel rated at 1,000 amps will consist of 250 kcmil THW copper conductors paralleled four times per phase. Determine the NE Code-minimum wire size of individual copper equipment-grounding conductors installed in each parallel run of rigid metal conduit.

Load-side equipment-grounding requirements: Ref. 250-79(e)
Wire size must be based on the 1,000-amp OCPD rating. Ref. 250-95
Wire size: No. 2/0 AWG copper Ref. T250-95

Code-challenge 9–18. The feeder circuit for a 3-phase, 4-wire, 208Y/277-volt lighting panel rated at 800 amps will consist of 400 kcmil THW aluminum conductors paralleled three times per phase. Determine the NE Code-minimum wire size of individual aluminum equipment-grounding conductors installed in each parallel run of rigid metal conduit.

Load-side equipment-grounding requirements: Ref. (A) ________________
Wire size must be based on the (B) ____________-amp OCPD rating.
Ref. (C) ________________
Wire size: (D) ____________ aluminum Ref. (E) ________________

9.13 VOLTAGE-DROP COMPENSATION

As discussed in Chapter 2 for circuit-conductor ampacities according to wire size, when a branch or feeder circuit will be subjected to either long distance (conductor length) or high amperage, the wire size of the circuit conductors must be increased to compensate for conductor voltage drop.

The maximum voltage drop allowed by the NE Code is 5% overall from point of supply to the connected load. Due to their lower amperage rating, branch-circuit conductors [refer to the fourth FPN to Sec. 210-19(a)] are normally limited to a maximum voltage drop of 3%. The service and feeder conductors [refer to the second FPN to Sec. 215-2(b)] are normally limited to a maximum of 2%.

When circuit conductors are adjusted in wire size to compensate for voltage drop, the third paragraph of Sec. 250-95 requires the equipment-grounding conductors to also be adjusted or up-scaled in wire size for the same compensation.

One method of up-scaling is to determine the number of wire sizes the circuit conductors had to be increased and increase the circular-mil area of the equipment-grounding conductor by the same number of wire sizes. This works fine for branch circuits rated up to 30 amps, simply because these circuits require a full-size equipment ground (reference T310-16 obelisk footnote and T250-95).

A better method of voltage-drop compensation is to compute the amount of increase in circular-mil area of the load-circuit conductor wire size as a ratio. Then apply the ratio in terms of circular-mil area to the wire size of the load-side equipment-grounding conductor, sized according to the OCPD rating of the circuit, to overrate or increase its wire size.

Because the wire size of both the circuit conductors and the equipment-grounding conductor can be taken as either AWG or kcmil, this method requires all the initial wire sizes to be converted over to circular-mil areas. Once the required compensated circular-mil area of the equipment-grounding conductor is computed, the area is then converted back to a standard wire size. As before, conversion from standard wire size to circular-mil area or vice versa is listed in Table 8 of Chapter 9. When the computed c.m.a. of the equipment-grounding conductor does not equal a standard wire size, the next higher standard wire size should be used.

For example, a 40-amp branch circuit, without compensation for higher ambient temperature or other loss of heat dissipation, requires a No. 8 AWG copper conductor. The 40-amp rating of the circuit OCPD requires a No. 10 AWG copper equipment-grounding conductor.

If exceptional length requires No. 4/0 AWG copper circuit conductors for the same circuit, Sec. 250-95 requires the No. 10 AWG copper equipment-grounding conductor to be increased in wire size by the same percentage as the No. 8 AWG was increased to the No. 4/0 AWG wire size. The overrate percentage cannot be effectively computed by dividing 4/0 by 8.

From Table 8, a No. 4/0 AWG conductor has a circular-mil area of 211,600 c.m. A No. 8 AWG conductor has a circular-mil area of 16,510 c.m. Dividing, the circuit wire size was increased by a factor of 12.8. The circular-mil area of the No. 10 AWG equipment-grounding conductor equals 10,380 c.m. Multiplying, the wire size of the equipment ground, compensated for the severe voltage drop, must have a circular-mil area of at least 132,864 c.m. Looking for the next higher circular-mil area value in Table 8, the compensated equipment-grounding conductor has to be a minimum wire size of No. 2/0 AWG.

Example 9–19. A 3-phase, 4-wire, 480Y/277-volt load center is rated at 60 amps. The non lighting load center is located 800 feet in wire length from the service equipment panel. Using THHN copper conductors, size the feeder conductors and the equipment-grounding conductor.

Wire size of the circuit conductors without voltage-drop compensation:

Wire terminal temperature limitations: 60°C — Ref. 110-14(c)(1), Ex. 1

Table-rated ampacity: 70 amps

Wire size: No. 4 AWG THHN copper — Ref. T310-16

Wire size of the circuit conductors with voltage-drop compensation:

Limit the voltage drop to 2% — Ref. 215-2(b), FPN No. 2

Computation:

$$R_{max} = \frac{1{,}000 \text{ ft} \times (0.02 \times 480 \text{ V})}{2 \times 800 \text{ ft} \times 60 \text{ A} \times 0.866}$$

$$R_{max} = 0.1155 \text{ ohms}$$

$$R_{table} = 0.0967 \text{ ohms}$$

Wire size: No. 2/0 AWG THHN copper Ref. Ch. 9, T8

Load-side equipment-grounding requirements: Ref. 250-79(e)

Wire size of the grounding conductor without voltage-drop compensation:
No. 10 AWG copper Ref. T250-95

Wire size of the grounding conductor with voltage-drop compensation:
Ref. 250-95

Conversion from AWG to circular-mil area: Ref. Ch. 9, T8

Circuit conductors without voltage-drop compensation:

No. 4 AWG = 41,740 c.m.

Circuit conductors with voltage-drop compensation:

No. 2/0 AWG = 133,100 c.m.

Equipment-grounding conductor without voltage-drop compensation:

No. 10 AWG = 10,380 c.m.

Circuit conductor c.m.a. overrate computation:

$$133{,}100 \text{ c.m.} \div 41{,}740 \text{ c.m.} = 3.189$$

Equipment-grounding conductor with voltage-drop compensation computation:

$$3.189 \times 10{,}380 \text{ c.m.} = 33{,}102 \text{ c.m.}$$

Conversion from circular-mil area to AWG: Ref. Ch. 9, T8

33,102 c.m. = (NHS) No. 4 AWG copper

Code challenge 9–19. The full-load current rating of a conveyor equipment assembly rated 3-phase, 208-volts is 80 amps. If the conveyor control panel is located 700 feet in wire length from the panel overcurrent-protection device, use THW copper conductors, and size both the branch-circuit conductors and the equipment-grounding conductor.

Wire size of the circuit conductors without voltage-drop compensation:

Wire terminal temperature limitations: (A) ____________ °C

Ref. (B) ____________________

Wire size: (C) ____________ THW copper Ref. (D) ____________________

Wire size of the circuit conductors with voltage-drop compensation:

Limit the voltage drop to (E) ____________ % Ref. (F) ____________________

Computation:

$$R_{max} = \frac{1{,}000 \text{ ft} \times [\_\_\_\_\_\_\_\_ \times (G) \_\_\_\_\_\_\_\_ \text{ V}]}{2 \times \_\_\_\_\_\_\_\_ \text{ ft} \times \_\_\_\_\_\_\_\_ \text{ A} \times (H) \_\_\_\_\_\_\_\_}$$

R_{max} = (I) ____________ ohms

R_{table} = (J) ____________ ohms

Wire size: (K) ____________ THW copper Ref. (L) ____________________

Load-side equipment-grounding requirements: Ref. (M) ____________________

Wire size of the grounding conductor without voltage-drop compensation:
(N) ____________ copper Ref. (O) ____________________

Wire size of the grounding conductor with voltage-drop compensation:
Ref. (P) ____________________

Conversion from AWG to circular-mil area: Ref. (Q) ____________________

Circuit conductors without voltage-drop compensation:

____________ AWG = (R) ____________ c.m.

Circuit conductors with voltage-drop compensation:

____________ AWG = (S) ____________ c.m.

Equipment-grounding conductor without voltage-drop compensation:

____________ AWG = (T) ____________ c.m.

Circuit conductor c.m.a. overrate computation:

____________ c.m. ÷ ____________ c.m. = (U) ____________

Equipment-grounding conductor with voltage-drop compensation computation:

____________ × ____________ c.m. = (V) ____________ c.m.

Conversion from circular-mil area to AWG: Ref. (W) ____________

____________ c.m. = (NHS) (X) ____________ copper

9.14 LOAD-CIRCUIT NEUTRAL CONDUCTORS

The busbar or terminal strip connected to the grounded neutral conductor in the service-equipment enclosure must be used as the common tie point for interconnection of the grounding-electrode system, the bonding grid of the structure, the equipment-grounding grid of the power distribution, and all the load-circuit neutral conductors.

Load-circuit neutral conductors are current-carrying circuit conductors that have been intentionally grounded by the grounding of the service neutral conductor. In reference to Secs. 250-23(a) and 250-61(b), these circuit conductors cannot be used as an equipment grounding or bonding means for the noncurrent-carrying metal parts of the power distribution on the load side of the service equipment. In other words, a feeder or branch-circuit neutral conductor must be both insulated and isolated from any type of equipment grounding or bonding on the load side of the service equipment.

If the load-circuit neutral conductor were mistakenly used as the equipment grounding/bonding means for an isolated metal enclosure on the load side of the service equipment, the opening of the grounded neutral conductor ahead of the equipment-grounding connection could put the metal enclosure at line potential and generate an unnecessary safety hazard to personnel.

For example, in Figure 9-20, a lighting fixture is supplied by a cable wiring method that does not contain an equipment-grounding conductor. The metal shroud (canopy) of the lighting fixture is purposely connected to the load-neutral conductor to attain a grounding means. The light is turned on. If the grounded neutral conductor were opened by a loose connection upstream from the light, or by an electrician working in the service panel who wanted to relocate its connection to the service-equipment grounded terminal strip, the light would go off, but the metal enclosure would be held at line potential.

Someone assumes the bulb is burned out, and in an attempt to change the light bulb, he or she touches the metal shroud. Because the shroud was grounded by connection to the load-neutral conductor that is now open, a severe electrical shock, and perhaps electrocution will occur. The grounded load-neutral conductor should never be used as the required equipment-grounding means.

All the noncurrent-carrying metal parts of the power distribution, including the metal enclosures of the connected load, must be effectively grounded, but their grounding connection can only be to the grounded service-neutral conductor through a separate, noncircuit conductor. This separate equipment-grounding conductor is not a current-carrying conductor except under fault overcurrent conditions.

The separate equipment-grounding conductor maintains the metal shroud of the lighting fixture (load, as just discussed), the metal housing of the motor (load), or the metal enclosure of a subpanel or motor control center (both are load centers) at ground potential (zero volts). The shock hazard is reduced, if not eliminated, because a difference in voltage potential cannot exist between the grounded surfaces of the building or other structure, where a person might be standing, and the metal enclosures of the power distribution that a person might touch.

If a fault condition should occur between an ungrounded circuit conductor and the metal enclosure, the separate equipment-grounding conductor provides a low-impedance return path for the current available from the faulted conductor back to the transformer winding such that the affected fuse or circuit breaker should quickly open to stop the current flow.

What is a rule without exception? In past NE Code editions, under certain conditions the metal frames (enclosure) of a dwelling-type electric range (stove) or clothes dryer could be grounded by direct connection to the grounded load-circuit neutral conductor. All the conditions listed in Sec. 250-60 had to be met.

In the 1996 Edition of the NE Code, the rules of Sec. 250-60 are still listed, but they apply only to existing electric range or clothes dryer branch circuits. All new instal-

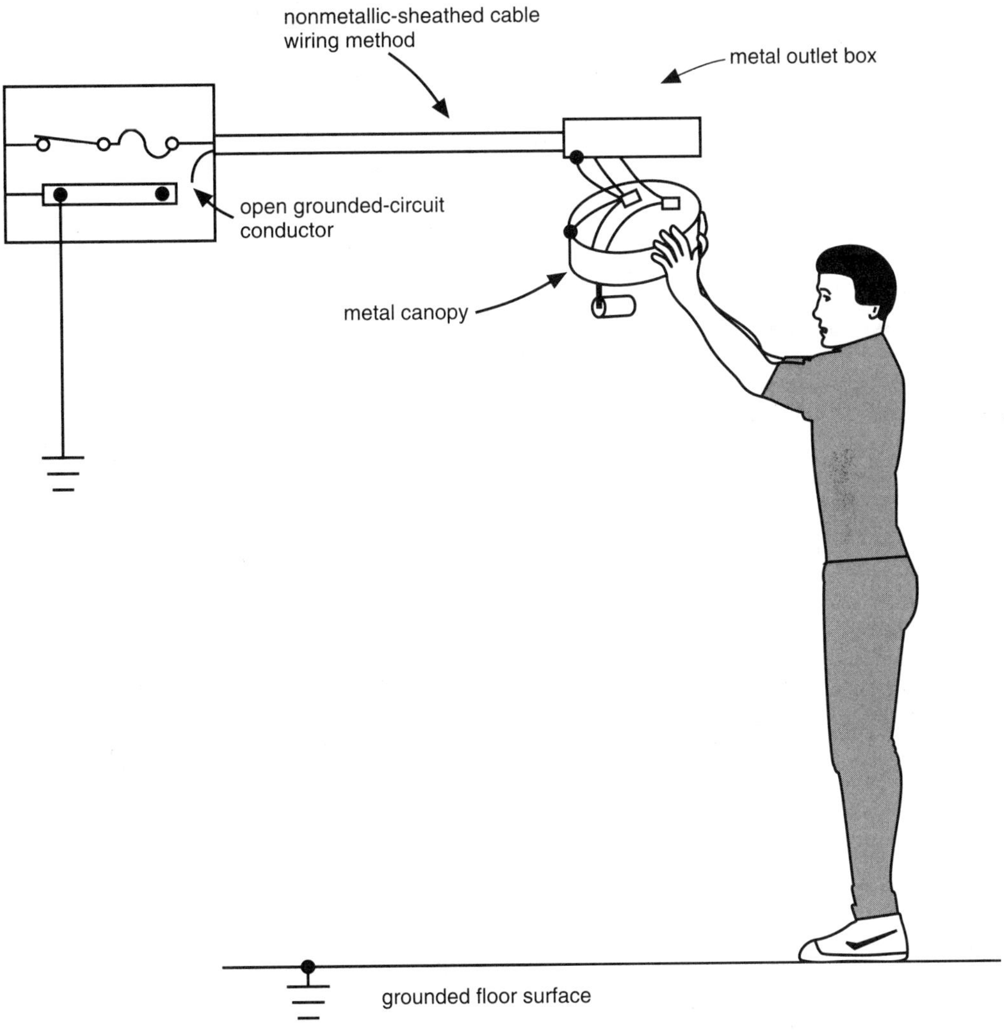

Fig. 9-20 A load-circuit neutral conductor can never be used as the equipment grounding means.

lations must comply with Secs. 250-57 and 250-59, which require a separate equipment-grounding conductor in both the branch-circuit supply circuit and the cord- and plug-connected receptacle for such appliances.

In reference to Sec. 250-60(a), the type of service is restricted to either a single-phase, 3-wire, 120/240-volt service, or a 3-phase, 4-wire, 208Y/120-volt service.

In addition to the right type of service supply, the grounding-neutral conductor is not allowed by Sec. 250-60(b) to be smaller than No. 10 AWG copper or No. 8 AWG aluminum (or c-c aluminum). The grounded multiwire branch-circuit neutral conductor in raceway-supplied appliances must be insulated. Subsection (c) will allow an SE cable feeder for the range with a covered, but not insulated, neutral conductor. When Romex is used, the neutral conductor must also be insulated.

The main point of this section requires the branch circuit to originate from the service equipment panel. If the branch-circuit OCPD for the range or clothes dryer originates in a subfed panel, the metal enclosure, as in the 1996 Edition of the NE Code, must be provided with a separate equipment-grounding means.

The drawback to the method of equipment grounding allowed by Sec. 250-60 is the location of the refrigerator near the range, or the washing machine near the clothes dryer. The metal frames of both of these appliances have to be grounded by a separate equipment-grounding means. If the metal frame of the electric clothes dryer is grounded by its branch-circuit neutral conductor, and the neutral prong of the dryer receptacle should fail, or the grounded load-neutral conductor should fall victim to a loose connection in the service panel, the dryer frame would be at line potential to the washing machine frame when the start button was pushed—a possible shock hazard.

The better load-side equipment-grounding system employs separate equipment-grounding conductors interconnected to all the metal enclosures throughout the power distribution system at one end, and connected common to the grounded service neutral busbar at the other end.

9.15 SEPARATELY DERIVED SYSTEMS

A separately derived system, such as a step-down transformer from a 3-phase, 4-wire, 480Y/277-volt system, to a 3-phase, 4-wire, 208Y/120-volt system (both wye configurations), must be treated as a service in terms of equipment grounding (Fig. 9-21). One end of all three secondary phase windings within the transformer is tied common in a wye configuration, and this common connection is bonded to the equipment-grounding system/bonding means (normally building steel) of the primary distribution.

Section 250-26 requires the transformer secondary grounding-electrode system because the derived voltage is one that operates in reference to earth ground (refer to Sec. 250-5).

The fourth-wire secondary common neutral conductor must be tied to this common grounded connection and run with the ungrounded secondary feeder conductors to the load center of the separately derived system.

The load center, normally a panel, must be provided with both a main disconnecting means and overcurrent (overload) protection for each of the ungrounded feeder conductors on the secondary circuit of the transformer. Article 384 governs the installation and use of panel and switch boards. Section 384-16(d) requires the panel to be provided with overcurrent protection on the secondary side of the transformer.

Article 240 governs the installation and use of OCPDs. The general rule of Sec. 240-21 requires the separately derived feeder conductors to be provided with overcurrent protection at the point of connection to the transformer secondary. Subsection (j) will allow the OCPD to be located as the main in the panel. Section 240-40 requires fuses to be provided with a disconnecting means (fused disconnect). Sections 240-80 and 240-81 will allow the circuit breaker to serve as both the OCPD and the means of panel disconnect.

The grounded busbar or terminal strip located in the load-center enclosure of this separately derived system serves as the common tie point (connection) to the separately derived grounded neutral conductor for all equipment bonding or grounding means (sup-

Fig. 9-21 Dry-type transformers are used to step down 480Y/277-volt lighting distributions to 208Y/120-volt lighting and receptacle distribution.

port equipment or building structure bonding) on both the supply and load sides of the separately derived system disconnecting means, for all grounding-electrode systems of the separately derived system within the building or other structure, and for all the separately derived load-circuit neutral conductors.

Section 250-26(b) requires the separately derived system grounding-electrode conductor to be sized in the same manner as the grounding-electrode conductors for the service equipment supplying the building or other structure in accordance with S & T 250-94.

The grounding electrode is required by Sec. 250-26(c) to be located in the same area as the transformer and panel of the separately derived system. The preferred grounding electrode (Fig. 9-22) is effectively grounded building steel. If the structure does not contain a bonding grid of building steel that is connected into the parent-voltage service equipment grounding-electrode system, the second choice is effectively grounded interior metal water pipe.

If effectively grounded (bonded to the parent grounding-electrode system), interior metal water pipe is not available, then Sec. 250 26(c) will allow any of the available or made electrodes listed in Secs. 250-81 and 83 (respectively) to be used as the required grounding electrode.

Caution here: the separately derived source cannot be connected to a separate grounding-electrode system that is isolated from (not connected to) the building bonding grid and the parent-voltage distribution grounding-electrode system.

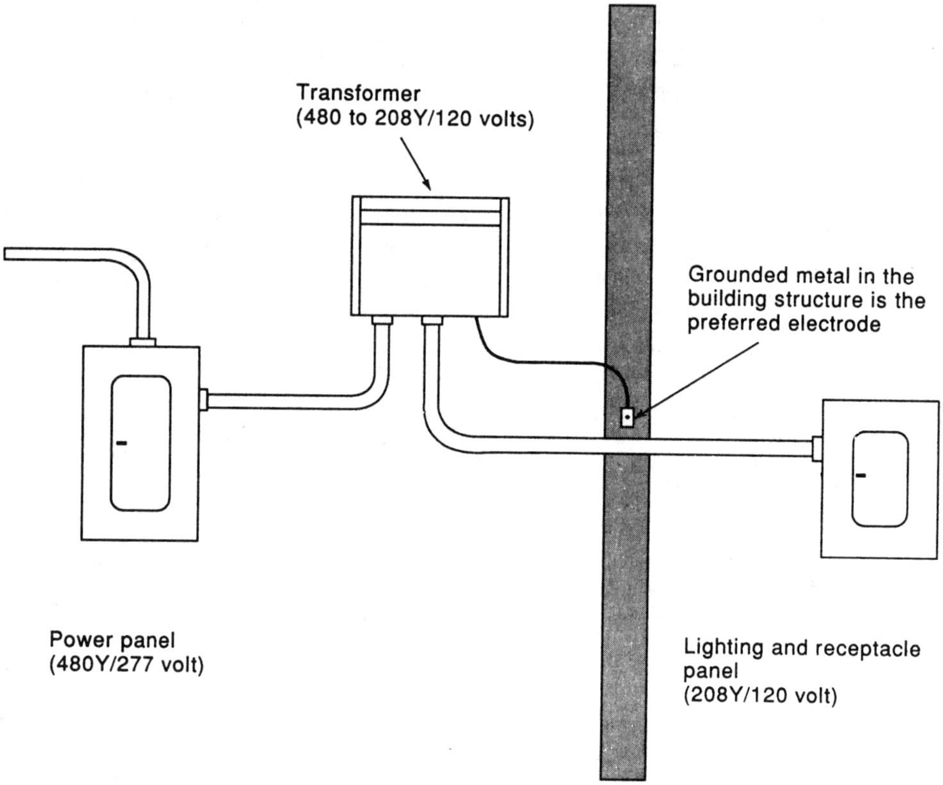

Fig. 9-22 Effectively grounded metal in the building structure is the preferred grounding electrode for a separately derived system.

Section 250-81 requires any and all grounding-electrode systems, either made or available in a given building or other structure, to be effectively bonded together to form the required service grounding-electrode system. If the separately derived source is connected to an independent grounding-electrode system (reference the second FPN to Sec. 250-86), a difference of potential could occur between the two separate grounding-electrode systems, or between the noncurrent-carrying metal parts used in the installation of the two systems.

If connection to the parent-service grounding-electrode system is not available in the immediate area of the transformer/panel location, the separately derived source can be grounded by means of the primary-circuit equipment-grounding conductor, which is a direct connection to the service grounding-electrode system.

Connected this way, the prevailing wire size of the primary feeder equipment-grounding conductor would be governed by the primary circuit OCPD rating and T250-95, and by the wire size of the secondary feeder-circuit conductors and T250-94. In reference to Sec. 310-15(c), the largest wire size, by respective computation, would automatically become NE Code minimum.

This same analogy must be applied to the equipment-grounding conductor in a mobile home feeder circuit. The interior metal water pipe or metal feed-in to the nonmetallic water piping distribution system is bonded to the equipment-grounding busbar in the panel inside the mobile home. The fourth-wire equipment-grounding conductor in this single-phase, 3-wire feeder is the only means for connecting or bonding the interior metal water pipe to the service equipment, which is located outside the mobile home on a pole or backboard, grounding-electrode system.

To meet the requirements of Sec. 250-80(a), the equipment-grounding conductor must be sized according to the wire size of the service conductors supplying the mobile home service and T250-94. To ground the noncurrent-carrying metal parts of the electrical power distribution within the mobile home, the same equipment-grounding conductor must be sized by the rating of the panel OCPD in the mobile home and T250-95.

The same philosophy holds true for data-processing equipment in the industrial or commercial arena. Although electrical noise on the equipment-grounding grid causes many problems in data processing, separately derived power systems for main-frame computers, programmable logic controllers (PLCs), or other types of data-processing equipment are not exempt from the bonding of the separately derived grounding-electrode system to the parent-voltage service grounding-electrode system.

Article 645 governs the supply and installation of data-processing equipment ("information technology equipment," to use the terminology of the 1996 NE Code). Unless the equipment is double insulated, Sec. 645-15 requires the interconnection of the grounding-electrode systems as just described.

For smaller units, where the separately derived system consists of a regulated power supply within the equipment, separately derived electrode system grounding is not required. The unit itself is grounded by the service grounding-electrode system through the equipment-grounding conductor in the branch-circuit supply.

Exception No. 4 to Sec. 250-74 will allow cord- and plug-connected information technology equipment to be supplied with a special receptacle that offers an isolated equipment-grounding terminal. The grounding terminal can be connected to an insulated equipment-grounding conductor extended back to the service neutral busbar of the power distribution without any other grounding connections (splice or taps) in any subfed panels or other load centers that are in series with the branch-circuit supply from the service equipment or the separately derived system to the receptacle.

In reference to the electrode grounding of a separately derived system within a building or other structure, a driven rod may be added at the location of the transformer and panel to meet the requirement of Sec. 250-26(c) for a grounding-electrode connection in the immediate vicinity of both. Section 250-91(c) would recognize this additional driven rod as "supplemental grounding." It must still be tied into the parent service grounding-electrode system through the equipment-grounding conductor of the transformer primary circuit.

The 120-volt control voltage derived from the higher power voltage delivered to a motor control center or other control system is also a separately derived system, but due to its limitations of available power from the secondary of the control transformer, a separate equipment grounding-electrode system does not have to be established with the control-transformer secondary.

This transformer is normally limited to single-phase operation. As shown in Figure 9-23, one end of the secondary winding is connected to the equipment-grounding system

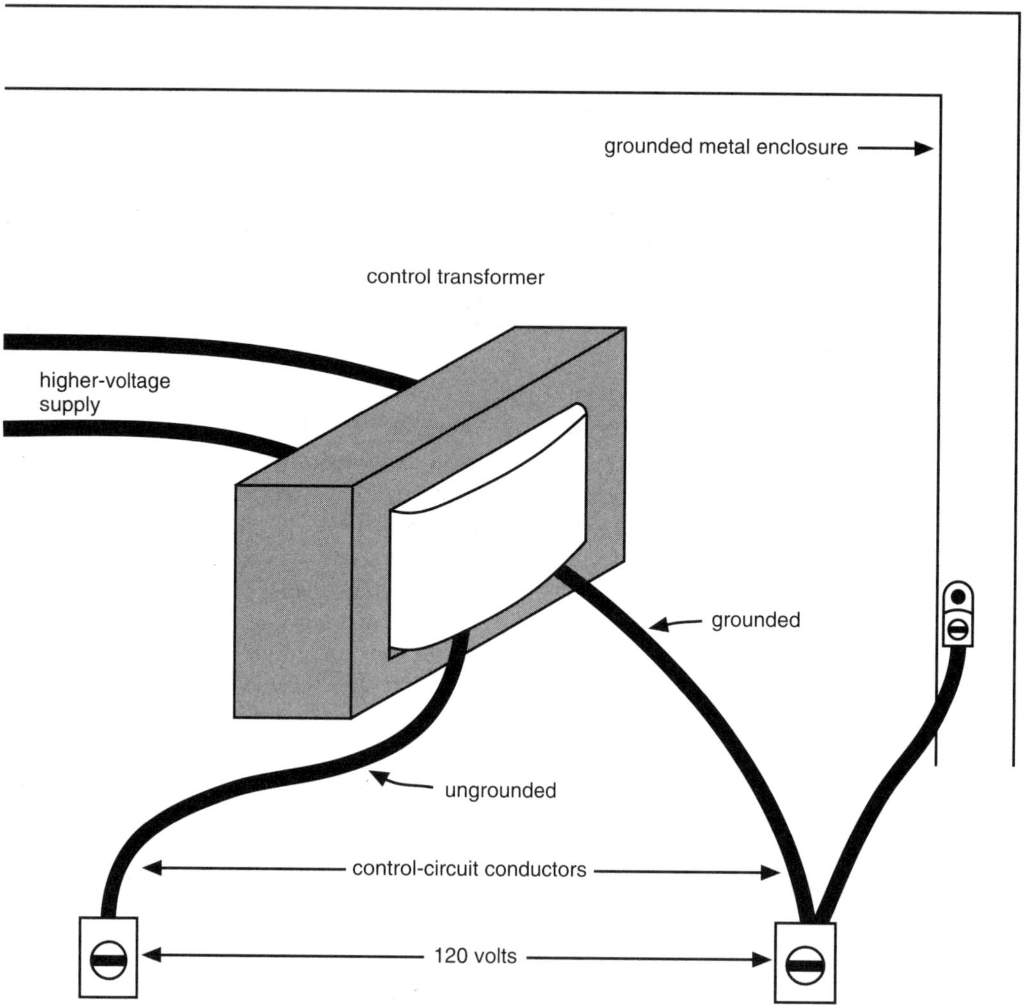

Fig. 9-23 The control transformer of process equipment is also a separately derived system that must be grounded.

of the primary distribution (normally within or to the metal enclosure of the control panel for the equipment) to provide a low-impedance return path for ground-fault overcurrent supplied by the control transformer. The unfused control-circuit return conductor (not to be considered as a neutral) is connected to this grounded end of the winding.

The other end of the winding must be fused because it supplies the ungrounded side of the control circuit. If one of the ungrounded control-circuit conductors should come in contact with any metal enclosure or other grounded surface either within the control cabinet, or located remote from the control center, high-magnitude ground-fault overcurrent will flow, opening the control fuse.

9.16 SINGLE SERVICE FOR MORE THAN ONE BUILDING

The rules governing the installation of a single service to a parent building, which will supply other buildings on the same premises, are found in Parts F and G of Art. 230 and Art. 225.

Section 230-70 requires that all buildings have a means to disconnect all the ungrounded conductors from the service-entrance conductors. The service-disconnecting means must be rated for service equipment and be suitable for its intended use (inside or outside, hazardous locations, etc.). The service-disconnecting means can be located either inside or outside the building but must be located at a readily accessible location nearest the point where the service-entrance conductors enter the building.

The service-disconnecting means for a single building (or for several buildings, as we shall see) can be a single main disconnect, or a group of two to six disconnecting means. Section 230-71(a) will allow the two to six service-disconnecting means to be contained in a single enclosure, or in a group of separate enclosures fed from a common gutter.

Service-entrance conductors as a general rule (Sec. 230-46) cannot be spliced. Exception No. 2 to Sec. 230-46 will allow a single set of service-entrance conductors contained in an auxiliary gutter to be tapped to supply the two to six service disconnects.

Section 230-90(a) requires that each ungrounded service conductor be provided with overload-overcurrent protection at its load end. Section 230-91(a) requires the overcurrent-protection device(s) to be either an integral part of (circuit breaker) or located adjacent to (fused knife switch) the service-disconnecting means.

The service-disconnecting means can be a fused knife switch or circuit breaker in an individual enclosure, or it can be the main breaker in a panel. The panel with main [Sec. 230-70(c)] would have to be rated as service equipment.

When a single service is used to supply more than one building, the overall volt-amp or current load in each respective building is used to determine whether each individual building can be supplied by a single branch circuit, or if a lighting-and-appliance or power distribution panel must be installed in each building.

The degree or amount of loading, and how each individual building is supplied (sub-panel feeder conductors or just a single branch circuit) will determine the grounding requirements for each individual building.

To meet the requirements of Sec. 240-21, the feeder or branch-circuit conductors must be provided overcurrent protection and a means of disconnect (Secs. 240-40 and 240-80) at the service point of the host (parent) building.

If the feeder circuit to the other building will be supplied from one of the two to six disconnects allowed by Secs. 230-71(a) and 230-72(a), it must be rated as service equipment.

The impedance of the increased wire length of the feeder conductors to the other building(s) limits the amount of fault overcurrent available to these buildings to the point that at the host building they are treated as additional services, but at the other building(s) they are treated as either feeders or branch circuits.

Article 225 governs the installation of outside feeder and branch circuits that would be used to supply the other buildings. Section 225-8(b) requires the feeder or branch circuit to have a second disconnecting means installed at each building or other structure load center at the point where the circuit conductors enter the building. For industrial installations, Exception No. 1 will allow the second service disconnecting means to be omitted (panel with main lug only).

Section 225-8(c) requires the second disconnecting means to be rated as service equipment because it is the central supply point for power in that building. Exception is made for garages or other outbuildings of dwelling services where 3- or 4-way lighting snap switches are used to control the other building loads.

The loads in the other building, controlled by the snap switches, must be limited to a single branch circuit. If the loads in the dwelling garage or other structure warrant a feeder circuit and secondary load center, this exception does not apply.

If for convenience a panel with a main circuit breaker is installed at the other building(s), effective selective overcurrent coordination is a must (Sec. 240-12), since overcurrent protection is provided at both ends of the feeder conductors.

If the volt-amp or current load in the other building(s) is such that only a small panel (limited number of branch circuits) is required, then the panel can be subfed from the service equipment panel of the parent building without the use of a gutter and the available two to six service-disconnecting means. The panel in the other building(s) would still have to be installed in the same manner (required disconnect) as was described for panels that were sourced from a group of two to six service-disconnecting means.

When the other building(s) is a warehouse or other building facsimile, and the lighting or other current loads are of low enough value (magnitude) that the building can be supplied by a single branch circuit, then the branch circuit can also originate in the service equipment panel of the parent or host building from a fused switch or circuit breaker.

9.17 GROUNDING FOR MORE THAN ONE BUILDING

If the two to six service-disconnecting means for the parent (host) and other supplied buildings are separate enclosures mounted on a common gutter box (auxiliary gutter), the grounded service neutral conductor contained in the gutter box, which is required by Sec. 250-23(b) to be tapped and run to each service enclosure, serves as the central bonding or tie point for the service grounding-electrode system.

If the panel or other load center in the host (of the service) or parent building is supplied as a subfed panel from its service-disconnecting means located outside the building with the other building(s) service-disconnecting means, the subfed panel must be supplied with a separate equipment-grounding conductor sized in accordance with S & T 250-95. Section 250-23(a) requires the grounded feeder neutral circuit conductor to be insulated and isolated from equipment grounding connections on the load side of the service-disconnecting means (past the service enclosure).

If the parent building service-equipment power panel is located inside and feeder or branch circuits are run back outside to the other building disconnecting means, the grounded neutral busbar in this panel can have common connections between the load-side equipment-grounding conductors and the grounded circuit neutral conductors that support feeder or branch circuits within the parent building.

Section 250-23(a) requires a service grounding-electrode system to be established at the parent building. Section 250-24(a) requires an additional grounding-electrode system to be established with the feeder or branch-circuit neutral (or other grounded-circuit conductor) in the disconnect or panel enclosure at all the other buildings (Fig. 9-24).

As the regrounding of the service neutral conductor at the building service-disconnecting means corrects the problem of voltage drop when the transformer is located outside the building, regrounding the feeder or branch-circuit neutral conductor at the other building serves the same purpose—it brings the earth in the immediate vicinity of the building and any grounded surfaces within the building up to the potential of the grounded neutral conductor in reference to earth ground for the grounding-electrode system at the host building.

Exception No. 2 to Sec. 250-24(a) does not require the regrounding of the circuit neutral conductor if a separate equipment-grounding conductor is extended with the circuit conductors. The grounding-electrode system of the other building must be established (connected) with (to) this separate equipment-grounding means.

The distance between the point of power supply and the farthest building, in terms of conductor wire length, can cause excessive voltage drop. The conductor wire length to this building includes both the wire length of the service drop or lateral to the parent building and the service drop or lateral from the parent building to the other building.

The higher resistance of undersized wire (in terms of voltage drop) is critical to branch-circuit installations of this type. The resistance of the conductors can be of such magnitude that it would restrict the fault overcurrent to what might be appear to the

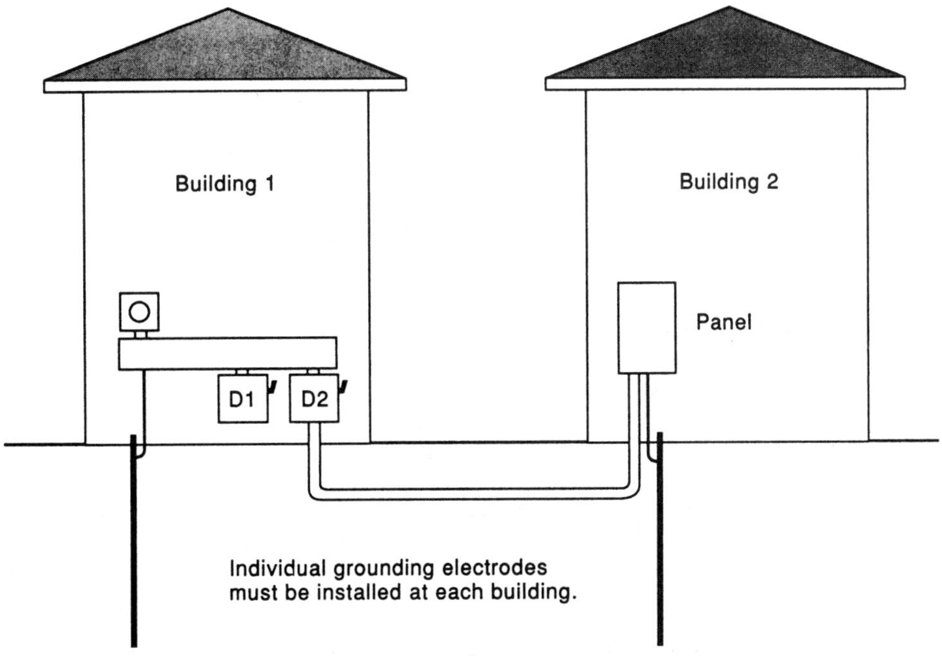

Fig. 9-24 Successive buildings supplied from a common service must have a separate service grounding-electrode system established at each building.

OCPD as normal load current (within the load-current rating of the fuse or circuit breaker), and the overcurrent-protection device would not open. Granted, this would be an extreme case of voltage drop, but any magnitude of voltage drop could delay the response of the branch-circuit overcurrent protection.

Any overcurrent-protection device (normally a fuse or CB) has inverse-time characteristics to some extent. Inverse time means that the elapsed time or speed of response to overcurrent by the device will be inversely proportional to the magnitude of the overcurrent. In other words, the higher the degree of overcurrent, the faster the overcurrent-protection device will open.

Connection of the feeder or branch-circuit equipment-grounding conductor to the grounding-electrode system in the building keeps the noncurrent-carrying metal parts and grounded surfaces of the building at a common potential with the faulted equipment or enclosure until the overcurrent-protection device in the building panel can clear the fault (open). This relieves or lessens the possibility of electrical shock to a person working in the other building.

Exception No. 1 to Sec. 225-8(b) omits the second disconnect requirement in the other building for industrial buildings under single management. When this option is taken, Sec. 250-24(c) requires the feeder circuit to contain the separate equipment-grounding conductor for common bonding of the other building grounding-electrode system. The circuit neutral or other grounded-circuit conductor [subsection (1)] must be isolated and insulated from any equipment or structure-grounding connections at the other building.

In all instances, whether service or main voltage distribution, separately derived systems of distribution or control, or power supply to more than one building, referencing the wiring system grounded circuit conductor with the equipment/structure grounding-electrode system will cause high-magnitude overcurrent to flow when a ground-fault condition occurs. The fuse or CB protecting the affected ungrounded circuit conductor will quickly blow or trip (see Chapter 7 of this text).

9.18 GFCI PROTECTION

Connecting one conductor in a distribution system to a grounding-electrode system and referencing or connecting the common bonding of all metal noncurrent-carrying parts or other surfaces within the structure that are considered to be grounded to that same grounding-electrode system is designed to provide a low-impedance path for ground-fault overcurrent when ungrounded circuit conductors come in contact with the metal parts or other grounded surfaces.

Although the response time is reduced considerably with this equipment-grounding system, several cycles of high-magnitude ground-fault overcurrent has to flow before the circuit breaker or fuse will respond (open) to the overcurrent condition. This high-magnitude ground-fault overcurrent can be deadly to a person if he or she is the connection between the ungrounded circuit conductor and the ground-return path to the grounding-electrode system.

Fortunately, body resistance for most people is high value as compared to circuit or load resistance in a distribution system; therefore, the flow of hazard current is kept at a low value (Ohm's law). This is the reason a person can get accidentally shocked without being electrocuted. But changes in body chemistry or dampness of the skin can reduce the body resistance to a much lower level. When this happens, the shock hazard due to ground-fault overcurrent can be increased to the point of possible electrocution, although the level of current flowing through the body would not be high enough to open the overcurrent-protection device.

To protect people against this type of hazard, ground-fault circuit interrupters (GFCIs) were developed to respond to ground-fault overcurrent at a preset level that is lower than what is required to stop the heart of the individual.

A GFCI protection device, which can be either a circuit breaker or a duplex receptacle, is designed to monitor the current flowing into and out of or back from a connected load (Fig. 9-25). A GFCI protection device will open the supply circuit when an imbalance of 5 milliamps occurs within the circuit. In other words, the current being conveyed to the load should be equal to the current returning from the load. If only part of the current returns, due to leakage current through faulty insulation, the presence of water in the equipment or conductor connections, or a person accidentally coming in contact with an ungrounded circuit conductor, the protective device will trip when the difference equals or exceeds five-thousandths of an ampere.

GFCI circuit breakers are available for both single-phase, phase-to-neutral and single-phase, phase-to-phase loads. Phase-to-neutral loads are supplied by a single-pole, 120-volt circuit breaker. Phase-to-phase loads (such as a circulating pump for a swimming pool) are supplied by a double-pole, 240-volt circuit breaker. GFCI receptacles are available only in the 125-volt rating.

GFCI is "people protection" and should be installed where persons are required to use electrical equipment around water or wet environments. Equipment that operates at the lower phase-to-neutral voltage and is used outside at ground level (the person using or exposed to the electrical equipment is in direct contact with the earth or ground) should be GFCI protected.

For dwelling structures [Sec. 210-8(a)], the NE Code is very specific as to which branch circuits (supplying selected areas) are to be GFCI-protected. All receptacles installed in bathrooms, all grade-level or floor-level accessible convenience receptacles located outside the structure, including garages, carports, or unfinished accessory buildings,

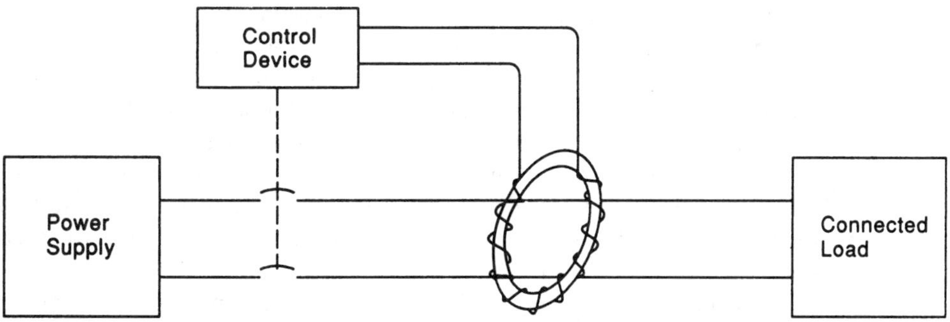

Fig. 9-25 The correct operation of a ground-fault circuit interrupter depends on the ac supply and return leads passing through a common toroidal coil, which functions as a current transformer.

and all convenience receptacles installed in underfloor crawlspaces or unfinished basements must be GFCI-protected. All countertop convenience receptacles located in the kitchen and any convenience receptacles located within six feet of the edge of a wetbar sink (measured radially from the edge) in any room must be GFCI-protected.

When a temporary service is installed on a construction site as allowed by Art. 305, all 125-volt, single-phase, 15- and 20-amp convenience receptacles are required by Sec. 305-6(a) to be ground-fault protected. The GFCI requirements can be avoided, if the rigorous test and follow-up procedures of Section 305-6(b) are practiced.

The importance of ground-fault circuit interrupters in commercial and industrial installations is not stressed by the NE Code as emphatically as it is for dwelling installations or temporary power for construction sites. Ground-fault protection for personnel, either in the form of GFCI circuit breakers (Sec. 215-9), or receptacles, should be installed on all neutral-load branch circuits where employees use hand tools or portable appliances in wet, damp, or grounded areas [refer to Sec. 210-8(b)].

If workbenches in plant shops or labs have grounded conductive tops or frames (such as steel, cast iron, or aluminum), convenience receptacle circuits supplying these work areas should be GFCI-protected. These are not NE Code requirements but good electrical practice for design and installation.

GFCI protection is also required for decorative lighting in fountains used for air-conditioning heat exchangers, in circuits supplying spas or hot tubs, and in circuits supplying therapeutic pools or tubs in health-care facilities. Article 680 lists the electrical installation requirements, including GFCI protection, equipment grounding, and structural bonding for swimming pools, fountains, spas and hot tubs, and similar installations that involve the use of water and wet locations.

9.19 GROUND-FAULT PROTECTION OF EQUIPMENT

The ground-fault circuit interrupter, which is "people protection," should not be confused with the ground-fault protection of equipment. Equipment ground-fault protection is required by Sec. 230-95 for services employing a solidly grounded wye, 3-phase, 4-wire distribution system rated at 1,000 amps or more, whose operating voltage is more than 150 volts to ground, but not more than 600 volts phase to phase.

The same requirements are placed on feeders with this type of rating by Sec. 215-10, and on the overcurrent protection of either services or feeders by Sec. 240-13.

Ground-fault circuit interrupters (GFCIs) and ground-fault protection (GFP) of equipment both sense fault current using current transformers. They differ mainly in their operating current levels. GFCI sensors are required to open the circuit when an imbalance of 5 milliamps is detected. Ground-fault protection for equipment is required by Sec. 230-95(a) to operate at an imbalance of 1,200 amps. The time delay is restricted to a maximum of one second if the imbalanced current (ground-fault overcurrent) can attain or exceed a magnitude of 3,000 amperes.

The requirement of ground-fault protection for equipment on this type of service can be avoided by using more than one service disconnect whose ratings are all less than 1,000 amps. Section 230-71(a) will allow up to six disconnects per service if they are grouped in any one location.

REVIEW QUESTIONS—PART ONE

Questions 1 and 2: In reference to blowing wind, the amount of charge that can accumulate on a building surface, and the rate at which it will discharge, is determined by the building materials used in its construction.

1. If the building is of wood construction, the number of free electrons that can be moved along the surface of the building is limited because dry wood will function as a good ____________.

2. If the building is of brick, block, or concrete construction, the number of electrons that can be moved by the wind is increased, as compared to dry wood, because masonry or other aggregate holds moisture, and water is easily ____________.

3. As defined in the NE Code, ____________ is the permanent joining of two or more metal parts to form a low-impedance electrical path that can safely conduct any current that is likely to be imposed.

4. As defined in the NE Code, ____________ is a low-impedance path into the earth's surface, either by direct penetration, or by direct connection to a bonding grid that has been effectively grounded through a grounding-electrode system.

5. The bonding grid of the metal parts for an in-ground swimming pool must be be mechanically interconnected with a minimum of wire size ____________ bare copper conductor.

6. The interconnection or bonding of all the metal parts contained within a building or other structure ensures that voltage gradients cannot be maintained between adjacent parts or surfaces. The static sources of electricity are still there, and the transfer of electrons still occurs, but the bonding grid provides a low-impedance return path, such that the static charge is quickly discharged on a continual basis.

True or False

7. The common bonding grid of a building or other structure effectively dissipates the energy of all the static sources except ____________. To correct this, the bonding grid within the structure, connected to a grounding-electrode system, forms a ____________ conductive path into the earth.

8. When the power transformer for the service of a building is located outside, the service neutral or other grounded-circuit conductor must be referenced to earth ground at both ends.

True or False

9. The lists of grounding electrodes that might be available in a given building or other structure include ______________________________, ______________________________, ______________________________, and a ____________________ electrode.

10. The lists of made grounding electrodes that an electrician may have to construct or install at a given building or other structure include ______________________________, ______________________________, and a ____________________ electrode.

11. Interior metal water pipe is required to be bonded to the service grounding-electrode system as if it were an electrode, because the interior metal water pipe is often used as a common connection (bus) when multiple grounding electrodes are to be connected to a single service.

True or False

12. Of all the available grounding electrodes, the metal underground water pipe is the only one that must be supplemented with at least one additional electrode.

True or False

13. To be used as the service grounding-electrode system, it is up to the installing electrician to ensure that the building steel is effectively bonded together and interconnected to a grounding-electrode system.

    True or False

14. If the rebar in the concrete footings of the building are to be used as the service grounding-electrode system, the rebar must be a minimum of ____________ feet long, at least ____________-inch in diameter, and encased in a minimum of ____________ inches of concrete. The rebar must be located near the ____________ of the concrete footing to ensure low-resistance to the earth.

15. Instead of rebar, ____________ feet of ____________ copper conductor can be encased in the concrete footing to serve as the required service grounding electrode.

16. A ground-ring service grounding electrode consists of a minimum of ____________ feet of copper conductor which is buried a minimum of ____________ feet deep. The conductor must be continuous and ____________ the building.

17. If approved by the local gas authority, underground metal gas pipe, either natural or bottled propane, can be used as the required service grounding electrode, provided at least 10 feet of unbroken pipe length is in direct contact with earth, and any coupling or other connections in-line between the grounding-electrode connection and the 10 feet of pipe is effectively jumpered around.

    True or False

18. Metal undeground pipe or tank systems which contain nonvolatile fluids can be used as the required service grounding-electrode system.

    True or False

19. Whether driven vertically, at a 45° angle, or bent and buried in a 1½ foot trench, an 8-foot rod or pipe can be used as the required service grounding electrode.

    True or False

20. A buried metal plate is suitable as a service grounding electrode if it offers a minimum of 2 square feet of area on each side for direct contact to the earth.

    True or False

21. Unless the installing electrician is willing to do a resistance check for 25 ohms or less on the installation of a made grounding electrode, a minimum of two separate made electrodes are required by the NE Code.

    True or False

22. If the driven rod is supplemented with a second driven rod, the two rods have to be a minimum of ____________ feet apart. The second rod should be ____________ back to the first.

23. Existing driven rods for lightning-arrestor grounding-electrode systems can be used as (in place of) the required service grounding electrode, provided the required resistance check for made electrodes is implemented before final inspection.

    True or False

24. Grounding-electrode conductors must be sized in accordance with S & T ____________, and connected by a method recognized by Sec. ____________.

25. For service or other circuit conductors to be paralleled, all the conductors paralleled to any one phase must be the ____________ wire size, which is at least ____________ AWG, have the same ____________ length, be of the ____________ conductor material, have the ____________ type of insulation, and be terminated in the same ____________.

REVIEW QUESTIONS—PART TWO

1. When ungrounded service conductors are paralleled, the size of the grounding-electrode conductor must be determined from the largest computed (summed or equivalent) overall area of all the conductors paralleled to a given phase.

   True or False

2. When the paralleling of the ungrounded service conductors yields an overall circular-mil area that is larger than 1,100 kcmil copper or 1,750 kcmil aluminum or c-c aluminum, the grounding-electrode conductor is never required to be larger than ____________copper or ____________aluminum or c-c aluminum.
3. Grounding-electrode conductors extended to either metal underground water pipe, other underground metal piping systems, or building steel must be full size as determined by S & T ___________. Grounding-electrode conductors extended to the other either available or made electrodes are sized by the Exception to Sec. ____________.
4. If the service power distribution for a building is a 3-wire, 575-volt delta, or other voltage delta system that is not referenced to ground, a grounding-electrode system does not have to be established for the noncurrent-carrying metal parts of the building or the power distribution equipment within the building.

   True or False

5. If a grounding-electrode conductor is installed in metal pipe for protection against physical damage, the NE Code requires the metal pipe to be made part of the grounding-electrode conductor by bonding at both ends.

   True or False

6. The grounded service neutral conductor of a building or other structure must be sized according to the maximum unbalanced load current that can occur due to varying load conditions and according to the maximum ground-fault overcurrent available from the transformer at the load-end of the ungrounded service conductors.

   True or False

7. When minimum current loading occurs, the reduced in wire size grounded service-neutral conductor must be at least equal in wire size to the required grounding-electrode conductor of the service.

   True or False

8. The equipment-grounding means for all the noncurrent-carrying metal parts of the distribution has to be established, regardless of whether or not the building or other structure has an established bonding grid and grounding-electrode system.

   True or False

9. When the circular-mil area of single ungrounded service conductors, or the total c.m.a. of ungrounded service conductors in parallel, exceeds ____________ copper, or ____________ aluminum or c-c aluminum, the c.m.a. by wire size of the grounded service-neutral conductor for these higher-amperage services must be at least equal to ⅛ the area of the largest c.m.a. by phase.

10. When grounded service-neutral conductors are paralleled, the smallest wire size allowed by the NE Code is ____________.

11. The interconnection of metal raceway and other metal enclosures used in the power distribution, which house (contain) individual insulated circuit conductors, can serve as the required equipment-grounding system for the metal enclosures of lighting fixtures, motors, or other load-side equipment, provided the raceway is connected to the grounded service-neutral conductor in the service equipment enclosure.

    True or False

12. Because the grounded service-neutral conductor is interconnected with the bonding grid and the grounding-electrode system of the building or other structure, any ungrounded fused conductor of the power distribution that has a potential difference with the grounded neutral conductor of either the service, feeder, or branch circuit, has the same potential difference with the grounded surfaces of the floor, walls, support beams of the building (building steel), and the metal raceway, metal boxes, or other metal enclosures that support the wiring distribution or the connected loads.

    True or False

13. On the line side of the service OCPD, equipment-grounding conductors or bonding jumpers are sized according to the largest wire size, or largest total circular-mil area of the ungrounded service-entrance conductors paralleled to one phase and Table ____________.

14. On the load side of the service OCPD, equipment-grounding conductors or bonding jumpers are sized according to the rating of the feeder or branch-circuit OCPD and Table ____________.

15. When the wire size of the largest ungrounded service-entrance conductor exceeds ____________ copper or ____________ aluminum or c-c aluminum, the c.m.a. by wire size of a single bonding jumper or other equipment-grounding conductor extended to any metal service enclosure must at least equal ⅛ the wire size of the largest ungrounded service-entrance conductor.

16. When load-circuit conductors are adjusted in wire size to compensate for voltage drop, the equipment grounding conductor must also be adjusted in wire size for the same compensation.

    True or False

17. As a general rule, load-circuit grounded neutral conductors can be used as the equipment-grounding or bonding means for the noncurrent-carrying metal parts of the power distribution on the load side of the service equipment.

    True or False

18. A separately derived system must be treated as a service in terms of equipment grounding.

    True or False

19. The grounding electrode is required to be located in the same area as the transformer and panel of the separately derived system.

    True or False

20. A separately derived source can be connected to a separate grounding-electrode system that is not connected to the building bonding grid or the parent-voltage distribution grounding-electrode system.

True or False

21. Because the interior metal water pipe is bonded to the equipment-grounding busbar in the panel inside the mobile home, the fourth-wire equipment grounding conductor in this single-phase, 3-wire feeder must be sized according to the wire size of the ungrounded service conductors supplying the mobile home service and Table ____________, and by the rating of the panel OCPD in the mobile home and Table ____________.

22. Unless the separately derived power supply for large-scale information technology equipment is double insulated, the system requires a grounding-electrode system that is interconnected to the building-service grounding-electrode system.

True or False

23. To help reduce noise problems on the distribution equipment-grounding system, information technology equipment can be supplied through a receptacle that offers an isolated equipment-grounding means connected to the equipment-grounding system only at the service equipment-grounding terminal or central grounding point on a separately derived system.

True or False

24. The service-disconnecting means for either a single or several buildings, can be a single main disconnect, or a group of two to six individual disconnecting means.

True or False

25. When other buildings will be supplied from the service of one building, each building must have an established grounding-electrode system.

True or False

REVIEW PROBLEMS

1. Determine the NE Code-minimum wire size of a copper grounding-electrode conductor run to the building steel grounding electrode, if a retail store will be supplied with a 150-amp service. Use THW copper for the service-entrance conductors.
2. A small manufacturing concern will be supplied by a 3-phase, 4-wire, 208Y/120-volt service rated at 175 amps. If the service-entrance will consists of XHHW-2 copper conductors installed in conduit run down the side of the building where the ambient temperature will reach 115°F, determine the NE Code-minimum wire size of the copper grounding-electrode conductor run to underground metal water pipe.
3. A large manufacturing concern will be supplied by a 3-phase, 4-wire, 480Y/277-volt service rated at 1,000 amps. If the service-entrance will consists of 250 kcmil THHN copper conductors paralleled four times per phase, determine the NE Code-minimum wire size of the copper grounding-electrode conductor run to effectively grounded building steel.
4. A service rated at 500 amp will be supplied with No. 3/0 AWG THW copper ungrounded service-entrance conductors paralleled three times per phase. Determine the NE Code-minimum wire size of the copper grounding-electrode conductor run to an underground metal tank farm.

5. A single-phase, 3-wire, 120/240-volt service rated at 200 amps will be supplied with No. 3/0 AWG THW copper ungrounded service-entrance conductors. If the service has only minimum phase-to-neutral loading, determine the NE Code-minimum wire size of the service-entrance neutral conductor for ground-fault ampacity requirements.
6. A service rated at 1,200 amps will be supplied with 350 kcmil THW aluminum ungrounded service-entrance conductors paralleled five times per phase. If the service has only minimum phase-to-neutral loading and all the parallel service-entrance conductors will be installed in a common gutter, determine the NE Code-minimum wire size of the single grounded service-entrance neutral conductor for ground-fault ampacity requirements.
7. A service rated at 500 amps will be supplied with 2,000 kcmil THW aluminum ungrounded service-entrance conductors. If the service has only minimum phase-to-neutral loading, determine the NE Code-minimum wire size of the grounded service-entrance neutral conductor for ground-fault ampacity requirements.
8. A service rated at 2,000 amps will be supplied with 600 kcmil THW copper ungrounded service-entrance conductors paralleled five times per phase. If the service has only minimum phase-to-neutral loading and all the parallel service-entrance conductors will be installed in a common gutter, determine the NE Code-minimum wire size of the single grounded service-entrance neutral conductor for ground-fault ampacity requirements.
9. A service rated at 1,400 amps will be supplied with 300 kcmil THHN copper ungrounded service-entrance conductors paralleled five times per phase. If the service has only minimum phase-to-neutral loading and all the parallel service-entrance conductors will be installed as sets in five parallel rigid conduits, determine the NE Code-minimum wire size of the individual, parallel service-entrance neutral conductors for ground-fault ampacity requirements.
10. A 3-phase, 4-wire, 240/120-volt, high-leg delta service rated at 300 amps will be supplied with 350 kcmil XHHW-2 copper ungrounded service-entrance conductors. Determine the NE Code-minimum wire size of the copper bonding jumper used to ground the service-equipment enclosure.
11. As allowed by the Note 3 table to T310-16, a 150-amp dwelling service can be supplied with No. 2/0 AWG aluminum SE cable. Determine the NE Code-minimum wire size of a copper bonding jumper run to ground the meter can.
12. A service rated at 1,200 amps will be supplied with 500 kcmil THW c-c aluminum ungrounded service-entrance conductors paralleled three times per phase. Determine the NE Code-minimum wire size of the c-c aluminum bonding jumpers extended one each to the three parallel conduits.
13. A service rated at 1,200 amps will be supplied with 500 kcmil THW c-c aluminum ungrounded service-entrance conductors paralleled three times per phase. Determine the NE Code-minimum wire size of a single c-c aluminum bonding jumper loop-connected to ground all three parallel conduits.
14. A service rated at 500 amps will be supplied with 2,000 kcmil THW c-c aluminum ungrounded service-entrance conductors. Determine the NE Code-minimum wire size of the c-c aluminum bonding jumper run to the service conduit.
15. A service rated at 3,000 amps will be supplied with 1,500 kcmil THW c-c aluminum ungrounded service-entrance conductors paralleled six times per phase. Determine the NE Code-minimum wire size of the c-c aluminum equipment-grounding conductor used to bond the metal gutter of the service.
16. Determine the NE Code-minimum wire size of a copper equipment-grounding conductor run to the enclosure of a subfed 600-amp panel in an industrial concern.
17. Determine the NE Code-minimum wire size of a copper equipment-grounding conductor used to ground a single wireway that contains circuits rated 15, 30, 40, and 60 amps.

18. The feeder circuit for a 3-phase, 4-wire, 480Y/277-volt lighting panel rated at 1,200 amps, will consists of 350 kcmil THW copper conductors paralleled four times per phase. Determine the NE Code-minimum wire size of individual copper equipment-grounding conductors installed in each parallel run of rigid metal conduit.

19. A 3-phase, 4-wire, 480Y/277-volt load center is rated at 100 amps. The nonlighting load center is located 900 feet in wire length from the service equipment panel. Using THHN copper conductors, size the feeder-circuit conductors and the equipment-grounding conductor.

Chapter 10

Multioutlet Branch Circuits

A branch circuit, as defined in Art. 100, is the circuit conductors between the last overcurrent-protection device and the outlet(s) or utilization equipment served. The feeder circuit or circuits are any circuit conductors between the final (last) overcurrent-protection device (supplying the branch circuit), and any overcurrent-protection device(s) up to the load side of the service or switchgear equipment.

Main feeders, having overcurrent protection at the switchgear, will normally supply either busduct (busways, Art. 364) or panels (panelboards, Art. 384). Feeders extending from the busduct or panel are overcurrent-protected at the busduct or panel and, in turn, supply other load centers for individual branch circuits.

To classify a branch circuit or a feeder circuit, the correct assigned voltage must be known. The voltage supplied will be determined by the type of load served. Is it a phase-to-neutral load, a single-phase, phase-to-phase load, a 2-phase load, or a 3-phase load? Section 220-2 assigns the nominal voltages to be used when making NE Code calculations from the power rating (VA or kVA) to the required current. If other voltages are specified in test questions or other examples, Sec. 220-2 states that those voltages should be used for the calculations instead of the voltages listed in the section.

10.1 BRANCH-CIRCUIT CLASSIFICATIONS

Branch circuits have two major divisions: multioutlet (general-purpose) and individual (dedicated or specific-purpose).

Multioutlet branch circuits supply mainly lighting and convenience receptacle loads. Convenience (duplex) receptacles installed in all inhabitable rooms of dwellings and in all rooms of nondwellings are supplied by multioutlet branch circuits.

Multioutlet branch circuits can supply appliances or other utilization equipment that is either fixed (fastened in place) or stationary (not easily moved) and is either supplied by cord-and-plug connection to a receptacle or by a fixed (permanent) wiring method.

Portable equipment is almost always supplied as cord-and-plug and fed from a receptacle. Appliances or other utilization equipment that have a medium-to-heavy power consumption are usually supplied by an individual branch circuit and either plug- and cord-connected through a single receptacle, or wired directly by a fixed wiring method.

Feeders, by definition, supply two or more branch circuits that can be either individual or multioutlet. Feeder conductor ampacity and installation requirements are governed by Art. 215. Feeder load calculations are governed by Part B of Art. 220.

The installation and loading allowances of multioutlet branch circuits are governed by Art. 210. This article lists maximum loads, number of required outlets, outlet ratings, branch-circuit classifications, and other installation requirements for multioutlet branch circuits. The installation and load allowances of individual branch circuits are governed by the particular article that governs the appliance, motor, or other utilization equipment supplied by the branch circuit. Section 210-2 lists most, if not all, of these articles. Another list (more detailed) is found in Sec. 240-2, overcurrent protection requirements.

To determine either the volt-amp (VA) rating, the kVA rating, or the ampere rating of an individual branch circuit, all the conditions of the pertaining article must be satisfied. Chapter 4 contains articles that govern equipment for general use. Chapter 6 contains articles that govern equipment for special use.

To determine the VA, kVA, or ampere (current) rating of a multioultet branch circuit, both Arts. 210 and 220 must be used. Article 220, Part A lists assigned values that must be used for lighting and receptacle outlets installed on multioutlet branch circuits. The minimum number of required branch circuits is required by Sec. 220-4(a) to be determined from the total computed load of lighting and other outlets required for the building or other structure, and from the size and rating of the branch circuits used. The determined number of branch circuits must have a combined rating at least equal to the total computed load, and no load on any individual multioutlet branch circuit (as determined by Sec. 210-22) can be exceeded.

Section 210-6 lists the voltage limitations put on all branch circuits (either individual or multioutlet) for structures and/or occupancies. The maximum operating voltage of lighting fixtures or power hand tools, portable appliances, or other hand-held utilization equipment in the home or similar structure is restricted by the occupancy limitations of Sec. 210-6(a) to a maximum of 120 volts.

Lighting equipment, either electric-discharge or mogul-base incandescent, is allowed by Sec. 210-6(c) to operate at 277 volts to ground (lower phase-to-neutral voltage of a 3-phase, 4-wire, 480-volt wye distribution), provided the fixtures are listed for this voltage.

Athletic field lighting, parking-lot lighting, tunnel lighting, or other types of high-intensity outside lighting fixtures are allowed to operate at 600 volts (phase to phase). Certain height restrictions apply. Section 210-6(d)(1)a. requires a minimum height of 22 feet for outside pole lighting. For roadway or other types of tunnel lighting, the minimum height for the higher voltage is reduced by Sec. 210-6(d)(1)b. to a minimum of 18 feet.

Part B of Art. 210 governs branch-circuit ratings. Section 210-19(a) requires branch-circuit conductors to have an ampacity at least equal to the maximum load connected to the branch circuit. The branch-circuit conductors for multioutlet branch circuits must also have an ampacity rating equal to the branch-circuit overcurrent-protection rating [Secs. 240-3 and 240-6(a)]. The branch-circuit rating, as outlined in Sec. 210-3, is classified according to the rating of the branch-circuit overcurrent-protection device. Section 210-19(c) requires branch-circuit conductors to be wire size No. 14 AWG or larger (note exceptions).

Section 210-20 lists three conditions that must be met in sizing the overcurrent-protection device (OCPD) (fuse or circuit breaker) for the branch-circuit conductors. Condition (1) protects the conductors according to their allowed ampacities as listed in Tables 310-16 through 310-19 with all applicable derates applied. Section 240-6(a) lists the standard sizes (ratings or settings) of overcurrent-protection devices. Section 240-3 requires the rated-load ampacity of the circuit conductors to be greater than or equal to the rating or setting of the OCPD.

Condition (2) allows the branch-circuit OCPD to be sized according to the current-load requirements (Sec. 210-2) of the specific equipment supplied. Condition (3), which is discussed in the next section, limits the rating of the circuit (limits the rating of its OCPD) according to what type of lighting or convenience receptacle loads are supplied.

10.2 MULTIOUTLET BRANCH CIRCUITS

Section 210-3 list the classifications of multioutlet branch circuits by the standard ratings of 15, 20, 30, 40, and 50 amperes. Section 210-23 lists the maximum permissible loads that can be connected to these multioutlet branch circuits. Both Secs. 210-3 and 210-23 exempt special-purpose branch circuits from these listed maximums by allowing individual branch circuits to be rated according to the load they supply.

Section 210-23(a) will allow 15- and 20-amp branch circuits to supply any combination of lighting and receptacle outlets and outlets intended for appliances or other utilization equipment. The rating of any one cord-and-plug connected appliance or other utilization equipment (receptacle supplied) is restricted to 80% of the branch-circuit rating. The full-load current (FLC) of a refrigerator supplied by a 15-amp multioutlet branch circuit cannot exceed 12 amps (80% of 15 amps = 12 amps).

As the section reads, two cord-and-plug connected appliances could be supplied by a 15- or 20-amp branch circuit as long as neither of their FLC ratings exceeds 80% of the branch-circuit rating. But this is not true. Section 220-4(a) states that the total connected load on a multioutlet branch circuit cannot exceed the maximum loads specified in Sec. 210-22. Section 210-22 forbids the maximum load rating (current rating) connected to any branch circuit from exceeding the branch-circuit rating.

The FLC or total load-current rating of a wall-mounted water cooler (equipment fastened in place) on a multioutlet branch circuit cannot exceed 50% of the branch-circuit rating, when lighting outlets and other utilization equipment, which are not fastened-in-place but are cord-and-plug connected, are also supplied.

This rule applies even if the water cooler is cord-and-plug connected. If two water coolers are installed on a multioutlet branch circuit rated at either 15 or 20 amps, their combined ratings or FLCs cannot exceed 50% of the branch-cirucuit rating, if other equipment and/or lighting outlets are also supplied by the same branch circuit. If other outlets are not served, and the branch circuit supplies only the water coolers, the 50% restriction is lifted and their combined full-load current ratings cannot exceed the current rating of the branch circuit.

Section 210-23(b) will not allow 30-amp multioutlet branch circuits to supply lighting outlets in a dwelling structure. Lighting fixtures equipped with heavy-duty lampholders can be installed on 30-amp branch circuits in structures other than dwellings. This section and Sec. 210-21(a) both restrict the supply and installation of regular fluorescent tube lighting to either 15- or 20-amp branch circuits.

The lampholders for fluorescent lamps or tubes are usually of the light- to medium-duty type. The branch-circuit conductors of electric-discharge lighting (fluorescent as well as other types) are actually connected to the ballast rather than the lampholders, but the higher wattage ratings required for lampholders to qualify for the heavy-duty type puts the 20-amp restriction on the ballast circuit as well.

Section 210-22(b) requires the total volt-amp load rating of fluorescent lighting fixtures to be based on the wattage rating of the lamps and on the power losses of the ballast. In other words, the total volt-amp load allowance must be based on the full fixture volt-amp rating and not just on the wattage ratings of the lamps alone.

Section 210-23(c) will allow 40- and 50-amp multioutlet branch circuits to supply fixed lighting units equipped with heavy-duty lampholders, infrared heating units, and other types of utilization equipment in any nondwelling occupancy. These circuits can supply cooking equipment that is fastened in place in any structure or occupancy.

Multioutlet branch circuits rated 50 amps or larger are not allowed by Sec. 210-23(d) to serve any lighting outlets.

The required conductor size, the required overcurrent protection rating, the types of equipment allowed, the receptacle ratings, the maximum loads, and the permissible loads for the various multioutlet (general-purpose) branch circuits are summarized in Table 210-24.

10.3 CONVENIENCE RECEPTACLES

Section 210-21 requires an outlet device (either a lighting fixture, lampholder, or a receptacle) to have an ampere rating at least equal to the load served. Subsection (b) governs the rating of receptacle outlets, and Sec. 210-21(b)(1) requires a single receptacle to have an amp rating equal to the branch-circuit ampere rating when the branch circuit (individual or dedictated) has no other outlets. A single receptacle installed on a 20-amp branch circuit dedicated to that receptacle outlet would have to be rated at 20 amps.

By definition (Art. 100), a single receptacle is a single contact device, mounted on a single strap (yoke), that is intended for the connection of a single attachment plug at an outlet. A duplex receptacle consists of two single contact devices installed on a common strap. A duplex receptacle is not a single receptacle, even though it is considered to be a single receptacle outlet. Convenience receptacles are also available in a triplex configuration (three receptacles on a single strap). This triplex configuration, normally referred to as "interchangeable devices," would also be a single receptacle outlet. The duplex is the more common in use of all three (Fig. 10-1).

An outlet is defined as a point on the wiring system from which power can be taken. A receptacle outlet is further defined as an outlet that contains one or more receptacles. By this definition, a single duplex receptacle, installed as the only receptacle outlet on a 20-amp, 120-volt branch circuit, would change the rating of the branch circuit from individual to multioutlet. As such, either a 15-amp or 20-amp duplex receptacle could be installed.

When installed on multioutlet branch circuits, individual receptacle ratings must meet the requirements of T210-21(b)(3). Only receptacles rated at 15 amps can be installed on a 15-amp multioutlet branch circuit. Receptacles rated at either 15 or 20 amps can be installed on a 20-amp multioutlet branch circuit. Only receptacles rated at 30 amps can be installed on a 30-amp multioutlet branch circuit. Receptacles rated at either 40 or 50 amps can be installed on a 40-amp multioutlet branch circuit, but only receptacles rated at 50 amps can be installed on a 50-amp multioutlet branch circuit.

The maximum load-ampere (current) rating of a cord- and plug-connected load supplied from a receptacle installed on a multioutlet branch circuit is governed by T210-

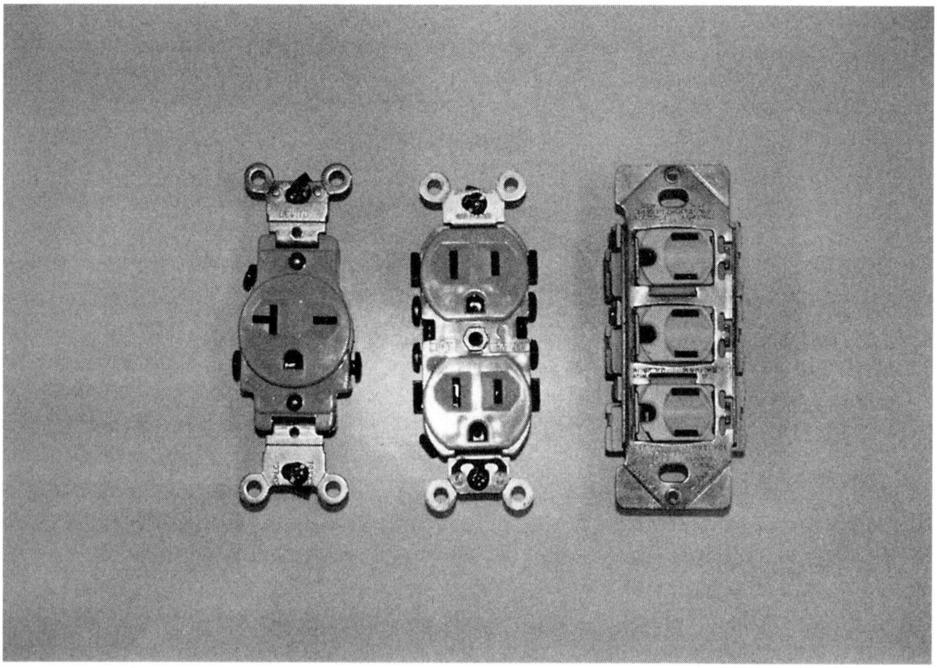

Fig. 10-1 All three of these plug-in receptacle straps represent a receptacle outlet as defined in Art. 100.

21(b)(2). The load on any receptacle, rated either 15, 20, or 30 amps cannot exceed 80% of the receptacle rating. The amperages listed by receptacle in T210-21(b)(2) correlate with Secs. 210-23(a) and (b).

Receptacles are also covered in Part L of Art. 410, along with cord connectors amd attachment plugs. Section 410-56(a) will not allow a receptacle installed for the attachment of portable cords to have a rating less than 15 amps at either 125 volts, or 250 volts. An NE Code-recognized receptacle as just described or a cord connector (the equivalent of a receptacle installed on the end of a flexible cord) cannot be suitable for use as a lampholder.

In years past it was not uncommon to install a keyless (porcelain base, screwshell) lighting fixture in a structure, and then install a plug adapter into the screwshell to be used as a receptacle outlet. In even earlier electrical installations, a twisted pair flexible cord was extended from a box and canopy in the ceiling and terminated in the middle of the room below "arm's reach" with a thumb-switch screwshell lampholder. These types of installation are not NE Code-recognized receptacle outlets.

Section 410-56(b) requires all receptacles rated at 20 amperes or less to be marked as CO/ALR if the receptacle can be connected directly to aluminum conductors. In the "copper crunch" of the 1960s, there was a giant push by the wiring industry to wire structures with aluminum wire. Aluminum wire is a good conductor of electricity, but it is more malleable (softer) than copper in a pressure connection. Also, aluminum has a higher expansion rate as compared to copper with temperature rise. Worst of all, aluminum has a tendency to react (galvanic action) to some of the other metals that are used as fasteners or connectors for what was previously a copper installation.

Branch circuits run with aluminum wire have a tendency to fail (due to loose connections that are a result of temperature expansion, galvanic action, or overtorqued connections) on receptacles that are not marked as CO/ALR. This label signifies that the connectors on the receptacle will accept either copper (CO) or aluminum (ALR) conductors. AL is the normal acronym for aluminum. The R suffixed to AL stands for Revised in this type of identification.

Receptacles, cord connectors, and attachment plugs are required by Sec. 410-56(h) to be manufactured in such a configuration (the prongs on the plugs and the squeeze connectors in the receptacle) that the receptacles or cord connectors will not accept attachment plugs or caps of a different voltage and/or current rating. Nongrounding-type receptacles and connectors must be manufactured in a discriminate configuration so as not to accept grounding-type plugs or caps.

Section 410-57 lists the installation requirements for receptacles in either damp or wet locations. The typical closed-cup cover installed on outdoor receptacles, which must be opened to plug in the equipment, is rated by Sec. 410-57(a) for only a damp location (Fig. 10-2). For an outdoor receptacle cover to qualify for use in a wet location [Sec. 410-57(b)], the receptacle must still be protected from the weather (falling rain) with the cord plug inserted into the receptacle (Fig. 10-3).

Section 410-57(c), which is new in the 1996 Edition of the NE Code, strictly forbids the installation of convenience receptacles in a bathtub or shower space. Whether GFCI-protected or not is not a qualifier in this ruling.

Section 410-57(d) requires the standpipes of floor receptacles to be installed high enough to prevent damage to the receptacles by floor-cleaning equipment.

Fig. 10-2 Outdoor receptacles must be protected from the weather.

Section 410-58 lists the construction requirements for grounding-type receptacles. Subsection (c) stipulates that a grounding terminal on a receptacle or plug shall be used for no other purpose except equipment grounding. Subsection (e) requires grounding-type plugs to be used only where equipment that requires grounding will be supplied by that plug.

Section 210-7 also governs the installation and use of receptacles and cord connectors. Subsection (a) requires all 15- and 20-amp convenience receptacles to be the grounding type. This subsection also correlates with Sec. 410-56(h) in requiring the grounding-type receptacles to be installed only on branch circuits that have the same voltage and current rating as the receptacle(s). Exception is noted in Sec. 210-7(a) for the 15-amp convenience (duplex) receptacles allowed by T210-21(b)(3) to be installed on branch circuits rated at 20 amps, and for the 50-amp receptacles which can be installed (same table) on 40-amp branch circuits.

Section 210-7(b) requires all grounding-type receptacles, except those mounted on portable generators (Exception No. 1), and replacement receptacles allowed by Sec. 210-7(d) (Exception No. 2) to have their grounding terminals effectively grounded. Subsection (c) describes the ways in which grounding-type receptacles can be grounded. The preferred way is by a separate equipment-grounding conductor. Section 210-7(c), FPN No. 1 refers the electrician to Sec. 250-91(b) for a list of acceptable equipment-grounding means.

The first FPN of Sec. 210-7(c) (not numbered) refers the electrician to Sec. 250-74, Excp. No. 4, which will allow an isolated equipment-grounding conductor, connected to the isolated grounding terminal on a special type of duplex (or single) receptacle, to be run all the way back to the service grounding-electrode system tie point in the panel or

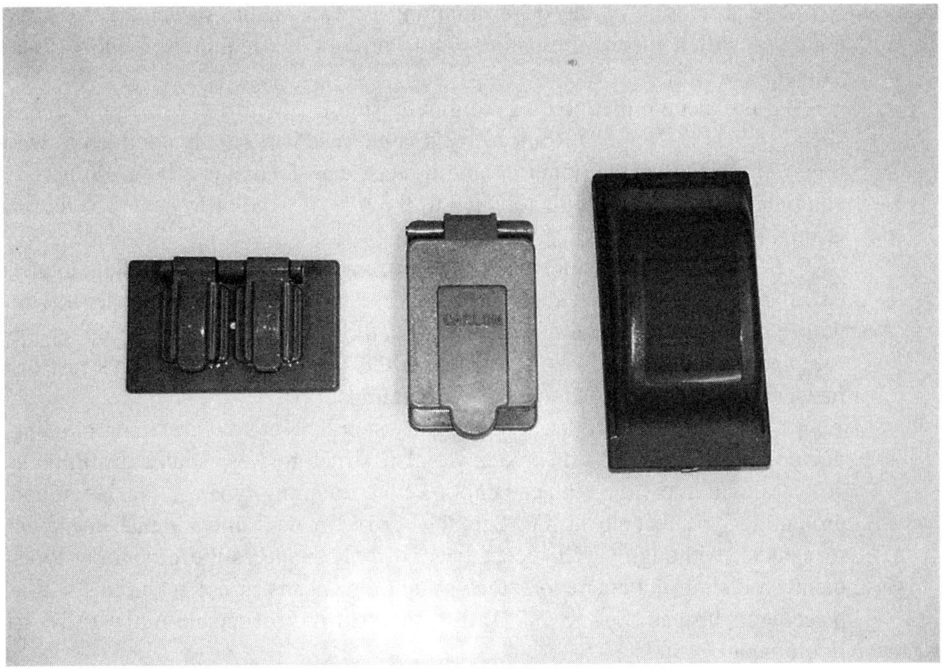

Fig. 10-3 Only the receptacle cover assembly on the right is rated as weatherproof.

switchgear (or the transformer grounding node for separately derived sources) without any other grounding connection or splice. This isolated (insulated) equipment-grounding means helps to reduce ac line noise on the supply branch circuit for computer and other electronic information technology equipment.

Section 210-7(d) addresses the problem of replacement for convenience receptacles in older structures. Not too many NE Code editions ago, grounding-type receptacles were not required for structures, so many nongrounding-type receptacles, especially in dwellings, are still in use. Sooner or later receptacles either fail or just wear out due to repetitive use. This section outlines how to replace these receptacles.

Replacement of nongrounding-type receptacles in an existing installation can be accomplished in many NE Code-correct ways. If the nongrounding-type receptacle was originally installed in a metal box, grounded by metal raceway, which would be typical of structures built of brick, block, or concrete, Sec. 210-7(d)(1) stipulates that replacement simply requires installing a grounding-type receptacle.

The general rule of Sec. 250-74 requires a bonding jumper to run between the grounded metal receptacle outlet box and the grounding terminal [green hexhead screw, Sec. 410-58(b)(1)] on the replacement grounding-type receptacle.

If the metal receptacle outlet box is surface mounted, Sec. 250-74, Excp. No. 1 will allow metal-to-metal contact of the yoke of the receptacle and the edge of the metal box to serve as the required equipment-grounding means. One of the cardboard retainers on the 6-32 mounting screws will have to be removed to facilitate the metal-to-metal contact.

If the metal receptacle outlet box is flush or recessed mounted, Sec. 250-74, Excp.

No. 2 requires one of the 6-32 mounting screws on the receptacle yoke (strap) to be equipped with a special grounding clip to provide the required grounding means (metal-to-metal contact).

If the metal outlet box was originally supplied with a 2-wire, grounding-type cable, such as No. 14/2 w/GND Romex or BX cable (use is normally confined to wooden structures), the splice of the homerun and downstream feeds must be made between the two equipment-grounding conductors and to the box, instead of to or on the receptacle screw terminal.

Section 250-114 stipulates that the arrangement of the grounding connections shall be in such a manner that disconnection or removal of the receptacle will not interfere with the continuity of the equipment grounding downstream of the outlet location. The receptacle would have to be grounded to the box in one of the ways just discussed for metal raceway–metal box receptacle grounding.

If the receptacle outlet was originally supplied by a 2-wire, nongrounding multiconductor cable (typical of the older wooden structures, especially dwellings), Sec. 210-7(d)(3)a. will allow the replacement of a nongrounding-type receptacle with another nongrounding-type receptacle. Or, Sec. 210-7(d)(3)b. will allow replacement with a GFCI receptacle. Although GFCI receptacles are only made as the grounding type, an equipment-grounding conductor or other grounding means is not required for this GFCI replacement. Instead, Sec. 210-7(d)(3)b. requires the receptacle outlet to be labeled "No Equipment Ground."

The operation of a GFCI-type receptacle does not depend on the equipment-grounding conductor to function under fault or other current-leakage conditions. Only the current-carrying conductors [the hot-leg (ungrounded) and the neutral (grounded)] are monitored on a standard 125-volt GFCI duplex receptacle. The operation is looking for what goes out (in terms of current), must come back. It doesn't matter to the GFCI operation where it went, if some of the current didn't come back. A GFCI receptacle is actually better "people protection" than a grounding-type receptacle with a low-impedance equipment-grounding means attached.

When several nongrounding-type receptacles supplied by a single circuit are to be replaced, the first one in the homerun to the panel can be replaced with a GFCI receptacle with feed-thru protection capability. All downstream receptacles can be replaced with grounding-type receptacles because they are GFCI-protected [Sec. 210-7(d)(3)c.]. An equipment-grounding means is not required for these protected receptacles, although each outlet must be labeled as "GFCI Protected" and "No Equipment Ground."

Section 210-7(d)(2) stipulates that any replacement receptacles, whether grounding or nongrounding-type, that are required by the 1996 Edition of the NE Code to be GFCI-protected (Sec. 210-8) will be replaced with GFCI receptacles, or else be protected by a GFCI circuit breaker. (Refer to Chapter 9, Sec. 9.18 in this text.)

If extensions are to be made to existing branch circuits that do not offer a grounding means (equipment-grounding conductor), Sec. 210-7(c), FPN No. 2 refers the electrician to Sec. 250-50. Only grounding-type convenience receptacles can be installed for additional outlets on an existing branch circuit.

A grounding means must be established for the added receptacle outlets. Section 250-50 (a) and (b), Exception will allow a separate equipment-grounding conductor to be added to the existing branch circuit, but instead of being routed back to the branch-circuit source (fuse or breaker service or subfed panel) with the original circuit conductors, the added conductor can be routed to the nearest accessible point on the existing grounding-electrode system for the older structure service.

This method can also be used for the replacement of nongrounding-type receptacles with grounding-type receptacles (adding the equipment-grounding conductor). Section 210-7(d)(1) refers to this same exception for providing a suitable grounding means for the receptacle outlet.

The drawbacks to running an additional equipment-grounding conductor are its size, how to fish down from the receptacle box in a concealed wall to the underside of the structure (normally a wooden structure with an underfloor crawlspace or unfinished basement), and how to protect the equipment-grounding conductor (which is often run the full length of the structure to access the service grounding-electrode system) from physical damage.

For a 15-amp multioutlet branch circuit, S & T 250-95 requires only a No. 14 AWG equipment-grounding conductor. A 20-amp branch circuit requires only a No. 12 AWG equipment-grounding conductor. These smaller conductors are not easily fished in the hollow (concealed chase) of a wall, especially an outside wall that is fully insulated (packed).

In past NE Code editions, these smaller additional equipment-grounding conductors could be run to the nearest effectively grounded metal water pipe (drinking water). But the fluent use of PVC and poly (polybutelene) nonmetallic water lines have made this almost nonexistent. (Changed to service grounding-electrode system in the 1993 edition of the NE Code.)

Section 250-92(c)(2) addresses the protection against physical abuse for these smaller added equipment-grounding conductors. The No. 14 or No. 12 AWG equipment-grounding conductor must be installed in accordance with Sec. 250-92(a), the protection of grounding-electrode conductors.

If the smaller conductor is effective, fished from the outlet location recessed in the wall to the unfinished basement or underfloor crawl space, Sec. 250-92(a) requires the conductor (smaller than wire size No. 8 AWG) to be installed in protective conduit or other raceway the length of its exposure. If metal raceway is employed, Sec. 250-92(b) requires the metal raceway to be bonded to the smaller equipment-grounding conductor at both ends (be made electrically continuous).

The easiest method of extensions of existing nongrounding-type branch circuits with additional convenience receptacle outlets is by adding nongrounding-type receptacles, or by installing a GFCI-type receptacle to protect downstream grounding-type receptacles.

10.4 LIGHTING OUTLETS

The general loading requirements of lighting outlets on multioutlet (general-purpose) branch circuits are found in Arts. 210 and 220. The general installation requirements of lighting fixtures, lampholders, and lamps are found in Art. 410. The installation of fixtures wires, their insulation, location, and ampacity requirements are governed by Art. 402. Flexible cords and cables used to supply lighting fixtures (as well as appliances and other utilization equipment) are governed by Art. 400. Switches used to control lighting outlets (as well as other electrical equipment) are governed by Art. 380.

Part B of Art. 410 governs the different locations in which lighting fixtures may be installed. The environmental atmosphere (structural air, inside or outside, and so on) will dictate the type of rating the fixture must have. Combustible material near the fixture location will dictate where and how it can be installed. And the type of fixture, whether it is a complete fixture, a lampholder, or just a pendant will also dictate where and how the fixture can be installed.

The type of mounting for the fixture (whether it is surface-mounted, recessed-mounted, or suspended on chain or cable), will also have bearing on its installation requirements. The type of lighting element, whether it is an incandescent lamp, a fluorescent tube, or some other type of electric-discharge lighting, and the element's operating voltage will control both the fixture location and its installation requirements.

A lampholder is just as the name implies, a holder for a lamp. Usually the lampholder is of the screwshell type, which may have a porcelain base that mounts over a lighting outlet box (often referred to as a keyless). The lampholder may be a brass screwshell base that mounts on top of a lamp stem. These are the common free-standing lamps used for decoration and lighting in both the home and office. This type of lamp usually contains a switch, either a key switch that must be turned or a pushbutton switch that is pushed from one side of the lampholder to the other. These brass screwshells can also be mounted on the end of a cord and be installed either as a pendant lampholder or as a portable lampholder. A lampholder can also be the pin type that is used with tubes in electric-discharge lighting.

Section 410-4(a) requires fixtures installed in either a wet or damp location to be identified (labeled or marked) as suitable for the installation. The lighting fixture must be installed in such a manner or location that water cannot enter or accumulate inside the fixture.

Fixtures installed in direct contact with the earth, masonry, or concrete, which are subject to water or other liquid accumulation, fixtures exposed to the weather and unprotected, fixtures mounted in areas such as dairies or car washes where walls or other areas are subject to direct spray or hosing, or lighting fixtures installed in or near water pools (Art. 680) such as swimming pools, therapeutic pools, and fountains, should all be rated for wet locations.

Lighting fixtures installed in protected areas such as roofed open porches, canopies, or marquees, and interior areas such as some basements, barns, or cold-storage warehouses, that are not in direct contact with either falling rain or run-off water, but are exposed to moisture and subsequent water buildup, need only be rated for damp locations.

Section 410-4(a) stipulates that a lighting fixture rated for a wet location can be installed in either a wet, damp, or dry location, but a fixture rated for a damp location can only be installed in either a damp or a dry location.

Lighting fixtures installed in corrosive locations are required by Sec. 410-4(b) to be labeled or otherwise marked to indicate them as suitable for the intended location.

Lighting fixtures can be installed within or beneath ducts and hoods of cooking equipment in nondwelling-type occupancies if all the conditions listed under Sec. 410-4(c) are met.

Section 410-4(d) will not allow pendant or other cord-connected fixtures, including hanging lighting fixtures, track lighting, or paddle fans to be installed within a perimeter measured three feet horizontally or eight feet vertically from the outer edge of a bathtub. This ruling does not apply to either recessed or surface-mounted fixtures that are installed over bathtubs or in shower stalls. Recessed or surface-mounted fixtures are required by Sec. 250-42(a) to be grounded, and by Sec. 410-4(b) to be approved for a damp location.

Section 410-8 lists the conditions under which a lighting fixture can be installed in a clothes closet. The required clearance depends on what type of fixture will be installed, and where it will be installed in reference to the clothes or other storage area. Section 410-8(c) will not allow keyless or pendant lampholders (not totally enclosed) to be installed in any clothes closet.

The storage space within a clothes closet as defined in Sec. 410-8(a), where any type of lighting fixture cannot be installed, is detailed in Figure 410-8 in the NE Code. The storage space, either hanging space from a rod or shelf space above the rod, is depicted to be three sided. From the front entry of the closet space to the left or right, along the side walls, and along the back wall, a minimum of 24 inches of storage space (no lighting fixtures) must be maintained up to a vertical height of 6 feet or the rod height, whichever is higher. Above the rod or 6-foot height is considered shelf space; the 24-inch storage space is reduced to only 12 inches on all three sides.

The 12-inch indent, as shown at the door opening in Figure 410-8, would only apply to walk-in closets five feet or wider across the door opening. The leading edge of a surface-mounted, totally enclosed, incandescent bulb fixture, which can be mounted either in the closet ceiling or on the wall space above the door [Sec. 410-8(d)(1)], must have a minimum clearance of twelve inches (one foot) from the nearest point or leading edge of the storage area (hanging clothes or shelf space).

The clearance is reduced to only six inches from the leading edge of either recessed incandescent lighting fixtures [Sec. 410-8(d)(3)], or fluorescent lighting fixtures that are either surface-mounted [Sec. 410-8(d)(2)] or flush-mounted [recessed, Sec. 410-8(d)(4)].

10.5 LIGHTING FIXTURE SUPPORTS

Part D of Art. 410 lists the support requirements of lighting fixtures. In general, Sec. 410-15(a) requires any fixture assembly that weighs more than 6 pounds, or is longer than 16 inches in any dimension, to be supported independently of a screwshell-type lampholder.

If the lighting fixture is to be supported by the outlet box in the ceiling or on the wall of a building or other structure, Sec. 410-16(a) requires the fixture to be secured to the box. The box in turn must be secured and supported as required by Secs. 370-23 and 370-27. (Box installation and support was discussed in Chapter 5, Sec. 5.7 of this text.) Any lighting fixture weighing in excess of 50 pounds must be supported independently of the outlet box from which it is supplied (Fig. 10-4).

If a paddle fan and lighting fixture assembly is to be mounted on an outlet box, Sec. 422-18(a) (considered to be an appliance) will allow the assembly to be supported by a ceiling outlet box that is installed and supported by the guidelines of Secs. 370-23 and 370-27. The weight of the fan, or the combined weight of a fan/light assembly cannot exceed 35 pounds [Sec. 422-18(b)].

The general rule of Sec. 370-27(c) will not allow a ceiling fixture (round or octagon) outlet box to be used as the only support for ceiling (paddle) fans. The Exception will allow boxes that are listed for the application to be used to mount the fan or the fan/light assembly.

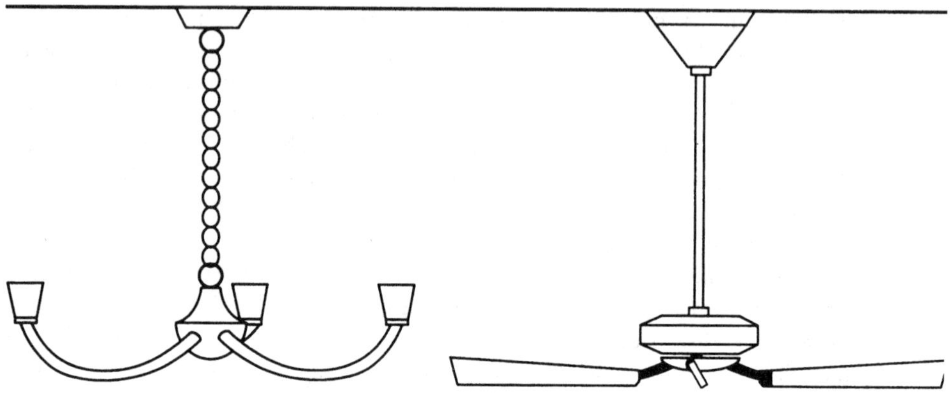

Fig. 10-4 Heavier lighting fixtures (exceeding 50 pounds) and paddle fans (exceeding 35 pounds) must be supported independently of the fixture outlet box.

Lay-in fixtures installed in the framing of suspended ceilings [Sec. 410-16(c)] can be supported by the ceiling framing if the fixture is secured to the frame by mechanical means such as bolts, screws, or rivets, or by clips that are identified for use with the framing members and the fixture. The framing members must be securely fastened to each other, and securely attached to the building structure at appropriate intervals.

Fluorescent lay-in fixtures installed in suspended ceilings cannot be supplied by flexible cords. Section 410-30(c)(1) requires the flexible cord to be visible for its entire length between the outlet box and the fixture. This type of fixture can be supplied with a fixture whip as allowed for flush and recessed fixtures. Section 410-67(c) will allow the whip to be between 4 and 6 feet in length.

Section 410-30(c) will allow fluorescent or other types of electric-discharge lighting fixtures that are supported by pipe or chain and suspended in open ceilings to be supplied with flexible cord, provided the cord is connected to the outlet box by a grounding-type attachment plug and receptacle. The cord can also be supplied by means of a busway plug.

The general rule of Sec. 410-31 states that lighting fixtures shall not be used as raceways for branch-circuit conductors. The exceptions cite fixtures that are listed for use as raceways and the conditions under which the branch-circuit conductors supplying the lighting fixture, as well as other lighting fixtures, can be run through the fixture.

The second paragraph of Sec. 410-31 stipulates that branch-circuit conductors which pass within 3 inches of an electric-discharge lighting ballast or ballast compartments must have an insulation rating as listed in this section. The types listed (including THW, refer to T310-13) have at least a 90°C (194°F) temperature rating.

10.6 FLUSH AND RECESSED LIGHTING FIXTURES

Part M of Art. 410 lists special requirements placed on flush and/or recessed lighting fixtures. Section 410-65(a) stipulates that fixtures shall not be installed adjacent to combustible materials, where such installation will expose the combustible material to temperatures in excess of 90°C or 194°F. Recessed fixtures are normally installed on hanger bars that suspend the fixture between two adjacent wooden ceiling joists. Recessed fixtures cannot be nailed or bolted directly to wooden joists or to wooden members (pearls or other horizontal braces/supports) that are nailed as a fixture support between the two wooden joists.

Section 410-65(c) requires recessed incandescent lighting fixtures to be thermally protected. A lighting fixture that is thermally protected normally has a solid-state device, sensitive to temperature, wired in series with the ungrounded fixture wire. The solid-state device is normally installed by the manufacturer in the forming shell or outer shroud of the recessed fixture (Fig. 10-5). When the temperature of the forming shell or outer shroud reaches a predetermined temperature, the solid-state device will open and remove power from the incandescent lamp(s). When the shell or outer shroud cools, the thermal protection will turn the lamps on automatically. Thermally protected ballasts supplying fluorescent lighting [Sec. 410-73(e)] function in the same way. The solid-state thermal device is mounted in the ballast.

Exceptions to Sec. 410-65(c) include (1) recessed incandescent fixtures installed in poured concrete, and (2) recessed incandescent fixtures whose construction design allows for heat dissipation equivalent to the thermal protection. Both of these recessed incandescent fixtures must be identified (listed and labeled) for their intended use.

Section 410-66(a) requires the recessed portions of lighting fixtures to be installed not less than ½ inch from wood ceiling joists or other combustible materials (Fig. 10-6). The Exception will allow a recessed incandescent fixture whose construction design allows for heat dissipation to be directly secured to the wood member. The fixture must be

Fig. 10-5 Recessed lighting fixtures are normally installed on adjustable hanger bars that suspend the fixture between two adjacent wooden ceiling joists.

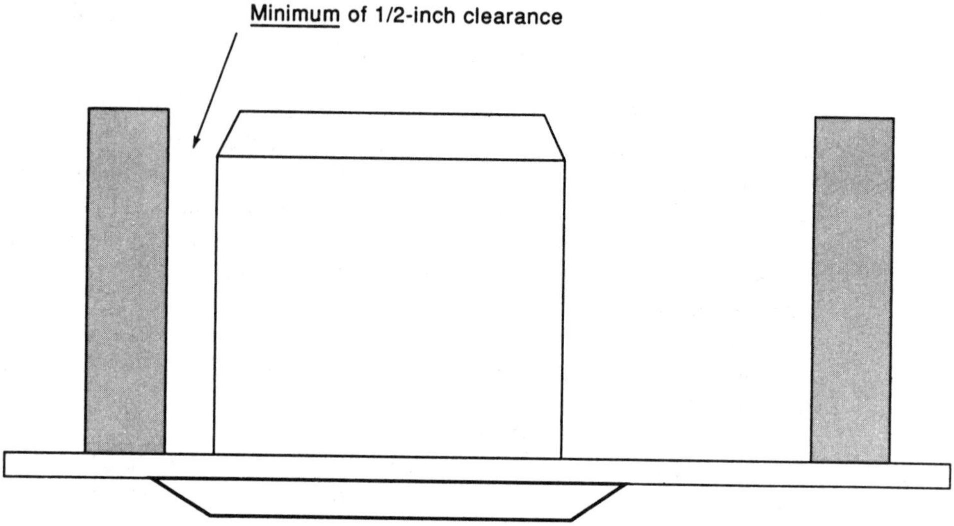

Fig. 10-6 Recessed portions of lighting fixtures must be installed with a minimum of 1/2-inch clearance from any combustible material.

identified as suitable for use with insulation being in direct contact with the fixture shell or shroud.

Unless the recessed fixture is identified for use with insulation in direct contact with the fixture (by exception), Sec. 410-66(b) will not allow thermal insulation to be installed either over or within 3 inches of any side of a recessed lighting fixture (Fig. 10-7).

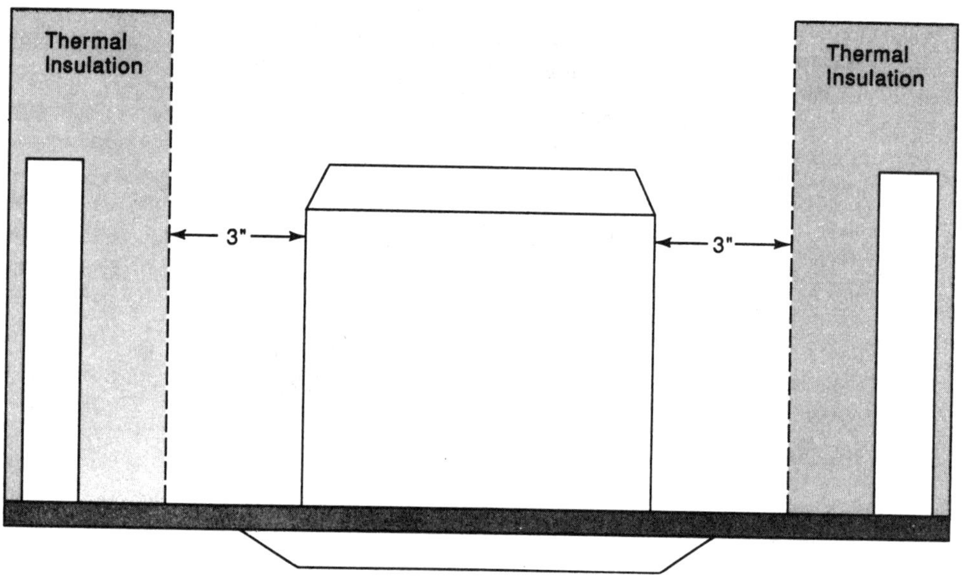

Fig. 10-7 Thermal insulation cannot be installed over or within three inches to the sides of the recessed portion of a lighting fixture.

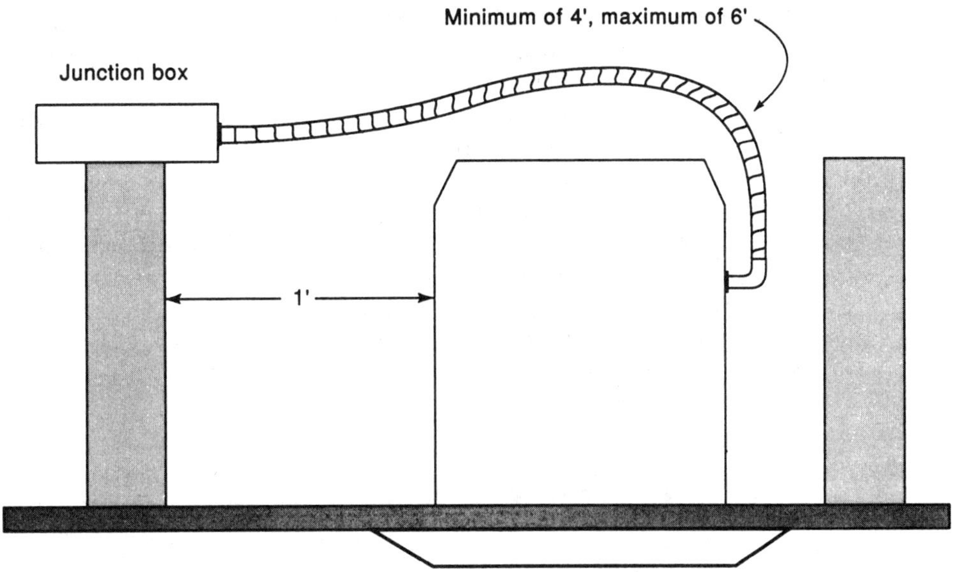

Fig. 10-8 Recessed incandescent lighting fixtures suffer from heat entrapment. The extra length of the fixture whip aids in heat dissipation.

For effective heat dissipation, Sec. 410-67(c) requires recessed fixtures to be supplied by tap conductors installed in a suitable (metal) raceway that is 4 to 6 feet long. The raceway must extend from the fixture shroud to an outlet box that is located not less than 1 foot from the fixture (Fig. 10-8).

10.7 FLUORESCENT AND OTHER ELECTRIC-DISCHARGE LIGHTING

Part P of Art. 410 lists the installation requirements for electric-discharge lighting that has an operating voltage (to the lamps) of 1,000 volts or less. Section 410-73(e) requires all fluorescent lighting fixtures to contain thermally protected ballasts. Replacement ballasts, installed in existing fixtures that originally contained ballasts not thermally protected, must also be thermally protected.

Section 410-75 will not allow electric-discharge lighting fixtures with an open-circuit (lamp) voltage greater than 300 volts to be installed in any residential occupancy, unless the fixture is so designed that the lampholders do not have any exposed live parts when the lamps are located in place, being installed, or being removed.

Section 410-79 requires snap switches used to control electric-discharge lighting to be installed according to their ratings as listed in Sec. 380-14. Subsection (a) of Sec. 380-14 requires ac general-use snap switches to have a rating at least equal to the full-load current rating of the lights regardless of whether the means of lighting is resistive, inductive [subsection (1)], or tungsten filament [subsection (2)]. Section 380-14(b)(2) restricts the full-load current rating of inductive lighting, including electric-discharge lighting, to a maximum of 50% of the rating of an ac/dc general-use snap switch.

Section 380-14(d) is new in the 1996 edition of the National Electrical Code. A 3-phase, 4-wire wye power distribution system rated at 600 volts has a single-phase voltage rating (line-to-neutral voltage) of 347 volts.

Snap switching for lighting, including 2-way, 3-way, and 4-way switching, and their installation requirements are discussed in Chapter 11 of this text.

If the fluorescent or other electric-discharge lighting fixtures are to be switched by the circuit breaker in the supply panel, Sec. 380-11 requires circuit breakers mounted in panels or other enclosures as described in Sec. 380-3 to have the correct number of poles (either single-pole, double-pole, or triple-pole, according to the way the lights are to be switched). Section 240-81 requires the circuit breaker, by way of its handle position, to clearly indicate whether the circuit is in the open (OFF) or closed (ON) position. Section 240-83(d) requires the circuit breaker to be approved for switching duty and to be marked with the identifying letters "SWD."

REVIEW QUESTIONS—PART ONE

1. A branch circuit consists of the conductors that extend between the last overcurrent-protection device within a building power distribution and the connected ____________.
2. A feeder circuit consists of the conductors that extend between the service equipment and/or overcurrent protection and the ____________ overcurrent-protection device supplying a connected load.
3. A ____________ - ____________ branch circuit can supply two or more lighting and/or receptacle outlets and is referred to as a multioutlet branch circuit.
4. An ____________ branch circuit can supply a single outlet or connection point (permanent wiring method) for a single appliance or other utilization equipment and is referred to as a dedicated or specific-purpose branch circuit.
5. The NE Code-assigned current or ampere ratings of multioutlet branch circuits are: ____________, ____________, ____________, ____________, and ____________ amperes.
6. The rating of any one cord- and plug-connected appliance or other type of utilization equipment cannot exceed ____________% of the branch-circuit rating if the multioutlet branch circuit is rated at either 15 or 20 amperes.
7. The total rating of one or more appliances or other utilization equipment fastened in place cannot exceed ____________% of the branch-cirucit rating if the multioutlet branch circuit is rated at either 15 or 20 amperes and lighting or receptacle outlets are not supplied.
8. If lighting or other receptacle outlets are also supplied, the total rating of one or more appliances or other utilization equipment fastened in place cannot exceed ____________% of the branch circuit rating if the multioutlet branch circuit is rated at either 15 or 20 amperes.
9. The required conductor size, the required overcurrent-protection rating, the types of equipment that can be supplied, the receptacle ratings, the maximum loads, and the permissible loads for various multioutlet (general-purpose) branch circuits are summarized in Table ____________.
10. An ____________ is defined as a point on the wiring system at which current is taken to supply some type of utilization equipment.
11. A single receptacle installed as a single outlet on an individual branch circuit must have an ampere rating at least equal to the ampere rating of the branch circuit.

    True or False

12. A duplex receptacle installed as a single receptacle outlet on a dedicated single-phase, two-wire, 20-amp, 120-volt branch circuit can have an ampere rating of either 15 or 20 amps.

True or False

13. A cord- and plug-connected load supplied from a 15-amp receptacle installed on a 20-amp multioutlet branch circuit is restricted to a maximum load of ____________ amperes.
14. A cord- and plug-connected load supplied from a 20-amp receptacle installed on a 20-amp multioutlet branch circuit is restricted to a maximum load of ____________ amperes.
15. Receptacles, cord connectors, and attachment plugs are required to be manufactured in such a configuration that the receptacles or cord connectors will not accept attachment plugs or caps of different voltage and/or current ratings. Nongrounding-type receptacles or cord connectors must be configured so as to not accept grounding-type plugs or caps.

True or False

16. In existing installations, nongrounding-type convenience (or duplex) receptacles can be replaced with either another nongrounding-type receptacle, a GFCI receptacle, or if an equipment-grounding means was provided with the original installation, by a grounding-type receptacle.

True or False

17. Lighting fixtures that will be installed in damp or wet locations must be identified as suitable for the intended location. The fixture must be designed and installed so that water cannot enter or accumulate within the fixture.

True or False

18. A lighting fixture rated for a damp location can be installed in either a wet or a damp location and be Code legal.

True or False

REVIEW QUESTIONS—PART TWO

1. A permanent lighting fixture cannot be installed over or within five feet horizontally to the water's edge (wall) of a swimming pool unless it is located at a minimum height of twelve feet above the maximum water level.

True or False

2. Unless installed a minimum of five feet above the maximum water level, a permanently installed lighting fixture located between five and ten feet of the perimeter (water's edge) of a swimming pool must be GFCI-protected.

True or False

3. Neither pendant or other cord-connected hanging lighting fixtures, track lighting, nor paddle fans, cannot be installed within ____________ feet horizontally or ____________ feet vertically from the rim (perimeter edge) of a bathtub.
4. Pendant or keyless-type lampholders, which do not have a protective globe installed over the bulb of the incandescent lighting fixture, can be installed in a clothes closet provided a mini-

mum of 24 inches is maintained between the fixture and the leading edge of the clothes or other storage space.

True or False

5. The installation of surface-mounted, totally enclosed, incandescent lighting fixtures must have a minimum of ____________ inches between the fixture and the leading edge of the clothes or storage shelf (whichever is closer) when installed in clothes closets.
6. The installation of recessed incandescent lighting fixtures, and either recessed or surface-mounted fluorescent lighting fixtures must have a minimum of ____________ inches of clearance between the fixture and the leading edge of the clothes or storage shelf (whichever is closer) when installed in clothes closets.
7. Any lampholder or other part of a lighting fixture assembly that weighs more than ____________ pounds, or is longer than ____________ inches in any dimension must be supported independently of a screwshell-type lampholder.
8. If a wall or ceiling box will be the sole means of support for a lighting fixture, the lighting fixture can weigh a maximum of ____________ pounds.
9. If an approved outlet box will be used as the sole means of support for a ceiling-mounted paddle fan or fan-light combination, the fan or fan-light assembly can weigh a maximum of pounds.
10. The framing members of a suspended ceiling can be used as the sole means of support for a lay-in type fluorescent lighting fixture, only if the fixture enclosure is mechanically secured to the framing members.

    True or False

11. Lay-in–type fluorescent lighting fixtures installed in suspended ceilings cannot be cord- and plug-connected and supplied by flexible cord because the cord must be ____________ for its entire length.
12. When fluorescent lighting fixtures or assemblies are approved (listed) for use as raceways, the contained branch-circuit conductors cannot pass within ____________ inches of the contained ballast(s), unless their insulations have a 90°C temperature rating.
13. The transformer ballasts of fluorescent or other types of electric-discharge lighting installed indoors must be thermally protected.

    True or False

14. The shells or shrouds of recessed lighting fixtures must be thermally protected.

    True or False

15. Unless a recessed lighting fixture is rated as suitable for use when in direct contact with thermal insulation, the forming shell or shroud cannot be installed within ____________ inch(es) of combustible materials such as wood ceiling joists.
16. Unless a recessed lighting fixture is rated as suitable for use when in direct contact with thermal insulation, thermal insulation cannot be installed over (covering) or within ____________ inch(es) of the lighting fixture forming shell or shroud.
17. If panel circuit breakers will be used to control (switch) fluorescent or other electric-discharge lighting branch circuits, the circuit breakers must be labeled (identified) as ____________ on their frame (molded case).

Chapter 11

Lighting Switches

The switch was discussed in the introduction to overcurrent protection in Chapter 7 as the device or component of the electrical circuit that provided directional control of the connected load.

A simple toggle switch is normally used to control the on/off status of electrical utilization equipment. This chapter will concentrate on two-, three-, and four-way switching of overhead or other general lighting in a room or other area of a building or other structure. The switching methods discussed can also be used to control different motor or heating loads supplied by the power distribution system within the building or other structure.

A knife switch is normally used for the disconnecting means of motors and other electrical utilization equipment. This disconnect is not intended as the directional control (on/off control) of the motor or other equipment. Rather, it is a safety disconnecting means, which will allow the service, repair, or replacement of the motor or other equipment, without exposing the electrician or other repairperson to the hazards of an unexpected startup (energization). The disconnecting means of electrical utilization equipment will be discussed in Chapter 12 of this text.

11.1 GENERAL LIGHTING

The safety disconnecting means for general overhead lighting is normally provided by the branch-circuit fused switch or circuit breaker at the lighting or power panel supply. According to the control and supply scheme, using the control switch as a disconnecting means for lighting circuits can be a safety hazard, simply because the branch-circuit supply, which is continuous power, is often terminated in the lighting outlet box for supply to the switch location.

In some applications, toggle switches used to control the general lighting, normally referred to as just "lighting switches" or "snap switches," are used to switch the supply circuit for convenience receptacles in a room. Controlled in this manner, the receptacles normally supply lamps that are used to light the room in place of overhead lighting fixtures.

11.2 INSTALLATION REQUIREMENTS

The installation and application of lighting switches, which includes voltage and current ratings, wiring identification, permitted location, switching arrangements, and methods of switching, are governed by Art. 380 of the NE Code.

One of the more important items concerning light switches is their accessibility. Section 380-8(a) requires all switches to be installed in a readily accessible location, and located such that the highest position of the handle is no more than 6 feet, 7 inches above the finished floorline (Fig. 11-1), base of a working platform, or finished grade when installed outside the building or other structure.

Article 380 governs all types of switches, not just lighting switches. The Exceptions to Sec. 380-8(a) allow greater operating heights for knife switches used as equipment disconnecting means. These switches can be operated by portable means such as hooksticks, ropes, or other means of extension. Greater heights are required when fusible plug-in disconnects are mounted on overhead busways, or when motor disconnects or other equipment isolating switches are rendered inaccessible to unqualified personnel by their height.

Section 380-8(b) lists the maximum operating voltages that are allowed between adjacent light switches. Snap switches are the primary way of controlling the lighting means in buildings or structures. Often several snap switches will be ganged at a central location such as a foyer or door entrance.

Ganging consists of removing the sideplates of metal construction boxes [regular device boxes as listed in T370-16(a)] and bolting the frames of the boxes together to create several (or a "gang") of switch openings. A two-gang switch box can also be constructed by using a 4-inch or 4 11/16-inch square metal box with a two-gang adapter (plaster ring) installed on its opening. When snap switches are mounted side by side in this manner (Fig. 11-2), Sec. 380-8(b) limits the voltage between two adjacent switches to a maximum of 300 volts (phase-to-phase or line voltage).

Three hundred volts between the switch conductors only presents a problem on the popular 480Y/277-volt, 3-phase, 4-wire distribution systems employing multiwire (Sec. 210-4) branch circuits. When adjacent switches are supplied by different phases (phases

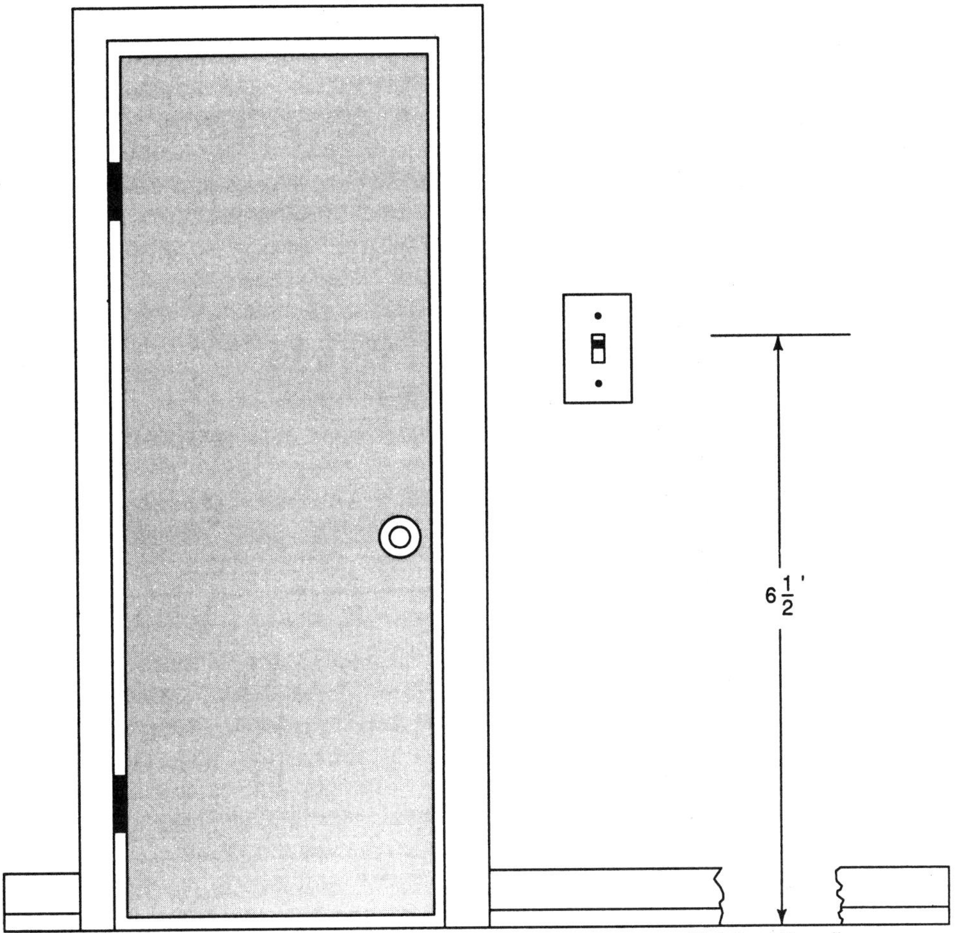

Fig. 11-1 Lighting switches must meet the same accessibility requirements of service disconnects and panel mains.

A, B, and C, or lines 1, 2, and 3), the voltage to ground from each phase is only 277 volts, but the voltage between the continuous-supply or switched-supply conductors of two adjacent switches is the phase-to-phase voltage of 480 volts. The arcing distance and therefore the safety hazard is increased still further on a 600Y/347-volt, 3-phase, 4-wire lighting power distribution. Section 380-8(b) does not prohibit the ganging of snap switches for these higher-voltage lighting circuits; it only requires that fixed barriers be installed between the switches.

Section 380-9 lists the requirements for faceplates installed on flush-mounted snap switches. Flush-mounted indicates the switch and its enclosure (box) are recessed in the wall or partition (Sec. 370-20). The faceplate is slightly raised as compared to the surface or face of the wall. In comparison, a surface-mounted switch would be installed in a box that is surface-mounted on the wall.

Section 380-9 lists the required thicknesses of faceplates, both metal and nonmetal.

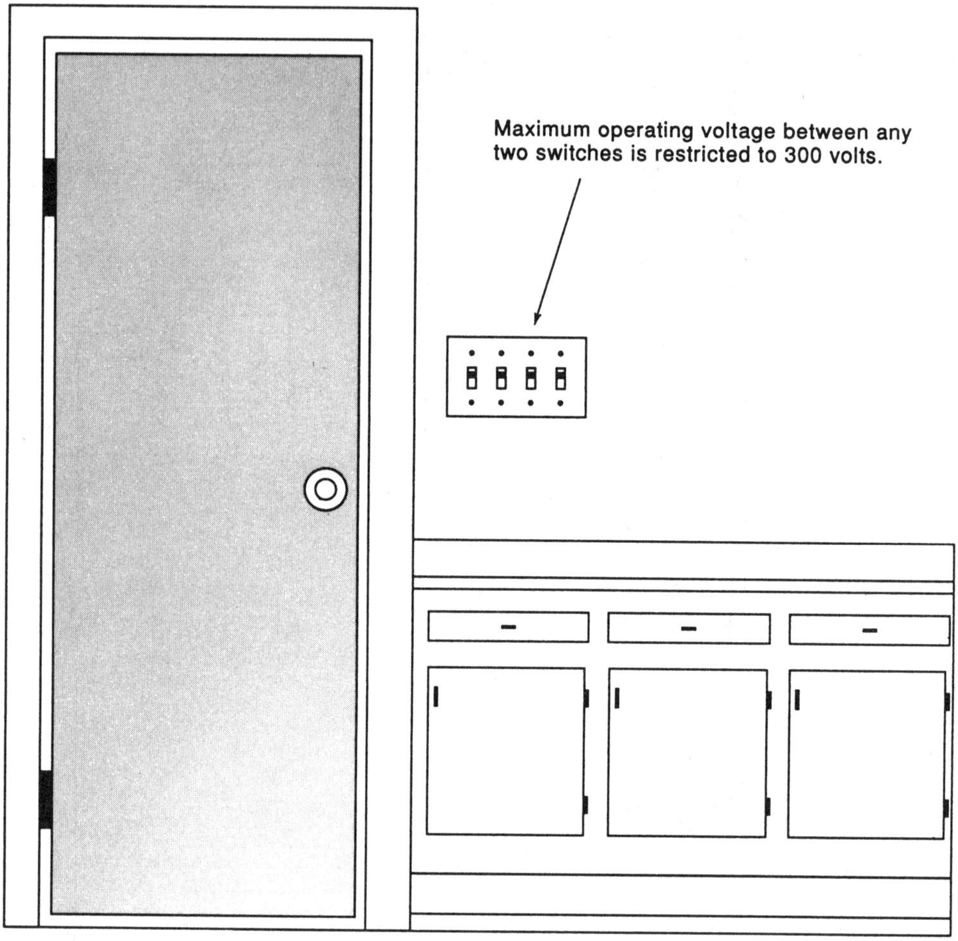

Fig. 11-2 Ganging of lighting switches on a 480Y/277-volt, 3-phase, 4-wire multiwire branch circuit requires fixed partitions (barriers) between the switches.

The faceplate must have a thickness, either by form or reinforcement, to provide adequate mechanical strength. The switch is supported by the box, but the faceplate must be of adequate strength to effectively close the wall opening containing the switch (Fig. 11-3).

Although not spelled out literally in an NE Code section, the faceplate for a switch installed in a surface-mounted box, such as a handy box, also has to effectively close the box opening. Section 370-18 addresses unused openings in a box, which applies specifically to knock-outs that have been removed. Between the reading of these two sections (370-18 and 380-9), the switch faceplate has to effectively close the wall opening and the box opening.

The faceplate for a handy box is tailored to the box opening (less than 1/16-inch overhang). The faceplate of switches installed in a 4-inch or 4 11/16-inch square, surface-mounted box is a raised cover with the same dimensions as the box.

Switch drops, or other switch supply/control schemes should be made with a wiring

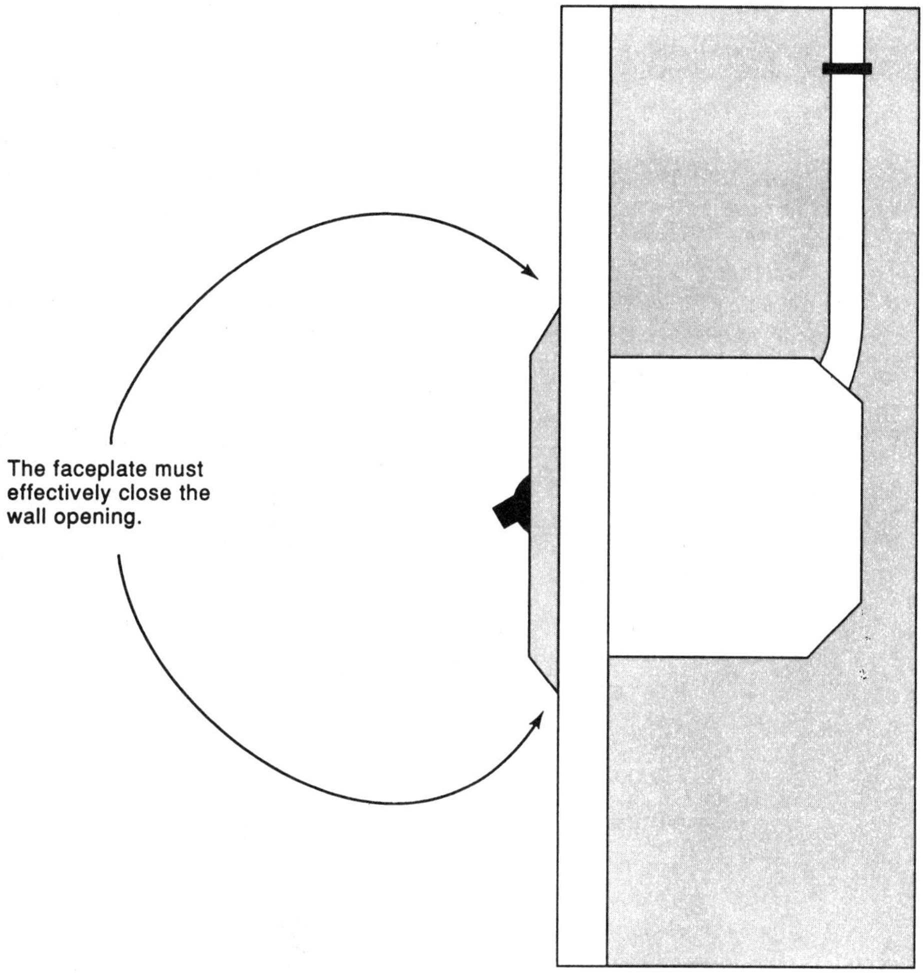

Fig. 11-3 Section 380-9 requires the switch faceplate to effectively close the wall opening containing the switch.

method (either conduit and individual insulated conductors or a multiconductor cable) that contains an equipment-grounding means. The switch assembly is mounted on a metal strap. Metal faceplates, or nonconductive faceplates like nylon or bakelite, are mounted and supported with metal machine screws threaded into the metal strap. If the ungrounded supply or switch-leg conductor should come in contact with the strap or plate, a shock hazard, at line potential, could be present on the faceplate or the screws. Proper grounding of the strap or the metal box that contains the switch would reduce the possiblilty of this shock hazard.

Section 380-9 requires switches that are installed in ungrounded metal boxes, which are accessible from grounded floors (concrete or tile floors are considered to be grounded surfaces) or other conductive surfaces, to be covered only with nonconductive, noncombustible faceplates.

Section 380-10(b) lists the mounting requirements of both flush-mounted and surface-mounted snap switches. If the box is mounted such that the front edge (face) is recessed from the finished surface of the wall (Fig. 11-4), the switch yoke (strap) must have plaster ears (plaster rings) on its ends, and the plaster ears must rest against the surface of the wall. If the front edge of the box is flush with the finished wall surface, or if the front edge protrudes from (extends past) the wall surface (Fig. 11-5, same requirement as for surface-mounted), the switch strap must rest against the front edge of the box. The extension plaster ears can be removed when the box is surface-mounted.

Almost all lighting switches are manufactured with plaster ears. These are the metal O-rings that are located at the four corners of the switch strap. The corner of the strap extending to the O-ring is normally crimped or indented to facilitate easy removal.

The switch must be supported either by the finished wall surface or by the front

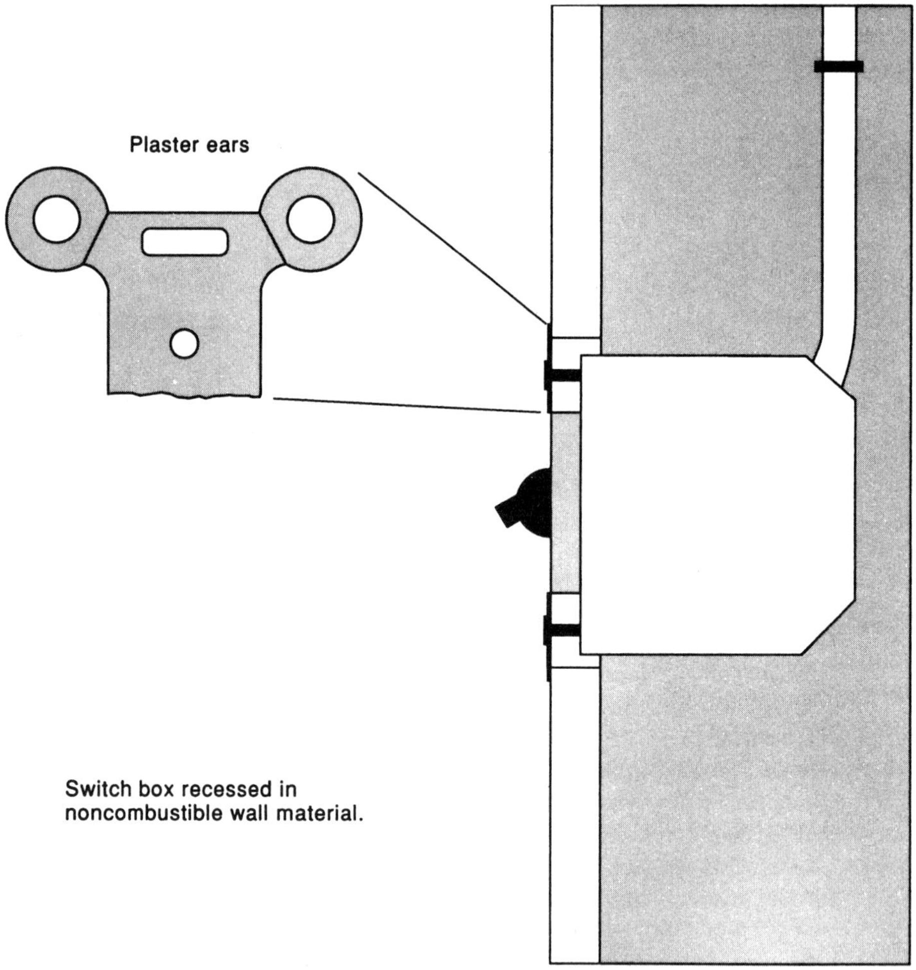

Fig. 11-4 When installed in recessed boxes, the switch strap must rest against the wall surface.

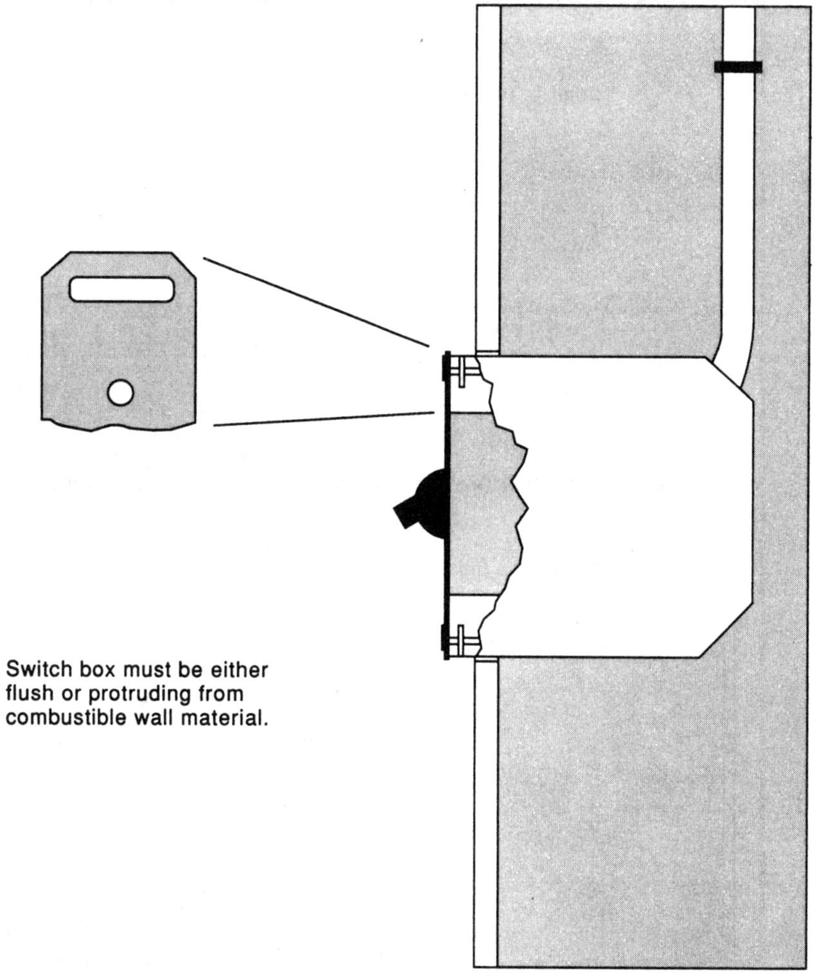

Fig. 11-5 Switch boxes installed in combustible wall material must be set either flush with the finished wall surface or protrude out from it.

edge of the outlet box. If the switch strap is not supported in this manner, the switch faceplate would be the only support for the switch. This would be an NE Code violation.

If the front edge of the box protrudes too far from the wall surface, the faceplate must still meet the requirements of Sec. 380-9: The faceplate must cover the entire wall opening and effectively close the opening. Oversized faceplates may be required.

For light switches installed in bathroom or kitchen walls that are partially tiled or paneled, the switch box would have to be set entirely in the tile or out of the tile (Fig. 11-6). The switch cannot be set straddling the coping of the tile, and either recessed in the tiled portion of the wall to be flush with the back wall finish, or raised in the outer wall finish to be flush with the tile. The switch box can be installed half-in/half-out of the tile portion of the wall if the box is set flush with the finish of the back wall and the coping of the tile is dressed around the switchplate.

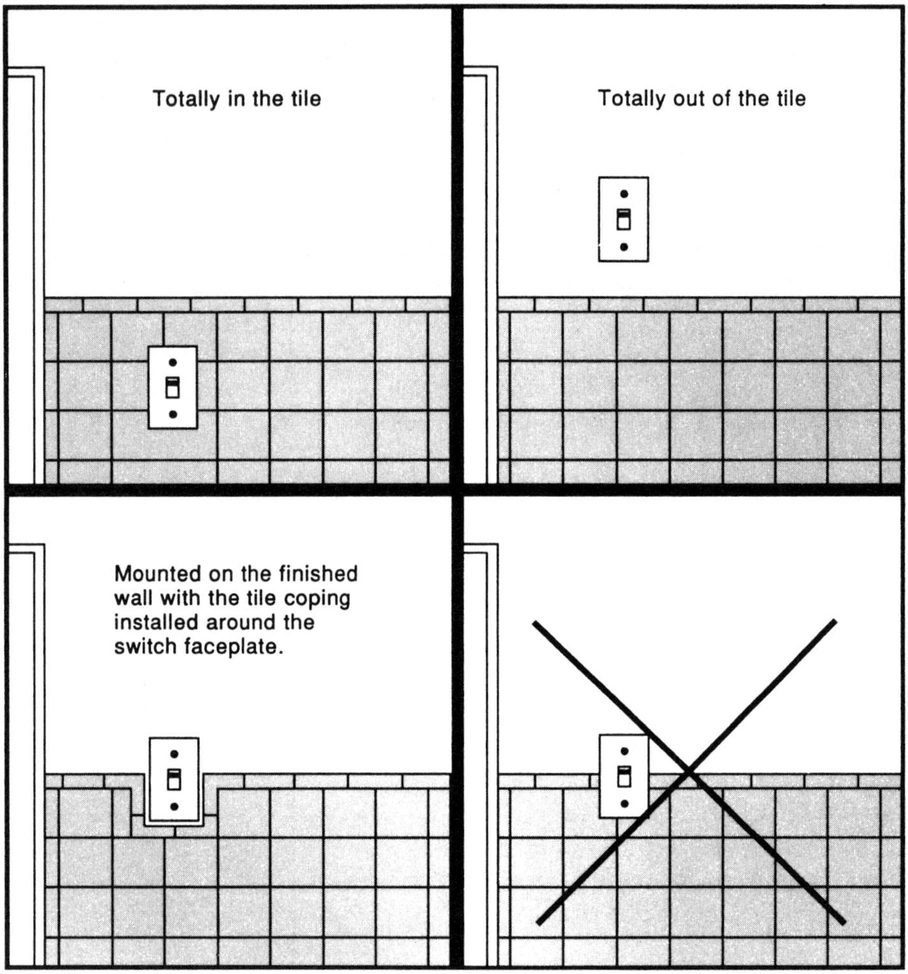

Fig. 11-6 When a wall is partially tiled or paneled, the light switch cannot be mounted half-and-half across the tile coping.

Article 370 lists the installation requirements for boxes and fittings. Section 370-18 requires all the unused openings in light switch boxes to be effectively closed (all unused knock-outs plugged—often referred to as uh-oh plugs). Section 370-20 requires wall or ceiling boxes recessed in combustible materials such as wood or fiberboard to be installed such that the front edge of the box is flush with the wall surface. The front edge of switch boxes installed in walls made of noncombustible materials such as concrete, tile, plaster, or sheetrock (plasterboard) can be recessed up to ¼ inch.

Section 370-21 requires the wall around the switch box to be effectively sealed by the finished installation. A maximum of ⅛-inch air gap or space is allowed between the edge of the box or fitting (plaster ring) and the cross-section of the wall surface. For new work in drywall (sheetrock), plaster, or tile construction, the wall finishing contractor normally plasters around the box. In panel, block, or concrete wall construction, the electrician is normally required to caulk or in some manner seal the openings around the switch box.

Switches and switch enclosures are required to be rated for the type of installation involved. Section 380-3 requires the switch to be externally operable and listed for its intended use. The box or other enclosure (refer to Arts. 370 and 373) must have adequate bending space and gutter space for the contained conductors, switches, and other devices in a common enclosure.

Section 380-4 requires switches installed in wet locations to be contained in weatherproof enclosures. Neither switches nor convenience receptacles [Sec. 410-57(c)] can be installed in a tub or shower space. Section 380-4 makes exception for switches that are factory-mounted within the tub or shower space and installed as part of a listed assembly.

Articles 500 through 503 cover electrical installations in hazard (classified) locations. Sections 501-6, 502-6, and 503-4 require switch enclosures installed in hazardous locations to have a rating equal to the class and division of the hazardous location.

11.3 SNAP SWITCH RATINGS

The ampere and voltage ratings of lighting snap switches is governed by Sec. 380-14. Two principal types of switches are ac general-use snap switches and ac/dc general-use snap switches.

Snap switches are normally rated at either 15 or 20 amperes with voltage ratings ranging from 120 to 300 volts. For ac lighting circuits employing ac snap switches, the load-current rating of the fixtures, either electric-discharge [fluorescent or other types—Sec. 380-14(a)(1)], or regular incandescent lighting [tungsten filament—Sec. 380-14(a)(2)] cannot exceed the ampere rating of the switch. In parallel, Sec. 210-23(a) restricts the supply of lighting fixtures assembled with light-duty rated screwshell or pin-type lampholders to either 15- or 20-amp multioutlet branch circuits.

The more common lighting circuits use an ac supply rated at either 120, 277, or 347 volts, but dc lighting circuits are sometimes employed. Section 380-14(b)(3) requires an ac/dc general-use snap switch used to control tungsten filament lighting (incandescent lighting) to be "T" rated. The total current load of the controlled lighting outlets (fixtures) cannot exceed the ampere rating of the switch.

11.4 SWITCHING ARRANGEMENTS

Switch connections are governed by Sec. 380-2. Subsection (b) stipulates that switches or circuit breakers shall not be installed in (cannot disconnect) the grounded-circuit conductor (neutral) of a circuit. Exceptions include (1) switching arrangements that disconnect all the circuit conductors (both grounded and ungrounded) simultaneously, and (2) switching configurations that assure first-make, last-break connection of the grounded-circuit conductor. For normal 2-, 3-, and 4-way switching arrays (arrangements) using regular snap-switches, switches are not allowed to be wired in series with the neutral or other grounded-circuit conductor.

11.5 CONDUCTOR IDENTIFICATION

Because all electrical circuits consist of a minimum of two wires, the supply and return of the electrical-power, lighting-circuit conductors must be insulated to protect the circuit from unwanted short-circuit or ground-fault overcurrents, and color-coded to identify their function (or side) within the circuit.

The two sides of the lighting circuit are the same two sides of supply and return in the branch circuit. The supply side is normally designated as the ungrounded side. The return side is normally designated as the grounded side of the circuit. The ungrounded side is fused or circuit-breaker protected at the point of supply (homerun) of the branch circuit. The grounded side has a solid connection (screw terminal) at the point of supply with the grounded service-neutral conductor and the grounding-electrode system of the building.

Equipment-grounding conductors or bonding jumpers also have a solid connection with the service neutral and the grounding-electrode system. These conductors are not required to be insulated because they are not current-carrying conductors under normal circuit operation. The lack of insulation identifies their function. When insulated, these equipment-grounding conductors must be green or green with one or more yellow stripes. The required colors of branch-circuit supply and lighting-circuit conductors are governed by Secs. 200-7, 210-5, and 310-12.

Ungrounded-circuit conductors [Sec. 310-12(c)], which include the supply conductors (switch feeds), the traveler conductors, and the switch-leg conductors (switch returns), can be any color except white, natural gray, or green. The grounded-circuit conductor [the neutral conductor in a multiwire branch circuit—Secs. 200-6, 210-5(a), and 310-12(a)] can be either white or natural gray.

If circuits are to be switched from more than one location, either 3-way or 4-way switching must be employed. A pair of ungrounded traveler conductors are used as the parallel switched conductors, which interconnect the 3-way and/or 4-way switches. Traveler conductors, which can be any color except white, natural gray, or green, are not required for two-way or simple on/off switching.

When installing lighting circuits, the conductor grouping requirements of Sec. 300-20(a) have to be honored at all times. When the homerun (branch-circuit supply from the panel or other load center) is to the lighting outlet location, the conduit or cable extended to the lighting outlet from the supply panel or other load center must contain both the ungrounded- and the grounded-circuit conductors (supply and return).

The metal outlet box must be grounded. If the box is supplied with metallic conduit, the conduit can serve as the required equipment-grounding means of the noncurrent-carrying metal parts of the lighting circuit. The safer raceway installation, which is NE Code-required when the raceway is nonmetallic or when a multiconductor wiring method is employed [Sec. 250-92(b)], is to run a separate equipment-grounding conductor with all circuit conductors to either the lighting outlets or the switching points within the circuit.

If the homerun is to the lighting outlet box and the switch drop, which consists of the ungrounded switch-feed (supply) and switch-leg (return) conductors, is fed from the lighting outlet box, the grounded-circuit conductor does not have to be run to the switch box; however, the equipment-grounding conductor does.

If the homerun is to the switch box, and the lighting outlet box is fed through the switch box, meaning the circuit supply is run from the panel to the switch box first, and the lighting load is fed from that switch box, then naturally the grounded-circuit conductor must be spliced and run through the switch box.

When switching arrangements utilize 3-way switches, the pair of travelers and either the ungrounded switch-feed conductor, the ungrounded switch-leg conductor, or the grounded-circuit conductor, depending on the switching configuration, must be contained in any cable, conduit, or tubing run between the two 3-way switches.

If one or more 4-way switches are installed between the two 3-way switches, and these switch boxes are in-line series-fed between the two 3-way switch boxes, a pair of traveler conductors and either the switch-feed conductor, the switch-leg conductor, or the grounded-circuit conductor, depending on the circuit configuration, must be contained in the cable, conduit, or tubing run to or from each 4-way switch box.

A drop from a junction box installed between two 3-way switch boxes to a 4-way switch box would only contain two sets (pairs) of traveler conductors and an equipment-grounding conductor. The switch-feed, switch-leg, or grounded-circuit conductor, one of which must be extended between the two 3-way switches, would not have to be run to and from the four-way switch in a single conduit run or cable installation.

Variations as to how and where the switches are located and wired are numerous. The standard rule of thumb is if power is supplied to a switch or a load, the return conductor for that power must be contained in the same raceway.

The requirement of "what goes out must come back in the same raceway" [Sec. 300-20(a)] stems from the inductive heating effect of surrounding ferrous metal surfaces when circuit conductors convey ac current. The flow of ac current along either the supply or return circuit conductor generates a varying magnetic field that surrounds the conductor.

When one circuit conductor without the other circuit conductor either passes through or is run along a metal surface, portions of that metal surface are submersed in this varying magnetic field. The varying magnetic field induces current flow in the metal surface.

The current flow in the metallic box or raceway is not routed to any connected load. As a result, the current travels as high-magnitude eddy currents through different parts of the metal surface. Eddy currents in a metal object or surface generate varying levels of heat in the metal. The generated heat of the outlet or device box or raceway has a detrimental effect on the single-circuit conductor's insulation.

When both sides of the circuit conductors supplying a load are installed (grouped together) in the same metal raceway or pass through the same hole in a metal box or other enclosure opening, the varying magnetic field of the supply conductor is cancelled by the generated field of the same current in the return conductor: The polarity of the fields are 180° out of phase between the length of the supply and the length of the load-return conductor.

Conductor grouping offers little difficulty in conduit or other piping raceway installations using individual, insulated conductors. The switch return conductors are normally a different color than the phase (hot or ungrounded) supply conductors. These are not NE Code requirements, as long as the colors used are not white, natural gray, or green, only good trade practice.

Problems tend to arise when the wiring method is multiconductor cable. The individually insulated conductors within the cable are color coded (colors are fixed) at the point of manufacture. For wiring methods such as BX or Romex cables, the insulated white conductor is in every cable run whether or not a grounded-circuit conductor (neutral) is required for the intended installation.

In a No. 12/3 w/GND (with ground) Romex cable, the manufactured colors are normally red, black, white, and a bare equipment-grounding conductor. In a No. 12/2 w/GND Romex cable, the red conductor is left out. Often the insulated white conductor contained in cable wiring methods is "used and abused" because it is not installed according to NE Code guidelines.

Rules governing the use of the white insulated conductor, which is NE Code-intended to be the grounded-circuit conductor, are found in Art. 200. The general rule of Sec. 200-7 stipulates that any conductor (either individual insulated and installed in raceway or contained in a multiconductor cable) that is white or natural gray in color shall not be used for any circuit purpose except as a current-carrying grounded-circuit conductor. Four exceptions are listed.

Exception No. 1 will allow an insulated white or natural gray conductor to be used as an ungrounded current-carrying conductor if it is reidentified at all terminations and at all accessible points.

In a wooden-frame dwelling structure, almost all branch circuits are wired with a multiconductor wiring method such as BX or Romex. A common installation is to supply a storage-type electric water heater with a 2-wire, 30-amp, 240-volt (or 208-volt) branch circuit. No. 10/2 w/GND Romex cable contains two insulated conductors, black and white, and a bare equipment-grounding conductor.

Section 200-7, Excp. No. 1 will allow the contained insulated white conductor to be used as the second ungrounded supply conductor for the water heater provided it is reidentified (Fig. 11-7) to a color other than natural gray or green at all points of accessibility: At the supply end, when connected to one terminal on the double-pole circuit breaker or fused switch, at the load end, when connected to the red insulated conductor in the water heater makeup enclosure, and at a junction box, should an in-line splice be required in the Romex (or BX) cable run to the water heater.

Section 200-7, Excp. No. 2 will allow the insulated white conductor contained in a multiconductor cable to be used as the switch feed (switch supply for either 2-way, 3-way, or 4-way switching) without being reidentified. This conductor will often be referred to as a "white-hot feed" (WHF). If the white conductor is to be used as a switch leg (switch return), it must be reidentified to a color other than natural gray or green.

Section 200-7, Excp. No. 3 will allow the insulated white conductor contained in a flexible cord to serve as an ungrounded current-carrying conductor without being reidentified if it is used to supply an appliance. Flexible cords are not a recognized wiring method for branch-circuit or lighting control-circuit wiring.

Section 200-7, Excp. No. 4 applies only to circuits that are rated less than 50 volts. If the circuit is not required by Sec. 250-5(a) to operate with reference to earth ground (to be grounded—supply conductors contain a grounded-circuit conductor), then the insulated white or natural gray conductor can be used as an ungrounded current-carrying circuit conductor.

The typical 2-wire doorbell circuit is an example of this exception. The doorbell or chimes circuit for a dwelling is normally supplied by a power-limited transformer rated at 50 volt-amps. The 12- or 24-volt secondary is not required to be referenced to earth ground by the NE Code; therefore, the contained insulated white conductor in

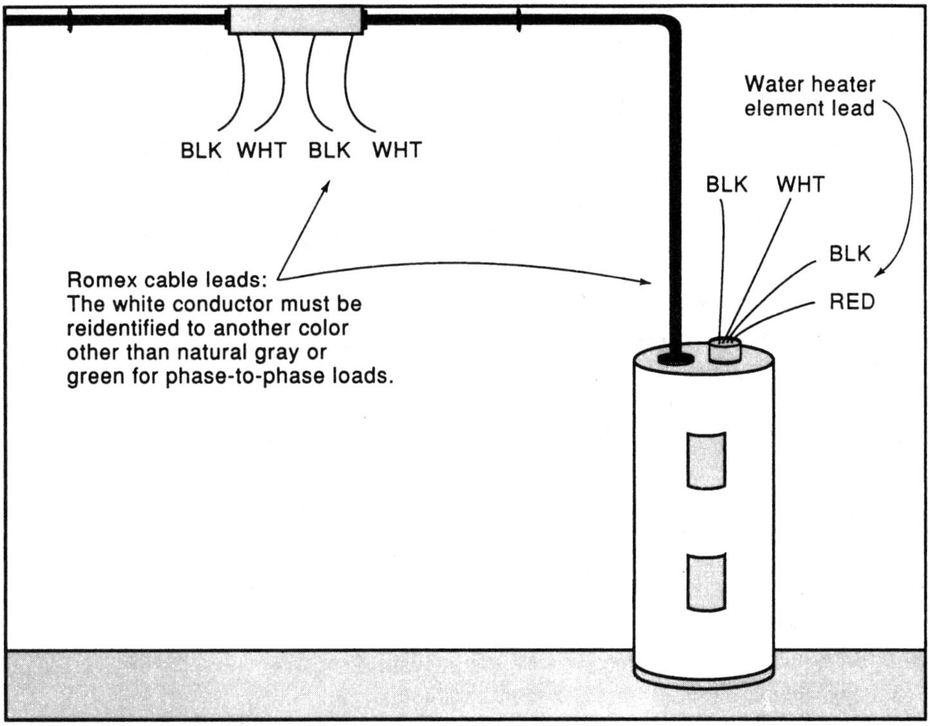

Fig. 11-7 When the white conductor of a multiconductor cable is used as an ungrounded phase conductor, it must be reidentified.

the covered, twisted-pair cable can be used for either side (supply or return) of the circuit.

For 2-way, 3-way, or 4-way switching arrays of overhead or receptacle-lamp room lighting using multiconductor cables (such as Romex or BX), Exceptions No. 1 and No. 2 to Sec. 200-7 are used to determine when the contained insulated white conductor must be reidentified. If the conductor is being used as the supply to a switching arrangement (switch feed) from a lighting outlet, it can be tied (spliced) to the ungrounded supply conductor (white-to-black) without being reidentified.

If the contained insulated white conductor is to be used as a switch leg (switch return, connection between the switch and the controlled load), it must be reidentified to a color other than natural gray or green. Normally, reidentification is accomplished with black electrical tape, since this tape is readily available on most tool pouches. Other colors of phase tape can be used, as well as painting and colored labels.

The NE Code does not restrict the white or natural gray conductor from being used as a traveler in the three-way and four-way switching arrangements, as long as it is reidentified by way of Sec. 200-7, Excp. No. 1. The switching arrangements of multiconductor cable wiring methods shown in this text will not use the white conductor for travelers, only for a switch supply (switch feed or white-hot feed) that is not reidentified, and for a switch leg that is reidentified.

For wiring diagrams of different switching arrangements, the color code of the respective conductors is normally indicated by the color abbreviated to three letters labeled (written) near the conductor. Common color abbreviations (acronyms) include: white, WHT; white-hot feed, WHF; black, BLK; red, RED; brown, BRN; blue, BLU; yellow, YEL; and lavender, LAV. Equipment-grounding conductors are normally not shown on lighting-circuit wiring diagrams since their presence is not required for the operation of the circuit (safety circuit—not a current-carrying conductor during normal circuit operation); therefore the acronym for green (GRN) was not specified.

For the purposes of this text, if multiconductor cable such as Romex or BX is employed in the switching arrangement, the multiconductor cable is indicated by an elliptical circle drawn around the respective conductors in a common cable.

Devices using standard symbols for switches, lighting outlets, or receptacle outlets are normally drawn inside a broken- or solid-line box. To honor the conductor grouping requirements of Sec. 300-20(a), all splices are drawn inside the broken or solid line to indicate splices within the box or other enclosure.

11.6 TERMINAL IDENTIFICATION

Section 200-10(a) requires any electrical device provided with terminals intended for the makeup to more than one ungrounded (phase) conductor (double-pole, double-throw switches), or for makeup to both an ungrounded phase (supply) conductor and the grounded-circuit conductor (neutral on a multiwire branch circuit), to have the respective terminals marked or so identified as to prevent cross connections.

In other words, a keyless or other screwshell-type lighting fixture (Fig. 11-8) must have screw terminals that are identified by distinguishing colors, paint, labels, or etching for both the ungrounded- and grounded-circuit conductors. Section 200-10(c) requires the wire terminal intended for the grounded-circuit conductor to be internally connected (point of manufacture) to the screwshell of the fixture.

If the lighting fixture is the screwshell type that uses fixture leads instead of terminals (Fig. 11-9), subsection (d) requires the lead attached to the screwshell of the lampholder to be a solid white- or natural gray-colored conductor that is easily distinguishable from the other lead connected to the center button of the screwshell (normally a black lead wire).

Section 200-10(b) requires the terminals on receptacles (Fig. 11-10), attachment plugs (caps), and cord connectors (equivalent to a receptacle on the end of a flexible cord) to be color identified. The color of the ungrounded (either switched or unswitched) terminal is not specified, although it cannot be identified as either white or green.

The terminal intended for the grounded-circuit conductor has to be identified with the screw color white or near white (substantially white or silver), or the word "white" or the letter "W" can be stamped, etched, molded, or printed in the area located adjacent to the terminal.

The terminal for the equipment-grounding conductor [Sec. 410-58(b) as applied to receptacles, and Sec. 250-119 in general] must either be a green-colored hexhead screw, nut, or pressure-type wire connector. The terminal or entrance hole for the equipment-grounding conductor can also be painted, etched, stamped, or in some manner identified with the word green, the letters "G" or "GR," or the ground symbol: ⏚

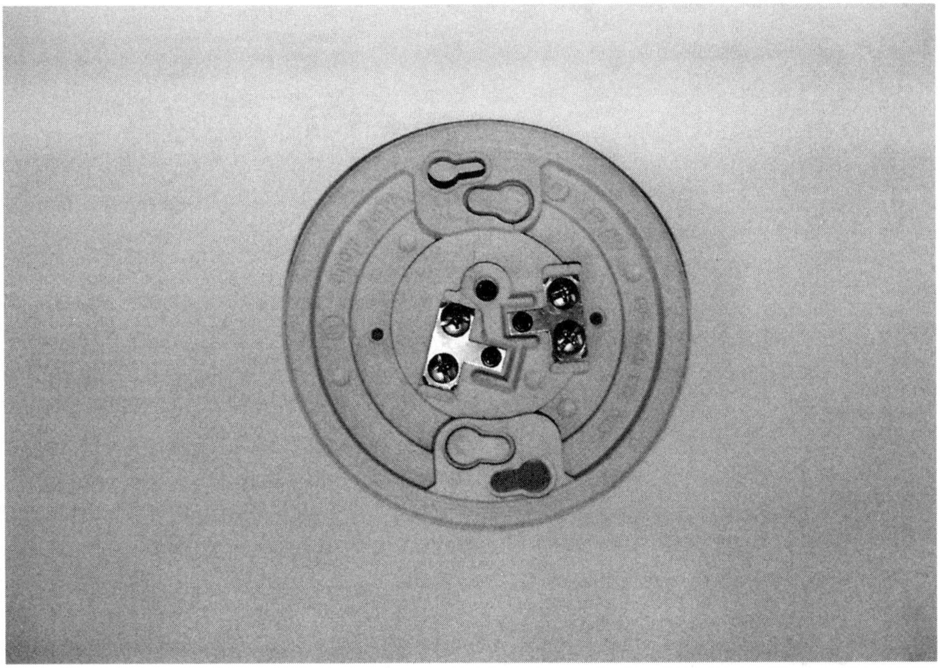

Fig. 11-8 The screw terminals on the base of a lighting fixture must be color coded to identify the grounded and ungrounded sides of the circuit.

11.7 TWO-WAY SWITCHING

Two-way or on/off switching (Fig. 11-11) indicates that the power to the lighting load is controlled from only one location and can be switched either "on" or "off." The ac general-use snap switch (light switch, Fig. 11-12) has two wire terminals. Both are on one side of the switch body (assembly), one at each of the two ends. The ungrounded supply of the homerun (continuous power) is connected to one terminal. Switched power, which must be extended to the lighting fixture as the circuit switch leg, is connected to the other terminal.

In terms of circuit operation, the ac general-use snap switch is not discriminatory. In other words, it makes no difference whether the switch feed is connected to the top terminal and the switch leg is connected to the bottom terminal, or vice versa.

There is, however, a top and a bottom on these switches. The words ON and OFF are either etched or embossed on the two flat sides (top and bottom) of the snap switch handle. When installed, the word OFF should be visible with the switch in the down position. The word ON should be visible with the handle in the up (highest) position. This is NE Code-required for all switches used as a disconnecting means (Sec. 380-7), and standard practice for 2-way (single location) light switching. Most practicing electricians install continuous power to the top terminal and switched power to the bottom terminal to parallel line-side and load-side wire terminations on circuit breakers or fused knife switch disconnecting means.

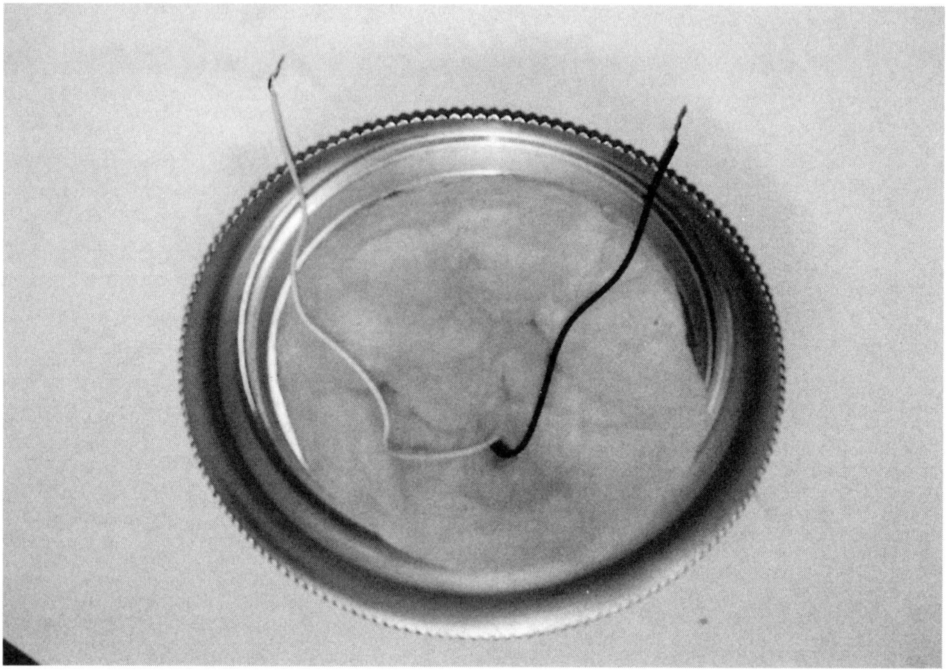

Fig. 11-9 Pigtail leads extending from the base of a lighting fixture must be color coded to identify the grounded and ungrounded sides of the circuit.

Power can be supplied to the switch from the homerun (homerun indicates the branch-circuit supply originating in the service power panel or other panel load center), or power can be supplied to the lighting outlet. Only two circuit conductors are required for either type of switching arrangement (array).

When the homerun is to the switching location, the continuous ungrounded supply conductor is connected to one terminal of the snap switch. The switch-leg conductor is connected to the other switch terminal and continues on with the grounded-supply conductor to the lighting outlet.

When the homerun is to the lighting-outlet location, the ungrounded supply is run as the switch feed with the switch leg of the circuit as the switch drop to the switching location. The grounded-supply conductor of the homerun is connected directly to the intended grounded terminal of the lighting fixture. The switch leg is connected to the intended ungrounded terminal of the lighting fixture.

Most light-switching installations begin with a drawing or wiring diagram of the intended switching arrangement. Wiring diagrams are important because they relate the number of conductors that must be installed in the raceway or contained in the multiconductor cable that connects the two points (the switching location and the lighting-outlet location) of the electrical distribution. The wiring diagram also helps to establish the colors of insulated circuit conductors that must be installed.

Fig. 11-10 The screw terminals on a convenience receptacle (duplex) must be color coded to identify the grounded and ungrounded sides of the circuit.

Example 11–1. In the wiring diagram of Figure 11-13, 2-way switching of a single lighting outlet is shown using single insulated, color-coded conductors installed in conduit with the following criteria:

1. The homerun is to the switch location.
2. The lighting outlet is fed (supplied) from the switch location.

At the source. The black ungrounded-supply conductor (phase conductor) is connected to the load terminal of the circuit breaker. Any color except white, natural gray, or green could have been used.

The white grounded-supply conductor is connected to the grounded busbar. White was used, although natural gray could have been used.

OFF ON

Fig. 11-11 A simple toggle switch is referred to as a two-way switch because it is either ON or OFF.

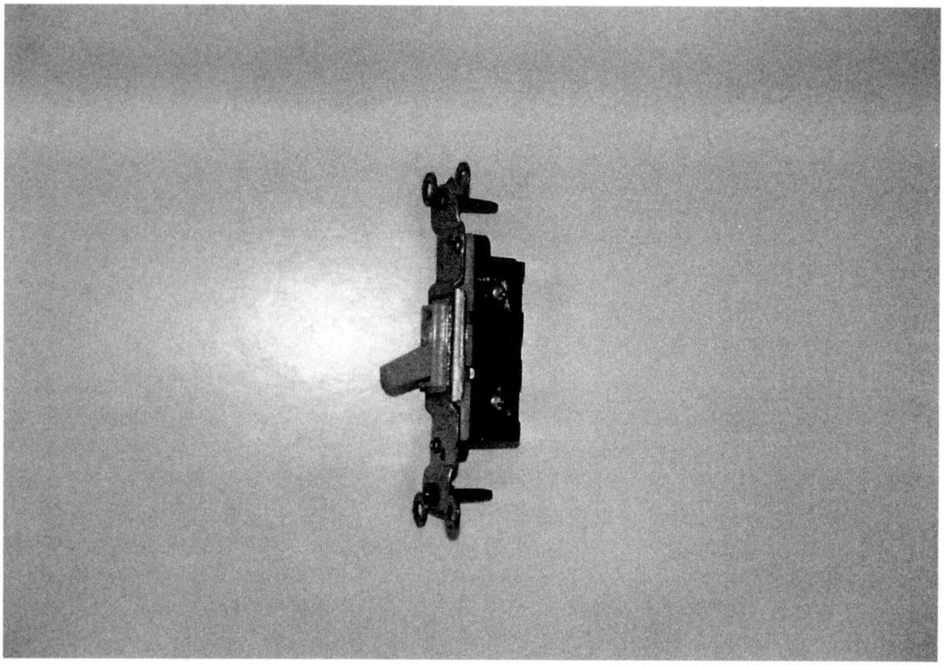

Fig. 11-12 The screw terminals two-way lighting switch are the same color because they are only on one side of the circuit.

At the switch. The black ungrounded-supply conductor is connected to one terminal on the 2-way switch. The red ungrounded switch-leg conductor is connected to the other terminal of the switch and extended to the lighting outlet.

The color of the ungrounded conductor is changed from black to red to indicate switched power, although black could continue on to the lighting outlet. Any color except white, natural gray, or green could be used for the ungrounded switch-leg conductor.

The white grounded-supply conductor is spliced to the white grounded-circuit conductor without connection to the switch and extended to the lighting outlet.

At the lighting outlet. The red ungrounded switch-leg conductor is connected to the brass screw terminal or black lead on the lighting fixture base (not shown).

The white grounded-circuit conductor is connected to the silver (light-colored) screw terminal or white lead on the lighting fixture base (not shown).

The makeup description given at the source, the control point, and the connected load is the final step in a wiring diagram of lighting circuits. Color coding of the branch-circuit supply and lighting-circuit conductors serves no useful purpose if it is not followed through with wire-to-device termination.

Code challenge 11–1. Complete the wiring diagram of Figure 11-14 using single insulated, color-coded conductors installed in conduit. The two lighting outlets should be controlled by the one switch, with the following criteria:

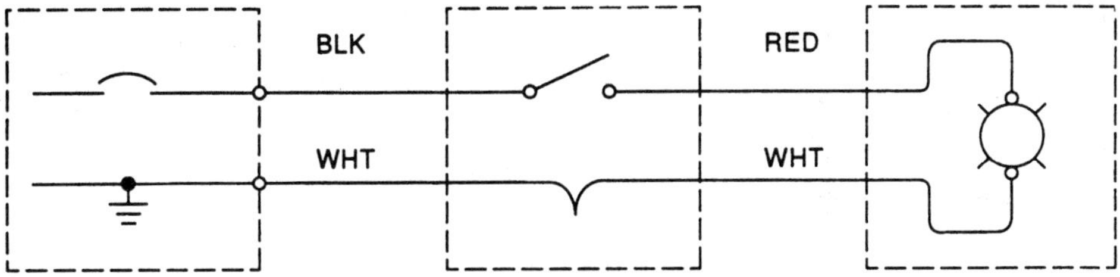

Fig. 11-13 Drawing of a two-way switching arrangement with power supplied to and the light fed from the switch enclosure.

1. The homerun is to the switch location.
2. The first lighting-outlet location is fed from the switch location.
3. The second lighting-outlet location is fed from the first lighting-outlet location.

At the source. The (A) ___________ (color) ungrounded-supply conductor is connected to the load terminal of the circuit breaker. The (B) ___________ grounded-supply conductor is connected to the grounded busbar.

At the switch. The (C) ___________ ungrounded-supply conductor is connected to one terminal on the 2-way switch. The (D) ___________ ungrounded switch-leg conductor is connected to the other terminal on the 2-way switch. The (E) _________ grounded-supply conductor is spliced to the (F) ___________ grounded-circuit conductor without connection to the switch, and extended to the first lighting outlet.

At the first lighting outlet. The (G) ___________ ungrounded switch-leg conductor is spliced and pigtailed. The pigtail is connected to the ungrounded terminal on the lighting fixture. The splice is extended as the (H) ___________ ungrounded switch-leg conductor to the second lighting outlet. The (I) ___________ grounded-circuit conductor is spliced and pigtailed. The pigtail is connected to the grounded terminal of the lighting fixture. The splice is extended as the (J) ___________ grounded-circuit conductor to the second lighting outlet.

At the second lighting outlet. The (K) ___________ ungrounded switch-leg conductor is connected to the ungrounded terminal on the lighting fixture. The (L) ___________ grounded-circuit conductor is connected to the grounded terminal on the light.

When two lighting outlets are controlled from a single switching location and power is supplied to one lighting outlet, with the switching location supplied from another lighting outlet, a third circuit conductor must be extended between the two lighting fixtures because continuous power must be supplied to the switch, and both the switched-power and the grounded-circuit conductors must be run between the lighting outlets.

Example 11–2. In the wiring diagram of Figure 11-15, 2-way switching is shown using color-coding of a multiconductor cable wiring method. The two lighting fixtures are controlled by one switch with the following criteria:

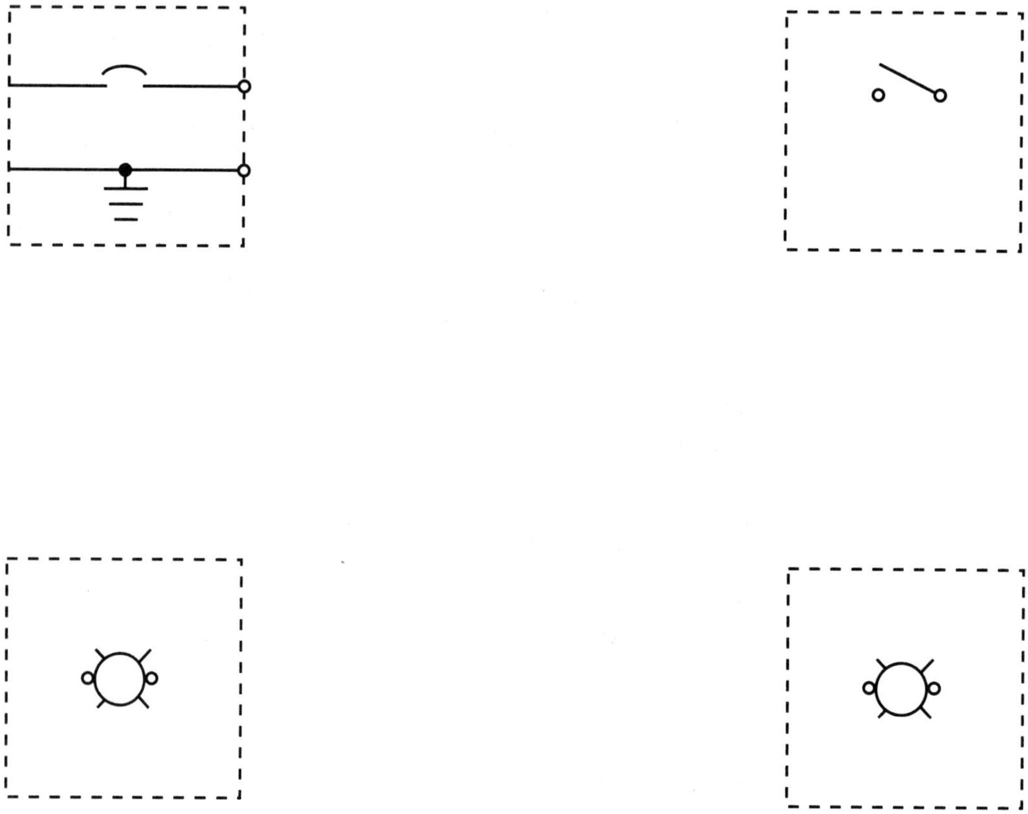

Fig. 11-14 Challenge drawing of two-way switching with power supplied to and one light fed from the switch enclosure. The second light should be fed from the first lighting outlet.

1. The homerun is to the first lighting-outlet location.
2. The second lighting-outlet location is fed from the first lighting-outlet location.
3. The switch drop is from the second lighting-outlet location.

At the source. The black ungrounded-supply conductor is connected to the load terminal on the circuit breaker. The white grounded-supply conductor is connected to the grounded busbar.

At the first lighting outlet. The black ungrounded-supply conductor is spliced to the black ungrounded-circuit conductor without connection to the lighting fixture, and extended to the second lighting outlet. The white grounded-supply conductor is spliced and pigtailed. The pigtail is connected to the grounded terminal on the lighting fixture. The splice is to the white grounded-circuit conductor extended to the other lighting outlet. The red ungrounded switch-leg conductor, extended from the other lighting outlet, is connected to the ungrounded terminal on the lighting fixture.

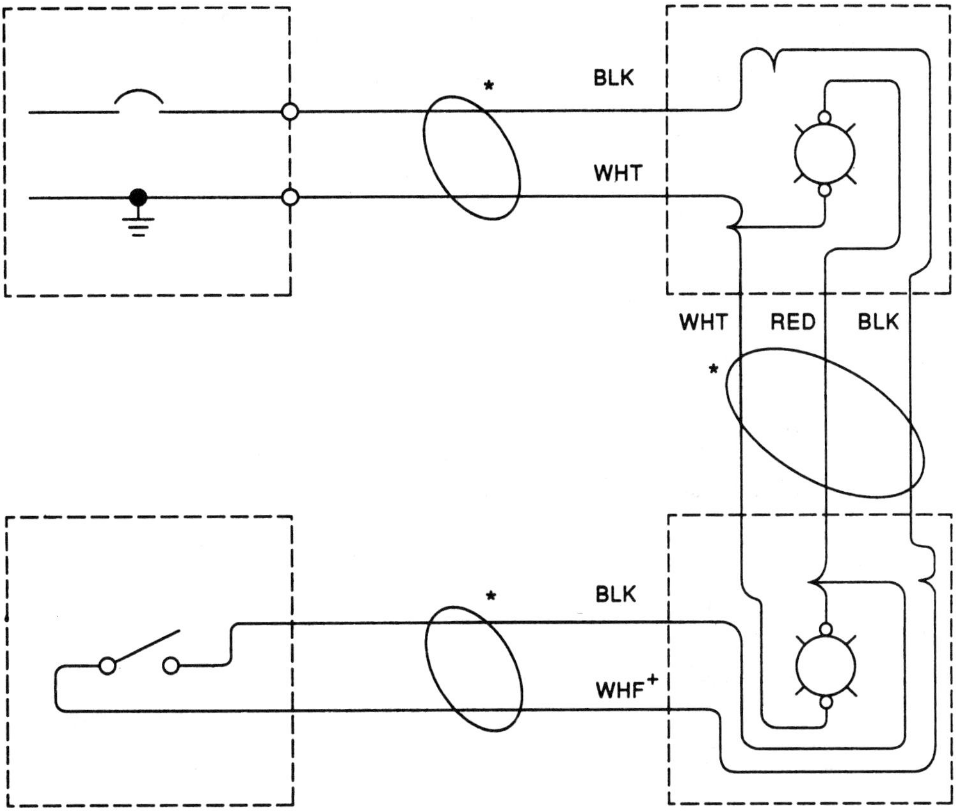

Fig. 11-15 When a switching arrangement consists of multiconductor cable, each cable must contain the correct number of conductors from point to point.

At the second lighting outlet. The black ungrounded-supply conductor is spliced to the ungrounded white-hot feed conductor in the switch drop multiconductor cable without reidentification of the white conductor. The white grounded-circuit conductor, extended from the first lighting outlet, is connected to the grounded terminal on the lighting fixture. The black ungrounded switch-leg conductor in the switch drop is spliced and pigtailed. The pigtail is connected to the ungrounded terminal on the lighting fixture. The splice is to the red ungrounded switch-leg conductor extended to the first lighting outlet.

At the switch location. The ungrounded white-hot feed conductor of the switch drop is connected to one terminal on the 2-way switch. The black switch-leg conductor is connected to the other terminal on the 2-way switch.

When the white insulated conductor in a multiconductor cable wiring method is used to supply the light switch, Sec. 200-7, Excp. No. 2 refers to this conductor as a white-hot feed—reidentification is not required. Had the white conductor in this example been used as the switch leg, Sec. 200-7, Excp. No. 1 requires the insulated conductor to be reidentified to another color (other than gray or green).

Code challenge 11–2. Complete the wiring diagram of Figure 11-16 using proper color coding of a multiconductor cable wiring method. The two lighting outlets should be controlled by the one switch, with the following criteria:

1. The homerun is to the first lighting outlet.
2. The switch location is fed from the first lighting outlet.
3. The second lighting outlet is fed from the switch location.

At the source. The (A) ____________ ungrounded-supply conductor is connected to the load terminal on the circuit breaker. The (B) ____________ grounded-supply conductor is connected to the grounded busbar.

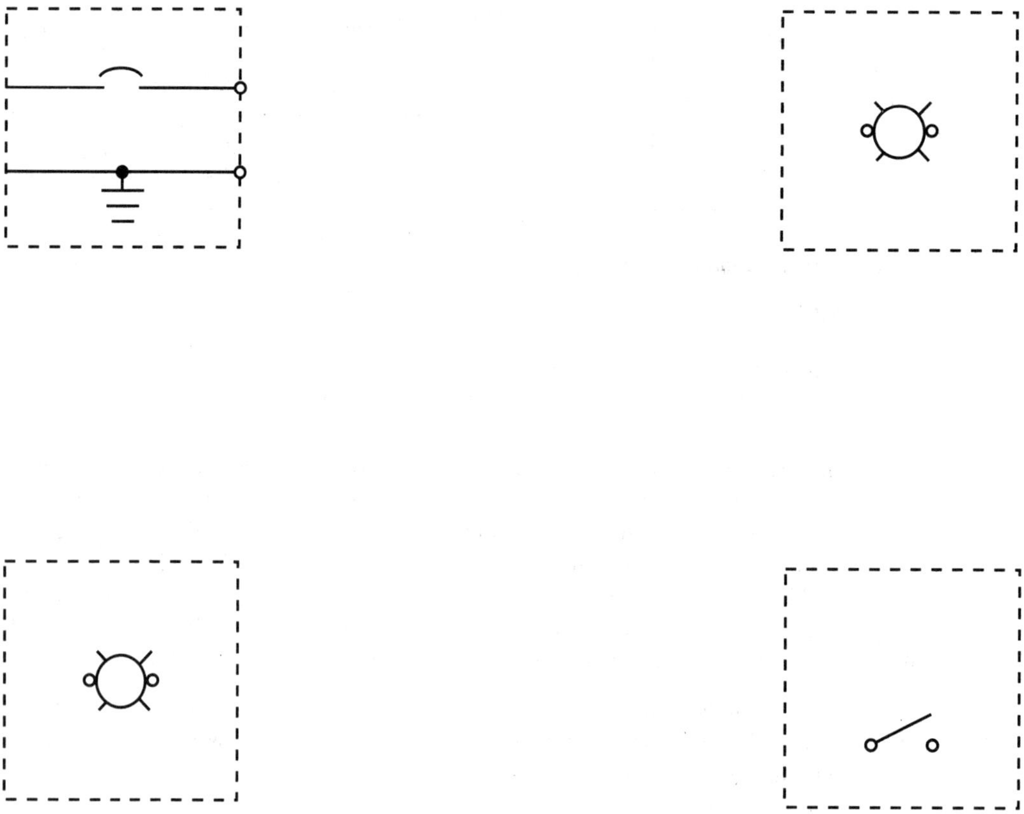

Fig. 11-16 Challenge drawing of two-way switching with the switch-leg feeding back with the homerun to the first lighting outlet.

At the first lighting outlet. The (C) ____________ ungrounded-supply conductor is spliced to the (D) ____________ ungrounded-circuit conductor without connection to the lighting fixture, and extended to the switch location. The (E) ____________ grounded-supply conductor is spliced and pigtailed. The pigtail is connected to the (F) ____________ terminal on the lighting fixture. The splice is extended as the (G) ____________ grounded-circuit conductor to the switch location. The (H) ____________ ungrounded switch-leg conductor, extended from the switching location, is connected to the (I) ____________ terminal on the lighting fixture.

At the switch. The (J) ____________ ungrounded-circuit conductor is connected to one terminal of the switch. The switch-leg conductor is spliced and pigtail-connected to the other switch terminal. The splice is to the (K) ____________ ungrounded switch-leg conductor extended to the first lighting outlet and to the (L) ____________ ungrounded switch-leg conductor extended to the second lighting outlet. The (M) ____________ grounded-circuit conductor extended from the first lighting outlet is spliced to the (N) ____________ grounded-circuit conductor without connection to the switch, and extended to the second lighting outlet.

At the second lighting outlet. The (O) ____________ ungrounded switch-leg conductor is connected to the (P) ____________ terminal of the lighting fixture. The (Q) ____________ grounded-circuit conductor is connected to the (R) ____________ terminal of the lighting fixture.

11.8 THREE-WAY SWITCHING

When lighting fixtures or other electrical outlets are to be controlled (switched) from two locations, two 2-way switches cannot be used. Instead, two 3-way switches are wired in series.

A 3-way switch, as shown in Figure 11-17, has a single-pole, double-throw switching arrangement in which power supplied to a common terminal can be switched (from one to the other) between two alternate switched terminals. Travelers are connected in parallel to the two switched terminals on one 3-way switch and extended to the other pair of switched terminals on the second 3-way switch.

The two common terminals and the pair of traveler conductors appear to separate the line (supply) and load terminals of a typical 2-way switch into two remote switching locations. As such, the 3-way switch connected to the homerun is often referred to as the

Fig. 11-17 A three-way switch does not have ON and OFF positions. Instead, the switching is between two alternate positions (traveler terminals) from a common supply.

line-side 3-way. The 3-way switch connected to the switch leg (switch return to the circuit) is often referred to as the load-side 3-way.

The handle of a 3-way switch is not marked with ON and OFF, as is the common 2-way lighting snap switch. Instead, the handle is left blank on both sides. The switching of a 3-way switch is between two alternate positions. Because these two switching positions supply a pair of travelers that connect to the second 3-way switch, the lights can be on with either or both switch handles in either the up or down position. The pattern depends on how the travelers are wired on initial installation, and which switch was used last to turn the lights on (or off).

At first glance, a 3-way switch looks like a 2-way switch except the body of the switch is wider. On one side of the 3-way switch, two screw terminals appear at both ends just like the 2-way. Only one screw terminal is located on the other side of the switch body at one end (Fig. 11-18).

The two screw terminals located directly across from each other on one end of the strap are the two switched terminals for the traveler conductors. The two traveler conductors are connected in parallel between the line-side and load-side 3-way switches. A pair of traveler conductors (as a rule) is never connected to a common side on a 3-way switch.

The single-screw terminal (only on one side at the other end) is the common wire terminal for the ungrounded-circuit supply conductor on the line-side 3-way switch, and the switch-leg terminal on the load-side 3-way switch.

Fig. 11-18 On a three-way, the pair of traveler terminals are located directly across from each other on one end of the strap. The common terminal is one located at the other end of the switch.

As with 2-way light switching using single insulated conductors, the color coding for ungrounded-circuit conductors used with 3-way switching is not restricted to any particular color, as long as the colors are not white, natural gray, or green (Sec. 310-12).

The traveler conductors are normally a different color from any switch-leg conductors contained in the circuit (Fig. 11-19). Both the travelers and the switch legs are normally a different color from the ungrounded supply or circuit conductors (switch or device feeds) contained in the circuit, especially when the circuit conductors are contained in a common conduit run, junction box, or switch enclosure. This is not an NE Code requirement, but a good electrical practice. Switch-feed conductors, switch-leg conductors, and traveler conductors are all ungrounded-circuit conductors.

Example 11–3. In the wiring diagram of Figure 11-20, 3-way switching is shown using color-coded insulated conductors in conduit. The lighting fixture is controlled by two switches with the following criteria:

1. The homerun is to the line-side 3-way switch location.
2. The load-side 3-way switch location is fed from the line-side 3-way switch location.
3. The lighting outlet is fed from the load-side 3-way switch location.

Fig. 11-19 Because one traveler terminal and the common terminal are located on the same side of the three-way switch, they are normally different colors to aid in correct wiring.

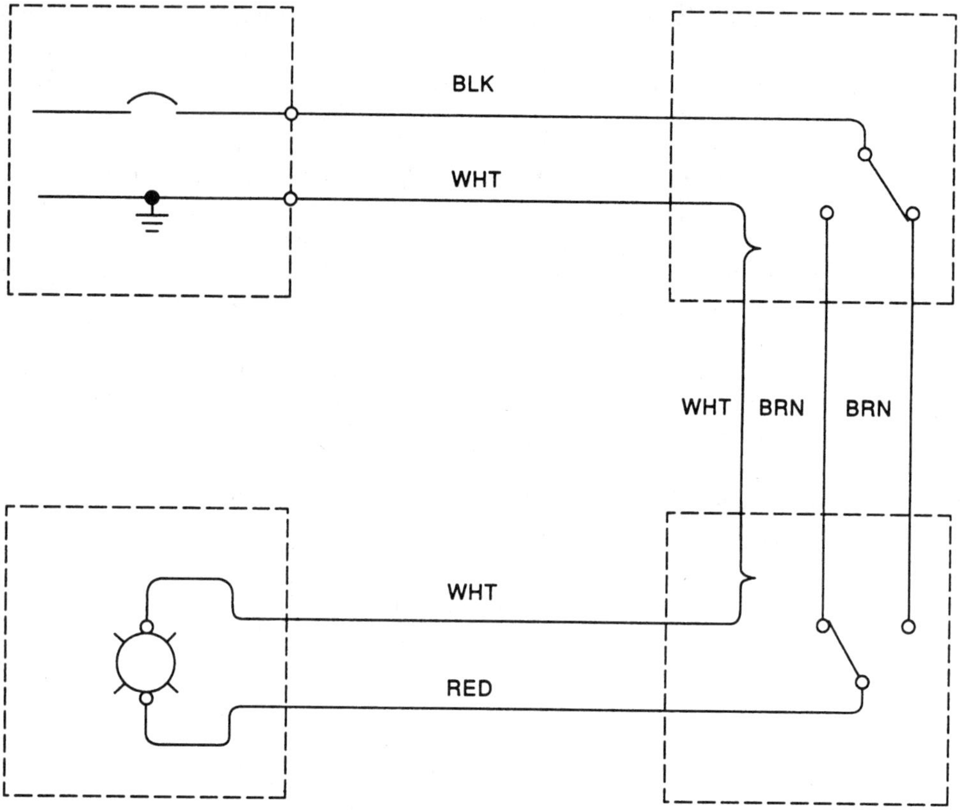

Fig. 11-20 When a lighting fixture is to be switched from two positions, a pair of travelers must connect the *two* three-way switches.

At the source. The black ungrounded-supply conductor is connected to the load terminal on the circuit breaker. The white grounded-supply conductor is connected to the grounded busbar.

At the line-side 3-way switch location. The black ungrounded-supply conductor is connected to the common terminal on the 3-way switch. The white grounded-supply conductor is spliced to the white grounded-circuit conductor without connection to the switch and extended to the second switching location. The two brown ungrounded-traveler conductors are connected to the switched terminals on the 3-way switch.

At the load-side 3-way switch location. The white grounded-circuit conductor is spliced without connection to the switch, and extended to the lighting outlet. The two brown ungrounded-traveler conductors are connected to the switched terminals on the 3-way switch. The red ungrounded switch-leg conductor is connected to the common terminal on the 3-way switch and extended to the lighting outlet.

At the lighting outlet. The red ungrounded switch-leg conductor is connected to the ungrounded terminal on the lighting fixture. The white grounded-circuit conductor is connected to the grounded terminal on the lighting fixture.

Code challenge 11–3. Complete the wiring diagram of Figure 11-21 using single insulated, color-coded conductors installed in conduit. The single lighting outlet should be controlled from two switch locations, with the following criteria:

1. The homerun is to the line-side 3-way switch location.
2. The lighting outlet is fed from the line-side 3-way switch location.
3. The load-side 3-way switch location is fed from the lighting outlet.

At the source. The (A) __________ ungrounded-supply conductor is connected to the load conductor on the circuit breaker. The (B) __________ grounded-supply conductor is connected to the grounded busbar.

At the line-side 3-way switch location. The (C) __________ ungrounded-supply conductor is connected to the (D) __________ terminal on the 3-way switch. The (E) __________ grounded-supply conductor is spliced to the (F) __________ grounded-circuit conductor without connection to the 3-way switch, and extended to the lighting outlet. The two (G) __________ ungrounded-traveler conductors are connected to the (H) __________ terminals on the 3-way switch.

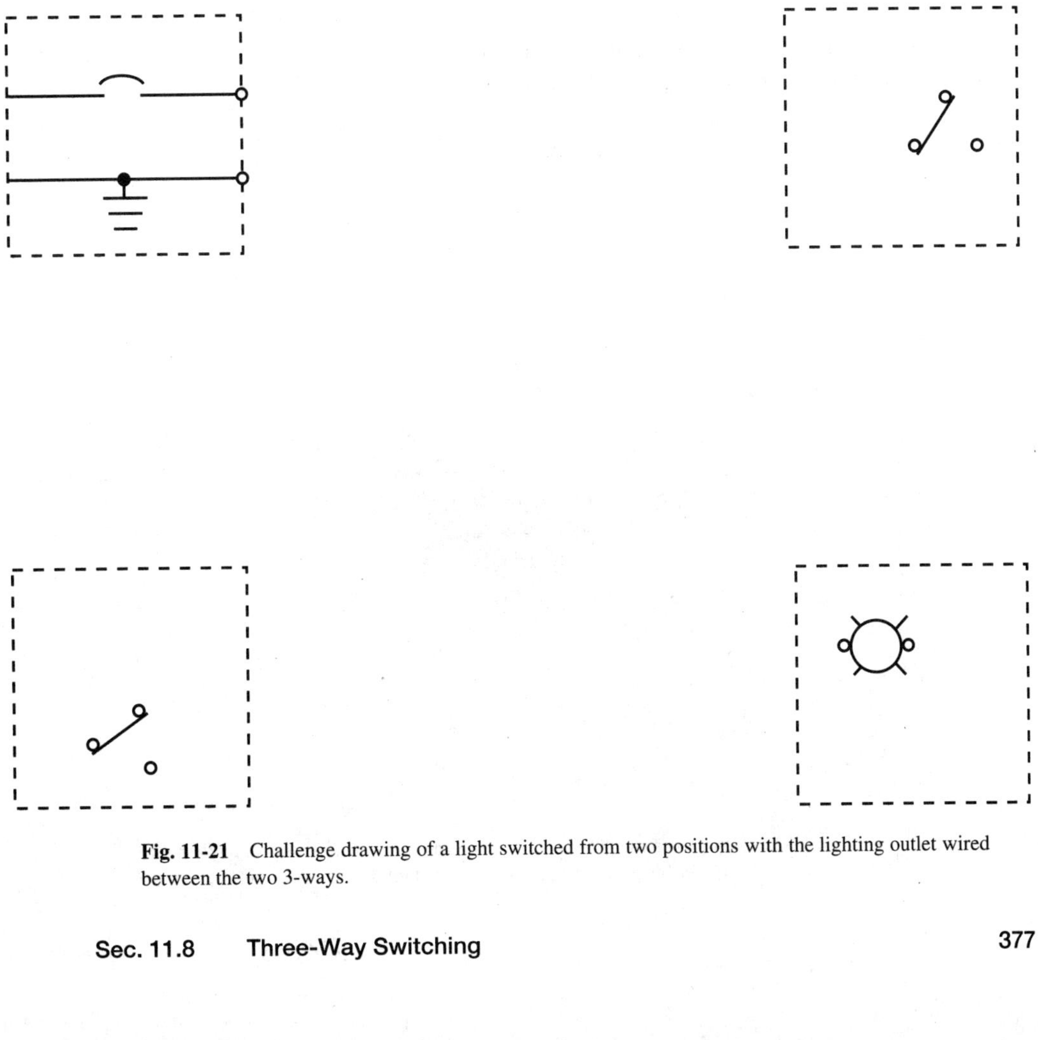

Fig. 11-21 Challenge drawing of a light switched from two positions with the lighting outlet wired between the two 3-ways.

At the lighting outlet. The (I) ____________ grounded-circuit conductor is connected to the grounded terminal on the lighting fixture. The two (J) ____________ ungrounded-traveler conductors are spliced without connection to the lighting fixture and extended to the load-side 3-way switch. The (K) ____________ ungrounded switch-leg conductor is connected to the ungrounded terminal of the lighting fixture.

At the load-side 3-way switch location. The (L) ____________ ungrounded switch-leg conductor is connected to the (M) ____________ terminal on the 3-way switch. The two (N) ____________ ungrounded-traveler conductors are connected to the (O) ____________ terminals on the 3-way switch.

11.9 SPLIT RECEPTACLES

On a standard duplex receptacle, breakaway tabs on both sides of the receptacle connect the electrical supply to the two single receptacles together. On a polarized duplex receptacle, the longer blade slot is the grounded-circuit terminal. The screws on this side of the receptacle are silver in color, as compared to the brass screws serving the ungrounded terminal (short blade slot) on the other side of the receptacle.

When a duplex receptacle is to be wired as a split receptacle, it is imperative that the breakaway tab on the grounded terminal not be removed. The breakaway tab on the ungrounded terminal (Fig. 11-22) must be removed to facilitate the split receptacle. In a

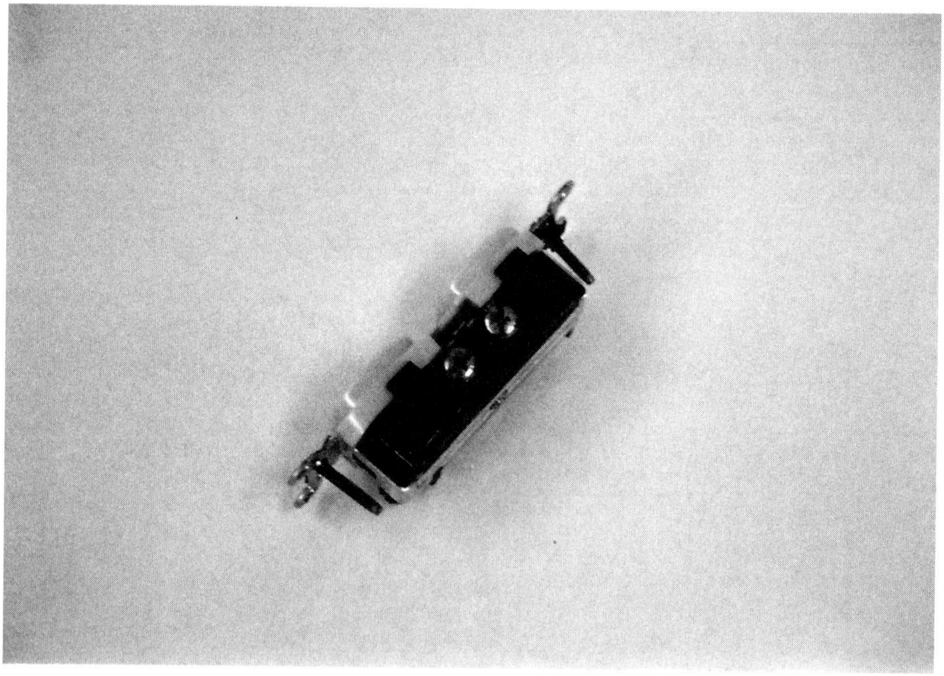

Fig. 11-22 In order to function as a split receptacle, the breakaway tab on the ungrounded side (short-blade slot–brass screws) of a duplex receptacle must be removed.

split-receptacle circuit, one receptacle of the duplex has a switched source (ungrounded, but switched supply). The other receptacle has an unswitched source (ungrounded supply). Remember, both receptacles in the duplex assembly share the grounded-terminal connection.

The NE Code makes no specific reference as to the relative location of the switched receptacle. The switched receptacle can be located on the top with the unswitched receptacle on the bottom or vice versa. Good trade practice would locate the switched receptacle on the bottom, such that constant (continuous or nonswitched) power would be available to cord-connected appliances or other utilization equipment without going under cords that are plugged in for switching convenience.

For example, in the great room of a dwelling, lighting is often furnished by decorative lamps. The lamps are plugged into the switched receptacle of a split duplex receptacle circuit. The switched receptacles are controlled by a wall switch. If the top receptacle is switched, a vacuum cleaner or carpet sweeper would have to be plugged in under the lamp cord. Or, if plugged into the top receptacle, the lighting circuit would have to be turned on in order to apply power to the cleaner or sweeper.

On a grounding-type duplex receptacle, the NE Code does not specify in which direction the polarity of the plug must be maintained. In some structures you may find the grounding terminal (banana plug or round hole located at the base of the tee formed with the two blade slots) located on top of the ungrounded and grounded terminals. In other structures, the grounding terminal may be located under the ungrounded and grounded terminal slots.

It is good trade practice to locate the grounding slot on top of the bladed terminals from a safety standpoint. If the receptacle is located behind a desk or some other type of furniture, and a conductive object like a screwdriver or letter opener is dropped on the plug-in cord of a lamp or other equipment, the attachment cap may be partially unplugged by the falling object.

When this occurs and the grounding terminal is on top, the grounding pin of the plug would be the first conductor the metal object comes in contact with. If a person should reach "in the blind" for the metal object, which could be in contact with the grounding pin of the attachment plug, no shock hazard exists.

If the configuration is reversed, with the grounding terminal of the duplex receptacle on the bottom, and the attachment cap (attachment plug) is partially unplugged by the fallen object, the person reaching "in the blind" could come in contact with a 120-volt source—a shock hazard.

The counter-argument for the grounding terminal's being located under the circuit terminals uses the delta configuration of the attachment plug as a vital means of support for the attachment plug. The blades and pin of the receptacle are held in place within the receptacle only by spring tension. The metal used to construct the slots is a metal strip that has been bowed inward from both sides of the slot. The bowed sections of the slot are compressed by the thickness of the inserted blade.

When plugged in, the spring tension of both blades tends to offer more support to the attachment plug and cord—more support as compared to just the single grounding slot's being installed on top. Both persuasions have their good and bad points.

When duplex receptacles are installed on their side, like the installation of a multioutlet assembly (plugmold), it is almost universally agreed that the grounded terminal, the longer slot, is installed over the ungrounded (shorter slot) terminal.

Example 11–4. In the wiring diagram of Figure 11-23, 3-way switching is shown using color coding of a multiconductor cable wiring method. The single split-receptacle outlet is controlled by two switches with the following criteria:

1. The homerun is to the receptacle outlet.
2. The load-side 3-way switch location is fed from the receptacle outlet.
3. The line-side 3-way switch location is fed from the load-side 3-way switch location.

At the source. The black ungrounded-supply conductor is connected to the load terminal on the circuit breaker. The white grounded-supply conductor is connected to the grounded busbar.

At the receptacle outlet. The black ungrounded-supply conductor is spliced and pigtailed. The pigtail is connected to the ungrounded terminal on the top receptacle. The splice is to the ungrounded white-hot feed (WHF) circuit conductor extended to the load-side 3-way switch location. The white grounded-supply conductor is connected to one of

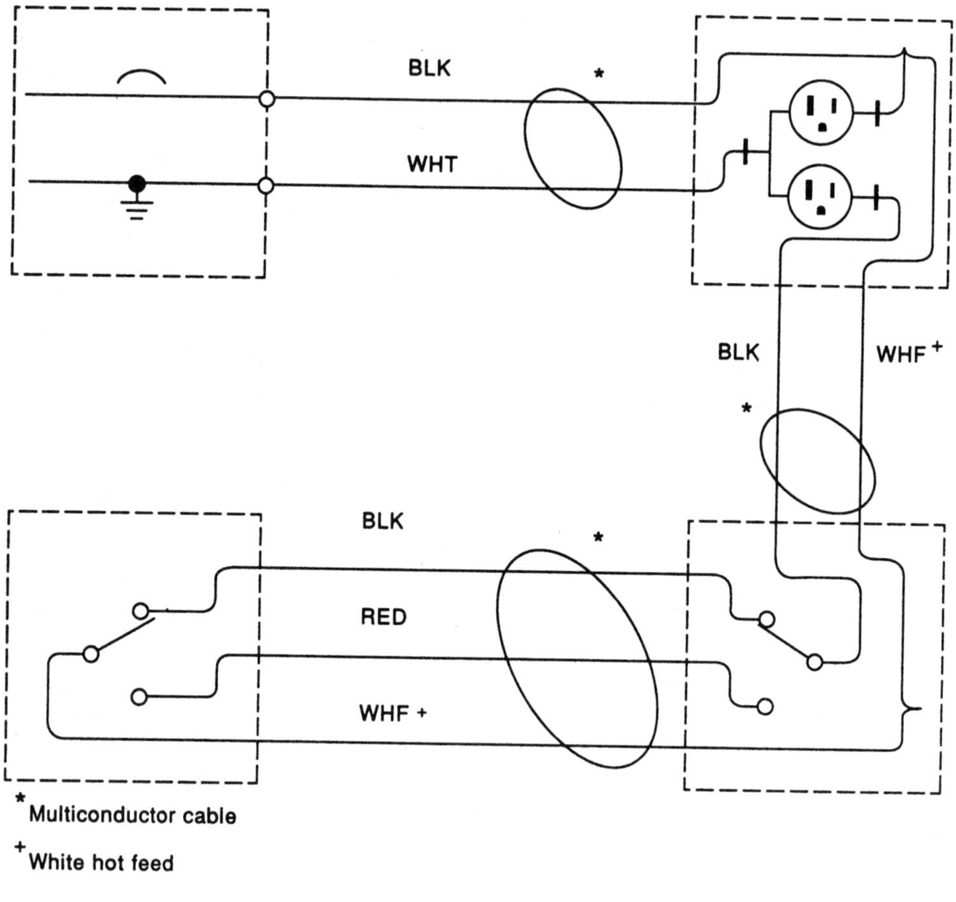

Fig. 11-23 Sections 200-7, 300-20(a), and 310-12 govern the installation of a split receptacle outlet supplied by a multiconductor wiring method.

the grounded terminals on the duplex receptacle. The black ungrounded switch-leg conductor, extended from the load-side 3-way switch location, is connected to the ungrounded terminal on the bottom receptacle.

At the load-side 3-way switch. The ungrounded WHF circuit conductor extended from the receptacle outlet is spliced to the ungrounded WHF circuit conductor without connection to the 3-way switch, and extended to the line-side 3-way switch. The black ungrounded switch-leg conductor is connected to the common terminal on the 3-way switch. The red and black ungrounded traveler conductors, extended from the line-side 3-way switch, are connected to the two switched terminals on the load-side 3-way switch.

At the line-side 3-way switch. The ungrounded WHF circuit conductor is connected to the common terminal on the 3-way switch. The red and black ungrounded-traveler conductors are connected to the two switched terminals on the 3-way switch.

In this example, the first 3-way switch in the supply of power from the receptacle outlet was used as the load-side 3-way switch, such that the contained white conductor in the multiconductor cable could be extended to the other 3-way switch as a white-hot feed without reidentification.

This type of installation practice minimizes confusion when the switches and receptacles are installed for makeup. On rough in, only the leads of the different colored conductors are exposed at each box location. If the white conductor had been swapped from the white-hot feed to a traveler conductor between the two 3-way switches, Sec. 200-7, Excp. No. 1 requires reidentification at both points of termination.

Code challenge 11–4. Complete the wiring diagram of Figure 11-24 using proper color coding of a multiconductor cable wiring method. The single split-receptacle outlet should be controlled by the two switches, with the following criteria:

1. The homerun is to the load-side 3-way switch location.
2. The line-side 3-way switch location is fed from the load-side 3-way switch location.
3. The split-receptacle outlet is fed from the load-side 3-way switch location.

At the source. The (A) ____________ ungrounded-supply conductor is connected to the load terminal on the circuit breaker. The (B) ____________ grounded-supply conductor is connected to the grounded busbar.

At the load-side 3-way switch. The (C) ____________ ungrounded-supply conductor is spliced to both the ungrounded (D) ____________ circuit conductor extended to the line-side 3-way switch and to the (E) ____________ ungrounded-supply conductor extended to the receptacle outlet. The (F) ____________ grounded-supply conductor is spliced to the (G) ____________ grounded-circuit conductor without connection to the switch and extended to the receptacle outlet. The (H) ____________ ungrounded switch-leg conductor is connected to the (I) ____________ terminal on the 3-way switch and extended to the receptacle outlet. The (J) ____________ and (K) ____________ ungrounded-traveler conductors are connected to the (L) ____________ terminals on the 3-way switch.

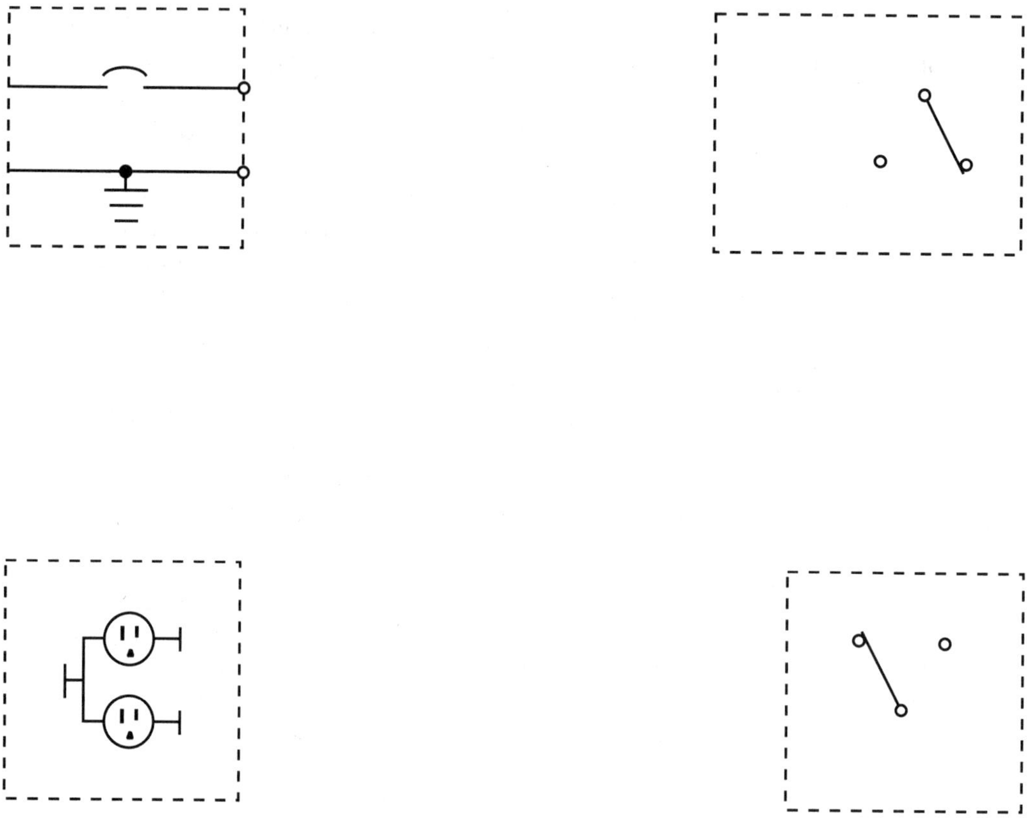

Fig. 11-24 Challenge drawing of a split receptacle with the homerun and the receptacle terminated to the same three-way switch location.

At the line-side 3-way switch. The ungrounded (M) ____________ circuit conductor is connected to the (N) ____________ terminal on the 3-way switch. The (O) ____________ and (P) ____________ ungrounded-traveler conductors are connected to the (Q) ____________ terminals on the three-way switch.

At the receptacle outlet. The (R) ____________ grounded-supply conductor is connected to one of the (S) ____________ terminals on the duplex receptacle. The (T) ____________ ungrounded-supply conductor is connected to the (U) ____________ terminal on the top receptacle. The (V) ____________ ungrounded switch-leg conductor is connected to the (W) ____________ terminal on the bottom receptacle.

11.10 FOUR-WAY SWITCHING

If switching is required at more than two locations, one or more 4-way switches must be installed in series between the two 3-way switches. Series arrays of 4-way switches cannot be installed for multiple-location switching without the use of two 3-ways. As the use of two 3-way switches splits the line and load terminals of a 2-way switch into two remote loca-

Fig. 11-25 Pairs of travelers in series with a four-way switch are either cross-connected or paralleled by the switching action.

tions, a 4-way switch splits the two 3-ways to generate a third remote switching location.

Four-way switches can only be inserted in series with the traveler conductors between two 3-ways. If four switching locations are required, the two 4-ways must be wired in series with each other between the two 3-ways. The switching arrangement has to begin and end with a 3-way switch. If a lighting fixture is to be switched from seven locations, five 4-way switches (7–2) and two 3-way switches would have to be wired in series.

A 4-way switch has four terminals, two on each side. The internal switching arrangement consists of two positions (Fig. 11-25). In one switch position, the terminals are crossed diagonally within the switch. In the other position, the travelers are merely paralleled through the switch. The switch is normally drawn on a wiring diagram in the crossed-diagonal position to avoid confusion.

The four screw terminals have two different colored pairs of terminals located one pair at each end of the switch (Fig. 11-26). One terminal is located on either side. The

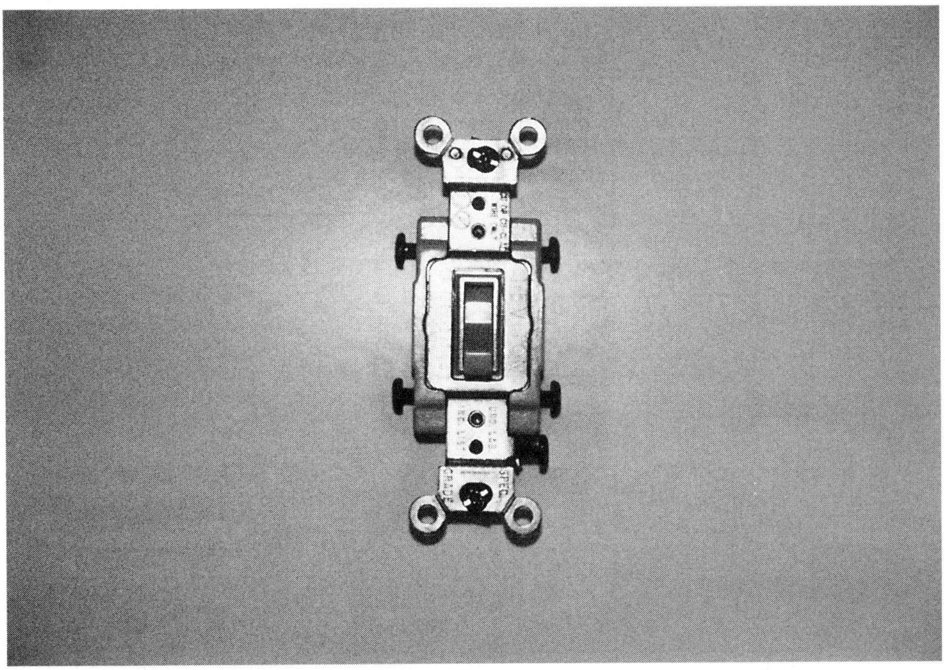

Fig. 11-26 A four-way lighting switch is a four-terminal device. One pair of travelers is connected across the switch to the terminals on either side. The other pair of travelers is connected in the same manner at the other end.

4-way switch is inserted (parallel to the strap) in series with two 3-way switches. The pair of traveler conductors from the line-side 3-way switch is connected at one end of the 4-way switch (straddling the strap, one traveler conductor on each side). The pair of traveler conductors from the load-side 3-way switch is connected at the other end of the strap in the same manner. A pair of traveler conductors is never connected common to one side (parallel to the strap) of a 4-way switch.

Example 11–5. In the wiring diagram of Figure 11-27, 4-way switching is shown using color-coded insulated conductors in conduit. The two switched receptacle outlets are controlled by three switches with the following criteria:

1. The homerun is to the first receptacle outlet.
2. The second receptacle outlet location is fed from the first receptacle outlet.
3. The load-side 3-way switch location is fed from the first receptacle outlet.
4. The 4-way switch location is fed from the load-side 3-way switch location.
5. The line-side 3-way switch location is fed from the 4-way switch location.

At the source. The black ungrounded-supply conductor is connected to the load terminal on the circuit breaker. The white grounded-supply conductor is connected to the grounded busbar.

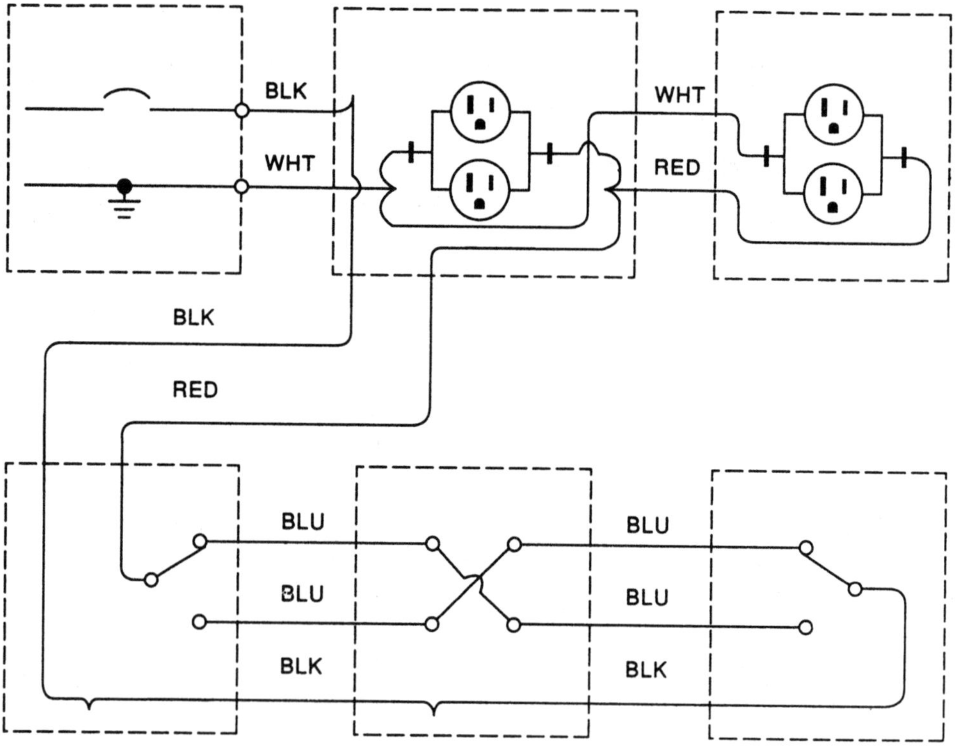

Fig. 11-27 Conductor grouping [Sec. 300-20(a)] and outlet box volume requirements (Sec. 370-6) are important concerns in this switching arrangement.

At the first receptacle outlet. The black ungrounded-supply conductor is spliced to the black grounded-circuit conductor without connection to the receptacle and extended to the load-side 3-way switch location. The white grounded-supply conductor is spliced and pigtailed. The pigtail is connected to one of the grounded terminals on the duplex receptacle. The splice is to the white grounded-circuit conductor extended to the other receptacle outlet. The red ungrounded switch-leg conductor is spliced and pigtailed. The pigtail is connected to one of the ungrounded terminals on the duplex receptacle. The splice is to the red ungrounded-circuit conductors extended to the other receptacle outlet.

At the second receptacle outlet. The red ungrounded switch-leg conductor is connected to one of the ungrounded terminals on the duplex receptacle. The white grounded-circuit conductor is connected to one of the grounded terminals on the duplex receptacle.

At the load-side 3-way switch. The black ungrounded-circuit conductors are spliced without connection to the switch, and extended to the 4-way switch location. The red ungrounded switch-leg conductor is connected to the common terminal on the 3-way switch. The two blue ungrounded-traveler conductors are connected to the switched terminals on the 3-way switch and extended to the 4-way switch.

At the 4-way switch. The black ungrounded-circuit conductors are spliced without connection to the four-way switch and extended to the line-side 3-way switch. The two blue ungrounded traveler conductors extended from the load-side 3-way switch are connected to one pair of switched terminals (at one end, parallel to the strap) on the 4-way switch. The two blue ungrounded-traveler conductors extended from the line-side 3-way switch are connected to the other pair of switched terminals) on the 4-way switch.

At the line-side 3-way switch. The two blue ungrounded-traveler conductors are connected to the switched terminals on the 3-way switch. The black ungrounded switch-feed conductor is connected to the common terminal on the 3-way switch.

The practicing electrician should note that a grounded-circuit conductor was not extended to the switches. Switches are control devices that must be wired in the ungrounded side of the circuit only. A white conductor is used to convey or connect the grounded side of the circuit.

Code challenge 11–5. Complete the wiring diagram of Figure 11-28 using single insulated, color-coded conductors installed in conduit. The two switched receptacle outlets should be should be controlled from three switch locations, with the following criteria:

1. The homerun is to the 4-way switch location.
2. The line-side 3-way switch location (on the left) is fed from the 4-way switch location.
3. The load-side 3-way switch location (on the right) is fed from the 4-way switch location.
4. The first receptacle outlet is fed from the 4-way switch location.
5. The second receptacle outlet is fed from the first receptacle outlet.

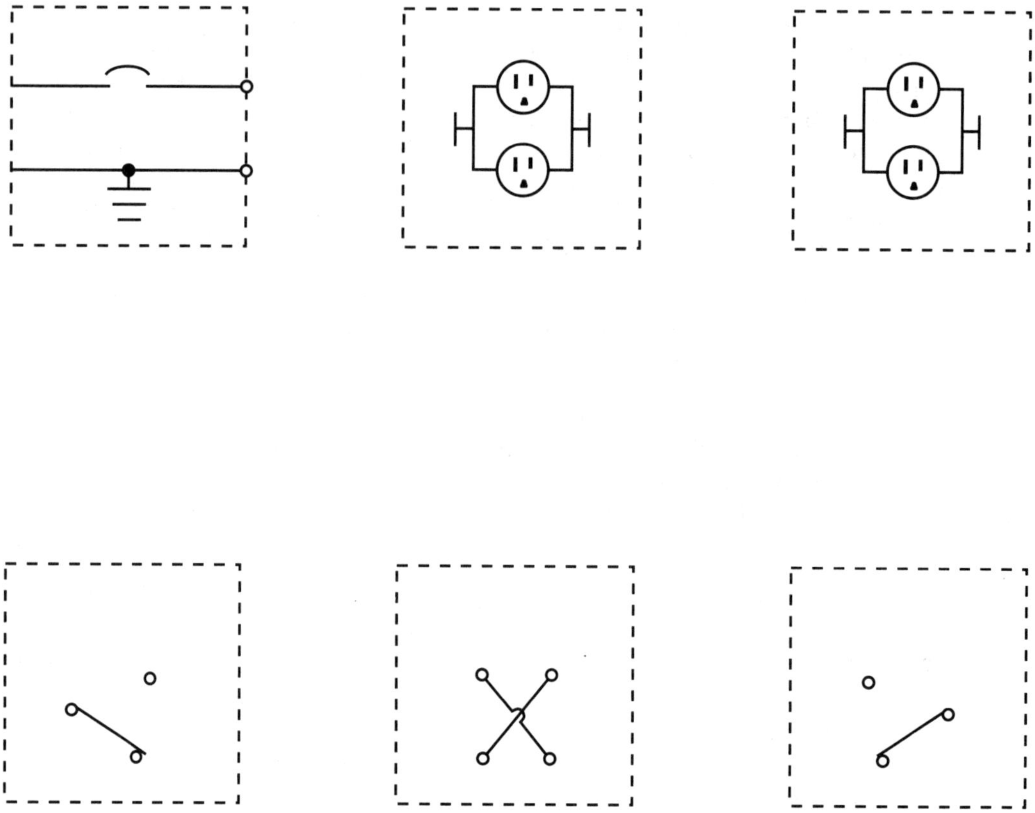

Fig. 11-28 Challenge drawing with both the homerun and receptacle feed terminated at the four-way switch location.

At the source. The (A) ____________ ungrounded-supply conductor is connected to the load terminal on the circuit breaker. The (B) ____________ grounded-supply conductor is connected to the grounded busbar.

At the 4-way switch. The (C) ____________ ungrounded-supply conductor is spliced to the (D) ____________ ungrounded switch-feed conductor without (E) ____________ to the 4-way switch, and extended to the line-side 3-way switch. The (F) ____________ grounded-supply conductor is spliced to the (G) ____________ grounded-circuit conductor without (H) ____________ to the 4-way switch, and extended to the first receptacle outlet. The (I) ____________ ungrounded switch-leg conductor from the load-side 3-way switch is spliced without (J) ____________ to the 4-way switch, and extended to the first receptacle outlet. The two (K) ____________ ungrounded traveler conduc-tors, extended from the line-side 3-way switch, are connected to one pair of (L) ____________ terminals on the 4-way switch. The two (M) ____________ ungrounded traveler conductors, extended from the load-side 3-way switch, are connected to the other pair of (N) ____________ terminals on the 4-way switch.

At the line-side 3-way switch. The (O) ____________ ungrounded switch-feed conductor is connected to the (P) ____________ terminal on the 3-way switch. The two (Q) ____________ ungrounded-traveler conductors are connected to the (R) ____________ terminals on the 3-way switch.

At the load-side 3-way switch. The (S) ____________ ungrounded switch-leg conductor is connected to the (T) ____________ terminal on the 3-way switch. The two (U) ____________ ungrounded-traveler conductors are connected to the (V) ____________ terminals on the 3-way switch.

At the first receptacle outlet. The (W) ____________ ungrounded switch-leg conductor is spliced and pigtailed. The pigtail is connected to the (X) ____________ terminal on the duplex receptacle. The splice is to the other (Y) ____________ ungrounded switch-leg conductor extended to the second receptacle outlet. The (Z) ____________ grounded-circuit conductor is spliced and pigtailed. The pigtail is connected to the (1A) ____________ terminal on the duplex receptacle. The splice is the (1B) ____________ grounded-circuit conductor extended to the second receptacle outlet.

At the second receptacle outlet. The (1C) ____________ ungrounded switch-leg conductor is connected to the (1D) ____________ terminal on the duplex receptacle. The (1E) ____________ grounded-circuit conductor is connected to the (1F) ____________ terminal on the duplex receptacle.

When used as either the grounded-circuit conductor, or as a WHF conductor (switch-feed) in lighting circuits, the white insulated conductor in a multiconductor cable, such as Romex or BX, can be without reidentification. If the white conductor is used as the switch-leg (switch return) for the circuit from a switching location, then it must be reidentified to a color other than gray or green.

Example 11–6. In the wiring diagram of Figure 11-29, 4-way switching is shown using color coding of a multiconductor cable wiring method. The single lighting outlet is controlled by three switches with the following criteria:

1. The homerun is to the 4-way switching location.
2. The lighting outlet is fed from the 4-way switch location.
3. The line-side (left-side) 3-way switch is fed from the 4-way switch location.
4. The load-side (right-side) 3-way switch is fed from the 4-way switch location.

At the source. The black ungrounded-supply conductor is connected to the load terminal on the circuit breaker. The white grounded-supply conductor is connected to the grounded busbar.

At the 4-way switch. The black ungrounded-supply conductor is spliced to the ungrounded WHF conductor without connection to the 4-way switch, and extended to the line-side 3-way switch. The white grounded-supply conductor is spliced to the white grounded-circuit conductor without connection to the 4-way switch, and extended to the lighting outlet. The white ungrounded switch-leg conductor, extended from the load-side

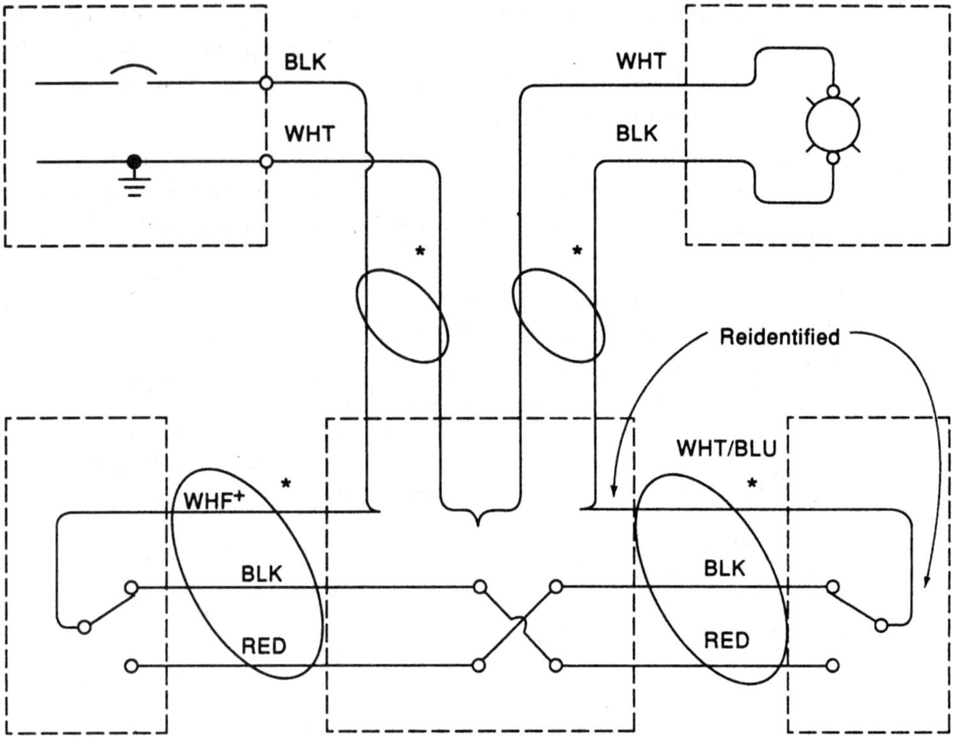

* Multiconductor cable

\+ White hot feed

Fig. 11-29 This wiring arrangement satisfies the requirements of Sec. 200-7, Exceptions No. 1 and No. 2, Secs. 300-20(a), and 310-12.

3-way switch, is reidentified to blue and spliced to the black ungrounded switch-leg conductor without connection to the 4-way switch. The black ungrounded switch-leg conductor is extended to the lighting outlet. The black and red ungrounded-traveler conductors, extended from the line-side 3-way switch, are connected to a pair of switched terminals on the 4-way switch. The black and red ungrounded-traveler conductors, extended from the load-side 3-way switch, are connected to the other pair of switched terminals on the 4-way switch.

At the line-side 3-way switch. The ungrounded WHF conductor is connected to the common terminal on the 3-way switch. The red and black ungrounded-traveler conductors are connected to the switched terminals on the 3-way switch.

At the load-side 3-way switch. The ungrounded white switch-leg conductor, reidentified to blue, is connected to the common terminal on the 3-way switch. The red and black ungrounded traveler conductors are connected to the switched terminals on the 3-way switch.

At the lighting outlet. The black ungrounded switch-leg conductor is connected to the ungrounded terminal on the lighting fixture. The white grounded-circuit conductor is connected to the grounded terminal on the lighting fixture.

Code challenge 11–6. Complete the wiring diagram of Figure 11-30 using proper color-coding of a multiconductor cable wiring method. The single lighting outlet should be controlled from three switch locations, with the following criteria:

1. The homerun is to the line-side 3-way switch location.
2. The 4-way switch location is fed from the line-side 3-way switch location.
3. The load-side 3-way switch location is fed from the 4-way switch location.
4. The lighting outlet is fed from the 4-way switch location.

At the source. The black (A) ____________-supply conductor is connected to the load terminal of the (B) ____________. The white (C) ____________-supply conductor is connected to the (D) ____________ busbar.

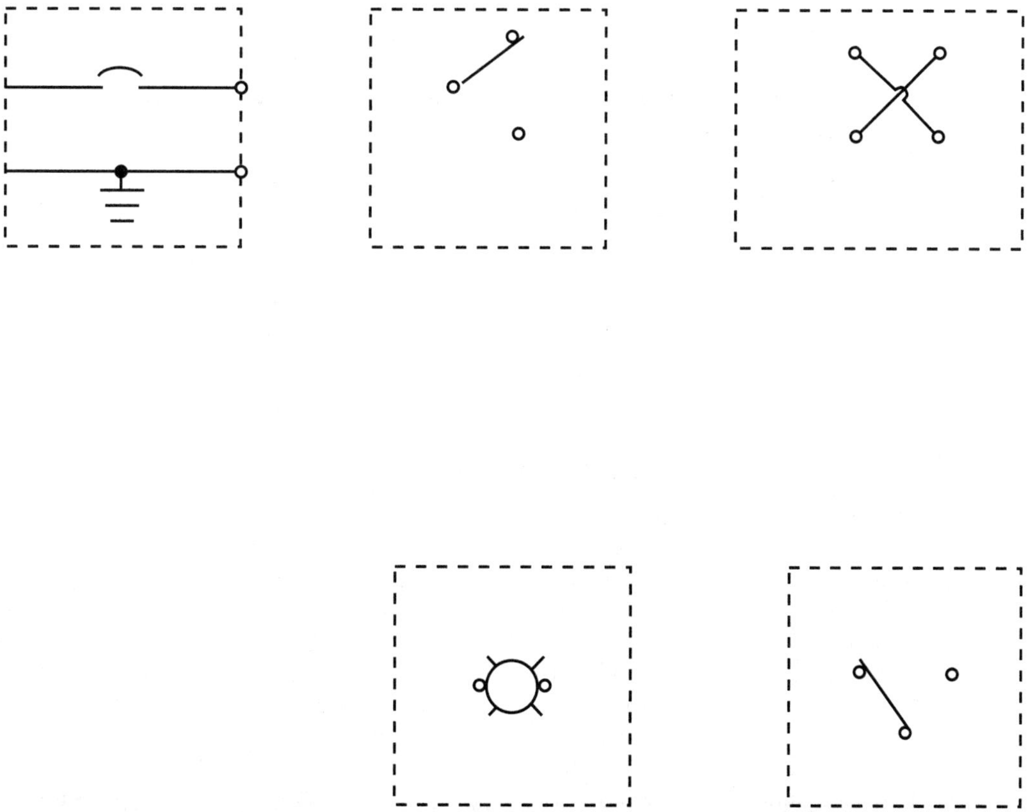

Fig. 11-30 Support drawing with the homerun to one of the three-way switch locations, and the light feed from the four-way switch location.

At the line-side 3-way switch. The black (E) ____________-supply conductor is connected to the (F) ____________ terminal on the 3-way switch. The black and red (G) ____________-traveler conductors are connected to the (H) ____________ terminals on the 3-way switch. The white (I) ____________-supply conductor is spliced to the white (J) ____________-circuit conductor without (K) ____________ to the 3-way switch, and extended to the 4-way switch.

At the 4-way switch. The white (L) ____________-circuit conductor from the line-side 3-way switch is spliced without (M) ____________ to the four-way switch, and extended to the lighting outlet. The black and red ungrounded-(N) ____________ conductors, extended from the line-side 3-way switch, are connected to one pair of (O) ____________ terminals on the 4-way switch. The black and red ungrounded-(P) ____________ conductors, extended from the load-side 3-way switch, are connected to the other pair of (Q) ____________ terminals on the 4-way switch. The ungrounded white switch-(R) ____________ conductor, extended from the load-side 3-way switch, is reidentifed to (S) ____________, and spliced to the black (T) ____________ switch-leg conductor. The black (U) ____________ switch-leg conductor is extended to the lighting outlet.

At the load-side 3-way switch. The ungrounded white switch (V)-____________ conductor is reidentified to (W) ____________, and connected to the (X) ____________ terminal on the 3-way switch. The black and red ungrounded (Y)-____________ conductors are connected to the (Z) ____________ terminals on the three-way switch.

At the lighting outlet. The black (1A) ____________ switch-leg conductor is connected to the (1B) ____________ terminal of the lighting fixture. The white (1C) ____________-supply conductor is connected to the intended (1D) ____________ terminal of the lighting fixture.

Switching arrangements or configurations can be made in various ways; the examples shown in this chapter were designed to give the practicing electrician a basic understanding of NE Code requirements for switching circuits.

REVIEW QUESTIONS

1. As a general rule, switches used to control or disconnect electrical power from a given load must be ____________ accessible.
2. As a general rule, switches used to control or disconnect electrical power from a given load must be located no more than ____________ feet, ____________ inches above the finished floor line, grade level, or working platform.
3. When snap (toggle) switches are ganged together to control more than one phase on a 480Y/277-volt multiwire lighting branch circuit, a fixed ____________ must be installed to isolate switches that control different phase power.
4. Metal construction (device) boxes can be ganged together by removing the sideplates and fastening one or more boxes together to attain the required number of ganged switches.

   True or False

5. Lighting switch faceplates installed in ungrounded metal boxes must be constructed of non-combustible, ____________ material.
6. Faceplates mounted over recessed-mounted lighting switches must effectively close the wall opening and ____________ against the wall surface.
7. When surface-mounted switches are used for lighting switches, the switch faceplate must effectively seal (close) the box.

   True or False

8. Switches installed in recessed boxes must be mechanically supported by the plaster ears on the switch resting against the ____________ surface.
9. Switches installed in surface-mounted boxes must be mechanically supported by the strap of the switch resting against the face of the ____________ .
10. Switches installed in recessed boxes that are flush with the finished wall surface, or project (standout) from the wall surface must be mechanically supported by the ____________ of the switch resting against the face of the box.
11. Device or outlet boxes installed in ceilings or walls fabricated of combustible material must be mounted so that the open edge (front) of the box is at least ____________ with the finished ceiling or wall surface, or project there from.
12. Device or outlet boxes installed in ceilings or walls fabricated of noncombustible material can be mounted such that the open edge (front) of the box is recessed no more than ____________-inch from the finished ceiling or wall surface.
13. For recessed lighting switch boxes installed in kitchen or bathroom walls that are partially paneled or tiled, the box must be located such that it is completely in the tile or paneling, completely out of the tile or paneling, or midway of the two wall surfaces with the tile or paneling coped or dressed around the box opening such that the faceplate completely closes the wall opening.

   True or False

14. Lighting switches installed in wet locations must have ____________ enclosures.
15. Lighting switches installed in hazardous locations must be contained in enclosures that are approved for the class and division of the hazardous location.

   True or False

16. As a general rule, switches or circuit breakers cannot be installed in series with the grounded-supply conductor of lighting circuits.

   True or False

17. Ungrounded-circuit conductors, whether switch feeds, switch legs, or travelers, can be any color except ____________, ____________ ____________, or ____________.
18. Neutrals or other grounded-circuit conductors wire size No. 6 AWG or smaller must be colored ____________ or ____________ ____________ along their entire length.
19. When covered or insulated, equipment-grounding conductors wire size No. 6 AWG or smaller must be colored either ____________ or ____________, with one or more ____________ stripes along their entire length.
20. When Romex, BX, or other multiconductor wiring methods are used to supply an appliance or other utilization equipment that requires a phase-to-phase (as compared to phase-to-neutral) voltage, the contained insulated white conductor can be reidentified to another color other than ____________ ____________ or ____________ at all points of accessibility and/or termination, and be used as an ungrounded circuit conductor.

21. When multiconductor wiring methods are used for switch drops, the contained insulated white conductor can be used as the switch-feed conductor without being reidentified. The conductor is referred to as a ____________-____________-____________.

22. When multiconductor wiring methods are used for switch drops to 3- and 4-way switching, and the contained insulated white conductor is used as the switch-return conductor (switch leg), it must be reidentified to a color other than ____________-____________ or ____________.

23. When single, insulated traveler conductors are installed in raceway between 3-way and 4-way switches, they can be any color except ____________, ____________-____________, or ____________.

24. A ____________-way switch has a common terminal on one end and a pair of alternate switch terminals on the other end. A ____________-way switch has two pairs of switched terminals, one set (pair) at end.

25. If a lighting or other electrical outlet is to be switched from more than two locations, one or more ____________-way switches must be installed in series between two ____________-way switches that start and finish the switching array.

REVIEW PROBLEMS

1. Draw a wiring diagram for the control of three lighting outlets from one switch location. Use single insulated, color-coded conductors installed in conduit with the following criteria:
   a. The homerun is to the first lighting outlet.
   b. The second lighting outlet is fed the first lighting outlet.
   c. The third lighting outlet is fed the second lighting outlet.
   d. The switching location is fed the third lighting outlet.
2. Draw a wiring diagram for the control of two lighting outlets from one switch location. Use proper color coding of a multiconductor cable wiring method with the following criteria:
   a. The homerun is to the first lighting outlet.
   b. The second lighting outlet is fed the first lighting outlet.
   c. The switching location is fed the first lighting outlet.
3. Draw a wiring diagram for the control of two lighting outlets from two switch locations. Use single-insulated, color-coded conductors installed in conduit with the following criteria:
   a. The homerun is to the first switching location.
   b. The first lighting outlet is fed the first switching location.
   c. The second lighting outlet is fed the first lighting outlet.
   d. The second switching location is fed the second lighting outlet.
4. Draw a wiring diagram for the control of two receptacle outlets from two switch locations. Use proper color coding of a multiconductor cable wiring method with the following criteria:
   a. The homerun is to the first receptacle outlet.
   b. The first switching location is fed from the first receptacle outlet.
   c. The second switching location is fed from the first switching location.
   d. The second receptacle outlet is fed from the second switching location.
5. Draw a wiring diagram for the supply and control of two split-receptacle outlets switched from two locations. Use single-insulated, color-coded conductors with the following criteria:
   a. The homerun is to the first receptacle outlet.
   b. The line-side switching location is fed from the first receptacle outlet.
   c. The second receptacle outlet is fed from the first receptacle outlet.
   d. The load-side switching location is fed from the second receptacle outlet.

6. Draw a wiring diagram for the control of two lighting outlets switched from three locations. Use proper color-coding of a multiconductor wiring method such as UF cable, with the following criteria:

a. The three switching locations are drawn left to right across the page.
b. The homerun is to the center switching location.
c. The line-side switching location (on the left) is fed from the center switching location.
d. The load-side switching location (on the right) is fed from the center switching location.
e. The first lighting outlet is fed from the load-side switching location.
f. The second lighting outlet is fed from the first lighting outlet.

Chapter 12

Individual Branch Circuits

A branch circuit that has two or more outlets is referred to as multioutlet or general purpose. An individual or single branch circuit dedicated to a single outlet, or to a single piece (unit) of equipment is referred to as specific purpose. According to their size, appliances, other utilization equipment, or motors may be supplied by either type of branch circuit.

12.1 APPLIANCES

An appliance, as defined in Art. 100, is utilization equipment such as clothes washers, dishwashers, electric ranges, or refrigerators that are used for household chores, food processing or preserving, or other functions in the home (dwelling). Utilization equipment, which has a much broader definition, converts electrical energy into mechanical, chemical, heating, lighting, or other forms of energy.

The installation of appliances as well as many other types of utilization equipment is governed by the requirements of Art. 422. Part B lists the general branch-circuit requirements for appliances. Section 422-4(a) stipulates that the rating of an individual branch circuit shall not be less than the nameplate rating of the appliance.

All appliances are required by Sec. 110-3(b) to be listed and labeled. The nameplate is used to identify the correct supply voltage, the number of required phases, the full-load current, and other pertinent information required for the safe installation and operation of the appliance.

Exception No. 2 to Sec. 422-4(a) requires the branch circuit supplying any appliance that can be operated continuously (draws full-load current without interruption for at least three hours, Art. 100) to be rated at 125% of the appliance's nameplate full-load current.

As indicated by the change in Sec. 220-3(a) in the 1996 edition of the NE Code, the rating of the branch circuit is referred to as its computed load, and as such, the 125% overrate of the nameplate full-load or rated-load current of the appliance or other utilization equipment is used to determine both the rating or setting of the branch-circuit overcurrent-protection device and the rated ampacity of the circuit conductors.

When appliances or other utilization-equipment loads are supplied by a multioutlet branch circuit, Sec. 422-4(b) requires the computed load of the branch circuit to be determined by the guidelines of Sec. 210-23.

For an individual appliance branch circuit, Sec. 422-5 requires the branch-circuit overcurrent-protection device to be sized according to the ampacity of the branch-circuit conductors supplying the appliance (Sec. 240-3). If the nameplate on the appliance specifies a "maximum branch-circuit overcurrent device rating," then this rating cannot be exceeded, even if the rated ampacity of the conductors (and Sec. 240-3) would allow a larger size or rating of the overcurrent-protection device.

12.2 DISCONNECTING MEANS

Part C of Art. 422 lists the general installation requirements for appliances. Part D governs the control and protection of appliances. Section 422-20 requires all appliances to have an individual disconnecting means that will disconnect all the ungrounded phase conductors to the appliance from their source.

For small appliances not rated over 300 volt-amps or ⅛ horsepower, Sec. 422-21(a) will allow the branch-circuit overcurrent-protection device to serve as the required disconnecting means. This means a branch-circuit fused switch or more commonly a circuit breaker can serve as the required disconnecting means.

For large appliances, such as water heaters that are rated over 300 volt-amps or ⅛ horsepower [Section 422-21(b)], a branch-circuit fused switch or circuit breaker (standard circuit breakers are marketed in sizes of 15 amps and larger) is allowed to serve as the required disconnecting means only if it is within sight of and readily accessible to the appliance.

The definition of within sight of ("In Sight From," Art. 100) requires the branch-circuit fused switch or circuit breaker to be within line of sight at a maximum of 50 feet from the appliance. For the disconnecting means to be readily accessible, Sec. 380-8(a) requires the handle of the knife switch or circuit breaker, when in its highest position, to be a maximum of 6 feet, 7 inches (2 meters) above the finished floorline.

If the panel (containing the fused switch or circuit breaker) is not readily accessible to the appliance, a second disconnect, located readily accessible, in sight of, and within 50 feet of the appliance is required (Fig. 12-1). The second disconnect is not required if the branch-circuit fused switch or circuit breaker is capable of being locked in the open position.

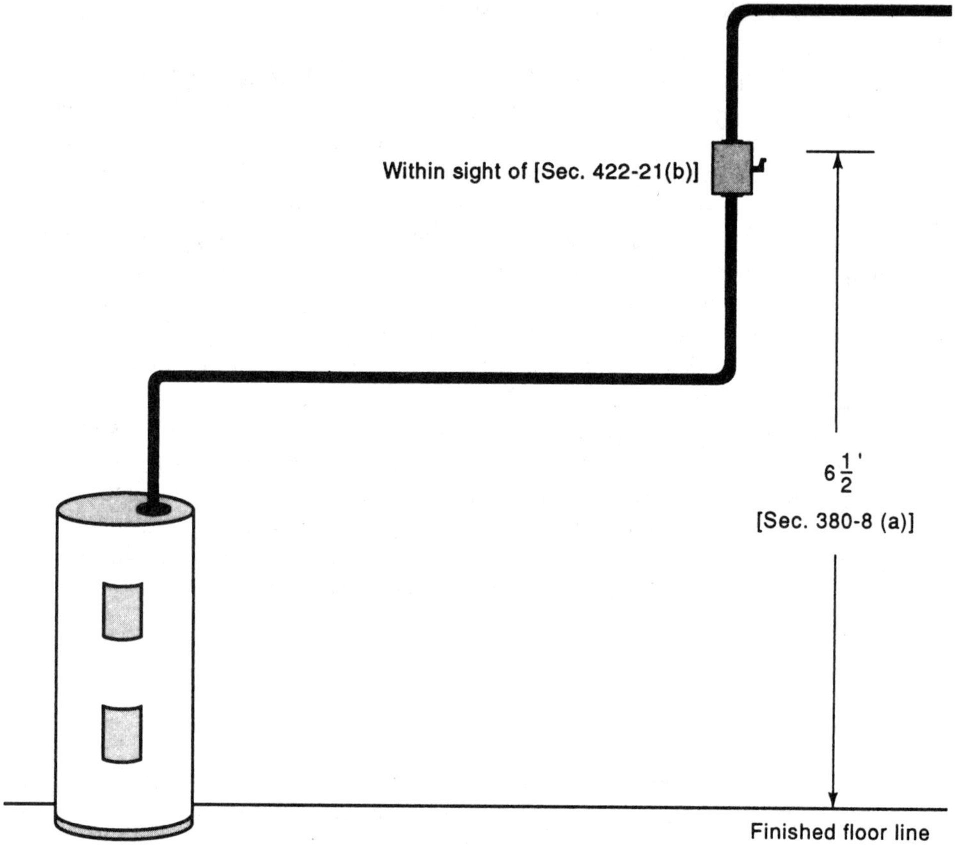

Fig. 12-1 Appliances such as water heaters are required by Sec. 422-21(b) to have a disconnecting means that is within sight of the appliance.

Section 422-22 lists cord- and plug-connected portable or stationary appliances that can use the separable plug and receptacle for the required disconnect. Equipment such as oil- or gas-fired furnaces are considered to be fixed utilization equipment that require a permanent wiring method for their electrical supply (Fig. 12-2). As a result, this type of fixed equipment must be supplied through a circuit interrupter or knife-switch disconnecting means and should not be cord- and plug-connected.

Section 422-25 lists installations where the unit switch on the appliance can serve as the required disconnecting means. The unit switch must disconnect all ungrounded conductors, and a second disconnecting means ahead of the appliance unit switch must be provided.

In a multifamily dwelling structure, such as an apartment or condominium complex, Sec. 422-25(a) requires the other (second) disconnecting means to be located within the individual dwelling unit that contains the appliance. For apartment complexes that have the electrical equipment for the individual apartments contained in a central electrical room, the other disconnect can be in this room (outside the dwelling unit), as long as it is on the same floor level as the dwelling unit that contains the appliance.

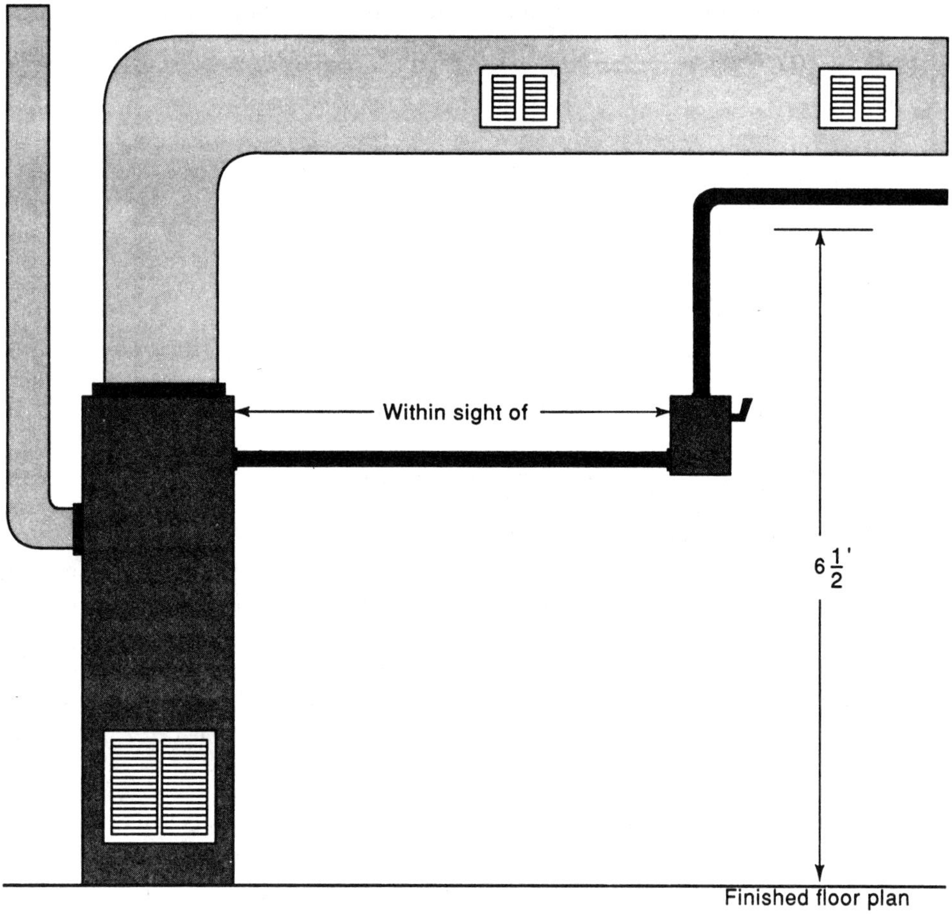

Fig. 12-2 A receptacle and attachment plug cannot be used as the required disconnecting means for fixed utilization equipment such as building furnaces.

Section 422-25(b) says basically the same thing about a two-family dwelling structure (duplex apartment). The other disconnecting means, which can be the individual branch-circuit supply or the service-disconnecting means (permitted to control lamps or other appliances), is allowed to be located either inside or outside the dwelling structure.

In reference to the service-disconnecting means, Sec. 230-70(a) requires the service equipment of any structure (dwelling or nondwelling) to be located at the nearest accessible point from finished grade or finished floorline either outside or immediately inside the entry wall of the service conductors (service-entrance conductors) into the building or other structure.

As with the other dwelling structures, Sec. 422-25(c) allows either the branch-circuit supply or the service-disconnecting means, located either inside or outside the structure, to serve as the other disconnecting means in a single-family dwelling.

For the unit switch on an appliance (or other utilization equipment) to serve as the disconnecting means required by Sec. 422-20 in a nondwelling occupancy or structure,

Sec. 422-25(d) will allow the branch-circuit fused switch or circuit breaker to serve as the second disconnecting means, only if the panel or other load center (service equipment or the service-disconnecting means) is readily accessible for servicing of the appliance.

The other disconnect would have to be in the immediate vicinity of the appliance. For example, an electric air dryer (dehumidifier) is located in an unfinished basement of the building, but the service equipment is located upstairs in the hall of the building. The other disconnect is not readily accessible. A second disconnect would be required at the location of the electric air dryer. The unit switch cannot be used as a disconnect, even if the other disconnect, located not readily accessible, is capable of being locked open [general requirement of Sec. 422-21(b)].

12.3 FIXED ELECTRIC SPACE HEATING EQUIPMENT

Part A of Art. 424 lists the general requirements for branch circuits supplying fixed electric space heating. Section 424-1 defines electric space heating equipment or units as fixed (permanently mounted and connected) utilization equipment used for heating the inside open spaces of a structure. This article does not apply to heating equipment such as resistance plate or wrap-around heaters, or induction coil-type heaters that are used in manufacturing processes.

Section 424-3(a) will allow individual (specific-purpose) branch circuits to supply any size (any current rating) of fixed electric space heating equipment. But, this section restricts a multioutlet branch circuit supplying only (dedicated to) fixed electric space heating equipment to branch circuits rated at 30 amps or less.

Section 424-3(b) requires the branch-circuit conductors and overcurrent-protection devices supplying fixed electric-space heating equipment containing resistance heating elements to be rated not less than 125% of nameplate full-load current (FLC).

Section 424-19 requires fixed electric space heating equipment to have a disconnecting means. The different subsections describe required disconnects according to type of installation and type of occupancy.

12.4 MOTORS

The installations and uses of motors are just as diversified as the many different kinds and horsepower ratings of motors. In general, motors or equipment that contains motor loads must be provided with a disconnecting means, supply conductors, and some type of overcurrent protection. Motors can be supplied either by a multioutlet or an individual branch circuit, with the individual (specific-purpose) being the more common.

The installation of electrical equipment used to supply, support, or control motors is governed by Art. 430 in the NE Code. Because motors draw high in-rush currents on starting, the branch-circuit overcurrent protection, the feeder overcurrent protection, and the branch or feeder conductors supplying a motor or several motors are sized in special ways. These special ways are different as compared to the guidelines of Secs. 210 and 220 for regular branch and feeder circuits.

If the motor is a hermetically sealed refrigerant motor-compressor (the motor is hermetically sealed, usually on the low-pressure side inside the assembly housing [crankcase] of the refrigerant compressor) used in air conditioning or other refrigeration equipment, the branch-circuit installation must meet the requirements of both Arts. 430 and 440.

12.5 MOTOR CURRENT AND VOLTAGE RATINGS

Other than fractional-horsepower, general-purpose motors, and specific-purpose motors rated greater than 500 horsepower, general-purpose motors have two separate ratings for full-load current. One rating is the full-load current rating specified on the motor nameplate. The other is an NE Code-specified value of rated full-load current as determined by Art. 430.

The full-load current rating for the same horsepower, voltage, and phase rating of a motor will vary among different manufacturers, as well as different trade lines or grades of motors from the same manufacturer. Motors of the same rating will vary in efficiency and power factor, as well as type of bearings, types of enclosures, and types of windings. All of the factors affect the full-load current rating of the motor.

When the power distribution for a building or manufacturing process is being designed and laid out, both the voltage and phase rating of the distribution and the required horsepower for a given task or process are known, but the type of motor is not known. The NE Code specifies the worst case (highest value) in terms of full-load current that can be expected according to horsepower, voltage, and phase rating for each motor. Section 430-6(a) requires the NE Code table-value full-load current rating of the motor to be used to size all the electrical support equipment for the motor, except individual running overload protection.

Even if the specific motor, in terms of manufacture and nameplate full-load current, is known at the time of design and initial installation, the power distribution would still have to be designed for worst case because the owner or operator of the motor could change manufacture and put in the worst-case (highest full-load current-rated) motor should replacement be required.

The worst-case motor full-load currents, as listed in Tables 430-147 through 150, must be used in computing the rated-load current of the service, the feeder circuit, the branch circuit, and all other electrical support or supply equipment such as the rating of overcurrent-protection devices, disconnects, and motor starters (mechanical contactors, or solid-state drive controllers).

Only the heaters in the motor starter or other types of separate (running) overload-overcurrent protection are to be sized by the motor nameplate full-load current rating because a motor, whatever its full-load current rating is, must be protected from overheating by separate overload protection that is tailored (sized) according to its individual full-load current rating.

If the motor is changed out for replacement with a motor with a different full-load current rating, the branch circuit (as well as the supply feeder, the service, and all other support equipment) remains unchanged, because it was sized for the worst case initially. The specific running overload-overcurrent protection would be the only thing in the electrical control and supply that would have to also be changed with motor replacement—again, tailored to the individual or specific motor rating.

The motor full-load current tables are located at the end of Art. 430. Table 430-147 lists the full-load currents (FLCs) of direct current (dc) motors according to both voltage rating (column) and their horsepower rating (row). Table 430-148 lists the full-load currents of single-phase alternating current (ac) motors in the same manner. Table 430-149 lists the full-load currents of 2-phase ac motors. Table 430-150 lists the full-load currents of 3-phase ac motors.

Example 12–1. Determine the table-value full-load current of a 208-volt, single-phase, ac motor rated at 5 hp:

For single-phase ac motors: Ref. T430-148

5-hp motor FLC = 30.8 amps

Code challenge 12–1. Determine the table-value full-load current (FLC) of a 230-volt, 3-phase ac motor rated at 10 hp.

For 3-phase ac motors: Ref. (A) ____________

10-hp motor FLC = (B) ____________ amps

The first paragraph of Tables 430-148, 430-149, and 430-150 stipulates that these table-value full-load currents are for motors designed with running speeds that are usual for belted motors with normal torque characteristics.

For ac torque motors, Sec. 430-6(b) stipulates that the rated-load current used to size the branch-circuit supply and support equipment, including separate motor-overload protection, shall be the nameplate locked-rotor current of the motor.

For motors supplied and controlled by ac variable-speed, variable-frequency drives or dc variable-voltage drives, or other multiple-speed motors (adjustable voltages and/or adjustable torques), Sec. 430-6(c) stipulates that the rated-load current used to size the disconnects, controllers, conductors, and overcurrent protection shall be based on the nameplate maximum operating current marked on the motor and/or controller.

The last sentence of Sec. 430-6(c) requires the motor FLC listed in Tables 430-149 or 430-150 to be increased to 150%, and this computed value to be used as the rated-load current for ac adjustable voltage or adjustable torque motors if the maximum operating current is not specified on the motor nameplate.

The student electrician should also note that T430-148 only lists single-phase motor FLCs for motors rated from fractional horsepower up to 10 hp. If a larger horsepower-rated, single-phase ac motor is to be installed, the motor nameplate full-load current will have to be used to size the disconnects, controllers, conductors, and overcurrent protection.

For larger dc motors (T430-147) or 2-phase ac motors (T430-149) rated in excess of 200 horsepower, the motor nameplate rated or full-load current must be used to size the disconnects, controllers, conductors, and overcurrent protection. The maximum of 500-horsepower rating for general-purpose motor applications given in the introduction only applies to 3-phase ac motors (T430-150).

12.6 MOTOR COMPRESSOR NAMEPLATE INFORMATION

Due to the variance in manufacture methods and materials, hermetically sealed refrigerant motor compressors also have a specific rated-load (full-load) current and a worst-case branch-circuit selection current. Both are listed on the equipment nameplate.

The branch-circuit selection current (Sec. 440-2, Def.) is used in the same manner as the NE Code table-value full-load current for regular motor installations to size all the electrical supply and support equipment of the hermetically sealed refrigerant motor compressor, except the separate overload-overcurrent protection.

The rated-load current (Sec. 440-2, Def.) is used to size the separate (running) overload protection. For some motor compressors, the nameplate will not specify a branch-circuit selection current. In this instance, the rated-load current is used to size all the support equipment. When specified on the nameplate, the branch-circuit selection current will always be greater than or equal to the rated-load current.

12.7 MOTOR DISCONNECT—LOCATION

Part I of Art. 430 governs the required disconnecting means for motors. Section 430-101 requires a disconnect for both the motor and its controller (Fig. 12-3). Section 430-109 will allow either a horsepower-rated switch, a circuit breaker, or a molded-case switch

Fig. 12-3 The disconnect for the motor and controller can be in the same enclosure as the controller. Several disconnects can be located in a common location.

(circuit interrupter, which looks like a circuit breaker, but does not offer any means of detecting and opening the circuit due to an overcurrent condition) to serve as the required disconnecting means for permanent (hard-wired) motors and their controllers.

When a motor is mounted on portable equipment, Exception No. 6 to Sec. 430-109 will allow the plug on the end of the supply cord, in conjunction with the supply receptacle to serve as the required disconnecting means. Other exceptions list various other means that can also serve as the motor disconnecting means.

If the motor circuit is limited to a single disconnecting means, Sec. 430-102(a) requires it to be located ahead of (on the line or supply side of) the motor controller (Fig. 12-4).

The single disconnect must be in sight from the controller location (Fig.12-5). As defined in Art. 100, in sight from indicates a straight line of sight, not around oblique corners of a wall, or obstructed view where machinery or ductwork systems are blocking the line of sight. Within line of sight also places a distance restriction on the disconnect location. The disconnect cannot be located more than 50 linear feet from the controller location.

The motor circuit may require more than one disconnecting means. Section 430-102(b) requires the single disconnect ahead of the controller to also be in sight from both the motor location and the driven machinery location. To meet the requirements of both sections, the single disconnect must be within 50 feet of the controller, within 50 feet of the motor, and within 50 feet of the farthest point of the driven machinery.

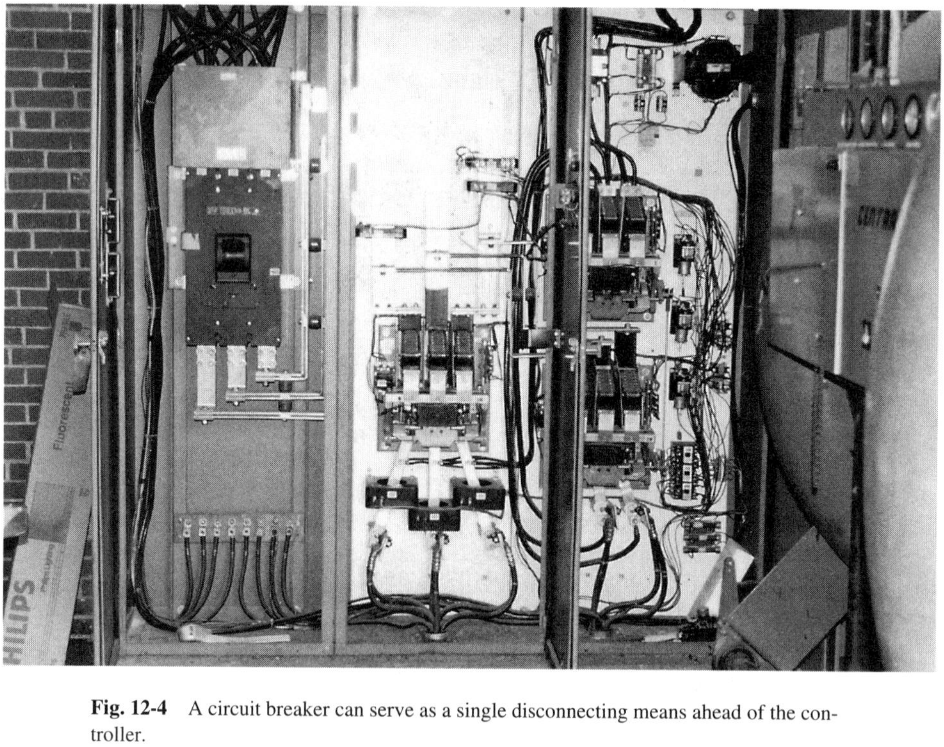

Fig. 12-4 A circuit breaker can serve as a single disconnecting means ahead of the controller.

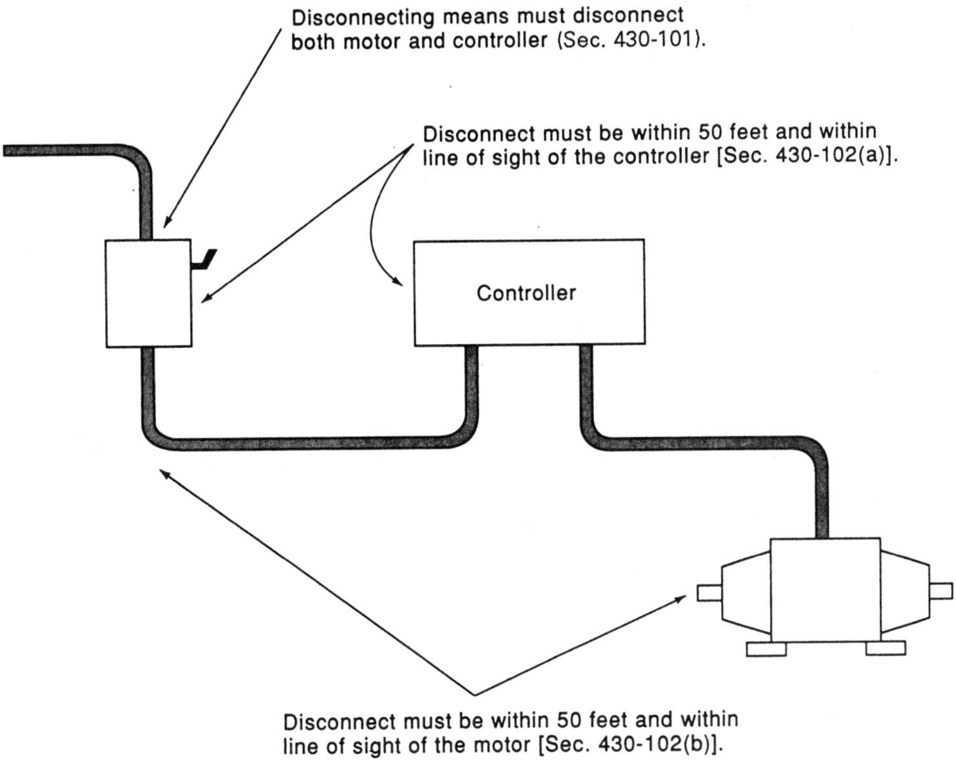

Fig. 12-5 A single disconnecting means can serve both motor and controller. The "50-foot" and "line of sight" are required by "in sight from" as defined in Art. 100.

If the motor is located around an oblique corner from the controller (motor starter) location as shown in Figure 12-6, then the location of the disconnect ahead of the controller cannot meet the in-sight-from requirements of Sec. 430-102(b). The motor and the driven machinery require a second disconnecting means that is in sight from both.

The linear measurement of 50 feet, as allowed by in sight from, is taken as the maximum distance the electrician or other maintenance person can visually verify that the disconnect handle is in the down (off) position as required by Sec. 430-104. Most knife switch-type disconnects normally have a red, orange, or otherwise bright-colored tip or plastic boot on the end of the handle as a visual aid. The on (up) or off (down) status of a circuit breaker or circuit interrupter (molded-case switches) is hard to verify (by sight) at 50 feet. Other safety codes, such as the mechanical code for refrigeration or gas equipment limits the distance, within line of sight, to a maximum of 6 feet.

As indicated in Figure 12-7, an enclosure for a combination starter (contains a knife switch disconnect, the motor-circuit fuses, and the motor starter) can probably be seen when located more than 50 feet from the motor or driven machinery location, but when the 50-foot limit is exceeded, a second disconnecting means is still required.

For some motor installations, an alternative to the second disconnecting means is found in the Exception to Sec. 430-102(b). If the disconnect ahead of the controller loca-

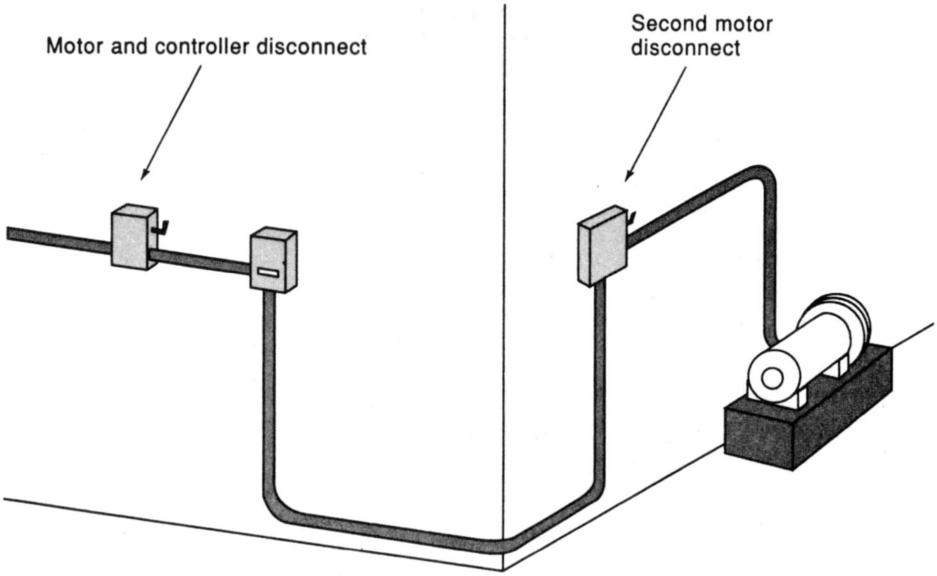

Fig. 12-6 Although a controller can be out of sight from the motor it serves, the disconnect cannot. A second disconnecting means is required.

tion is individually capable of being locked in the open position, the requirements of Sec. 430-102(b) can be disregarded (the motor or driven machinery can be located anywhere in relation to the controller and its disconnect location without a second disconnecting means located at the motor).

For rooftop motor-driven equipment, where the motor controllers and required line-side disconnects are located on the production floor or mezzanine (raised platform) inside the production area, this exception can create a safety hazard—especially when more than one motor is located on the roof, and more than one motor-disconnecting means are at a common location inside the building: What if you mistakenly lock out (lock in the open position) the wrong disconnect?

When you approach the motor on the roof and it is running, you will realize your mistake immediately. But what if the motor is cycled off by process control, and when you are in the middle of disconnecting the motor leads from the motor-circuit conductors in the terminal housing, the process cycles the motor on—a definite electrical shock hazard.

Instead of disconnecting the motor electrically—you could be in the process of changing drive belts or drive chains, or removing the housing on an in-line spring coupling between the motor and the driven machinery, when the motor cycled on. Physical injury is almost guaranteed should the motor start after you have locked it out (or at least think you have).

The better design, given this scenario, is to locate a second knife switch disconnect immediately adjacent to the motor location and lock it out if you are going to work on the motor or the driven machinery.

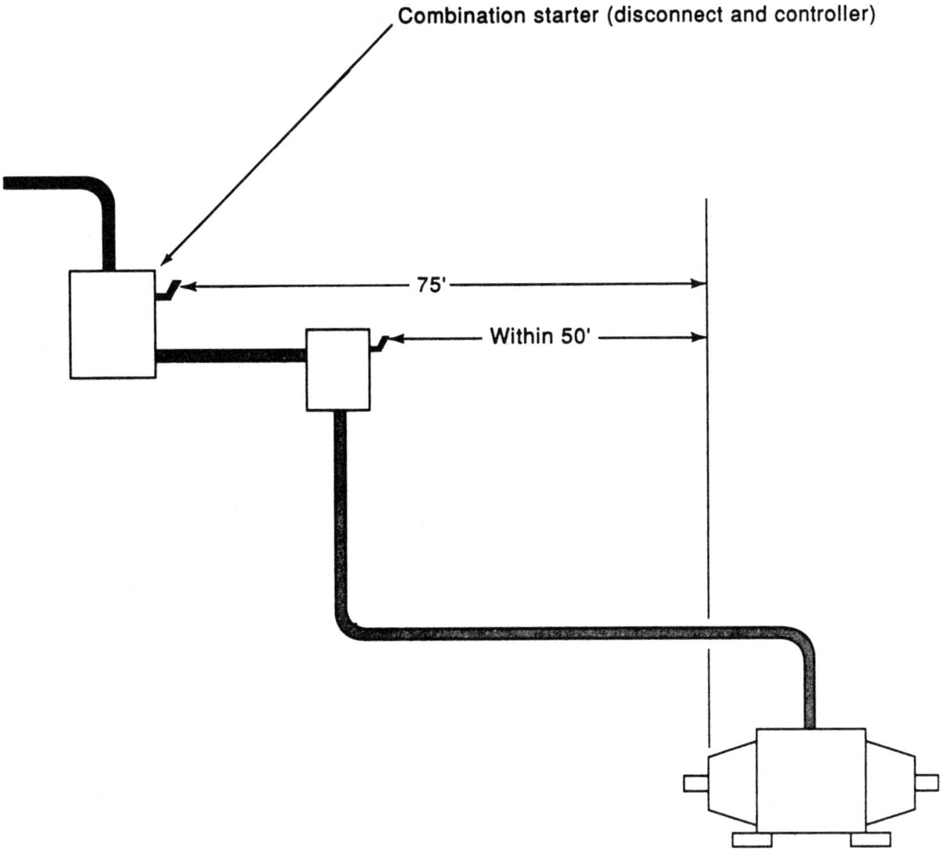

Fig. 12-7 To meet the "in sight from" requirements of Art. 100, the motor-disconnecting means must be located within 50 feet and in line of sight of the motor.

A typical industrial installation is shown in Figure 12-8. High-ampacity busduct is run above the supplied equipment in a high-bay manufacturing area at 25 feet or higher above the floorline. A plug-in fused disconnect is installed and the circuit is piped down (raceway) to the controller location. In this situation, a second disconnecting means, either ahead of the controller or motor (preferably ahead of the motor for servicing) is required.

When the single disconnect ahead of the controller (capable of being locked open) is the only disconnecting means for the motor circuit, Sec. 430-107 requires it to be readily accessible. When the motor is provided a second disconnecting means, Sec. 430-107 will allow the disconnect ahead of the controller to be accessible but does not require it to be readily accessible (due to its height in Fig. 12-8), provided the disconnect ahead of the motor (and driven machinery) is readily accessible.

As with the disconnecting means for appliances or other utilization equipment, to qualify as readily accessible, Sec. 380-8(a) limits the operating height of the disconnect handle in the on (up) position to a maximum of 6 feet, 7 inches (2 meters) above the deck on a working platform (mezzanine), above the finished floorline, or above finished grade when located outside.

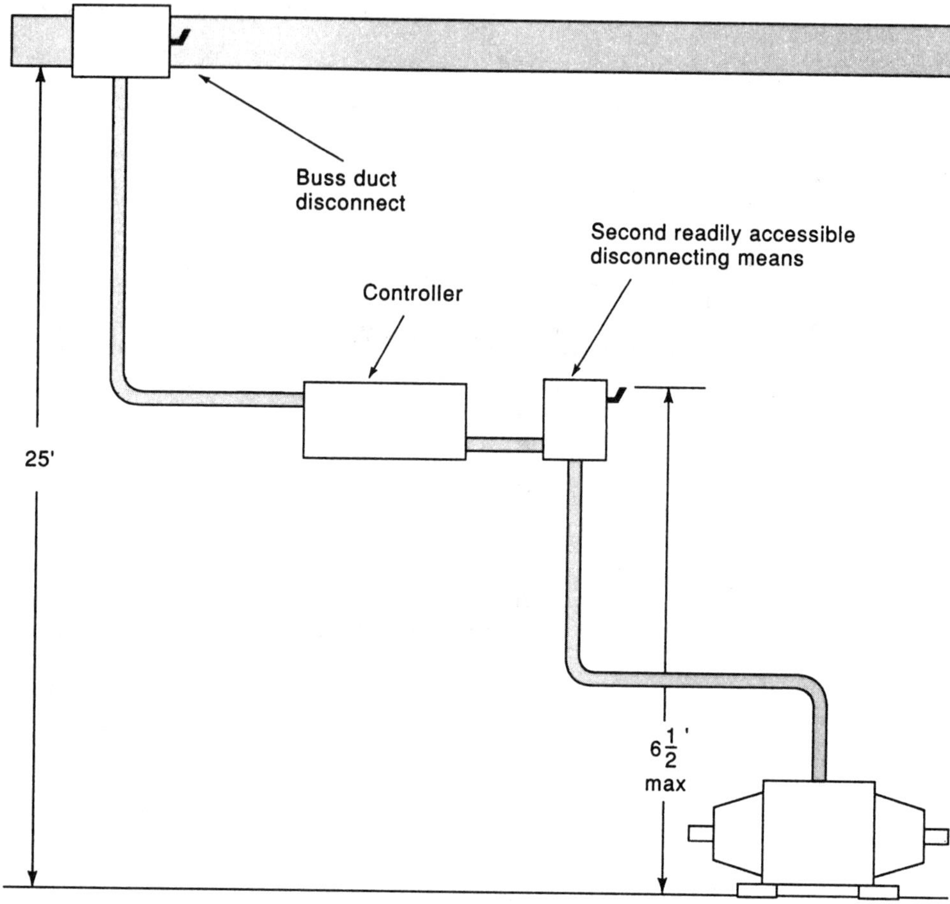

Fig. 12-8 A motor must have a *readily accessible* disconnecting means. The restriction of 6 feet 7 inches (2 meters) is imposed by Sec. 380-8(a).

Article 100 further defines readily accessible as capable of being reached quickly, without a person having to climb over obstacles such as stored boxes or located equipment (transformer enclosure) or resort to portable means such as chairs or ladders.

For the motor-circuit disconnect to qualify as readily accessible, Sec. 430-107 also requires the disconnect to be freely operable. In other words, the motor-circuit disconnect cannot be located behind locked doors, or inside locked cabinets or other enclosures that are sealed closed by special screws, bolts, or other fasteners. These types of fasteners, which are often employed to protect the equipment against vandalism, normally require special tools for opening.

The disconnect definitely cannot be locked in the on position. In an emergency situation, such as someone's hand caught in operating machinery, the motor disconnect is one thing you need to find, get to, and open as soon as possible.

Rules governing the location, access, and operation of disconnects for hermetically sealed refrigerant motor compressors, air conditioning, or other refrigeration equipment are found in Part B of Art. 440. Section 440-14 (Fig. 12-9) requires the disconnecting

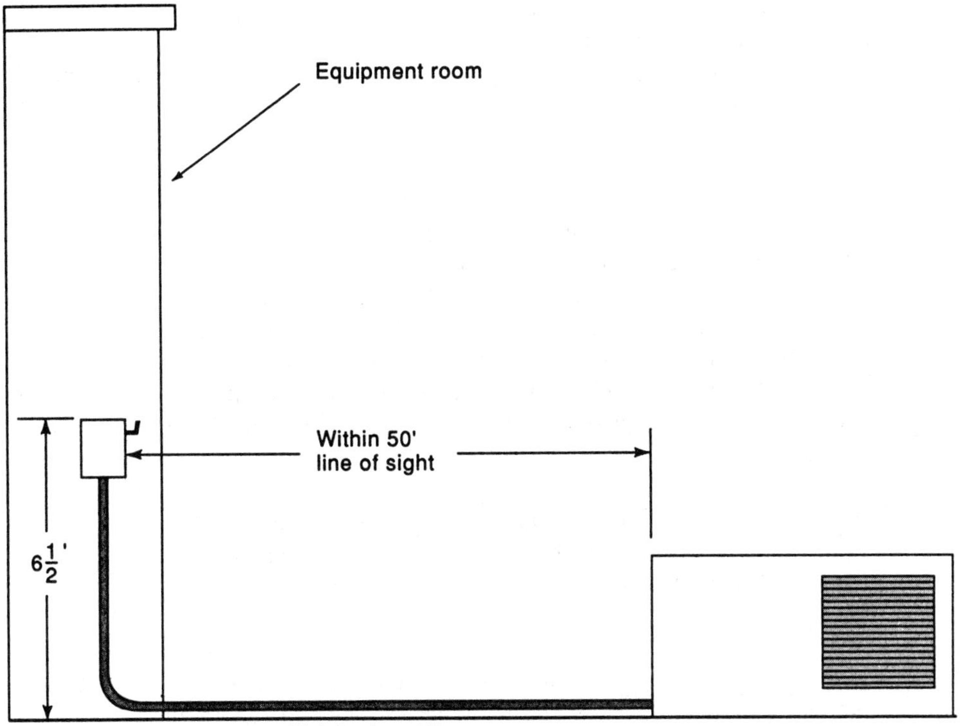

Fig. 12-9 Air-conditioning and refrigeration equipment are required to have a disconnecting means that is both within sight of and readily accessible.

means to be located within sight from and readily accessible to all air-conditioning and refrigeration equipment.

12.8 MOTOR DISCONNECT—RATING

A disconnecting means for a motor circuit must be dual-rated. Section 430-109 requires the switch or circuit breaker to be horsepower rated. Section 430-110(a) requires the switch or circuit breaker to have a minimum full-load (rated-load) current rating.

The horsepower rating for the standard motor is the easier of the two. A motor rated at 30 horsepower requires a disconnecting means rated at a minimum of 30 horsepower. The disconnect also has to have the same voltage rating and the same phase rating as the motor and its supply. It would be impossible to disconnect all the supply conductors to a motor rated 3-phase, 3-wire with a single pole (single-line) switch.

Most switch-type (knife switch) disconnecting means have different horsepower ratings (different values) for the type of fuse overcurrent protection employed. When instant-type (fast acting) will be used as the motor overcurrent protection, the horsepower rating is normally less as compared to the horsepower rating for a motor protected with time-delay fuses.

Section 430-110(a) sets the ampere rating (full-load or rated-load current rating) of the motor-disconnecting means at a minimum of 115% of the full-load current(s) of the motor or motors served. Section 440-12(a)(1) requires the disconnecting means for a single hermetically sealed refrigerant motor compressor to be rated at 115% of the rated-load current, or the branch-circuit selection current when specified on the motor compressor nameplate.

Horsepower-rated knife switches, circuit breakers, or molded-case switches will normally have nameplate full-load current ratings at assigned voltage ratings. To serve as the motor or motor compressor disconnecting means, the 115% overrate computation of the table-value full-load current or the branch-circuit selection current (when specified) cannot exceed the disconnect nameplate full-load current rating.

It seems reasonable that the full-load current ratings of knife switch disconnects or circuit interrupters (molded-case switches) would follow the standard current ratings of available overcurrent-protection devices (fuses or circuit breakers), but this is not the case. The more common full-load current ratings of switch-type disconnecting means include:

30 amps	400 amps	1,400 amps
60 amps	600 amps	1,600 amps
100 amps	800 amps	1,800 amps
200 amps	1,200 amps	2,000 amps

Example 12–2. Determine the nameplate full-load current rating of a fused knife switch used as the disconnecting means for a 30-hp, 230-volt, 3-phase squirrel-cage induction motor.

Motor full-load current rating: Ref. T430-150

Motor FLC = 80 amps

Knife switch overrate computation: Ref. 430-110(a)

115% of 80 amps = 92 amps

Knife switch rating: Ref. Other sources

Next higher standard (NHS) size: 100 amps

The knife switch-disconnecting means in this example must have a minimum full-load current rating of 100 amperes (next higher standard size) and a minimum horsepower rating of 30 hp at 230 volts, 3-phase to meet all NE Code requirements.

The horsepower rating of the disconnecting means varies according to the number of phases and according to the motor-circuit voltage. Both are a direct indication of full-load and locked-rotor current conditions: The higher the horsepower rating, the higher the full-load or locked-rotor current rating. The higher the voltage rating for a given horsepower, the lower the full-load or locked-rotor current rating. For the same number of phases, a switch-type disconnecting means will normally have different horsepower ratings for different voltage ratings.

Code challenge 12–2. Determine the nameplate full-load current rating of a fused knife switch used as the disconnecting means for a 50-hp, 460-volt, 3-phase wound-rotor induction motor.

Motor full-load current rating: Ref. (A) ____________

Motor FLC = (B) __________ amps

Knife switch overrate computation: Ref. (C) ____________

(D) __________ % of __________ amps = (E) __________ amps

Knife switch full-load current rating: Ref. Other sources
(NHS): (F) __________ amps

A hermetically sealed refrigerant motor compressor does not have a horsepower rating specified on the nameplate, as do standard motors. Nonetheless, the knife switch, circuit breaker, or circuit interrupter disconnecting means still has to have the dual rating to meet the requirements of Sec. 430-109.

Section 440-12(a) places three minimum ratings on the required disconnecting means of a hermetically sealed motor compressor. Two are based on the rated-load current or branch-circuit selection current when specified. The third is based on the locked-rotor current rating of the motor compressor.

As the term implies, the locked-rotor current is the line value of circuit current the motor windings will draw when the rotor is mechanically held at standstill with power applied.

A motor or motor compressor will draw its nameplate rated-load or full-load current only when operating at rated voltage and rated frequency, and the rotor is turning at its baseline speed under full-load (torque) conditions. If the motor is underloaded, the motor-circuit will draw less current than the full-load rating and still run at baseline speed. If the motor is overloaded, the motor circuit will draw current in excess of the full-load rating trying to maintain baseline speed. Normally, the rotor is turning at less than baseline speed when running against a mechanical overload.

The turning speed of the rotor controls the motor-circuit impedance, which by Ohm's law controls the value of current drawn by the motor. On initial startup, the motor circuit momentarily draws locked-rotor current, which is normally about six times the rated-load current specified on motor nameplate (rms values). As the rotor starts to turn, the value of line current decreases rapidly. When the rpm of the turning rotor attains baseline speed, the value of line current drops to the load level (not necessarily the full-load current rating).

If the motor becomes overloaded, due to either binding in the coupling to the driven machinery (straight couplings, chains, gears, or belts), to bad or failing bearings (under or over lubricated) within the motor or driven machinery, or due just to a torque overload of the driven machinery, the motor will draw overload overcurrent. When the motor becomes severely overloaded (loss of rotor speed), the rotor will stop turning and the higher value of locked-rotor current will be drawn by the motor circuit until something opens (disconnects or disables) the supply circuit.

If that something is the motor-circuit disconnect switch, it must be capable of clearing (interrupting) these higher values of overload overcurrents (as compared to the much higher values of fault overcurrent), and maintaining the open (extinguishing the arc) without self destructing.

In terms of full-load current rating, Sec. 440-12(a)(1) requires the rated-load current or branch-circuit selection current when specified to be overrated to 115%. If the overrate computation yields a nonstandard size or rating of switch or circuit breaker, then the next higher standard size must be installed to meet NE Code requirements.

In terms of horsepower rating, Sec. 440-12(a)(2) requires both the nameplate full-load current rating and the nameplate locked-rotor current rating to be converted into a horsepower rating. In comparison, the largest equated horsepower rating becomes NE Code minimum for the hermetically sealed refrigerant motor compressor in question.

For the full-load current rating, the rated-load current or the branch-circuit selection current when specified (without the 115% overrate) is equated to horsepower by either Table 430-148, 430-149, or 430-150, according to the phase rating of the motor compressor.

The motor FLC tables are technically used in reverse: The full-load current rating of the hermetically sealed motor compressor is equated to horsepower by the motor FLC values listed in the respective table according to voltage and phase rating. When the rated-load or branch-circuit selection current does not correspond to a standard listing of motor FLC, the next higher standard size (horsepower rating) is NE Code minimum.

The nameplate locked-rotor current (LRC) rating of the motor compressor is equated to horsepower by selecting a locked-rotor current according to the phase-rating and voltage-rating columns in either T430-151(a) or (b) that is greater than or equal to the nameplate LRC rating. The selected locked-rotor current identifies by row heading the required horsepower rating of the disconnect.

The method that identifies the largest horsepower rating is the minimum rating of the disconnecting means allowed by the NE Code in Sec. 440-12(a)(2).

Example 12–3. Determine both the full-load current rating and the horsepower rating of a circuit breaker used as the disconnecting means for a single-phase, 230-volt, hermetically sealed refrigerant motor compressor that has a nameplate rated-load current of 28 amps, a branch-circuit selection current of 32 amps, and a locked-rotor current (LRC) of 184 amps.

Motor compressor full-load current determination: Ref. 440-12(a)

Motor FLC = branch-circuit selection current = 32 amps

Circuit breaker full-load overrate computation: Ref. 440-12(a)(1)

115% of 32 amps = 36.8 amps

Circuit breaker full-load current rating: Ref. 240-6(a)

NHS: 40 amps

Circuit breaker horsepower rating determination: Ref. 440-12(a)(2)
FLC to horsepower rating conversion: Ref. T430-148

Motor FLC = (NHS) 40 amps: 7½ hp

LRC to horsepower rating conversion: Ref. T430-151(a)

Motor LRC = (NHS) 240 amps: 7½ hp

NE Code-minimum horsepower rating:
Compare for larger when the two ratings are not equal

Tables 430-151 and 152 were revised in the 1996 edition of the NE Code to reflect the current trends in the motor manufacturing concern. Table 430-151, which lists the locked-rotor current of the motor according to the voltage rating, has been broken down into Tables 430-151(a) and 430-151(b). The use of T430-151(a) is limited to single-phase standard motors or hermetically sealed refrigerant motor compressors. Table 430-151(b) must be used when determining the locked-rotor currents of either 2- or 3-phase motors or motor compressors.

The columns in T430-151(b) have been divided into two columns under each voltage rating for either 2- or 3-phase motors. The split in current ratings, mainly for locked-rotor currents, came about due to innovation in the manufacture and operation of motors. Recently manufactured motors are classified as "design" motors. The locked-rotor current listings in T430-151(b) depend on whether the motor is rated as design B, design C, design D, or design E. Technically, which value of listed locked-rotor current to use depends solely on whether or not it is a design E motor.

In prior NE Code editions, motors were classified by "Code letter" according to the type of winding employed in the manufacture of the stator and rotor circuits. Different winding techniques change the rotor-circuit impedance, and sometimes the stator-circuit impedance. In addition to the turning speed of the rotor discussed earlier, changes in motor-circuit impedance for the same phase, voltage, and horsepower rating will yield a corresponding change in both the motor's full-load curent rating and its locked-rotor current rating.

Code letter classification is still required on the motor nameplate in the 1996 NE Code edition by Sec. 430-7(a)(8). The Code letter, in conjunction with T430-7(b), can be used to determine the amount of stored energy the motor must draw from the line on startup, or give back to the line on shut down. This information is critical when determining withstand (interrupt) ratings of overcurrent-protection equipment such as fuses and circuit breakers.

For determination of the locked-rotor current in T430-151(b), or the rating of the branch-circuit motor-circuit fuses or circuit breaker short-circuit, ground-fault protection according to the overrate percentages in T430-152, the Code letters are no longer used to control the allowed overrate percentages for these devices (overcurrent-protection ratings are discussed in Chapters 13 and 14 of this text). Instead, the design letter, and whether it is a design E motor or one of the other design-letter motors, is used to determine either the locked-rotor current rating or the higher fault protection overrate of the branch-circuit fuses or circuit breaker.

The design E motor is described as an energy efficient motor. It has lower heat losses (I^2R losses) as compared to the standard NEMA design B, C, or D motors. It also has higher values of in-rush (starting) current and locked-rotor current as compared to the other 3-phase ac motors. The values listed in T430-151 for locked-rotor currents of design B, C, and D are about the same values found in prior NE Code edition tables. The addition is for the new design E, energy-efficient motors. As specified in the fine print note to Sec. 430-7(a)(9), motor design definitions can be found in NEMA MG1-1993 standards and in the ANSI/IEEE 100-1992 standards.

As a general rule, the locked-rotor current rating of an ac induction motor, other than design E, is equal to 600% or six times the motor full-load current rating as found in either Table 430-148, 430-149, or 430-150. The locked-rotor current rating is equal to 800% or eight times the table-value full-load current rating when the motor is a design E motor rated at 3 hp or more.

Code challenge 12–3. Determine the both full-load current rating and the horsepower rating of a circuit breaker used as the disconnecting means for a 3-phase, 208-volt hermetically sealed refrigerant motor compressor that has a nameplate rated-load current of 68 amps, a branch-circuit selection current of 75 amps, and a locked-rotor current of 402 amps.

Motor compressor full-load current determination: Ref. (A) ________________

Motor FLC = (B) ____________ amps

Circuit breaker full-load overrate computation: Ref. (C) ________________

(D) ____________ % of ____________ amps = (E) ____________ amps

Circuit breaker full-load current rating: Ref. (F) ________________
NHS: (G) ____________ amps

Circuit breaker horsepower rating determination: Ref. (H) ________________
FLC to horsepower rating conversion: Ref. (I) ________________

Motor FLC = (NHS) (J) ____________ amps: ____________ hp

LRC to horsepower rating conversion: Ref. (K) ________________

Motor LRC = (NHS) (L) ____________ amps: ____________ hp

NE Code-minimum horsepower rating:

Compare for (M) ____________: (N) ____________ hp

If two or more motors or motor compressors or one or more motors or motor compressors and additional loads (such as power or lighting loads) are to be supplied from a common feeder, and this combined load is connected to the feeder through a single disconnect, Secs. 430-110(c) and 440-12(b) require the switch or circuit breaker disconnecting means to be dual-rated—for both the full-load current rating and the horsepower rating.

The disconnecting means [Secs. 430-110(c)(1) and 440-12(b)(1)] must be treated as if it is controlling a single motor whose full-load current rating is equal to the summation of all the current loads supplied at their rated-load (full-load) condition.

As with a single-motor disconnect, Secs. 430-110(c)(2) and 440-12(b)(2) require the total combination-load current to be overrated to 115% to determine the NE Code-minimum full-load current rating.

The summation of all the full-load currents must include the rated-load current or branch-circuit selection current when specified for all the motor compressors in the group, the table-value motor FLCs of all standard motors in the group, plus the rated-load current (full-load current) of all the other nonmotor loads supplied by the single disconnect.

For the horsepower rating, Secs. 430-110(c)(1) and 440-12(b)(1)a. require the horsepower rating of the feeder or individual branch-circuit disconnecting means for the combination load to be equated by Tables 430-148 through 430-150 according to the summation of the full-load current rating of all the loads without the 115% overrate.

Besides the full-load current horsepower rating, Secs. 430-110(c)(1) and 440-12(b)(2)b. requires the single disconnect to have a locked-rotor current horsepower rating.

The locked-rotor current rating of the circuit breaker or switch-type disconnecting means, according to horsepower rating [Table 430-151(a) or 430-151(b)], must be greater than or equal to the sum of the locked-rotor currents of all the motor and motor-compressor loads, plus the sum of the full-load currents of all the nonmotor loads. The motor locked-rotor current ratings are determined according to horsepower, phase, and voltage rating, as well as the design (design E or other for 3-phase) by either Table 430-151(a) or 430-151(b). The motor-compressor locked-rotor current ratings must be determined by individual nameplate listings.

Example 12–4. Determine the NE Code-minimum full-load current rating and horsepower rating of a single knife switch-disconnecting means for a 480Y/277-volt, 3-phase, 4-wire feeder that supplies the following loads:

Overhead lighting rated at 20 amps.

1 Transformer rated at 5 kVA, 3-phase, 480-208Y/120 volts for receptacle loads and control voltage

2 Hermetically sealed refrigerant motor compressors that have nameplate rated-load currents of 21 amps, nameplate branch-circuit selection currents of 23 amps, and nameplate locked-rotor currents of 130 amps each

2 Design B water-pump motors rated at 10 hp, 3-phase, 460 volts each

2 Design B condensor-coil fan motors rated at 5 hp, 3-phase, 460 volts each

Full-load current determination/computation:

Overhead lighting: (given) 20 amps

Transformer FLC: (primary circuit)

$$\frac{1000 \times 5\text{ kVA}}{1.732 \times 480\text{ volts}} = 6.0\text{ amps}$$

Motor compressors: motor FLC = 23 amps — Ref. 440-2, Def.
Motor FLCs: — Ref. T430-150
5 hp: 7.6 amps
10 hp: 14 amps

Locked-rotor current determination:
Motor compressors: (given) motor LRC = 130 amps
Motor LRCs: — Ref. T430-151(b)
5 hp: 46 amps
10 hp: 81 amps

Disconnect full-load current computation:

Overhead lighting:		20.0 amps
Transformer:		6.0 amps
Motor-compressors:	2 × 23 amps =	46.0 amps
Pump motors:	2 × 7.6 amps =	15.2 amps
Fan motors:	2 × 14 amps =	28.0 amps
Total full-load current:		115.2 amps

NE Code-minimum knife switch disconnect full-load current rating:
Overrate computation: Ref. 430-110(c)(2) — Ref. 440-12(b)(2)

115% of 115.2 amps = 132.5 amps

Rating: (NHS) 200 amps — Ref. Other sources

Knife-switch horsepower rating for the full-load current summation:
Ref. 430-110(c)(1) — Ref. 440-12(b)(1)a.
Table-value FLC = 124 amps: 100 hp — Ref. T430-150

Disconnect full-load current computation at locked rotor:

Overhead lighting:		20 amps
Transformer:		6 amps
Motor compressors:	2 × 130 amps =	260 amps
Pump motors:	2 × 46 amps =	92 amps
Fan motors:	2 × 81 amps =	162 amps
Total full-load current:		540 amps

Knife switch horsepower rating for the locked-rotor current summation:
Ref. 430-110(c)(1) — Ref. 440-12(b)(1)b.
Table-value LRC = 543 amps: 75 hp — Ref. T430-151(b)

NE Code-minimum knife switch disconnect horsepower rating:
Compare for larger: 100 horsepower

Although Art. 440 pertains primarily to the branch-circuit requirements of hermetically sealed refrigerant motor compressors, chiller units, or other air-conditioning or refrigeration equipment, Sec. 440-11 requires the guidelines of Sec. 440-12 to be used for the feeder disconnect computations.

Code challenge 12–4. Determine the NE Code-minimum full-load current rating and horsepower rating of a single knife switch-disconnecting means for a 460-volt 3-phase, 3-wire delta feeder that supplies the following loads:

1 Transformer rated at 10 kVA, 3-phase, 460-240/120 volts for control voltage, overhead lighting, and receptacle loads

2 Hermetically sealed refrigerant motor compressors that have nameplate rated-load currents of 27 amps, nameplate branch-circuit selection currents of 30 amps, and nameplate locked-rotor currents of 165 amps each

2 Design E condensor-coil fan motors rated at 7½ hp, 3-phase, 460 volts each

2 Design E water-pump motors rated at 15 hp, 3-phase, 460 volts each

Full-load current determination/computation:

Transformer FLC: (primary circuit)

$$\frac{\text{(A) \_\_\_\_\_\_\_\_} \times \text{\_\_\_\_\_\_\_\_ kVA}}{\text{(B) \_\_\_\_\_\_\_\_} \times \text{\_\_\_\_\_\_\_\_ volts}} = \text{(C) \_\_\_\_\_\_\_\_ amps}$$

Motor compressors: Ref. (D) ____________

Motor FLC = (E) ________ amps

Motor FLCs: Ref. (F) ____________

7.5 hp: (G) ________ amps

15 hp: (H) ________ amps

Locked-rotor current determination:

Motor compressors: (given) motor LRC = (I) ________ amps

Motor LRCs: Ref. (J) ____________

7.5 hp: (K) ________ amps

15 hp: (L) ________ amps

Disconnect full-load current computation:

Transformer:				________ amps
Motor compressors:	(M) ________	× ________	amps =	________ amps
Fan motors:	(N) ________	× ________	amps =	________ amps
Pump motors:	(O) ________	× ________	amps =	________ amps
Total full-load current:			(P)	________ amps

NE Code-minimum knife switch disconnect full-load current rating:

Overrate computation: Ref. (Q) ________ Ref. (R) ________

(S) ________ % of ________ amps = (T) ________ amps

Rating: (NHS) (U) ________ amps Ref. Other sources

Knife switch horsepower rating for the full-load current summation:

Ref. (V) ______________ Ref. (W) ______________

Table-value FLC = __________ amps: (X) __________ hp

Ref. (Y) __________

Disconnect full-load current computation at locked rotor:

Transformer:				______	amps
Motor compressors:	(Z ______	× ______	amps =	______	amps
Fan motors:	(1A) ______	× ______	amps =	______	amps
Pump motors:	(1B) ______	× ______	amps =	______	amps
			(1C)	______	amps

Knife switch horsepower rating for the locked-rotor current summation:

Ref. (1D) ______________ Ref. (1E) ______________

Table-value LRC = __________ amps: (1F) __________ hp

NE Code-minimum knife switch disconnect horsepower rating:

Compare for (1G) __________: (1H) __________ hp

Section 440-12(e) states that when either the rated-load current or the locked-rotor current computation requires a disconnect rated in excess of 100 horsepower, the disconnecting means is allowed by Sec. 430-109, Excp. No. 5 to be an isolating switch that is plainly marked "Do not operate under load." An isolating switch, as defined in Art. 100, has no interrupt rating. Its sole purpose is to isolate an electric circuit from its power source after the load has already been disconnected by some other means.

Sections 430-110(c)(3) and 440-12(c) state that the locked-rotor current of small motors and/or motor compressors not covered in Table 430-151(a) or 430-151(b) can be assumed to be six times the nameplate full-load current. The locked-rotor current of 3-phase, design E ac induction motors can be assumed to be eight times the nameplate full-load current.

If the locked-rotor current rating is not specified on the nameplate of multi-motor/combination-load equipment such as large chiller units, the nameplate-rated full-load current should be converted to equivalent horsepower for the rating of the chiller-disconnecting means.

12.9 DRIVE DISCONNECT—RATING

The examples in the previous section apply to knife switches, circuit interrupters, or circuit breakers that supply and disconnect motors operating at baseline speeds. When speed control or a lower starting torque is required, ac or dc motors are often supplied through some form of drive controller instead of an electromechanical (across-the-line) motor starter.

Rules governing the speed or reduced-voltage startup of a dc motor are not very explicit in the NE Code. Other than the overrates found in Sec. 430-22(a), Excp. No. 2 for the rated ampacity of motor-circuit conductors supplying dc motors, there are no specific regulations that pertain to dc motor controllers or disconnecting means.

A disconnecting means located ahead of an SCR (dc) controller would have to be sized according to the general requirements of Secs. 430-109 and 430-110. The disconnect would have to be dual rated: For the full-load current rating, the knife switch, circuit interrupter (molded-case switch), or circuit breaker would have to be sized at a minimum of 115% of the maximum motor full-load rating specified on the nameplate of the SCR controller. The horsepower rating of the disconnecting means would have to be at least equal to the nameplate horsepower rating specified on the controller.

Because of the fixed frequency of the utility supply (60 Hz in the United States and 50 Hz in European countries), not too many years ago, speed control of a 3-phase ac induction motor was limited to some form of resistance placed in the rotor circuit of a specially wire-wound rotor type of ac induction motor. Then, as now, the wire-wound induction motor is made to function as both a motor and a transformer, which reduces the amount of torque (and speed) available to the connected load by the turning rotor. The transformer power is wasted as dissipated heat from the resistors.

Another type of reduced speed control, which is actually a reduced torque by reduced voltage control for a 3-phase ac induction motor, is the wye-delta starter/controller. When a 3-phase ac induction motor is operated in a wye winding configuration, the line voltage is reduced by the square root of 3 when presented as the individual (phase) winding voltage. Since the motor windings are impedance rated for the higher delta line voltage, the reduced voltage reduces the winding current.

The reduction of winding current generates reduced torque from the turning rotor, and as a result, the driven machinery will turn slower when compared to the same motor windings wired in a delta winding configuration. In the delta winding configuration, the voltage is returned to the line value. Because the motor winding is impedance rated for the higher line voltage, both the winding voltage and current are increased by the square root of 3. For a motor that drives a high-torque load, this would appear as a two-speed control of the motor.

Recent developments in the use of solid-state (electronic) devices have brought about the use of variable-speed, variable-frequency ac drive controllers to where the speed of any ac induction motor can be adjusted or controlled by changing the frequency of the supply voltage to the motor.

The speed of an ac induction motor is based on the number of pole pairs in the stator circuit and the frequency of the ac supply voltage. By formula, the synchronous speed of an ac induction motor can be computed:

$$S = 120 \times f \div P$$

Where:

S = synchronous speed of the revolving magnetic field in rpms (revolutions per minute)
f = frequency in Hertz (cycles per second) of the ac supply
P = number of poles (magnetic pole faces) within the stator circuit

The magnetic poles in the structure of a stator assembly for an ac induction motor will always be an even number, often referred to as pole pairs. The poles represent the north and south poles of the magnetic field caused to rotate within the stator assembly. Each north pole has to have an accompanying south pole to make one pole pair.

Supplied with a 60-Hz ac source, the synchronous speed of a 2-pole induction motor (1 pole pair) equals 120 × 60 ÷ 2: 3,600 rpms. Because the ac power supply of the rotor circuit is by induction (transformer action) from the revolving stator magnetic field, there is a difference between the synchronous speed of the ac induction motor and the actual baseline speed of the turning rotor. The difference between these two speeds is referred to as the slip of the motor. The percentage of slip is normally 2% or less, such that the baseline speed of this 2-pole ac induction motor is about 3,550 rpms.

Operating at 60 Hertz or cycles per second, the rotating magnetic field inside the stator assembly of a two-pole ac induction motor makes one full revolution per cycle. With 60 cycles per second, the rotating magnetic field will make 3,600 revolutions in one minute, hence 3,600 rpms.

A four-pole ac induction motor has a synchronous speed of 1,800 rpms (120 × 60 ÷ 4). Its baseline speed is normally 1,750 rpms. Operating at the same 60-Hz supply frequency, the rotating magnetic field inside the typical four-pole ac induction motor makes only one-half revolution per cycle: 3,600 cycles per minute × ½ revolution per cycle = 1,800 rpms.

An increase in the number of poles (at the point of motor manufacture) yields a corresponding decrease in the synchronous and baseline speeds of the rotor, but it generates a corresponding increase in torque. Without getting into a full motor-operations course, a four-pole ac induction motor introduces a second magnetic field, which is operating to aid the single magnetic field that was available when the motor was only in a two-pole configuration.

If the frequency can be increased with the increase in the number of pole pairs, the loss in speed due to an increase in poles will not affect the baseline speed of the motor. By the same token, the required slip of the ac induction motor can be compensated for by only a slight increase in frequency such that the running baseline speed of the rotor in the typical 3-phase ac induction motor can be set to the synchronous speed of the incoming ac supply.

To change the frequency of the ac supply from the utility requires rectification of the ac power supply into a dc power supply, and then with solid-state firing circuitry, which more or less pulses the now steady-state dc voltage, supply the motor windings with an almost sinusoidal (ac) waveform at a different frequency.

According to controller-supply frequency, the standard four-pole ac induction motor, which has a traditional rotor baseline speed of 1,750 rpms, can be caused to turn at speeds less than or greater than 1,750 rpms. At higher frequencies, the speed is limited by the structural strength of the rotor assembly. Higher centrifugal forces are placed on the rotor at higher rpms. Given a high enough spindle speed, the rotor could fly apart. Doubling the ac supply frequency (from 60 to 120 Hertz) is a common application. Some ac drives will allow the frequency to be increased to 420 Hertz, which is seven times the ac supply frequency.

When the motor is operated at subfrequencies (frequencies lower than the ac supply), the magnitude of the voltage applied to the motor windings must also be reduced or adjusted. The primary element of circuit impedance in a motor circuit when an ac voltage is applied is the inductive reactance of both the rotor and stator circuits (the rotor-circuit impedance being reflected by transformer action into the stator circuit). Inductive reactance is computed by the formula,

$$X_L = 6.28 \times f \times L$$

Where:

X_L = the inductive reactance
6.28 = 2 pi (the electrical degrees in a full cycle)
f = the supply-circuit frequency

A decrease in the supply frequency to lower the baseline speed will also decrease the motor-circuit impedance. Unless the supply-circuit voltage is reduced, the motor will overheat due to an increase in the motor-circuit current.

As with dc speed controllers, rules governing the speed or reduced-voltage startup of an ac induction motor with the distribution supply and the motor-circuit supply (from the controller) operating at the same phase are not very explicit in the NE Code. In other words, there are no explicit rules governing the disconnecting means of a 3-phase supply to an ac drive controller, which in turn will supply a 3-phase motor at the same line voltage.

A disconnecting means located ahead of an ac drive controller would have to be sized according to the general requirements of Secs. 430-109 and 430-110. The disconnect would have to be dual rated: For the full-load current rating, the knife switch, circuit interrupter (molded-case switch), or circuit breaker would have to be sized at a minimum of 115% of the maximum motor full-load rating specified on the nameplate of the solid-state controller. The horsepower rating of the disconnecting means would have to be at least equal to the nameplate horsepower rating specified on the controller.

Since the ac drive controller converts the ac distribtuion supply into a dc supply, and then pulses this dc supply to appear as a sinsoidal ac supply for the motor circuit, it is a very easy task to supply an ac drive with a single-phase ac source and supply and speed control a 3-phase ac motor. In NE Code terms, this type of ac drive controller is referred to as a "phase converter."

The installation and use of phase converters is governed by Art. 455 of the NE Code. For the disconnect rating, the overrate to use is determined by the intended load of the phase converter (the use of a phase converter is not limited to just the control and supply of a motor load). If the rating of the connected load for the phase converter is not known at the time of design of the electrical installation, or at the time of actual installation, Sec. 455-8(c) requires the knife switch, circuit interrupter, or circuit breaker to have a full-load current rating greater than or equal to 115% of the full-load current rating specified on the nameplate of the phase converter.

When the phase converter is installed for a specific motor load, the exceptions to Sec. 455-8(c) are used to size the phase converter-disconnecting means.

In parallel to Sec. 430-102, the phase converter is the controller when it supplies a motor load. Section 455-8(a) requires the disconnecting means to be located readily accessible on the line side (supply side) of the motor controller. In parallel to Sec. 430-109, Sec. 455-8(b) will allow the phase converter-disconnecting means to be either a knife switch, a circuit interrupter (molded-case switch), or a circuit breaker.

When a 3-phase ac motor is supplied from a single-phase ac source, the difference in the two line currents for the same voltage is the square root of 3. The full-load current rating for a 3-phase ac motor of a given horsepower and voltage must be determined from T430-150 of the NE Code. To compute the single-phase supply line current, the 3-phase motor FLC must be increased to 173% (multiplied by the square root of 3).

As will be discussed in Chapter 13, motors are normally allowed to operate in a running overload condition up to 25% of their rated-load or full-load current rating. When the single-phase converter supplies a 3-phase motor, the 125% must be applied to the adjusted single-phase current, such that the 173% is multiplied by the 125%. The single-phase current is now equal to 216% of the determined (NE Code table-value) 3-phase motor FLC.

To align with the general overrate requirements of a motor disconnect found in Sec. 430-110(a), a motor disconnect must have a full-load current rating greater than or equal to 115% of the full-load current rating of the motor. The 216% overrate of the three-phase current must be further increased by the 115% overrate, such that the single-phase line current, for a specific motor load, must be equal to a minimim of 250% (115% of the already 216% overrate) of the 3-phase full-load current rating.

For the full-load current rating, Exception No. 1 to Sec. 455-8(c) allows the single-phase disconnect ahead of phase converter to be rated at a minimum of 250% of the computed full-load 3-phase current to be supplied.

Example 12–5. A 3-phase, 115-volt, design B pump motor rated at 2 horsepower will be supplied from a single-phase, 115-volt source through a phase converter. Determine the full-load ampere rating of a single-pole circuit breaker installed as the disconnecting means ahead of the converter.

Three-phase full-load current determination: Ref. T430-150

Motor FLC = 13.6 amps

Single-phase FLC overrate computation: Ref. 455-8(c), Excp. No. 1

250% of 13.6 amps = 34 amps

Circuit breaker full-load current rating: Ref. 240-6(a)
NHS: 35 amps

The full-load current rating of a knife switch or circuit interrupter (molded-case switch) would be computed in the same manner, although the standard sizes of these two types of disconnecting means do not follow the standard ratings of overcurrent-protection devices as listed in Sec. 240-6(a). As given in an earlier section, the next higher standard size of knife switch or circuit interrupter for this example is 60 amperes.

Code challenge 12–5. A 3-phase, 230-volt, design B fan motor rated at 7½ horsepower will be supplied from a single-phase, 230-volt source through a phase converter. Determine the full-load ampere rating of a double-pole circuit breaker installed as the disconnecting means ahead of the converter.

Three-phase full-load current determination: Ref. (A) ____________________

Motor FLC = (B) ____________ amps

Single-phase FLC overrate computation: Ref. (C) ____________________

(D) ____________ % of ____________ amps = ____________ amps

Circuit breaker full-load current rating: Ref. (E) ____________________

NHS: (F) ____________ amps

If the motor-circuit voltage (phase converter output voltage) is higher than the single-phase ac input voltage, like a 120-volt source is used to supply a 3-phase, 460-volt (480-volt) ac motor, Sec. 455-8(c), Ex. 1(b) requires the single-phase full-load current, which is already equal to 250% of the 3-phase motor full-load current, to be further increased by the ratio of the higher output voltage to the lower input voltage.

Example 12–6. A 3-phase, 230-volt, design B pump motor rated at 5 horsepower will be supplied from a single-phase, 115-volt source through a phase converter. Determine the full-load ampere rating of a double-pole circuit breaker installed as the disconnecting means ahead of the converter.

Three-phase full-load current determination: Ref. T430-150

Motor FLC = 15.2 amps

Single-phase FLC overrate computation: Ref. 455-8(c), Excp. No. 1

250% of 15.2 amps = 38 amps

Output/input voltage ratio overrate computation: Ref. 455-8(c), Excp. No. 1

$$\frac{230 \text{ volts}}{115 \text{ volts}} \times 38 \text{ amps} = 76 \text{ amps}$$

Circuit breaker full-load current rating: Ref. 240-6(a)

NHS: 80 amps

In addition to a phase converter, when the output voltage is higher than the input voltage, the ac drive controller is also being used as a transformer to step up the supply voltage. As will be demonstrated in Chapter 15, when the voltage is stepped up by a transformer, the current is stepped down. When computed from the load back to the single-phase source, if the voltage is doubled, it will take twice the value of the single-phase current to drive the higher-voltage 3-phase load.

Code challenge 12–6. A 3-phase, 460-volt, design B fan motor rated at 15 horsepower will be supplied from a single-phase, 240-volt source through a phase converter. Determine the full-load ampere rating of a double-pole circuit breaker installed as the disconnecting means ahead of the converter.

Three-phase full-load current determination: Ref. (A) ____________________

Motor FLC = (B) ___________ amps

Single-phase FLC overrate computation: Ref. (C) ___________

(D) ___________ % of ___________ amps = ___________ amps

Output/input voltage ratio overrate computation: Ref. (E) ____________________

$$\frac{\_\_\_\_\_\_\_\_\_\_\ \text{volts}}{\_\_\_\_\_\_\_\_\_\_\ \text{volts}} \times \_\_\_\_\_\_\_\_\_\_\ \text{amps} = \text{(F)}\ \_\_\_\_\_\_\_\_\_\_\ \text{amps}$$

Circuit breaker full-load current rating: Ref. (G) ____________________
NHS: (H) ___________ amps

In comparsion to a motor disconnect as sized by Sec. 430-110 for the standard motor application or Sec. 440-12(a) for a hermetically sealed refrigerant motor compressor, which both require the knife switch, circuit interrupter, or circuit breaker to be dual rated for both full-load current and horsepower, Exceptions No. 1 and No. 2 to Sec. 455-8(c) allow the disconnect to be rated either in terms of full-load current, or in terms of horsepower, neither require the dual rating.

When a phase converter supplies a single-motor load, and a single-phase horsepower-rated disconnect is to be installed ahead of the phase converter, Sec. 455-8(c), Excp. No. 2 requires the locked rotor-current rating of the 3-phase motor as determined by T430-151(b) (according to either design E or other design motor) to be overrated to 200% to determine the NE Code-minimum single-phase locked-rotor current rating of the disconnect.

The disconnect must then be sized in accordance with T430-151(a) for a single-phase horsepower rating. The table-listed value of single-phase locked-rotor current, according to horsepower and voltage rating, must be greater than or equal to the computed locked-rotor current of the phase converter.

Example 12–7. A 3-phase, 230-volt, design E ac induction motor rated at 5 horsepower will be supplied with a single-phase, 230-volt source through a phase converter. Determine the horsepower rating of the single-phase knife-switch disconnecting means.

Three-phase locked-rotor current determination: Ref. T430-151(b)

Motor LRC = 122 amps

Single-phase LRC overrate computation: Ref. 455-8(c), Excp. No. 2

200% of 122 amps = 244 amps

Knife switch horsepower rating: Ref. T430-151(a)

Table-value LRC = 300 amps: 10 hp

Code challenge 12–7. A 3-phase, 230-volt design E ac induction motor rated at 3 horsepower will be supplied with a single-phase, 230-volt source through a phase converter. Determine the horsepower rating of the single-phase knife switch-disconnecting means.

Three-phase locked-rotor current determination: Ref. (A) ______________

Motor LRC = (B) __________ amps

Single-phase LRC overrate computation: Ref. (C) ______________

(D) __________ % of __________ amps = (E) __________ amps

Knife switch horsepower rating: Ref. (F) ______________

Table-value LRC = (G) __________ amps: (H) __________ hp

When a phase converter supplies several motors, or several motors and nonmotor loads, and a single-phase horsepower-rated disconnect is to be installed ahead of the phase converter, Sec. 455-8(c), Excp. No. 2 requires the summation of all the 3-phase nonmotor and motor full-load currents as determined by T430-150, other than the largest motor FLC, to be added to the locked-rotor current rating of the largest motor as determined by T430-151(b) (according to design E or other design motor) to be overrated to 200% for a total single-phase locked-rotor current rating.

Example 12–8. One 3-phase, 230-volt, design B ac induction motor rated at 3 horsepower and two 3-phase, 230-volt, design B ac induction motors rated 2 horsepower will be supplied with a single-phase, 230-volt source through a phase converter. Determine the horsepower rating of the single-phase circuit interrupter-disconnecting means.

Three-phase locked-rotor current determination: Ref. T430-151(b)

3-hp motor LRC = 64 amps

Other three-phase motor FLC determination: Ref. T430-150

2-hp motor FLC = 6.8 amps

Three-phase current computation: Ref. 455-8(c), Excp. No. 2

Largest motor LRC: 64.0 amps
Other motor FLC: 2 × 6.8 amps = 13.6 amps
Total: 77.6 amps

Single-phase LRC overrate computation: Ref. 455-8(c), Excp. No. 2

200% of 77.6 amps = 155.2 amps

Circuit interrupter horsepower rating: Ref. T430-151(a)

Table-value LRC = 168 amps: 5 hp

Rating the single-phase disconnect for the phase converter in terms of horsepower is limited to low-value horsepower-rated 3-phase ac motors. The locked-rotor current rating of the largest single-phase disconnect, which is rated at 10 horsepower, is only 300 amps.

Code challenge 12–8. One 3-phase, 230-volt, design C ac induction motor rated at 7½ horsepower and two 3-phase, 230-volt, design E ac induction motors rated at 3 horsepower will be supplied with a single-phase, 230-volt source through a phase converter. Determine the horsepower rating of the single-phase circuit interrupter-disconnecting means.

Three-phase locked-rotor current determination: Ref. (A) ____________

(B) ____________ hp motor LRC = ____________ amps

Other three-phase motor FLC determination: Ref. (C) ____________

(D) ____________ hp motor FLC = ____________ amps

Three-phase current computation: Ref. (E) ____________

Largest motor LRC: (F) ____________ amps
Other motor FLC: ____________ × (G) ____________ amps = ____________ amps
Total: (H) ____________ amps

Single-phase LRC overrate computation: Ref. (I) ____________

(J) ____________ % of ____________ amps = (K) ____________ amps

Circuit interrupter horsepower rating: Ref. (L) ____________

Table-value LRC = ____________ amps: (M) ____________ hp

As with the full-load current computations required by Exception No. 1 to Sec. 455-8(c), if the motor-circuit voltage (phase-converter output voltage) is higher than the single-phase ac input voltage, the single-phase locked-rotor current, which is already equal to 200% of the adjusted 3-phase motor locked-load current must be further increased by the ratio of the higher output voltage to the lower input voltage to determine the single-phase locked-rotor current horsepower rating.

12.10 MOTOR CONTROLLERS

In some applications the knife switch or other disconnecting means is used as the means of control for a motor. In other instances, switches, similar in operation to the lighting snap switches discussed in Chapter 11, are used in a secondary, lower-voltage control circuit to direct the operation of a relay, contactor, or motor starter, or as used with ac or dc drive controllers, the amount of power that will be applied to the motor terminals.

Relays, contactor (power-rated relays), and motor starters (contactors furnished with running overload protection for the motor—Fig. 12-10) are all electromechanical devices, in which a solenoid is activated by an electrical signal and through a generated magnetic field, mechanically (electromechanically) closes one or more sets of switching contacts that are held open either by gravity or spring tension, or a combination of both.

The higher voltage- and current-rated conductors connected to the line and load sides of the power contacts (switches) on the relay, contactor, or starter are the motor-

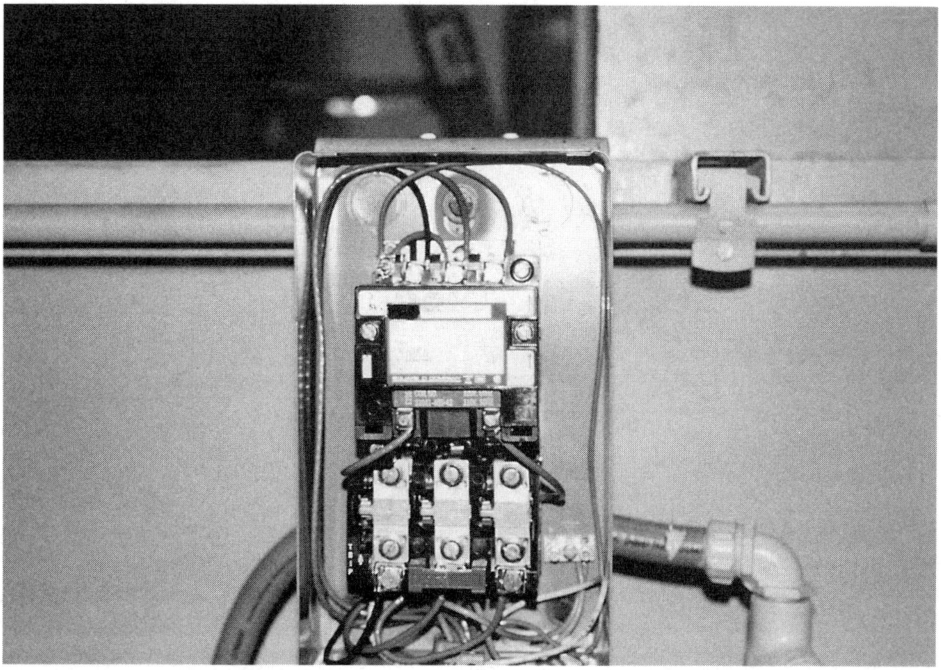

Fig. 12-10 The elements of the overload relay are series wired with the motor on the load side of the power contacts of the starter.

circuit conductors. The lower voltage- and current-rated signaling conductors, which through switches are used to supply power to the solenoids of the relay, contactor, or starter, are the motor-control circuit conductors.

Part F of Art. 430 governs the installation of motor-control circuits. Part G governs the installation of motor controllers. Part H governs the installation of motor-control centers (Fig. 12-11).

In general [Secs. 430-81 (a) & (b)], the controller for a stationary, hard-wired motor rated greater than ⅛ horsepower is a contactor that serves to start and stop the motor [Sec. 430-82(a)] by opening all the supply conductors to the motor. The control contactor, often called an automatic (magnetic) starter, or a manual (nonmagnetic) starter, is rated in terms of horsepower, as well as the full-load current it is capable of interrupting.

The starter is rated by the National Electrical Manufacturers Association (NEMA) in terms of sizes. The NEMA size defines a range of horsepower ratings the starter is capable of controlling. NEMA starter sizes range from "00" (double ought) up to size No. 8. The horsepower ratings are based on both the voltage rating of the motor and whether the motor is wired as single-phase or polyphase. The contactor shown in Figure 12-12 is rated NEMA size 2.

Section 430-82(a) requires the starter or controller to be capable of interrupting the stalled or lock-rotor current of the motor in conjunction with starting and stopping the

Fig. 12-11 Motor control centers contain the disconnect, overcurrent protection, and control means (starters) of several motors in a common enclosure.

Fig. 12-12 Starter contactors are rated in NEMA sizes from double-ought (00) up to size 8. The starter picutred is NEMA size 2.

motor. Except for 3-phase, design E motors, which are normally computed at 800%, the locked-rotor current rating of an ac induction motor is equal up to 600% of the motor's rated full-load current. Section 430-83 lists the required horsepower and voltage ratings of the motor controllers.

Unless the controller has a dual function as both controller and disconnect, Sec. 430-84 states that the controller need not open all the supply conductors, only enough to stop the motor. This would constitute one conductor for a dc or single-phase ac motor, and two conductors for either a 2-phase or 3-phase ac motor.

If the controller, as designated by Sec. 430-111, is allowed to serve as the required motor-disconnecting means, Sec. 430-84, Exception requires the multipole fused knife-switch, circuit interrupter (molded-case switch), or circuit breaker to open all motor supply conductors in the stopping of the motor.

If one of the motor supply conductors (line conductors) is a grounded-circuit (current-carrying) conductor, Sec. 430-85 is in agreement with Sec. 430-111 in stipulating that the grounded-circuit conductor cannot be opened by the controller unless all the motor-circuit conductors are opened simultaneously. As a general rule, no switches [Sec. 380-2(b)] or circuit overcurrent-protection devices (Sec. 240-22) are allowed in any circuit or supply conductor (current-carrying conductor) that is intentionally grounded.

12.11 MOTOR CONTROL CIRCUITS

Section 430-71 defines motor control-circuit conductors as signaling conductors that carry only signal current that directs the performance of the starter or motor controller. In other words, these conductors convey only the power supplied to the solenoids of control relays, power contactors, or motor starters, and indicating (status) lights. Motor control-circuit conductors do not convey motor-circuit (supply) current.

Motor-control circuits are generally rated at 120 volts or less. When these circuits control higher-voltage motors, the control voltage is normally generated from a control (signaling) transformer whose secondary will maintain the lower 120-volt control voltage. The elements of a control circuit and their overcurrent protection must include all the equipment and/or devices concerned with the function of the circuit. In conjunction with the conductors, the source of energy for the control circuit, the contactor operating coil, all switches that govern the energization of the coil, and even the raceway containing the signaling conductors are all part of the motor-control circuit.

Article 725 [refer to 430-72(a)] governs the use, protection, and installation of control circuits, and breaks them down into three basic classes. The classes, listed as either class 1, class 2, or class 3, are governed by the energy available to the circuit. Class 1 power circuits normally have no power limitations but must have overcurrent protection. Class 2 and class 3 signaling circuits are "inherently limited" by their transformer source and normally do not require overcurrent protection in their secondaries.

Class 2 and class 3 circuits are considered as safe from a fire initiation standpoint (Sec. 725-2, Def.). Class 2 circuits are also considered as safe from a shock or electrocution standpoint. A doorbell signaling transformer is an example of a class 2 control circuit. A step-down voltage transformer for an intercom is an example of a class 3 control circuit.

Most motor-control circuits are class 1 signaling circuits because of the relatively high energy required to operate the coils. Section 725-21(b) restricts the rating of a class 1 signaling circuit to 600 volts or less with no limitation on the power output. Class 1 signaling circuits must be wired within the guidelines of Secs. 725-21 through 725-29. Section 725-23 requires separate overcurrent protection for class 1 signaling circuits. Section 725-24 gives the allowed location of the required protection.

When several motors are controlled from a central location (motor control center), the motor control-circuit voltage is normally derived from a control transformer. With this type of installation, the motor control voltage is not derived or supplied from the load side of any of the individual motor disconnects.

If motors are installed individually in discrete locations, the control voltage is normally taken from the supply side (line side) of the motor controller. If the motor supply voltage is greater than 120 volts to ground, a signaling transformer (Fig. 12-13) is normally connected to the supply side of the motor controller, and the secondary derived voltage is taken at the lower control voltage (120 volts).

Section 430-72(a) states that motor control circuits are not to be considered as branch circuits, and when tapped from the supply side of the motor starter (controller), Sec. 430-72(b) will not allow either the branch-circuit short-circuit, ground-fault protection (overcurrent protection) of the motor, or the supplementary overcurrent protection for the control-circuit conductors to exceed the values listed in T430-72(b).

Fig. 12-13 The lower-value control voltage of a motor circuit is often derived through a transformer from the higher-voltage line supply to the motor circuit.

Table 430-72(b), column A requires control-circuit conductors that fall under this category to be protected according to their ampacities listed in T310-16. Separate or supplementary motor control-circuit overcurrent protection for wire sizes No. 16 or No. 18 AWG must be sized in accordance with Sec. 725-23 for class 1 control (signaling) circuits, and in accordance with Sec. 725-41 for either class 2 or class 3 control circuits.

The overrates of overcurrent protection as compared to the ampacities of the motor control-circuit conductors in column B of Table 430-72(b) are much higher than the corresponding ratings listed in Column A. The basic rule of column A must allow for the motor control-circuit conductors to be run remote from the starter enclosure. The remote control station may be outside of the building or structure from where the starter is located.

The higher values of protection allowed by column B take into consideration that the control-circuit conductors do not leave the starter enclosure [Sec. 430-72(b), Excp. No. 1]. This restricts the hazard of the control-circuit conductors being exposed to physical damage. Confinement to the starter enclosure also reduces the safety hazard to personnel from exposed motor control-circuit conductors that have been subjected to physical damage.

In the basic rule (column A), a No. 14 AWG THW copper control-circuit conductor would have to be protected from overcurrent at a maximum value of 20 amps (T430-72, column A Note to T310-16).

By Sec. 430-72(b), Excp. No. 1 and column B of T430-72(b), the same No. 14

AWG THW copper control-circuit conductor, contained solely in the starter enclosure, can have overcurrent protection rated at a maximum of 100 amperes. The 100-amp protection is actually the rating of the motor branch-circuit short-circuit, ground-fault protection (discussed in Chapter 13).

If the installation of the motor-control circuit meets the requirements of Sec. 430-72(b), Excp. No. 1, but the motor's branch-circuit short-circuit, ground-fault protection rating exceeds the values listed in column B of Table 430-72(b), then the motor control-circuit conductors must have separate or supplementary overcurrent protection as sized by the basic rule in column A.

Section 430-72(b), Excp. No. 2 governs the installation of a control circuit for a magnetic starter (or other controller) in which stop and start switches and/or other pilot devices are located outside the starter enclosure, and the motor control circuit is tapped from the supply side (line side) of the motor starter. The motor branch-circuit short-circuit, ground-fault overcurrent protection rating is restricted to the values listed in column C of Table 430-72(b), or the motor control-circuit conductors must be protected by separate overcurrent protection sized in accordance with column A.

Section 430-72(b), Excp. No. 3 and Sec. 430-72(c) govern the use of signaling transformers to derive the motor control-circuit voltage. The general rule of Sec. 430-72(c) requires the transformer to have overcurrent protection sized in accordance with Art. 450 [Sec. 450-3(b)]. The control transformer does not have to be supplied by tapping the line side of the motor starter. The control transformer does not have to be located in the motor-starter enclosure.

If the motor control-circuit conductors are the secondary conductors of a single-phase, two-wire (single-voltage) transformer, the control-circuit conductors can be protected by the transformer's primary overcurrent protection. The size of the transformer's overcurrent protection is restricted by Table 30-72(b). The restriction depends on whether the installation of the motor-control circuit conductors falls under the basic rule of column A, the Exception of column B, or the Exception of column C.

In many industrial installations a central motor-control cabinet (control center) will have a separate transformer used to derive the control voltage. With this application, the transformer will have primary overcurrent protection that is sized in accordance with Sec. 450-3. The control-circuit wiring is normally sized by "company policy" at No. 12 AWG copper. The secondary overcurrent protection must be sized by Sec. 725-21 (class 1—unlimited power supply) according to the smallest manufacturer-installed wiring on the starters contained within the motor-control center [Sec. 430-72(a)].

12.12 CONTROL-CIRCUIT WIRING DIAGRAMS

If a practicing electrician is at his or her point of training that a journeyman's license is at hand, motor control-circuit wiring diagrams should be part of his or her acquired talents. The more common motor-control diagrams include a 2-wire motor-control circuit, a 3-wire motor-control circuit, and an "H-O-A" or hands-off-automatic motor-control circuit.

A 2-wire control circuit can be a manual control circuit in which a motor-rated switch is used to control the motor directly. This switch is normally a fractional-horsepower or small-motor control method. The run or stop status of the motor is controlled in the same manner as the snap switches were used to control the on or off status of the overhead lighting fixtures in Chapter 11. In most instances, the switching is confined to a single location. Figure 12-14 shows some of the common symbols used for motor control-circuit switches.

For larger motor installations using 2-wire motor control, a selector switch with contact block (Fig. 12-15), or a simple snap switch (toggle-type light switch) is used to control the motor starter. The switch engages or disengages the starter contactor by control power applied to the starter solenoid coil (Fig. 12-16). The starter coil acts as a solenoid or mechanical plunger that mechanically engages contacts in the motor circuit and supplies power to the motor.

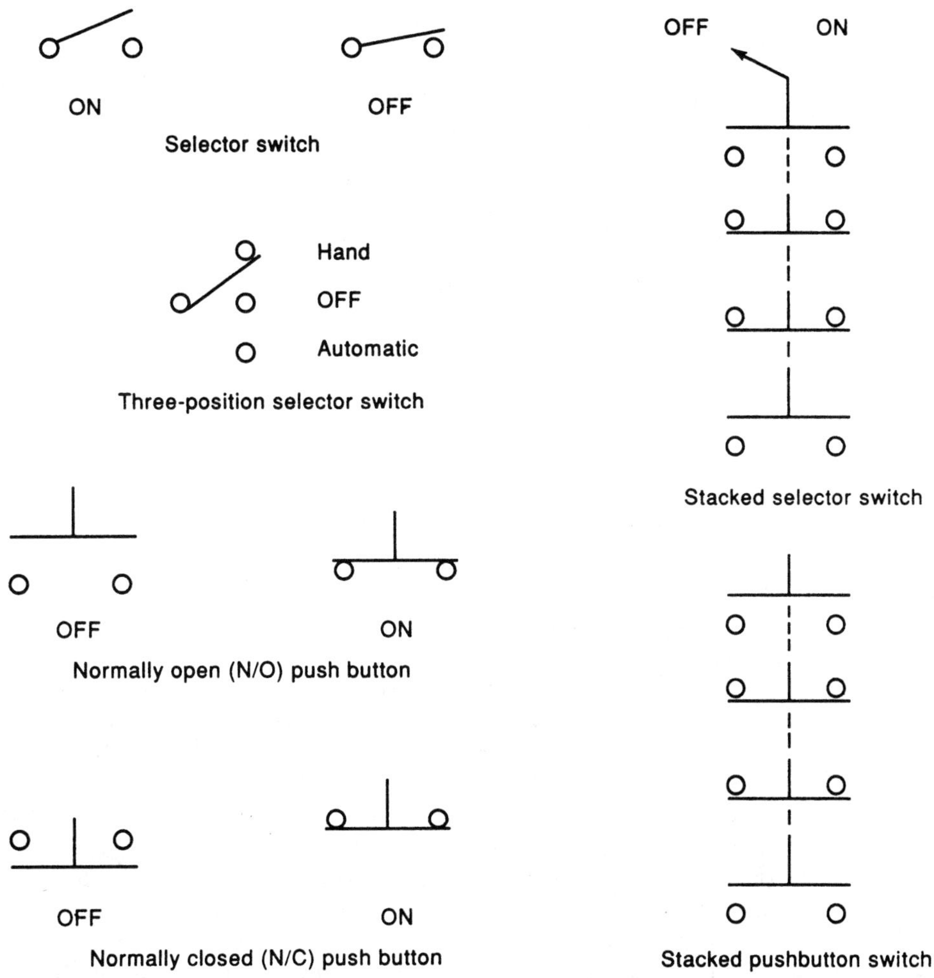

Fig. 12-14 Motor control circuits normally use either selector switches or pushbutton switches to direct the operation of the motor controller.

Fig. 12-15 Both pushbutton (momentary) switch operators (left) and selector (maintain) switch operators are used in motor-control circuits.

A 2-wire control circuit must use a select or maintain-type switch which ensures power in the control circuit is constantly supplied or maintained to the starter solenoid coil through that switch. A pushbutton switch, also referred to as a momentary switch, cannot be used with a 2-wire control circuit. When a momentary pushbutton switch is installed in a 2-wire control circuit, power is supplied to the motor starter coil only when the button is pushed. Releasing the pushbutton removes power from the coil, and the motor drops out.

Running overload-overcurrent protection for the motor windings, the motor-circuit conductors, and the branch-circuit supply conductors is provided by an overload relay mounted on the starter. The control circuit is wired in series with the contacts on the overload relay (Fig. 12-17). If for some reason the motor goes into mechanical overload and starts drawing current higher than its full-load current rating, and higher than the load-current rating of the heating elements in the overload relay (normally set at 125% of the motor nameplate FLC), the heater element(s) will respond to the mechanical overload by opening N/C contacts which will disable the control circuit.

Disabling or opening the control circuit causes the motor starter to open the power contacts in the motor circuit: Power is removed from the motor circuit. The control circuit will be disabled by opening of the overload-relay control contacts even though the selector or on/off switch is still made or closed.

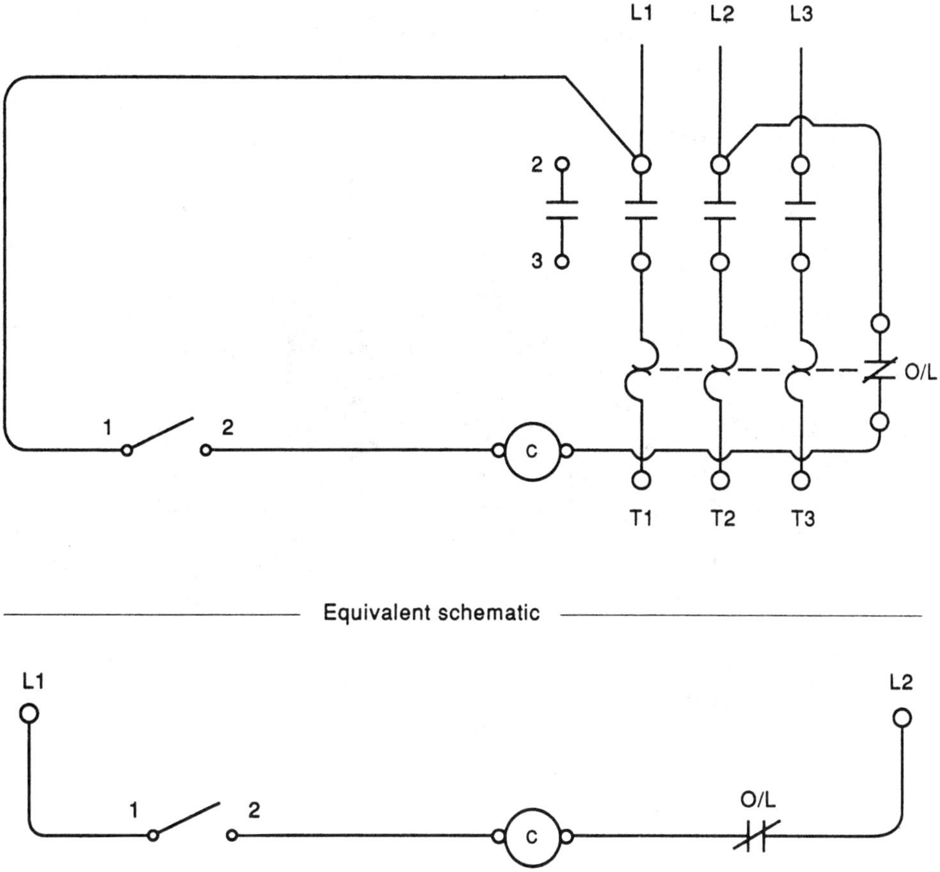

Fig. 12-16 A two-wire motor-control circuit uses a select or maintain-type switch to apply power to or remove power from the contactor coil.

A 3-wire control circuit, as shown in Figure 12-18, uses the same type of magnetic motor starter and overload relay protection; it differs in the type of switches used. Instead of selector switches that maintain the position they are switched to, a 3-wire control circuit utilizes momentary pushbutton switches.

Pushbutton switches are normally momentary switches held in one of two positions by spring tension. Activating the switch by the push of the finger changes the status of the switch to the other position. When the finger is removed, the switch contacts spring-return to the previous (nonactivated) position. These switches also use contact blocks mounted to the back of the switch operator for the actual switching action (Fig. 12-19).

The contacts on a pushbutton switch can be wired as normally open (N/O) or as normally closed (N/C) switches. Normally open or normally closed indicates the status of the switch at rest. At rest indicates the status of the control circuit when none of the buttons are being pushed (no human intervention).

Fig. 12-17 The control circuit of a motor is wired in series with the normally open/held-closed switch contacts on a separate overload relay.

A 2-wire control circuit requires only one switch and two wires. A 3-wire control circuit requires two switches and three wires. The control circuits get their names from the number of conductors that must be extended to the switching station from the starter. For 2-wire control, one conductor out to the single maintain-type switch and another conductor back from the single switch is a total of two conductors.

As with overhead light switching, discussed in Chapter 11, one conductor in the 2-wire control is the continuous power supply conductor from the control-voltage source. The return conductor is the switch-leg conductor, which is connected to the line side or ungrounded side of the solenoid coil on the electromechanical starter. The other side of the coil is referred to as the grounded or common side. Another circuit conductor is extended between this terminal and the return side (grounded side) of the supply transformer or other control-voltage source.

For 3-wire control, the same supply and switch-leg conductors are extended to the switching station. The continuous power is connected to the line side of the N/C stop pushbutton. The switch-leg power is connected to the load side of the N/O start pushbutton. The two switches must be wired in series.

Since both switches are momentary, switched power is applied to the solenoid in the starter only when the start pushbutton switch is closed (pushed) with the stop switch remaining closed. This is often referred to as a jog function of the motor because the seal-

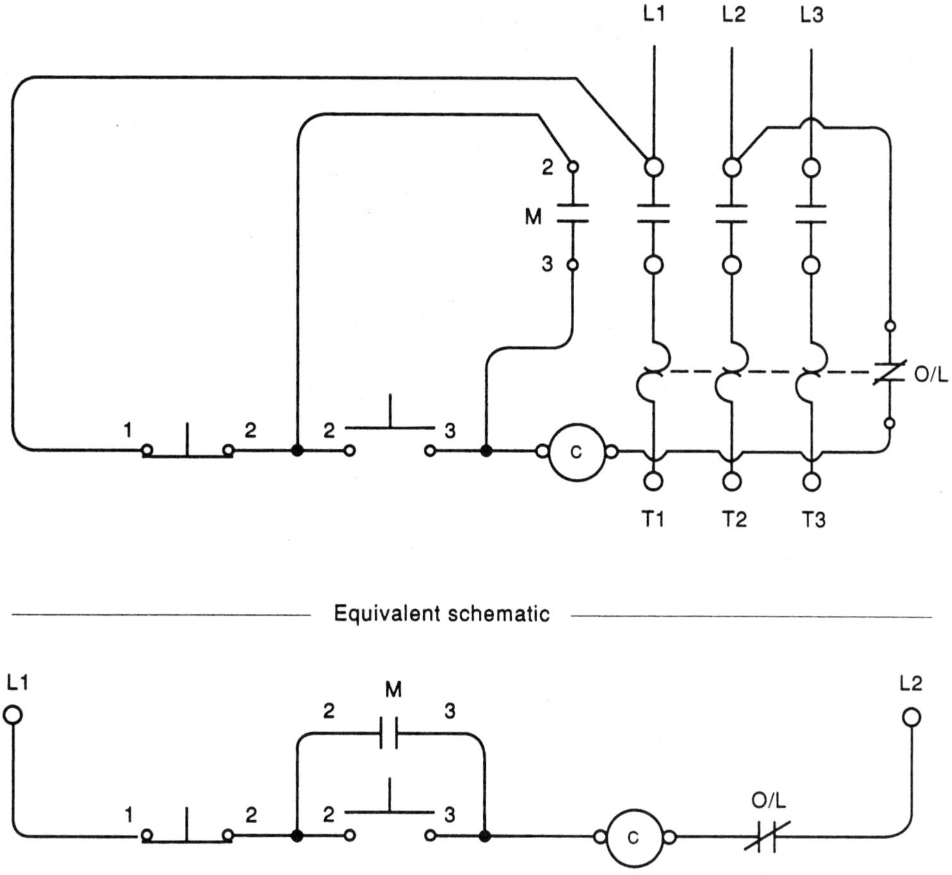

Fig. 12-18 A three-wire motor control circuit uses both normally open (N/O) and normally closed (N/C) pushbutton switches to control the contactor coil.

in circuit is disabled. The seal-in circuit requires the third conductor, which is extended from the starter location to the switch location, hence the name 3-wire control.

The seal-in circuit consists of a set of N/O auxiliary contacts that are mounted on the power contactor of the motor starter. These contacts close when the power contacts supplying the motor close, and open when these power contacts open. The N/O auxilary contacts must be wired in parallel to the N/O start switch. Momentarily closing the start switch momemtarily applies power to the starter solenoid. The closing of the N/O auxilary contacts with the closing of the power contacts parallels the momentarily closed start switch and seals in the motor in a run condition. The start switch can be released (opened) without stopping the motor.

To wire the N/O auxilary contacts (also referred to as either holding, maintaining, or sealing contacts) in parallel to the N/O start pushbutton switch, the third conductor from the switching location is connected to one side of the N/O auxiliary contacts, the

Fig. 12-19 Separate switch contact blocks in either N/O, N/C, or a combination of N/O and N/C can be mounted on either type of switch operator (pushbutton or selector).

other side of the N/O auxiliary contacts must be connected common with the switch leg at the line side of the motor starter solenoid coil.

When the stop button is pushed, power is removed from the control circuit because this N/C pushbutton is forced into an open condition. When power is removed from the starter coil (de-energized) the contactor drops out, and the seal-in auxiliary contacts returns to their N/O status. When the stop button is released, the motor remains disabled (stopped) because both the start switch and the seal-in contacts are open. If the N/C contacts on the overload relay should open due to motor overload, the control circuit is disabled in the same manner as with a 2-wire control.

For NE Code exams or testing at the journeyman's level, the practicing electrician may be asked to complete a 3-wire wiring diagram for a motor-control circuit whose control voltage is derived from the line side of the motor starter through a transformer (Fig. 12-20). In this type control circuit, a 2-wire transformer is used to stepdown the usually higher motor supply voltage to the common 120-volt motor-control voltage. The transformer has both a 2-wire primary and a 2-wire secondary.

The transformer primary is wired across supply lines L1 and L2 on the motor starter. The transformer secondary has to complete a closed loop with the stop pushbutton switch, the start pushbutton switch, the starter solenoid coil, and the N/C control contacts

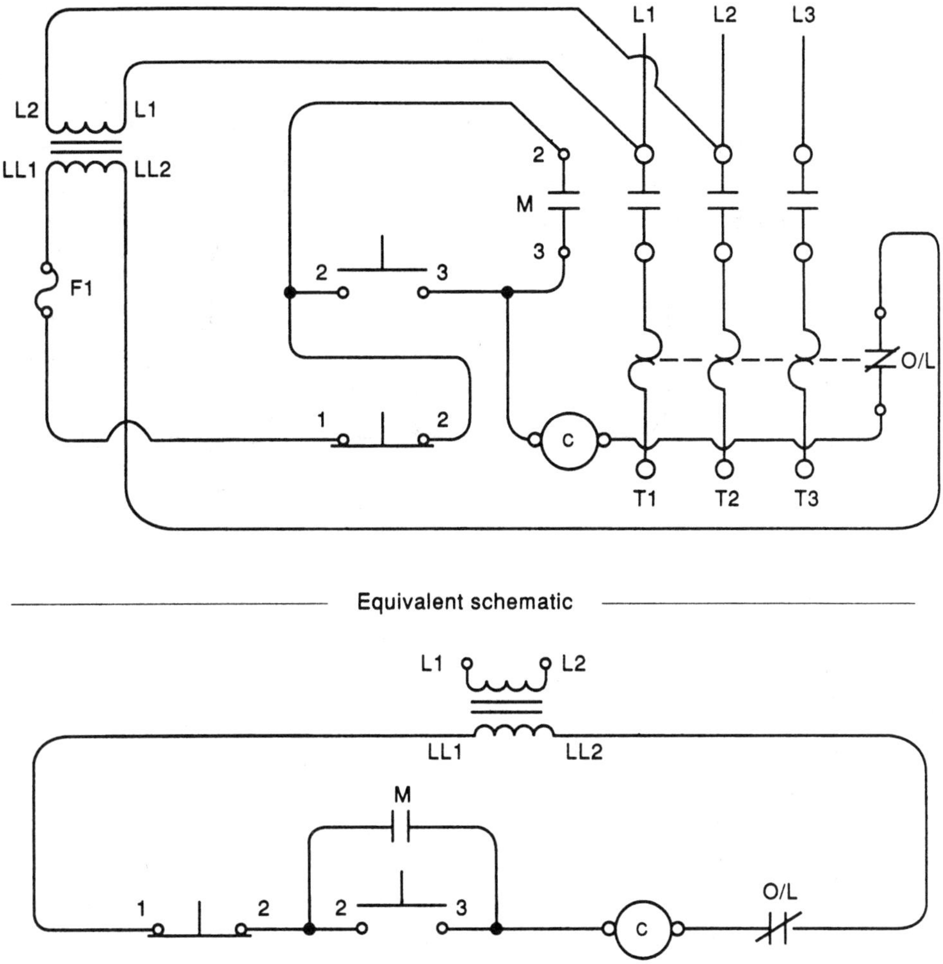

Fig. 12-20 When motors are supplied with voltages rated higher than 120 volts to ground, a transformer is normally used to derive the control voltage.

of the overload relay wired in series. The overload-relay contacts are drawn as a set of N/C contacts and identified with the acronym O/L.

As with all 3-wire control circuits, the auxiliary contacts, shown as a fourth set of N/O contacts beside the power contacts in the motor circuit, must be wired in parallel with the N/O contacts on the start pushbutton switch.

A hands-off-automatic or H-O-A control circuit, as shown in Figure 12-21, uses a three-position selector switch that offers two parallel control circuits for the motor starter coil. In the OFF postion (normally the center position of the selector switch), both parallel control circuits are disabled (no power applied). In the HANDS control position, power is supplied to the motor starter coil without any interrupt except for the overload-

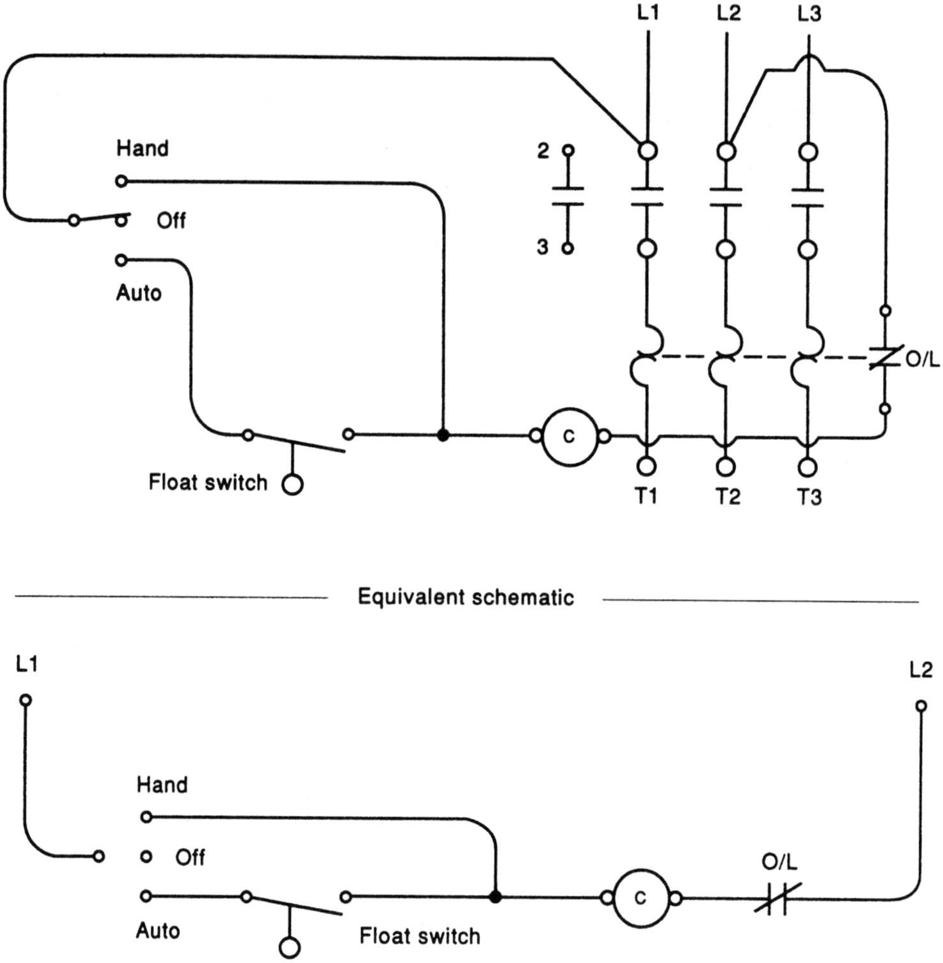

Fig. 12-21 Sump pumps are normally controlled by H-O-A switches. When automatic is selected, the float switch starts and stops the pump according to pre determined liquid levels.

relay control contacts. In the AUTO position, power is applied to some type of pilot device, which in turn controls the status of the starter solenoid.

The pilot device can be a float switch on a sump pump. When the water in the sump reaches a predetermined level, the float closes the N/O switch contacts and applies power to the starter solenoid. The closing of the N/O power contacts within the starter applies power to the motor circuit. The turning rotor of the motor applies mechanical power to the pump, and water is removed from the sump. The overload protection is in both parallel control circuits. Should the motor draw overload current, the control circuit will be disabled and power will be removed from the pump motor.

Pilot devices (Fig. 12-22) can be many types of switches including float switches,

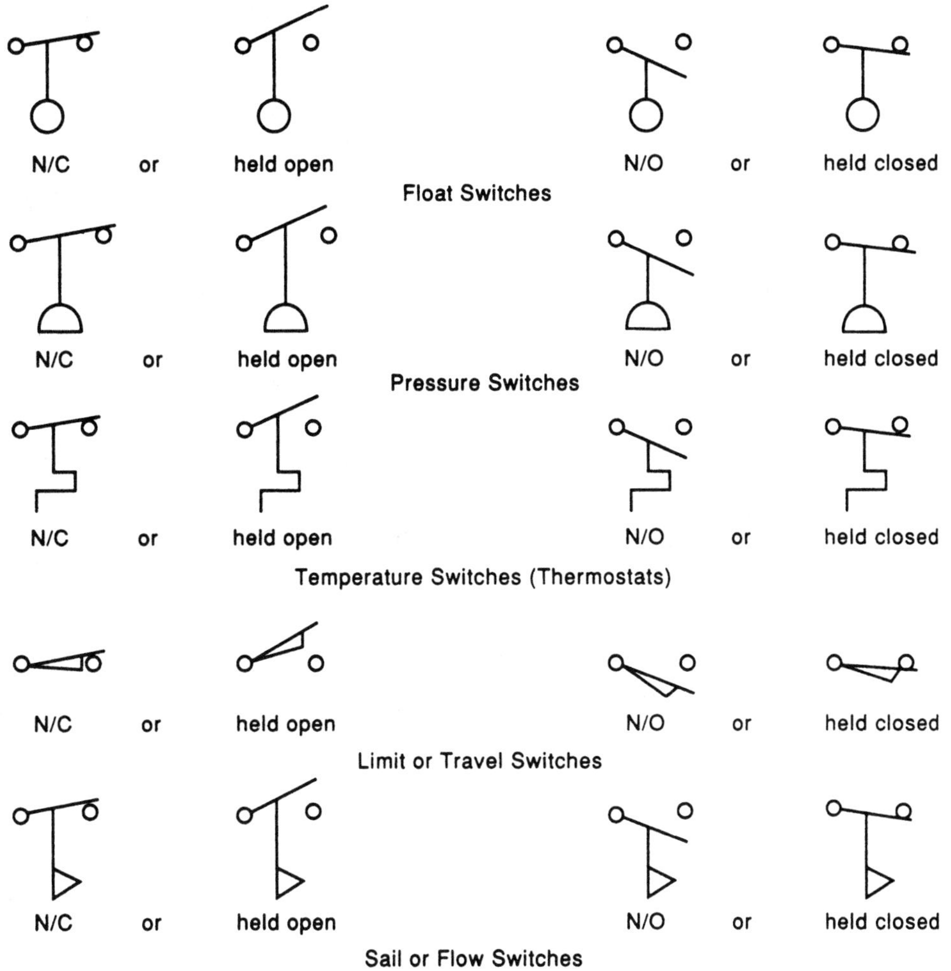

Fig. 12-22 Pilot devices sense the material being moved by a motor drive train. Motor power is *not* conveyed through the pilot device.

pressure switches, temperature switches (thermostats), limit (or travel) switches, and sail (or flow) switches.

REVIEW QUESTIONS

1. A specific-purpose or ____________ branch circuit is dedicated to a single piece of equipment or single receptacle outlet.

2. Appliances are utilization equipment manufactured in standard sizes, which are normally designed to aid in the many chores around the home.

True or False

3. The installation of appliances on multioutlet branch circuits must meet the requirements of Sec. ___________ .

4. The rating of a branch-circuit used to supply a continuous-load appliance must be overrated to ___________ % of the nameplate rating of the appliance.

5. All appliances must have a means to disconnect all ungrounded (phase) conductors from the electrical source.

   True or False

6. The rating of both the circuit conductors and overcurrent protection of a branch circuit supplying fixed electric space heating must be overrated to ___________ % of the nameplate full-load current.

7. All equipment installed as fixed electric space heating must have a means to disconnect all the ungrounded (phase) conductors from the electrical source.

   True or False

8. Other than fractional-horsepower, general-purpose motors, and specific-purpose motors rated in excess of 200 horsepower (500 horsepower for 3-phase), the NE Code table-value full-load current must be used to size all electrical supply and support equipment except separate individual overload protection.

   True or False

9. The nameplate-rated full-load current of any motor must be used to size individual separate overload (running overload) protection.

   True or False

10. When hermetically sealed refrigerant motor compressors have two nameplate ratings of full-load current, the rated-load current will always be larger than the branch-circuit selection current.

    True or False

11. The branch-circuit selection current, when listed on the nameplate of a hermetically sealed refrigerant motor compressor, must be used to size all electrical supply and support equipment except separate (running) overload protection.

    True or False

12. The nameplate rated-load current of any hermetically sealed refrigerant motor compressor must be used to size individual running overload protection.

    True or False

13. The NE Code table-value full-load current ratings of general-purpose motors are found in Tables ___________ through ___________.

14. A disconnecting means must be located in the circuit on the line side of any motor controller.

    True or False

15. A disconnecting means must be located in the circuit on the line side of any controller for a hermetically sealed refrigerant motor compressor.

    True or False

16. One or more disconnecting means may be required when the motor is located remote from its controller, because a disconnecting means required for the controller, and a disconnecting means required for the motor or motor-driven machinery (which can be the same disconnect), must be located within line-of-sight, and within 50 feet of the equipment it serves.

    True or False

17. If the disconnecting means ahead of the motor controller is capable of being locked in the open position, this disconnect can be more than 50 feet from, and not visible from the motor or motor-driven machinery, and be NE Code legal.

    True or False

18. The disconnecting means required for a condensor, chiller, air conditioner, or other refrigeration unit that contains one or more hermetically sealed motor compressors must be located within line of sight, and within 50 feet of the equipment it serves.

    True or False

19. If a motor and its controller are supplied through a single- to three-phase converter, a disconnecting means must be located ahead of, within line of sight, and within 50 feet of the phase converter.

    True or False

20. When a single disconnecting means serves a phase converter, controller, and motor it must be readily accessible. If more than one disconnect is required to be NE Code legal, only one disconnecting means, preferably the one at the motor, must be readily accessible.

    True or False

21. A motor controller can be a manual switch that interrupts the supply circuit to the motor. When used as the disconnecting means, the manual switch must disconnect all the motor-circuit conductors from the branch-circuit supply conductors.

    True or False

22. When a motor-control circuit uses maintain-type switches, the control-loop is referred to as a ____________-wire control.

23. When a motor control circuit uses momentary-type switches, the control-loop is referred to as a ____________-wire control.

24. If a control circuit uses momentary-type switches, the N/O start switch must be wired in ____________ with a set of N/O auxiliary contacts mounted on the motor-starter contactor.

25. Whether the control circuit uses momentary- or maintain-type switches, and whether the motor can be controlled manually (by hand) or automatically by pilot devices (sensing switches), all control circuits must be wired in series with a set of N/O contacts that are held closed against spring tension on an overload-control relay. When the motor is mechanically overloaded, the overload-relay contacts return to their normally open status, disabling first the control circuit and then the motor supply circuit.

    True or False

REVIEW PROBLEMS

1. Determine the NE Code table-value full-load current of a 2-phase, 460-volt ac motor rated at 200 horsepower.
2. Determine the nameplate full-load current rating of a fused knife switch used as the disconnecting means for a 3-phase, 460-volt, 60 hp wound-rotor ac induction motor.
3. Determine both the full-load current rating and the horsepower rating of a circuit breaker used as the disconnecting means for a 3-phase, 460-volt, hermetically sealed refrigerant motor compressor that has a nameplate rated-load current of 80 amps, a branch-circuit selection current of 85 amps, and a locked-rotor current (LRC) of 440 amps.
4. Determine the NE Code-minimum full-load current rating and horsepower rating of a single knife switch-disconnecting means for a 575-volt, 3-phase, 3-wire feeder that supplies the following loads:
   - 1 Transformer rated at 12.5 kVA, 3-phase, 575-208Y/120 volts for receptacle loads and control voltage
   - 2 Hermetically sealed refrigerant motor compressors which have nameplate rated-load currents of 8 amps, nameplate branch-circuit selection currents of 10 amps, and nameplate locked-rotor currents of 48 amps each
   - 2 Design D condensor-coil fan motors rated at 10 hp, 3-phase, 575 volts each
   - 2 Design D water-pump motors rated at 15 hp, 3-phase, 575 volts each
5. A 3-phase, 208-volt design D blower motor rated at 3 horsepower will be supplied with a single-phase, 208-volt source through a phase converter. Determine the full-load ampere rating of a single-pole circuit breaker installed as the disconnecting means ahead of the converter.
6. A 3-phase, 208-volt design E fluoroveyor motor rated at 10 horsepower will be supplied with a single-phase, 115-volt source through a phase converter. Determine the full-load ampere rating of a single-pole circuit breaker installed as the disconnecting means ahead of the converter.
7. A 3-phase, 208-volt design D blower motor rated at 7½ horsepower will be supplied with a single-phase, 208-volt source through a phase converter. Determine the horsepower rating of the single-phase knife switch disconnecting means.
8. One 3-phase, 230-volt, design E ac induction motor rated at 3 horsepower and two 3-phase, 230-volt, design E ac induction motors rated 1½ horsepower will be supplied with a single-phase, 230-volt source through a phase converter. Determine the horsepower rating of the single-phase circuit interrupter-disconnecting means.

Chapter 13

Motors—Part I

Any electrical circuit consists of three basic components: a power supply, a connected load, and the circuit conductors that connect the two. The amount of current that will flow from the power supply to the connected load is governed by the magnitude of the applied voltage, the number of phases in the supply, and the magnitude of the load resistance, or in ac circuits, the load impedance.

Because the circuit conductors are in series with the connected load, the circuit current is further impeded by the wire resistance of the conductors. Sized correctly, according to their rated ampacity and the prevailing environmental conditions of operation (elevated operating temperature and voltage drop considerations—refer to Chapter 2 of this text), the conductor resistance is considered negligible. The load impedance is considered to be the circuit impedance.

For a resistive load, such as the heating elements in a storage-type water heater or tungsten-filament, incandescent lighting, the load resistance is fixed at the point of manufacture. The rated full-load current is drawn every time the load is turned on (power is applied).

The impedance of a motor is governed by both its manufactured operating characteristics and the inertia or other resistance to the turning motion of its rotor (by the connected load). The motor's impedance, and therefore its load current, can vary from almost infinite impedance and near zero amps, when operated in a no-load condition, to a very low value of circuit impedance, and as much as six times its rated-load current (eight times for design E, energy-efficient motors), when the motor is either started or operating in a full-load condition and suddenly mechanically overloaded to the point of stalling. When a motor stalls, it is forced into a locked-rotor condition.

The value or amount of line current drawn by the motor falls between these two extremes according to the torque requirements of the connected load. When the motor draws current in excess of its full- or rated-load current, as specified on the motor nameplate, the motor is said to be in overload.

Overcurrent due to overload of a motor will cause the circuit conductors to overheat. Extensive overheating will cause conductor insulation failure, which will cause the overcurrent condition to escalate from an overload to a fault overcurrent condition. High-magnitude fault overcurrent, whether short-circuit or ground-fault, will normally destroy both the circuit conductors and the windings of the motor if not stopped almost instantly.

The size or rating of the overcurrent-protection device for the motor circuit must be low enough in value to provide adequate fault protection, but high enough in value to allow the motor to operate with varying load conditions from no load to full load, and in some instances to a certain amount or degree of running overload, yet protect the circuit conductors from extensive overheating.

13.1 LOAD CHARACTERISTICS

Motor circuits require special provisions to allow the motor to be operated most efficiently according to the conditions of use. Motors draw high in-rush current, equivalent to its locked-rotor current rating, on starting. Once the inertia of the load is overcome and the rotor of the motor begins to turn, the load current diminishes. Once the rotor attains baseline or running speed, the amount of current drawn is dependent upon the amount of energy required to keep the load moving.

The resistance of the driven load to the turning of the rotor within the motor is governed by weight, friction, and the type of work being performed by the load equipment.

The weight of the driven load is determined by its mass or density and the pull of gravity. The blade assembly of a small exhaust fan in a dwelling kitchen does not have much weight. Both the motor's in-rush starting current and its full- or running-load current would be of relatively low value. The fan is only moving air from the inside to the outside of the structure.

In comparison, on starting, the drive motor for a belt-conveyor assembly used to tranport cases of material out to a loading dock could draw high-value in-rush current for a relatively long period of time to bring the conveyor belt up to line speed. Once line speed is attained, the value of motor full-load current is determined by the length of the belt, its angle of inclination, the individual weight of the cases, and the number of cases being conveyed.

The friction of the bearing surfaces, within the motor and the driven equipment also resist the turning of the rotor. The amount of friction is governed by the type of bearing and lubricant employed in both the motor and the driven equipment.

The type of work performed by the driven equipment also governs the amount of current drawn by the motor. The small exhaust fan would have a low value of full-load current that would remain relatively constant. The belt-drive motor would have a relatively high value of full-load current that would fluctuate according to the number of cases being conveyed. If the exhaust fan were replaced with an air compressor, the value of in-rush starting current and full-load current of the air-compressor motor could equal or even exceed the current drawn by the conveyor-belt motor.

Motor circuits are normally designed to allow the motor to operate in a continuous overload condition. Operated in overload, the motor will draw current in excess of nameplate-rated full-load current. The magnitude of allowed overload overcurrent and its duration are normally governed by the construction and other operating characteristics of the motor, which are established at the point of manufacture. Remember, the motor will overheat when overloaded. If allowed to get too hot, the motor-winding insulation will fail.

13.2 OVERCURRENT PROTECTION

As discussed in Chapter 7, overcurrent drawn by an electrical circuit can be the result of either overload, short circuit, or ground fault. The branch circuit overcurrent-protection device (OCPD) can be the only means of protection, but normally a motor is provided with a separate means of overload-overcurrent protection.

To protect the motor against mechanical overload protects the circuit against electrical overload. Thermal devices, such as solder-dash pots, bimetallic strips, or thermal resistors (standard symbol—heating element in Fig. 13-1), current transformers, electronic sensing devices, or other equipment that is sensitive to (responds to) the magnitude of current drawn by the motor, are normally used to provide the separate or additional means of overload-overcurrent protection for the motor circuit.

Only the branch-circuit fuse or circuit breaker can provide the necessary short-circuit or ground-fault protection required for the motor and its supply circuit conductors. Separate overload protection is normally referred to as the motor's running overload protection (Fig. 13-2).

13.3 RUNNING OVERLOAD PROTECTION

Individual motor and branch-circuit protection against overload overcurrent is governed by the requirements of Part C in Art. 430 of the NE Code. Motors can be protected at a value of overload overcurrent that is less than their full-load current rating, but normally the motor is allowed to operate in an overload condition. The rating of the separate overload device is set at 100% plus some percentage of the motor's full-load current rating. The value or percentage of overrate is determined by the conditions of use and by Secs. 430-32, 430-33, 430-34, 430-42, and 430-55 of the NE Code.

Section 430-6(a) requires the rating or setting of the running overload protection, which can be a separate overload relay, a thermal protector integral with (contained in) the motor windings, or a solid-state (electronic) control device located in the motor controller, to be sized in accordance with the actual motor full-load current as listed on the individual nameplate.

Overload relays installed on the load side of the control contactor (Fig. 13-3) are by

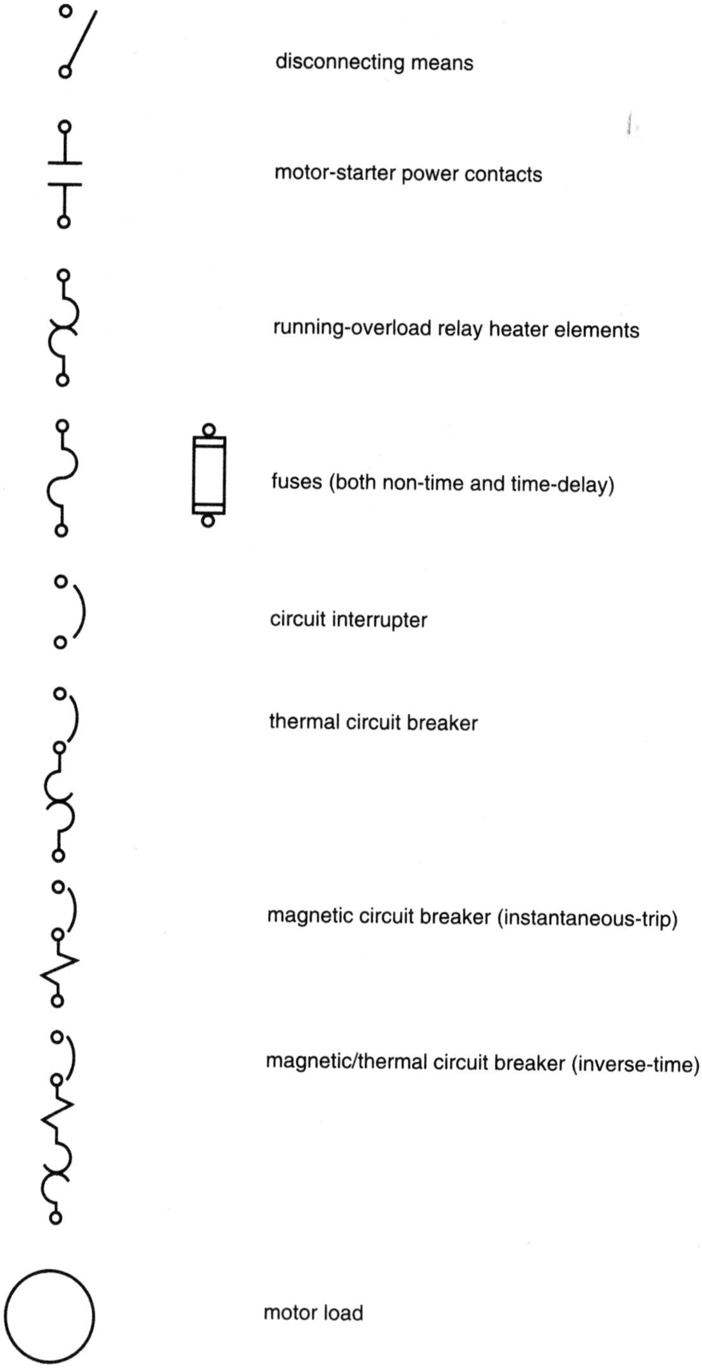

Fig. 13-1 Some of the common electrical symbols used in the one-line drawing of a motor supply.

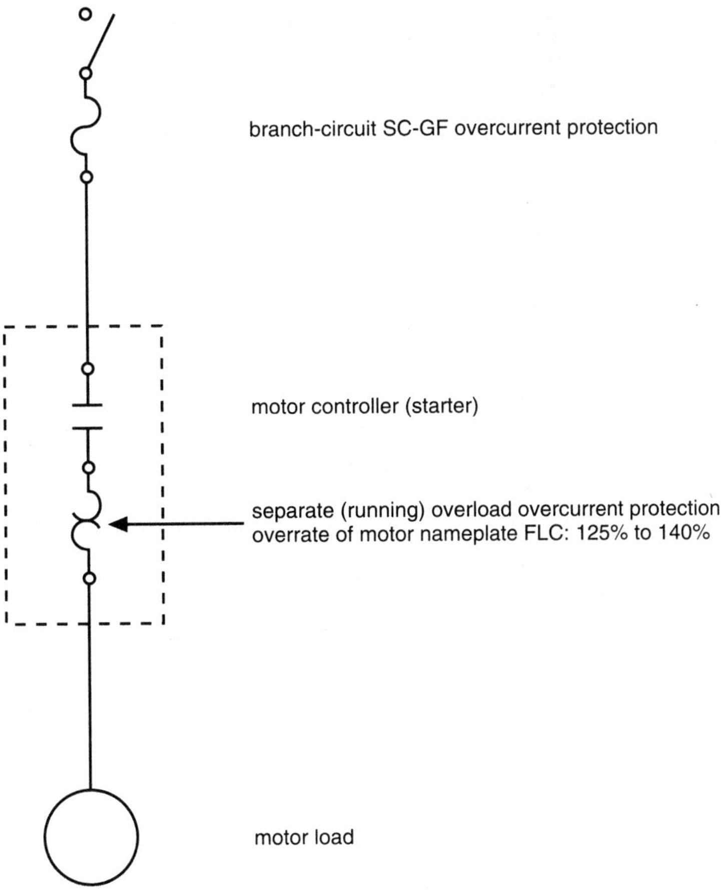

Fig. 13-2 As allowed by Sec. 430-32, the overcurrent protection for a motor is normally split with the branch-circuit OCPD providing the fault protection and a separate relay providing the overload protection.

far the most common type of running overload protection. The allowed percentage of overrate of the motor full-load current (FLC) is determined by the type of motor and the guidelines of Secs. 430-32 and 430-34.

The 125% overrate found in Sec. 430-32(a)(1) for intergral-horsepower motors, and Sec. 430-32(c)(1) for fractional-horsepower motors, is the norm of the electrical trade when sizing overload protection, although both sections stipulate that the protection device rating for any motor that does not have at least one of the special ratings (increased service factor or limited temperature rise) shall be restricted to a maximum of 115% overrate for separate overload protection.

Example 13–1. A 3-phase, 230-volt, 25-hp squirrel-cage induction motor has a service fator of 1.15 and a nameplate full-load current rating of 64 amps. Determine the full-load current rating of the heaters installed in the separate overload relay.

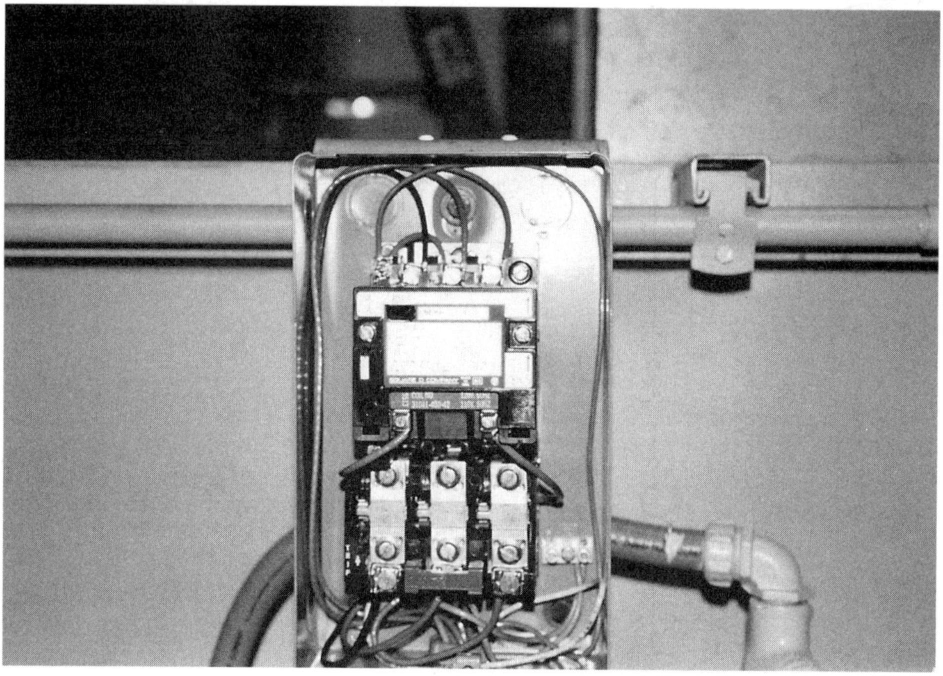

Fig. 13-3 A motor starter consists of an overload relay wired on the load side of the power contactor.

Motor full-load current rating determination: Ref. 430-6(a)
Motor FLC = 64 amps

Starter-heater overrate computation: Ref. 430-32(a)(1)

125 % of 64 amps = 80 amps

Standard sizes of overload-relay heating elements are not listed in the NE Code. Instead, standard heater ratings must be determined by the manufacturer of the motor starter (or other motor controller). In most instances a heater table, which lists the load current range of the heater identified by a model number, is furnished with the starter.

Code challenge 13–1. A 3-phase, 230-volt, 20-hp squirrel-cage induction motor has a service fator of 1.15 and a nameplate full-load current rating of 51 amps. Determine the full-load current rating of the heaters installed in the separate overload relay.

Motor full-load current rating determination: Ref. (A) ____________

Motor FLC = (B) ____________ amps

Starter-heater overrate computation: Ref. (C) ____________

____________ % of ____________ amps = (D) ____________ amps

Motors that drive high-inertia loads, such as large flywheel assemblies, belt or chain conveyors, or pumps started under load, normally have long acceleration periods (extended starting times) to attain their baseline (running) speed. Running overload protection overrated to the percentages allowed by Sec. 430-32 may heat to the point of tripping out (opening) during a normal start. Section 430-34 allows the percentages to be increased from 115% to 130% for general-purpose motors. For special-rated motors, the 125% overrate can be increased up to 140%.

The overrate percentages listed in Sec. 430-32 can be labeled as the normal or standard overrates for separate running overload protection. When identified as such, the higher percentages listed in Sec. 430-34 should be labeled as NE Code maximum. Any higher rating or setting of the overload relay heater's current rating could not be considered as the required running overload protection.

Example 13–2. A 3-phase, 460-volt, 200-hp wound-rotor induction motor is used to drive a rock crusher. If the motor has a service factor (SF) of 1.15, and a nameplate rated-load current (FLC) of 235 amps, determine the NE Code-maximum full-load current rating of the heaters installed in a separate overload relay.

Motor full-load current rating determination: Ref. 430-6(a)
Motor FLC = 235 amps

Starter-heater overrate computation: Ref. 430-34

140 % of 235 amps = 329 amps

In comparison to Example 13-1, the described load indicated a high-inertia, long-acceleration type start. The maximum rating was also asked for in the question, hence the 140% overrate.

Code challenge 13–2. A 3-phase, 208-volt, 100 hp wound-rotor induction motor is used to drive a belt conveyor for crushed coal. If the motor has a service factor (SF) of 1.15, and a nameplate rated-load current of 270 amps, determine the NE Code-maximum full-load current rating of the heaters installed in a separate overload relay.

Motor full-load current rating determination: Ref. (A) ____________
Motor FLC = (B) ____________ amps

Starter-heater overrate computation: Ref. (C) ____________

____________ % of ____________ amps = (D) ____________ amps

The ac induction motor within a hermetically sealed refrigerant motor compressor can also be protected against running overload by a separate sensing device that is wired in series with the branch-circuit fuses or circuit breaker. Branch-circuit requirements for this type of motor load are found in Art. 440. Section 440-52 lists the same type of devices and allowed overrates for separate overload protection for motor compressors as described for general motor installations in Secs. 430-32 and 430-34.

Instead of full-load current, rated-load current is normally listed on the nameplate of the hermetically sealed refrigerant motor compressor. The procedure shown in these last two examples would be used to determine the current rating of the heater elements in an overload relay for the motor compressor.

13.4 CIRCUIT CONDUCTORS

Section 240-3(f) requires both the branch-circuit conductors (extended from the panel or other load center to the motor controller) and the motor-circuit conductors (extended from the controller to the motor) to be protected against overcurrent in accordance with their rated ampacity (Fig. 13-4). Since the rating of the separate overload heaters is based

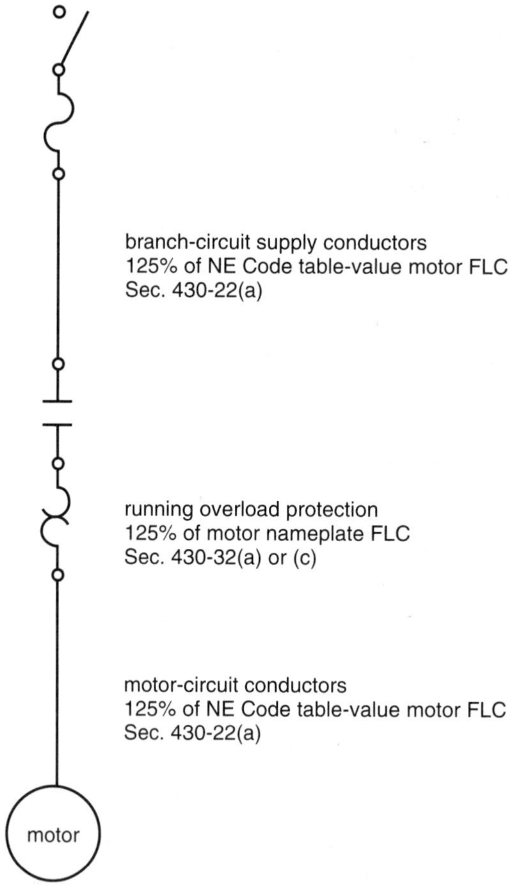

Fig. 13-4 The rated ampacity of the circuit conductors for a motor must equal or exceed the motor's overload protection rating.

on an overrate of the motor full-load current, the rated ampacity of the circuit conductors, according to wire size and any derates imposed by operating at elevated temperatures or loss of heat dissipation, must be determined according to the percentage of overload allowed.

Sections 430-6(a), 430-32, and 440-52 (motor compressors) require the overload-relay heaters to be rated or sized according to the nameplate full-load or rated-load current. In comparison, the branch-circuit or motor-circuit conductors are not sized according to these nameplate values.

As outlined in Chapter 12 on the rating of motor-circuit disconnects, a motor branch circuit is sized according to the horsepower requirements of the driven load. The same type of motor (horsepower, frame size, voltage ratings, etc.) is produced by several manufacturers. The nameplate full-load current of individual motors may vary among different manufacturers according to how each motor was constructed and the materials used.

The running overload heaters (from the previous section) must be sized according to the nameplate full-load current to tailor the circuit overload-overcurrent protection to the specific motor. Section 430-6(a) requires the circuit conductors; the disconnects; the short-circuit, ground-fault overcurrent-protection devices; and any other support or supply equipment for the motor to be sized for the worst-case or highest full-load current of any motor available for installation or replacement.

Tables 430-147 through 430-150 in the NE Code list the worst-case full-load currents of different motors according to the type of power supply (either ac or dc), and according to the voltage, phase (ac), and horsepower rating.

Part B of Art. 430, and Part D of Art. 440 lists the guidelines for sizing motor-circuit conductors.

Although the current rating of the overload relay heaters can range from 115 to 140% of the nameplate full-load current of the motor, Secs. 430-22(a) and 440-32 set the overrate of both the branch-circuit and motor-circuit conductors, for across-the-line (full-voltage) starting, at 125% of the NE Code table-value full-load current.

Example 13–3. A single-phase, 208-volt, 10-hp wound-rotor induction motor is used to drive a sump pump. If the motor has a limited surface temperature rise of not over 40°C, and a nameplate rated-load current of 50 amps, determine the NE Code-standard full-load current rating of the heaters installed in a separate overload relay, and the NE Code-minimum wire size of the THHN copper conductors used for both the branch-circuit and motor-circuit supplies.

Starter-heater overload-current rating determination:

Motor FLC = 50 amps — Ref. 430-6(a)

Overrate computation: — Ref. 430-32(a)(1)

$$125\% \text{ of } 50 \text{ amps} = 62.5 \text{ amps}$$

Circuit-conductor full-load current rating determination: Ref. 430-6(a)

Motor FLC = 55 amps — Ref. T430-148

Circuit-conductor ampacity overrate computation: Ref. 430-22(a)

125% of 55 amps = 68.8 amps

Circuit-conductor wire size determination:
Wire terminal temperature limitation: 60°C Ref. 110-14(c)(1), Excp. 1
Table-rated ampacity: 70 amps Ref. T310-16
Wire size: No. 4 AWG THHN copper

The THHN insulation is listed in T310-16 in the 90°C temperature column (for either copper, aluminum, or c-c aluminum). Section 110-14(c) requires the rated ampacity of the wire (by wire size) to be determined according to the lowest temperature rating of any device in the circuit. Since the temperature rating of the supply or load terminals for the motor circuit was not specified, Sec. 110-14(c)(1) required worst-case table-rated ampacities for NE Code-minimum: The rated ampacity was selected from the 60°C temperature column. Exception No. 1 to Sec. 110-14(c)(1) allowed the higher temperature-rated insulation.

Code challenge 13-3. A single-phase, 230-volt, 7½-hp wound-rotor induction motor is used to drive the circulating pump for a swimming-pool filtration system. If the motor has a limited surface temperature rise of not over 40°C, and a nameplate rated-load current of 38 amps, determine the NE Code-standard full-load current rating of the heaters installed in a separate overload relay, and the NE Code-minimum wire size of the XHH copper conductors use for both the branch-circuit and motor-circuit supplies.

Starter-heater overload-current rating determination:
Motor FLC = (A) ___________ amps Ref. (B) ___________

Overrate computation: Ref. (C) ___________

___________ % of ___________ amps = (D) ___________ amps

Circuit-conductor full-load current rating determination: Ref. (E) ___________

Motor FLC = (F) ___________ amps Ref. (G) ___________

Circuit-conductor ampacity overrate computation: Ref. (H) ___________

___________ % of ___________ amps = (I) ___________ amps

Circuit-conductor wire size determination:
Wire terminal temperature limitation: (J) ___________ °C Ref. (K) ___________
Table-rated ampacity: (L) ___________ amps Ref. (M) ___________
Wire size: (N) ___________ XHH copper

Due to the variable starting requirements of hermetically sealed refrigerant motor compressors, the NE Code does not offer a table listing of worst-case full-load current ratings for motor compressors. Instead, two values of load current are specified on the nameplate. Section 440-2 defines the higher load-current rating (worst case) as the "branch-circuit selection current." When specified on the nameplate, this section requires the branch-circuit selection current to be used in the same manner as the NE Code table-value full-load currents are used for general motor applications.

The branch-circuit selection current will always be greater than or equal to the rated-load current of the hermetically sealed refrigerant motor compressor. The rated-load current is defined in this same section as the current drawn when the motor compressor is operated at rated load, rated voltage, and rated frequency. The value would therefore change if one or more of these parameters vary from the nameplate listing, hence a higher value of branch-circuit selection current.

For the supply of a single hermetically sealed refrigerant motor compressor, Sec. 440-32 requires the rated ampacity of the branch-circuit and motor-circuit conductors, determined by wire size and the terminal temperature limitations of Sec. 110-14(c), to be equal to a minimum of 125% of the nameplate branch-circuit selection current specified on the motor-compressor nameplate.

Example 13–4. The nameplate rated-load current of a 3-phase, 460-volt, 10-ton hermetically sealed refrigerant motor compressor is 12 amps. The nameplate branch-circuit selection current rating is 15 amps. Determine the NE Code-standard full-load current rating of the heaters installed in a separate overload relay, and the NE Code-minimum wire size of the THW copper conductors used for both the branch-circuit and motor-circuit supplies.

Starter-heater overload-current rating determination:
Motor FLC = 12 amps — Ref. 440-52(a)(1)

Overrate computation: (NE Code standard) — Ref. 430-32(a)(1)

125 % of 12 amps = 15 amps

Circuit-conductor full-load current rating determination: — Ref. 440-32

Motor FLC = 15 amps

Circuit-conductor ampacity overrate computation: — Ref. 440-32

125% of 15 amps = 18.8 amps

Circuit-conductor wire size determination:
Wire terminal temperature limitation: 60°C — Ref. 110-14(c)(1), Excp. 1

Table-rated ampacity: 20 amps — Ref. T310-16
Wire size: No. 14 AWG THW copper

In reference to the obelisk footnote to T310-16, the maximum rating of an overcurrent-protection device for No. 14 AWG THW copper is 15 amperes. This restriction technically applies to 15-, 20,- and 30-amp multioutlet branch circuits sized in accordance with Sec. 210-23(a), when lighting, utilization equipment, and receptacle loads will be supplied from a single multioutlet branch circuit. The receptacles represent unknown loads.

If we elect to apply the restriction of the footnote, the No. 14 AWG copper conductors are protected against overload by the setting of the overload-relay heaters at 15 amps.

Sec. 440-52(a)(1) will allow the heaters in the separate overload relay, which protects a hermetically sealed refrigerant motor compressor, to be set at a maximum of 140%. The NE Code standard of 125% was applied in the example in accordance with Sec. 430-32(a)(1).

Code challenge 13–4. The nameplate rated-load current of a 3-phase, 208-volt, 10-ton hermetically sealed refrigerant motor compressor is 27 amps. The nameplate branch-circuit selection current rating is 30 amps. Determine the NE Code-standard full-load current rating of the heaters installed in a separate overload relay, and the NE Code-minimum wire size of the RHW copper conductors used for both the branch-circuit and motor-circuit supplies.

Starter-heater overload-current rating determination:

Motor FLC = (A) ___________ amps   Ref. (B) ___________

Overrate computation: (NE Code standard)   Ref. (C) ___________

___________ % of ___________ amps = (D) ___________ amps

Circuit-conductor full-load current rating determination: Ref. (E) ___________

Motor FLC = (F) ___________ amps

Circuit-conductor ampacity overrate computation: Ref. (G) ___________

___________ % of ___________ amps = (H) ___________ amps

Circuit-conductor wire size determination:

Wire terminal temperature limitation: (I) ___________ °C Ref. (J) ___________

Table-rated ampacity: (K) ___________ amps   Ref. (L) ___________

Wire size: (M) ___________ THW copper

When the heaters in a separate overload relay are sized at 130 to 140 percent of the motor nameplate full-load current to allow successful, repetitive starting of high-inertia, long-acceleration loads, Secs. 430-32 and 430-34, as shown in Fig. 13-5, are technically in NE Code violation of Sec. 240-3 if the rated ampacity of the circuit conductors is overrated only to the 125% computation of Sec. 430-22(a): The conductors can overheat due to the allowed running overload.

Of course, the 130 to 140% is applied to the motor-nameplate full-load current, which is normally of a lower value than the table-value full-load current the 125% is applied to, but the conductors can still overheat due to overload overcurrent and their insulation eventually fail.

The better electrical design will allow for the possibility of the overload heaters be-

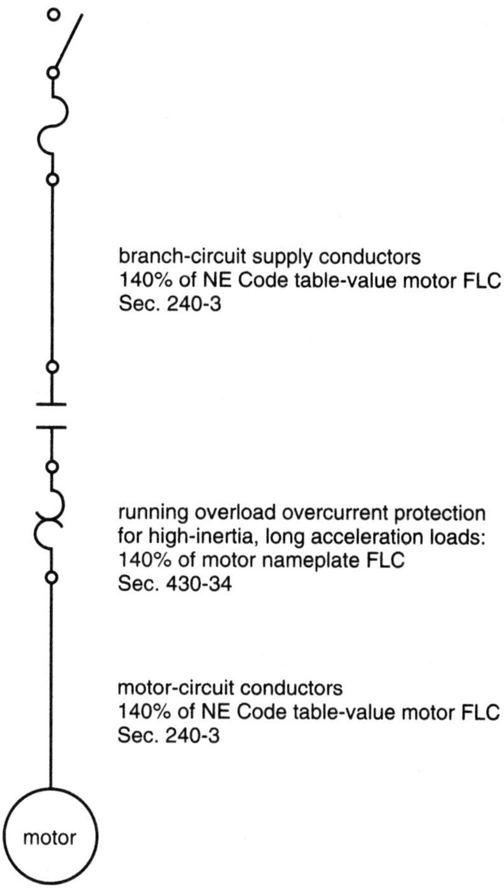

Fig. 13-5 When motors drive high-inertia, long-acceleration loads, the requirements of Sec. 240-3 supercedes the requirements of Sec. 430-22(a).

ing set at the NE Code-maximum of 140%, and disregard Sec. 430-22(a) in favor of Sec. 240-3: The 140% overrate should be applied to the table-value FLC rating to determine the minimum ampacity rating of the circuit conductors.

Example 13–5. A 2-phase, 575-volt, 40-hp ac induction motor will be used to drive a coal-chute conveyor that can be started under load. If the motor has a service factor of 1.15, and a nameplate full-load current rating of 32 amps, determine the NE Code-maximum full-load current rating of the heaters installed in a separate overload relay, and the design-minimum wire size of the XHHW aluminum conductors used for both the branch-circuit and motor-circuit supplies.

Starter-heater overload-current rating determination:

Motor FLC = 32 amps — Ref. 430-6(a)

Overrate computation: — Ref. 430-34

$$140\% \text{ of } 32 \text{ amps} = 44.8 \text{ amps}$$

Circuit-conductor full-load current rating determination: Ref. 430-6(a)
Motor FLC = 36 amps Ref. T430-149

Circuit-conductor ampacity overrate computation: (design) Ref. 240-3

140% of 36 amps = 50.4 amps

Circuit-conductor wire size determination:
Wire terminal temperature limitation: 60°C Ref. 110-14(c)(1), Excp. 1
Table-rated ampacity: 55 amps Ref. T310-16
Wire size: No. 4 AWG XHHW aluminum

For the ampacity overrates of the circuit conductors, Sec. 240-3 overrides the overrate minimum of Sec. 430-22(a) by the 140% overrate of the circuit conductor overload-overcurrent protection allowance of Sec. 430-34.

Code challenge 13–5. A 2-phase, 230-volt, 25 hp ac induction motor will be used to drive a box conveyor than can be started under load. If the motor has a service factor of 1.15, and a nameplate full-load current rating of 55 amps, determine the NE Code-maximum full-load current rating of the heaters installed in a separate overload relay, and the design-minimum wire size of the RHW-2 c-c (copper-clad) aluminum conductors use for both the branch-circuit and motor-circuit supplies.

Starter-heater overload-current rating determination:
Motor FLC = (A) ____________ amps Ref. (B) ____________

Overrate computation: Ref. (C) ____________

____________ % of ____________ amps = (D) ____________ amps

Circuit-conductor full-load current rating determination: Ref. (E) ____________
Motor FLC = (F) ____________ amps Ref. (G) ____________

Circuit-conductor ampacity overrate computation: (design) Ref. (H) ____________

____________ % of ____________ amps = (I) ____________ amps

Circuit-conductor wire size determination:

Wire terminal temperature limitation: (J) ____________ °C Ref. (K) ____________
Table-rated ampacity: (L) ____________ amps Ref. (M) ____________
Wire size: (N) ____________ RHW-2 c-c aluminum

When the motor is rated dc, a means of rectification is required to convert the ac branch-circuit supply into a dc motor supply. The 125% overrate of Sec. 430-22(a) applies to the branch-circuit conductors supplying the drive controller. Exception No. 2 to

Sec. 430-22(a) lists the overrate percentages required for the motor-circuit conductors (extended between the controller and the motor) according to the type of rectification employed.

When a 3-phase ac motor is supplied through a variable-voltage, variable-frequency (variable-speed) drive controller from a 3-phase source, both the branch-circuit supply conductors and the motor-circuit conductors are sized according to the 125% overrate of Sec. 430-22(a).

If the ac drive controller is supplied from a single-phase source and in turn supplies a 3-phase motor, only the motor-circuit conductors on the load side of the phase converter are sized at the 125% overrate of Sec. 430-22(a). The single-phase line-side branch-circuit supply conductors of the phase converter ac drive must be sized according to the requirements of Sec. 455-6.

If the motor is rated as a "duty cycle" motor, the overrate percentage for both the branch-circuit supply and motor-circuit (load) conductors must be determined according to the type of duty service of the motor (row heading) and the minute rating (column heading) of the motor in T430-22(a), Exception.

If the motor is rated as multispeed, the industrial version of the common 3-speed floor fan found in the average home, the second paragraph to Sec. 430-22(a) requires the general 125% overrate to be applied to the highest full-load current rating found on the motor nameplate.

Examples of sizing the branch-circuit and motor-circuit conductors for these special types of motors will be shown with the development of the branch-circuit short-circuit, ground-fault protection requirements in Chapter 14 of this text (Motors—Part II).

For general motor applications, when more than one motor or hermetically sealed refrigerant motor compressor will be supplied by a branch circuit, feeder, or service, Sec. 430-24 requires the rated ampacity of the circuit conductors to be greater than or equal to the 100% rating of all loads supplied, plus a 25% overrate applied to the largest motor FLC of any motor supplied.

Example 13–6. Determine the NE Code-minimum wire size of THW copper conductors used to supply two 20-hp and three 30-hp ac induction motors, supplied from a 3-phase, 3-wire, 575-volt delta distribution.

Motor full-load current rating determinations: Ref. T430-150

20 hp: 22 amps — 30 hp: 32 amps

Circuit-conductor ampacity overrate computation: Ref. 430-24

20 hp:	2×22 amps =	44 amps
30 hp:	3×32 amps =	96 amps
Overrate:	25% of 32 amps =	8 amps
Total:		148 amps

Circuit-conductor wire size determination:

Wire terminal temperature limitation: 75°C — Ref. 110-14(c)(2)

Table-rated ampacity: 150 amps — Ref. T310-16

Wire size: No. 1/0 AWG THW copper

If the feeder circuit in this example will also supply nonmotor loads, such as lighting, heating, or convenience receptacle loads, the full-load current of these loads will have to be computed in accordance Art. 220. For lighting loads, Secs. 220-10(a) and 220-10(b) (computed load) require an overrate to 125% to compensate for continuous load. For heating loads, Sec. 220-15 requires the fixed electric space heating loads to be computed in reference to Art. 424. For convenience receptacle loads, Sec. 220-3(c)(6) requires a minimum load assignment of 180 volt-amps per receptacle outlet.

Code challenge 13–6. Determine the NE Code-minimum wire size of RHW c-c (copper-clad) aluminum conductors used to supply four 10-hp and six 15-hp 3-phase ac induction motors supplied from a 3-phase, 4-wire, 480Y/277-volt power distribution.

Motor full-load current rating determinations: Ref. (A) ____________
10 hp: (B) ____________ amps 15 hp: (C) ____________ amps

Circuit-conductor ampacity overrate computation: Ref. (D) ____________
10 hp: ____________ × ____________ amps = (E) ____________ amps
15 hp: ____________ × ____________ amps = (F) ____________ amps
Overrate: (G) ____________ % of ____________ amps = ____________ amps
Total: (H) ____________ amps

Circuit-conductor wire size determination:
Wire terminal temperature limitation: (I) ____________ °C Ref. (J) ____________
Table-rated ampacity: (K) ____________ amps Ref. (L) ____________
Wire size: (M) ____________ RHW c-c aluminum

13.5 COMBINED OVERCURRENT PROTECTION

An electric motor is not required by the National Electrical Code to be protected against a running overload condition by a separate overcurrent-protection device (OCPD). The branch-circuit fuses or circuit breaker, which are normally overrated to allow successful, repetitive starting, yet provide the necessary short-circuit, ground-fault overcurrent protection, can be sized or rated to provide the necessary overload-overcurrent protection as well. When used as full overcurrent protection for the motor circuit (short-circuit, ground-fault, and overload), Sec. 430-55 refers to this as combined overcurrent protection, and restricts the rating of the branch-circuit OCPD to the overrates allowed by Sec. 430-32 (Fig. 13-6).

As before for separate overload-overcurrent protection ratings, the computed rating of the fuses or circuit breaker for this type of installation should be based on the motor nameplate FLC. If the motor is changed out and replaced with another motor that has a higher or lower FLC rating, the rating of the branch-circuit fuses or circuit breaker will have to be changed accordingly.

Example 13–7. The blower motor for a central vacuum system of a building is rated 3 hp, 230 volts, single-phase. If the motor has a nameplate rated-load current of 15

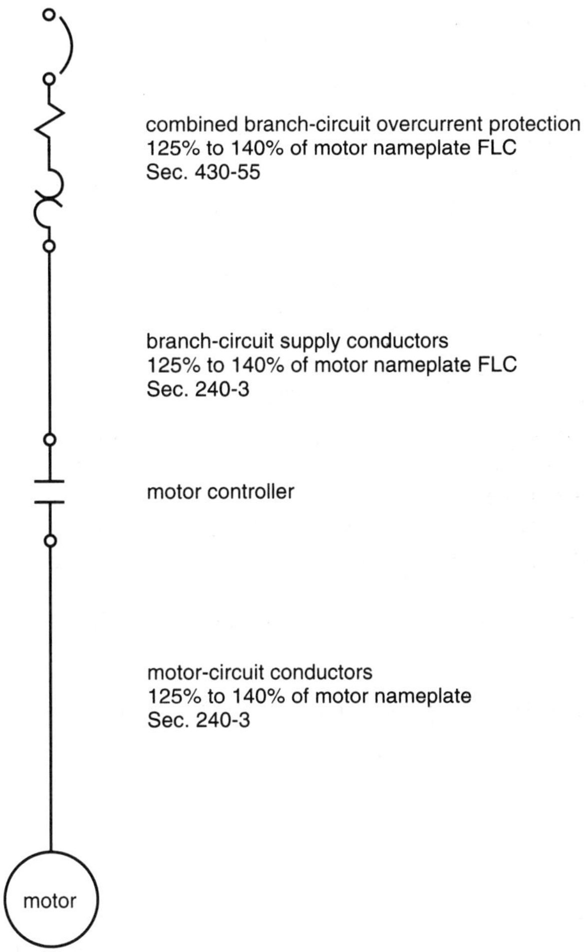

Fig. 13-6 When the overcurrent protection for a motor is combined in the branch-circuit OCPD, the overload overpercentages of Sec. 430-32 are NE Code maximum.

amperes and a service factor of 1.15, determine the NE Code-maximum rating of time-delay fuses installed as the combined overcurrent protection for the motor branch circuit.

Motor full-load current rating determination: Ref. 430-6(a)
Motor FLC = 15 amps

Fuse overrate computation for combined protection: Ref. 430-55
125% of 15 amps = 18.8 amps Ref. 430-32(a)(1)

Standard OCPD ratings: Ref. 240-6(a)
NHS: 20-amp time-delay fuses Ref. (NHS) 430-34

The next higher standard (NHS) size of fuse for the combined overcurrent protection rating, as listed in Sec. 240-6(a), was allowed by the reference to Sec. 430-34 in the second paragraph of Sec. 430-32(a)(1). The window of overrate for the overload-overcurrent protection rating of a motor circuit is between 125% and 140%. When the 125% is applied to the branch-circuit OCPD rating, the NHS should be selected for nonstandard computations. When the 140% overrate yields a nonstandard size or rating of fuse or circuit breaker, the next lower standard (NLS) size must be selected for the rating of combined overcurrent protection.

Although the circuit conductors were not sized in this example, the NE Code rules demonstrated in the previous examples still apply. The branch-circuit conductors would have to be sized at a minimum overrate of 125% of the NE Code table-value full-load current rating for this motor to meet the requirements of Secs. 430-6(a) and 430-22(a).

Code challenge 13–7. The exhaust fan motor located in the wall near the roof line of a high-bay manufacturing building is rated 3 hp, 115 volts, single-phase. If the motor has a nameplate rated-load current of 31 amperes and a service factor of 1.15, determine the NE Code-maximum setting of an inverse-time circuit breaker installed as the combined overcurrent protection for the motor branch circuit.

Motor full-load current rating determination: Ref. (A) ____________

Motor FLC = (B) ________ amps

CB overrate computation for combined protection:

Ref. (C) ________ Ref. (D) ____________

________ % of ________ amps = (E) ________ amps

Standard OCPD ratings: Ref. (F) ____________

NHS: (G) ________ -amp CB Ref. (NHS) (H) ________

This method of using inverse-time circuit breakers or time-delay fuses as the combined overcurrent protection is often used on small equipment when an ac general-use snap switch or a horsepower-rated selector switch is used as the motor controller. Combined overcurrent protection is also used on portable attachment plug and receptacle-supplied equipment where the branch-circuit overcurrent protection serves as the motor running-overload protection.

When the next higher standard size of branch-circuit OCPD above the 125% computation of Sec. 430-32(a)(1) exceeds the 140% cut-off of Sec. 430-34, the branch-circuit fuses or circuit breaker cannot serve as the combined motor-circuit overcurrent protection. The branch circuit must be treated as a normal motor circuit with additional (separate) overload protection, such as thermal cutouts on the motor or supplemental fuses wired in-line (in series with) to the motor must be installed to keep the overload-overcurrent protection rating within the range of 125 to 140%.

Larger motors, which will normally be permanent or "hard-wired," will have either a manual or electromagnetic starter that serves as the controller. These starters will normally have an overload relay wired in series (the heater elements) with the motor on the load side of the starter contactor.

With this type of controller, the overcurrent protection is broken up into short-circuit, ground-fault overcurrent protection, which is provided by the fuses or circuit breakers on the line side (supply side) of the starter (at the origin of the branch circuit), and overload-overcurrent protection, which is provided by a separate overload relay on the load side of the starter.

The requirements of Secs. 430-32, 430-34, and 430-55 apply to motors in general applications. Sections 430-33 and 430-42 apply to motor installations that have a specific purpose, or the rating of the overload protection is modified according to either the load served, or to the horsepower rating of the motor.

13.6 BRANCH-CIRCUIT SHORT-CIRCUIT GROUND-FAULT PROTECTION

When motors are provided with separate, individual running overload-overcurrent protection, the protection against the high-magnitude overcurrents of short-circuit and ground-fault (SC-GF) conditions is provided at both the branch-circuit and feeder levels by either fuses or circuit breakers.

Since motors draw high-magnitude in-rush currents on starting, especially for high-inertia, long-acceleration period loads, the branch-circuit fuses or circuit breaker will normally have high-value load-current ratings or settings as compared to the rated full-load current of the motors. These higher values are necessary to allow repetitive starting without nuisance tripping or opening (fuses) of the branch-circuit protection.

As discussed in Chapter 7, fuses and circuit breakers have both an overload (full-load) rating and a fault (interrupt) rating. The overload-overcurrent rating is determined by the load rating of the fuse or circuit breaker. The fault or interrupt rating, which is determined by the design, construction, and operating characteristics of the protection device, is normally several times higher than the load rating (within a range of 5,000 to 200,000 amperes).

The load rating of the motor branch-circuit fuses or circuit breaker is purposely set at a higher rating than the separate running overload protection current rating to allow the operation of these series-fed overcurrent-protection devices to be selective (the fuse or circuit breaker responds to fault overcurrents, not to running overload overcurrents). The separate overload protection is set within a range of 125 to 140% of the individual motor full-load current. The branch-circuit fuses are rated above these overload percentages with in a range of 175 to 225%. Circuit breakers are set within a range of 250 to 400%.

As with the branch-circuit or motor-circuit conductors, the overrate computations for the fuses or circuit breaker are applied to the NE Code table-value of motor full-load current. The overrate for running overload protection is applied to the individual motor nameplate full-load current.

Part D of Art. 430 governs the selection of motor branch-circuit fault protection (short circuit or ground fault). Section 430-52(b) requires the rating of the branch-circuit fuses or the setting of the circuit breaker to be capable of carrying the starting current of the motor (without opening—nuisance tripping).

The in-rush starting current of a motor can be as much as 600% of the motor's FLC (800% for 3-phase, design E motors). If the motor has a successful start, this high-magnitude in-rush overcurrent is of short duration. The branch-circuit protection and the individual motor overload protection must have some type of built-in time delay that will allow these harmless in-rush starting currents to pass without opening the motor supply circuit. By the same token, the time delay cannot be too long—failure to start will draw high-magnitude overcurrents indefinitely.

Section 430-52(c)(1) requires the branch-circuit fuses or circuit breaker to have a rating or setting determined by the overrate percentages listed in T430-152. The table, listed under Part M and located at the end of Art. 430, is a grid of columns and rows. The row headings list the types of motors for which the branch-circuit overcurrent protection is to be sized. The other four columns list the overpercentages allowed, according to the type of overcurrent-protection device to be installed.

In determining the type of motor, several questions must be answered. Is the motor ac or dc? If the motor is dc, then the allowed overpercentages or overrates will be found listed according to the type of overcurrent-protection device selected under the last main-row heading.

If the motor is ac, then several more questions have to be answered. What is the phase rating of the motor? If the motor is rated 3-phase, what is the design letter rating of the motor? Is the motor a standard squirrel-cage induction motor or a wound-rotor induction motor? If the motor is rated as wound rotor, is it rated synchronous (no slip) or standard induction (a low percentage of slip)? The answers to all these questions will determine the maximum overrate of branch-circuit SC-GF protection allowed for a particular motor.

The first row of T430-152 lists the percentages allowed according to the type of OCPD for all single-phase ac motors, regardless of their voltage rating. The second row heading, which contains subheadings for the next three rows, list the overrate percentages for either 2-phase or 3-phase (or other polyphase, special-purpose) ac motors.

For 2-phase ac motors, the percentages are listed according to "squirrel-cage: other than design E." For 3-phase motors, the overrate percentage to use depends on whether the motor is not rated as design E (same percentages as 2-phase), or it is rated as design E (separate row of overrate percentages).

The use of Sec. 430-52(c) and T430-152 to determine the overrate percentages allowed for the SC-GF overcurrent protection is limited to an individual branch circuit supplying a single motor. Section 430-52(c) cannot be used to size the branch-circuit overcurrent protection unless separate overload protection is provided for the motor.

As with the overrates allowed by Secs. 430-32 and 430-34, Sec. 430-52(c)(1) offers both a normal or standard overrate of the branch-circuit fuse or circuit breaker, and for high-inertia, long-acceleration loads, an NE Code-maximum overrate to allow repetitive starting without nuisance opening of the branch-circuit overcurrent-protection device.

For standard overrates, the overpercentages are determined by T430-152, according to the type of motor and the type of overcurrent-protection device. The fuse rating or circuit breaker setting is chosen from the list of standard device ratings listed in Sec. 240-6(a). When the overrate computation yields a nonstandard size of fuse or circuit breaker, Exception No. 1 to Sec. 430-52(c)(1) will allow the next higher standard size or rating to be used.

For high-inertia, long-acceleration loads, Exception No. 2 to Sec. 430-52(c)(1) lists the NE Code-maximum overrate percentages of the branch-circuit OCPD according to the type of overcurrent protection. When the overrate computation, using the percentages found in either subsection a, b, c, or d of the exception, yields a nonstandard rating of fuses or setting of a circuit breaker, the next lower standard size must be used.

If the motor branch circuit will be protected with time-delay fuses (Fig. 13-7), the normal or standard overrate is listed in T430-152 at either 150 or 175% according to the type of motor installed. The NE Code-maximum value of overrate for time-delay fuses is listed in Sec. 430-52(c)(1), Excp. No. 2(b) at 225%.

The rated ampacity of the branch-circuit supply or motor-circuit conductors and the resulting wire size is not governed by these higher ratings or settings of motor-circuit short-circuit, ground-fault protection. As mentioned earlier, to use these higher overrate values of branch-circuit overcurrent protection requires separate running overload protection for the motor.

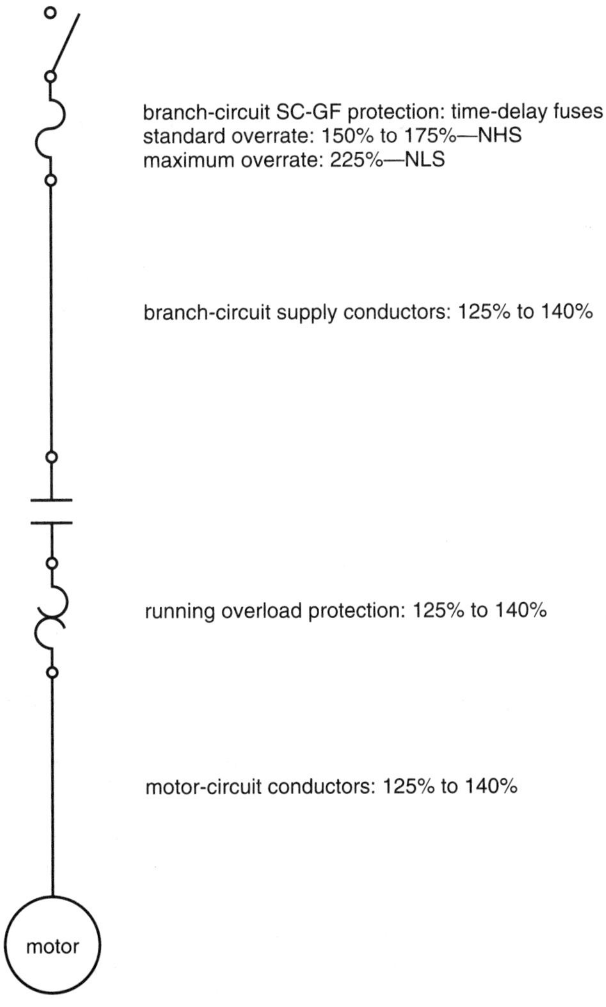

Fig. 13-7 The window of the ratings for time-delay fuses protecting a single motor is from 150% to 225%.

As discussed in Section 13.4, Sec. 430-22(a) requires the rated ampacity of both the branch-circuit supply and motor-circuit conductors to be computed at a minimum of 125% of the NE Code table-value motor FLC to parallel the allowed overrates for the overload protection listed in Sec. 430-32.

For high-inertia, long-acceleration motor loads, Sec. 240-3 requires the rated-load ampacity of the branch-circuit supply and motor-circuit conductors to be computed at a miminimum of 140% to parallel the allowed overrates of Sec. 430-34.

The standard overrate as found in T430-152 for the branch-circuit SC-GF overcurrent-protection ratings correspond to the standard overload percentages listed in 430-32. The rated ampacity of the branch-circuit and motor-circuit conductors should be computed at a minimum of 125% of the motor FLC as required by Sec. 430-22(a).

For high-inertia, long-acceleration starting, the NE Code-maximum overrate of Sec. 430-52(c)(1), Excp. No. 2 corresponds to the maximum overload percentages allowed by Sec. 430-34. For time-delay fuses, the table-value motor FLC should be overrated to 225%, and their rating selected according to the listings in Sec. 240-6(a) (with the NLS for nonstandard computations).

The minimum rated ampacity of the branch-circuit supply and motor-circuit conductors should be computed at a minimum of 140% of the motor FLC as required by Sec. 240-3. Using T310-16, the selected wire size of the circuit conductors must have a table-rated ampacity greater than or equal to the computed minimum rated-load ampacity.

Example 13–8. For a 3-phase, 60-hp, 460-volt, design B ac induction motor with a nameplate-rated limited temperature rise of not over 40°C, determine:

1. Both the standard and the NE Code-maximum rating of time-delay fuses used to provide the branch-circuit SC-GF protection of the motor
2. The NE Code-minimim wire size of the branch-circuit and motor-circuit conductors for each size of time-delay fuse selected using THW copper

Motor full-load current rating determination: Ref. T430-150

Motor FLC = 77 amps

SC-GF protection rating determinations: Ref. 430-52(c)(1)

Standard overrate computation: Ref. T430-152

175% of 77 amps = 134.8 amps

Standard OCPD ratings: Ref. 240-6(a)

NHS: 150 amps Ref. (NHS) 430-52(c)(1), Excp. 1

NE Code-maximum overrate computation: Ref. 430-52(c)(1), Excp. 2(b)

225% of 77 amps = 173.2 amps

NE Code-maximum OCPD rating: Ref. 240-6(a)

NLS: 150 amps Ref. (NLS) 430-52(c)(1), Excp. 2(b)

Standard circuit-conductor ampacity overrate computation:
Ref. 430-32(a)(1) Ref. 430-22(a)

125% of 77 amps = 96.3 amps

Standard circuit-conductor wire size determination:
Wire terminal temperature limitation: 60°C Ref. 110-14(c)(1), Excp. 1
Table-rated ampacity: 110 amps Ref. T310-16
Wire size: No. 1 AWG THW copper

NE Code-maximum circuit-conductor ampacity overrate computation:
Ref. 430-34 Ref. 240-3

140% of 77 amps = 107.8 amps

NE Code-maximum circuit-conductor wire size determination:
Wire terminal temperature limitation: 75°C Ref. 110-14(c)(2)
Table-rated ampacity: 115 amps Ref. T310-16
Wire size: No. 2 AWG THW copper

Either the No. 1 or the No. 2 AWG THW copper conductors connected to the load side of the fuseholder containing 150-amp time-delay (TD) fuses may appear as an NE Code violation when installed in the industrial plant or other manufacturing concern, but the practicing electrician must remember this is a motor supply circuit. Both the branch-circuit and motor-circuit conductors are protected within their rated ampacity by separate overload-overcurrent devices, whose maximum rating (at 140%) will not exceed the 110- or 115-amp rating of the conductors.

Because of the wire terminal temperature limitations of Sec. 110-14(c) and the magic cutoff of 100 amperes, increasing the 125% overrate in this example to 140% pushed the overload protection rating of the circuit conductors past the 100-amp cutoff of Sec. 110-14(c)(1).

Using the 75°C insulation-temperature column to determine the rated ampacity of the circuit conductors as allowed by Sec. 110-14(c)(2) actually resulted in a smaller wire size for a computed higher value of current. But this is NE Code legal.

Should the electrician elect to stay with the 60°C insulation temperature column, the overrate to 140% for the running overload protection would increase the wire size of the circuit conductors back to No. 1 AWG THW copper.

Code challenge 13–8. For a 3-phase, 25-hp, 575-volt, design E ac induction motor with a nameplate-rated limited temperature rise of not over 40°C, determine:

1. Both the standard and the NE Code-maximum rating of time-delay fuses used to provide the branch-circuit SC-GF protection of the motor
2. The NE Code-minimim wire size of the branch-circuit and motor-circuit conductors for each size of time-delay fuse selected using XHHW copper

Motor full-load current rating determination: Ref. (A) ______________
Motor FLC = (B) __________ amps

SC-GF protection rating determinations: Ref. (C) ______________
Standard overrate computation: Ref. (D) ______________

__________ % of __________ amps = (E) __________ amps

Standard OCPD ratings: Ref. (F) ______________
NHS: (G) __________ amps Ref. (NHS) (H) ______________

NE Code-maximum overrate computation: Ref. (I) ______________

__________ % of __________ amps = (J) __________ amps

NE Code-maximum OCPD rating: Ref. (K) ______________
NLS: (L) __________ amps Ref. (NLS) (M) ______________

Standard circuit-conductor ampacity overrate computation:
Ref. (N) __________ Ref. (O) ______________

__________ % of __________ amps = (P) __________ amps

Standard circuit-conductor wire size determination:
Wire terminal temperature limitation: (Q) ____°C Ref. (R) ______________
Table-rated ampacity: (S) __________ amps Ref. (T) ______________
Wire size: (U) __________ XHHW copper

NE Code-maximum circuit-conductor ampacity overrate computation:
Ref. (V) __________ Ref. (W) ______________

__________ % of __________ amps = (X) __________ amps

NE Code-maximum circuit-conductor wire size determination:
Wire terminal temperature limitation: (Y) ____°C Ref. (Z) ______________
Table-rated ampacity: (1A) __________ amps Ref. (1B) ______________
Wire size: (1C) __________ XHHW copper

If the motor branch circuit will be protected by nontime fuses (Fig. 13-8), the standard overrate is listed in T430-152 at either 150% or 300%, depending on the type of motor. The NE Code-maximum value of overrate is listed in Sec. 430-52(c)(1), Excp. No. 2(a) at 400%. If the standard overrate computation is greater than 600 amps, Excp. No. 2(d) sets the NE Code-maximum overrate at 300% instead of 400%.

Example 13–9. Determine the standard and maximum ratings of nontime fuses used to provide the branch-circuit short-circuit, ground-fault protection for a single-phase, 208-volt, ac squirrel-cage induction motor rated at 10 hp.

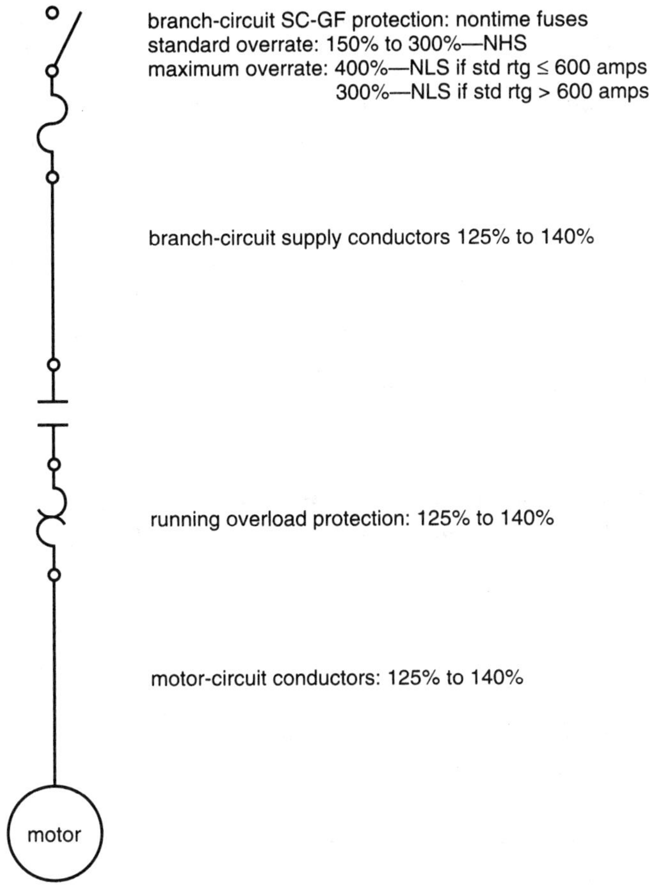

Fig. 13-8 The window of the ratings for non-time fuses protecting a single motor is from 150% to 400%.

Motor full-load current rating determination: Ref. T430-148
Motor FLC = 55 amps

SC-GF protection rating determinations: Ref. 430-52(c)(1)
Standard overrate computation: Ref. T430-152
300% of 55 amps = 165 amps

Standard OCPD ratings: Ref. 240-6(a)
NHS: 175 amps Ref. (NHS) 430-52(c)(1), Excp. 1

NE Code-maximum overrate computation: Ref. 430-52(c)(1), Excp. 2(a)

400% of 55 amps = 220 amps

NE Code-maximum OCPD rating: Ref. 240-6(a)
NLS: 200 amps Ref. (NLS) 430-52(c)(1), Excp. 2(a)

The branch-circuit and motor-circuit conductors for this motor installation would be sized in the same manner as the previous example: The rated-ampacity of the circuit conductors should be computed at 125% if the fuses are sized at 175 amps, and at 140% if the instant (nontime-delay) fuses are sized at 200 amps.

Code challenge 13–9. Determine the standard and maximum ratings of non-time fuses used to provide the branch-circuit short-circuit, ground-fault protection for a single-phase, 200-volt, ac squirrel-cage induction motor rated at 7 1/2 hp.

Motor full-load current rating determination: Ref. (A) ____________

Motor FLC = (B) ________ amps

SC-GF protection rating determinations: Ref. (C) ____________
Standard overrate computation: Ref. (D) ____________

________ % of ________ amps = (E) ________ amps

Standard OCPD ratings: Ref. (F) ____________
NHS: (G) ________ amps Ref. (NHS) (H) ____________

NE Code-maximum overrate computation: Ref. (I) ____________

________ % of ________ amps = (J) ________ amps

NE Code-maximum OCPD rating: Ref. (K) ____________
NLS: (L) ________ amps Ref. (NLS) (M) ____________

If the motor branch circuit will be protected by an inverse-time circuit breaker (Fig. 13-9), the standard overrate is listed in T430-152 at either 150% or 250%, depending on the type of motor. The NE Code-maximum value of overrate is listed in Sec. 430-52(c)(1), Excp. No. 2(c)(1) at 400% for motors rated at 100 amps or less (FLCs), and 300% [Excp. No. 2(c)(2)] for motors rated in excess of 100 amps.

Example 13–10. Determine the standard and maximum settings of an inverse-time circuit breaker used for the branch-circuit short-circuit, ground-fault overcurrent protection for a 2-phase, 230-volt, ac squirrel-cage induction motor rated at 20 hp.

Motor full-load current rating determination: Ref. T430-149
Motor FLC = 47 amps

SC-GF protection rating determinations: Ref. 430-52(c)(1)
Standard overrate computation: Ref. T430-152

250% of 47 amps = 117.5 amps

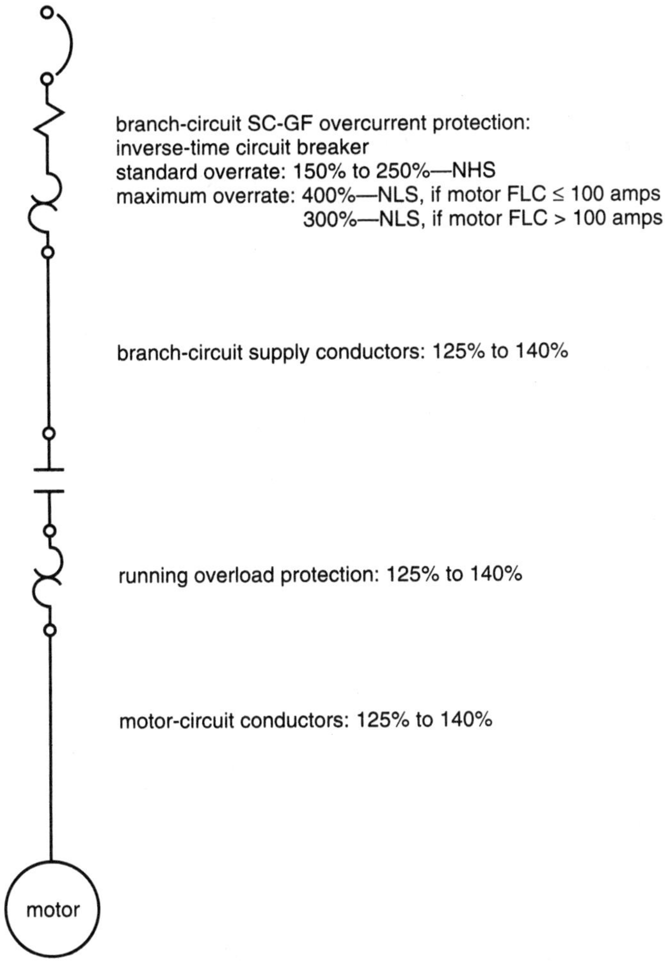

Fig. 13-9 The window of the ratings for an inverse-time circuit breaker protecting a single motor is from 150% to 400%.

Standard OCPD ratings:	Ref. 240-6(a)
NHS: 125 amps	Ref. (NHS) 430-52(c)(1), Excp. 1

NE Code-maximum overrate computation:	Ref. 430-52(c)(1), Excp. 2(c)(1)

$$400\% \text{ of } 47 \text{ amps} = 188 \text{ amps}$$

NE Code-maximum OCPD rating:	Ref. 240-6(a)
NLS: 175 amps	Ref. (NLS) 430-52(c)(1), Excp. 2(c)(1)

For horsepower ratings that yield table-value motor FLCs in excess of 100 amperes, the standard 250% overrate of T430-152 with the next higher standard setting of CB as allowed by Sec. 430-52(c)(1), Excp. No. 1, and the limited 300% maximum overrate imposed by Sec. 430-52(c)(1), Excp. No. 2(c)(2) with the next lower standard setting as listed in Sec. 240-6(a) will normally equate to the same setting or rating of inverse-time circuit breaker.

Code challenge 13–10. Determine the standard and maximum settings of an inverse-time circuit breaker used for the branch-circuit short-circuit, ground-fault overcurrent protection for a 2-phase, 575-volt, ac squirrel-cage induction motor rated at 100 hp.

Motor full-load current rating determination: Ref. (A) ____________
Motor FLC = (B) ____________ amps

SC-GF protection rating determinations: Ref. (C) ____________
Standard overrate computation: Ref. (D) ____________

____________ % of ____________ amps = (E) ____________ amps

Standard OCPD ratings: Ref. (F) ____________
NHS: (G) ____________ amps Ref. (NHS) (H) ____________

NE Code-maximum overrate computation: Ref. (I) ____________

____________ % of ____________ amps = (J) ____________ amps

NE Code-maximum OCPD rating: Ref. (K) ____________
NLS: (L) ____________ amps Ref. (NLS) (M) ____________

If the motor branch circuit will be protected by an instantaneous-trip circuit breaker (Figure 13-10), the normal overrate is listed in T430-152 at 250% for dc motors, and 800% for all ac motors except design E. Design E motors have a standard overrate of 1,100% because of their higher value of in-rush starting current.

The NE Code-maximum values of overrate for instantaneous-trip circuit breakers are listed in Sec. 430-52(c)(3), Excp. No. 1. For 3-phase design B, C, or D ac induction motors which drive high-inertia, long-acceleration loads, the 800% can be increased to 1,300%. For 3-phase design E ac induction motors which drive high-inertia, long-acceleration loads, the 1,100 in T430-152 can be increased to a maximum of 1,700%.

There is no allowed increase over the table values for synchronous or wound rotor ac induction motors, nor dc motors when an instantaneous-trip circuit breaker is used as the short-circuit, ground-fault overcurrent protection. The increase to the 1,300% for 3-phase ac induction motors other than design E and 1,700% for 3-phase design E motors is only allowed after engineering evaluation of the proposed installation.

As discussed in Chapter 7, an instantaneous-trip circuit breaker uses a magnetic-trip latch to clear the circuit in the event of an overcurrent. It does not contain a separate device with time-delayed response to momentary, low-level overload. Its overload and interrupt thresholds are one and the same. The response rating of the instantaneous-trip circuit breaker must be set above the locked-rotor current rating of the motor (above 600 to 800% for design E).

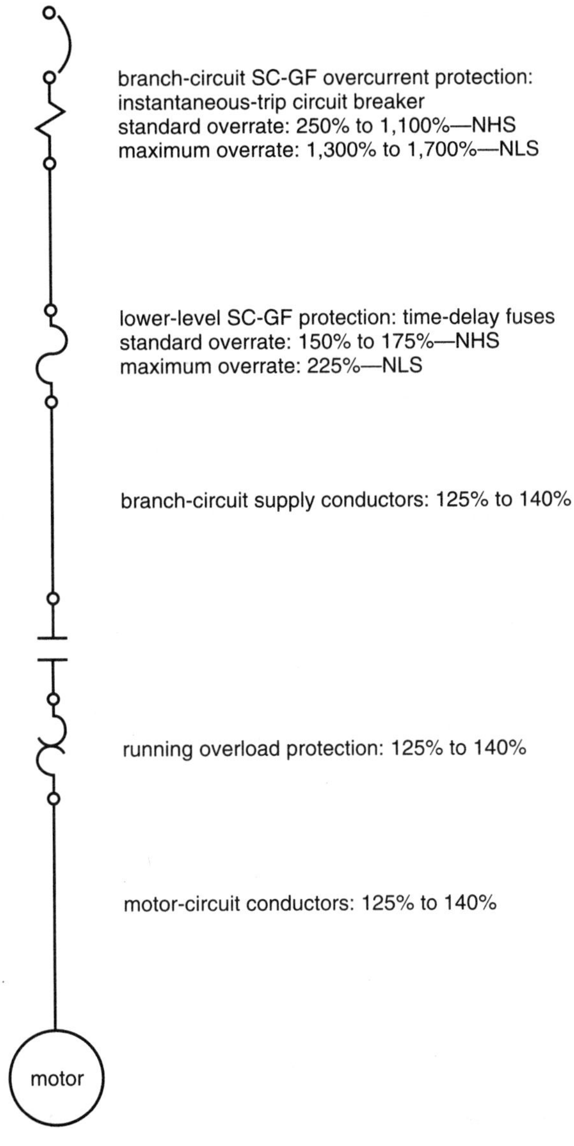

Fig. 13-10 Because of the high-value overrates allowed, an instantaneous-trip circuit breaker protecting a single motor must be supplemented with fuse protection.

The instantaneous-trip circuit breaker can only be used as supplemental, high-level fault protection for the motor circuit. Section 430-52(c)(3) will allow an instantaneous-trip circuit breaker to be used as the single-motor branch-circuit short-circuit, ground-fault protection only if its setting is adjustable and part of a combination controller which contains coordinated overload and additional, lower-level short-circuit, ground-fault protection in each circuit conductor.

In other words, the instantaneous-trip circuit breaker can only be used in conjunction with either time-delay or nontime fuses and a separate overload relay. The relay provides the necessary low-level overload-overcurrent protection. The fuses provide the low-level short-circuit, ground-fault overcurrent protection. The CB provides the high-level short-circuit, ground-fault overcurrent protection.

Example 13–11. If a 3-phase, 460-volt, 60-hp, design E ac squirrel-cage induction motor has a service factor of 1.15 and a nameplate full-load current rating of 74 amperes, determine:

1. The standard and maximum settings of an instantaneous-trip circuit breaker used for the high-level branch-circuit SC-GF overcurrent protection
2. The standard and maximum ratings of time-delay fuses used for the low-level branch-circuit SC-GF overcurrent protection
3. The standard and maximum settings of thermal heaters installed in a separate overload relay
4. The standard and maximum wire size of the XHHW copper branch-circuit and motor-circuit conductors

Motor FLC determination for the branch-circuit SC-GF protection and the rated ampacity of the circuit conductors:

77 amps Ref. 430-6(a) Ref. T430-150

Motor FLC determination for the thermal overload heaters to be installed in a separate overload relay:

74 amps Ref. 430-6(a)

Branch-circuit SC-GF protection setting determinations for the instantaneous-trip circuit breaker:

Ref. 430-52(c)(3)

Standard overrate computation: Ref. T430-152

$$1{,}100\% \text{ of } 77 \text{ amps} = 847 \text{ amps}$$

Standard OCPD ratings: Ref. 240-6(a)

NHS: 1,000 amps Ref. (NHS) 430-52(c)(3), Excp. 1

NE Code-maximum overrate computation: Ref. 430-52(c)(3), Excp. 1

$$1{,}700\% \text{ of } 77 \text{ amps} = 1{,}309 \text{ amps}$$

NE Code-maximum OCPD rating: Ref. 240-6(a)

NLS: 1,200 amps Ref. (NLS) 430-52(c)(3), Excp. 1

Branch-circuit SC-GF protection setting determinations for the TD fuses:

Ref. 430-52(c)(1)

Standard overrate computation: Ref. T430-152

175% of 77 amps = 134.8 amps

Standard OCPD ratings: Ref. 240-6(a)
NHS: 150 amps Ref. (NHS) 430-52(c)(1), Excp. 1

NE Code-maximum overrate computation: Ref. 430-52(c)(1), Excp. 2(b)

225% of 77 amps = 173.3 amps

NE Code-maximum OCPD rating: Ref. 240-6(a)
NLS: 150 amps Ref. (NLS) 430-52(c)(1), Excp. 2(b)

Starter-heater load-current overrate computations:
Standard rating: Ref. 430-32(a)(1)

125% of 74 amps = 92.5 amps

NE Code-maximum rating: Ref. 430-34

140% of 74 amps = 103.6 amps

Standard circuit-conductor ampacity overrate computation:
Ref. 430-32(a)(1) Ref. 430-22(a)

125% of 77 amps = 96.3 amps

Standard circuit-conductor wire size determination:
Wire terminal temperature limitation: 60°C Ref. 110-14(c)(1), Excp. 1
Table-rated ampacity: 110 amps Ref. T310-16
Wire size: No. 1 AWG XHHW copper

NE Code-maximum circuit-conductor ampacity overrate computation:
Ref. 430-34 Ref. 240-3

140% of 77 amps = 107.8 amps

NE Code-maximum circuit-conductor wire size determination:
Wire terminal temperature limitation: 75°C Ref. 110-14(c)(2), Excp. 1
Table-rated ampacity: 115 amps Ref. T310-16
Wire size: No. 2 AWG XHHW copper

Code challenge 13–11. If a 230-volt, 3-phase, 50-hp, design E ac squirrel-cage induction motor has a service factor of 1.15 and a nameplate full-load current rating of 126 amperes, determine:

1. The standard and maximum settings of an instantaneous-trip circuit breaker used for the high-level branch-circuit SC-GF overcurrent protection
2. The standard and maximum ratings of time-delay fuses used for the low-level branch-circuit SC-GF overcurrent protection
3. The standard and maximum settings of thermal heaters installed in a separate overload relay
4. The standard and maximum wire size of the RHW-2 copper branch-circuit and motor-circuit conductors.

Motor FLC determination for the branch-circuit SC-GF protection and the rated ampacity of the circuit conductors:

(A) ____________________ amps Ref. (B) ____________________
Ref. (C) ____________________

Motor FLC determination for the thermal overload heaters to be installed in a separate overload relay:

(D) ____________ amps Ref. (E) ____________________

Branch-circuit SC-GF protection setting determinations for the instantaneous-trip circuit breaker: Ref. (F) ____________________

Standard overrate computation: Ref. (G) ____________________

____________ % of ____________ amps = (H) ____________ amps

Standard OCPD ratings: Ref. (I) ____________________
NHS: (J) ____________ amps Ref. (NHS) (K) ____________________

NE Code-maximum overrate computation: Ref. (L) ____________________

____________ % of ____________ amps = (M) ____________ amps

NE Code-maximum OCPD rating: Ref. (N) ____________________
NLS: (O) ____________ amps Ref. (NLS) (P) ____________________

Branch-circuit SC-GF protection setting determinations for the TD fuses:
Ref. (Q) ____________________

Standard overrate computation: Ref. (R) ____________________

____________ % of ____________ amps = (S) ____________ amps

Standard OCPD ratings: Ref. (T) ____________________
NHS: (U) ____________ amps Ref. (NHS) (V) ____________________

NE Code-maximum overrate computation: Ref. (W) ____________________

____________ % of ____________ amps = (X) ____________ amps

NE Code-maximum OCPD rating: Ref. (Y) ____________

NLS: (Z) ____________ amps Ref. (NLS) (1A) ____________

Starter heater load-current overrate comutations:

Standard rating: Ref. (1B) ____________

____________ % of ____________ amps = (1C) ____________ amps

NE Code-maximum rating: Ref. (1D) ____________

____________ % of ____________ amps = (1E) ____________ amps

Standard circuit-conductor ampacity overrate computation:

Ref. (1F) ____________ Ref. (1G) ____________

____________ % of ____________ amps = (1H) ____________ amps

Standard circuit-conductor wire size determination:

Wire terminal temperature limitation: (1I) ____________ °C

Ref. (1J) ____________

Table-rated ampacity: (1K) ____________ amps Ref. (1L) ____________

Wire size: (1M) ____________ RHW-2 copper

NE Code-maximum circuit-conductor ampacity overrate computation:

Ref. (1N) ____________ Ref. (1O) ____________

____________ % of ____________ amps = (1P) ____________ amps

NE Code-maximum circuit-conductor wire size determination:

Wire terminal temperature limitation: (1Q) ____________ °C

Ref. (1R) ____________

Table-rated ampacity: (1S) ____________ amps Ref. (1T) ____________

Wire size: (1U) ____________ RHW-2 copper

Section 440-22 governs the rating or setting of the branch-circuit short-circuit, ground-fault protection for hermetically-sealed refrigerent motor-compressors. For an individual motor compressor (Fig. 13-11), the rating of either time-delay or nontime fuses, or of an inverse-time circuit breaker are all set at a standard of 175% and a maximum of 225%. These overrates are parallel to the overrates allowed for time-delay fuses in general motor applications [T430-152 and Sec. 430-52(c)(1), Excp. No. 2(b)].

The motor compressor must be provided with separate running overload protection sized in reference to Sec. 440-52. The rated ampacity of the branch-circuit supply and motor-circuit conductors must be determined by Sec. 440-32 when the standard 125% overload overrate is used in conjunction to the 175% SC-GF overrate setting, and by Sec. 240-3 when the 140% overload overrate is used in conjunction with the 225% SC-GF overrate setting (for starting under load against head pressure).

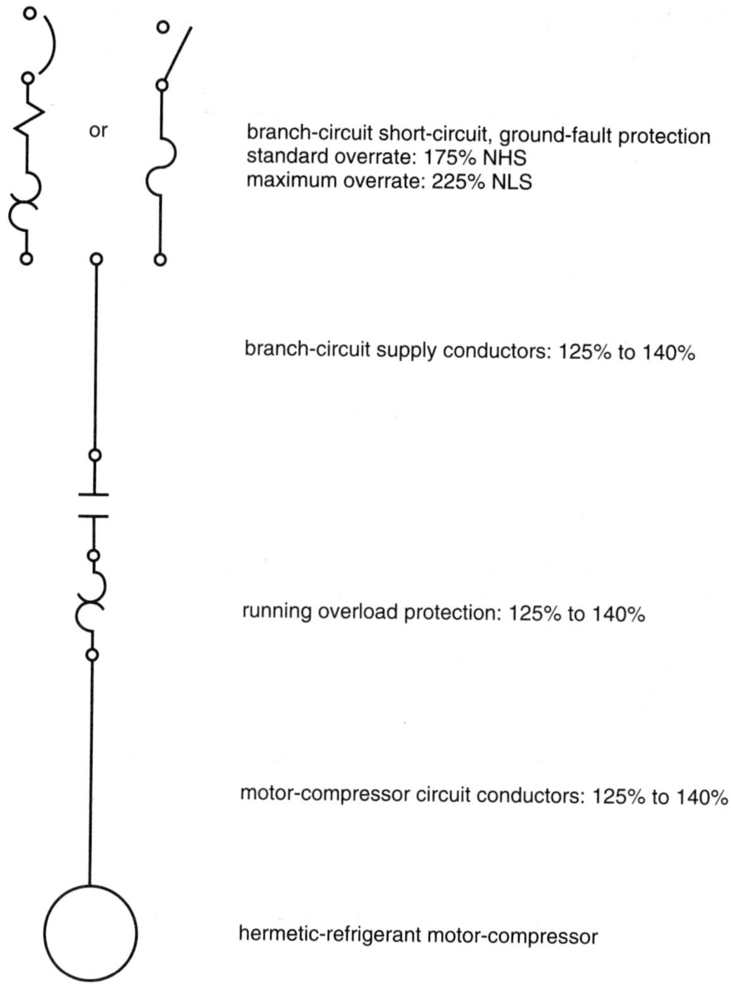

Fig. 13-11 The window of branch-circuit SC-GF protection for a single motor compressor is 175% to 225%, regardless of the type of OCPD used.

The overrate for the motor-circuit running overload protection must be applied to the rated-load current of the hermetically sealed motor compressor as stated on the nameplate. The branch-circuit selection current, when specified on the nameplate, must be used for both the overrate of the SC-GF overcurrent protection and the overrate of the rated-ampacity of the circuit conductors.

Example 13–12. If a hermetically sealed refrigerant motor-compressor has a nameplate rated-load current of 80 amperes and a branch-circuit selection current rating of 88 amperes, determine:

1. The standard and maximum ratings of time-delay fuses used for the branch-circuit SC-GF overcurrent protection

2. The standard and maximum settings of thermal heaters installed in a separate overload relay
3. The standard and maximum wire size of the RHW copper-clad aluminum branch-circuit and motor-circuit conductors

Motor FLC determination for the branch-circuit SC-GF protection and the rated ampacity of the circuit conductors:
88 amps Ref. 440-22(a) Ref. 440-32

Motor FLC determination for the thermal overload heaters to be installed in a separate overload relay:
80 amps Ref. 440-52(a)(1)

Branch-circuit SC-GF protection setting determinations for the TD fuses:
Ref. 440-22(a)

Standard overrate computation:

175% of 88 amps = 154 amps

Standard OCPD ratings: Ref. 240-6(a)
NHS: 175 amps Ref. (NHS) 440-22(a)

NE Code-maximum overrate computation:

225% of 88 amps = 198 amps

NE Code-maximum OCPD rating: Ref. 240-6(a)
NLS: 175 amps Ref. (NLS) 440-22(a)

Starter heater overload-current overrate computations:
Standard rating: Ref. 430-32(a)(1)

125% of 80 amps = 100 amps

NE Code-maximum rating: Ref. 440-52(a)(1)

140% of 80 amps = 112 amps

Standard circuit-conductor ampacity overrate computation:
Ref. 430-32(a)(1) Ref. 440-32

125% of 88 amps = 110 amps

Standard circuit-conductor wire size determination:
Wire terminal temperature limitation: 75°C Ref. 110-14(c)(2)
Table-rated ampacity: 120 amps Ref. T310-16
Wire size: No. 1/0 AWG RHW c-c aluminum

NE Code-maximum circuit-conductor ampacity overrate computation:
Ref. 440-52(a)(1) Ref. 240-3

140% of 88 amps = 123.2 amps

NE Code-maximum circuit-conductor wire size determination:
Wire terminal temperature limitation: 75°C Ref. 110-14(c)(2)
Table-rated ampacity: 135 amps Ref. T310-16
Wire size: No. 2/0 AWG RHW c-c aluminum

Instead of the standard overrate of the branch-circuit short-circuit, ground-fault overcurrent protection for the motor compressor being located in one part of the NE Code and the maximum overrate value being located in another part, both are found in Sec. 440-22(a) for the individual single hermetically sealed motor compressor. The standard overrate of 175% allowed the next higher size because a range was given up to the 225%. The next lower size was specified and referenced by the same section for the maximum overrate of 225% because an exception that would allow the next higher size was not listed.

Code challenge 13–12. If a hermetically sealed refrigerant motor compressor has a nameplate rated-load current of 54 amperes and a branch-circuit selection current rating of 59 amperes, determine:

1. The standard and maximum settings of an inverse-time circuit breaker used for the branch-circuit SC-GF overcurrent protection
2. The standard and maximum settings of thermal heaters installed in a separate overload relay
3. The standard and maximum wire size of the THW aluminum branch-circuit and motor-circuit conductors

Motor FLC determination for the branch-circuit SC-GF protection and the rated ampacity of the circuit conductors:
(A) ____________ amps Ref. (B) ____________________
Ref. (C) ____________________

Motor FLC determination for the thermal overload heaters to be installed in a separate overload relay:
(D) ____________ amps Ref. (E) ____________________

Branch-circuit SC-GF protection setting determinations for the inverse-time circuit breaker: Ref. (F) ____________________

Standard overrate computation:

____________ % of ____________ amps = (G) ____________ amps

Standard OCPD ratings: Ref. (H) ____________________
NHS: (I) ____________ amps Ref. (NHS) (J) ____________________

NE Code-maximum overrate computation:

____________ % of ____________ amps = (K) ____________ amps

NE Code-maximum OCPD rating: Ref. (L) ____________________
NLS: (M) ____________amps Ref. (NLS) (N) ____________________

Starter heater overload-current overrate computations:
Standard rating: Ref. (O) ____________________

____________ % of ____________ amps = (P) ____________amps

NE Code-maximum rating: Ref. (Q) ____________________

____________ % of ____________ amps = (R) ____________ amps

Standard circuit-conductor ampacity overrate computation:
Ref. (S) ____________________ Ref. (T) ____________________

____________ % of ____________ amps = (U) ____________ amps

Standard circuit-conductor wire size determination:
Wire terminal temperature limitation: (V) ____________ °C
Ref. (W) ____________________
Table-rated ampacity: (X) ____________ amps Ref. (Y) ____________________
Wire size: (Z) ____________ THW aluminum

NE Code-maximum circuit-conductor ampacity overrate computation:
Ref. (1A) __________________________ Ref. (1B) ____________________

____________ % of ____________ amps = (1C) ____________ amps

NE Code-maximum circuit-conductor wire size determination:
Wire terminal temperature limitation: (1D) ____________ °C
Ref. (1E) ____________________
Table-rated ampacity: (1F) ____________ amps Ref. (1G) ____________________
Wire size: (1H) ____________ THW aluminum

These standard and maximum overrates of the branch-circuit fuses or circuit breaker are the ceiling or highest values allowed by the National Electrical Code. The NE Code does not require that all motors or motor compressors be protected at these higher ratings or settings of overcurrent protection. Many factors govern the actual size of overcurrent-protection device selected. The main consideration should be the type of driven load and its torque requirements (under both starting and running conditions).

The wire size of the branch-circuit conductors is fixed at the 125% overrate of the motor FLC (140% for good design and high-inertia loads) plus any derates that might apply due to the prevailing conditions (higher or lower ambient temperatures, more than three conductors in a raceway, or voltage-drop considerations) that hamper conductor heat dissipation. The size of the branch-circuit and motor-overcurrent protection needs only to be high enough to ensure repetitive motor starts without nuisance tripping.

The full-load currents of motors with horsepower ratings greater than 200 horsepower (500 for 3-phase) will not be found in Tables 430-147 through 430-150. The requirement of Sec. 430-6(a) to use the table-value full-load motor current for sizing the motor disconnect, controllers, conductors, and overcurrent protection is void for these larger horsepower-rated motors. Instead, the motor nameplate full-load current is used to size all the supply and control equipment for the motor.

13.7 FEEDER SHORT-CIRCUIT GROUND-FAULT PROTECTION

When several motors are supplied from a common feeder circuit, the motor-feeder short-circuit, ground-fault (SC-GF) protection device ratings are determined by the guidelines and requirements of Sec. 430-62.

For a specific motor load (Fig. 13-12), the overrate rating or setting (above total motor FLC) of feeder overcurrent protection, as outlined in Sec. 430-62(a), allows for the high-value in-rush starting currents of the motors as it simultaneously provides SC-GF protection for both the motors and the feeder conductors.

Both motor and feeder-conductor overload-overcurrent protection is provided on an individual motor basis at the branch-circuit level by the separate overload relays (or other overload-overcurrent protection devices) installed in each motor controller (starter).

The rating or setting of the feeder-circuit fuses or circuit breaker must be based on the summation of the largest standard rating of branch-circuit fuses or circuit breaker computed for the general motor application (T430-152) or the motor-compressor [Sec. 440-22(a)], and the full-load current rating (100%) of all the other motors supplied.

Most of Art. 440 governs the branch-circuit requirements for hermetically sealed refrigerant motor compressors. Section 440-7 requires either the rated-load current (FLC) of the motor-compressor, or (Sec. 440-7, Excp.) the branch-circuit selection current of the motor-compressor when specified on the nameplate to be treated as the full-load current of a single motor when determining:

1. The minimum ampacity of motor feeder and/or branch-circuit conductors by Sec. 430-24
2. The maximum rating or setting of the branch-circuit SC-GF protection for motors in accordance Secs. 430-53 (b) and (c)
3. The maximum rating or setting of the feeder SC-GF protection in accordance with Sec. 430-62(a)

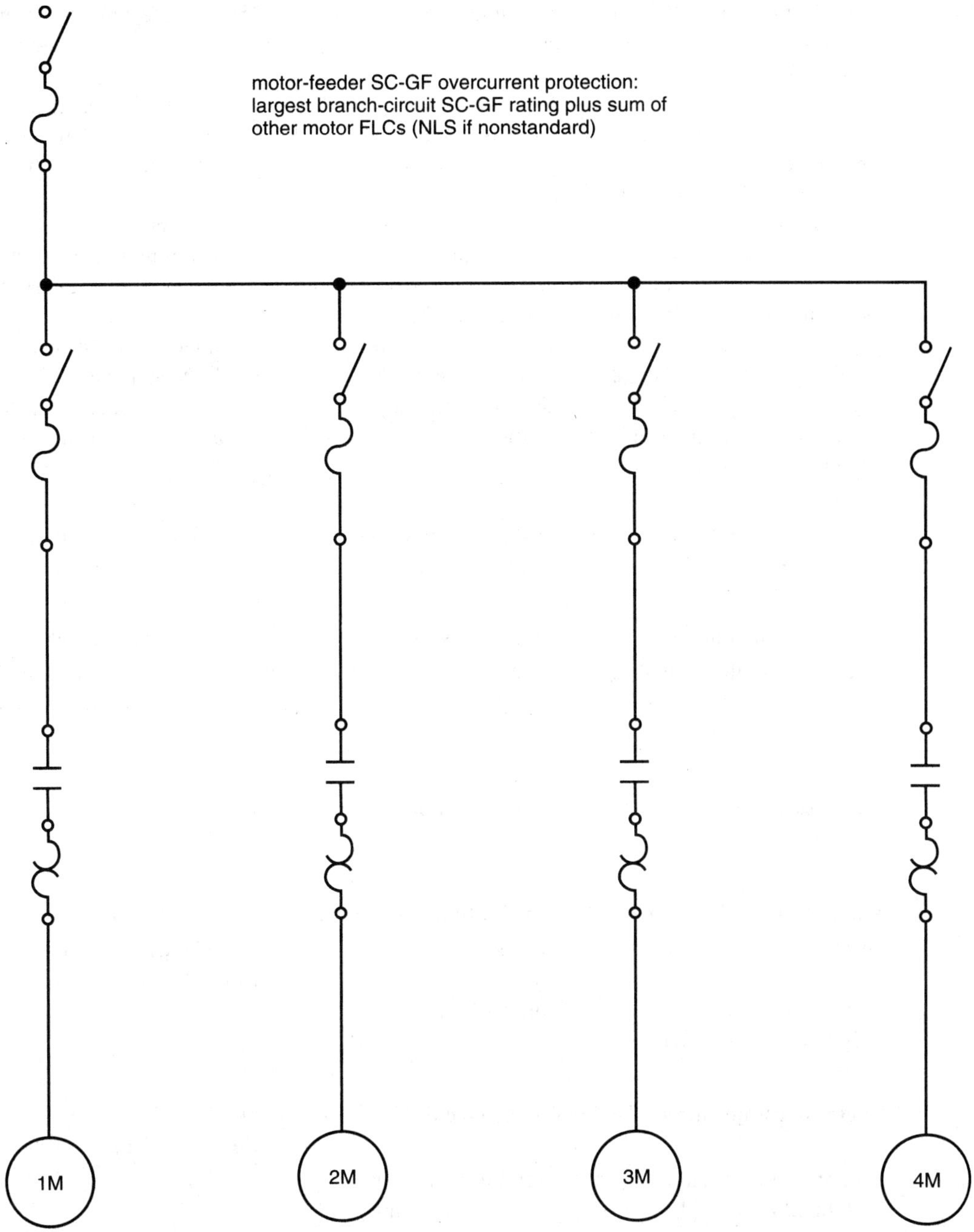

Fig. 13-12 The rating of the motor feeder SC-GF protection is determined by Sec. 430-62(a).

Section 440-22(a) allows the same overrate value of 175% for the branch-circuit SC-GF protection of a motor compressor, regardless of the type of overcurrent protection employed.

If the computation of the feeder SC-GF protection overrate yields a nonstandard size of overcurrent-protection device as listed in Sec. 240-6(a), Sec. 430-62(a) requires the next lower standard size of fuses or circuit breaker to be installed for the feeder short-circuit, ground-fault overcurrent protection.

Section 430-24 requires the ampacity of the feeder-circuit conductors, determined according to selected wire size, to be greater than or equal to the summation of all the motor currents with all the motors operating at the full-load condition, with the largest motor in the maximum allowed overload condition.

In other words, the rated ampacity of the feeder-circuit conductors must at least equal 100% of the full-load current of all the motors supplied by the feeder, plus 25% of the largest motor FLC. By good design, if the largest motor is allowed to operate in the NE Code-maximum overload condition of 140%, then the overrate should be taken at 40%, instead of the 25%.

Example 13–13. Two 20-hp and two 40-hp, 3-phase, 460-volt design B ac squirrel-cage induction motors are to be supplied from a common feeder. Determine:

1. The maximum size of time-delay fuses for the motor-feeder SC-GF protection
2. The minimum wire size of the feeder conductors; use single insulated XHHW-2 copper conductors installed in IMC conduit, which passes through an ambient air temperature of 125°F

Motor full-load current rating determinations: Ref. T430-150

20 hp: 27 amps
40 hp: 52 amps

Motor-feeder SC-GF protection overrate determinations: Ref. 430-62(a)

Overrate computations for the largest motor branch-circuit SC-GF protection:
Ref. T430-152

20 hp: 175% of 27 amps = 47.3 amps
40 hp: 175% of 52 amps = 91 amps

Overrate computations for the feeder-circuit SC-GF protection:
Ref. 430-62(a)

Largest branch-circuit protection rating:	91 amps
20 hp: 2 × 27 amps =	54 amps
40 hp:	52 amps
Total:	197 amps

NE Code-maximum OCPD rating: Ref. 240-6(a)

NLS: 175 amps Ref. (NLS) 430-62(a)

Circuit-conductor ampacity overrate computation: Ref. 430-24

20 hp:	2 × 27 amps =	54 amps
40 hp:	2 × 52 amps =	104 amps
Overrate:	25% of 52 amps =	13 amps
Total:		171 amps

Circuit-conductor wire size determination:

Higher ambient temperature overrate computation: Ref. T310-16,CF

$$171 \text{ amps} \div 0.76 = 225 \text{ amps}$$

Table-rated ampacity: 225 amps

Wire size: No. 3/0 AWG XHHW-2 copper Ref. T310-16

Wire terminal temperature limitation: 75°C Ref. 110-14(c)(2), Excp. 1

Table-rated ampacity: 175 amps

Wire size: No. 2/0 AWG XHHW-2 copper Ref. T310-16

NE Code-minimumn wire size: Ref. 310-15(c)

Compare for larger: No. 3/0 AWG XHHW-2 copper

Some NE Code sources, such as Example No. 8 in Part B of Chapter 9 in the NE Code, deem that Exception No. 1 to Sec. 430-53(c)(1) is also allowed by the requirements of Sec. 430-62(a) such that the 91 amps in the example should be increased to the next higher standard size of 100 amps of fuse or circuit breaker for the summation in Sec. 430-62(a). This would make the maximum overrate equal to 206 amps. The next lower standard size will increase to 200 amps.

In some instances, the use of the next higher size for the largest branch-circuit SC-GF rating will not yield a higher size in the feeder SC-GF rating. Since the wording of Sec. 430-62(a) indicates that the percentage overrate computation allowed by T430-152 is to be taken as the maximum, in this text we will not use the NHS for the largest branch-circuit rating determination.

Code challenge 13–13. Two 30-hp and three 40-hp, 3-phase, 230-volt design D ac squirrel-cage induction motors are to be supplied from a common feeder. Determine:

1. The maximum size of time-delay fuses for the motor-feeder SC-GF protection
2. The minimum wire size of the feeder conductors; use single insulated THHN aluminum conductors installed in rigid metal conduit, which passes through an ambient air temperature of 115°F.

Motor full-load current rating determinations: Ref. (A) ____________________

30 hp: (B) ____________ amps 40 hp: (C) ____________ amps

Motor-feeder SC-GF protection overrate determinations:
Ref. (D) ____________________

Overrate computations for the largest motor branch-circuit SC-GF protection:
Ref. (E) ____________________

30 hp: ____________ % of ____________ amps = (F) ____________ amps
40 hp: ____________ % of ____________ amps = (G) ____________ amps

Overrate computations for the feeder-circuit SC-GF protection:
Ref. (H) ____________________

Largest branch-circuit protection rating: ____________ amps
30 hp: ____________ × ____________ amps = ____________ amps
40 hp: (I) ____________ × ____________ amps = ____________ amps
Total: (J) ____________ amps

NE Code-maximum OCPD rating: Ref. (K) ____________________
NLS: (L) ____________ amps Ref. (NLS) (M) ____________________

Circuit-conductor ampacity overrate computation:
Ref. (N) ____________________

30 hp: ____________ × ____________amps = ____________ amps
40 hp: ____________ × ____________amps = ____________ amps
Overrate: (O) ____________ % of ____________ amps = ____________ amps
Total: (P) ____________ amps

Circuit-conductor wire size determination:

Higher ambient temperature overrate computation:
Ref. (Q) ____________________

____________ amps ÷ ____________ = (R) ____________ amps

Table-rated ampacity: (S) ____________ amps
Wire size: (T) ____________ THHN aluminum Ref. (U) ____________________

Wire terminal temperature limitation: (V) ____________ °C ____________Ref. (W)
Table-rated ampacity: (X) ____________ amps
Wire size: (Y) ____________ THHN aluminum Ref. (Z) ____________________

NE Code-minimumn wire size: Ref. (1A)____________________
Compare for (1B) ____________: (1C) ____________ THHN aluminum

In both the Example and the Code challenge, the largest-rated branch-circuit short-circuit, ground-fault protection device corresponded to the largest motor supplied by the feeder. This won't always be the case.

The overrate percentage allowed by T430-152 is determined by both the type of motor and the type of overcurrent-protection device that will be employed. Because a mixture of different types of motors can be supplied by a common feeder, the difference in percentages could cause a lower horsepower-rated motor to have a higher rating of branch-circuit SC-GF protection as compared to a higher horsepower-rated motor. This is why it is important, as shown in both the Example and the Code challenge, to compute the individual branch-circuit SC-GF overcurrent-protection rating for each motor.

Example 13–14. Four 3-phase, 230-volt, ac induction motors are to be supplied by a common feeder. Two of the motors are rated design B squirrel-cage at 15 hp. The other two motors are rated wound-rotor at 25 hp. Determine the maximum rating of the feeder SC-GF protection using nontime fuses.

Motor full-load current rating determinations: Ref. T430-150
15 hp: 42 amps 25 hp: 68 amps

Motor-feeder SC-GF protection overrate determinations: Ref. 430-62(a)
Overrate computations for the largest motor branch-circuit SC-GF protection:
Ref. T430-152

15 hp: 300% of 42 amps = 126 amps
25 hp: 150% of 68 amps = 102 amps

Overrate computations for the feeder-circuit SC-GF protection:
Ref. 430-62(a)

Largest branch-circuit protection rating:	126 amps
15 hp:	42 amps
25 hp: 2 × 68 amps =	136 amps
Total:	304 amps

NE Code-maximum OCPD rating: Ref. 240-6(a)
NLS: 300 amps Ref. (NLS) 430-62(a)

The overrate computation of the feeder protection device is not based on the largest or greatest horsepower-rated motor, but on the largest standard branch-circuit protection rating of any motor in the group. Both dc (constant voltage) motors and ac wound-rotor induction motors have lower overrate percentages when compared to either single-phase or polyphase ac squirrel-cage induction motors. If an instantaneous-trip circuit breaker will be used for the feeder SC-GF protection, 3-phase design E squirrel-cage induction motors have a higher overrate percentage value (1,100%) than any other type of motor listed in T430-152.

Code challenge 13–14. Five 3-phase, 460-volt, ac induction motors are to be supplied by a common feeder. Three of the motors are rated design D squirrel-cage at 60 hp. The other two motors are rated wound-rotor at 75 hp. Determine the maximum rating of the feeder SC-GF protection using an inverse-time circuit breaker.

Motor full-load current rating determinations: Ref. (A) ____________
60 hp: (B) ____________ amps 75 hp: (C) ____________ amps

Motor-feeder SC-GF protection overrate determinations:
Ref. (D) ____________

Overrate computations for the largest motor branch-circuit SC-GF protection:
Ref. (E) ____________

60 hp: ________ % of ________ amps = (F) ________ amps
75 hp: ________ % of ________ amps = (G) ________ amps

Overrate computations for the feeder-circuit SC-GF protection:
Ref. (H) ____________

Largest branch-circuit protection rating: ________ amps
60 hp: (I) ________ × ________ amps = ________ amps
75 hp: ________ × ________ amps = ________ amps
Total: (J) ________ amps

NE Code-maximum OCPD rating: Ref. (K) ____________
NLS: (L) ________ amps Ref. (NLS) (M) ____________

The overcurrent protection should be "like protection." In other words, the branch-circuit protection could be fuses or circuit breakers. If the feeder protection is specified as fuses, then the overrates for the largest normal branch-circuit protection requirement of Sec. 430-62(a) should be determined from all the branch circuits being protected by fuses, regardless of what type of branch-circuit protection is actually installed. If the feeder protection is going to be an inverse-time circuit breaker, then the largest standard branch-circuit SC-GF overcurrent protection rating should be determined from the overrates for an inverse-time circuit breaker.

Example 13–15. Two 3-phase, design E squirrel-cage induction motors rated at 10 hp, two 3-phase wound-rotor ac induction motors rated at 20 hp, and one 3-phase, 30-ton hermetically sealed refrigerant motor compressor will be supplied by a common 460-volt feeder. Determine:

1. The standard and maximum ratings of an instantaneous-trip circuit breaker protecting the individual branch circuits of the two 10-hp motors
2. The standard and maximum ratings of time-delay fuses also protecting the individual branch circuits of the two 10-hp motors
3. The standard and maximum ratings of nontime fuses protecting the individual branch circuits of the two 20-hp motors
4. The standard and maximum ratings of time-delay fuses protecting the branch circuit of the motor compressor for a nameplate rated-load current of 24 amps and a branch-circuit selection current of 27 amps
5. The maximum setting of the inverse-time circuit breaker protecting the motor feeder

Motor full-load current rating determinations: Ref. T430-150
10 hp: 14 amps 20 hp: 27 amps
Motor-compressor: 27 amps Ref. 440-22(a)

1. **For the 10-hp motors: Individual branch-circuit SC-GF protection setting determinations for the instantaneous-trip CB:**

Ref. 430-52(c)(3)

Standard overrate computation: Ref. T430-152

1,100% of 14 amps = 154 amps

Standard OCPD ratings: Ref. 240-6(a)
NHS: 175 amps Ref. (NHS) 430-52(c)(3), Excp. 1

NE Code-maximum overrate computation: Ref. 430-52(c)(3), Excp. 1

1,700% of 14 amps = 238 amps

NE Code-maximum OCPD rating: Ref. 240-6(a)
NLS: 225 amps Ref. (NLS) 430-52(c)(3), Excp. 1

2. **For the 10-hp motors: Individual branch-circuit SC-GF protection rating determinations for the TD fuses:**

Ref. 430-52(c)(1)

Standard overrate computation: Ref. T430-152

175% of 14 amps = 24.5 amps

Standard OCPD ratings: Ref. 240-6(a)
NHS: 25 amps Ref. (NHS) 430-52(c)(1), Excp. 1

NE Code-maximum overrate computation: Ref. 430-52(c)(1), Excp. 2(b)

225% of 14 amps = 31.5 amps

NE Code-maximum OCPD rating: Ref. 240-6(a)
NLS: 30 amps Ref. (NLS) 430-52(c)(1), Excp. 2(b)

3. **For the 20-hp motors: Individual branch-circuit SC-GF protection setting determinations for the nontime fuses:**

Ref. 430-52(c)(1)

Standard overrate computation: Ref. T430-152

150% of 27 amps = 40.5 amps

Standard OCPD ratings: Ref. 240-6(a)
NHS: 45 amps Ref. (NHS) 430-52(c)(1), Excp. 1

NE Code-maximum overrate computation: Ref. 430-52(c)(1), Excp. 2(a)

400% of 27 amps = 108 amps

NE Code-maximum OCPD rating: Ref. 240-6(a)
NLS: 100 amps Ref. (NLS) 430-52(c)(1), Excp. 2(a)

4. For the 30-ton motor-compressor branch-circuit SC-GF protection rating determinations for the TD fuses:
Ref. 440-22(a)

Standard overrate computation: 175% of 27 amps = 47.3 amps
Standard OCPD ratings: Ref. 240-6(a)
NHS: 50 amps Ref. (NHS) 440-22(a)

NE Code-maximum overrate computation: 225% of 27 amps = 60.8 amps
NE Code-maximum OCPD rating: Ref. 240-6(a)
NLS: 60 amps Ref. (NLS) 440-22(a)

5. Motor-feeder SC-GF protection overrate determinations using an inverse-time CB:
Ref. 430-62(a)

Overrate computations for the largest motor branch-circuit SC-GF protection:
Ref. T430-152 Ref. 440-22(a)

10 hp:	250% of 14 amps = 35.0 amps
20 hp:	150% of 27 amps = 40.5 amps
Motor-compressor:	175% of 27 amps = 47.3 amps

Overrate computations for the feeder-circuit SC-GF protection:
Ref. 430-62(a)

Largest branch-circuit protection rating:	47.3 amps
10 hp: 2 × 14 amps =	28.0 amps
20 hp: 2 × 27 amps =	54.0 amps
Total	129.3 amps

NE CODE-maximum OCPD rating: Ref. 240-6(a)
NLS: 125 amps Ref. (NLS) 430-62(a)

Code challenge 13–15. Two 3-phase, design B squirrel-cage induction motors rated at 25 hp, two 3-phase wound-rotor ac induction motors rated at 30 hp, and one 3-phase, 40-ton hermetically sealed refrigerant motor compressor will be supplied by a common 208-volt feeder. Determine:

1. The standard and maximum ratings of an instantaneous-trip circuit breaker protecting the individual branch circuits of the two 25-hp motors

2. The standard and maximum ratings of time-delay fuses also protecting the individual branch circuits of the two 25-hp motors
3. The standard and maximum ratings of nontime fuses protecting the individual branch circuits of the two 30-hp motors
4. The standard and maximum ratings of time-delay fuses protecting the branch circuit of the motor compressor for a nameplate rated-load current of 87 amps and a branch-circuit selection current of 90 amps
5. The maximum setting of the inverse-time circuit breaker protecting the motor feeder

Motor full-load current determinations: Ref. (A) ____________
25 hp: (B) ____________ amps 30 hp: (C) ____________ amps
Motor-compressor: (D) ____________ amps Ref. (E) ____________

1. **For the 25-hp motors: Individual branch-circuit SC-GF protection setting determinations for the instantaneous-trip CB:**
Ref. (F) ____________

Standard overrate computation: Ref. (G) ____________

____________ % of ____________ amps = (H) ____________ amps

Standard OCPD ratings: Ref. (I) ____________
NHS: (J) ____________ amps Ref. (NHS) (K) ____________

NE Code-maximum overrate computation: Ref. (L) ____________

____________ % of ____________ amps = (M) ____________ amps

NE Code-maximum OCPD rating: Ref. (N) ____________
NLS: (O) ____________ amps Ref. (NLS) (P) ____________

2. **For the 25-hp motors: Individual branch-circuit SC-GF protection rating determinations for the TD fuses:**
Ref. (Q) ____________

Standard overrate computation: Ref. (R) ____________

____________ % of ____________ amps = (S) ____________ amps

Standard OCPD ratings: Ref. (T) ____________
NHS: (U) ____________ amps Ref. (NHS) (V) ____________

NE Code-maximum overrate computation: Ref. (W) ____________

____________ % of ____________ amps = (X) ____________ amps

NE Code-maximum OCPD rating: Ref. (Y) ____________________
NLS: (Z) ____________ amps Ref. (NLS) (1A) ____________________

3. **For the-30 hp motors: Individual branch-circuit SC-GF protection setting determinations for the nontime fuses:**

Ref. (1B) ____________________

Standard overrate computation: Ref. (1C) ____________________

____________ % of ____________ amps = (1D) ____________ amps

Standard OCPD ratings: Ref. (1E) ____________________
NHS: (1F) ____________ amps Ref. (NHS) (1G) ____________________

NE Code-maximum overrate computation: Ref. (1H) ____________________

____________ % of ____________ amps = (1I) ____________ amps

NE Code-maximum OCPD rating: Ref. (1J) ____________________
NLS: (1K) ____________ amps Ref. (NLS) (1L) ____________________

4. **For the 40-ton motor-compressor branch-circuit SC-GF protection rating determinations for the TD fuses:**

Ref. (1M) ____________________

Standard overrate computation:

____________ % of ____________ amps = (1N) ____________ amps

Standard OCPD ratings: Ref. (1O) ____________________
NHS: (1P) ____________ amps Ref. (NHS) (1Q) ____________________

NE Code-maximum overrate computation:

____________ % of ____________ amps = (1R) ____________ amps

NE Code-maximum OCPD rating: Ref. (1S) ____________________
NLS: (1T) ____________ amps Ref. (NLS) (1U) ____________________

5. **Motor-feeder SC-GF protection overrate determinations using an inverse-time CB:**

Ref. (1V) ____________________

Overrate computations for the largest motor branch-circuit SC-GF protection:
Ref. (1W) ____________________ Ref. (1X) ____________________
25 hp: (1Y) __________ % of __________ amps = __________ amps
30 hp: (1Z) __________ % of __________ amps = __________ amps
Motor-compressor: (2A) __________ % of __________ amps = __________ amps

Overrate computations for the feeder-circuit SC-GF protection:

Ref. (2B) ________________

Largest branch-circuit protection rating:	(2C) ____________	amps
25 hp:	____________	amps
30 hp: (2D) ____________ × ____________ amps =	____________	amps
Motor-compressor:	____________	amps
Total:	(2E) ____________	amps

NE Code-maximum OPCD rating: Ref. (2F) ________________

NLS: (2G) ____________ amps Ref. (NLS) (2H) ________________

Sometimes a large QMB (quick-make-and-break) panel is installed with an ampere rating much greater than the total amperage of a group of motors actually supplied from that panel. Sometimes feeders with ampacities greater than a group of motors actually supplied are pulled to a common supply point for "add-ons" or modifications of machinery in the future.

Section 430-62(b) states that when larger-ampacity feeders are installed for reserved capacity, the size (setting or rating) of the feeder overcurrent-protection device can be based on the ampacity of the feeders instead of the actual connected motor load. For this type of installation, Sec. 430-24 would be disregarded. The rated ampacity of the feeder conductors would have to be at least equal to the feeder overcurrent-protection rating or setting.

Many times at the design stage of a proposed electrical installation, the actual connected motor loads for a motor control center or a central panel are determined, and then the next available standard size panel or other type of control center is selected. The rating or setting of the feeder overcurrent protection, and the rated ampacity of the feeder-circuit conductors are then selected to meet or equal the rating of the panel or control center.

13.8 FEEDER SHORT-CIRCUIT GROUND-FAULT PROTECTION—COMBINATION LOADS

When a predominantly motor feeder circuit will also supply power and lighting loads, the feeder short-circuit, ground-fault protection must be sized in accordance with Sec. 430-63.

The feeder loads are divided into two groups: motor and nonmotor. The motor loads are computed in accordance with Art. 430, Secs. 430-52 through 430-62. The nonmotor loads are computed in accordance with Art. 220, Sec. 220-10(b).

The computation of the motor current depends on whether only a single motor or several motors are to be supplied from the common motor feeder. By reference to Sec. 430-52 in Sec. 430-63, if the motor load is a single motor (Fig. 13-13), the motor contribution to the motor-feeder SC-GF protection overrate is computed as the NE Code-maximum of the branch-circuit SC-GF overcurrent-protection rating or setting as determined by either Excp. No. 2 to Sec. 430-52(c)(1), or by Excp. No. 1 to Sec. 430-52(c)(3) if the feeder will be protected by an instantaneous-trip circuit breaker. If the motor load is several motors, the motor contribution is computed by the guidelines of Sec. 430-62(a).

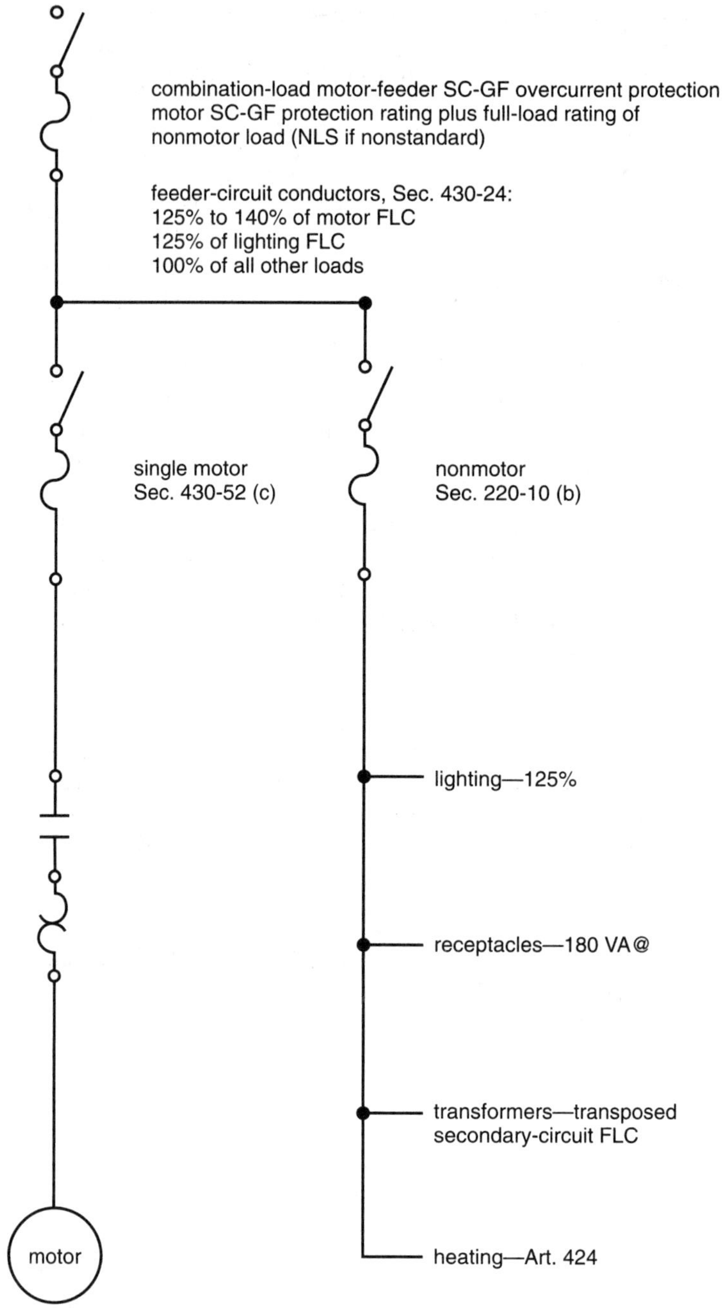

Fig. 13-13 The rating of the combination motor feeder SC-GF protection supplying a single motor is determined by Sec. 430-63.

The computation of the nonmotor load current for the motor-feeder overcurrent protection overrate must follow the guidelines of Sec. 220-10(b): The continuous full-load currents must be overrated to 125% and added to the noncontinuous full-load currents to determine the nonmotor contribution to the motor-feeder SC-GF protection overrate.

To satisfy the requirements of Sec. 430-63, the two subtotals of motor and nonmotor overrate currents are added together. The resulting total is the maximum size or rating of short-circuit, ground-fault overcurrent protection allowed for the motor feeder. If the summation results in a nonstandard size or rating of fuses or circuit breaker, then the next lower standard size as listed in Sec. 240-6(a) must be used.

Example 13–16 (single motor). A 3-phase, 4-wire, 208Y/120-volt, 150-amp panel located in an equipment room will supply the following loads:

1. Overhead lighting rated at 20 amps
2. Convenience receptacles rated at 10 amps
3. A 3-phase, 208-volt, design B ac induction motor rated at 25 hp

Determine:

1. For a high-inertia, long-acceleration load, the maximum rating of time-delay fuses for the branch-circuit SC-GF overcurrent protection of the motor
2. The maximum setting of the motor-feeder SC-GF overcurrent protection using an inverse-time circuit breaker
3. The minimum wire size of the feeder-circuit conductors using THHN copper

Motor full-load current: 74.8 amps — Ref. T430-150

1. Branch-circuit SC-GF protection rating determination using TD fuses:

NE Code-maximum overrate computation: Ref. 430-52(c)(1), Excp. 2(b)

225% of 74.8 amps = 168.3 amps

NE Code-maximum OCPD rating: Ref. 240-6(a)

NLS: 150 amps — Ref. (NLS) 430-52(c)(1), Excp. 2(b)

2. Feeder-circuit SC-GF protection rating determination using an inverse-time circuit breaker:

Overrate computation: Ref. 430-63

Motor load: Ref. 430-53(c)(1), Excp. No. 2(c)(1)

400% of 74.8 amps = 299.2 amps

Nonmotor loads: Ref. 220-10(b)

Lighting: 125% of 20 amps =	25.0 amps
Receptacles:	10.0 amps
Total:	334.2 amps

NE Code-maximum OCPD rating: Ref. 240-6(a)
NLS: 300 amps Ref. (NLS) 430-63

3. **The minimum wire size of the feeder-circuit conductors using THHN copper.**

Ampacity overrate computation: Ref. 430-24

Motor: Ref. 430-34 Ref. 240-3

140% of 74.8 amps = 104.7 amps

Nonmotor: Ref. 220-10(b)

Lighting: 125% of 20 amps =	25.0 amps
Receptacles:	10.0 amps
Total:	139.7 amps

Circuit-conductor wire size determination:

Wire terminal temperature limitation: 75°C Ref. 110-14(c)(2), Excp. 1
Table-rated ampacity: 150 amps Ref. T310-16
Wire size: No. 1/0 AWG THHN copper

Although the branch-circuit SC-GF protection for the motor in this example was specified as time-delay fuses, the feeder SC-GF protection was sized for the percentages allowed for an inverse-time circuit breaker. As required by Sec. 430-62(a), when computing the overrate for the feeder short-circuit, ground-fault protection device, selection of the largest branch-circuit SC-GF protection must be based on "like protection" of the feeder.

In reference to Sec. 430-24, the rated-ampacity of the feeder-circuit conductors must be at least equal to the sum of the the motor load at the full-load condition, plus a 25% overrate of the largest motor FLC, plus the nonmotor full-load current, as overrated by Sec. 220-10(b) for continuous load.

Because the branch-circuit SC-GF protection overrate of the single motor load was computed at the maximum as allowed by Sec. 430-52(c)(1), Excp. No. 2(C)(1), the 140% overrate for the motor running-overload protection (Sec. 430-34) was assumed. Instead of the 25% minimum overrate allowed by Sec. 430-24, the 140% overrate of the motor load was applied to the minimum rated ampacity of the circuit conductors to meet the requirements of Sec. 240-3.

Code challenge 13–16 (single motor). A 3-phase, 4-wire, 208Y/120-volt, 225-amp panel located in an equipment room will supply the following loads:

1. Overhead lighting rated at 25 amps
2. Convenience receptacles rated at 15 amps
3. A 3-phase, 208-volt, design C ac induction motor rated at 40 hp

Determine:

1. For a high-inertia, long-acceleration load, the maximum rating of time-delay fuses for the branch-circuit SC-GF overcurrent protection of the motor
2. The maximum setting of the motor-feeder SC-GF overcurrent protection using an inverse-time circuit breaker
3. The minimum wire size of the feeder-circuit conductors using XHHW-2 copper

Motor full-load current: (A) ____________ amps Ref. (B) ____________________

1. **Branch-circuit SC-GF protection rating determination using TD fuses:**

NE Code-maximum overrate computation: Ref. (C) ____________________

____________ % of ____________ amps = (D) ____________amps

NE Code-maximum OCPD rating: Ref. (E) ____________________
NLS: (F) ____________ amps Ref. (NLS) (G) ____________________

2. **Feeder-circuit SC-GF protection rating determination using an inverse-time circuit breaker:**

Overrate computation: Ref. (H) ____________________
Motor load: Ref. (I) ____________________
(J) ____________ % of ____________ amps = ____________ amps
Nonmotor loads: Ref. (K) ____________________
Lighting: (L) ____________ % of ____________ amps = ____________ amps
Receptacles: ____________ amps
Total: (M) ____________ amps

NE Code-maximum OCPD rating: Ref. (N) ____________________
NLS: (O) ____________ amps Ref. (NLS) (P) ____________________

3. **The minimum wire size of the feeder-circuit conductors using XHHW-2 copper.**

Ampacity overrate computation: Ref. (Q) ____________________
Motor load: Ref. (R) ____________ Ref. (S) ____________________
(T) ____________ % of ____________ amps = ____________ amps
Nonmotor loads: Ref. (U) ________________
Lighting: (V) ____________ % of ____________ amps = ____________ amps
Receptacles: ____________ amps
Total: (W) ____________ amps

Standard circuit-conductor wire size determination:
Wire terminal temperature limitation: (X) ____________° C
Ref. (Y) ____________________
Table-rated ampacity: (Z) __________ amps Ref. (1A) ____________________
Wire size: (1B) ____________ XHHW-2 copper

When several motors are supplied from a 3-phase feeder, the higher voltage of 480 or 575 volts is normally used to take advantage of lower full-load current ratings for relatively high horsepower ratings (Fig. 13-14). The draw back to these higher distribution voltages is the requirement for a transformer to separately derive the required 120-volt supply for convenience receptacles and incandescent lighting.

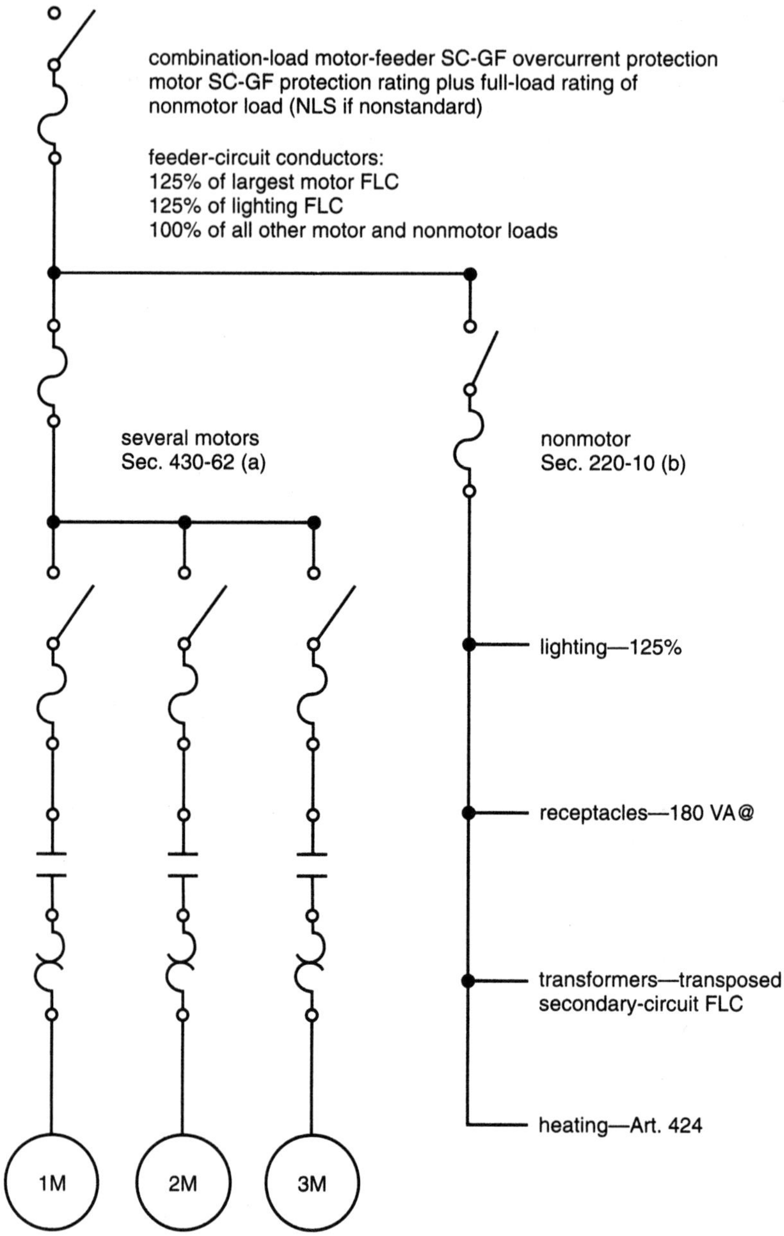

Fig. 13-14 The rating of the combination motor feeder SC-GF protection supplying several motors is determined by Sec. 430-63.

To meet NE Code-minimum requirements when a 3-phase transformer is used to supply secondary lighting and receptacle loads from a predominantly motor feeder, the voltage ratio of secondary-to-primary voltage can be applied to the secondary full-load current rating to compute the primary full-load current at the rated-load condition.

For some applications, especially for future growth or modification of the separately derived system, the 3-phase power formula is used with the VA or kVA rating of the transformer to compute the maximum primary full-load current that will be drawn when the transformer is supplying its maximum load rating (power rating). Both ways are acceptable, although the second method technically is not NE Code minimum.

Example 13–17 (several motors, Fig. 13-14). For a 3-phase, 4-wire, 480Y/277-volt, motor-feeder, using nontime fuses, determine the maximum size of feeder short-circuit, ground-fault protection allowed for the following loads:

1. 100 amps of fluorescent lighting at 277 volts
2. A 12-kVA, 3-phase, 480Y–208Y/120-volt transformer supplying 20 amps of convenience receptacle loads
3. Three 25-hp, 460-volt, 3-phase air handlers rated design B squirrel cage
4. Four 50-hp, 460-volt, 3-phase water pumps rated design B squirrel cage
5. A hermetically sealed motor compressor with nameplate rated-load current 93 amps, and a branch-circuit selection current of 98 amps

Full-load current determinations:

Receptacle transformer:

$$208 \text{ volts} \div 480 \text{ volts} \times 20 \text{ amps} = 8.67 \text{ amps}$$

Motors: Ref. T430-150

25 hp: 34 amps 50 hp: 65 amps

Motor-compressor: 98 amps Ref. 440-7, Excp.

Combination motor-feeder SC-GF protection rating determination using nontime fuses:

Ref. 430-63

Several-motor feeder SC-GF overcurrent protection overrate allowances:

Ref. 430-62(a)

Overrate computations for the largest single-motor branch-circuit SC-GF protection, using nontime fuses:

Motors: Ref. T430-152

25 hp: 300% of 34 amps = 102 amps

50 hp: 300% of 65 amps = 195 amps

Motor-compressor: Ref. 440-22(a)

$$175\% \text{ of } 98 \text{ amps} = 171.5 \text{ amps}$$

Overrate computation for the several-motor feeder SC-GF protection: Ref. 430-62(a)

Largest branch-circuit protection overrate:	195 amps
25 hp: 3 × 34 amps =	102 amps
50 hp: 3 × 65 amps =	195 amps
Motor-compressor:	98 amps
Total:	590 amps

Overrate computation of the nonmotor full-load current: Ref. 220-10(b)

Lighting: 125% of 100 amps =	125.0 amps
Receptacles:	8.7 amps
Total:	133.7 amps

Overrate summation for the combination motor-feeder SC-GF fault protection: Ref. 430-63

Motor:	590.0 amps
Nonmotor:	133.7 amps
Total:	723.7 amps

NE Code-maximum OCPD rating: Ref. 240-6(a)

NLS: 700 amps Ref. (NLS) 430-63

Code challenge 13–17 (several motors). For a 3-phase, 3-wire, 440-volt motor feeder, using time-delay fuses, determine the maximum size of feeder short-circuit, ground-fault protection allowed for the following loads:

1. 75 amps of fluorescent lighting at 254 volts
2. A 12-kVA, 3-phase, 440–208Y/120-volt transformer supplying 28 amps of convenience receptacle loads
3. Three 40-hp, 460-volt, 3-phase air handlers rated design B squirrel cage
4. Four 75-hp, 460-volt, 3-phase water pumps rated design B squirrel-cage
5. A hermetically sealed motor compressor with nameplate rated-load current 74 amps, and a branch-circuit selection current of 76 amps

Full-load current determinations:

Receptacle transformer:

__________ volts ÷ __________ volts × __________ amps = (A) __________ amps

Motor: Ref. (B) ______________

40 hp: (C) __________ amps 75 hp: (D) __________ amps

Motor-compressor: (E) __________ amps Ref. (F) ______________

Combination motor-feeder SC-GF protection rating determination using time-delay fuses:

Ref. (G) ____________________

Several-motor feeder SC-GF overcurrent protection overrate allowances:

Ref. (H) ____________________

Overrate computations for the largest single-motor branch-circuit SC-GF protection, using time-delay fuses:

Motors: Ref. (I) ____________________

40 hp: ____________ % of ____________ amps = (J) ____________ amps

75 hp: ____________ % of ____________ amps = (K) ____________ amps

Motor-compressor: Ref. (L) ____________________

____________ % of ____________ amps = (M) ____________ amps

Overrate computation for the several-motor feeder SC-GF protection:

Ref. (N) ____________________

Largest branch-circuit protection overrate:			____________	amps
40 hp: (O) ____________	× ____________	amps =	____________	amps
75 hp: (P) ____________	× ____________	amps =	____________	amps
Motor-compressor:			____________	amps
Total:		(Q)	____________	amps

Overrate computation of the nonmotor full-load current:

Ref. (R) ____________________

Lighting:	(S) ____________	% of ____________	amps = ____________	amps
Receptacles:			____________	amps
Total:			(T) ____________	amps

Overrate summation for the combination motor-feeder SC-GF fault protection:

Ref. (U) ____________________

Motor:		____________	amps
Nonmotor:		____________	amps
Total:	(V)	____________	amps

NE Code-maximum OCPD rating: Ref. (W) ____________________

NLS: (X) ____________ amps Ref. (NLS) (Y) ____________________

When a chiller unit is supplied by a feeder circuit with other motor and nonmotor loads, the nameplate values of "minimum wire size" (Sec. 440-35) and "maximum fuse size" [Sec. 440-22(b)] are used "as is" for the conductor ampacity computations of Sec. 430-24 or Sec. 430-25, and for the feeder overcurrent computations of Sec. 430-62(a) or Sec. 430-63.

When determining the required ampacity of the feeder conductors for a nonmotor continuous load such as nondwelling lighting, the 125% overrate, required by Sec. 220-10(b) must be applied to both the rating of the feeder fuses or setting of the circuit breaker and the rated ampacity of the feeder-circuit conductors.

Example 13–18. For a 3-phase, 4-wire, 480Y/277-volt, motor feeder supplying the following loads:

1. 30 amps of fluorescent lighting at 277 volts
2. A 7½-kVA, 3-phase, 480Y–208Y/120-volt transformer supplying 15 amps of convenience receptacle loads
3. One 5-hp, 460-volt, 3-phase air compressor rated design C squirrel cage
4. Two 20-hp, 460-volt, 3-phase air handlers rated design C squirrel cage
5. Four 25-hp, 460-volt, 3-phase water pumps rated design C squirrel cage
6. A chiller unit with nameplate specifications of No. 2 AWG THW copper as the minimum supply conductor wire size, and 175-amp time-delay fuses as the maximum disconnect fuse size.

Determine:

1. The maximum setting of an inverse-time circuit breaker used for the feeder SC-GF overcurrent protection
2. The minimum wire size of the feeder-circuit conductor using XHHW-2 copper conductors

Full-load current determinations:
Receptacle transformer:

$$208 \text{ volts} \div 480 \text{ volts} \times 15 \text{ amps} = 6.5 \text{ amps}$$

Motors: Ref. T430-150
5 hp: 7.6 amps
20 hp: 27 amps
25 hp: 34 amps
Chiller Unit:
Supply conductor rated-ampacity: Ref. 440-35
Wire terminal temperature limitations: 60°C
Ref. 110-14(c)(1), Excp. 1
No. 2 AWG THW copper: 95 amps Ref. T310-16
Overcurrent protection rating: Ref. 440-22(c)
Fuse size: (given) 175 amps

1. The maximum setting of an inverse-time circuit breaker used for the feeder SC-GF overcurrent protection.

Ref. 430-63

Several-motor feeder SC-GF overcurrent protection overrate allowances:
Ref. 430-62(a)

Overrate computations for the largest single-motor branch-circuit SC-GF protection, using an inverse-time circuit breaker:

Motors: Ref. T430-152

5 hp: 250% of 7.6 amps = 19.0 amps
20 hp: 250% of 27.0 amps = 67.5 amps
25 hp: 250% of 34.0 amps = 85.0 amps

Chiller unit: 175 amps Ref. 440-22(b)

Overrate computation for the several-motor feeder SC-GF protection:
Ref. 430-62(a)

Largest branch-circuit protection overrate:	175.0 amps
5 hp:	7.6 amps
20 hp: 2 × 27 amps =	54.0 amps
25 hp: 4 × 34 amps =	136.0 amps
Total:	372.6 amps

Overrate computation of the nonmotor full-load current: Ref. 220-10(b)

Lighting: 125% of 30 amps =	37.5 amps
Receptacles:	6.5 amps
Total:	44.0 amps

Overrate summation for the combination motor-feeder SC-GF fault protection:
Ref. 430-63

Motor:	372.6 amps
Nonmotor:	44.0 amps
Total:	416.6 amps

NE Code-maximum OCPD rating: Ref. 240-6(a)
NLS: 400 amps Ref. (NLS) 430-63

2. The minimum wire size of the feeder-circuit conductore using XHHW-2 copper conductors.

Overrate computations: Ref. 430-24

Chiller unit:	95.0 amps
Motor-load:	
5 hp:	7.6 amps
20 hp: 2 × 27 amps =	54.0 amps
25 hp: 4 × 34 amps =	136.0 amps
Largest motor: 25% of 34 amps =	8.5 amps
Nonmotor load: Ref. 220-10(b)	44.0 amps
Total:	345.1 amps

Feeder conductor wire size determination:

Wire terminal temperature limitations: 75°C — Ref. 110-14(c)(2), Excp. 1
Table-rated ampacity: 380 amps — Ref. T310-16
Wire size: 500 kcmil XHHW-2 copper

In this example, the 175-amp fuses protecting the chiller unit, which is rated as multimotor/combination-load equipment, was used as the largest branch-circuit short-circuit ground-fault overcurrent-protection rating. But the 95-amp rating of the No. 2 AWG copper conductors specified as minimum on the nameplate was not considered in the 25% overrate of the largest motor to size the motor-feeder conductors.

As a rule multimotor/combination-load equipment is not considered when making the comparisons required by either Sec. 430-24 or Sec. 430-62(a), but the reference made to Sec. 440-22(a) by Sec. 430-62(a), allows the 175-amp fuse rating of the chiller unit to be considered in the determination of the largest branch-circuit short-circuit, ground-fault protection rating in this example.

The rating of the 175 fuses for the SC-GF overcurrent protection of the chiller unit was determined by Sec. 440-22(b). Because it is multimotor/combination-load equipment, the fuse rating was determined by adding the full-load currents of one or two fan motors, a signaling transformer, and perhaps a small crankcase resistive heating load to the 175% overrate of the contained hermetically sealed refrigerant motor compressor branch-circuit selection current.

For the largest motor overrate requirement of Sec. 430-24, the branch-circuit selection current of the motor-compressor cannot be gleaned out of the 95-amp rating of the No. 2 AWG circuit conductors specified on the nameplate. As a result, the 25% largest motor overrate was applied to the largest motor given in the example.

Code challenge 13–18. For a 3-phase, 4-wire, 460Y/265-volt motor feeder supplying the following loads:

1. 30 amps of fluorescent lighting at 265 volts
2. A 7 1/2-kVA, 3-phase, 460Y–208Y/120-volt transformer supplying 18 amps of convenience receptacle loads
3. One 3-hp, 460-volt, 3-phase air compressor rated design B squirrel cage
4. Two 30-hp, 460-volt, 3-phase air handlers rated design B squirrel cage
5. Four 40-hp, 460-volt, 3-phase water pumps rated design B squirrel cage
6. A chiller unit with nameplate specifications of No. 1 AWG THW copper as the minimum supply conductor wire size, and 150-amp time-delay fuses as the maximum disconnect fuse size

Determine:

1. The maximum rating of time-delay fuses used for the feeder SC-GF overcurrent protection
2. The minimum wire size of the feeder-circuit conductor using XHH aluminum conductors

Full-load current determinations:

Receptacle transformer:

__________ volts ÷ __________ volts × __________ amps = (A) __________ amps

Motors: Ref. (B) ________________

3 hp: (C) __________ amps

30 hp: (D) __________ amps

40 hp: (E) __________ amps

Chiller Unit:

Supply conductor rated-ampacity: Ref. (F) ________________

Wire terminal temperature limitations: (G) __________ °C

Ref. (H) ________________

No. 1 AWG THW copper: (I) __________ amps

Overcurrent protection rating: Ref. (J) ________________

Fuse size: (given) (K) __________ amps

1. The maximum rating of time-delay fuses used for the feeder SC-GF overcurrent protection.

Ref. (L) ________________

Several-motor feeder SC-GF overcurrent protection overrate allowances:

Ref. (M) ________________

Overrate computations for the largest single-motor branch-circuit SC-GF protection, using time-delay fuses:

Motors: Ref. (N) ________________

3 hp: (O) __________ % of __________ amps = __________ amps

30 hp: (P) __________ % of __________ amps = __________ amps

40 hp: (Q) __________ % of __________ amps = __________ amps

Chiller unit: (R) __________ amps Ref. (S) ________________

Overrate computation for the several-motor feeder SC-GF protection:

Ref. (T) ________________

Largest branch-circuit protection overrate: (U) __________ amps

3 hp: __________ amps

30 hp: (V) __________ × __________ amps = __________ amps

40 hp: (W) __________ × __________ amps = __________ amps

Total: (X) __________ amps

Overrate computation of the nonmotor full-load current:

Ref. (Y) ________________

Lighting: (Z) __________ % of __________ amps = __________ amps

Receptacles: __________ amps

Total: (1A) __________ amps

Overrate summation for the combination motor-feeder SC-GF fault protection:

Ref. (1B) ______________

Motor: ______________ amps
Nonmotor: ______________ amps
Total: (1C) ______________ amps

NE Code-maximum OCPD rating: Ref. (1D) ______________
NLS: (1E) ______________ amps Ref. (NLS) (1F) ______________

2. **The minimum wire size of the feeder-circuit conductor using XHH aluminum conductors.**

Overrate computations: Ref. (1G) ______________
Chiller unit: (1H) ______________ amps
Motor-load:
3 hp: ______________ amps
30 hp: (1I) ______________ × ______________ amps = ______________ amps
40 hp: (1J) ______________ × ______________ amps = ______________ amps
Largest motor: (1K) ______________ % of ______________ amps = ______________ amps
Nonmotor load: Ref. (1L) ______________ ______________ amps
Total: (1M) ______________ amps

Feeder conductor wire size determination:
Wire terminal temperature limitations: (1N) ______________ °C
Ref. (1O) ______________
Table-rated ampacity: (1P) ______________ amps Ref. (1Q) ______________
Wire size: (1R) ______________ XHH aluminum

Chapter 14 will continue the discussion on motors, concentrating on motor feeder tap rules and special task motors such as variable-speed drive controllers.

REVIEW QUESTIONS

1. Because the impedance of a motor is governed by both its manufactured operating characteristics and the inertia or other resistance to the motion of its connected load, the motor-load current can vary from almost infinite impedance and zero amps, when operated in a no-load condition, to a very low value of circuit impedance, and as much as six times its rated-load current, when the motor is either started, or forced into a locked-rotor condition.

   True or False

2. Overcurrent drawn by a motor circuit can be the result of either ______________, ______________, or ______________.
3. To protect a motor against mechanical overload, protects the circuit against electrical overload.

   True or False

4. Only the branch-circuit fuse or circuit breaker can provide the necessary ___________ or ___________ overcurrent protection required for the motor and its supply circuit conductors. Separate overload protection for a motor is normally referred to as ___________ overload-overcurrent proection.

5. The rating or setting of the separate overload protection device for a motor must be determined by overrating the NE Code table-value full-load current of the individual motor.

   True or False

6. For general motor applications, the standard overrate percentage for the sensing elements in a separate overload relay is either ___________ % or ___________ %, according to the type of motor or its ratings. The NE Code-maximum overrate percentage for running overload protection is either ___________ % or ___________ %.

7. The motor, hermetically sealed within a refrigerant motor compressor, can be protected against running overload by a separate sensing device, provided the rating or setting of the sensing device does not exceed ___________ % of the motor compressor's rated-load current.

8. The rated-ampacity of any circuit conductor must be greater than or equal to the rating of the overload-overcurrent protection device protecting the circuit.

   True or False

9. The circuit conductors, the disconnects, the SC-GF overcurrent-protection devices, and any other support or supply equipment for the motor must be sized according to the table-value full-load current listed in the NE Code.

   True or False

10. For hermetically sealed refrigerant motor compressors, the rating or setting of the running overload-protection device is determined by overrating the rated-load current listed on the nameplate. The rating or setting of the circuit conductors, the disconnects, the SC-GF overcurrent protection devices, or any other support or supply equipment for the motor-compressor must be determined by overrating the branch-circuit selection current listed on the nameplate.

    True or False

11. The branch-circuit selection current will always be greater than or equal to the rated-load current of the hermetically sealed refrigerant motor compressor.

    True or False

12. The better electrical design will allow for the possibility of the overload heaters in a motor starter being set at the NE Code maximum, and apply the ___________ % overrate to the table-value FLC rating of the motor to determine the minimum ampacity rating of the circuit conductors.

13. When more than one motor is supplied by a branch-circuit, feeder, or service, the rated ampacity of the circuit conductors must be greater than or equal to the ___________ % rating of all loads supplied, plus ___________ % overrate applied to the largest motor FLC of any motor supplied.

14. When used as the overload-overcurrent protection, the rating of the branch-circuit fuses or circuit breaker, already used as the necessary SC-GF overcurrent protection of the motor, is restricted to the overrates listed in Secs. ___________ and ___________.

15. When motors are provided with separate, individual running overload protection, the protection against the high-magnitude currents of short-circuit and ground-fault conditions is provided at the branch-circuit level by either fuses or circuit breakers.

    True or False

16. When overrating the branch-circuit fuses or circuit breaker for the standard rating of the motor branch-circuit SC-GF overcurrent protection and a nonstandard size or rating is determined, the next ____________ (lower or higher) standard size or rating can be installed.

17. When overrating the branch-circuit fuses or circuit breaker for the NE Code-maximum rating of the motor branch-circuit SC-GF overcurrent protection and a nonstandard size or rating is determined, the next ____________ (lower or higher) standard size or rating must be installed.

18. If the motor branch circuit will be protected against SC-GF overccurent with time-delay fuses, the standard overrate is either ____________ % or ____________ % according to the type of motor installed. The NE Code-maximum value of overrate is ____________ %.

19. If the motor branch circuit will be protected against SC-GF overcurrent with nontime fuses, the standard overrate is either ____________ % or ____________ % according to the type of motor installed. The NE Code-maximum value of overrate is ____________ %, provided the standard overrate did not yeild a fuse rating in excess of 600 amps.

20. If the motor branch circuit will be protected against SC-GF overcurrent with nontime fuses, the standard overrate is either ____________ % or ____________ % according to the type of motor installed. The NE Code-maximum value of overrate is restricted to ____________ %, if the standard overrate fuse rating exceeds of 600 amps.

21. If the motor branch circuit will be protected against SC-GF overcurrent by an inverse-time circuit breaker, the standard overrate is either ____________ % or ____________ % according to the type of motor installed. The NE Code-maximum value of overrate is restricted to ____________ %, if the motor FLC is 100 amps or less. The NE Code-maximum value of overrate is restricted to ____________ %, if the motor FLC is over 100 amps.

22. If the motor branch circuit will be protected against SC-GF overcurrent by an instantaneous-trip circuit breaker, the standard overrate is either ____________ %, ____________ %, or ____________ % according to the type of motor installed. The NE Code-maximum value of overrate is restricted to either ____________ % or ____________ % according to the type of motor installed.

23. For motor branch-circuit overcurrent protection, an instantaneous-trip circuit breaker can only be used as supplemental, high-level SC-GF overcurrent protection for the motor circuit. The instantaneous-trip circuit breaker can be used only if its setting is adjustable, and it is part of a combination controller that contains coordinated overload and additional, lower-level SC-GF overcurrent protection in each circuit conductor.

    True or False

24. The rating or setting of the branch-circuit SC-GF overcurrent protection for a hermetically sealed refrigerent motor compressor, is set at a standard overrate of ____________ % and a NE Code-maximum of ____________ % regardless of the type of fuse or circuit breaker to be used.

25. When several motors are supplied from a common feeder circuit, the motor-feeder SC-GF overcurrent protection device rating must be based on the summation of the largest standard rating of branch-circuit fuses or circuit breaker computed for the general-application motor or the hermetically sealed refrigerant motor compressor and the full-load current rating (100%) of all the other motors supplied.

    True or False

26. If the computation of the motor-feeder SC-GF overcurrent protection overrate yields a nonstandard size of overcurrent-protection device, the next lower standard size (NLS) of fuses or circuit breaker must be installed for the feeder SC-GF protection.

    True or False

27. When large-ampacity feeders are installed for reserved capacity of a motor feeder, the size of the feeder overcurrent-protection device can be based on the ampacity of the feeders instead of the actual connected motor load. For this type of installation, the ampacity of the feeder conductors would have to be at least equal to the rating or setting of the feeder overcurrent-protection device.

True or False

28. When a predominantly motor-feeder circuit will also supply power and lighting loads, the rating or setting of the feeder SC-GF overcurrent-protection device must be determined in accordance with Sec. ____________.

29. When a chiller unit is installed on a feeder circuit with other motor and nonmotor loads, the nameplate values of "minimum wire size" and "maximum fuse size" are used "as is" for the conductor ampacity overrate computations and for the feeder SC-GF overcurrent protection overrate computations.

True or False

REVIEW PROBLEMS

1. A 3-phase, 575-volt, 20-hp squirrel-cage induction motor has a nameplate full-load current rating of 20 amps. If the motor is rated with a limited temperature rise of not over 40°C, determine the full-load current rating of the heaters installed in the separate overload relay.
2. A 3-phase, 230-volt, 75-hp wound-rotor induction motor is used to drive a wood shredder. If the motor has a service factor (SF) of 1.15 and a nameplate rated-load current of 187 amps, determine the NE Code-maximum full-load current (FLC) rating of the heaters installed in a separate overload relay.
3. A single-phase, 230-volt, 5-hp wound-rotor induction motor is used to drive a sump pump. If the motor has a service factor (SF) of 1.15, and a nameplate rated-load current (FLC) of 23 amps, determine the NEC Code-maximum full-load current rating of the heaters installed in a separate overload relay, and the NE Code-minimum wire size of the XHHW copper conductors used for both the branch-circuit and motor-circuit supplies.
4. The nameplate rated-load current of a 3-phase, 230-volt, 10-ton hermetically sealed refrigerent motor compressor is 23 amps. The nameplate branch-circuit selection current rating is 27 amps. Determine the standard full-load current rating of the heaters installed in a separate overload relay, and the NE Code-minimum wire size of the RHW copper conductors used for both the branch-circuit and motor-circuit supplies.
5. A 2-phase, 230-volt, 20-hp ac induction motor will be used to drive a blocked-ice conveyor than can be started under load. If the motor has a service factor of 1.15, and a full-load current rating of 44 amps, determine the NE Code-maximum full-load current rating of the heaters installed in a separate overload relay, and the design-minimum wire size of the THW copper conductors used for both the branch-circuit and motor-circuit supplies.
6. Determine the NE Code-minimum wire size of THW aluminum conductors used to supply two 30-hp and three 20-hp ac induction motors, supplied from a 3-phase, 3-wire, 480-volt delta distribution.
7. The blower motor for a steam take-away fan on the roof of a building is rated 1½ hp, 115 volts, single-phase. If the motor has a nameplate rated-load current of 18 amperes, and a service fac-

tor of 1.15, determine the NE Code-maximum size (rating) of time-delay fuses that can be installed as the combined overcurrent protection for the motor branch circuit.

8. For a 3-phase, 20-hp, 200-volt, design B ac induction motor with a nameplate-rated limited temperature rise of not over 40°C, determine:
   a. Both the standard and the NE Code-maximum ratings of time-delay fuses used to provide the branch-circuit SC-GF overcurrent protection of the motor
   b. The NE Code-minimim wire size of the branch-circuit and motor-circuit conductors for each size of time-delay fuse selected using RHW aluminum

9. Determine the standard and maximum ratings of nontime fuses used to provide branch-circuit SC-GF overcurrent protection for a single-phase, 230-volt, squirrel-cage induction motor rated at 5 hp.

10. Determine the standard and maximum settings of an inverse-time circuit breaker used for the branch-circuit SC-GF overcurrent protection for a 3-phase, 460-volt, squirrel-cage induction motor rated at 7½ hp.

11. If a 3-phase, 460-volt, design E ac induction motor rated at 40 hp, has a service fator of 1.15 and a nameplate full-load current rating of 48 amperes, determine:
   a. The standard and maximum settings of an instantaneous-trip circuit breaker used for the high-level branch-circuit SC-GF overcurrent protection
   b. The standard and maximum ratings of time-delay fuses used for the low-level branch-circuit SC-GF overcurrent protection
   c. The standard and maximum settings of thermal heaters installed in a separate overload relay
   d. The standard and maximum (design) wire size of the THWN copper branch-circuit and motor-circuit conductors

12. If a hermetically sealed refrigerant motor compressor has a nameplate rated-load current of 63 amperes and a branch-circuit selection current rating of 68 amperes, determine:
   a. The standard and maximum ratings of time-delay fuses used for the branch-circuit SC-GF overcurrent protection
   b. The standard and maximum settings of thermal heaters installed in a separate overload relay
   c. The standard and maximum wire size of the RHW copper-clad aluminum branch-circuit and motor-circuit conductors

13. Two 15-hp and two 25-hp, 3-phase, 208-volt design C squirrel-cage induction motors are to be supplied from a common feeder. Determine:
   a. The maximum size of time-delay fuses for the motor-feeder SC-GF overcurrent protection
   b. The minimum wire size of the feeder conductors using single-insulated THWN copper conductors installed in IMC conduit, which passes through an ambient air temperature of 120°F

14. Five 3-phase, 460-volt, squirrel-cage induction motors are to be supplied by a common feeder. Two of the motors are rated design C at 20 hp. The other three motors are rated wound rotor at 25 hp. Determine the maximum rating of the feeder SC-GF protection using nontime fuses.

15. Two 3-phase, design B squirrel-cage induction motors rated at 7½ hp, two 3-phase wound-rotor ac induction motors rated at 10 hp, and one 3-phase, 10-ton hermetically sealed refrigerant motor compressor will be supplied by a common 208-volt feeder. Determine:
   a. The standard and maximum ratings of an instantaneous-trip circuit breaker protecting the individual branch circuits of the two 7½-hp motors
   b. The standard and maximum ratings of time-delay fuses also protecting the individual branch circuits of the two 7½ hp-motors
   c. The standard and maximum ratings of nontime fuses protecting the individual branch circuits of the two 10-hp motors

d. The standard and maximum ratings of time-delay fuses protecting the branch circuit of the motor-compressor for a nameplate rated-load current of 36 amps and a branch-circuit selection current of 40 amps
e. The maximum setting of the inverse-time circuit breaker protecting the motor feeder

16. A 3-phase, 4-wire, 208Y/120-volt, 100-amp panel located in an equipment room will supply the following loads:
a. Overhead lighting rated at 10 amps
b. Convenience receptacles rated at 9 amps
c. A 3-phase, 208-volt, design B, ac induction motor rated at 20 hp

Determine:
a. For a high-inertia, long-acceleration load, the maximum rating of time-delay fuses for the branch-circuit SC-GF protection of the motor
b. The maximum setting of the motor-feeder SC-GF overcurrent protection using an inverse-time circuit breaker
c. The minimum wire size of the feeder-circuit conductors using THW-2 copper

17. For a 3-phase, 4-wire, 460Y/265-volt motor feeder, using nontime fuses, determine the maximum size of feeder SC-GF overcurrent protection allowed for the following loads:
a. 65 amps of fluorescent lighting at 265 volts
b. A 7½ kVA, 3-phase, 460Y–208Y/120-volt transformer supplying 15 amps of convenience receptacle loads
c. Three 15-hp, 460-volt, 3-phase air handlers rated design B squirrel-cage
d. Four 20-hp, 460-volt, 3-phase water pumps rated design B squirrel-cage
e. A hermetically sealed motor compressor with nameplate rated-load current of 31 amps and a branch-circuit selection current of 36 amps

18. For a 3-phase, 4-wire, 480Y/277-volt, motor-feeder supplying the following loads:
a. 20 amps of fluorescent lighting at 277 volts
b. A 12-kVA, 3-phase, 480Y–208Y/120-volt transformer supplying 20 amps of convenience receptacle loads
c. One 2-hp, 460-volt, 3-phase air compressor rated design D squirrel cage
d. Two 10-hp, 460-volt, 3-phase air handlers rated design D squirrel cage
e. Four 15-hp, 460-volt, 3-phase water pumps rated design D squirrel cage
f. A chiller unit with nameplate specifications of No. 1 AWG THW copper as the minimum supply conductor wire size, and 125-amp time-delay fuses as the maximum disconnect fuse size

Determine:
a. The maximum setting of an inverse-time circuit breaker used for the feeder SC-GF overcurrent protection
b. The minimum wire size of the feeder-circuit conductor using XHH c-c aluminum conductors

Chapter 14

Motors—Part II

The discussion on motors actually started in Chapter 12 with the required disconnecting means for electrical utilization equipment. For general motor applications (typical running speeds and torque requirements), Chapter 13 demonstrated the means available to provide overcurrent protection for the motor circuit, plus the NE Code requirements for motor-circuit conductor ampacity according to wire size.

Chapter 14 will discuss other motor circuits which have special application as compared to what has already been discussed. Feeder tap circuits, more than one motor on a branch circuit, duty cycle motors, variable speed dc drives, and variable speed ac drives will be examined.

14.1 MOTOR FEEDER TAPS

The term "circuit tap" describes a supply circuit of lower ampacity (smaller wire size) for a piece of equipment, such as a load center or a unit control center, or an individual load, such as a motor load, that is connected or "tapped" to a higher ampacity (larger wire size) feeder circuit without an overcurrent-protection device being installed at the point of tap.

In other words, the overcurrent protection for the tap conductors is being split in the same manner as for motor loads: The higher-ampacity conductors are protected at their point of power supply by a fuse or circuit breaker sized according to the rated ampacity of the larger feeder-circuit conductors. In a tap circuit, the feeder-circuit fuses or circuit breaker serve as the required short-circuit, ground-fault (SC-GF) overcurrent protection for the lower-ampacity tap-circuit conductors. The overload-overcurrent protection for the tap conductors is provided separately (or additionally) at the load end of the tap.

The general tap rules for all phases of electrical installations (motor and nonmotor loads at the branch-circuit or feeder-circuit level) are listed in the subsections of Sec. 240-21. The general rule of Sec. 240-21(a) requires an overcurrent-protection device, wired in series with each ungrounded (phase) conductor [Sec. 240-20(a)], to be installed at the point where the ungrounded conductors of either the feeder or branch circuit receive their supply. Section 240-21(h) refers the student electrician to Sec. 430-28 for motor feeder taps, and to Sec. 430-53 for motor branch-circuit taps.

Again, a tap consists of a conductor of lower ampacity connected to a protected conductor of higher ampacity without overcurrent protection installed at the point of the tap. The tap conductor is normally overcurrent-protected according to its rated ampacity at the point of termination.

For motor feeders, Sec. 430-28 lists two tap rules: The 10-foot tap rule and the 25-foot tap rule. Special provisions must be met for each one.

To use the 10-foot tap rule (Fig. 14-1), the overall length of the motor-circuit conductors between the point of tap and connection to the branch-circuit overcurrent-protection device cannot exceed 10 feet. The tap conductors must be totally enclosed in raceway or gutter space within the motor controller to protect them from physical damage. The rating or setting of the feeder overcurrent-protection device cannot be greater than ten times (1,000%) the rated ampacity of the tap conductors.

Example 14–1 (10-foot tap rule). A 3-phase, 230-volt, design B, ac induction motor rated at 20 hp will be supplied by conductors tapped to a larger feeder circuit without overcurrent protection at the point of tap. For the following conditions, determine if this is an NE Code-legal installation:

1. The No. 3 AWG THW copper tap conductors, enclosed in electrical metallic tubing (EMT), are a total of 8 feet in length between the point of tap and the point of termination to a circuit breaker.
2. The inverse-time CB, which provides the necessary branch-circuit SC-GF overcurrent protection for the motor circuit, is rated at 175 amps.
3. The feeder circuit, which consists of 500 kcmil THW copper conductors, is protected at the point of supply by time-delay fuses rated at 500 amps.

Motor full-load current: 54 amps — Ref. T430-150

Tap-conductor wire size justification:

Conductor ampacity overrate computation: — Ref. 430-22(a)

$$125\% \text{ of } 54 \text{ amps} = 67.5 \text{ amps}$$

Minimum wire size determination:

Terminal temperature limitation: 60°C — Ref. 110-14(c)(1), Excp. 1

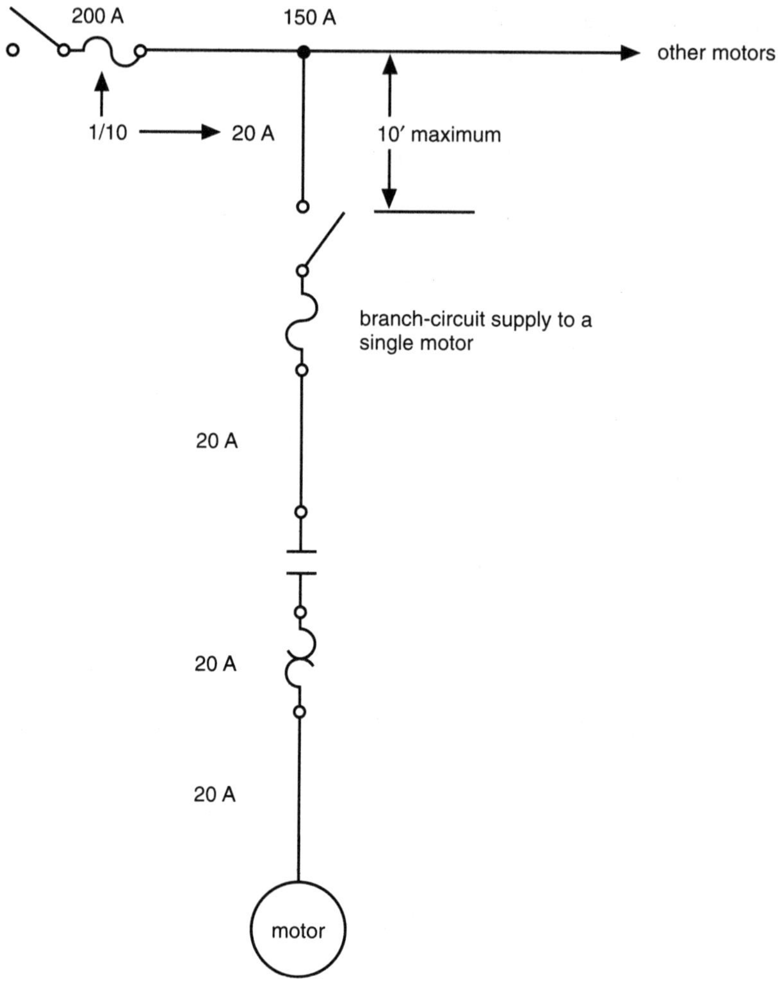

Fig. 14-1 The 10-foot motor tap rule allows the branch-circuit disconnect to be installed remote from the connection to the feeder bus.

Table-rated ampacity: 70 amps Ref. T310-16
Wire size: No. 4 AWG THW copper

Single-motor branch-circuit SC-GF overcurrent protection overrate computations, using an inverse-time CB:

Standard overrate computation: Ref. 430-52(c)(1)
Ref. T430-152

250% of 54 amps = 135 amps

Standard OCPD ratings: Ref. 240-6(a)
NHS: 150 amps Ref. (NHS) 430-52(c)(1), Excp. 1

NE Code-maximum overrate computation: Ref. 430-52(c)(1), Excp. 2(c)(1)

400% of 54 amps = 216 amps

NE Code-maximum OCPD rating: Ref. 240-6(a)
NLS: 200 amps Ref. (NLS) 430-52(c)(1), Excp. 2(c)(1)

Maximum feeder OCPD rating, using the 10-foot tap rule:
Given wire size: No. 3 AWG THW copper
Terminal temperature limitation: 60°C Ref. 110-14(c)(1), Excp. 1
Table-rated ampacity: 85 amps Ref. T310-16
Overrate computation: Ref. 430-28

1,000% of 85 amps = 850 amps

The NE Code requirements using the 10-foot tap rule have been met.

By reference to Part B of Art. 430 in Sec. 430-28, the restrictions imposed by the 10-foot tap rule in this example include all NE Code rules pertaining to a single-motor branch-circuit supply. The maximum length of 8 feet does not justify the overall installation.

To qualify the No. 3 AWG THW copper tap conductors, the NE Code-minimum wire size was computed according to the standard overrate requirement of Sec. 430-22(a). The fact that the given tap wire size was larger than the NE Code-minimum wire size of a single-motor branch circuit checks OK for the 10-foot tap. If the specified tap wire size had been No. 6 AWG THW copper, the other computations would not have been warranted: All NE Code motor-circuit requirements have to be met.

Since the operating parameters of the motor were not specified, both the standard and NE Code-maximum settings of the branch-circuit inverse-time circuit breaker were computed. The intended 175-amp rating of the tap installation fell between the two extremes, which checked OK for the 10-foot tap rule.

And finally, the rated ampacity of the tap-circuit conductors has to be a minimum of one-tenth the ampacity (current) rating or setting of the feeder SC-GF overrcurrent protection. Instead of dividing the 500-amp rating of the feeder protection by 10, the example took 1,000% (10 times) of the rated ampacity of the circuit conductors. As long as this computation at least equals the feeder overcurrent-protection device rating or setting, all the conditions of the 10-foot rule have been met.

Code challenge 14–1 (10-foot tap rule). A 3-phase, 460-volt, design E, ac induction motor rated at 50 hp will be supplied by conductors tapped to a larger feeder circuit without overcurrent protection at the point of tap. For the following conditions, determine if this is an NE Code-legal installation:

1. The No. 2 AWG THW copper tap conductors, enclosed in EMT, are a total of 9 feet in length between the point of tap and the point of termination to a circuit breaker.

2. The inverse-time CB, which provides the necessary branch-circuit SC-GF overcurrent protection for the motor circuit, is rated at 225 amps.
3. The feeder circuit, which consists of 600 kcmil THW copper conductors, is protected at the point of supply by time-delay fuses rated at 600 amps.

Motor full-load current: (A) ____________ amps Ref. (B) ____________

Tap-conductor wire size justification:

Conductor ampacity overrate computation: Ref. (C) ____________

(D) ____________ % of ____________ amps = ____________ amps

Minimum wire size determination:

Terminal temperature limitation: (E) ______ °C Ref. (F) ____________

Table-rated ampacity: (G) ____________ amps Ref. (H) ____________

Wire size: (I) ____________ THW copper

Single-motor branch-circuit SC-GF overcurrent protection overrate computations, using an inverse-time CB:

Ref. (J) ____________

Standard overrate computation: Ref. (K) ____________

(L) ____________ % of ____________ amps = ____________ amps

Standard OCPD ratings: Ref. (M) ____________

NHS: (N) ____________ amps Ref. (NHS) (O) ____________

NE Code-maximum overrate computation: Ref. (P) ____________

(Q) ____________ % of ____________ amps = ____________ amps

NE Code-maximum OCPD rating: Ref. (R) ____________

NLS: (S) ____________ amps Ref. (NLS) (T) ____________

Maximum feeder OCPD rating, using the 10-foot tap rule:

Given wire size: (U) ____________ THW copper

Terminal temperature limitation: (V) ______ °C Ref. (W) ____________

Table-rated ampacity: (X) ____________ amps Ref. (Y) ____________

Overrate computation: Ref. (Z) ____________

(1A) ____________ % of ____________ amps = ____________ amps

The NE Code requirements using the 10-foot tap rule (1B) ____________ (have or have not) been met.

The 10-foot tap length for the motor circuit can be increased to 25 feet if the tap-conductor rated ampacity (according to wire size) is increased to at least ⅓ the rated ampacity of the feeder-circuit conductors (Fig. 14-2). The rules of totally enclosed in raceway or other suitable protection from physical damage and terminated to the individual motor branch-circuit overcurrent-protection device still apply.

The rated ampacity of the tap conductors in the 10-foot tap rule must be at least equal to ⅒ the rating or setting of the larger feeder-circuit SC-GF overcurrent-protection rating or setting (CB). In comparison, the rated ampacity of the tap conductors in the 25-foot tap rule must be at least equal to ⅓ the rated-ampacity of the larger feeder-circuit conductors.

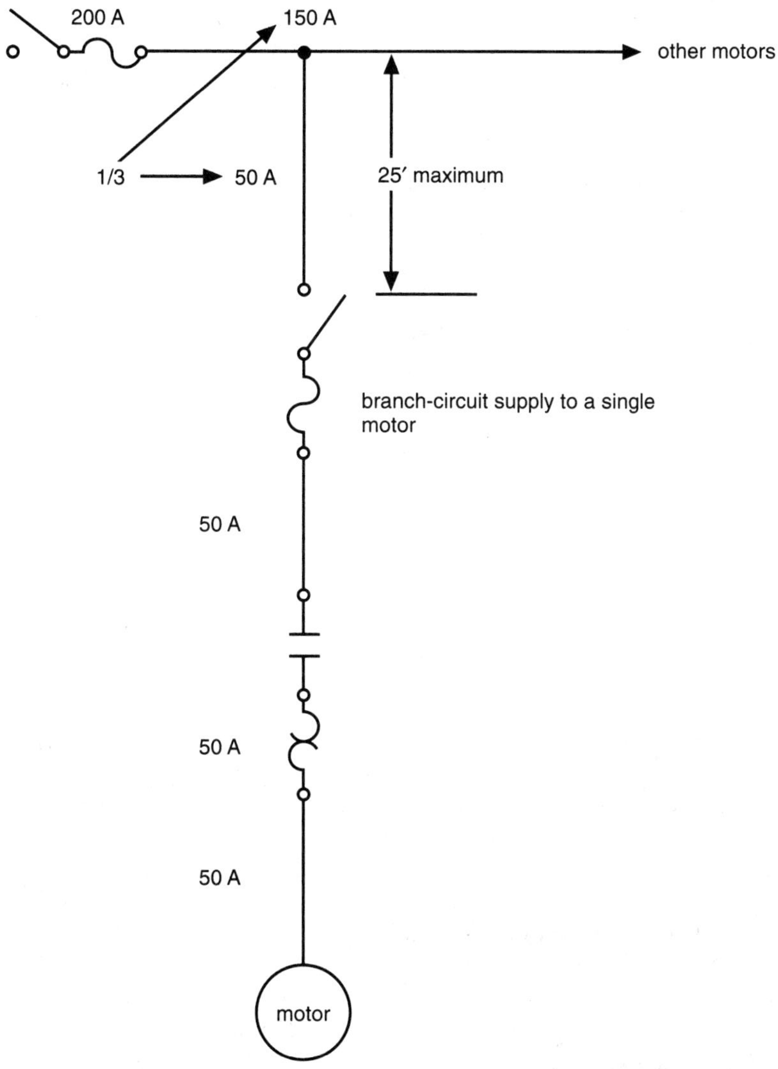

Fig. 14-2 The 25-foot motor tap rule allows the branch-circuit disconnect to be installed remote from the connection to the feeder bus.

Example 14–2 (25-foot tap rule). A 3-phase, 208-volt, design E, ac induction motor rated at 50 hp will be supplied by conductors tapped to a larger feeder circuit without overcurrent protection at the point of tap. For the following conditions, determine if this is an NE Code-legal installation:

1. The tap consists of a multiconductor cable that contains three No. 4/0 AWG THW copper conductors and a No. 1 AWG THW copper equipment-grounding conductor. The multiconductor cable, which has a total length of 22 feet and is routed to be protected from physical damage, is terminated to a fused knife-switch disconnect.
2. The time-delay fuses in the knife switch, which provide the necessary branch-circuit SC-GF overcurrent protection for the motor circuit, are rated at 300 amps.
3. The feeder circuit, which consists of 1,000 kcmil THW copper conductors, is protected at the point of supply by an inverse-time circuit breaker rated at 1,000 amps.

Motor full-load current: 143 amps — Ref. T430-150

Tap-conductor wire size justification:

Conductor ampacity overrate computation: — Ref. 430-22(a)

125% of 143 amps = 178.8 amps

Minimum wire size determination:

Terminal temperature limitation: 75°C — Ref. 110-14(c)(2)
Table-rated ampacity: 200 amps — Ref. T310-16
Wire size: No. 3/0 AWG THW copper

Single-motor branch-circuit SC-GF overcurrent protection overrate computations, using time-delay fuses:

Standard overrate computation: — Ref. 430-52(c)(1), Ref. T430-152

175% of 143 amps = 250.3 amps

Standard OCPD ratings: — Ref. 240-6(a)
NHS: 300 amps — Ref. (NHS) 430-52(c)(1), Excp. 1

NE Code-maximum overrate computation: — Ref. 430-52(c)(1), Excp. 2(b)

225% of 143 amps = 321.8 amps

NE Code-maximum OCPD rating: — Ref. 240-6(a)
NLS: 300 amps — Ref. (NLS) 430-52(c)(1), Excp. 2(b)

Maximum feeder OCPD rating, using the 25-foot tap rule:

Given tap-conductor wire size: No. 4/0 AWG THW copper
Terminal temperature limitation: 75°C — Ref. 110-14(c)(2)
Table-rated ampacity: 230 amps — Ref. T310-16

Given feeder-conductor wire size: 1,000 kcmil THW copper
Terminal temperature limitation: 75°C — Ref. 110-14(c)(2)
Table-rated ampacity: 545 amps — Ref. T310-16
Overrate computation: — Ref. 430-28

300% of 230 amps = 690 amps

The NE Code requirements using the 25-foot tap rule have been met.

As with the 10-foot tap rule, the rated ampacity of the No. 4/0 AWG copper tap conductors in the multiconductor cable exceeded the NE Code-minimum single motor-circuit conductor rated ampacity as computed by Sec. 430-22(a).

The rated ampacity of the No. 4/0 AWG tap conductors had to also equal at least ⅓ the rated ampacity of the main feeder-circuit conductors. As computed in the example, taking 300% of the rated-ampacity of the tap conductors actually computed the maximum rated-load ampacity of the main feeder conductors. The actual 545-amp rated-load ampacity of the 1,000 kcmil THW copper feeder conductors did not exceed the 690 amps allowed. This made it an NE Code-legal tap installation.

The 230-amp rated ampacity of the tap conductors could have been divided by the 545-amp rating of the main feeder conductors. As long as the ratio equals or exceeds ⅓ (0.333 or 33⅓%), the tap installation will still be NE Code legal.

Code challenge 14–2 (25-foot tap rule). A 3-phase, 230-volt, design B, ac induction motor rated at 75 hp will be supplied by conductors tapped to a larger feeder circuit without overcurrent protection at the point of tap. For the following conditions, determine if this is an NE Code-legal installation:

1. The tap consists of a multiconductor cable that contains three 250 kcmil RHW copper conductors and a No. 1/0 AWG THW copper equipment-grounding conductor. The multiconductor cable, which has a total length of 11 feet and is routed to be protected from physical damage, is terminated to a fused knife switch disconnect.
2. The time-delay fuses in the knife switch, which provides the necessary branch-circuit SC-GF overcurrent protection for the motor circuit, are rated at 400 amps.
3. The feeder circuit, which consists of 1,500 kcmil RHW copper conductors, is protected at the point of supply by an inverse-time circuit breaker rated at 1,000 amps.

Motor full-load current: (A) ____________ amps Ref. (B) ____________________
Tap-conductor wire size justification:
Conductor ampacity overrate computation: Ref. (C) ____________________

(D) ____________ % of ____________ amps = ____________ amps

Minimum wire size determination:
Terminal temperature limitation: (E) ______ °C Ref. (F) ____________________
Table-rated ampacity: (G) ____________ amps Ref. (H) ____________________
Wire size: (I) ____________ RHW copper

Single-motor branch-circuit SC-GF overcurrent protection overrate computations, using time-delay fuses:

Ref. (J) ________________

Standard overrate computation: Ref. (K) ________________

(L) ____________ % of ____________ amps = ____________ amps

Standard OCPD ratings: Ref. (M) ________________

NHS: (N) ____________ amps Ref. (NHS) (O) ________________

NE Code-maximum overrate computation: Ref. (P) ________________

(Q) ____________ % of ____________ amps = ____________ amps

NE Code-maximum OCPD rating: Ref. (R) ________________

NLS: (S) ____________ amps Ref. (NLS) (T) ________________

Maximum feeder OCPD rating, using the 25-foot tap rule:

Given tap-conductor wire size: (U) ____________ RHW copper

Terminal temperature limitation: (V) _____ °C Ref. (W) ________________

Table-rated ampacity: (X) ____________ amps Ref. (Y) ________________

Given feeder-conductor wire size: (Z) ____________ RHW copper

Terminal temperature limitation: (1A) ____ °C Ref. (1B) ________________

Table-rated ampacity: (1C) ____________ amps Ref. (1D) ________________

Overrate computation: Ref. (1E) ________________

(1F) ____________ % of ____________ amps = ____________ amps

The NE Code requirements using the 25-foot tap rule (1G) ____________ (have or have not) been met.

For all practical purposes, the 10-foot motor-feeder tap rule of Sec. 430-28 parallels the 10-foot tap rule for general circuit installations found in Sec. 240-21(b). The 25-foot motor-feeder tap rule of Sec. 430-28 parallels the 25-foot tap rule for general circuit installations found in Sec. 240-21(c).

The 100-foot motor-feeder circuit tap rule found in the Exception to Sec. 430-28, which is a modification of the 25-foot tap rule, parallels the general high-bay manufacturing area 100-foot tap rule found in Sec. 240-21(e). To keep the conduit or other raceway run visible along its entire length, the 25-foot maximum applies to the horizontal span in both 100-foot tap rules. The tap must be made a minimum of 30 feet above the finished floor line. The total vertical drop cannot exceed 75 feet. Other rules such as no splices and minimum wire sizes of the tap conductors also apply.

14.2 COMBINATION LOADS—SINGLE BRANCH CIRCUIT

Sections 430-53 (a), (b), and (c) lists the rules that govern three separate situations or installation conditions where several motors and/or other loads are to be supplied by a single multioutlet branch circuit.

Section 430-53(a) lists the conditions under which two or more motors, rated at one horsepower or less, can be supplied by a multioutlet branch circuit that has no other loads. The conditions, as shown in Figure 14-3, are best examined by the following example:

Example 14–3. Assuming that the motors have individual overload protection, and that each individual motor controller has a full-load current rating of at least 20 amps, can two ⅙-hp and one ¼-hp, single-phase, 115-volt, squirrel-cage ac induction motors be supplied by a single 20-amp branch-circuit rated at 120 volts?

Do the motors have separate, individual overload protection? Yes.

Motor full-load currents: Ref. T430-150

⅙ hp: 4.4 amps ¼ hp: 5.8 amps

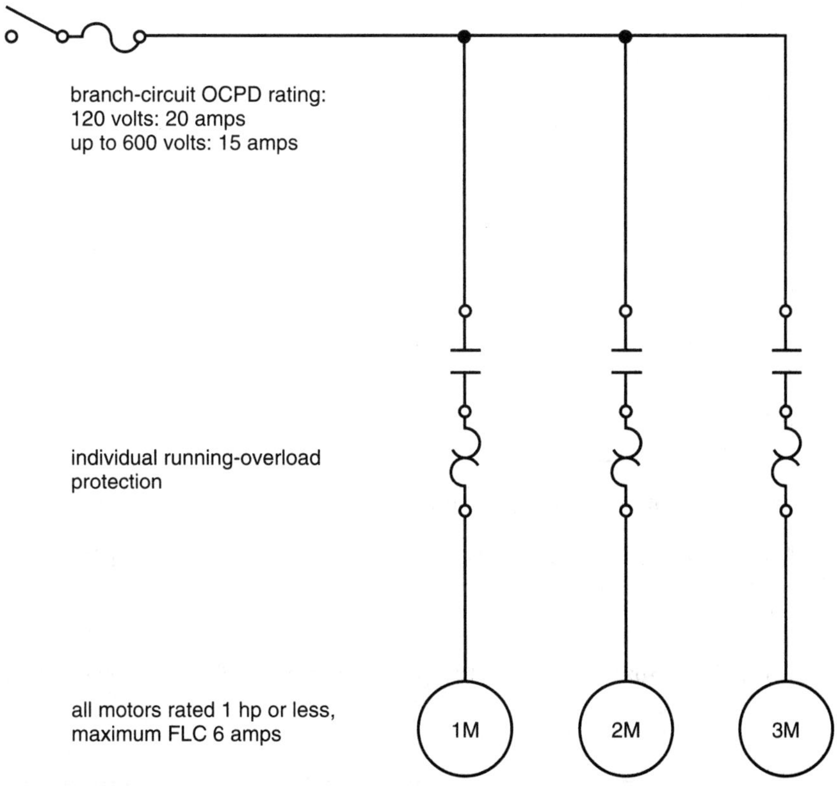

Fig. 14-3 Several motors supplied from a single branch circuit are allowed by Sec. 430-53(a)(1).

Does any motor full-load current exceed 6 amps? No.

Does the branch-circuit SC-GF overcurrent-protection rating exceed any motor controller rating in the group? No.

Does the summation of the full-load current of all the motors exceed the branch-circuit rating?

⅙ hp: 2 × 4.4 amps =	8.8 amps
¼ hp:	5.8 amps
Total:	14.6 amps

No.

The NE Code requirements for several fractional horsepower-rated motors supplied by an individual branch circuit have been met.

Ref. 430-53(a)

According to the motor FLCs listed in T430-148 for single-phase ac motors rated at 115 volts, the supply of several fractional-horsepower motors from a single branch circuit is limited to a maximum of ¼ horsepower because of Sec. 430-53(a)(1): The motor FLC cannot exceed 6 amps.

If the circuit voltage is changed to the 230-volt rating, the horsepower rating in T430-148 increases to ½ horsepower, but the branch-circuit protection rating in Sec. 430-53(a) decreases from 20 to a maximum of 15 amperes.

The requirements of Sec. 430-53(a) are better suited for 3-phase motors with FLCs as listed in T430-150. Except for the lower 115-volt rating, several motors rated up to 1 hp can be supplied by a single branch circuit that is overcurrent protected at a maximum of 15 amps.

Code challenge 14–3. Assuming the motors have individual overload protection, and that each individual motor controller has a full-load current-rating of at least 15 amps, can three ½-hp and two ¾-hp, 3-phase, 208-volt, squirrel-cage ac induction motors be supplied by a single 15-amp branch-circuit rated at 208 volts?

Do the motors have separate, individual overload protection?

(A) ____________ (Yes or No)

Motor full-load currents: Ref. (B) ________________

½ hp: (C) ____________ amps ¾ hp: (D) ____________ amps

Does any motor full-load current exceed 6 amps? (E) ____________ (Yes or No)

Does the branch-circuit SC-GF overcurrent protection rating exceed any motor controller rating in the group?

(F) ____________ (Yes or No)

Does the summation of the full-load current of all the motors exceed the branch-circuit rating?

½ hp: (G) ____________ × ____________ amps = ____________ amps

¾ hp: (H) ____________ × ____________ amps = ____________ amps

Total: (I) ____________ amps

(J) ____________ (Yes or No)

The NE Code requirements for several fractional-horsepower–rated motors supplied by an individual branch circuit (K) ____________ (have or have not) met.

Ref. (L) ______________________

If integral-horsepower motors are to be supplied from a single branch circuit (Fig. 14-4), or if motor and nonmotor loads are to be supplied from a single branch circuit, the motor loads must be computed by the guidelines of Sec. 430-53(b). The nonmotor loads must be computed by the guidelines of Sec. 220-3(a). Section 430-53(a) would not be applicable since only motor loads of one horsepower or less are cited in that section.

Three principle conditions must be met: The rating of the branch-circuit OCPD cannot exceed the maximum rating or setting of the branch-circuit SC-GF protection device as computed by Exception No. 2 to Sec. 430-52(c)(1) for the smallest motor supplied. Each motor must have individual overload protection (can be thermal protection integral with the motor). The branch-circuit OCPD should not open under the most severe loading situation.

Example 14–4. Two 2-hp, and one 3-hp, 3-phase, 230-volt squirrel-cage induction motors will be supplied by a single 3-phase, 230-volt branch circuit. If the motors are so interlocked that they cannot be started simultaneously, if each motor has individual

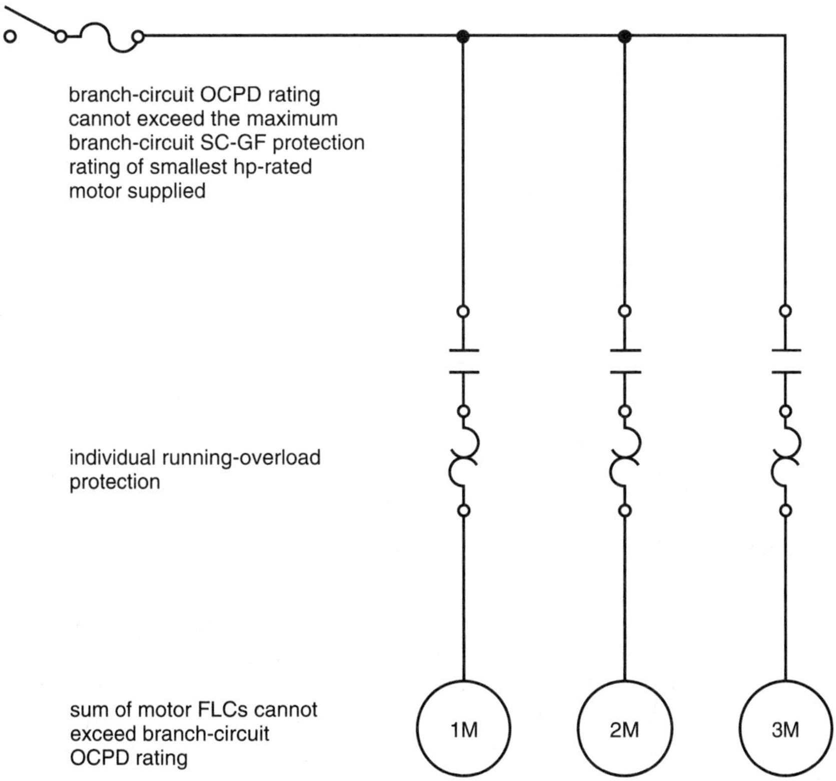

Fig. 14-4 For integral horsepower-rated motors supplied by a single branch circuit, the smallest motor has to be protected from fault overcurrent [Sec. 430-53(b)].

overload protection, and if the branch circuit is protected by a 3-phase, inverse-time circuit breaker rated at 25 amps, would this be an NE Code-legal installation?

Motor full-load currents: Ref. T430-150

2 hp: 6.8 amps   3 hp: 9.6 amps

Does the total full-load current of all the motors exceed the branch-circuit rating?

2 hp: 2 × 6.8 amps =	13.6 amps
3 hp:	9.6 amps
Total:	23.2 amps

No.

Does the branch-circuit SC-GF overcurrent-protection rating exceed the NE Code-maximum protection rating for the smallest motor in the group?

2-hp, inverse-time circuit breaker:

Overrate computation: Ref. 430-52(c)(1), Excp. No. 2(c)(1)

400% of 6.8 amps = 27.2 amps

NE Code-maximum rating: Ref. 240-6(a)

25 amps

No.

The NE Code requirements for several integral-horsepower–rated motors supplied by an individual branch circuit have been met.

Ref. 430-53(b)

As a rule, to use Sec. 430-53(b), the rated horsepower of the motors have to be close, such as a 2-hp motor and a 3-hp motor supplied from the same branch circuit. If the horsepower ratings are far apart, such as a 2-hp and a 7½ hp motor supplied by the same branch circuit, the low value of the branch-circuit fuse rating or circuit breaker setting required to protect the smaller motor against fault conditions will normally open under even moderate load conditions.

The most severe load condition, other than all loads being connected when the circuit is energized (started simultaneously), would be all loads operating in the full-load condition and startup of the largest motor under load.

Code challenge 14–4. Two 5-hp, and one 71/2-hp, 3-phase, 460-volt squirrel-cage induction motors are to be supplied by a single 3-phase, 460-volt branch circuit. If the motors are so interlocked that they cannot be started simultaneously, if each motor has individual overload protection, and if the branch circuit is protected by a 3-phase, inverse-time circuit breaker rated at 30 amps, would this be an NE Code-legal installation?

Motor full-load currents: Ref. (A) ____________________

5 hp: (B) ____________ amps   7½ hp: (C) ____________ amps

Does the total full-load current of all the motors exceed the branch-circuit rating?

5 hp: ____________ × ____________ amps = ____________ amps
7½ hp: ____________ amps
Total: (D) ____________ amps
(E) ____________ (Yes or No)

Does the branch-circuit SC-GF overcurrent-protection rating exceed the NE Code-maximum rating for the smallest motor in the group:

(F) ____________ hp, inverse-time circuit breaker:

Overrate computation: Ref. (G) ____________

(H) ____________ % of ____________ amps = ____________ amps

NE Code-maximum rating: Ref. (I) ____________
(J) ____________ amps
(K) ____________ (Yes or No)

The NE Code requirements for several integral-horsepower–rated motors supplied by an individual branch circuit (L) ____________ (have or have not) been met.
Ref. (M) ____________

14.3 COMBINATION-LOAD EQUIPMENT—SINGLE BRANCH CIRCUIT

Section 430-53(c) governs the maximum rating or setting of the overcurrent protection device of multimotor and/or combination-load equipment installed on an individual branch circuit (Fig. 14-5). The bulk of this section deals with the field assembly of multi-motor and/or combination-load equipment.

Machinery completely assembled with more than one motor would be considered as only multimotor equipment. Machinery completely assembled with a motor, inspection lights, and strip heating or other support equipment would be considered as combination-load equipment. Machinery that contains more than one motor and other support equipment would be considered as multimotor/combination-load equipment.

In field assembly of multimotor and/or combination-load equipment, the branch-circuit SC-GF overcurrent-protection device, the controller, and the individual (separate) overload-overcurrent protection device for each motor must all be listed (by a local authority-recognized listing agency) for group use, and be provided with instructions (by the manufacturer) for use with each other. If circuit breakers are used for the branch-circuit protection, Sec. 430-53(c)(3) will only allow the inverse-time type.

In field assembly, the rating or setting of the branch-circuit overcurrent-protection device is computed at the NE Code-maximum overrates [Sec. 430-52(c), Excp. No. 2] for the largest motor in the group, plus the sum of the other motor FLCs and/or the total current load of the rest of the circuit.

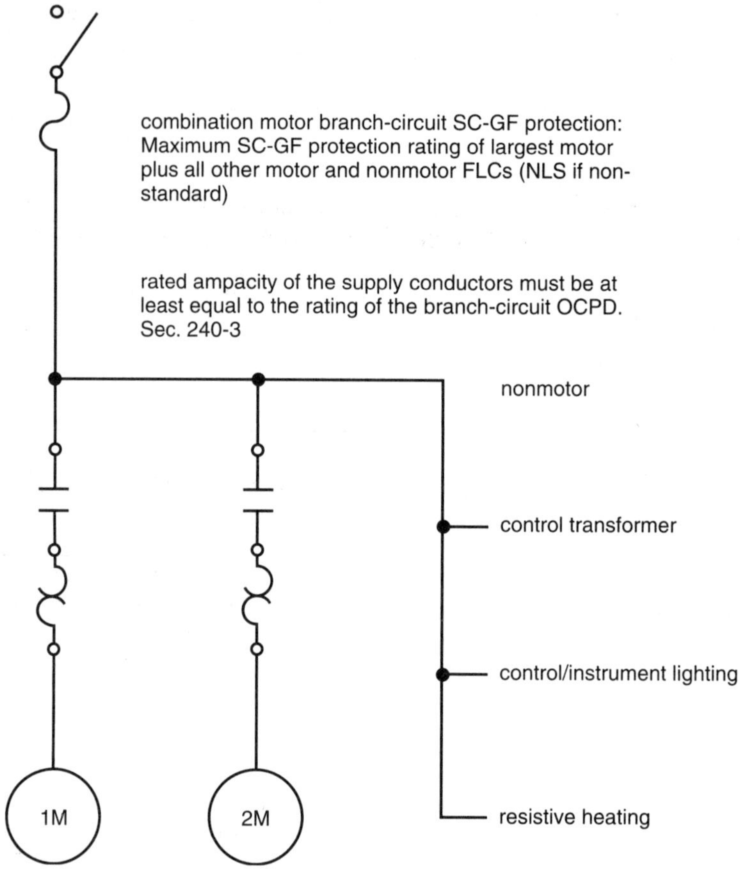

Fig. 14-5 Sec. 430-53(c) governs the installation of multimotor/combination-load equipment on a single branch circuit.

When the computed full-load current does not correspond to a standard size of fuse or setting of a circuit breaker as listed in Sec. 240-6(a), Sec. 430-53(c)(4) will allow the next higher standard size to be used. The next higher standard size is allowed only when the branch-circuit supply conductors have an ampacity rating greater than the computed value of overcurrent protection.

Example 14–5. The computed rating of the branch-circuit overcurrent-protection device for a field assembly of multimotor/combination-load equipment is equal to 61 amps. If No. 6 AWG THW copper conductors are used to supply this branch-circuit load, can the branch-circuit fuses or inverse-time circuit breaker be rated at 70 amps?

Table-rated ampacity determination of the branch-circuit conductors:

Wire terminal temperature limitations: 60°C — Ref. 110-14(c)(1), Excp. 1
No. 6 AWG THW copper: 55 amps — Ref. T310-16

Next higher standard size of fuse or circuit breaker above conductor rated ampacity:

60 amps Ref. 240-6(a)

The NE Code requirements for the field assembly of multimotor/combination-load equipment on an individual branch circuit have not been met. The 70-amp branch-circuit overcurrent-protection rating cannot be used.

Ref. 430-53(c)(4)

The disqualifier in this example is the wire terminal temperature limitations of Sec. 110-14(c). The table-rated ampacity of No. 6 AWG THW copper as listed in the 75°C temperature column of T310-16 is 65 amps. At this point, the installation is NE Code legal.

Since the wire terminal temperature rating of either the circuit breaker or the terminal strip in the equipment (could be wire nuts) was not specified, Sec. 110-14(c)(1), Excp. 1 allows the 75°C rated THW copper wire to be used as the branch-circuit supply for this equipment, but limits the rated ampacity of No. 6 AWG copper to the ampacity listed in the 60°C temperature column of T310-16. The next higher rating or setting of branch-circuit overcurrent protection above the 55-amp rating, as listed in Sec. 240-6(a), is only 60 amps. The 70-amp circuit breaker rating would be in NE Code violation of Sec. 430-53(c)(4).

The requirements of Sec. 430-53(c)(4) are in parallel to Sec. 240-3(b): As long as the rated ampacity of the circuit conductors exceeds the computed rated-load (full-load) current, but neither correspond to a standard rating or setting of overcurrent-protection device, the next higher standard rating or setting (size) can be installed without a further increase in the rated ampacity (wire size) of the circuit conductors.

Code challenge 14–5. The computed rating of the branch-circuit overcurrent-protection device for a field assembly of multimotor/combination-load equipment is equal to 83 amps. If No. 3 AWG THW copper conductors are used to supply this branch-circuit load, can the branch-circuit fuses or inverse-time circuit breaker be rated at 100 amps?

Table-rated ampacity determination of the branch-circuit conductors:

Wire terminal temperature limitations: (A) ____________ °C

Ref. (B) ____________________

No. 3 AWG THW copper: (C) __________ amps Ref. (D) ____________________

Next higher standard size of fuse or circuit breaker above conductor rated ampacity:

(E) ____________ amps Ref. (F) ____________________

The NE Code requirements for the field assembly of multimotor/combination-load equipment on an individual branch circuit (G) ____________ (have or have not) been met. The 100-amp branch-circuit overcurrent-protection rating (H) ____________ (can or cannot) be used.

Ref. (I) ____________________

When a single hermetically sealed refrigerant motor compressor and other loads are supplied by a single branch-circuit, Sec. 440-22(b) requires the rating or setting of the branch-circuit overcurrent-protection device to be sufficient (large enough) to handle the starting current of the equipment. Two conditions are cited:

Condition (1) is the procedure to be followed when the motor compressor is the largest load connected to the circuit.

Condition (2) is the procedure to be followed when the motor compressor is not the largest load connected to the circuit.

With either condition, the rating or setting of the multimotor and/or combination-load equipment branch-circuit overcurrent-protection device must also meet the requirements of Sec. 430-53.

When the motor compressor is the largest branch-circuit load, the rating of the branch-circuit short-circuit, ground-fault protection can be computed at 225% [NE Code maximum, parallel to Sec. 430-53(c), using the overrate percentages found in Sec. 440-22(a)] of the motor compressor's largest nameplate current rating (either rated-load current or branch-circuit selection current, whichever is greater), plus the sum of the largest nameplate current rating of any other motor compressors, the sum of the table-rated FLCs of any motors (Tables 430-147 through 430-150), and the sum of full-load currents [Sec. 220-3(a)] for any additional nonmotor loads supplied by the individual branch circuit.

The rated-load ampacity of the branch-circuit supply conductors is governed by the requirements of Sec. 440-34. As always (by reference to Secs. 440-32 and 440-33), the rated-load ampacity of the circuit conductors, according to wire size, must be greater than or equal to 100% of the connected loads at the full-load (rated-load) condition, plus a 25% overrate of the largest motor or motor compressor FLC.

Example 14–6. A single-phase, 2-wire, 240-volt branch circuit will supply a multimotor/combination-load equipment that consists of the following loads:

1 Hermetically sealed refrigerant motor-compressor with a nameplate rated-load current rating of 14 amps
2 Condensor fans rated at ½ hp each
2 Water pumps rated at 1 hp each
1 Resistive-type strip heater with a full-load current rating of 5 amps
1 Control transformer with a primary full-load current rating of 3 amps

Determine:

1. The NE Code-maximum rating of the branch-circuit SC-GF overcurrent protection, using time-delay fuses
2. The NE Code-minimum wire size of the branch-circuit supply conductors, using THHN copper

Motor full-load current determinations: Ref. T430-148

½ hp: 4.9 amps 1 hp: 8 amps

Overrate computation of the branch-circuit SC-GF overcurrent protection rating: Ref. 440-22(b)(1)

Motor-compressor:	225% of 14 amps =	31.5 amps
½ hp:	2 × 4.9 amps =	9.8 amps
1 hp:	2 × 8 amps =	16.0 amps
Strip heat:		5.0 amps
Control transformer:		3.0 amps
Total:		65.3 amps

Standard rating of the time-delay fuses: Ref. 240-6(a)

NLS: 60 amps Ref. (NLS) 440-22(b)(1)

Rated-ampacity overrate computation of the branch-circuit supply conductors: Ref. 440-34

Motor-compressor:		14.0 amps
½ hp:	2 × 4.9 amps =	9.8 amps
1 hp:	2 × 8 amps =	16.0 amps
Strip heat:		5.0 amps
Control transformer:		3.0 amps
Largest motor:	25% of 14 amps =	3.5 amps
Total:		51.3 amps

NE Code-minimum wire size determination:

Terminal temperature limitation: 60°C Ref. 110-14(c)(1), Excp. 1

Table-rated ampacity: 55 amps Ref. T310-16

Wire size: No. 6 AWG THHN copper

The rated-load ampacity of the No. 6 AWG THHN copper conductors, as found in the 90°C temperature column of T310-16 is 75 amps. This rating exceeds both the full-load current computations in the example, as well as the 60-amp branch-circuit overcurrent-protection device rating. But, Sec. 110-14(c)(1) mandated the 60°C temperature column rating of only 55 amps because higher temperature wire terminal ratings were not specified.

The better electrical installation, which is by design instead of NE Code-minimum, is to size the branch-circuit conductors according to the general requirement of Sec. 240-3. The rated-load ampacity of the circuit conductors, as determined by the 60°C temperature column of T310-16 according to wire size, should be greater than or equal to the 60-amp rating or setting of the branch-circuit overcurrent-protection device: No. 4 AWG THHN copper.

Code challenge 14–6. A single-phase, 2-wire, 208-volt branch circuit will supply a multimotor/combination-load equipment that consists of the following loads:

1 Hermetically sealed refrigerant motor compressor with a nameplate rated-load current rating of 12 amps and a branch-circuit selection current of 14 amps.
2 Water pumps rated at ¾ hp each
2 Condensor fans rated at ¼ hp each
1 Resistive-type strip heater with a full-load current rating of 4 amps
1 Control transformer with a primary full-load current rating of 1½ amps

Determine:

1. The NE Code-maximum rating of the branch-circuit SC-GF overcurrent protection, using time-delay fuses
2. The NE Code-minimum wire size of the branch-circuit supply conductors, using XHH copper

Motor full-load current determinations: Ref. (A) ____________

¼ hp: (B) ________ amps ¾ hp: (C) ________ amps

Overrate computation of the branch-circuit SC-GF overcurrent protection rating:

Ref. (D) ____________

Motor-compressor: (E) ______ % of ______ amps = ______ amps
¼ hp: ______ × ______ amps = ______ amps
¾ hp: ______ × ______ amps = ______ amps
Strip heat: ______ amps
Control transformer: ______ amps
Total: (F) ______ amps

Standard rating of the time-delay fuses: Ref. (G) ____________

NLS: (H) ________ amps Ref. (NLS) (I) ____________

Rated-ampacity overrate computation of the branch-circuit supply conductors:

Ref. (J) ____________

Motor-compressor: = ______ amps
¼ hp: ______ × ______ amps = ______ amps
¾ hp: ______ × ______ amps = ______ amps
Strip heat: ______ amps
Control transformer: ______ amps
Motor-compressor: (K) ______ % of ______ amps = ______ amps
Total: (L) ______ amps

NE Code-minimum wire size determination:

Terminal temperature limitation: (M) ____ °C Ref. (N) ____________
Table-rated ampacity: (O) ________ amps Ref. (P) ____________
Wire size: (Q) ________ XHH copper

If the hermetically sealed refrigerant motor compressor is not the largest load connected to the branch circuit, then the motor compressor is treated as just another motor load. The branch-circuit overcurrent protection is sized in accordance with Sec.

430-53(c)(4) when other motor loads are supplied, and in accordance with Sec. 240-3 when other nonmotor loads are supplied.

If the machinery is factory-assembled as multimotor and/or combination-load equipment, the requirements of Sec. 430-53(c) are met by the manufacturer. The project engineer must meet other NE Code requirements to correctly supply and protect this type of equipment.

For branch-circuit or feeder conductors, Secs. 430-25 and 440-35 require their rated-load ampacity to be at least equal to the minimum-circuit ampacity as listed on the multimotor and/or combination-load equipment nameplate.

For the branch-circuit SC-GF overcurrent protection, Sec. 430-54 stipulates that the maximum rating of the branch-circuit protection shall not exceed the maximum protection rating listed on the nameplate of multimotor and/or combination-load equipment. This is especially true for equipment such as chiller units or condensor units that contain a hermetically sealed refrigerant motor compressor.

In reference to Sec. 110-3(b), if the nameplate of listed multimotor and/or combination-load equipment specifies "fuses," then only fuses can be used as the overcurrent protection for the equipment. On the other hand, if the nameplate specifies "overcurrent-protection rating," then either fuses or circuit breakers can be installed.

Unless the multimotor/combination-load equipment is the major portion of the motor-feeder load, the maximum branch-circuit protection rating on the nameplate of multimotor and/or combination-load equipment would be used at 100% in the motor-feeder SC-GF overcurrent-protection calculations of Secs. 430-62 and 430-63.

14.4 DUTY-CYCLE MOTOR CIRCUITS

The requirements of Sec. 430-22(a) and the 125% overrate for the circuit-conductor ampacity, apply to all motors in general applications. Motors normally run continuously under varying load conditions. Motors designed for a specific task or purpose may only run intermittently or for short periods of time. An electric garage door opener or a freight elevator are examples of specific-task or duty-cycle motor applications.

Specific-task motors normally have duty cycles in which the motors run for a period of time, and then they are at rest (off) for a (longer) period of time (time to cool down). Section 430-22(a), Excp. No. 1 alters the 125% overrate requirement when a motor is used with duty-cycle service.

Table 430-22(a) Exception establishes the percentages of motor full-load current the motor supply conductor ampacities must be equal to. The overrate or underrate percentage to be used is dependent on both the type of duty cycle and on the minute rating of the motor. The duty cycles are defined in Art. 100.

Duty-cycle motors are normally operated in an overload condition. Section 430-33 omits the overload requirements of Part C in Art. 430. The rating of the motor overcurrent protection is limited only to the standard rating of branch-circuit SC-GF overcurrent protection computed by the overrates allowed by T430-152.

Example 14–7. A 3-phase, 460-volt, design B ac induction motor rated at 50 hp is used for periodic duty on a roll-conveyor assembly in a pasteboard box factory. If the motor is rated for continuous duty, determine:

1. The NE Code-maximum rating of the motor SC-GF overcurrent protection, using time-delay fuses
2. The NE Code-minimum wire size of motor branch-circuit conductors, using THHN copper

Motor full-load current: 65 amps Ref. T430-150

Single-motor branch-circuit SC-GF overcurrent protection overrate computations:

Ref. 430-33 Ref. T430-152

175% of 65 amps = 113.8 amps

Standard ratings of time-delay fuses: Ref. 240-6(a)

NLS: 110 amps Ref. (NLS) 430-33

Conductor ampacity overrate computation for a continuous-duty rated motor used for periodic duty:

Ref. T430-22(a), Excp.

140% of 65 amps = 91 amps

NE Code-minimum conductor wire size determination:

Wire terminal temperature limitation: 60°C Ref. 110-14(c)(1), Excp. 1

Table-rated ampacity: 95 amps Ref. T310-16

Wire size: No. 2 AWG THHN copper

Duty-cycle motors are normally identified on the nameplate as "minute-rated" motors. According to T430-22(a) Exception, the motor can be 5-minute rated, 15-minute rated, 30-minute-rated, or 60-minute rated.

The second paragraph of Sec. 430-33 stipulates that any motor application is considered to be continuous duty, unless the nature of the driven machinery (apparatus) is such that the motor cannot operate continuously (three hours or more—nonstop, Art. 100) under any condition of use.

Code challenge 14–7. A 3-phase, 208-volt, design B ac induction motor rated at 30 hp is used for varying duty on a fluoroveyor assembly in a plastics factory. If the motor is only 30-minute rated, determine:

1. The NE Code-maximum rating of the motor SC-GF overcurrent protection, using an inverse-time circuit breaker
2. The NE Code-minimum wire size of motor branch-circuit conductors, using RHW copper

Motor full-load current: (A) ____________ amps Ref. (B) ____________

Single-motor branch-circuit SC-GF overcurrent protection overrate computations:

Ref. (C) ____________ Ref. (D) ____________

(E) ____________ % of ____________ amps = ____________ amps

Standard rating of time-delay fuses: Ref. (F) ____________

NLS: (G) ____________ amps Ref. (NLS) (H) ____________

Conductor ampacity overrate computation for a 30-minute rated motor used for varying duty:

Ref. (I) ____________

(J) ____________ % of ____________ amps = ____________ amps

NE Code-minimum conductor wire size determination:

Wire terminal temperature limitation: (K) ________ °C Ref. (L) ____________

Table-rated ampacity: (M) ____________ amps Ref. (N) ____________

Wire size: (O) ____________ RHW copper

14.5 MULTIPLE-SPEED MOTOR CIRCUITS

Some ac induction motors are wound as consequent-pole motors where both ends (beginning and ending leads) of the stator windings are brought out of the motor. Through the use of relays or contactors, the windings can be interconnected to change the pole configuration of the motor. As discussed in Sec. 12.9 of Chapter 12, for a motor drive controller-disconnecting means, changing the pole configuration changes the baseline (running) speed of the motor, hence the name multispeed motors.

For the branch-circuit SC-GF overcurrent protection, the overrate percentages are determined by T430-152 for standard-load operation, and by Sec. 430-52(c)(1), Excp. No. 2 for motors that drive high-inertia, long-acceleration loads. Section 430-52(c)(4) requires the overrates to be applied to the smallest winding protected. The smallest winding would reflect the lowest winding full-load current rating indicated on the motor nameplate according to speed rating.

Separate overload protection, located on the load side of the controller, is required by Sec. 430-32(a)(1) for integral-horsepower-rated multispeed ac motors and Sec. 430-32(c)(1) for fractional-horsepower-rated multispeed ac motors to be installed for each winding connection. The overrate values of the overload heaters in a separate overload relay are the same as for any motor: 125% standard; 140% NE Code maximum. In correlation to Sec. 430-52(c)(4), the allowed overrate should be applied to the lowest winding full-load current rating indicated on the motor nameplate according to speed rating.

The second paragraph of Sec. 430-22(a) requires the 125% overrate applied to the branch-circuit conductors on the line side of the controller to be based on the highest winding full-load current rating indicated on the motor nameplate according to speed rating.

<u>Example 14–8.</u> A single-phase, 208-volt, 5-hp, 2-speed ac induction motor has winding full-load current ratings of 24 amps and 32 amps for the respective speeds. Determine:

1. The NE Code-maximum rating of the motor branch-circuit SC-GF overcurrent protection, using nontime fuses

2. The NE Code-minimum wire size of motor branch-circuit conductors, using XHHW copper

Branch-circuit SC-GF overcurrent-protection overrate computations for a two-speed motor:

Ref. 430-52(c)(1)
Motor FLC: 24 amps — Ref. 430-52(c)(4)
Overrate computation: — Ref. T430-152

300% of 24 amps = 72 amps

Standard rating of nontime fuses: — Ref. 240-6(a)
NHS: 80 amps — Ref. (NHS) 430-52(c)(1), Excp. 1

Branch-circuit conductor ampacity overrate computation for a two-speed motor:
Motor FLC: 32 amps — Ref. 430-22(a)
Overrate computation:

125% of 32 amps = 40 amps

Branch-circuit conductor wire size determination:
Wire terminal temperature limitation: 60°C — Ref. 110-14(c)(1), Excp. 1
Table-rated amapcity: 40 amps — Ref. T310-16
Wire size: No. 8 AWG XHHW copper

The type of driven load was not specified in this example. The maximum percentages of overrate for the branch-circuit SC-GF overcurrent protection of a motor circuit should not be implemented unless the type of connected (driven) load is determined to be a high-inertia, long-acceleration type load.

Code challenge 14–8. A 3-phase, 460-volt, 50-hp, 2-speed ac induction motor has winding full-load current ratings of 58 amps and 64 amps for the respective speeds. Determine:

1. The NE Code-maximum rating of the motor branch-circuit SC-GF overcurrent protection, using an inverse-time circuit breaker
2. The NE Code-minimum wire size of motor branch-circuit conductors, using THHW copper

Branch-circuit SC-GF overcurrent protection overrate computations for a two-speed motor:

Ref. (A) ____________________
Motor FLC: (B) ____________ amps — Ref. (C) ____________________
Overrate computation: — Ref. (D) ____________________

(E) ____________ % of ____________ amps = ____________ amps

Standard ratings of inverse-time circuit breakers: Ref. (F) ________________
NHS: (G) ____________ amps Ref. (NHS) (H) ________________

Branch-circuit conductor ampacity overrate computation for a two-speed motor:
Motor FLC: (I) ____________amps Ref. (J) ________________
Overrate computation:

(K) ____________ % of ____________ amps = ____________ amps

Branch-circuit conductor wire size determination:
Wire terminal temperature limitation: (L) ____ °C Ref. (M) ________________
Table-rated ampacity: (N) ____________ amps Ref. (O) ________________
Wire size: (P) ____________ THHW copper

14.6 WOUND-ROTOR SPEED CONTROL

Another form of ac motor speed control uses a wound-rotor induction motor equipped with a slip-ring brush assembly that offers field connection to the rotor-circuit windings in the motor terminal housing. The windings in the rotor circuit of a standard squirrel-cage ac induction motor, whether cast or wire-wound, is purposely short-circuited on the rotor assembly, such that no field connections to the rotor circuit are available in the motor terminal housing. In both types of motors, stator winding lead connections must be provided in the motor terminal housing.

For speed control or reduced line current (stator current) yet full torque on starting, the wound-rotor induction motor is forced to function as both a motor and a transformer. The motor action applies torque to the rotor. The transformer action applies ac power to a resistive load bank connected in the secondary (rotor circuit) of the motor. (Transformers will be discussed in Chapters 15 and 16 of this text.)

Without going too deep into motor theory, the amount of power available from the stator circuit of the wound-rotor induction motor is determined by the frequency, voltage, and horsepower ratings at the point of manufacture. Maximum horsepower is available when the rotor circuit is short circuited (as with the standard squirrel-cage induction motor) causing maximum rotor current flow. When resistance is placed in series with the rotor windings, the rotor-circuit current is reduced, which in turn reduces the strength of the rotor field. The slip of the motor increases, lowering the speed of the rotor.

Speed (or torque) is controlled by the amount of resistance in the rotor circuit. The resistance is normally staged in the form of a resistor bank, meaning full resistance is placed in the rotor circuit on starting. As the rotor begins to speed up, the resistance is shorted out (short-circuited, paralleled, or shunted) in successive closures of multiple power contactors, until all the resistance is removed, and the rotor is turning at baseline speed (determined by the number of poles in the motor; refer to Chapter 12, Sec. 12.9).

For speed control, different values (normally referred to as steps) of resistance are left in the rotor circuit (Fig. 14-6). The motor may be rated for operation at two, three, or four different speeds, although the step speed should never be less than 50% of baseline speed at the slowest speed setting.

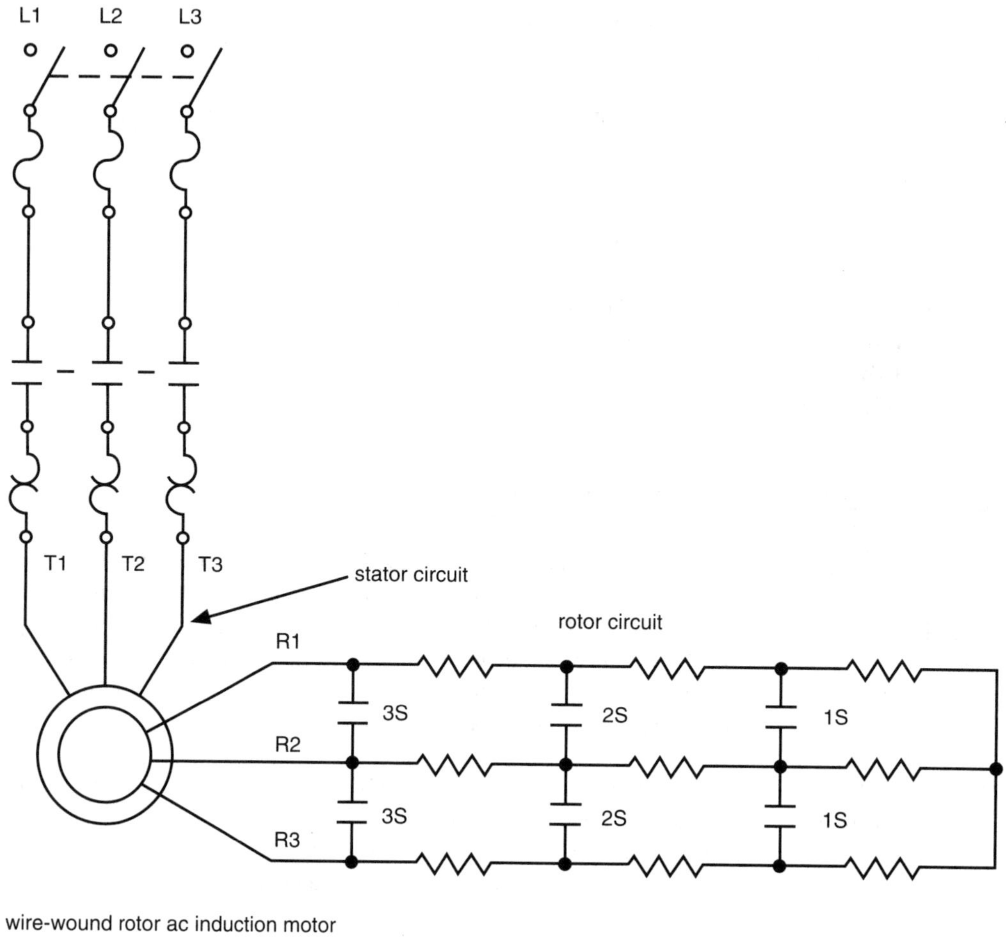

Fig. 14-6 Speed control of a wire-wound rotor ac induction motor is accomplished by making the motor function as a transformer.

In terms of branch-circuit SC-GF protection and the wire size of the branch-circuit supply conductors, the motor circuit for wound-rotor induction motors is sized in the same manner as general-application motor circuits.

Section 430-22(a) requires the branch-circuit supply conductors to the motor controller, and the motor-circuit supply conductors to the stator circuit to be sized at a minimum of 125% of the NE Code table-value motor full-load current to correlate with the allowed overload percentages of Sec. 430-32. When the wound-rotor induction motor drives high-inertia, long-acceleration loads, Sec. 240-3 requires the same circuit conductors to be sized at a minimum of 140% overrate to correlate with the allowed overpercentages of Sec. 430-34.

There are no special provisions listed in Sec. 430-52(c) for wound-rotor ac induction motors (other than the low value of percentages listed in T430-152). The standard overrate percentages are found in T430-152, according to the type of branch-circuit

SC-GF overcurrent protection selected for the motor circuit. The NE Code-maximum overrate percentages are found in Excp. No. 2 to Sec. 430-52(c)(1).

For separate overload protection (standard versus the Exception), Sec. 430-32(d) will allow the secondary rotor circuit (resistance bank) of the wound rotor ac induction motor to be protected against running overload by the overload heaters (or other separate overload-detection devices) located on the load side of the motor starter (controller), which is the line side (supply side) of the stator circuit.

In other words, the branch-circuit, motor-circuit supply for a wound-rotor induction motor is computed and installed in the same manner as for general-application motor circuits. Additional circuitry is required for the secondary rotor circuit. The added concern is the rated ampacity of the secondary circuit conductors as outlined in Sec. 430-23.

When the motor is rated as continuous duty—in other words, speed control is part of the manufacturing process—Sec. 430-23(a) requires the rated ampacity of the wound-rotor secondary-circuit conductors to be computed at a minimum of 125% of the maximum (full-load) secondary rotor current as specified on the motor nameplate.

When the wound-rotor induction motor is minute-rated as a duty-cycle motor, Sec. 430-23(b) requires the rated ampacity of the wound-rotor secondary-circuit conductors to be computed at the minimum overrate percentages listed in Table 430-22(a) Exception. As with continuous-duty rated wound-rotor induction motors, the overrate percentage selected must be applied to the maximum (full-load) secondary rotor current as specified on the motor nameplate.

When the resistor load bank will be located separate from the motor controller, the rated ampacity and resulting wire size of the secondary rotor-circuit conductors must be determined according to the duty classification of the resistors and the overrate or underrate percentages listed in T430-23(c). The percentage to use depends on whether the resistors are rated for starting duty, meaning no speed control—just a ramp-type start to line speed; intermittent duty—mainly starting, but some speed control; or continuous-duty—starting and speed control.

Example 14–9. A 3-phase, 460-volt, wound-rotor, ac induction motor rated at 200 hp has a nameplate rotor-circuit current FLC rating of 230 amps. If the resistors in the secondary load bank are only rated for "light starting duty," determine the NE Code-minimum wire size of the secondary rotor-circuit conductors using THW copper.

Secondary rotor-circuit conductor ampacity overrate computation:

Ref. T430-23(c)

35% of 230 amps = 80.5 amps

Secondary rotor-circuit conductor wire size determination:

Wire terminal temperature limitation: 60°C — Ref. 110-14(c)(1), Excp. 1
Table-rated amapcity: 85 amps — Ref. T310-16
Wire size: No. 3 AWG THW copper

Again, the branch-circuit and motor-circuit supply conductors, the branch-circuit SC-GF protection, and the motor-circuit running overload protection would be sized in the same manner as motors for general applications in this example. The secondary rotor-circuit conductor wire size is an add-on to the procedures demonstrated in Chapter 13 of this text.

Code challenge 14–9. A 3-phase, 230-volt, wound-rotor, ac induction motor rated at 100 hp has a nameplate rotor-circuit current FLC rating of 242 amps. If the resistors in the secondary load bank are only rated for "light intermittent duty," determine the NE Code-minimum wire size of the secondary rotor-circuit conductors using THHN copper.

Secondary rotor-circuit conductor ampacity overrate computation:

Ref. (A) ______________

(B) __________ % of __________ amps = __________ amps

Secondary rotor-circuit conductor wire size determination:

Wire terminal temperature limitation: (C) _____ °C Ref. (D) ______________

Table-rated ampacity: (E) __________ amps Ref. (F) ______________

Wire size: (G) __________ THHN copper

14.7 WYE-DELTA STARTING

When large integral horsepower motors drive high-inertia, long-acceleration loads, a wye-delta starting method is often employed to reduce the torque applied to the driven load on across-the-line startup (full-voltage starting).

Wye-delta starting is limited to 3-phase power distribution systems and a special type of delta-wound motor that has the beginning and ending of all three motor windings (one for each phase) brought out of the motor assembly for field connection to the starter load conductors in the motor terminal housing.

When a 3-phase motor is wye connected, the values of line current and phase or winding current are equal. The values of line voltage and winding voltage differ by the square root of 3 (1.732). The line voltage is 173% (1.732) greater than the winding voltage. Said another way, the winding or phase voltage is only 58% (57.7%) of the line voltage value (discussed in greater detail in Chapter 15, Transformers).

Voltage and current ratios trade places when switching from wye to delta. When a 3-phase motor is delta connected, the values of line voltage and phase or winding voltage are equal. The delta line current is 173% (square root of 3) greater that the single-phase winding current. As with the wye voltage ratio, the delta winding or phase current is equal to 58% (57.7%) of the line current value.

According to T430-150, a 3-phase, 460-volt, delta-wound ac induction motor rated at 50 hp will draw a line current of 65 amps when operated at full load. In the delta connection, the winding current is equal to 58% of the 65 amps or 37.7 amps. Dividing the winding voltage by the winding current, the winding impedance: 460 volts ÷ 37.7 amps = 12.2 ohms.

If the motor winding leads are connected for a wye start, the motor winding, rated at 460 volts, will only receive 267 volts. Remember in the wye configuration, the winding voltage is equal to 58% of the line voltage. Dividing the winding voltage by the winding impedance: 267 volts ÷ 12.2 ohms = 21.9 amps. The motor is starting with only one-third of the normal line current (21.9 amps ÷ 65 amps). Reduced line current means reduced torque on starting.

Once the rotor has attained about one-half to two-thirds of its running speed, a change in the control contactors switches the motor from a wye start to a delta run condition. The motor will run at line speed, and if operating at full load in our example, the line current will be 65 amps.

If the motor is heavily loaded, the rotor may never attain the one-half running speed due to the loss in torque. More often the control scheme uses a timed-delay start which will automatically change the motor winding connections from wye to delta, regardless of the value of attained rotor speed.

The decrease in torque can be computed at the square of the voltage loss. Decreasing from 460 volts to 267 volts, the voltage decrease was to 58%. This value squared is equal to 33.6% or one-third (normalizing the values)—the same percentage as in the reduction of line current between a wye start and a delta run of the motor.

In terms of sizing the branch-circuit supply to a wye-delta motor circuit, the SC-GF overcurrent protection, disconnecting means, and supply conductors are sized in the same manner as for motors of general application shown in Chapter 13. The separate overload protection and the motor-circuit conductors are sized in a different manner.

A wye-delta starting method requires a special assembly of control power contactors which can be used to start the motor in a wye configuration, and once started, convert the motor connections into a delta configuration (Fig. 14-7). This is why six motor leads instead of the normal three are extended from the motor controller to the motor terminal housing.

The line values of motor full-load current are present in the supply circuit only up to the line side of the power contactors (Fig. 14-8). From the power contactors to the motor (in the motor circuit), only one-third of the motor full-load current is being conveyed in the wye starting configuration. When changed over to a delta run configuration, the motor circuit current is increased to only 58% of the motor nameplate full-load current.

Using across-the-line electromechanical motor starters, a single overload relay with three separate overload heaters are required with the wye-delta starting scheme. For the size of the overload heaters, the delta winding current, which is 58% of the motor FLC line current, is overrated according to the percentages found in Sec. 430-32 for standard motor operation (125%), and by Sec. 430-34 for high-inertia, long-acceleration load starting (140% maximum).

Section 430-6(a) requires the overrate percentage of the running overload protection (starter heaters) to be applied to the actual motor full-load current as specified on the individual motor nameplate. The 50 hp motor in our discussion has a worst-case FLC rating (NE Code table value) of 65 amps. If the nameplate FLC for this motor is 60 amps, the overload heater current, before allowed overrate, would have to be computed at 58%. The standard operation 125% overrate, or the NE Code-maximum 140% overrate would be applied to a motor FLC current of only 34.8 amps. This modification for wye-delta starting is allowed by the third paragraph of Sec. 430-32(a)(1).

Section 430-6(a) requires the branch-circuit supply conductors for general motor

Fig. 14-7 A wye-delta starting method requires a special assembly of three power contactors that can start the motor in a wye configuration and switch it to run in delta configuration.

applications to be sized according to the worst-case FLC rating as listed in the NE Code. The second paragragh of Sec. 430-22(a) (new in the 1996 edition of the NE Code) will allow both the branch-circuit supply conductors and the motor-circuit conductors to be sized according to the motor nameplate FLC.

The better design method is to size the branch-circuit supply conductors according to the worst-case motor FLC and to size the motor-circuit conductors according to the nameplate FLC. Sized in this manner, the branch-circuit supply conductors will not have to be resized with the change and replacement with a motor of another manufacture (different motor FLC).

The Exception to Sec. 430-22(a) will allow the motor-circuit conductors to be sized at the same current value as the overload heaters.

Example 14–10. If a 3-phase, 460-volt, 200-hp, design B, delta-wound ac induction motor, which has a service factor of 1.15 and a nameplate full-load current rating of 232 amperes, will be supplied with a wye-delta starting method, determine:

1. The standard and maximum ratings of time-delay fuses used for the branch-circuit SC-GF overcurrent protection
2. The standard and maximum ratings of thermal heaters installed in a separate overload relay

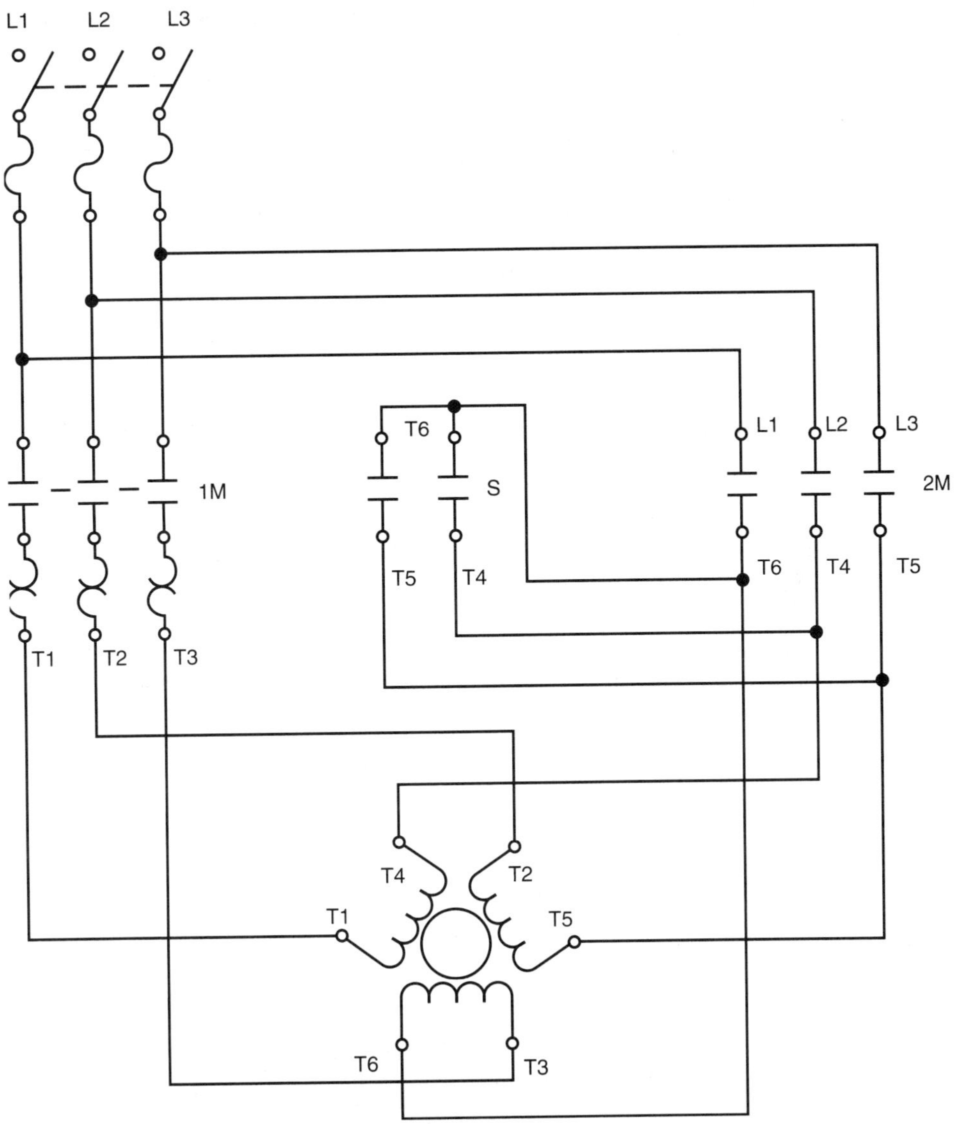

Fig. 14-8 Typical wiring diagram of a wye-delta motor-starter circuit.

3. The standard and maximum wire sizes of the XHHW copper branch-circuit supply conductors
4. The standard and maximum wire sizes of the XHHW copper motor-circuit load conductors

Motor FLC determination for the branch-circuit SC-GF protection and the rated ampacity of the circuit conductors:

240 amps Ref. 430-6(a) Ref. T430-150

Motor FLC determination for the thermal overload heaters to be installed in a separate overload relay:

232 amps — Ref. 430-6(a)

Branch-circuit SC-GF protection setting determinations for the TD fuses:

Ref. 430-52(c)(1)

Standard overrate computation: Ref. T430-152

175% of 240 amps = 420 amps

Standard OCPD ratings: Ref. 240-6(a)

NHS: 450 amps — Ref. (NHS) 430-52(c)(1), Excp. 1

NE Code-maximum overrate computation: Ref. 430-52(c)(1), Excp. 2(b)

225% of 240 amps = 540 amps

NE Code-maximum OCPD rating: Ref. 240-6(a)

NLS: 500 amps — Ref. (NLS) 430-52(c)(1), Excp. 2(b)

Starter heater load-current determinations:

Derate computation for wye-delta starting: Ref. 430-32(a)(1)

58% of 232 amps = 134.6 amps

Standard rating computation: Ref. 430-32(a)(1)

125% of 134.6 amps = 168.3 amps

NE Code-maximum rating computation: Ref. 430-34

140% of 134.6 amps = 188.4 amps

Branch-circuit conductor ampacity determinations:

Standard overrate computation: Ref. 430-32(a)(1) Ref. 430-22(a)

125% of 240 amps = 300 amps

Standard wire size determination:

Wire terminal temperature limitation: 75°C — Ref. 110-14(c)(2), Excp. 1

Table-rated ampacity: 310 amps — Ref. T310-16

Wire size: 350 kcmil XHHW copper

NE Code-maximum overrate computation:

Ref. 430-34 Ref. 240-3

140% of 240 amps = 336 amps

NE Code-maximum wire size determination:
Wire terminal temperature limitation: 75°C — Ref. 110-14(c)(2), Excp. 1
Table-rated ampacity: 380 amps — Ref. T310-16
Wire size: 500 kcmil XHHW copper

Motor-circuit conductor ampacity determinations:
Derate computation for wye-delta starting: — Ref. 430-22(a), Excp.

58% of 240 amps = 139.2 amps

Standard overrate computation: Ref. 430-32(a)(1) Ref. 430-22(a)

125% of 139.2 amps = 174 amps

Standard wire size determination:
Wire terminal temperature limitation: 75°C — Ref. 110-14(c)(2), Excp. 1
Table-rated ampacity: 175 amps — Ref. T310-16
Wire size: No. 2/0 AWG XHHW copper

NE Code-maximum overrate computation:
Ref. 430-34 Ref. 240-3

140% of 139.2 amps = 194.9 amps

NE Code-maximum wire size determination:
Wire terminal temperature limitation: 75°C — Ref. 110-14(c)(2), Excp. 1
Table-rated ampacity: 200 amps — Ref. T310-16
Wire size: No. 3/0 AWG XHHW copper

Code challenge 14–10. If a 3-phase, 460-volt, 150-hp, design B, delta-wound ac induction motor, which has a nameplate limited surface temperature rise of 40°C and a full-load current rating of 172 amperes, will be supplied with a wye-delta starting method, determine:

1. The standard and maximum ratings of nontime fuses used for the branch-circuit SC-GF overcurrent protection
2. The standard and maximum settings of thermal heaters installed in a separate overload relay
3. The standard and maximum wire size of the THHN copper branch-circuit conductors
4. The standard and maximum wire size of the THHN copper motor-circuit conductors

Motor FLC determination for the branch-circuit SC-GF protection and the rated ampacity of the circuit conductors:
(A) ____________ amps Ref. (B) ______________ Ref. (C) ______________

Motor FLC determination for the thermal overload heaters to be installed in a separate overload relay:

(D) ____________ amps Ref. (E) ________________

Branch-circuit SC-GF protection setting determinations for the nontime fuses:

Ref. (F) ________________

Standard overrate computation: Ref. (G) ________________

____________ % of ____________ amps = (H) ____________ amps

Standard OCPD ratings: Ref. (I) ________________

NHS: (J) ____________ amps Ref. (NHS) (K) ________________

NE Code-maximum overrate computation: Ref. (L) ________________

____________ % of ____________ amps = (M) ____________ amps

NE Code-maximum OCPD rating: Ref. (N) ________________

NLS: (O) ____________ amps Ref. (NLS) (P) ________________

Starter heater load-current determinations:

Derate computation for wye-delta starting: Ref. (Q) ________________

(R) ____________ % of ____________ amps = ____________ amps

Standard rating computation: Ref. (S) ________________

____________ % of ____________ amps = (T) ____________ amps

NE Code-maximum rating computation: Ref. (U) ________________

____________ % of ____________ amps = (V) ____________ amps

Branch-circuit conductor ampacity determinations:

Standard overrate computation:

Ref. (W) ________________ Ref. (X) ________________

____________ % of ____________ amps = (Y) ____________ amps

Standard wire size determination:

Wire terminal temperature limitation: (Z) ____________ °C

Ref. (1A) ________________

Table-rated ampacity: (1B) ____________ amps Ref. (1C) ________________

Wire size: (1D) ____________ THHN copper

NE Code-maximum overrate computation:

Ref. (1E) ____________________ Ref. (1F) ______________

____________ % of ____________ amps = (1G) ____________ amps

NE Code-maximum wire size determination:

Wire terminal temperature limitation: (1H) ____________ °C

Ref. (1I) ______________

Table-rated ampacity: (1J) ____________ amps Ref. (1K) ______________

Wire size: (1L) ____________ THHN copper

Motor-circuit conductor ampacity determinations:

Derate computation for wye-delta starting: Ref. (1M) ______________

____________ % of ____________ amps = (1N) ____________ amps

Standard overrate computation:

Ref. (1O) ____________________ Ref. (1P) ______________

____________ % of ____________ amps = (1Q) ____________ amps

Standard wire size determination:

Wire terminal temperature limitation: (1R) ____________ °C

Ref. (1S) ______________

Table-rated ampacity: (1T) ____________ amps Ref. (1U) ______________

Wire size: (1V) ____________ THHN copper

NE Code-maximum overrate computation:

Ref. (1W) ____________________ Ref. (1X) ______________

____________ % of ____________ amps = (1Y) ____________ amps

NE Code-maximum circuit-conductor wire size determination:

Wire terminal temperature limitation: (1Z) ____________ °C

Ref. (2A) ______________

Table-rated ampacity: (2B) ____________ amps Ref. (2C) ______________

Wire size: (2D) ____________ THHN copper

14.8 DC DRIVE CONTROLLERS

Until the recent development of variable-frequency, reduced-voltage controllers, which can be used to vary the speed of a standard ac squirrel-cage induction motor, speed control could only be accomplished using either rheostats or SCR voltage controllers and dc motors, or resitive load banks and wound-rotor ac induction motors.

Variable speed dc drives (motor and its controller—Fig. 14-9) are still widely used in industry due to the greater torque available from dc motors as compared to ac motors.

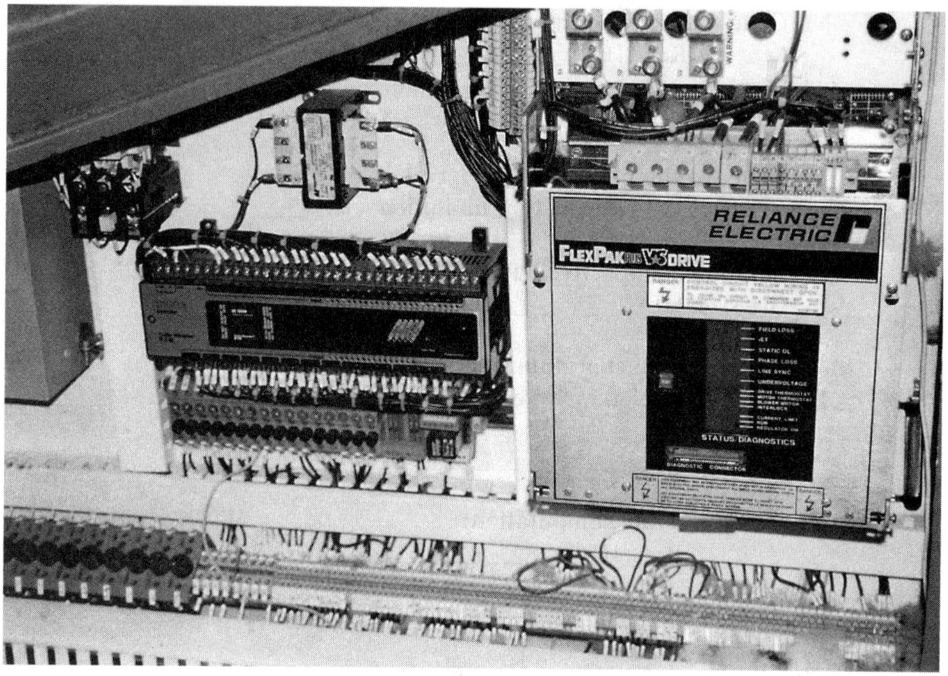

Fig. 14-9 Variable voltage SCR controllers are used to vary the speed of a dc motor. (Courtesy of Reliance Electric)

When a single dc motor is supplied from an individual branch circuit, the branch circuit technically ends at the line side of the dc controller. The motor-circuit conductors extend from the load side of the controller to the supply terminals of the motor.

As a rule, the industrial plant is supplied with a 3-phase ac power distribution system, which must be rectified at the controller to dc power for supply to the dc motor. In comparison to ac power within a circuit, which is always changing in magnitude and periodically reversing direction (governed by the supply frequency), dc current flows in only one direction within the circuit. The level of dc current may be steady state (constant-voltage level), or the dc current may be pulsating (voltage level varies in amplitude according to the type of filtration on the rectifier or according to the type of controller).

The old method of controlling the speed of a dc motor is to place a large (high wattage rating) variable resistance (power rheostat) in series with the armature (rotary circuit) of the dc motor on the load side of the rectifier power supply. Full torque and baseline speed are accomplished when full voltage is applied to the armature circuit of a dc motor. When resistance is added to the armature circuit by the operator turning the rheostat, some of the armature voltage is dropped across this series resistance. The armature receives lower voltage, hence less rotor torque and lower speed.

Instead of resistance, which causes many problems in terms of heat dissipation (the high-value armature current passes through the rheostat; the loss of voltage is dissipated as heat), an "SCR" controller (silicon-controlled rectifier) is used as an electronic or solid-state switch (no moving parts) to switch on the rectified ac to dc power at various angles (firing angles) or degrees in the rise and fall of each alternation in the sinusoidal ac waveform.

The dc current from an SCR controller is pulsating. Average values of the dc current and voltage govern the torque and speed of the armature in the dc motor.

In terms of supply and support equipment, NE Code rules or regulations concerning dc motors apply to both the old-type rheostat controller and the SCR drive controller.

The worst-case average-value armature currents for dc motors are found in T430-147 in the NE Code. These values of current are normally used in the design stage of the structure power distribution when nameplate values of dc controllers or motors are not known, only the horsepower and voltage ratings.

In reference to Sec. 430-52(d), the rating of the branch-circuit overcurrent protection for a dc motor is governed by the nameplate rating of fuses or circuit breaker specified on the drive controller. In the absence of nameplate values, the worst-case values listed in the table are often used on the test or licensing exam, or at the design and layout stage of the power distribution.

Dc motors, especially those supplied by circuits that implement speed control, do not have the high-value in-rush starting currents that are typical of ac induction motors with full-voltage (across-the-line) starting; therefore, the overrate percentage listed in T430-152 is only 150% for fuses or an inverse-time circuit breaker. The overrate of an instantaneous-trip circuit breaker, which must have supplemental (lower-level) branch-circuit SC-GF overcurrent protection is rated at only 250%.

The SCR controller can be viewed as unit equipment. If the nameplate has a specified maximum fuse size or other overcurrent protection device rating, then Secs. 430-52(c)(5) and 430-54 will not allow the branch-circuit SC-GF overcurrent protection rating or setting to exceed this specified value.

If the nameplate does not specify a maximum branch-circuit overcurrent-protection rating, then the SCR controller, as with a dc rheostat controller, must be provided with branch-circuit SC-GF overcurrent protection sized in accordance with Sec. 430-52(c) and T430-152.

Dc motors supplied through an SCR controller are normally thermally protected by thermistors (special resistors whose value will change with a rise in temperature) or some type of temperature sensing device located (embedded) in the motor windings.

In the terminal housing of the motor, two control-circuit conductors are identified with either the letters "TP" or just "P" indicating thermal protection. These sensing conductors, normally the smallest in wire size when compared to either the field (stator circuit) or armature (rotor circuit) leads, must be extended to the drive controller to provide thermal overload protection for the dc motor.

The control loop of the motor thermal protection interrupts the control circuit of the SCR controller in much the same manner as the overload heaters in the electromechanical starter opens a set of control contacts to interrupt control power supplied to the starter solenoid. Section 430-32(a)(3), in correlation with Sec. 430-39, recognizes this type of thermal protection to be used in lieu of the starter heaters in the typical load-side starter overload relay. The overrates listed in Sec. 430-32(a)(1) are coordinated by the motor manufacturer, such that the standard 125% overrate and the maximum 140% overrate (Sec. 430-34) of the motor FLC for the rating of the running overload protection do not apply.

When the speed of the dc motor is controlled by a rheostat (old method of control), the dc motor must also be protected from running overload. The motor can be internally thermally protected as with the SCR controller, or the motor circuit can be protected by a separate overload relay. For the latter, the overrate percentages found in Sec. 430-32 do apply.

The branch-circuit conductors supplying either type of dc controller are sized by Sec. 430-22(a) at 125% of the worst-case NE Code table-value dc motor FLC.

The motor-circuit conductors are sized by Exception No. 2 to Sec. 430-22(a). The overrate percentage is determined according to the type of ac to dc rectification employed (Fig. 14-10). For single-phase, half-wave rectification (only one alternation of the ac power), Sec. 430-22(a), Excp. No. 2(a) requires the rated ampacity of the motor-circuit conductors to be overrated to 190% of the NE Code table value of dc motor FLC.

For single-phase, full-wave rectification (both alternations of the ac power), Sec. 430-22(a), Excp. No. 2(b) requires the rated-ampacity of the motor-circuit conductors to be overrated to only 150% of the NE Code table value of dc motor FLC.

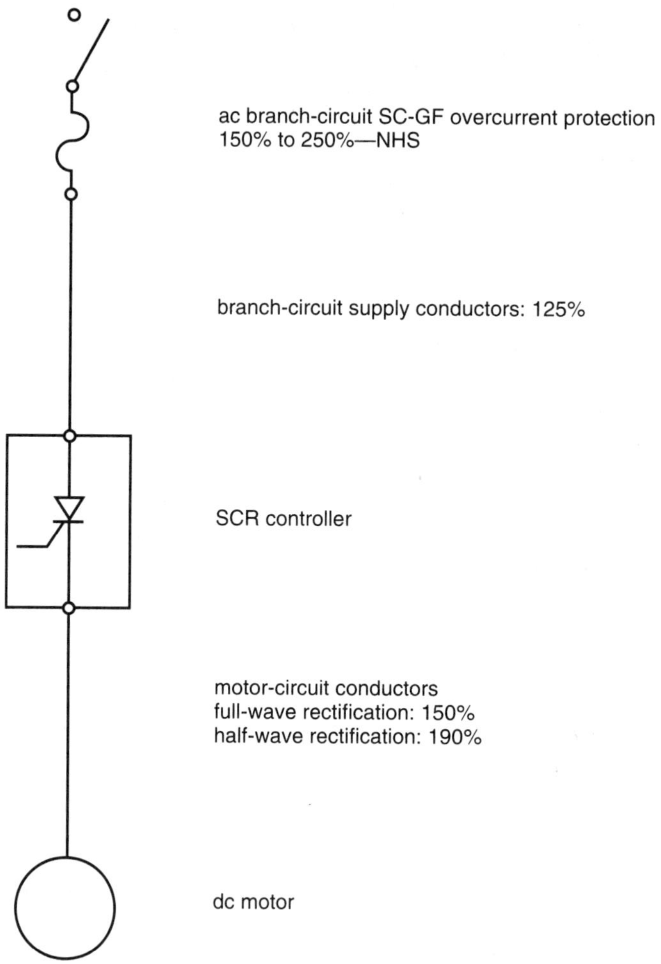

Fig. 14-10 The type of rectification is used to determine the required overrate of the circuit conductors supplying a dc motor.

Example 14–11. A 20-hp, 240-volt armature, dc series motor is supplied and controlled by a single-phase, full-wave, solid-state rectifier. Determine:

1. The standard rating of the branch-circuit SC-GF overcurrent protection, using time-delay fuses
2. The minimum wire size of the branch-circuit conductors, using THWN copper
3. The minimum wire size of the motor-circuit conductors, using THWN copper

Motor full-load current: 72 amps — Ref. T430-147

Branch-circuit SC-GF protection setting determinations for the TD fuses: Ref. 430-52(c)(1)

Standard overrate computation: Ref. T430-152

150% of 72 amps = 108 amps

Standard OCPD ratings: Ref. 240-6(a)

NHS: 110 amps — Ref. (NHS) 430-52(c)(1), Excp. 1

Branch-circuit conductor ampacity determinations:

Overrate computation: Ref. 430-39 — Ref. 430-22(a)

125% of 72 amps = 90 amps

Wire size determination:

Wire terminal temperature limitation: 60°C — Ref. 110-14(c)(1), Excp. 1

Table-rated ampacity: 95 amps — Ref. T310-16

Wire size: No. 2 AWG THWN copper

Motor-circuit conductor ampacity determinations:

Overrate computation: Ref. 430-22(a), Excp. 2(b)

150% of 72 amps = 108 amps

Wire size determination:

Wire terminal temperature limitation: 75°C — Ref. 110-14(c)(2), Excp. 1

Table-rated ampacity: 110 amps — Ref. T310-16

Wire size: No. 1 AWG THWN copper

NE Code-maximum values of branch-circuit SC-GF overcurrent protection were not computed in this example, simply because of the speed control of the dc motor offered by an SCR controller. High-value, in-rush starting currents for high-inertia, long-acceleration loads, are not prolonged in this type of motor circuit because of the control limits on the motor supply current.

The same analysis can be applied to the branch-circuit supply conductors. Their NE Code-minimum rated ampacity was computed at the standard running overload-overcurrent protection overrate of only 125%. The main difference for the motor-circuit conductors, in comparison for general application of ac motor circuits, is the 190% overrate of the motor-circuit conductors for half-wave rectification and the 150% overrate for full-wave rectification.

Code challenge 14–11. A 50-hp, 500-volt armature, dc series motor is supplied and controlled by a single-phase, half-wave, solid-state rectifier. Determine:

1. The standard rating of the branch-circuit SC-GF overcurrent protection, using nontime fuses
2. The minimum wire size of the branch-circuit conductors, using THHN copper
3. The minimum wire size of the motor-circuit conductors, using THHN copper

Motor full-load current: (A) ____________ amps Ref. (B) ____________________

Branch-circuit SC-GF protection setting determinations for the nontime fuses:

Ref. (C) ____________________

Standard overrate computation: Ref. (D) ____________________

(E) ____________ % of ____________ amps = ____________ amps

Standard OCPD ratings: Ref. (F) ____________________

NHS: (G) ____________ amps Ref. (NHS) (H) ____________________

Branch-circuit conductor ampacity determinations:

Overrate computation: Ref. (I) ____________ Ref. (J) ____________________

(K) ____________ % of ____________ amps = ____________ amps

Wire size determination:

Wire terminal temperature limitation: (L) ____________ °C

Ref. (M) ____________________

Table-rated ampacity: (N) ____________ amps Ref. (O) ____________________

Wire size: (P) ____________ THHN copper

Motor-circuit conductor ampacity determinations:

Overrate computation: Ref. (Q) ____________________

(R) ____________ % of ____________ amps = ____________ amps

Wire size determination:

Wire terminal temperature limitation: (S) ____________ °C

Ref. (T) ____________________

Table-rated ampacity: (U) ____________ amps Ref. (V) ____________________

Wire size: (W) ____________ THHN copper

14.9 AC DRIVE CONTROLLERS

Other than the use of multispeed motors, speed control of an ac induction motor is normally accomplished through the use of either a static inverter or a variable-frequency/reduced-voltage drive controller.

When a 3-phase power supply is rectified to a dc power supply and then through a series of firing circuits for transistors, the dc power is converted into a pulsating dc source at 3-phase, the pulse waveform of the dc is made to appear as a near sinusoidal ac signal within a range of frequencies. The phase converter is being used only for speed control.

The branch-circuit supply and support equipment and the motor circuit for this type of motor controller would be sized and protected in the same manner as a general-application motor circuit as outlined in Chapter 13. The only difference is that the allowed NE Code-maximum values of the branch-circuit SC-GF overcurrent protection, running overload protection, and circuit conductors would not be warranted, since as with a dc drive controller, the in-rush starting current to the stator circuit is limited by the controller.

The same type of ac variable-frequency/reduced-voltage drive controller can be supplied with a single-phase source, and through rectification and firing circuitry of the transistors, the output can be a generated near sinusoidal ac single-phase, 2-phase, or 3-phase waveform at variable frequencies and voltages (Fig. 14-11).

Fig. 14-11 When a variable-frequency/reduced-voltage solid-state drive controller is used to change the phase of the supply to match the phase of the motor, it is referred to as a phase converter by the NE Code.

The National Electrical Code refers to this second type of ac supply and speed control of the motor-driven equipment as a "phase converter." Installation and protection guidelines are listed in Art. 455.

When a phase converter is installed to supply either an unknown load or a nonmotor load, Sec. 455-7 requires the rating or setting of the branch-circuit overcurrent-protection device to be overrated to 125% of the full-load or rated-load current specified on the phase-converter nameplate. Section 455-6 requires the same 125% overrate for the computation of the minimum rated ampacity of the single-phase branch-circuit supply conductors.

If a 3-phase motor is supplied through a phase converter from a single-phase electrical source at the same voltage rating (Fig. 14-12), Excp. No. 1 to Sec. 455-7 requires the rating or setting of the branch-circuit overcurrent-protection device to be overrated to 250% of the worst-case 3-phase motor full-load or rated-load current as listed in T430-150.

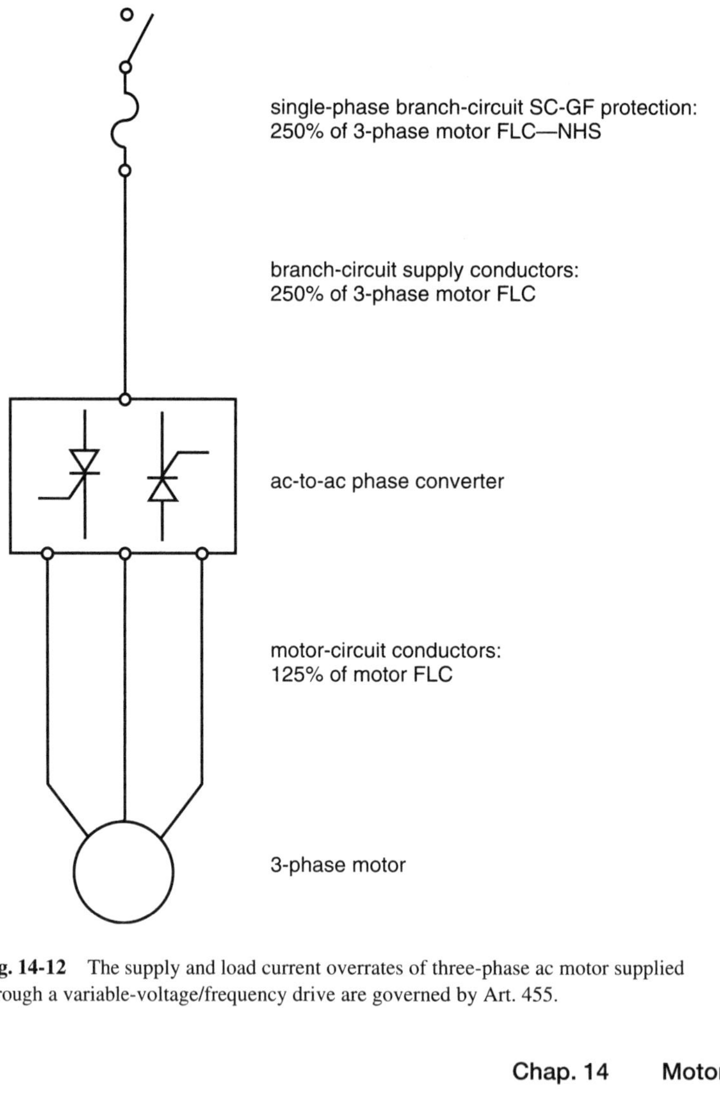

Fig. 14-12 The supply and load current overrates of three-phase ac motor supplied through a variable-voltage/frequency drive are governed by Art. 455.

According to this exception, "not greater than" indicates the 250% computation is the NE Code-maximum rating of the fuses or circuit breaker. If the computation yields a nonstandard size (setting or rating) as listed in Sec. 240-6(a), then the next lower standard size should be used. Exception No. 2 will allow the next higher standard size, rating, or setting of the fuse or circuit breaker to be installed, provided the rated ampacity of the circuit conductors is at least equal to the 250% overrate required by Sec. 455-6 Exception. Section 455-7, Excp. No. 2 parallels Sec. 240-3(b).

Justification for the 250% overrate of the 3-phase full-load current for the computation of the minimum single-phase supply current can be found in Chapter 12, Sec. 12.9 of this text—Drive Disconnect Ratings).

The Exception to Sec. 455-6 requires the rated ampacity of the single-phase branch-circuit supply conductors (on the input or line side of the converter) to also be at least equal to the 250% overrate of the 3-phase motor full-load current.

The 3-phase, motor-circuit conductors (on the output or load side of the converter) are overrated by the general requirements of Sec. 430-22(a) at a minimum of 125% of the 3-phase motor full-load current.

Example 14–12. A 10-hp, 230-volt, 3-phase ac motor will be supplied through a single- to three-phase static converter from a 230-volt, single-phase, ac source.

Determine:

1. The standard rating of the branch-circuit SC-GF overcurrent protection, using non-time fuses
2. The minimum wire size of the branch-circuit conductors, using THW copper
3. The minimum wire size of the motor-circuit conductors, using THW copper

Motor full-load current: 28 amps — Ref. T430-150

Phase converter—motor load branch-circuit SC-GF overcurrent protection determination:

Overrate computation: — Ref. 455-7, Excp. No. 1

250% of 28 amps = 70 amps

Standard rating of nontime fuses: 70 amps — Ref. 240-6(a)

Branch-circuit conductor ampacity determinations:

Overrate computation: — Ref. 455-6, Excp.

250% of 28 amps = 70 amps

Wire size determination:

Wire terminal temperature limitation: 60°C — Ref. 110-14(c)(1), Excp. 1

Table-rated ampacity: 70 amps — Ref. T310-16

Wire size: No. 4 AWG THW copper

Motor-circuit conductor ampacity determinations:

Overrate computation: Ref. 430-22(a)

125% of 28 amps = 35 amps

Wire size determination:

Wire terminal temperature limitation: 60°C Ref. 110-14(c)(1), Excp. 1

Table-rated ampacity: 40 amps Ref. T310-16

Wire size: No. 8 AWG THW copper

Separate running overload protection was not sized in this example. As with an SCR drive controller for a dc motor, the static inverter uses other forms of motor overload protection instead of the typical heater assembly in a separate motor-starter overload relay.

Code challenge 14–12. A 15-hp, 230-volt, 3-phase ac motor will be supplied through a single- to three-phase static converter from a 230-volt, single-phase, ac source. Determine:

1. The standard rating of the branch-circuit SC-GF overcurrent protection, using an inverse-time circuit breaker
2. The minimum wire size of the branch-circuit conductors, using RHW aluminum
3. The minimum wire size of the motor-circuit conductors, using RHW aluminum

Motor full-load current: (A) ____________ amps Ref. (B) ____________

Phase converter—motor load branch-circuit SC-GF overcurrent protection determination:

Overrate computation: Ref. (C) ____________

(D) ____________ % of ____________ amps = ____________ amps

Standard OCPD ratings: Ref. (E) ____________

NHS: (F) ____________ amps Ref. (NHS) (G) ____________

Branch-circuit conductor ampacity determinations:

Overrate computation: Ref. (H) ____________

(I) ____________ % of ____________ amps = ____________ amps

Wire size determination:

Wire terminal temperature limitation: (J) ____________ °C

Ref. (K) ____________

Table-rated ampacity: (L) ____________ amps Ref. (M) ____________

Wire size: (N) ____________ RHW aluminum

Motor-circuit conductor ampacity determinations:

Overrate computation: Ref. (O) ________________

(P) __________ % of __________ amps = __________ amps

Wire size determination:

Wire terminal temperature limitation: (Q) __________ °C

Ref. (R) ________________

Table-rated ampacity: (S) __________ amps Ref. (T) ________________

Wire size: (U) __________ RHW aluminum

If the 3-phase output voltage of the static phase converter is greater than the single-phase input voltage rating, the 250% overrate of both the single-phase branch-circuit supply conductors and overcurrent-protection devices must be further overrated or increased by a percentage determined by taking the ratio (division) of the output voltage to (by) the input voltage. Refer to 455-6, Excp. for the conductors, and 455-7 for the branch-circuit SC-GF overcurrent protection.

Example 14–13. A 3-phase, 460-volt, 5-hp ac motor will be supplied by a single-phase, 208-volt source through a phase converter. **Determine:**

1. The standard rating of the branch-circuit SC-GF overcurrent protection, using time-delay fuses
2. The minimum wire size of the branch-circuit conductors, using RHW copper
3. The minimum wire size of the motor-circuit conductors, using RHW copper

Motor full-load current: 7.6 amps Ref. T430-150

Phase converter—motor load branch-circuit SC-GF overcurrent protection determination:

Ref. 455-7, Excp. No. 1

Overrate computation:

250% of 7.6 amps = 19 amps

Output-to-input voltage overrate computation:

$$\frac{460 \text{ volts}}{208 \text{ volts}} \times 19 \text{ amps} = 42 \text{ amps}$$

Standard rating of nontime fuses: Ref. 240-6(a)

NHS: 45 amps Ref. (NHS) 455-7, Excp. No. 2

Branch-circuit conductor ampacity determinations: Ref. 455-6, Excp.

Overrate computation:

250% of 7.6 amps = 19 amps

Output-to-input voltage overrate computation:

$$\frac{460 \text{ volts}}{208 \text{ volts}} \times 19 \text{ amps} = 42 \text{ amps}$$

Wire size determination:

Wire terminal temperature limitation: 60°C Ref. 110-14(c)(1), Excp. 1
Table-rated ampacity:55 amps Ref. T310-16
Wire size: No. 6 AWG RHW copper

Motor-circuit conductor ampacity determinations:m

Overrate computation: Ref. 430-22(a)

125% of 7.6 amps = 9.5 amps

Wire size determination:

Wire terminal temperature limitation: 60°C Ref. 110-14(c)(1), Excp. 1
Table-rated ampacity: amps Ref. T310-16
Wire size: No. 14 AWG RHW copper

The motor-circuit conductors in this example only have to be rated for a minimum ampacity of 9.5 amps. In T310-16, wire size No. 14 AWG copper is the smallest wire size listed with a rated ampacity in the 60°C insulation temperature column. The table-rated ampacity of the 60°C temperature column is mandated by the wire terminal temperature limitations of Sec. 110-14(c).

Code challenge 14–13. A 15-hp, 575-volt, 3-phase ac motor will be supplied through a single- to three-phase static converter from a 230-volt, single-phase, ac source. **Determine:**

1. The standard rating of the branch-circuit SC-GF overcurrent protection, using non-time fuses
2. The minimum wire size of the branch-circuit conductors, using XHH copper
3. The minimum wire size of the motor-circuit conductors, using XHH copper

Motor full-load current: (A) ____________ amps Ref. (B) ____________________

Phase converter—motor load branch-circuit SC-GF overcurrent protection determination:

Ref. (C) ____________________

Overrate computation:

(D) ____________ % of ____________ amps = ____________ amps

Output-to-input voltage overrate computation:

$$\frac{\underline{\qquad\qquad} \text{ volts}}{\underline{\qquad\qquad} \text{ volts}} \times \underline{\qquad\qquad} \text{ amps} = \text{(E)} \underline{\qquad\qquad} \text{ amps}$$

Standard OCPD ratings: Ref. (F) ________________
NHS: (G) __________ amps Ref. (NHS) (H) ________________

Branch-circuit conductor ampacity determinations: Ref. (I) ________________
Overrate computation:

(J) __________ % of __________ amps = __________ amps

Output-to-input voltage overrate computation:

$$\frac{\_\_\_\_\_\_\_\_\_\_ \text{ volts}}{\_\_\_\_\_\_\_\_\_\_ \text{ volts}} \times \_\_\_\_\_\_\_\_\_\_ \text{ amps} = \text{(K) } \_\_\_\_\_\_\_\_\_\_ \text{ amps}$$

Wire size determination:
Wire terminal temperature limitation: (L) __________ °C
Ref. (M) ________________
Table-rated ampacity: (N) __________ amps Ref. (O) ________________
Wire size: (P) __________ XHH copper

Motor-circuit conductor ampacity determinations:
Overrate computation: Ref. (Q) ________________

(R) __________ % of __________ amps = __________ amps

Wire size determination:
Wire terminal temperature limitation: (S) __________ °C
Ref. (T) ________________
Table-rated ampacity: (U) __________ amps Ref. (V) ________________
Wire size: (W) __________ XHH copper

Other special-purpose motors such as small clock or timer motors, or shaded-pole motors used for exhaust fans in kitchens, bathrooms, and so forth were not discussed in this chapter. These types of motors are considered to be low-current, high-impedance motors that do not require special overrates in terms of SC-GF protection of circuit conductor supply.

Similar to a limited-power control source as discussed in Chapter 12 with motor control circuits, these high-impedance motors will not draw high values of overcurrent during a locked-rotor condition. Overcurrent protection for these motors is normally limited to the rating of the fuse or circuit breaker protecting the 15- or 20-amp, 120-volt multioutlet branch circuit from which they are supplied.

REVIEW QUESTIONS

1. The general tap rules for all phases of electrical installations are listed in the subsections of Sec. __________.

2. The general tap rules for motor-feeder circuits are found in the subsections of Sec. ____________.

3. A tap consists of a conductor of lower ampacity connected to a protected conductor of higher ampacity without overcurrent protection installed at the point of the tap. The tap conductor is normally overcurrent-protected according to its rated ampacity at the point of termination.

   True or False

4. For a motor-feeder tap, the rated-ampacity of the tap conductors in the 10-foot tap rule must be at least equal to ____________ the rating or setting of the larger feeder-circuit SC-GF overcurrent-protection rating (fuse) or setting (CB).

5. For a motor-feeder tap, the rated-ampacity of the tap conductors in the 25-foot tap rule must be at least equal to ____________ the rated-ampacity of the larger feeder circuit conductors.

6. The installation requirements of several motors and/or other loads installed on a single branch circuit are listed in Sec. ____________.

7. When a single hermetically sealed refrigerant motor compressor and other loads are supplied by a single branch circuit, the rating or setting of the branch-circuit protection device must be sufficient (large enough) to handle the starting current of the equipment.

   True or False

8. Branch-circuit conductors supplying motors designed for a specific task or duty cycle are sized according to the overrate percentages listed in Table ____________.

9. Duty-cycle motors are normally operated in an overload condition. The rating of the motor overcurrent protection is limited only to the standard rating of branch-circuit, short-circuit, ground-fault protection computed by the overrates allowed by Table ____________.

10. For multispeed, ac induction motors, the 125% overrate applied to the branch-circuit conductors on the line side of the controller must be based on the smallest winding full-load current rating indicated on the motor nameplate according to speed rating.

    True or False

11. For multispeed, ac induction motors, the overrate applied to the branch-circuit SC-GF overcurrent protection of the motor must be based on the smallest winding full-load current rating indicated on the motor nameplate according to speed rating.

    True or False

12. When a wound-rotor ac induction motor is rated as continuous-duty, the rated ampacity of the wound-rotor secondary-circuit conductors must be computed at a minimum of ____________ % of the maximum (full-load) secondary rotor current as specified on the motor ____________.

13. When a wound-rotor ac induction motor is minute-rated as a duty-cycle motor, the rated-ampacity of the wound-rotor secondary-circuit conductors must be computed at the minimum overrate percentages listed in Table ____________.

14. When the resistor load bank for a wound-rotor ac induction motor will be located separate from the motor controller, the rated ampacity and resulting wire size of the secondary rotor-circuit conductors must be determined according to the duty classification of the resistors and the overrate or underrate percentages lised in Table ____________.

15. Wye-delta starting is limited to 3-phase power distribution systems and a special type of delta-wound motor that has six motor winding leads (the beginning and ending of all 3-phase motor windings) brought out of the motor assembly for field connection to the starter load conductors in the motor terminal housing.

    True or False

16. When a 3-phase motor is wye connected, the values of winding current and line current are ____________ . The value of the winding voltage is only ____________ % of the line voltage value.

17. When a 3-phase motor is delta connected, the values of winding voltage and line voltage are ____________ . The value of the winding current is only ____________ % of the line current value.

18. Using across-the-line electromechanical motor starters, an overload relay with three separate overload heaters are required with the wye-delta starting scheme. For the size of the overload heaters, the delta winding current, which is ____________ % of the motor FLC line current, is overrated according to the percentages found in Sec. ____________ for standard motor operation, and by Sec. ____________ for high-inertia, long-acceleration load starting.

19. The branch-circuit conductors supplying an SCR controller for a dc motor must be overrated to ____________ % of the NE Code table-value dc motor FLC.

20. The motor-circuit conductors, extended between an SCR controller and a dc motor, must be overrated to either ____________ % or ____________ % of the NE Code table-value motor FLC. The percentage to use depends on whether the ac-to-dc rectification is full-wave or half-wave.

21. In reference to Question 20, the motor branch-circuit short-circuit, ground-fault protection device must be sized or rated according to the overrate percentages found in Sec. ____________ and Table ____________ .

22. When a phase converter is installed to supply either an unknown load or a nonmotor 3-phase load from a single-phase source, the rated ampacity of the branch-circuit supply conductors and the rating or setting of the branch-circuit overcurrent-protection device must be overrated to ____________ % of the full-load or rated-load current specified on the phase converter nameplate.

23. When a phase converter is installed to supply a specific 3-phase motor load from a single-phase source at the same voltage rating, the rated ampacity of the branch-circuit supply conductors and the rating or setting of the branch-circuit overcurrent-protection device must be overrated to ____________ % of the NE Code table-value full-load current of the motor.

24. If the 3-phase output voltage of a phase converter supplying a motor load is greater than the single-phase input voltage rating, the overrate of both the single-phase branch-circuit supply conductors and overcurrent protection devices must be further overrated or increased by a percentage determined by taking the ____________ of the output voltage to the input voltage.

REVIEW PROBLEMS

1. A 3-phase, 200-volt, design B, ac induction motor rated at 20 hp will be supplied by conductors tapped to a larger feeder circuit without overcurrent protection at the point of tap. For the following conditions, determine if this is an NE Code-legal installation:
   a. The No. 2 AWG THW copper tap conductors, enclosed in IMC, are a total of 6 feet in length between the point of tap and the point of termination by a 3-phase fused disconnect which contains time-delay fuses.
   b. The time-delay fuses, which provide the necessary branch-circuit SC-GF overcurrent protection for the motor circuit, are rated at 110 amps.
   c. The feeder circuit, which consists of 400 kcmil THW copper conductors, is protected at the point of supply by time-delay fuses rated at 350 amps.

2. A 3-phase, 230 volt, design E, ac induction motor rated at 30 hp will be supplied by conductors tapped to a larger feeder circuit without overcurrent protection at the point of tap. For the following conditions, determine if this is an NE Code-legal installation:
   a. The tap consists of a multiconductor cable that contains three No. 1 AWG THW copper conductors and a No. 6 AWG THW copper equipment-grounding conductor. The multiconductor cable, which has a total length of 15 feet and is routed to be protected from physical damage, is terminated to a fused knife switch disconnect
   b. The time-delay fuses in the knife switch, which provide the necessary branch-circuit SC-GF overcurrent protection for the motor circuit, are rated at 175 amps
   c. The feeder circuit, which consists of 600 kcmil THWN copper conductors, is protected at the point of supply by an inverse-time circuit breaker rated at 800 amps
3. Assuming the motors have individual overload protection, and that each individual motor controller has a full-load current rating of at least 15 amps, can two ¼-hp and one ½-hp, 3-phase, 115-volt, squirrel-cage ac induction motors be supplied by a single 20-amp branch-circuit rated at 120 volts?
4. Two 3-hp and one 5-hp, 3-phase, 460-volt squirrel-cage induction motors are to be installed on a single 3-phase, 460-volt, branch-circuit. If the motors are so interlocked that they cannot be started simultaneously, if each motor has individual overload protection, and if the branch circuit is protected by a 3-phase, inverse-time circuit breaker rated at 25 amps, would this be an NE Code-legal installation?
5. The computed rating of the branch-circuit overcurrent protection for a field assembly of multimotor/combination-load equipment is equal to 92 amps. If No. 2 AWG XHHW copper conductors are used to supply this branch-circuit load, can the branch-circuit fuses or inverse-time circuit breaker be rated at 100 amps?
6. A single-phase, 2-wire, 230-volt branch circuit will supply a multimotor/combination-load equipment that consists of the following loads:

   1 Hermetically sealed refrigerant motor compressor with a nameplate rated-load current rating of 10 amps
   2 Condensor fans rated at ¼ hp each
   2 Water pumps rated at ½ hp each
   1 Resistive-type strip heater with a full-load current rating of 3 amps
   1 Control transformer with a primary full-load current rating of 2 amps

   Determine:
   a. The NE Code-maximum rating of the branch-circuit SC-GF overcurrent protection, using time-delay fuses
   b. The NE Code-minimum wire size of the branch-circuit supply conductors, using THWN copper
7. A 3-phase, 230-volt, design C ac induction motor rated at 20 hp is used for intermittent duty on a roll-conveyor assembly in a canned-goods factory. If the motor is rated for continuous duty, determine:
   a. The NE Code-maximum rating of the motor SC-GF overcurrent protection, using nontime fuses
   b. The NE Code-minimum wire size of motor branch-circuit conductors, using XHHW-2 copper
8. A single-phase, 230-volt, 3-hp, 2-speed ac induction motor has winding full-load current ratings of 16 amps and 19 amps for the respective speeds.

Determine:

a. The NE Code-maximum rating of the motor branch-circuit SC-GF overcurrent protection, using time-delay fuses
b. The NE Code-minimum wire size of motor branch-circuit conductors, using THW copper

9. A 3-phase, 460-volt, wound-rotor, ac induction motor rated at 100 hp has a nameplate rotor-circuit current FLC rating of 118 amps. If the resistors in the secondary load bank are rated for "extra-heavy starting duty," determine the NE Code-minimum wire size of the secondary rotor-circuit conductors using THW copper.

10. If a 3-phase, 460-volt, 125-hp, design B, delta-wound ac induction motor, which has a service factor of 1.15 and a nameplate full-load current rating of 148 amperes, will be supplied with a wye-delta starting method, determine:

a. The standard and maximum ratings of time-delay fuses used for the branch-circuit SC-GF overcurrent protection
b. The standard and maximum ratings of thermal heaters installed in a separate overload relay
c. The standard and maximum wire sizes of the XHHW aluminum branch-circuit supply conductors
d. The standard and maximum wire sizes of the XHHW aluminum motor-circuit load conductors

11. A 5-hp, 180-volt armature, dc series motor is supplied and controlled by a single-phase, half-wave, solid-state rectifier. Determine:

a. The standard rating of the branch-circuit SC-GF overcurrent protection, using time-delay fuses
b. The minimum wire size of the branch-circuit conductors, using XHHW copper
c. The minimum wire size of the motor-circuit conductors, using XHHW copper

12. A 2-hp, 115-volt, 3-phase ac motor will be supplied through a single- to three-phase static converter from a 115-volt, single-phase, ac source. Determine

a. The standard rating of the branch-circuit SC-GF overcurrent protection, using time-delay fuses
b. The minimum wire size of the branch-circuit conductors, using RHW copper
c. The minimum wire size of the motor-circuit conductors, using RHW copper

13. A 3-phase, 230-volt, 10-hp ac motor will be supplied by a single-phase, 200-volt source through a phase converter. Determine:

a. The standard rating of the branch-circuit SC-GF overcurrent protection, using an inverse-time circuit breaker
b. The minimum wire size of the branch-circuit conductors, using RHW copper
c. The minimum wire size of the motor-circuit conductors, using RHW copper

Chapter 15

Transformers—Part I

As discussed in Chapter 2 with the rated-load ampacity of the circuit conductors, increasing the wire size to compensate for conductor length in terms of resistance and resulting voltage drop is limited due to physical size and sheer weight of the conductors. Another way to reduce the voltage losses along the length of a circuit conductor—or worse yet, cross-country transmission lines—is to reduce the amount of current that must be conveyed. Transformers are used to step up the voltage and step down the load current where distance is a problem between the supply and the load.

A transformer can also be used to limit the magnitude of fault overcurrent (short-circuit or ground-fault overcurrent) available to a connected load, such as a secondary panel, motor-control center, or other type of load center, from a large switchgear or unit substation.

The amount of fault overcurrent available at the load terminals of a switchgear is determined by the length of service conductors between the connection to the utility source and the unit substation, the impedance rating of the contained transformer, and the load-current rating of the secondary switchgear. When a secondary load center is located in the immediate vicinity of the switchgear, the available fault overcurrent, which is already limited (as compared to the infinite utility bus by the unit substation transformer) can still be relatively high due to the short conductor length.

The impedance of a second isolation transformer, which includes both the primary winding resistance, the inductive load resistance of the iron core, and the reflected load impedance of the secondary circuit, are used to limit the available fault overcurrent within the interrupt rating (operating limits) of the fuses or circuit breaker protecting the load circuit and connected utilization equipment.

For signaling, remote control, or other low-power, low-voltage control or sensing circuits, even higher-impedance control (signaling) transformers are used to limit the rated-load current of the transformer primary circuit to a value well below the load-current rating of the supply branch-circuit fuse(s) or circuit breaker. Signaling transformers were discussed in Chapter 12, Sec. 12.11, Motor Control Circuits.

The installation, use, and protection of both large power transformers and lower power-rated control transformers are governed by Art. 450 in the NE Code.

15.1 PRINCIPLE OF TRANSFORMATION

Line wires in the cross-country transmission of electrical power, or circuit conductors within the building (or other structure) power distribution that conduct ac sinusoidal current, have a varying magnetic field generated perpendicular to the length of the conductor that is also sinusoidal. If a conductor in a second circuit, which can be mechanically and electrically isolated and insulated from the line circuit (primary circuit), is immersed in the associated sinusoidal magnetic field, the principle of induction causes a voltage to be induced in the second (secondary) circuit. The relative motion between the magnetic field and this second conductor is provided by the sinusoidal magnetic field. This is referred to as transformer action.

As shown in Fig. 15-1, the simplest construction of a transformer consists of two seperate coils of wire (windings) wrapped around a common iron core. When an ac voltage is applied to the primary winding (coil) of the transformer, a concentrated, alternating

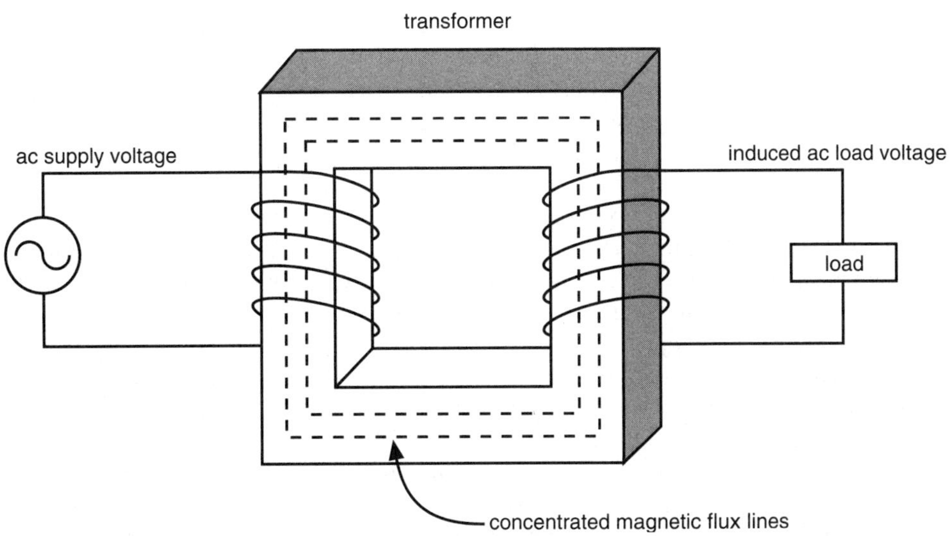

Fig. 15-1 Transfer of ac power in either an alternator, motor, or transformer depends on magnetic coupling and voltage induced by opposing magnetic fields.

magnetic field is generated in the iron core. The generated magnetic field causes an alternating voltage to be induced in the secondary coil (winding). If a load is connected to the secondary winding, current is caused to flow. Hence, the voltage and current of the primary supply circuit have been "transformed."

Technically, magnetic lines of force, often referred to as "flux lines," form a concentric pattern around the outer surface of the conductor that fans out perpendicular to the overall length of the conductor. The concentric flux lines are located side by side along the total length of the primary circuit. The heaviest concentration of flux lines is in the center, closest to the surface of the conductor. As the flux lines fan out from the surface of the conductor, the air offers high resistance (reluctance) to the flow of the flux lines, diminishing the strength of the magnetic field.

Wrapping the length of the respective circuit conductors in a spiral coil tends to concentrate the number of flux lines around the body of the coil. The number of flux lines at any cross section along the surface of the coil increases as lines from adjacent wraps add and concentrate the magnetic field as it travels across the surface, perpendicular to the parallel run of the conductors.

The magnetic field is further concentrated in the iron core, which serves as a good conductor of flux lines. A better core construction would envelop the windings as much as possible in an iron container to create multiple, complete flux paths.

The core does not have to be constructed of iron or other ferrous metal for transformer action to take place. In communication or other amplifier circuits, the core of the transformer may be constructed of a hollow tube made out of cardboard. This would be referred to as an "air-core" transformer. For signaling, control, or power transformers, the iron core provides a concentrated magnetic field path for better coupling (power transfer) of the two circuits.

The primary circuit of the transformer is the source of electrical power for both the transformer and its connected load. The secondary circuit of the transformer is the supply circuit (or power supply) for the connected load (from the transformer).

In the primary circuit, the coil of wire creates an inductive load that restricts the flow of primary current. Looking at the primary circuit only, the single coil of wire with an ac source serves as a reactor, which, except for wire resistance (heat losses), can only store the electrical energy. The main restriction of primary-circuit current is reflected impedance from the secondary circuit of the transformer.

The greatest restriction (highest value of impedance) reflected into the primary circuit will occur when the secondary circuit is open (no connected load). If the transformer were ideal, without secondary-circuit current flow there should be no primary-circuit current flow. But no transformer is ideal. A low value of primary-circuit current flows with the secondary circuit open.

This open-secondary primary-circuit current is referred to as the transformer's "excitation current." This current supports the continual reversal of the magnetic poles (core losses, dissipated as heat). As the sinusoidal waveform produced by the alternator (generator of the ac power) periodically reverses the direction of current flow along the length of the primary winding, the flux lines in the concentrated magnetic field of the iron core are also caused to reverse their direction of flow, such that the poles of the created magnet in the iron core are continually changing from north to south, south to north, and north to south in sinusoidal fashion.

The iron core is also a good conductor of electricity. If the iron core is a solid cast of iron, the relative motion of the sinusoidal magnetic field will cause stray currents to flow at random across the iron core (in reference to the flow of the flux lines). These stray currents, referred to as "eddy currents," cannot be channeled either in or out of the core structure to do useful work. Instead, their presence creates additional heat in the core structure.

The effect of eddy current is reduced to a bare-bones minimum by constructing the iron core of individual thin sheets of steel that is allowed to oxidize before being assembled. The oxidation serves as an insulator to the flow of eddy currents. The iron core is assembled by riveting several of the sheets together (laminated) to form the required cross section of the flux path.

The excitation current contributes to the energy expended due to eddy current losses and to the continual demagnetization and remagnetization of the iron core (field reversal—often referred to as hysterisis). The expended electrical energy of the excitation current is referred to as iron core losses (I^2R losses) in the transformer, which are normally dissipated as heat.

For power transformers, the excitation current is such a low value when compared to the secondary full-load current rating that the excitation current or transformer losses are ignored. The power rating for the transformer is considered the same for the primary and secondary circuits.

Without a connected load (complete path), voltage is available from the secondary leads or terminals of the transformer, but secondary current in an open circuit cannot flow—again, the greatest value of reflected impedance.

The least restriction (lowest value of impedance) reflected into the primary circuit of the transformer will occur when the two ends of the secondary winding are shorted together. In essence, the transformer becomes its own load. The restriction to secondary current flow is limited only to the resistance of the wire that makes up the secondary winding and the magnetic losses of the iron core.

Short-circuit current (secondary shorted) is referred to as fault overcurrent. When the transformer circuit is in a fault condition, the high-magnitude overcurrent generates excessive heat in the conductors, which can cause the conductor insulation to fail. The associated magnetic field of the fault overrcurrent also puts mechanical stress on the conductors, especially at their terminations, and the framing of the iron core. The stress comes from the "motor action" (inductive reaction) between the conductors and the metal enclosures that protect the conductors.

15.2 CIRCUIT CONNECTIONS

As with lighting circuits discussed in Chapter 11, transformer circuits are normally drawn using standard symbols. The introductory transformer circuit of Fig. 15-1 is redrawn in Fig. 15-2, using the standard symbol for a transformer. The two parallel bars between the two winding symbols indicate a transformer with an iron core.

The two open circles in the indicated circuit conductors represent either wire terminals on some sort of terminal block, or short sections of wire called "field leads," which are the ends of the windings within the transformer. Because a single terminal block or all the field leads of both transformer circuits are contained in the single terminal housing,

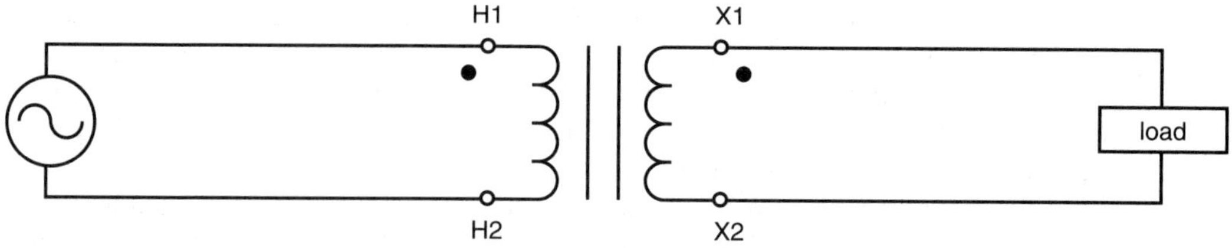

Fig. 15-2 Standard electrical symbols showing the ac supply, a transformer and its connected load in an electrical schematic.

the winding leads have to be identified to distinguish between the beginning and ending of a particular winding. The respective circuit windings have to be further identified to distinguish the higher-voltage winding leads from the lower-voltage winding leads.

The capital letter H is used to identify the higher-voltage winding. The capital letter X is used to identify the lower-voltage winding. The "in" (beginning or start) of the winding is identified with the number 1. The "out" (ending or finish) of the same winding is identified with the number 2. The leads or terminals identified as H1 and H2 are the in and out connections of the higher-voltage winding. The leads or terminals identified as X1 and X2 are the in and out connections of the lower-voltage winding.

If a single-phase transformer has two primary windings (dual-winding transformer), the in of the second winding will be identified as H3. The out will be identified as H4. If the single-phase transformer also has two lower-voltage windings, the in and out of the second winding will be identified as X3 and X4. The numbers help to identify the polarity of the transformer windings. Traditionally, the ins will always be odd numbers and the outs will always be even numbers. The in and out of a single winding will always be two consecutive numbers, odd then even.

The letter X was chosen as the insignia for the lower-voltage winding instead of the letter L (as H indicates "higher," it seems reasonable that L would indicate "lower"), simply because the capital letter L is used to identify line values, such as line 1 (L1) and line 2 (L2).

The two solid dots adjacent to the H1 and X1 terminals are also used in establishing the polarity of the two windings. As drawn, H1 and X1 are "in-phase." Ac current is sinusoidal. One alternation is positive in reference to a zero-volt baseline. The second alternation is negative in reference to the same baseline. The solid dots indicate that X1 in the secondary circuit will go positive when H1 goes positive and negative when H1 goes negative.

For buck or boost transformations, or for correct polarity in a 3-phase bank formation (wye or delta), the dots ensure that the transformer leads will be interconnected correctly (more about this as the different transformer connections are introduced and discussed).

As drawn, the transformer in Figure 15-2 is used in a step-down configuration. The higher line voltage is being stepped down to a lower working voltage.

15.3 TRANSFORMER RATIOS

The construction of a transformer, the two windings, primary and secondary, permanently wound on a stationary iron core, will not allow a true dc voltage (from a storage battery) to be transformed (voltage stepped up, stepped down, or isolated).

The three principal components of induction include a conductor, a magnetic field, and relative motion between the two. Ac voltage provides the relative motion within a transformer with its continual change in amplitude and periodic reversal in polarity (direction of current flow). True dc has a constant level of voltage available and a constant level of current flow that is governed by Ohm's law and the total circuit resistance.

The student electrician should note that a transformer will convey pulsating dc voltages and currents produced either by an oscillator in an electronic circuit or derived through recitification of an ac circuit. Steady-level dc, as supplied by a storage-cell battery, applied to the primary of a transformer cannot generate a secondary voltage.

The power rating of a transformer is equal to the product of primary-circuit voltage and current or secondary-circuit voltage and current. The magnitude of currents flowing in the respective circuits of the transformer are governed by the magnitude of the load current connected to the secondary circuit. The power rating of the transformer indicates the maximum amount of power that can be drawn by the load without seriously overheating the transformer.

The voltage rating of a transformer or its windings is determined by the number of turns or wraps of wire around the iron core. Wrapping or coiling the wire effectively creates an inductive reactor (coil) in either the primary or secondary circuit. Instead of rating the respective coils of wire in terms of inductive reactance, the coils (windings) are rated in terms of voltage.

The voltage rating is an indication of the length of wire used to construct the coil of the winding. The total length is not as important as the number of wraps (around the core) required per volt. The voltage ratio and the turns ratio are directly proportional.

By formula:

$$\frac{N_{pri}}{N_{sec}} = \frac{V_{pri}}{V_{sec}} \quad \text{or}: \ N_{pri} \times V_{sec} = N_{sec} \times V_{pri}$$

where:

N( ) = (pri) primary and (sec) secondary number of turns

V( ) = (pri) primary and (sec) secondary voltage ratings

For step-down transformer configurations, the turns or voltage ratio (to say one says the other) is normally specified as a multiple of one. If the number of turns in the primary-circuit winding equals 2,400 wraps of wire and the number of turns in the secondary-circuit winding equals 60 wraps of wire, the turns ratio is 2,400 to 60. Since both numbers can be reduced (divided) by 60, the turns ratio is equal to 40-to-1 (normally written 40:1).

The voltage ratio is stated in the same manner: If a single winding transformer (one winding in the primary and one in the secondary) is used to step down a supply voltage of 480 volts to a sensing (control) voltage of only 12 volts, the turns ratio and voltage ratio are also 40-to-1: $480 \div 12 = 40$.

If the transformer primary winding has a high-value voltage rating, the coil will consist of a relatively high number of turns or wraps of small wire (wire size as compared to the secondary circuit). This winding, because of the small cross-sectional area of the wire, will also have a relatively low-value current rating.

If the transformer secondary winding has a low-value voltage rating, the coil will consist of a relatively low number of turns or wraps of larger wire (wire size as compared to the primary circuit). This winding, because of the large cross-sectional area of the wire, will also have a relatively high-value current rating.

Because the voltage ratio and turns ratio of a transformer are directly proportional, the turns ratio per se is rarely used in transformer power or circuit value computations.

Technically, the primary of a transformer can be either winding. The secondary can be either winding. The primary winding is connected to the power source that is being transformed (either step-down for a lower-circuit voltage, step-up for a higher-circuit voltage, or merely transformer isolation for the same circuit voltage). The secondary winding is connected to the load to which the transformer is supplying power.

If the transformer is used in step-up configuration, the ratio is stated in reverse. If the 12 volts in the previous discussion is the supply voltage and 480 volts is the transformed (secondary) load voltage, the ratio is stated as 1-to-40 (1:40).

In most plant distributions or control processes using electrical power, the transformer primary will be connected to the higher voltage. The secondary is used to derive the lower working voltage for the connected load.

Because the power rating (product of voltage and current) for a transformer is the same for both the primary and secondary circuits (ignoring core losses), in step-down voltage configurations, the available current is stepped up. In step-up voltage configurations, the available current is stepped down.

As with the number of turns and voltage ratios, the primary and secondary circuit currents also have a ratio. Because there is only one power rating for a transformer, the current ratio is inversely proportional to the voltage ratio.

By formula:

$$\frac{A_{sec}}{A_{pri}} = \frac{V_{pri}}{V_{sec}} \quad \text{or: } V_{pri} \times A_{pri} = V_{sec} \times A_{sec}$$

where:

A( ) = (pri) primary and (sec) secondary current ratings

V( ) = (pri) primary and (sec) secondary voltage ratings

Example 15–1. If the full-load power rating of the secondary circuit of a transformer is given at 240 volts and 104 amps, determine the primary-circuit full-load current rating for a primary-circuit voltage rating of 575 volts.

$$V_{pri} \times A_{pri} = V_{sec} \times A_{sec}$$

$$A_{pri} = \frac{V_{sec} \times A_{sec}}{V_{pri}}$$

$$A_{pri} = \frac{240 \text{ volts} \times 104 \text{ amps}}{575 \text{ volts}}$$

$$A_{pri} = 43.4 \text{ amps}$$

The example gave the secondary voltage and full-load current ratings of this power transformer. If the primary values of voltage and current had been given, the same format would have been used to compute the secondary-circuit full-load current.

When the voltage ratio of a power transformer is 1-to-1, the current ratio is also 1-to-1. Computations of the other circuit current, when either primary or secondary values are given, is not required. The transformer is being used for isolation of the primary and secondary circuits (often used for noise abatement in computer or other data-processing supply systems, and for limiting fault overcurrent available to the load circuit). The primary current is equal to the secondary current. The primary voltage is equal to the secondary voltage.

Code challenge 15–1. If the full-load power rating of the secondary circuit of a transformer is given at 230 volts and 164 amps, determine the primary-circuit full-load current rating for a primary-circuit voltage rating of 480 volts.

$$A_{pri} = \frac{\text{(A) \_\_\_\_\_\_\_\_\_\_ volts} \times \text{\_\_\_\_\_\_\_\_\_\_ amps}}{\text{(B) \_\_\_\_\_\_\_\_\_\_ volts}}$$

$$A_{pri} = \text{(C) \_\_\_\_\_\_\_\_\_\_ amps}$$

The voltage and current ratios, which are inversely proportional to each other, are math variations of the power formula. The primary-circuit current just computed will be drawn only when the secondary full-load current, according to the transformer's power rating, is drawn by a connected load.

The actual secondary load current of the transformer may be well under its full-load capability (200 amps available, but only a 60-amp load is connected to the transformer). The magnitude of the secondary-circuit current is determined by the secondary-circuit voltage rating and the resistance (or ac impedance) of the connected load. The higher the secondary-circuit resistance, the lower the load current, and vice versa. The actual primary-circuit current value will always be in direct proportion to the actual secondary load current (neglecting transformer losses).

Example 15–2. In the previous example problem, if the actual secondary-circuit load current is 73 amps, determine the actual primary-circuit current.

$$A_{pri} = \frac{240 \text{ volts} \times 73 \text{ amps}}{575 \text{ volts}}$$

$$A_{pri} = 30.5 \text{ amps}$$

The power rating of the transformer, or the actual value of power being supplied to the connected load in this example, is the product of the secondary voltage times the secondary current. Since the power being transformed has to be equal in each respective circuit, dividing the secondary actual power rating by the primary-circuit voltage rating computes (in one combined step) the actual primary load current (again neglecting transformer losses).

Code challenge 15–2. In the previous Code challenge problem, if the actual secondary-circuit load current is only 67 amps, determine the actual primary-circuit current.

$$A_{pri} = \frac{\text{(A) \_\_\_\_\_\_\_\_\_\_ volts} \times \text{\_\_\_\_\_\_\_\_\_\_ amps}}{\text{(B) \_\_\_\_\_\_\_\_\_\_ volts}}$$

$$A_{pri} = \text{(C) \_\_\_\_\_\_\_\_\_\_ amps}$$

The ratios of voltage, current, and number of turns will hold true for any transformer, whether it is rated as a small signaling transformer (such as a common doorbell supply), a small power transformer (as shown in the example), or a large power transformer (often found in unit substations).

15.4 OVERCURRENT PROTECTION

As stated in the introduction, Art. 450 governs transformer installations. Because the transformer structure has no moving parts, it is a very rugged piece of electrical utilization equipment. Transformer circuits are set up in the same manner as motor circuits. Although good design would warrant a power rating greater than or equal to the full-load power requirements of the intended secondary loads, transformers are normally allowed to operate in an overload condition.

Like the motor, the transformer will overheat when it is overloaded. The amount of allowed overload is often determined by the amount of ventilation or forced cooling available to the transformer enclosure. Some transformers may have vented enclosures and be air cooled. Others may have sealed enclosures and be oil cooled. If the transformer is heavily overloaded, the winding insulation may fail due to the intense heat buildup.

In parallel to motor circuits, the overcurrent protection for a transformer circuit can be split with transformer overload protection installed in the secondary circuit, and short-circuit, ground-fault overcurrent protection installed in the primary circuit.

As discussed in the service requirements for a building or other structure in Chapter 8, a fuse or circuit breaker installed in the secondary circuit of a transformer can provide full overcurrent protection (short-circuit, ground-fault, and overload-overcurrent protection) for the connected load, but only overload-overcurrent protection for the transformer located on its line or supply side.

In parallel to the combined overcurrent protection of a motor circuit, when the transformer is provided with primary-circuit overcurrent protection only, the fuse or circuit breaker must be sized or rated to provide overload-overcurrent protection to the transformer as well as the necessary fault-overcurrent protection.

The high value of initial energization current to the primary circuit of a transformer, especially to a large power-rated transformer, can be of a longer duration as compared to the in-rush starting current of a motor circuit, simply because the predominant secondary load of a transformer might well be motor loads. The primary-circuit impedance of a transformer is predominantly the reflected secondary impedance. Until this reflected impedance is established, the transformer will draw high-value primary-circuit current.

The allowed overrates of transformer overcurrent-protection devices are found in Secs. 450-3(a) and 450-3(b) in the National Electrical Code. Which section to use, and as a result what value of overrate is allowed, is determined by the circuit voltage ratings of the transformer. If both the primary- and secondary-circuit voltage ratings are not over 600 volts nominal (600 volts or less), then the overrates are determined according to Sec. 450-3(b). If any circuit voltage rating is over 600 volts, then the overrates are determined according to Sec. 450-3(a).

Figure 15-3 shows a single line drawing of a transformer circuit provided with "combined" primary overcurrent protection. Because both circuits are rated at less than 600 volts, Sec. 450-3(b)(1) sets the overrate of the primary full-load current of the transformer to a maximum of 125%.

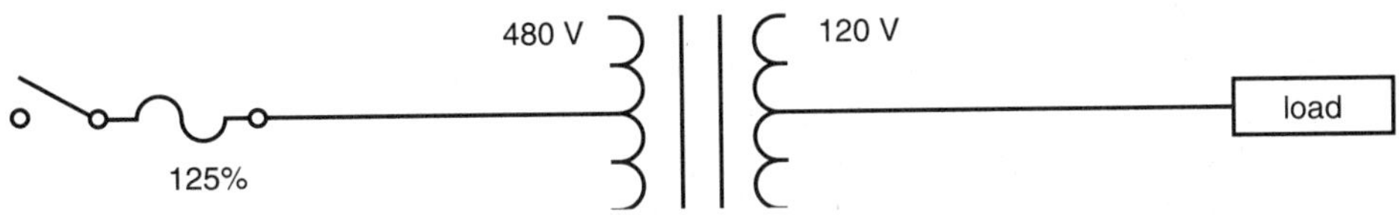

Fig. 15-3 Single-line drawing of a transformer circuit provided with *combined* (primary circuit only) overcurrent protection.

If the primary full-load current computation, determined by dividing the power rating of the transformer by the primary-circuit voltage rating, yields a nonstandard rating of a fuse, or setting of a circuit breaker, Excp. No. 1 to Sec. 450-3(b)(1) will allow the next higher standard size as listed in Sec. 240-6(a) to be installed.

If the computed primary full-load current is less that 9 amps, the 125% overrate can be increased to 167% with the next lower standard size of OCPD for nonstandard computations. If the computed primary full-load current is less than 2 amps, the overrate can be increased to a maximum of 300%.

When a feeder or branch circuit supplies two or more transformers, Excp. No. 2 to Sec. 450-3(b)(1) will allow the omission of individual transformer overcurrent protection if the rating of the feeder fuse or circuit breaker does not exceed the allowed overrate.

When a transformer will be energized with its secondary loads connected, similar to power being restored to your home when a lightning strike has caused a utility power outage, the initial in-rush energization current may be of such magnitude that the fuse or circuit breaker set at the 125% allowed overrate of combined primary-circuit overcurrent protection may open when power is applied.

Figure 15-4 shows the allowed overrates when the transformer overcurrent protection is split between the primary and secondary circuits. As allowed by Sec. 450-3(b)(2),

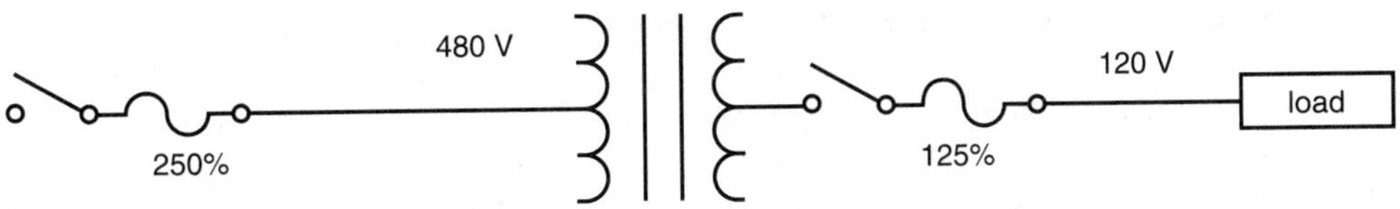

Fig. 15-4 Single-line drawing of a transformer circuit provided with *split* (primary and secondary circuit) overcurrent protection.

the 125% overrate for transformer overload-overcurrent protection is in the secondary circuit. The overrate of the primary SC-GF (short-circuit, ground-fault) overcurrent-protection device is increased to 250% to allow either a higher value or longer duration of inrush energization current, or both, to pass without opening the primary OCPD (overcurrent protection device).

Again, the respective circuit currents are computed by dividing the power rating of the transformer by the respective circuit voltages. When the overrates are applied to the currents and the computed current values do not correspond to standard ratings of fuses or circuit breakers as listed in Sec. 240-6(a), use of the next higher standard size is limited to the secondary-circuit overload-overcurrent protection only.

In parallel to Sec. 450-3(b)(1), Excp. No. 1, Exception to Sec. 450-3(b)(2), will allow an overrate up to 167% if the secondary computed load current is less than 9 amps. Like the primary-circuit overcurrent-protection device rating, when the 167% overrate yields a nonstandard size or rating, the next lower standard size or rating must be installed.

Examples of these required computations will accompany the drawings of different transformer circuit connections presented in both this chapter and Chapter 16.

When either or both circuits of the transformer will operate at over 600 volts, the overrate percentages are determined according Sec. 450-3(a). The allowed overrates are listed in either T450-3(a)(1) or T450-3(a)(2)b. What value of overrate to use is determined according to not only the voltage rating of the respective circuits, but also according to the impedance rating of the transformer, the type of overcurrent protection device to be installed, and to the location of the transformer. This seems like a lot of conditions and is the main reason that Sec. 450-3(b) was discussed first.

Except for the overrates allowed for the secondary circuit when operating at 600 volts or less, the two tables are identical. Whether to use the 125% overrate for the secondary-circuit overload-overcurrent protection, or the 250% overrate when the secondary-circuit voltage is rated at 600 volts or less, is decided by the heading of T450-3(a)(2)b: Does the transformer installation qualify as a "supervised location"?

The common pole-mounted power transformer for the service drop or lateral to a small business or a single-family dwelling, which is furnished and maintained by the serving utility, would qualify as a supervised location by virtue of its height. As a general rule, service-call electricians do not have access to either the supply or load terminals on the utility transformer.

A unit substation located in an industrial plant in a separate equipment room would also qualify as a supervised location, especially if the plant production is continuous (24-hour facility), and plant maintenance electricians or other qualified people are the only ones allowed access to the transformer vault.

On the other hand, the same unit substation located in an equipment room at an indoor mall or large department store may not qualify as a supervised location, simply because maintenance, service, or other repair of the electrical equipment would be by outside contractors who may not be familiar with the higher-voltage equipment.

As shown in Figure 15-5, the primary fuse is allowed to be overrated to a maximum of 300% of the primary full-load current, regardless of whether the transformer is located in a supervised or unsupervised location.

Whether to use the next higher standard size of fuse or circuit breaker when the

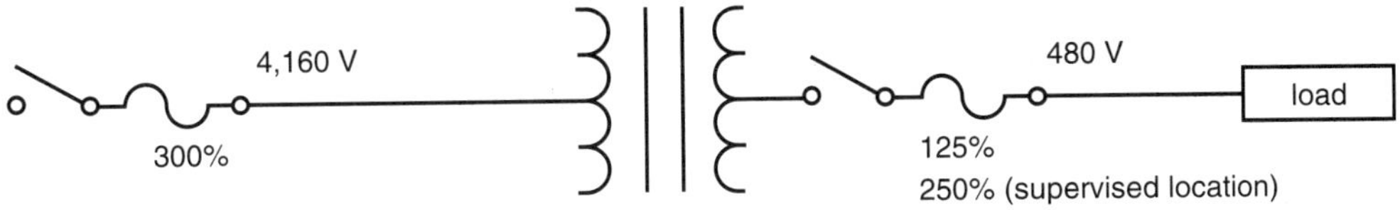

Fig. 15-5 When the primary circuit is rated over 600 volts, the fuse overrate is increased to 300%.

overrate computation yields a nonstandard size or rating of OCPD depends on whether or not the transformer is located in a supervised location. If the 125% overrate of the secondary overload-overcurrent protection for lower voltage systems is used (indicating a nonsupervised location), then Sec. 450-3(a)(1), Excp. No. 1 will allow the next higher standard size or rating of OCPD for both the primary- and secondary-circuit protection devices.

If the 250% overrate of the secondary overload-overcurrent protection for lower-voltage systems is used (indicating a supervised location), the overrate percentages listed in T450-3(a)(2)b. are maximum values because Sec. 450-3(a)(2)b. does not have a listed exception allowing the NHS.

The nameplate-rated impedance of the transformer, which is used to compute the maximum value of short-circuit or ground-fault overcurrent available at the secondary terminals of the transformer, controls the amount of overrate allowed for either the fuse or the circuit breaker protecting the secondary circuit when the secondary-circuit voltage rating is rated greater than 600 volts. The nameplate impedance only affects the overrate percentage of the inverse-time circuit breaker in the primary circuit.

The transformer impedance is expressed as a percentage on the nameplate. The percentage will range from 1 to a maximum of 10%. As shown in Figure 15-6, when the nameplate impedance is less than or equal to 6%, the secondary-circuit fuse overrate is limited to a maximum of 250% with the NHS if the computation yields a nonstandard rating [Sec. 450-3(a)(1), Excp. No. 1]. If the nameplate impedance is greater than 6% but not greater than 10%, the overrate is limited to 225% with the NHS.

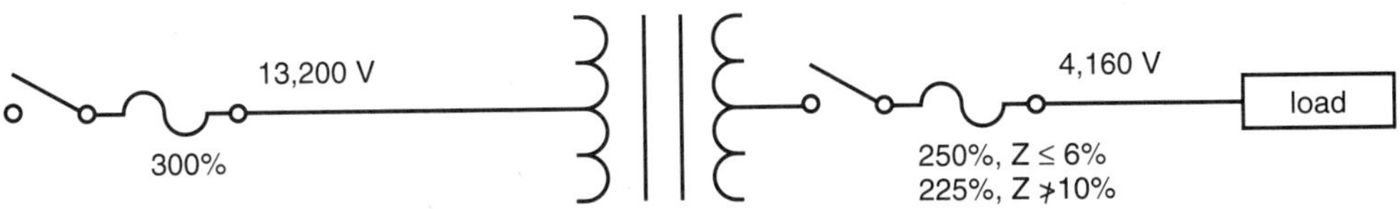

Fig. 15-6 The nameplate impedance (Z) of a transformer affects the overrate percentages of the secondary fuse for secondary circuits rated over 600 volts.

As shown in Figure 15-7, when both transformer circuits are protected by inverse-time circuit breakers, the nameplate impedance will affect the overrate percentages in both the primary and secondary circuits. When the nameplate impedance does not exceed 6%, the inverse-time circuit breaker providing the primary-circuit SC-GF overcurrent protection, can be overrated to 600% of the primary FLC. The inverse-time circuit breaker, providing the secondary overload-overcurrent protection for a circuit operating at over 600 volts, can be overrated to 300% of the secondary FLC. If the nameplate impedance exceeds 6%, the overrates are reduced to 400% and 250%, respectively.

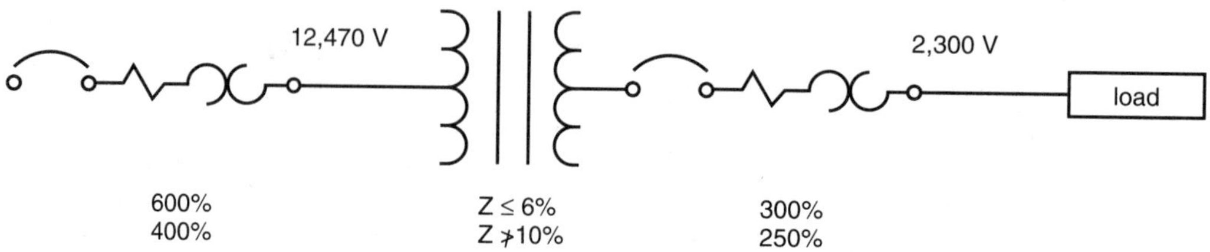

Fig. 15-7 When high-voltage transformer circuits are protected by inverse-time circuit breakers, the nameplate impedance will affect the overrate percentages.

As shown in Figure 15-8, if a transformer operating at not over 600 volts in either circuit is thermally protected with sensing devices embedded in the windings, Sec. 450-3(b)(2) will allow the primary fuse or circuit breaker to be overrated to either the 600% (six times) or 400% (four times) as allowed for circuit breakers by the 450-3(a) tables, according to the nameplate rated impedance.

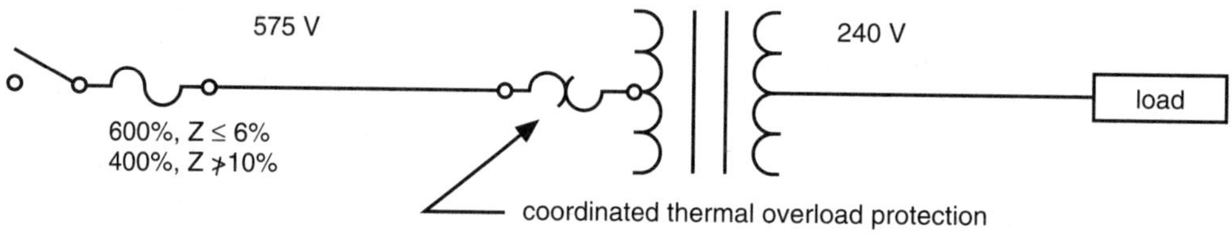

Fig. 15-8 For low-voltage transformer circuits, thermal overload protection wired in series with the primary circuit can be used in lieu of secondary overload protection.

The coordinated thermal-overload protection, which must interrupt the primary-circuit current and be installed at the point of manufacture, provides the necessary overload-overcurrent protection for the transformer. The higher percentages of overrate of the OCPD in the primary circuit, higher as compared to the 250% allowed by the first paragraph of Sec. 450-3(b)(2), provides only SC-GF overcurrent protection for the transformer.

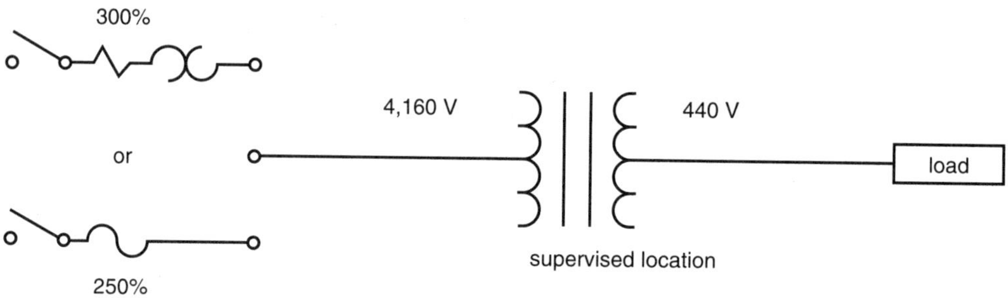

Fig. 15-9 Combined primary-circuit overcurrent protection is also allowed for transformers operating over 600 volts.

Section 450-3(b)(1) allowed the primary overcurrent-protection device to serve as combined overcurrent protection (short circuit, ground fault, and overload) for a transformer operating at not over 600 volts in either the primary or secondary circuit. As shown in Figure 15-9, Sec. 450-3(a)(2)a. will allow the primary OCPD to serve as the allowed combined overcurrent protection of a transformer for higher circuit voltages. The use of combined overcurrent protection is limited to supervised locations. The overrate for a fuse is limited to 250%. The overrate for electronic-actuated fuses or inverse-time circuit breakers is increased to 300%. Exception No. 1 will allow the NHS for either type of OCPD.

15.5 SHORT-CIRCUIT OVERCURRENT

The overrate percentages of the fuses or circuit breakers listed in Tables 450-3(a)(1) and 450-3(a)(2)b. are based on the values or percentages of nameplate-rated transformer impedance. The impedance rating is listed as a percentage that will vary among different types of transformers between 1 and 10%.

As introduced in Sec. 7.3 of Chapter 7 with the discussion of overcurrent, the greatest amount of short-circuit overcurrent available is at the shorted secondary terminals of a transformer (load-side of the fuse—Fig. 15-10). The interrupt rating of a fuse or circuit breaker connected to the load-side terminals has to be greater than or equal to this maximum overcurrent if the circuit will be cleared (opened) when forced into a faulted condition.

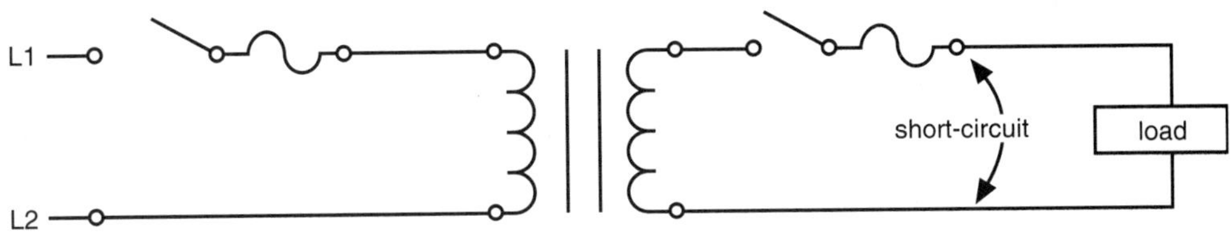

Fig. 15-10 The maximum short-circuit overcurrent from a transformer occurs at the load terminals of the windings and the fuse.

To compute the maximum short-circuit overcurrent the transformer is capable of delivering, the percent impedance should be converted to a decimal fraction and divided into 1. The product of this ratio times the secondary full-load current equals the maximum short-circuit overcurrent rating. By formula:

$$A_{sc} = \frac{1}{\text{decimal - value impedance}} \times A_{\text{sec FLC}}$$

Example 15–3. If a power transformer has a nameplate-rated impedance of 5% and a secondary full-load current rating of 600 amps, determine the maximum value of short-circuit current available at the secondary load terminals.

$$A_{sc} = \frac{1}{0.05} \times 600 \text{ amps} = 12{,}000 \text{ amps}$$

If only the kVA rating and voltage ratings of the transformer are known, the power formula can be used to compute the full-load secondary current. Which power formula to use depends on whether the transformer circuit is rated single-phase, 2-phase, or 3-phase.

Code challenge 15–3. If a power transformer has a nameplate-rated impedance of 6% and a secondary full-load current rating of 750 amps, determine the maximum value of short-circuit current available at the secondary load terminals.

$$A_{sc} = \frac{(A)\ \_\_\_\_\_\_\_\_\_\_}{(B)\ \_\_\_\_\_\_\_\_\_\_} \times \_\_\_\_\_\_\_\_\_\_ \text{ amps} = (C)\ \_\_\_\_\_\_\_\_\_\_ \text{ amps}$$

15.6 TRANSFORMER CIRCUIT CONDUCTORS

The high-value overrates of transformer overcurrent protection as introduced in Sec. 15.4 and listed in Secs. 450-3(a) & (b) in the NE Code are intended to protect the transformer windings. Article 240 lists the installation and sizing requirements for overcurrent-protection devices. Section 240-3 stipulates that circuit conductors, whether primary or secondary, must be protected against overcurrent in accordance with their rated ampacities. Section 240-3(i) is very direct in indicating that the overrates allowed by Art. 450 do not protect the circuit conductors.

As outlined in Sec. 13.4 of Chapter 13 with the discussion of motor-circuit supply conductors, in order for the circuit conductors to be protected against overcurrent in accordance with their rated ampacity, the table-rated ampacity listed according to wire size in the ampacity tables referenced by Sec. 310-15(a) (normally T310-16 up to 2,000 volts), should be determined according to the rating of the fuses or the setting of the circuit breaker providing the overload-overcurrent protection for the transformer.

For transformer circuits operating at 600 volts or less, the overload-overcurrent protection is set at 125% of the transformer FLC by Sec. 450-3(b)(1) for combined overcurrent protection (primary circuit only), and by Sec. 450-3(b)(2) for split protection (overload protection in secondary circuit; SC-GF protection in the primary). Both sections have exceptions that will allow the next higher standard size when the computation yields a nonstandard rating of OCPD. The rated-ampacity of the circuit conductors should be based on the selected rating of the fuse or circuit breaker and not just the 125% overrate of the respective circuit currents because of this NHS allowance.

Example 15–4. A power transformer is used to step down 480 volts to 208 volts. If the secondary full-load current rating is equal to 96 amps, determine:

1. The maximum rating of the time-delay fuses installed in the primary supply circuit for the combined overcurrent protection of the transformer
2. The minimum rating of the primary-circuit supply conductors, using THHN copper
3. The minimum rating of the secondary-circuit load conductors, using THHN copper

NE Code-maximum rating determination of the primary circuit fuses:
Primary FLC computation:

$$(208 \text{ volts} \div 480 \text{ volts}) \times 96 \text{ amps} = 41.6 \text{ amps}$$

Overrate computation: Ref. 450-3(b)(1)

125% of 41.6 amps = 52 amps

Standard ratings of TD fuses: Ref. 240-6(a)
NHS: 60 amps Ref. (NHS) 450-3(b)(1), Excp. 1

NE Code-minimum wire size of the primary-circuit conductors:
Wire terminal temperature limitations: 60°C Ref. 110-14(c)(1), Excp. 1
Table-rated ampacity: 70 amps Ref. T310-16
Wire size: No. 4 AWG THHN copper

NE Code-minimum wire size of the secondary-circuit conductors:
Secondary protection rating FLC computation:

(480 volts ÷ 208 volts) × 60 amps = 138.4 amps

Wire terminal temperature limitations: 75°C Ref. 110-14(c)(2), Excp. 1
Table-rated ampacity: 150 amps Ref. T310-16
Wire size: No. 1/0 AWG THHN copper

Computed at just the 125% overrate of the secondary FLC, which was only 96 amps, the secondary circuit conductors would only be sized for a rated ampacity of 120 amps. According to the 75°C temperature column of T310-16, the secondary-circuit conductors would only have to be wire size No. 1 AWG THHN copper instead of the computed No. 1/0. The difference of 8 amps between the 125% overrate of the primary protection (52 amps) and the rating of the primary-circuit fuses makes the No. 1 AWG secondary circuit conductors an NE Code violation of Sec. 240-3(i).

How to determine the rated-ampacity of the circuit conductors is discussed in Chapter 2. The rated ampacity and resulting wire size must be computed according to the prevailing conditions that hamper the rate of heat dissipation along the length of the circuit, as well as the wire terminal temperature limitations of Sec. 110-14(c).

Code challenge 15–4. A power transformer is used to step down 575 volts to 240 volts. If the secondary full-load current rating is equal to 120 amps, determine:

1. The maximum rating of the time-delay fuses installed in the primary supply circuit for the combined overcurrent protection of the transformer
2. The minimum rating of the primary-circuit supply conductors, using XHHW copper
3. The minimum rating of the secondary-circuit load conductors, using XHHW copper

NE Code-maximum rating determination of the primary circuit fuses:
Primary full-load current computation:

(__________ volts÷ __________ volts) × __________ amps = (A) __________ amps

Overrate computation: Ref. (B)________________

(C) __________ % of __________ amps = __________ amps

Standard ratings of TD fuses: Ref. (D)________________
NHS: (E) __________ amps Ref. (NHS) (F)________________

NE Code-minimum wire size of the primary-circuit conductors:
Wire terminal temperature limitations: (G) __________ °C
Ref. (H)________________
Table-rated ampacity: (I) __________ amps Ref. (J) ________________
Wire size: (K) __________ XHHW copper

NE Code-minimum wire size of the secondary-circuit conductors:
Secondary protection rating FLC computation:

(__________ volts÷ __________ volts) × __________ amps = (L) __________ amps

Wire terminal temperature limitations: (M) __________ °C
Ref. (N)________________
Table-rated ampacity: (O) __________ amps Ref. (P) ________________
Wire size: (Q) __________ XHHW copper

When the transformer overcurrent protection is split between primary and secondary OPCDs, the setting of the secondary circuit breaker or the rating of the secondary fuses must be projected into the primary circuit according to the voltage ratio to size the primary-circuit conductors to meet the requirements of Sec. 240-3.

Example 15–5. A power transformer is used to step down 575 volts to 360 volts. If the secondary full-load current rating is equal to 65 amps, determine:

1. The maximum rating of the inverse-time circuit breaker installed in the secondary load circuit as the overload-overcurrent protection of the transformer
2. The maximum rating of time-delay fuses installed in the primary supply circuit as the SC-GF overcurrent protection of the transformer
3. The minimum rating of the secondary-circuit load conductors, using THHN aluminum
4. The minimum rating of the primary-circuit supply conductors, using THHN aluminum

NE Code-maximum rating determination of the secondary-circuit CB:
Overrate computation: Ref. 450-3(b)(2)

125% of 65 amps = 81.3 amps

Standard ratings of inverse-time CB: Ref. 240-6(a)
NHS: 90 amps Ref. (NHS) 450-3(b)(2), Excp.

NE Code-maximum rating determination of the primary-circuit fuses:
Primary FLC computation:

$$(360 \text{ volts} \div 575 \text{ volts}) \times 65 \text{ amps} = 40.6 \text{ amps}$$

Overrate computation: Ref. 450-3(b)(2)

$$250\% \text{ of } 40.6 \text{ amps} = 101.7 \text{ amps}$$

Standard ratings of TD fuses: Ref. 240-6(a)
NLS: 100 amps Ref. (NLS) 450-3(b)(2)

NE Code-minimum wire size of the secondary-circuit conductors:
Wire terminal temperature limitations: 60°C Ref. 110-14(c)(1), Excp. 1
Table-rated ampacity: 100 amps Ref. T310-16
Wire size: No. 1/0 AWG THHN aluminum

NE Code-minimum wire size of the primary-circuit conductors:
Primary protection rating FLC computation:

$$(360 \text{ volts} \div 575 \text{ volts}) \times 90 \text{ amps} = 56.3 \text{ amps}$$

Wire terminal temperature limitations: 60°C Ref. 110-14(c)(1), Excp. 1
Table-rated ampacity: 65 amps Ref. T310-16
Wire size: No. 3 AWG THHN aluminum

As in the previous example, if the computed primary current of 40.6 amps had been overrated to the 125% allowed for the secondary overload-overcurrent protection rating or setting, the primary-circuit wire size would have been based on 50.8 amps, which in the 60°C aluminum column requires only a No. 4 AWG conductor—again, an NE Code violation of Sec. 240-3.

Code challenge 15–5. A power transformer is used to step down 600 volts to 340 volts. If the secondary full-load current rating is equal to 74 amps, determine:

1. The maximum rating of the inverse-time circuit breaker installed in the secondary load circuit as the overload-overcurrent protection of the transformer
2. The maximum rating of time-delay fuses installed in the primary supply circuit as the SC-GF overcurrent protection of the transformer
3. The minimum rating of the secondary-circuit load conductors, using XHHW copper-clad (c-c) aluminum
4. The minimum rating of the primary-circuit supply conductors, using XHHW c-c aluminum

NE Code-maximum rating determination of the secondary-circuit CB:

Overrate computation: Ref. (A)_______________

(B) __________ % of __________ amps = __________ amps

Standard ratings of inverse-time CBs: Ref. (C)_______________

NHS: (D) __________ amps Ref. (NHS) (E) _______________

NE Code-maximum rating determination of the primary circuit fuses:

Primary full-load current computation:

(__________ volts÷ __________ volts) × __________ amps = (F) __________ amps

Overrate computation: Ref. (G)_______________

(H) __________ % of __________ amps = __________ amps

Standard ratings of TD fuses: Ref. (I) _______________

NLS: (J) __________ amps Ref. (NLS) (K) _______________

NE Code-minimum wire size of the secondary-circuit conductors:

Wire terminal temperature limitations: (L) __________ °C

Ref. (M) _______________

Table-rated ampacity: (N) __________ amps Ref. (O)_______________

Wire size: (P) __________ XHHW c-c aluminum

NE Code-minimum wire size of the primary-circuit conductors:

Primary protection rating FLC computation:

(__________ volts÷ __________ volts) × __________ amps = (Q) __________ amps

Wire terminal temperature limitations: (R) __________ °C

Ref. (S) _______________

Table-rated ampacity: (T) __________ amps Ref. (U)_______________

Wire size: (V) __________ XHHW c-c aluminum

For power transformers with circuit voltages operating at over 600 volts, the NE Code-minimum rated ampacity of the respective circuit conductors would be determined (computed) in the same manner as shown in these examples. Whether to use a 125%, 225%, 250%, or 300% overrate of the transformer FLC rating depends on the voltage ratings of the secondary circuits and whether the installation is rated as a supervised or unsupervised transformer location.

For circuit voltages rated in excess of 2,000 volts, the rated ampacities and wire sizes listed in T310-16 are null and void. Section 310-15(a) references Tables 310-67 through 310-84 for circuit voltages within the range of 2,001 to 35,000 volts.

15.7 SINGLE-PHASE POWER TRANSFORMERS

When a single-phase source is applied to a transformer, whether it has only one winding in each circuit (primary and secondary), two windings in each circuit, or one winding in one circuit and two windings in the other circuit, it is still a single-phase transformer (Fig. 15-11).

Fig. 15-11 Single-phase dry-type transformers are often wall mounted for lower voltage supply of control or other process equipment.

The product of line values of voltage and current in the supply or load circuit of a single-phase transformer, as used in the examples for wire sizing the transformer circuit conductors in the previous section, is referred to as the apparent power of the service, feeder, or branch circuit that supplies the transformer.

For single-phase transformer circuits, the primary line voltage is equal to the primary winding or phase voltage. The primary line current is equal to the primary winding or phase current. The secondary line voltage is equal to the secondary winding or phase voltage. And the secondary line current is equal to the secondary winding or phase current. In formula,

$$V_{line} = V_{phase} \text{ (winding)}$$

and

$$A_{line} = A_{phase} \text{ (winding)}$$

for both the primary and secondary circuits.

The ratio of primary-to-secondary voltage is equal (directly proportional) to the turns ratio for a single-phase transformer. The ratio of primary-to-secondary current is inversely proportional (reciprocal) to both the turns and voltage ratios.

If the power and voltage ratings of the single-phase transformer are known, the current rating for full-load requirements can be determined by solving the power formula for the line current of the respective circuit. In formula,

$$\text{Power}_{\text{VA}} = \text{V}_{\text{line}} \times \text{A}_{\text{line}}$$

and

$$\text{A}_{\text{line}} = \text{Power}_{\text{VA}} \div \text{V}_{\text{line}}$$

For larger or higher power-rated transformers, the power rating is normally specified in terms of kVA instead of just VA (volt-amps) to keep the length of the number to just three or four digits. The power formula of line values is modified by dividing by 1,000 to read the power rating in kVA.

$$\text{Power}_{\text{kVA}} = (\text{V}_{\text{line}} \times \text{A}_{\text{line}}) \div 1{,}000$$

and

$$\text{A}_{\text{line}} = (\text{Power}_{\text{kVA}} \times 1{,}000) \div \text{V}_{\text{line}}$$

Figure 15-12 shows the wiring diagram of a single-phase, single-winding transformer used to step down a 240-volt distribution to a 120-volt source. The actual load current of the secondary circuit, which is equivalent to the secondary line current, is used to determine the primary line current value. The transformer could supply a full-load current within its rating of 100 amps: (12 kVA × 1,000) ÷ 120 volts = 100 amps.

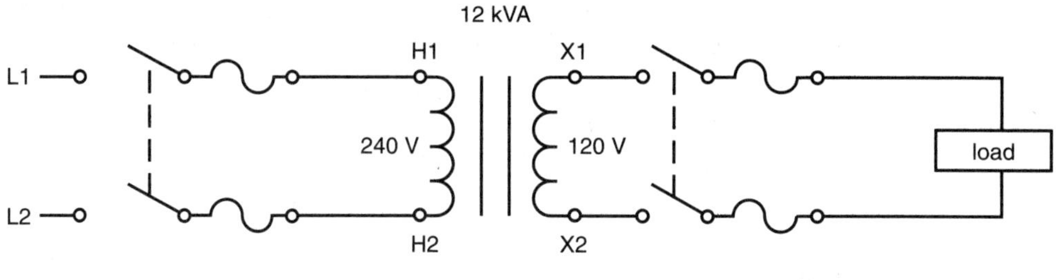

Fig. 15-12 Wiring diagram of a single-phase, single-winding power transformer.

Technically, the secondary load current could be rated up to 125 amps and be NE Code legal. The secondary OCPD rating is allowed by Sec. 450-3(b)(2) to be rated at a maximum of 125% of the secondary 100-amp FLC rating. Example 15-6 will demonstrate the NE Code allowances and requirements for both the overcurrent protection and the circuit conductors.

Example 15–6. The power transformer shown in Figure 15-12 is used to step down 240 volts to 120 volts. For the 12-kVA power rating of the transformer, determine:

1. The maximum rating of the inverse-time circuit breaker installed in the secondary load circuit as the overload-overcurrent protection of the transformer

2. The maximum rating of time-delay fuses installed in the primary supply circuit as the SC-GF overcurrent protection of the transformer
3. The minimum wire size of the secondary-circuit load conductors, using THW copper
4. The minimum wire size of the primary-circuit supply conductors, using THW copper

NE Code-maximum rating determination of the secondary-circuit CB:
Secondary FLC computation:

$$(12 \text{ kVA} \times 1{,}000) \div 120 \text{ volts} = 100 \text{ amps}$$

Overrate computation: Ref. 450-3(b)(2)

$$125\% \text{ of } 100 \text{ amps} = 125 \text{ amps}$$

Standard ratings of inverse-time circuit breakers: Ref. 240-6(a)

125 amps

NE Code-maximum rating determination of the primary-circuit fuses:
Primary FLC computation:

$$(12 \text{ kVA} \times 1{,}000) \div 240 \text{ volts} = 50 \text{ amps}$$

Overrate computation: Ref. 450-3(b)(2)

$$250\% \text{ of } 50 \text{ amps} = 125 \text{ amps}$$

Standard ratings of TD fuses: 125 amps Ref. 240-6(a)

NE Code-minimum wire size of the secondary-circuit conductors:
Wire terminal temperature limitations: 75°C Ref. 110-14(c)(2)
Table-rated ampacity: 130 amps Ref. T310-16
Wire size: No. 1 AWG THW copper

NE Code-minimum wire size of the primary-circuit conductors:
Primary protection rating FLC computation:

$$(120 \text{ volts} \div 240 \text{ volts}) \times 125 \text{ amps} = 62.5 \text{ amps}$$

Wire terminal temperature limitations: 60°C Ref. 110-14(c)(1), Excp. 1
Table-rated ampacity: 70 amps Ref. T310-16
Wire size: No. 4 AWG THW copper

Due to the full-load current ratings and the allowed overrates of Sec. 450-3(b)(2), the ratings of both the secondary circuit breaker protection and the primary-circuit fuse protection were standard sizes, such that the next higher standard size allowed by Sec.

450-3(b)(2), Excp. for the secondary rating was not required. Neither was the next lower standard size required for the primary protection for the lack of an exception to this same section.

Code challenge 15–6. For the power transformer circuit shown in Figure 15-12, change the power rating to 20 kVA and determine:

1. The maximum rating of the inverse-time circuit breaker installed in the secondary load circuit as the overload-overcurrent protection of the transformer
2. The maximum rating of time-delay fuses installed in the primary supply circuit as the SC-GF overcurrent protection of the transformer
3. The minimum wire size of the secondary-circuit load conductors, using RHW aluminum
4. The minimum wire size of the primary-circuit supply conductors, using RHW aluminum

NE Code-maximum rating determination of the secondary-circuit CB:
Secondary FLC computation:

(__________ kVA × 1,000) ÷ __________ volts = (A) __________ amps

Overrate computation: Ref. (B) __________

(C) __________ % of __________ amps = __________ amps

Standard ratings of inverse-time CBs: Ref. (D) __________
NHS: (E) __________ amps Ref. (NHS) (F) __________

NE Code-maximum rating determination of the primary circuit fuses:
Primary full-load current computation:

(__________ volts÷ __________ volts) × __________ amps = (G) __________ amps

Overrate computation: Ref. (H) __________

(I) __________ % of __________ amps = __________ amps

Standard ratings of TD fuses: Ref. (J) __________
NLS: (K) __________ amps Ref. (NLS) (L) __________

NE Code-minimum wire size of the secondary-circuit conductors:
Wire terminal temperature limitations: (M) __________ °C
Ref. (N) __________
Table-rated ampacity: (O) __________ amps Ref. (P) __________
Wire size: (Q) __________ RHW aluminum

NE Code-minimum wire size of the primary-circuit conductors:
Primary protection rating FLC computation:

(__________ volts÷ __________ volts) × __________ amps = (R) __________ amps

Wire terminal temperature limitations: (S) ___________ °C
Ref. (T) ____________________
Table-rated ampacity: (U) ___________amps Ref. (V) ____________________
Wire size: (W) ___________ RHW aluminum

The use of the single-phase, single-winding transformer, as shown in the diagram of Figure 15-12 is confined to the specific voltage ratings shown. A more versatile single-phase, dual-winding transformer used to step the 240-volt supply down to a 120-volt operations or control voltage is shown in the circuit diagram of Figure 15-13. The dual-winding rating is limited to the primary circuit only. Both 240-volt windings are magnetically coupled to the single secondary 120-volt winding.

One 240-volt primary-circuit winding is identified by H1 and H2. The second 240-volt winding is identified by H3 and H4. The single secondary winding is identified by the lower-voltage insignia of X1 and X2.

The transformer is wired correctly by paralleling the higher-voltage windings to the 240-volt, single-phase supply voltage. If this transformer has the same 12-kVA power rating as the single-winding transformer shown in Figure 15-11, each 240-volt winding is only capable of 6 kVA—half of the total power rating.

If the secondary-circuit power requirement is less than the 6 kVA available from one winding, either 240-volt primary winding is capable of driving or generating 120-volts in the secondary circuit when energized, but the other 240-volt winding cannot be left unconnected and open, nor can it be shorted (leads tied together) without causing a serious electrical safety hazard.

Because of the single 120-volt winding in the secondary, the use of the single-phase, dual-primary winding transformer, as shown in the wiring diagram of Figure

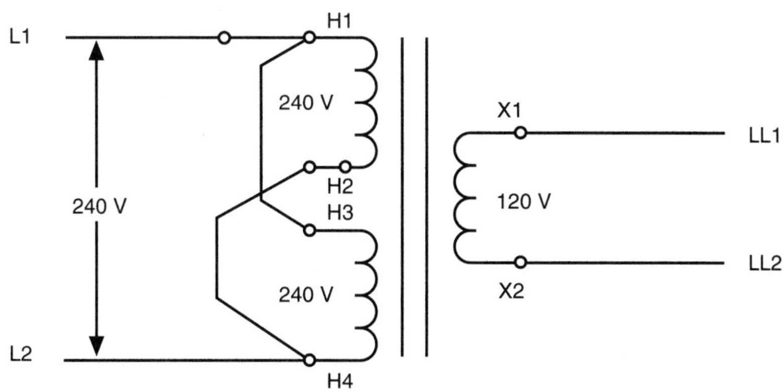

Fig. 15-13 A single-phase dual-winding transformer can be used to step down 240 volts to 120 volts.

15-13 is confined to the specific voltage ratings shown. A more versatile single-phase, dual-winding transformer, as shown in wiring diagram of Figure 15-14, has dual windings in both the primary and secondary circuits. The dual windings in the primary can be wired in series to transform a 480-volt supply down to either a 120-volt or 240-volt source, according to whether the secondary windings are wired in parallel (120 volts) or series (240 volts). Both 240-volt windings are magnetically coupled to the 120-volt winding.

As with the transformer shown in Figure 15-13, one 240-volt primary-circuit winding is identified by H1 and H2. The second 240-volt winding is identified by H3 and H4. In the secondary circuit of Figure 15-14, one 120-volt secondary-circuit winding is identified by X1 and X2. The second 120-volt winding is identified by X3 and X4.

The transformer is wired correctly by series wiring the two 240-volt windings across the 480-volt, single-phase supply voltage. The secondary windings are paralleled to generate the 120-volt load source.

According to the pattern of either series or parallel winding connections, this transformer can also be wired to step down a 480-volt supply to a 240-volt source (high voltage series, low voltage series), to step down a 240-volt source to a 120-volt source (high voltage parallel, low voltage parallel), to isolate a 240-volt source from a 240-volt load (high voltage parallel; low voltage series), to step up a 120-volt source to a 240-volt supply (low voltage parallel, high voltage parallel), to step up a 120-volt source to a 480-volt supply (low voltage parallel, high voltage series), to step up a 240-volt source to a 480-volt supply (low voltage series, high voltage series).

A common application of a single-phase power transformer in the industrial concern is found in process heating. The 480-volt plant distribution voltage is stepped down to 240 volts for supply to exposed process heating elements.

The overrate percentages of overcurrent protection for these transformers are found in Sec. 450-3(b). In most applications, the transformer is provided with split overcurrent

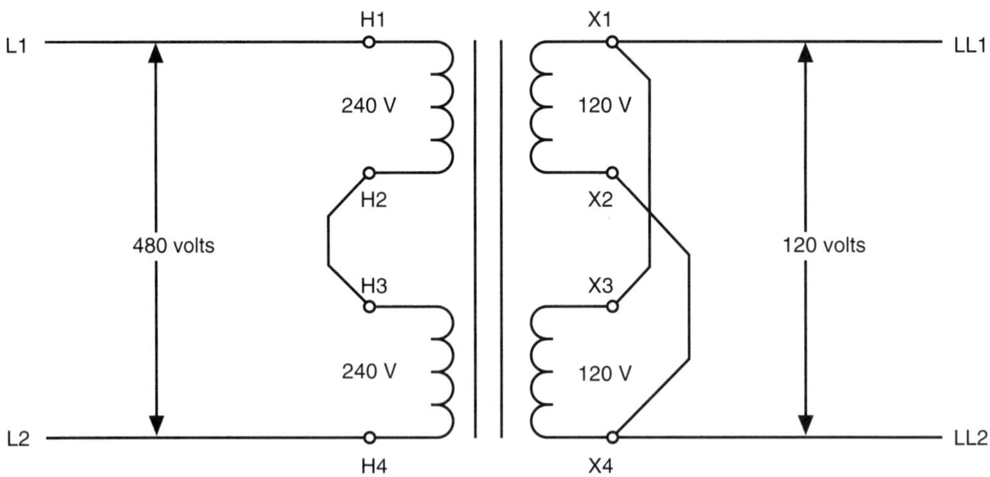

Fig. 15-14 A single-phase dual-winding transformer can be used to step down 480 volts to 120 volts.

protection (in both the primary and secondary circuits). Primary feeder is specified in both subsections [(b)(1) and (b)(2)] because a single feeder circuit protected against overcurrent at its supply end is allowed to supply and protect two or more of these power transformers, provided that each transformer is individually protected against overload-overcurrent on its secondary side.

The overrate percentages for these smaller power transformers are not as great as was cited for the utility-type distribution transformer in Sec. 450-3(a). As specified in Sec. 450-3(b)(2), the transformer can be purposely overloaded in terms of secondary load current by these allowed percentages. The better design criteria sizes the transformer power rating at least equal to, if not greater than, the computed full-load current rating. NE Code minimum allows the overload overrate.

In reference to Sec. 240-3, the rated-load ampacity of the circuit conductors is determined by the rating or setting of the secondary overload-overcurrent protection device rating. The secondary-circuit condcutors are sized directly according to the rating of the fuses or setting of the circuit breaker. The minimum rated-load ampacity of the primary-circuit conductors can be computed by multiplying the secondary to primary voltage ratio times the secondary overcurrent-protection device rating or setting.

When the 125% overrate of the secondary full-load current does not correspond to a standard rating of a fuse or setting of a circuit breaker, the exception to Sec. 450-3(b)(2) will allow the next higher standard size, as listed in Sec. 240-6(a), to be used.

With this in mind, a 75-kVA transformer used to step down a 480-volt feeder circuit to a 240-volt feeder circuit could supply a 400-amp load and be NE Code legal. By the power formula, the secondary full-load current rating of this transformer is only (75 ÷ 0.24) 312.5 amps. Increased to 125% equals only 390.6 amps. With the exception of the next higher standard size, a 400-amp CB or switched fuses could be installed for the secondary-circuit overload-overcurrent protection.

In the primary circuit, the full-load rating is only (75 ÷ 0.48) 156.3 amps. Increased to 250%, the maximum fuse or CB rating is 390.6 amps. But this is maximum. There is no exception that will allow the next higher standard size or rating of overcurrent protection device. The primary SC-GF overcurrent-protection rating of the transformer would be limited to either 350-amp fuses or a 350-amp circuit breaker.

Example 15–7. If a single-phase power transformer is rated 460/230 volts at 30 kVA, determine:

1. The maximum setting of the secondary circuit breaker
2. The maximum rating of the primary-circuit time-delay fuses
3. The minimum wire size of the secondary-circuit conductors, using THHW-2 copper
4. The minimum wire size of the primary-circuit conductors, using THHW-2 copper

NE Code-maximum setting of the secondary circuit breaker determination:

Secondary FLC computation:

$$(30 \text{ kVA} \times 1{,}000) \div 230 \text{ volts} = 130.4 \text{ amps}$$

Overrate computation: Ref. 450-3(b)(2)

125% of 130.4 amps = 163 amps

Standard OCPD ratings: Ref. 240-6(a)
NHS: 175 amps Ref. (NHS) 450-3(b)(2), Excp. No. 1

NE Code-maximum setting of the primary-circuit fuses determination:
Full-load current computation:

(30 kVA × 1,000) ÷ 460 volts = 65.2 amps

Overrate computation: Ref. 450-3(b)(2)

250% of 65.2 amps = 163 amps

Standard OCPD ratings: Ref. 240-6(a)
NLS: 150 amps Ref. (NLS) 450-3(b)(2)

NE Code-minimum wire size of the secondary-circuit conductors:
Wire terminal temperature limitations: 75°C Ref. 110-14(c)(2), Excp. No. 1
Table-rated ampacity: 175 amps Ref. T310-16
Wire size: No. 2/0 AWG THHW-2 copper

NE Code-minimum wire size of the primary-circuit conductors:
Primary FLC computation:

(230 volts ÷ 460 volts) × 175 amps = 87.5 amps

Wire terminal temperature limitations: 60°C Ref. 110-14(c)(1), Excp. No. 1
Table-rated ampacity: 95 amps Ref. T310-16
Wire size: No. 2 AWG THHW-2 copper

The rated ampacity of the secondary-circuit conductors exceeded 100 amperes. Section 110-14(c)(2), Excp. No. 1 allowed the use of the 90°C temperature-rated wire but required the table-rated ampacities listed in the 75°C temperature column to be used to size the wire.

In the primary circuit, the 175-amp rating of the secondary inverse-time circuit breaker was projected into the primary circuit by the inverse of the voltage ratio. The computation yielded a primary FLC less than 100 amperes. Section 110-14(c)(1), Excp. No. 1 allowed the use of the 90°C temperature-rated wire but required the table-rated ampacities listed in the 60°C temperature column to be used to size the wire.

Code challenge 15–7. If a single-phase power transformer is rated 575/240 volts at 37.5 kVA, determine:

1. The maximum setting of the secondary-circuit breaker
2. The maximum rating of the primary-circuit time-delay fuses
3. The minimum wire size of the secondary-circuit conductors, using XHHW aluminum
4. The minimum wire size of the primary-circuit conductors, using XHHW aluminum

NE Code-maximum setting of the secondary-circuit breaker determination:
Secondary FLC computation:

(__________ kVA × __________) ÷ __________ volts = (A) __________ amps

Overrate computation: Ref. (B)__________

(C) __________ % of __________ amps = __________ amps

Standard OCPD ratings: Ref. (D) __________
NHS: (E) __________ amps Ref. (NHS) (F) __________
NE Code-maximum rating determination of the primary-circuit fuses:
Primary FLC computation:

(__________ kVA × __________)÷ __________ volts = (G) __________ amps

Overrate computation: Ref. (H) __________

(I) __________ % of __________ amps = __________ amps

Standard OCPD ratings: Ref. (J) __________
NLS: (K) __________ amps Ref. (NLS) (L) __________

NE Code-minimum wire size of the secondary-circuit conductors:
Wire terminal temperature limitations: (M) ___ °C Ref. (N) __________
Table-rated ampacity: (O) __________ amps Ref. (P) __________
Wire size: (Q) __________ XHHW aluminum

NE Code-minimum wire size of the secondary-circuit conductors:
Primary FLC computation:

(__________ volts÷ __________ volts) × __________ amps = (R) __________ amps

Wire terminal temperature limitations: (S) ___ °C Ref. (T) __________
Table-rated ampacity: (U) __________ amps Ref. (V) __________
Wire size: (W) __________ XHHW aluminum

Almost all residential installations (one- and two-family dwellings) and small-scale commercial or industrial installations that have low power requirements will be supplied by a single-phase service. The most common installation is a single-phase power distribution transformer (Fig. 15-15) that has a single primary winding and two secondary windings. The primary winding can be connected in either a 3-phase wye or delta connection with other supply transformers for single-phase services within the utlity's distribution system.

The two secondary windings are both rated at 120 volts. The two windings are wired in series at the transformer with the center tap (connection between the two windings) connected to a grounding-electrode system. The service, established at the secondary connections of the power transformer, is rated as a single-phase, 3-wire, 120/240-volt distribution (Fig. 15-16). The lower neutral voltage of 120 volts is available from either phase A (LL1) or phase B (LL2) to the neutral conductor. The higher single-phase, phase-to-phase voltage of 240 volts is available only to loads that are connected between phases A and B (the two ungrounded supply conductors).

The transformer can supply 120-volt and 240-volt loads simultaneously. The load is said to be balanced when the same value (magnitude) of 120-volt load amperage is connected to each ungrounded phase conductor. The 240-volt loads, which utilize the full winding voltage, are balanced automatically.

The transformer primary normally consists of a single winding rated at 7,200 volts. The single winding is connected as single-phase (line-to-neutral) in a 3-phase, 4-wire wye primary distribution system.

In terms of voltage and current ratios, the single-phase power distribution transformer, which is either pad-mounted or pole-mounted some distance (the length across

Fig. 15-15 A single-phase, three-wire 120/240-volt, power transformer supply for a residential service.

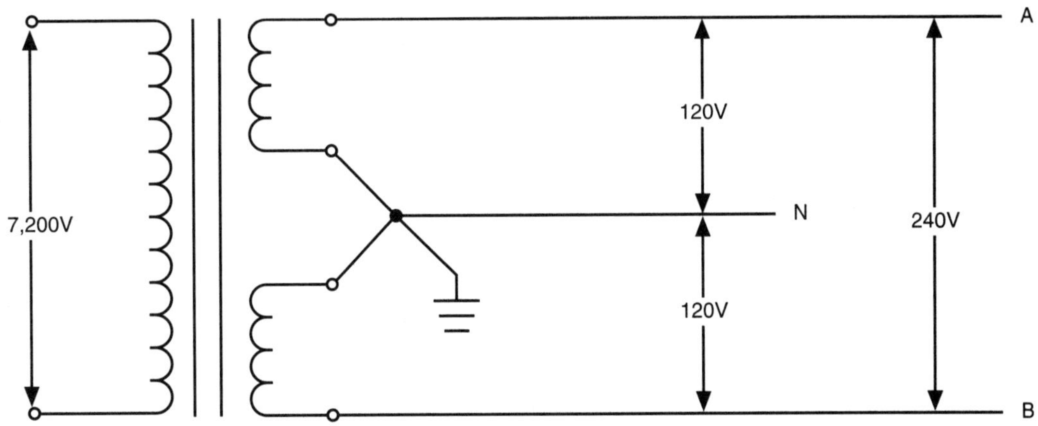

Fig. 15-16 Voltage relationships of the three supply lines from a single-phase residential service power transformer.

the yard) from the side of the dwelling structure, functions like any transformer. The kVA rating of the transformer is determined by the service full-load computations as determined by Art. 220 of the NE Code (refer to Chapter 17 of this text).

Example 15–8. For a single-phase, 3-wire, 120/240-volt, 200-amp service supply to a dwelling structure through a transformer with a 7,200-volt primary, determine:

1. The voltage ratio of the power transformer
2. The kVA rating of the transformer
3. The primary full-load current rating of the transformer

Voltage ratio computation:

$$7{,}200 \text{ volts} \div 240 \text{ volts} = 30{:}1$$

Power rating computation:

$$(240 \text{ volts} \times 200 \text{ amps}) \div 1{,}000 = 48 \text{ kVA}$$

Primary full-load current computation:

$$(240 \text{ volts} \times 200 \text{ amps}) \div 7{,}200 \text{ volts} = 6.67 \text{ amps}$$

The turns ratio was computed at 30:1 for the secondary line-to-line (phase-to-phase) voltage of 240 volts. The primary full-load current equals $6\frac{2}{3}$ amps. Each 120-volt secondary winding is coupled to the 7,200-volt primary with a turns ratio of 60:1. When one 120-volt winding is fully load at 200 amps, the primary full-load current is equal to only $3\frac{1}{3}$ amps.

Use of the power formula to determine the required VA or kVA rating of the transformer assumes the lower-voltage, phase-to-neutral loads are balanced (equal magnitude of current from phase A to the common neutral as there is from phase B). When the neutral loads are unbalanced, the transformer must be sized for the largest value of unbalanced neutral load current connected to any one phase, plus the current of the connected 240-volt, phase-to-phase, single-phase loads.

Code challenge 15–8. For a single-phase, 3-wire, 115/230-volt, 150-amp service supply to a dwelling structure through a transformer with a 7,200-volt primary, determine:

1. The voltage ratio of the power transformer
2. The kVA rating of the transformer
3. The primary full-load current rating of the transformer

Voltage ratio computation:

__________ volts ÷ __________ volts = (A) __________

Power rating computation:

(__________ volts × __________ amps) ÷ (B) __________ = (C) __________ kVA

Primary full-load current computation:

(__________ volts × __________ amps) ÷ __________ volts = (D) __________ amps

The overcurrent protection for this type of power transformer is allowed to be split between the primary and secondary circuits. Because the primary single-phase voltage is rated over 600 volts, the rating of the primary and secondary OCPDs must be determined according to the overrates found in Sec. 450-3(a) and the accompanying tables.

Good design of the power distribution system within a building or other structure normally sizes the transformer with a power rating at least equal to the full-load current of the secondary circuit as computed in the previous example. Utility tranformers, located outside on a pole or in a vault, are normally not sized in the same manner.

The service transformer would be utility-owned and therefore a supervised installation (not just anybody works on or with utility transformers). The overrate percentages allowed are listed in T450-3(a)(2)b. The primary circuit operates above 600 volts. The secondary operates at less than 600 volts. The selected overrate depends on the nameplate-rated impedance of the transformer.

When the transformer is provided with secondary overload-overcurrent protection, Sec. 450-3(a)(2)b. allows individual primary overcurrent protection for the transformer to be omitted. The 7,200-volt, single-phase trunk line of the utility can be fused at its point of origin to protect two or more transformers against fault overcurrent. The ampere rating of the higher-voltage fuse cannot exceed 300% of the primary full-load current rating.

The power transformers can be purposely undersized and meet NE Code requirements. For a supervised installation, the 200-amp service in Example 15-8 could be supplied by a transformer with a secondary full-load current rating of only 80 amps: 200 amps ÷ 2.5 (250%) = 80 amps. By the power formula, the kVA rating of the transformer need only be 19.2 kVA: 240 volts × 80 amps ÷ 1,000.

The circuit conductors of the power transformer must be sized according to the maximum-load current that will be drawn or supplied by the transformer. This is based on the rating or setting of the secondary overload-overcurrent fuses or circuit breaker.

For the 200-amp service, if the supply to the dwelling is in the form of a triplexed service-drop cable, the secondary conductors, installed in free air, can be sized by T310-17. Assuming RHW insulation, the service drop can consist of No. 1/0 AWG RHW copper or No. 2/0 AWG RHW aluminum or c-c aluminum conductors.

The service-entrance conductors, either single insulated conductors installed in conduit or SE cable, can be sized by Note 3 to Ampacity Tables 310-16 through 310-19 (because it is a single-phase dwelling supply) at either No. 2/0 AWG RHW copper, or No. 4/0 AWG RHW aluminum or c-c aluminum. For a nondwelling service of the same rating, T310-16 would require either No. 3/0 AWG RHW copper or 250 kcmil RHW aluminum or c-c aluminum for the same service-entrance conductrors.

The minimum rated-load ampacity of the primary feeder conductors (the utility trunk line) can be determined by applying the voltage ratio of secondary to primary voltages to the secondary full-load current rating. For the 200-amp service: 240 volts ÷ 7,200 volts × 200 amps = 6.67 amps.

The rated-load ampacity of the primary-circuit conductors (line conductors) to the transformer would have to be at least 6.67 amps at 7,200 volts. If the primary trunk lines supply more than one transformer, the primary trunk line fuse rating cannot exceed the 300% rating allowed by T450-3(a)(2)b.

If the transformer is purposely overloaded to the 250% maximum allowed in the secondary circuit by T450-3(a)(2)b., then 19.2 kVA ÷ 7,200 volts = 2.67 amps. The 300% overrate must be applied to this 2.67 amps to correctly size the primary fuses. The maximum rating of the primary fuses ([Sec. 240-6(a), Excp.] is only 8 amps.

The primary trunk line conductors of the utility are normally bare. The separation in the air is used as the high-voltage insulation. As such, the ampacity tables found in Art. 310 do not apply. The voltage rating determines the minimum air separation. The higher the voltage, the wider the separation. Bare conductor ampacities are listed in the NE Code in Appendix B, Table B-310-4. The smallest wire size listed is No. 8 AWG. No. 8 copper is good for 98 amps. No. 8 aluminum is good for 76 amps. Steel wire, which is common for the span strength of these trunk line circuits, is not listed.

In reference to Sec. 90-2(b)(5), the ampere rating, size, and installation requirements of these conductors, because they are utility owned, do not come under the jurisdiction of the NE Code. The example was expanded to demonstrate methods required to meet NE Code requirements.

Example 15–9. If a building will be supplied with a 400-amp, 120/240-volt, single-phase, 3-wire service, determine:

1. The minimum kVA rating of the transformer based on the panel rating, if the NE Code-maximum setting of the secondary overcurrent-protection device is employed for an unsupervised installation

2. The maximum rating of a trunk line fuse for a primary supply voltage of 7,200 volts
3. The minimum wire size of the triplexed service drop using RHW aluminum conductors
4. The minimum wire size of RHW aluminum conductors in conduit extended as the service entrance to the service-disconnecting means

NE Code-minimum power transformer kVA rating determination:

Secondary FLC computation: Ref. T450-3(a)(1)

$$400 \text{ amps} \div 1.25 = 320 \text{ amps}$$

Power rating computation:

$$240 \text{ volts} \times 320 \text{ amps} \div 1{,}000 = 76.8 \text{ kVA}$$

NE Code-maximum rating determination of the primary-circuit fuses:

Primary FLC computation:

$$(240 \text{ volts} \div 7{,}200 \text{ volts}) \times 320 \text{ amps} = 10.7 \text{ amps}$$

Overrate computation: Ref. T450-3(a)(1)

$$300\% \text{ of } 10.7 \text{ amps} = 32.1 \text{ amps}$$

Standard OCPD ratings: Ref. 240-6(a)

NHS: 35 amps Ref. (NHS) 450-3(a)(1), Excp. No. 1

NE Code-minimum wire size of the triplexed service-drop conductors:

Wire terminal temperature limitations: 75°C Ref. 110-14(c)(2)

Table-rated ampacity: 425 amps Ref. T310-17

Wire size: 400 kcmil RHW aluminum

NE Code-minimum wire size of the service-entrance conductors:

Wire terminal temperature limitations: 75°C Ref. 110-14(c)(2)

Table-rated ampacity: 425 amps Ref. T310-16

Wire size: 900 kcmil RHW aluminum

The Note 3 ampacity table to the 0- to 2,000-volt ampacity tables (notes following T310-19 in the NE Code) could not be used to reduce the wire sizes listed in this example. The service was to a nondwelling structure.

Table 310-17 allowed a much smaller wire size for the service drop conductors as compared to the service-entrance (SE) conductors sized by T310-16. The difference stems from the rate of heat dissipation for the two different sets of conductors. Table 310-17 list the ampacities allowed for circuit conductors suspended in free air (extra cooling). The SE conductors must be contained in an overall sheath (multiconductor cable, or else be installed in conduit or other raceway) to protect them from physical damage (heat entrapment).

Code challenge 15–9. If a building will be supplied with a 300-amp, 120/240-volt, single-phase, 3-wire service, determine:

1. The minimum kVA rating of the transformer based on the panel rating, if the NE Code-maximum setting of the secondary overcurrent protection device is employed for an unsupervised installation
2. The maximum rating of a trunk line fuse for a primary supply voltage of 7,200 volts
3. The minimum wire size of the triplexed service drop, using RHW copper conductors
4. The minimum wire size of RHW copper conductors in conduit extended as the service entrance to the service-disconnecting means

NE Code-minimum power transformer kVA rating determination:
Secondary FLC computation: Ref. (A) ________________

________ amps ÷ (B) ________ = (C) ________ amps

Power rating computation:

________ volts × ________ amps ÷ ________ = (D) ________ kVA

NE Code-maximum rating determination of the primary-circuit fuses:
Primary FLC computation:

(________ volts÷ ________ volts) × ________ amps = (E) ________ amps

Overrate computation: Ref. (F) ________________

(G) ________ % of ________ amps = ________ amps

Standard OCPD ratings: Ref. (H) ________________
NHS: (I) ________ amps Ref. (NHS) (J) ________________

NE Code-minimum wire size of the triplexed service-drop conductors:
Wire terminal temperature limitations: (K) ____ °C Ref. (L) ________________
Table-rated ampacity: (M) ________ amps Ref. (N) ________________
Wire size: (O) ________ RHW copper

NE Code-minimum wire size of the service-entrance conductors:
Wire terminal temperature limitations: (P) ____ °C Ref. (Q) ________________
Table-rated ampacity: (R) ________ amps Ref. (S) ________________
Wire size: (T) ________ RHW copper

Often electricians mistakenly think the single-phase, 3-wire, 7,200-volt–120/240-volt transformer is a 2-phase transformer because of the phase notation of A and B, but this is not the case. For a transformer to be rated as 2-phase, the two phases must have electrical time displacement (electrical degrees) between them. Phases A and B of this power distribution transformer are generated simultaneously from a single-phase source, which means the two 120-volt windings (wired in series) are in phase with each other.

Another confusing point about this type of 3-wire, single-phase service is the cancelling effect of the neutral current between the two phases. When only one phase has a neutral-current load, only one winding of the transformer is serving as a power supply for the connected load. This is true for either winding.

When both 120-volt windings are connected to phase-to-neutral loads, the two loads are basically put in series with or across a 240-volt power source. The current, which by Ohm's law is constant in a series circuit, is restricted by the total impedance of the 240-volt load (Fig. 15-17). But the neutral serves as a 120-volt reference such that neither load voltage can exceed 120 volts. The lower-impedance load will draw more current than the 120-volt circuit with the higher-impedance load. The difference attained by subtracting the current of the higher-impedance load from the current of the lower-impedance load is the unbalanced current that must be supplied from the respective 120-volt winding through the neutral conductor.

If the neutral conductor between the center-tap connection and the connected 120-volt loads should become open for any reason, the 120-volt reference, as well as the unbalanced neutral current being supplied to the lower-impedance load, would be lost. Serious voltage disturbances are the result.

When this happens, the total line current, supplied by the ungrounded phase conductors A and B, is reduced or consumed by the total impedance of the two phase-to-neutral loads which are now connected in series across a 240-volt source. The voltage drop across the two series impedances (voltage applied to each series load) will divide

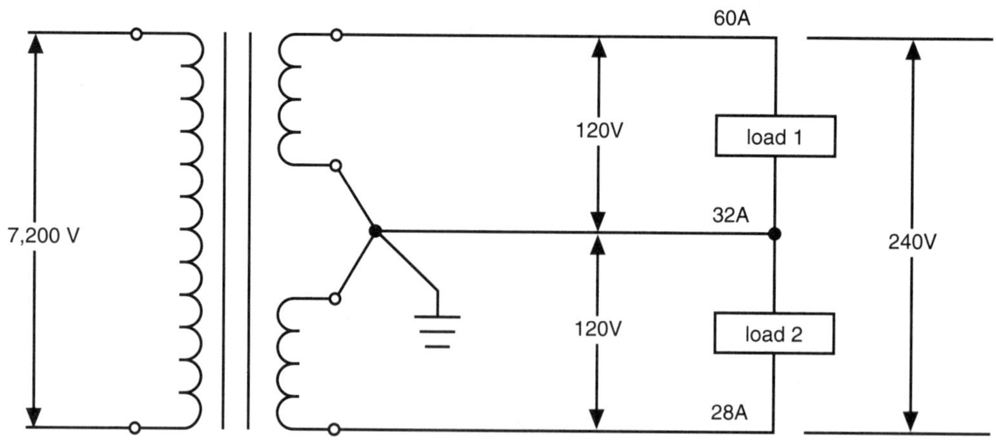

Fig. 15-17 The neutral conductor in the secondary circuit of a single-phase, three-wire power transformer conveys only the unbalanced load current between the two ungrounded phase conductors.

proportionately to the magnitude of the individual impedances (load resistances). In other words, the lower the load resistance, the lower the voltage now applied to that load will be. The low-impedance loads will suffer from undervoltage, and the high-impedance loads will suffer from overvoltage. In either instance, most electrical equipment, which operates at a rated voltage, will fail.

Section 220-22 will allow the neutral conductor on a single-phase, 3-wire system or a 3-phase, 4-wire system to be smaller than the ungrounded phase conductors. The feeder neutral conductor must have a rated ampacity at least equal to the maximum unbalanced load as determined by Sec. 220-22. If the neutral conductor is a service conductor, then its size must also meet the ground-fault return requirements of Sec. 250-23(b). As a general NE Code rule, there is no reduction allowed for the grounded neutral (circuit for two-wire) conductor in a smaller branch-circuit supply.

15.8 TWO-PHASE POWER TRANSFORMERS

Besides the common single-phase and 3-phase (Sec. 15.9) electrical power distribution systems, some areas of the country operate on a 2-phase system.

Two-phase distribution systems utilize two single-phase sources that are 90 electrical degrees out-of-phase with each other. Two-phase power can be transformed either by two individual single-phase transformers or by two separate windings mechanically displaced (offset) 90° from each other along the confines of a single iron core.

In either type of construction, the rating of the transformer bank (for two individual transformers) is the sum of the VA or kVA rating of each individual winding. In other words, if two single-phase transformers rated at 50 kVA each are banked in a 2-phase transformation, the total rating of the transformer bank is 2 × 50 kVA = 100 kVA.

In terms of phase (coil or individual winding) voltage and current, if both transformers have the same primary and secondary voltage and power ratings, then the total bank 2-phase power rating is equal to

$$\text{Power}_{\text{VA}} = 2 \times \text{V}_{\text{phase}} \times \text{A}_{\text{phase}}$$

On a 4-wire system as illustrated in Figure 15-18, each transformer is connected to its own individual single-phase source. The two phases do not interact with each other. As with single-phase transformation, the phase currents are equal to the line currents, and the phase voltages are equal to the line voltages, such that the power formula is

$$\text{Power}_{\text{VA}} = 2 \times \text{V}_{\text{line}} \times \text{A}_{\text{line}}$$

When the full-load current of an intended secondary load for a 2-phase transformer is known, the line values of voltage and current are simply plugged into the power formula to determine or compute the required power rating of the transformer.

Power ratings of 2-phase transformers are normally given or computed in terms of kVA. To convert the volt-amp rating to a kilovolt-amp rating, the 2-phase power formula is divided by 1,000. In formula,

$$\text{Power}_{\text{kVA}} = \frac{2 \times \text{V}_{\text{line}} \times \text{A}_{\text{line}}}{1{,}000}$$

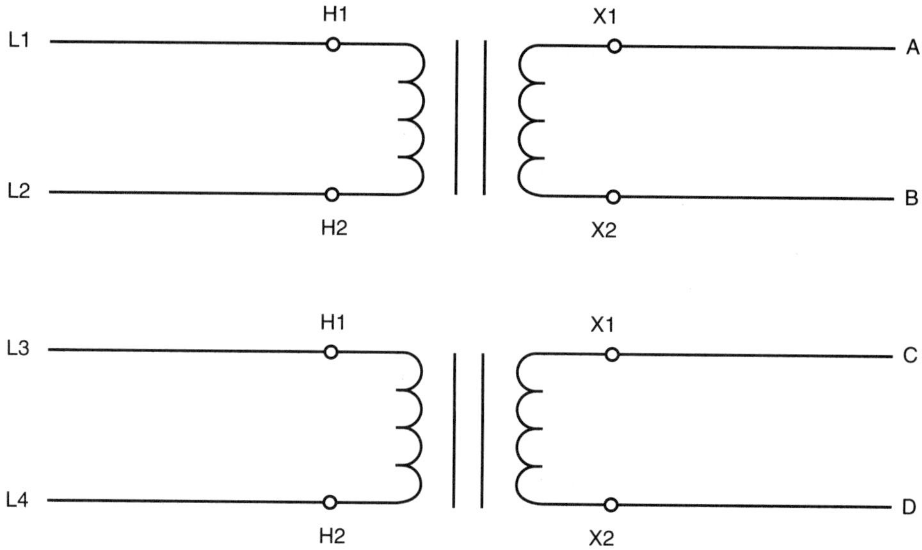

Fig. 15-18 On a two-phase power distribution, the two transformers are sourced from two single-phase sources that are 90 electrical degrees apart.

Example 15–10. A 2-phase transformer has secondary voltage and full-load current ratings of 480 volts and 45 amps. Determine the kVA power rating of the transformer.

$$(2 \times 480 \text{ volts} \times 45 \text{ amps}) \div 1{,}000 = 43.2 \text{ kVA}$$

Before the division by 1,000, the power rating of the motor equaled 43,200 volt-amps. Large values of power are more commonly expressed in kVA.

Code challenge 15–10. A 2-phase transformer has secondary voltage and full-load current ratings of 240 volts and 53 amps. Determine the kVA power rating of the transformer.

$$\frac{([A]\underline{\qquad\qquad} \times \underline{\qquad\qquad} \text{ volts} \times \underline{\qquad\qquad} \text{ amps})}{1{,}000} = (B)\underline{\qquad\qquad} \text{ kVA}$$

This power distribution can be set up as either a 4-wire or a 5-wire system with voltage ratings of 120 volts, 240 volts, 480 volts, 575 volts, or where high-value horsepower-rated motors are supplied, at 2,300 volts.

Two-phase motors and other 2-phase loads are rated as 4-wire. The service for this type of distribution would be rated as 5-wire to establish a ground-fault return path along the service neutral conductor and to establish a common bonding/grounding-electrode system between the power distribution and the conductive building surfaces.

As shown in Figure 15-19, the grounded service-neutral conductor in a 2-phase, 5-wire power distribution is the unbalanced load-current return conductor for both single-phase transformers. Available with operating voltages of either 480/240 or 240/120, the single grounded service neutral is wired common to any of the four 240- or 120-volt secondary windings of the two transformers.

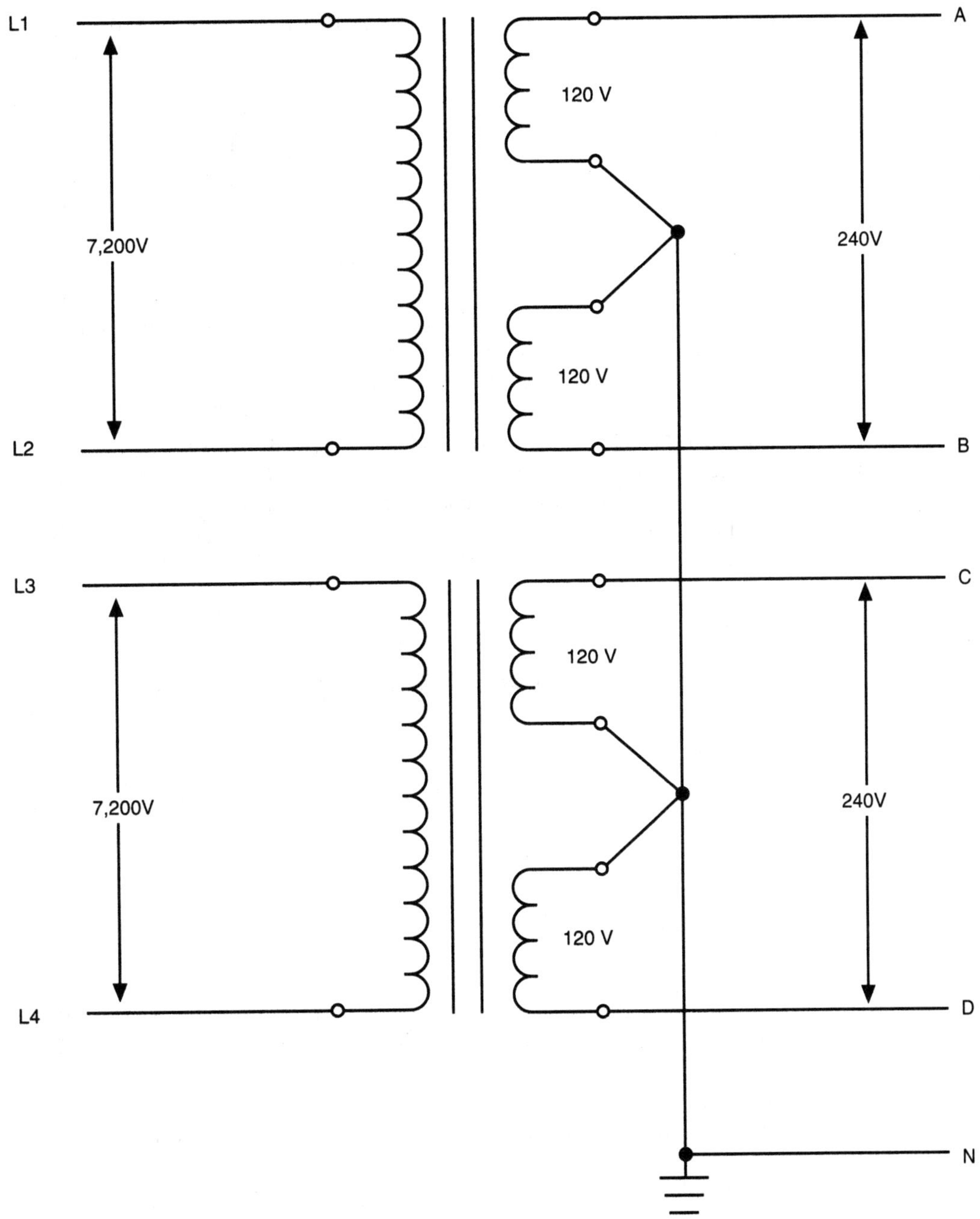

Fig. 15-19 In a two-phase, five-wire power distribution system, the fifth wire is the grounded circuit conductor that is common to both transfomers and all four windings.

As with single-phase systems, use of the power formula to determine the required VA or kVA rating of the 2-phase power transformer bank assumes that the lower-voltage, phase-to-neutral loads are balanced. When the neutral loads are unbalanced, each transformer must be sized for the largest value of unbalanced current it will supply (phase with the highest value of neutral load current), plus the current of the connected 240-volt, phase-to-phase, single-phase, and 2-phase loads. If the bank consists of a single 2-phase transformer on a common iron core, then the bank rating must be based on twice the largest rating as determined for the individual transformers in a 2-phase bank.

Because the phases do not effectively interact with each other, as with a 3-phase system (next section), the maximum unbalanced neutral current load of any of the four ungrounded phase conductors (reference to the common neutral) is computed, and this value, as determined by Sec. 220-22 in the NE Code, is increased to 140% to compensate for the loading from the second-phase neutral loads.

Here is another way to look at it: The neutral of either single-phase system is allowed to be downsized to 70% of the connected loads to the transformer. If the two phases do not interact, then the 140% overrate is in essence the summation of each respective 70% allowance for the maximum unbalanced load current.

In terms of overcurrent protection and circuit conductors, both are sized in the same manner as previously outlined for single-phase transformers. The procedures will be repeated in the 3-phase power computations.

15.9 THREE-PHASE POWER TRANSFORMERS

Three-phase power, which is the more common power distribution system, can be transformed either by three individual single-phase transformers (separate iron cores, separate containers [enclosures]) or by three separate windings mechanically displaced (offset) 120° from each other (between adjacent phase windings) along the confines of a single iron core. The connection of the individual in and out winding leads will be interconnected in either a wye or a delta configuration.

As shown in Figure 15-20, the 3-phase wye transformation is simply three transformer windings wired in parallel with their H1 leads connected one each to supply conductors L1, L2, and L3. The H2 leads are tied common to each other. In the wye secondary, the X2 leads of all three transformer windings are tied common. The X1 leads are connected, one each, to load conductors LL1, LL2, and LL3 (or phase A, phase B, and phase C). The X2 common connection can be grounded (tied to a grounding-electrode system) to create a 3-phase, 4-wire, neutral distribution.

As shown in Figure 15-21, the 3-phase delta is simply a closed-loop circuit in which the three transformer windings are wired in series. In the primary, the H2 lead of one transformer winding is connected to the H1 lead of a second transformer winding. The H2 lead of the second transformer winding is connected to the H1 lead of a third transformer winding. The H2 lead of the third transformer winding is connected to the H1 lead of the first transformer winding. Supply lines L1, L2, and L3 are connected, one each, to the three H1–H2 connections.

In the delta secondary, the X2 lead of one transformer winding is connected to the X1 lead of a second transformer winding. The X2 lead of the second transformer winding is con-

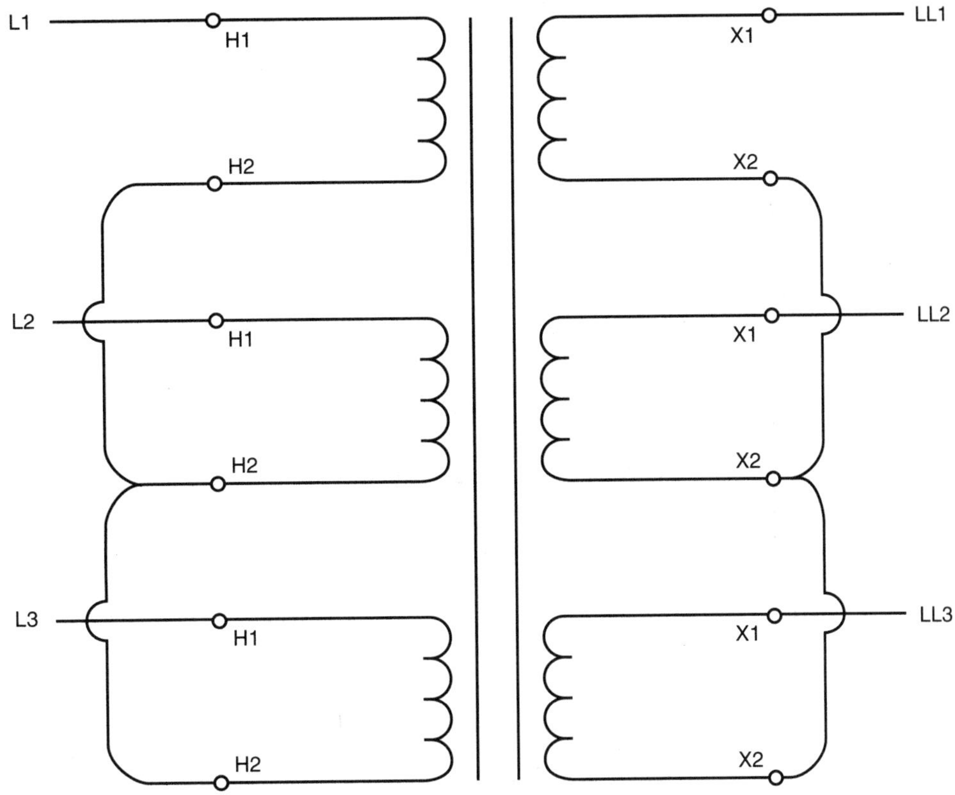

Fig. 15-20 In a wye-to-wye three-phase transformation, the primary and secondary coils appear connected in *parallel* by the one-end common connection.

nected to the X1 lead of a third transformer winding. The X2 lead of the third transformer winding is connected to the X1 lead of the first transformer winding. The secondary load conductors, LL1, LL2, and LL3 are connected, one each, to the three X1–X2 connections.

Deltas and wyes can be carried through or mixed between the primary and secondary connections. In other words, the 3-phase transformer can be connected as delta-delta, wye-wye, delta-wye, or wye-delta. Two transformers cannot be connected to transform 3-phase power in a wye connection. An open-delta configuration in the primary requires an open-delta configuration in the secondary. This is often referred to as a V–V connection (discussed in Sec. 15.11).

Individual single-phase transformers cannot be haphazardly wired together to form a 3-phase transformer bank. To prevent suicidal currents to the transformers, all of the following requirements must be met:

1. All three transformers in the bank must have the same or equal kVA rating.
2. All three transformers must be of the same construction (either core or shell).
3. All three transformers must have identical turn ratios, and the same high- and low-voltage ratings.

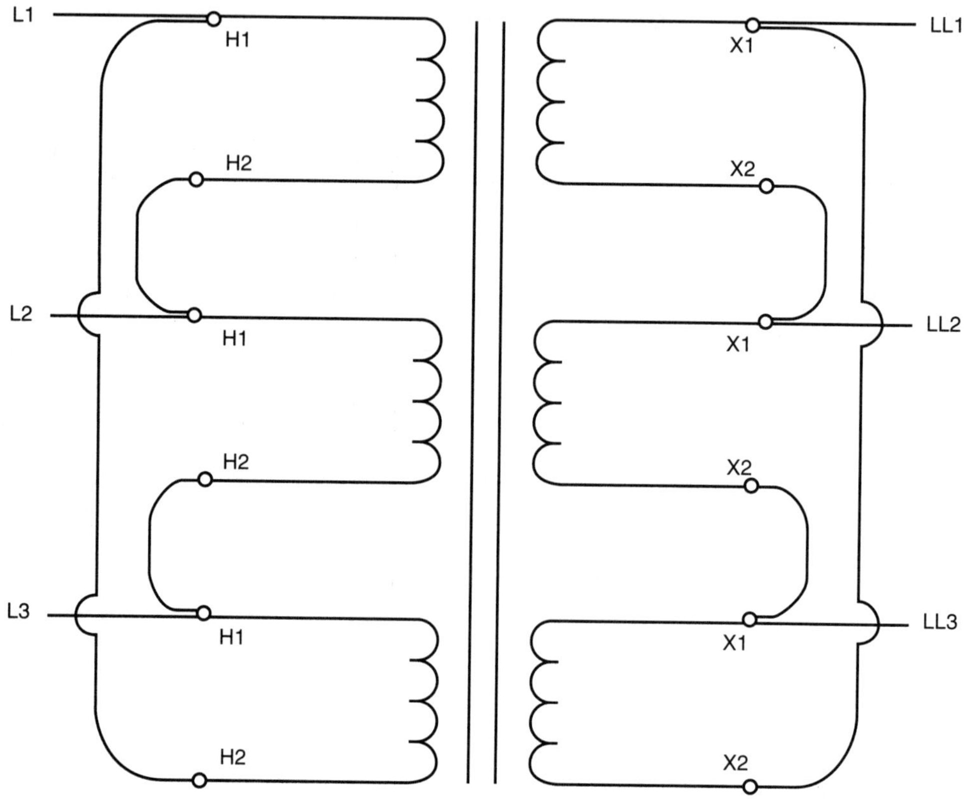

Fig. 15-21 In a delta-to-delta three-phase transformation, the windings in both the primary and secondary circuits appear as *closed loops.*

4. All three transformers must have identical resistances, reactances, and impedances, and, above all, the same polarities.

When 3-phase electrical transformations are accomplished using 3-phase transformers, all of these concerns are taken care of at the point of manufacture. The main worry then is only whether the transformation should be wye or delta.

In either type of construction and in either type of wiring configuration (wye or delta), the rating of the transformer bank (for three individual transformers) is the sum of the VA or kVA rating of each individual single-phase transformer. In other words, if three single-phase transformers rated at 75 kVA each are banked in a 3-phase transformation, the total rating of the transformer bank is 75 kVA + 75 kVA + 75 kVA = 225 kVA.

In terms of phase (coil or individual winding) voltage and current, if all three transformers have the same primary and secondary voltage and power ratings, then the total bank 3-phase power rating is equal to

$$\text{Power}_{\text{VA}} = 3 \times \text{V}_{\text{phase}} \times \text{A}_{\text{phase}}$$

Although in single- or two-phase transformation the phase (winding) currents are equal to the line currents and the phase voltages are equal to the line voltages, in 3-phase transformation, due to either the wye or delta wiring configuration, these values are not equal.

For a 3-phase distribution system, the power formula for line values of voltage and current (as measured with a voltmeter and an ammeter) in either the primary or secondary circuit is multiplied by the square root of 3 (1.732), as compared to just three when phase values are used. In formula,

$$\text{Power}_{VA} = 1.732 \times V_{line} \times A_{line}$$

As with 2-phase power transformers, the power computation of 3-phase transformers, especially when using line values, are normally stated in terms of kVA. In formula,

$$\text{Power}_{kVA} = \frac{1.732 \times V_{line} \times A_{line}}{1,000}$$

When the full-load current of an intended secondary load for a 3-phase transformer is known, the line values of voltage and current are simply plugged into the power formula to determine or compute the required power rating of the transformer.

Example 15–11. A 3-phase transformer has secondary voltage and full-load current ratings of 460 volts and 15 amps. Determine the kVA power rating of the transformer.

$$(1.732 \times 460 \text{ volts} \times 15 \text{ amps}) \div 1,000 = 12.0 \text{ kVA}$$

The actual computed load in this example equals 11.9508 kVA. In terms of thousands of volt-amps (kVA), the 50.8 volt-amps appears insignifcant; therefore, the power rating was rounded to only three-place accuracy.

Code challenge 15–11. A 3-phase transformer has secondary voltage and full-load current ratings of 575 volts and 53 amps. Determine the kVA power rating of the transformer.

$$\frac{([A]\,\_\_\_\_\_\_\_\_ \times \_\_\_\_\_\_\_\_ \text{ volts} \times \_\_\_\_\_\_\_\_ \text{ amps})}{1,000} = (B)\,\_\_\_\_\_\_\_\_ \text{ kVA}$$

Often the next higher standard size or power rating of a 3-phase transformer is selected and the respective circuit full-load or actual-load currents must be computed. By algebraic manipulation, either 3-phase power formula can be solved for the respective line current.

$$A_{line} = \text{Power VA} \div (1.732 \times V_{line})$$

or

$$A_{line} = (\text{Power kVA} \times 1,000) \div (1.732 \times V_{line})$$

Example 15–12. A 3-phase, 2,000-kVA transformer will be used to step down 13,200 volts to 480 volts. Determine both the primary and secondary full-load line currents.

Primary full-load line current computation:

$$(2{,}000 \text{ kVA} \times 1{,}000) \div (1.732 \times 13{,}200 \text{ volts}) = 87.5 \text{ amps}$$

Secondary full-load current computation:

$$(2{,}000 \text{ kVA} \times 1{,}000) \div (1.732 \times 480 \text{ volts}) = 2{,}406 \text{ amps}$$

If the primary-circuit voltage in this example is expressed as 13.2 kV, then the product of 1.732 times 13.2 kV, divided into 2,000 kVA also equals 87.5 amps. Division in the power formula must be by "like terms." The k's would cancel when computed in this manner. By the same token, 2,000 kVA divided by the product of 1.732 times 0.48 volts also equals 2,406 amps in the secondary computations.

Code challenge 15–12. A 3-phase, 2,500-kVA transformer will be used to step down 18,000 volts to 575 volts. Determine both the primary and secondary full-load line currents.

Primary full-load line current computation:

(__________ kVA × 1,000)÷(1.732 × __________ volts) = (A) __________ amps

Secondary full-load current computation:

(__________ kVA × 1,000)÷(1.732 × __________ volts) = (B) __________ amps

As with dual-voltage, single-phase systems that employ a common grounded neutral conductor (120/240, single-phase, 3-wire), the use of the power formula to compute either line currents or the required VA or kVA power rating of a 3-phase, 4-wire wye distribution system assumes that the lower-voltage, phase-to-neutral loads are balanced.

When the neutral loads are not balanced, the required rating of each transformer must be determined by the phase values of voltage and current. Each transformer must then be sized for the sum of the largest value of unbalanced neutral load current it will supply, plus the current of the connected phase-to-phase, single-phase loads, plus the current of the connected 3-phase loads.

If the transformer consists of a single 3-phase transformer on a common iron core, then the bank rating must be based on three times the largest rating as determined for the individual transformers in a 3-phase bank.

The same problem can occur on a single-voltage delta system. In a 3-phase, 3-wire delta configuration, the line voltage and phase or winding voltage are equal, such that a second, lower-value voltage is not available. A delta transformer configuration can supply single-phase, phase-to-phase loads (line-to-line loads) or 3-phase loads. When supplying single-phase loads, the volt-amp rating connected between any two lines must be as near as possible equal. As with the wye configuration, the 3-phase, 3-wire loads are automatically balanced when presented to the transformer windings.

The three individual windings in either the primary or secondary of the 3-phase transformer bank must be interconnected with each other and their respective line conductors in either a delta or wye configuration to transform the ac power. Both configurations are used in electrical power distribution. The operational difference between the two configurations lies in the difference between the line values of voltage and current as compared to the phase or coil values of voltage and current for the respective windings.

15.10 THREE-PHASE WYE POWER

When the 3-phase transformer leads are terminated in a wye configuration as shown in Figure 15-22, the start of each respective winding is connected directly to one of the line conductors, such that all of the line current of that phase passes through the coil of that single winding, and

$$A_{line} = A_{phase(winding)}$$

The line voltage between any two of the three supply lines in the wye transformer configuration is applied as a total voltage drop across two of the three windings wired in series, such that

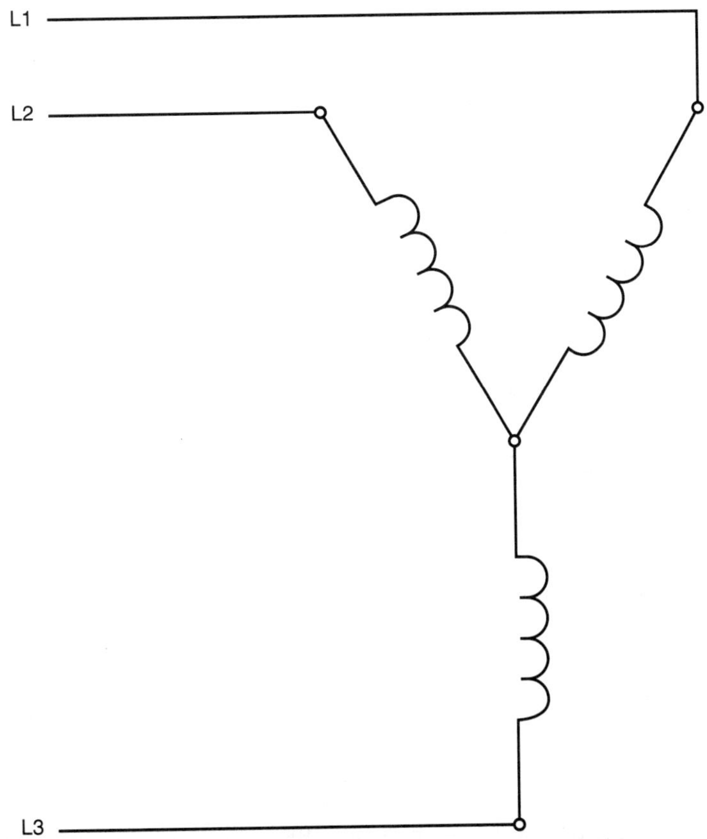

Fig. 15-22 In a wye configuration, the line and winding currents are equal (single path).

$$V_{line} = 1.732 \times V_{phase(winding)}$$

or

$$V_{phase} = V_{line} \div 1.732$$

Because of the 120° shift between the two line voltages, the phase voltages add vectorially (phasor quantities), not algebraically. If the phase voltages were added algebraically, the line voltage would be equal to two times the phase voltage.

Example 15–13. A 3-phase, 3,325.5-kVA transformer will be used to step down 18.0 kV to 480 volts in a wye-to-wye wiring configuration. Determine:

1. The primary line voltage
2. The primary full-load line current
3. The primary phase voltage
4. The primary full-load phase current
5. The secondary line voltage
6. The secondary full-load line current
7. The secondary phase voltage
8. The secondary full-load phase current

Primary line voltage: 1,000 × 18.0 kV = 18,000 volts
Primary full-load line current computation:

$$3{,}325.5 \text{ kVA} \div (1.732 \times 18.0 \text{ kV}) = 106.7 \text{ amps}$$

Primary phase voltage computation:

$$18{,}000 \text{ volts} \div 1.732 = 10{,}393 \text{ volts}$$

Primary full-load phase current = primary line current = 106.7 amps
Secondary line voltage: (given) 480 volts
Secondary full-load line current computation:

$$(3{,}325.5 \text{ kVA} \times 1{,}000) \div (1.732 \times 480 \text{ volts}) = 4{,}000 \text{ amps}$$

Secondary phase voltage computation:

$$480 \text{ volts} \div 1.732 = 277 \text{ volts}$$

Secondary full-load phase current = secondary line current = 4,000 amps

Because of the vector sum of the winding voltages in a 3-phase, 4-wire wye configuration, the line voltage is 173% (square root of 3) greater than the single-phase winding voltage of any one of the transformers. When computing the winding voltage in this example, the respective voltages were divided by the square root of 3. Technically, the single-phase winding voltage can be computed by taking 58% (57.7%, the reciprocal of 1.732) of the line-voltage value.

Code challenge 15–13. A 3-phase, 2,500-kVA transformer will be used to step down 12.5 kV to 208 volts in a wye-to-wye wiring configuration. Determine:

1. The primary line voltage
2. The primary full-load line current
3. The primary phase voltage
4. The primary full-load phase current
5. The secondary line voltage
6. The secondary full-load line current
7. The secondary phase voltage
8. The secondary full-load phase current

Primary line voltage: ____________ × ____________ kV = (A) ____________ volts
Primary full-load line current computation:

____________ kVA÷([B] ____________ × ____________ kV) = (C) ____________ amps

Primary phase voltage computation:

____________ volts ÷ ____________ = (D) ____________ volts

Primary full-load phase current = (E) ____________ amps
Secondary line voltage = (F) ____________ volts
Secondary full-load line current computation:

(____________ kVA × 1,000) ÷ (1.732 × ____________ volts) = (G) ____________ amps

Secondary phase voltage computation:

____________ volts ÷ ____________ = (H) ____________ volts

Secondary full-load phase current = (I) ____________ amps

When a unit substation, which consists of the fused high-voltage side, the 3-phase transformer assembly, and the secondary service equipment power panel (fused disconnect and busbars for further dissemination of power), is installed in an industrial plant, the supply power from an outside switchyard is normally medium voltage and the power distribution within the plant is normally 600 volts or less. If the service is 3-phase, 4-wire, wye-to-wye connected, then both the primary and secondary circuits are in wye configuration. The line and phase values of voltage and current would be computed as in this previous example.

In terms of the rating or setting of the circuit overcurrent-protection devices or the rated ampacity of the circuit conductors, only the line values are required.

The overrate of the secondary overload-overcurrent protection device must be determined in accordance with whether the unit substation installation will be in a supervised or unsupervised location. The difference between the two is in the comparison of the overrate percentages listed in the two tables referenced by Sec. 450-3(a).

If the location is supervised (only persons qualified will work on the equipment), T450-3(a)(2)b. will allow the secondary fuse or circuit breaker rating (for less than 600 volts) to be set at a maximum of 250% of the rated full-load current of the transformer. If the unit substation location cannot qualify as supervised, the overrate is limited by T450-3(a)(1) to only 125% of the secondary full-load current rating of the transformer. The other overrate percentages, listed according to percent of nameplate impedance, the voltage rating (over 600 volts), and the type of protection (either fuses or circuit breakers) are the same in either table.

In comparison to the delta wiring configuration, the wye configuration is the more common 3-phase power distribution system because of the dual voltage rating and the ease of establishing a ground-fault return path between the equipment-grounding system/building bonding grid and the fourth-wire service neutral conductor.

When determining the rating or setting of the circuit overcurrent-protection devices or the rated ampacity of the circuit conductors, the relationship between winding (phase) values and line values is not really a concern because only the line values are required.

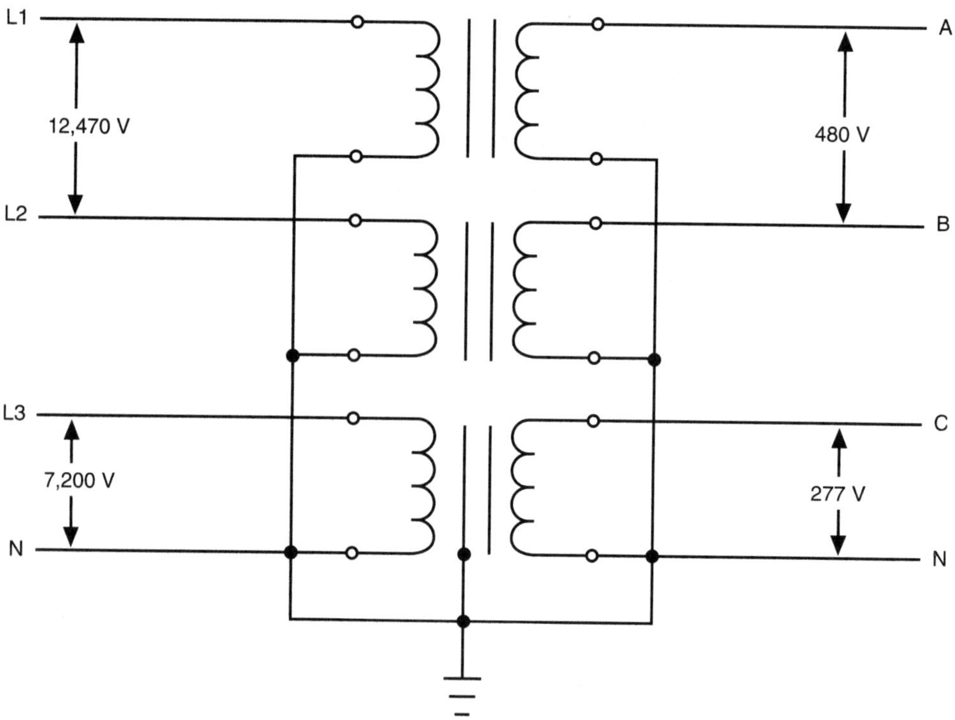

Fig. 15-23 In a wye configuration, the line voltage is 58% *greater* than the winding voltage.

Example 15–14. Figure 15-23 shows the relationship of the single-phase and 3-phase voltages in a wye-to-wye transformation. If this is the wiring diagram of a 3-phase, 4-wire, 2,000-kVA unit substation that will be installed for the power distribution within a building at an unsupervised location, for a primary supply voltage of 12,470Y/7,200 volts and a secondary service voltage of 480Y/277 volts, determine:

1. The maximum setting of the secondary inverse-time circuit breaker
2. The minimum wire size of the secondary conductors supplying the switchgear, using XHHW-2 copper paralleled eight times
3. The maximum rating of the primary disconnect fuses
4. The minimum wire size of the bonding jumper connecting the metal enclosure of the switchgear to the neutral busbar at the transformer secondary.

NE Code-maximum rating determination of the secondary circuit breaker:
Secondary FLC computation:

$$(2{,}000 \text{ kVA} \times 1{,}000) \div (1.732 \times 480 \text{ volts}) = 2{,}406 \text{ amps}$$

Overrate computation: Ref. T450-3(a)(1)

$$125\% \text{ of } 2{,}406 \text{ amps} = 3{,}008 \text{ amps}$$

Standard OCPD ratings: Ref. 240-6(a)
NHS: 4,000 amps Ref. (NHS) 450-3(a)(1), Excp. 1

NE Code-minimum wire size of the secondary service conductors:
Parallel-conductor amperage computation: Ref. 310-4

$$4{,}000 \text{ amps} \div 8 = 500 \text{ amps}$$

Wire terminal temperature limitations: 75°C Ref. 110-14(c)(2), Excp. 1
Table-rated ampacity: 520 amps Ref. T310-16
Wire size: 900 kcmil XHHW-2 copper

NE Code-maximum rating determination of the primary service fuses:
Primary FLC computation:

$$(2{,}000 \text{ kVA} \times 1{,}000) \div (1.732 \times 12{,}470 \text{ volts}) = 92.6 \text{ amps}$$

Overrate computation: Ref. T450-3(a)(1)

$$300\% \text{ of } 92.6 \text{ amps} = 277.8 \text{ amps}$$

Standard OCPD ratings: Ref. 240-6(a)
NHS: 300 amps Ref. (NHS) 450-3(a)(1), Excp. 1

NE Code-minimum wire size of the bonding jumper connecting the metal enclosure of the switchgear to the neutral busbar at the transformer secondary.

Parallel-conductor c.m.a. computation: Ref. 310-4

$$8 \times 900 \text{ kcmil} = 7{,}200 \text{ kcmil}$$

Copper bonding-jumper c.m.a. computation: Ref. 250-79(d)

$$12.5\% \text{ of } 7{,}200 \text{ kcmil} = 900 \text{ kcmil}$$

The accuracy employed in the computation of the secondary full-load current in this example yielded a few amps over the intended 2,400-amp rating of this unit substation. Carried into the 125% overrate allowed by T450-3(a)(1), the extra amps yielded an OCPD rating just over 3,000 amps. Using the next higher standard size of circuit breaker, as allowed by the exception to Sec. 450-3(a)(1) to attain a 4,000-amp service rating seems to be overshooting the intent of the NE Code.

If the secondary FLC rating were taken at 2,400 amps nominal, the overrate computation would have yielded a standard circuit breaker setting of 3,000 amps. This seems to be more along the lines of NE Code requirements.

The paralleling of the secondary-circuit conductors, was actually a review from Chapter 2, simply because T310-16 does not list any single-circuit conductor with a rated ampacity of 4,000 amps.

Sizing the line-side equipment bonding jumper extended to the switchgear enclosure was a review from Chapter 9. The wire size of the bonding jumper must at least equal the wire size of the grounding-electrode conductor as determined by T250-94. When the computed c.m.a. of the ungrounded service conductors exceeds the table cutoff of 1,100 kcmil copper, the c.m.a. of the single copper bonding jumper must equal a minimum of 12.5 % of the ungrounded service conductor c.m.a. (connected to any one phase) to meet the line-side equipment-grounding requirements of Sec. 250-79(d).

Because 12.5% equals ⅛, and the ungrounded service conductors were paralleled eight times, a full size equipment bonding jumper (same wire size as the paralleled ungrounded service conductors) was required.

Code challenge 15–14. A 3-phase, 4-wire, 576.4-kVA unit substation will be installed for the power distribution within a building at an unsupervised location. For a primary supply voltage of 12,470Y/7,200 volts and a secondary service voltage of 208Y/120 volts, determine:

1. The maximum setting of the secondary inverse-time circuit breaker
2. The minimum wire size of the secondary conductors supplying the switchgear, using THHN copper paralleled six times
3. The maximum rating of the primary disconnect fuses
4. The minimum wire size of the bonding jumper connecting the metal enclosure of the switchgear to the neutral busbar at the transformer secondary

NE Code-maximum rating determination of the secondary circuit breaker:

Secondary FLC computation:

(__________ kVA × 1,000) ÷ (1.732 × __________ volts) = (A) __________ amps

Overrate computation: Ref. (B) ________________

__________ % of __________ amps = (C) __________ amps

Standard OCPD ratings: (D) __________ amps Ref. (E) ________________

NE Code-minimum wire size of the secondary service conductors:

Parallel-conductor amperage computation: Ref. (F) ________________

__________ amps ÷ (G) __________ = __________ amps

Wire terminal temperature limitations: (H) __________ °C

Ref. (I) ________________

Table-rated ampacity: (J) __________ amps Ref. (K) ________________

Wire size: (L) __________ THHN copper

NE Code-maximum rating determination of the primary service fuses:

Primary FLC computation:

(__________ kVA × 1,000) ÷ (1.732 × __________ volts) = (M) __________ amps

Overrate computation: Ref. (N) ________________

__________ % of __________ amps = (O) __________ amps

Standard OCPD ratings: Ref. (P) ________________

NHS: (Q) __________ amps Ref. (NHS) (R) ________________

NE Code-minimum wire size of the bonding jumper connecting the metal enclosure of the switchgear to the neutral busbar at the transformer secondary.

Parallel-conductor c.m.a. computation: Ref. (S) ________________

__________ × __________ kcmil = (T) __________ kcmil

Copper bonding-jumper c.m.a. computation: Ref. (U) ________________

__________ % of __________ kcmil = (V) __________ kcmil

The secondary circuit of a 3-phase wye-connected transformer can be either 3-wire or 4-wire. A 3-wire wye configuration, like a 3-wire delta-connected transformer, is a single-voltage system.

Industrial process heating often requires very high concentrations of heat. One way to generate such large amounts of heat is to force high magnitude current to flow along copper-alloy busbars. For example, a 3-phase, 50-kVA transformer with a 480-volt primary will draw 60 amps under full-load current conditions. Ignoring iron-core losses, if the transformer supplies a resistive heating bank, 50 kW of heating is available. If the secondary voltage is reduced to only 12 volts, and the copper busbars are assembled in a wye configuration, the secondary load current is (480 volts ÷ 12 volts) × 60 amps = 2,400 amps.

It would be almost impossible to control the 2,400 amps at 12 volts with any type of switching means, whether solid-state devices or electromechanical contactors (power relays). But the 60-amps in the primary circuit is easily controlled by either.

In terms of combined overcurrent protection, Sec. 450-3(b)(1) will allow the transformer and secondary heating circuit to be protected in the primary circuit with a fuse or circuit breaker set at a maximum of 125% of the primary FLC. When the 125% overrate does not correspond to a standard size of fuse or circuit breaker, Exception No. 1 will allow the next higher standard size to be installed.

With the NE Code-maximum overrate of the primary overcurrent-protection rating, the 50-kVA, 3-phase transformer could supply a maximum heating load of 62.5 kW.

Example 15–15. A 3-phase, 3-wire, 75-kVA, wye-connected power transformer with voltage ratings of 480 volts and 6 volts will supply a process heating unit. If the secondary 6-volt circuit will consist of wye-connected copper-alloy busbars, and the control and overcurrent protection will be installed in the primary circuit only, determine:

1. The maximum rating of the primary disconnect fuses
2. The minimum wire size of the primary-circuit conductors, using THHN copper
3. The secondary full-load current rating, based on the rating of the primary-circuit fuses

NE Code-maximum rating determination of the primary fuses:
Primary FLC computation:

$$(75 \text{ kVA} \times 1{,}000) \div (1.732 \times 480 \text{ volts}) = 90.2 \text{ amps}$$

Overrate computation: Ref. 450-3(b)(1)

$$125\% \text{ of } 90.2 \text{ amps} = 112.8 \text{ amps}$$

Standard OCPD ratings: Ref. 240-6(a)
NHS: 125 amps Ref. (NHS) 450-3(b)(1), Excp. 1

NE Code-minimum wire size of the primary circuit conductors:
Wire terminal temperature limitations: 75°C Ref. 110-14(c)(2), Excp. 1
Table-rated ampacity: 130 amps Ref. T310-16
Wire size: No. 1 AWG THHN copper

Secondary full-load current computation with primary protection overrate:

$$(480 \text{ volts} \div 6 \text{ volts}) \times 125 \text{ amps} = 10{,}000 \text{ amps}$$

Code challenge 15–15. A 3-phase, 3-wire, 66-kVA, wye-connected power transformer with voltage ratings of 460 volts and 12 volts will supply a process heating unit. If the secondary 12-volt circuit will consist of wye-connected copper-alloy busbars, and the control and overcurrent protection will be installed in the primary circuit only, determine:

1. The maximum rating of the primary disconnect fuses
2. The minimum wire size of the primary-circuit conductors using RHW copper
3. The secondary full-load current rating, based on the rating of the primary-circuit fuses

NE Code-maximum rating determination of the primary fuses:
Primary full-load line current computation:

(__________ kVA × 1,000) ÷ (1.732 × __________ volts) = (A) __________ amps

Overrate computation: Ref. (B) __________

(C) __________ % of __________ amps = __________ amps

Standard OCPD ratings: Ref. (D) __________
NHS: (E) __________ amps Ref. (NHS) (F) __________

NE Code-minimum wire size of the primary-circuit conductors:
Wire terminal temperature limitations: (G) __________ °C
Ref. (H) __________
Table-rated ampacity: (I) __________ amps Ref. (J) __________
Wire size: (K) __________ THHN copper

Secondary full-load current computation with primary protection overrate:

(__________ volts ÷ __________ volts) × __________ amps = (L) __________ amps

When used as the service supply to a power panel within a building or other structure, the 3-phase, 4-wire wye offers the advantage of 3-phase power distribution, as it simultaneously permits lighting to be connected between any of the three phase conductors and the common neutral conductor (the fourth wire). Typical voltages for a 3-phase, 4-wire wye include 208Y/120, 440Y/254, 460Y/265, 480Y/277, and 600Y/347. If the lower-voltage 208Y/120 is used for the structure distribution, both lighting and general-purpose receptacle loads (duplex receptacles) can be connected as single-phase loads between any phase conductor and the common neutral conductor.

The 3-phase, 4-wire wye is easily balanced by ensuring that the current load connected between any one-phase conductor and the neutral conductor is equal to the single-phase, phase-to-neutral current loads on the other two phase conductors. Three-phase connected loads are automatically balanced. Single-phase, phase-to-phase loads (connected between any two of the three ungrounded phase conductors) are balanced when the current loads connected are equal, regardless of to which two phase conductors the loads are connected.

Sections 215-10, 230-95, and 240-13 require additional ground-fault overcurrent protection (in addition to the ungrounded conductor supply fuses or circuit breaker) to be installed on all solidly grounded wyes that are rated at 1,000 amps or more, and do not exceed 600 volts phase-to-phase, but do exceed 150 volts phase-to-ground. This ground-fault overcurrent protection requirement was imposed by the NE Code because of the excessive number of switchgear burndowns occurring when one or more legs (phases) of heavy or high current on this type distribution faulted (shorted) to ground.

The need or requirement for ground-fault protection can be avoided by installing multiple feeder fused disconnects that are all rated at less than 1,000 amperes or by eliminating the neutral conductor.

15.11 THREE-PHASE DELTA POWER

As shown in Figure 15-24, when the transformer leads (beginning and ending leads or terminals of each respective phase windings) of a 3-phase transformer are connected in a delta configuration, each winding is connected directly across two of the three supply lines, such that

$$V_{line} = V_{phase(winding)}$$

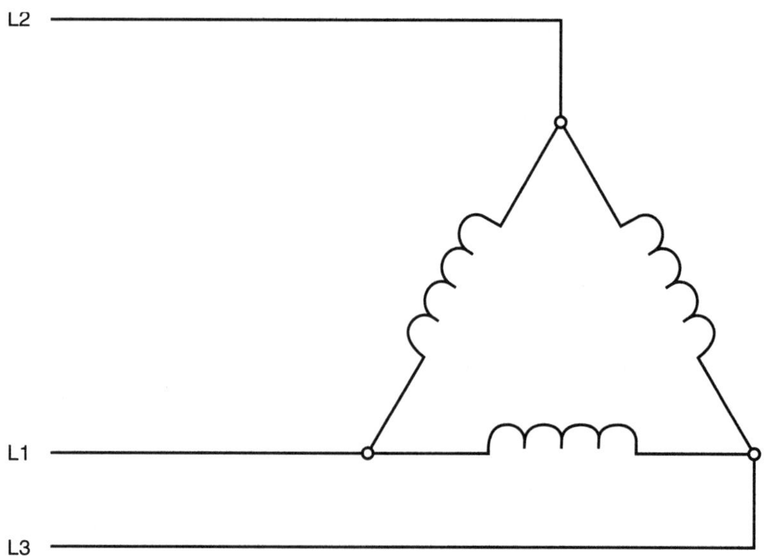

Fig. 15-24 In a delta configuration, the line and winding voltages are equal (one winding wired across any two lines).

The line current from any one of the three supply conductors in the delta transformer configuration has two paths it can take, because current flowing from one line into the transformer will actually flow out through two different windings to both of the other lines, such that

$$A_{line} = 1.732 \times A_{phase(winding)}$$

or

$$A_{phase} = A_{line} \div 1.732$$

Because of the 120° phase shift between any two adjacent line currents, the phase currents add vectorially (phasor quantities), not algebraically. If the two phase currents were added algebraically, the line current would be equal to two times the phase current.

Example 15–16. A 3-phase, 66.5-kVA transformer will be used to step down 13.2 kV to 240 volts in a delta-to-delta wiring configuration. Determine:

1. The primary line voltage
2. The primary full-load line current
3. The primary phase voltage
4. The primary full-load phase current
5. The secondary line voltage
6. The secondary full-load line current
7. The secondary phase voltage
8. The secondary full-load phase current

Primary line voltage: 1,000 × 13.2 kV = 13,200 volts
Primary full-load line current computation:

$$66.5 \text{ kVA} \div (1.732 \times 13.2 \text{ kV}) = 2.91 \text{ amps}$$

Primary phase voltage = primary line voltage = 13,200 volts
Primary full-load phase current computation:

$$2.91 \text{ amps} \div 1.732 = 1.68 \text{ amps}$$

Secondary line voltage: (given) 240 volts
Secondary full-load line current computation:

$$(66.5 \text{ kVA} \times 1{,}000) \div (1.732 \times 240 \text{ volts}) = 160 \text{ amps}$$

Secondary phase voltage = secondary line voltage = 240 volts
Secondary full-load phase current computation:

$$160 \text{ amps} \div 1.732 = 92.4 \text{ amps}$$

Because of the vector sum of the winding currents in a 3-phase, 3-wire delta configuration, the line current is 173% (square root of 3) greater than the single-phase winding current of any one of the transformers. When computing the winding current in this example, the respective currents were divided by the square root of 3. Technically, the single-phase winding current can be computed by taking 58% (57.7%, the reciprocal of 1.732) of the line-current value.

Code challenge 15–16. A 3-phase, 150-kVA transformer will be used to step down 18 kV to 480 volts in a delta-to-delta wiring configuration. Determine:

1. The primary line voltage
2. The primary full-load line current
3. The primary phase voltage
4. The primary full-load phase current
5. The secondary line voltage
6. The secondary full-load line current
7. The secondary phase voltage
8. The secondary full-load phase current

Primary line voltage: ____________ × ____________ kV = (A) ____________ volts

Primary full-load line current computation:

____________ kVA ÷ (____________ × ____________ kV) = (B) ____________ amps

Primary phase voltage = (C) ____________ volts

Primary full-load phase current computation:

____________ amps ÷ ____________ = (D) ____________ amps

Secondary line voltage = (E) ____________ volts

Secondary full-load line current computation:

(________ kVA × ________) ÷ (________ × ________ volts) = (F) ________ amps

Secondary phase voltage = (G) ____________ volts

Secondary full-load phase current computation:

____________ amps ÷ ____________ = (H) ____________ amps

As the wye-configured unit substation discussed in Example 15-14, the 3-phase delta configuration can also serve as the point of power supply for distribution within a building or other structure (service equipment or a unit substation). A common 3-phase transformation that was used in older commercial and industrial installations is the delta-to-delta (Fig. 15-25). The rules of Sec. 450-3 apply for the overrates of the primary and secondary circuits overcurrent protection.

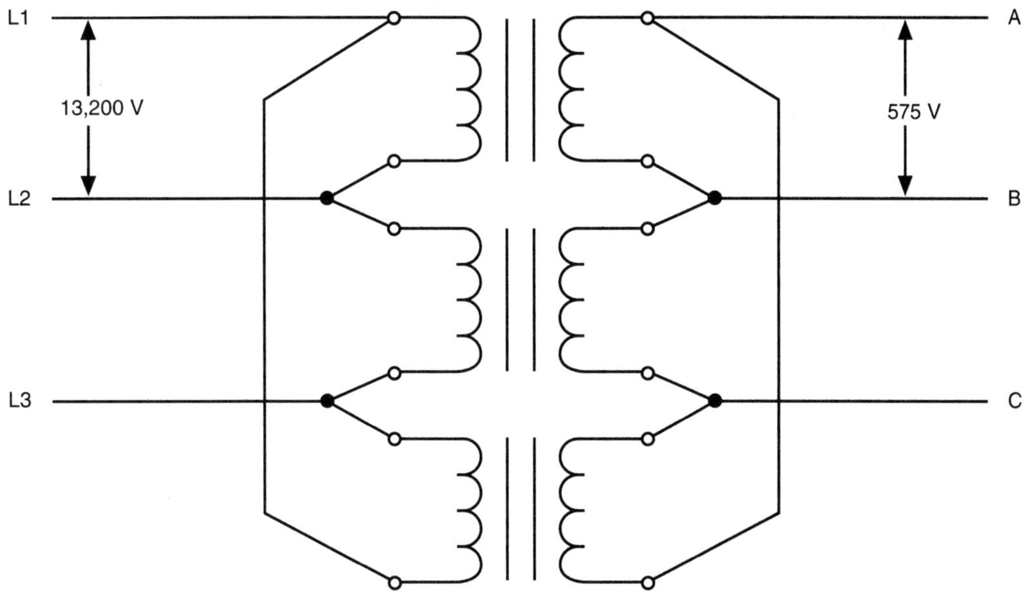

Fig. 15-25 A three-phase, three-wire floating delta distribution is a single-voltage system that is not referenced to ground.

Example 15–17. For a 3-phase, 1,600-amp switchgear supplied as a 3-wire delta with a primary supply voltage of 13.2 kV and a secondary service/power distribution voltage of 575 volts, determine:

1. The minimum kVA rating of the transformer within the unit substation if the NE Code-maximum setting of the secondary overcurrent protection device is employed for an unsupervised installation
2. The minimum wire size of the secondary conductors supplying the switchgear, using THW copper paralleled four times
3. The maximum rating of the primary disconnect fuses

NE Code-minimum power transformer kVA rating determination:
Secondary FLC computation: Ref. T450-3(a)(1)

$$1{,}600 \text{ amps} \div 1.25 = 1{,}280 \text{ amps}$$

Power rating computation:

$$(1.732 \times 575 \text{ volts} \times 1{,}280 \text{ amps}) \div 1{,}000 = 1{,}275 \text{ kVA}$$

NE Code-minimum wire size of the secondary service conductors:
Parallel-conductor amperage computation: Ref. 310-4

$$1{,}600 \text{ amps} \div 4 = 400 \text{ amps}$$

Wire terminal temperature limitations: 75°C — Ref. 110-14(c)(2)
Table-rated ampacity: 420 amps — Ref. T310-16
Wire size: 600 kcmil THW copper

NE Code-maximum rating determination of the primary service fuses:
Primary FLC computation:

(575 volts ÷ 13,200 volts) × 1,280 amps = 55.8 amps

Overrate computation: — Ref. T450-3(a)(1)

300% of 55.8 amps = 167.3 amps

Standard OCPD ratings: — Ref. 240-6(a)
NHS: 175 amps — Ref. (NHS) 450-3(a)(1), Excp. 1

This type of 3-phase connection (Fig. 15-25) is sometimes referred to as a "floating delta" since the secondary is not referenced to ground. The service grounding-electrode system must still be established for the noncurrent-carrying metal parts of the distribution system (metal raceways, boxes, and other enclosures), but there is no electrical connection from the delta secondary to it.

Code challenge 15–17. For a 3-phase, 2,400-amp switchgear supplied as a 3-wire delta with a primary supply voltage of 18 kV and a secondary service/power distribution voltage of 440 volts, determine:

1. The minimum kVA rating of the transformer within the unit substation if the NE Code-maximum setting of the secondary overcurrent-protection device is employed for an unsupervised installation
2. The minimum wire size of the secondary conductors supplying the switchgear, using RHW aluminum paralleled eight times
3. The maximum rating of the primary disconnect fuses

NE Code-minimum power transformer kVA rating determination:
Secondary FLC computation: — Ref. (A) ____________________

____________ amps ÷ ____________ = (B) ____________ amps

Power rating computation:

(1.732 × ____________ volts × ____________ amps)÷1,000 = (C) ____________ kVA

NE Code-minimum wire size of the secondary service conductors:
Parallel-conductor amperage computation: — Ref. (D) ____________________

____________ amps÷ ____________ = (E) ____________ amps

Wire terminal temperature limitations: (F) ___________ °C
Ref. (G) ____________________

Table-rated ampacity: (H) ___________ amps Ref. (I) ____________________

Wire size: (J) ___________ RHW aluminum

NE Code-maximum rating determination of the primary service fuses:

Primary FLC computation:

_________ volts ÷ _________ volts) × _________ amps = (K) _________ amps

Overrate computation: Ref. (L) ____________________

(M) ___________ % of ___________ amps = ___________ amps

Standard OCPD ratings: Ref. (N) ____________________

NHS: (O) ___________ amps Ref. (NHS) (P) ____________________

One advantage of a floating-delta power distribution is that the equipment will not single-phase (opening of one phase in a 3-phase supply) when any one phase faults to ground. Remember, in a 3-phase, 3-wire floating-delta secondary none of the phases are referenced to ground. Therefore, there is no return path to the power source (transformer). The respective winding does not see a low-resistance ground-fault path, so no high-magnitude current, other than normal load current, will flow. The equipment will continue to run.

In a 3-phase, 4-wire wye distribution, all three phase windings in the transformer are referenced or tied to earth ground in such a manner that a ground-fault condition on any phase conductor will place a direct short across the affected phase winding. The overcurrent protection for that feeder or branch-circuit will open, and that part of the distribution system will go into a single-phase condition.

A 3-phase motor will not start if any supply phase is open (in a single-phase condition). Three-phase motors that are running with light loads and suffer the loss of one phase will normally continue to run, but will draw excessive current in the other two legs (phases) to create its own third phase. The excessive current will cause heat buildup in the motor windings; the motor will suffer a torque loss and normally trip out on overload.

Although not specifically required by the NE Code, most secondary switchgear delta connections supplied as a floating delta have ground-fault monitors (usually amber lights) that detect and give an alarm of a phase conductor that has faulted to ground. Any one phase faulted to ground poses no real problem or shock hazard to personnel. The problem arises when one delta phase conductor is grounded in one part of the structure or plant's distrubution system, and a second phase faults to ground in another part of the distribution system. The building structure now becomes a low-resistance path or load that can almost represent a "dead short" between the two grounded phases.

Since the floating delta does not offer a fourth-wire neutral conductor, only 3-phase and/or single-phase, phase-to-phase loads can be supplied from the switchgear. The lower phase-to-neutral voltages for lighting, receptacles, and control purposes must be generated as separately derived systems by additional transformers at secondary load centers.

One commonly used transformer configuration for this purpose in commercial and industrial installations is the delta-wye transformation (Fig. 15-26). The secondary of this system is a 3-phase, 4-wire wye. The fourth wire is the grounded neutral conductor which carries only the unbalanced current of the three phase or ungrounded supply conductors. The neutral conductor and the common tie-point of the secondary wye are referenced (tied) to earth ground through a grounding-electrode system.

Example 15–18. If a 3-phase power transformer rated 37.5 kVA is used to derive a 3-phase, 4-wire, 208Y/120-volt supply from a 3-phase, 3-wire, 575-volt delta feeder circuit (Fig. 15-23), determine:

1. The maximum setting of the secondary circuit breaker
2. The maximum rating of the primary-circuit time-delay fuses
3. The minimum wire size of the secondary-circuit conductors, using THHW-2 copper
4. The minimum wire size of the primary-circuit conductors, using THHW-2 copper

NE Code-maximum setting of the secondary circuit breaker determination:
Secondary FLC computation:

$$(37.5\ \text{kVA} \times 1{,}000) \div (1.732 \times 208\ \text{volts}) = 104.1\ \text{amps}$$

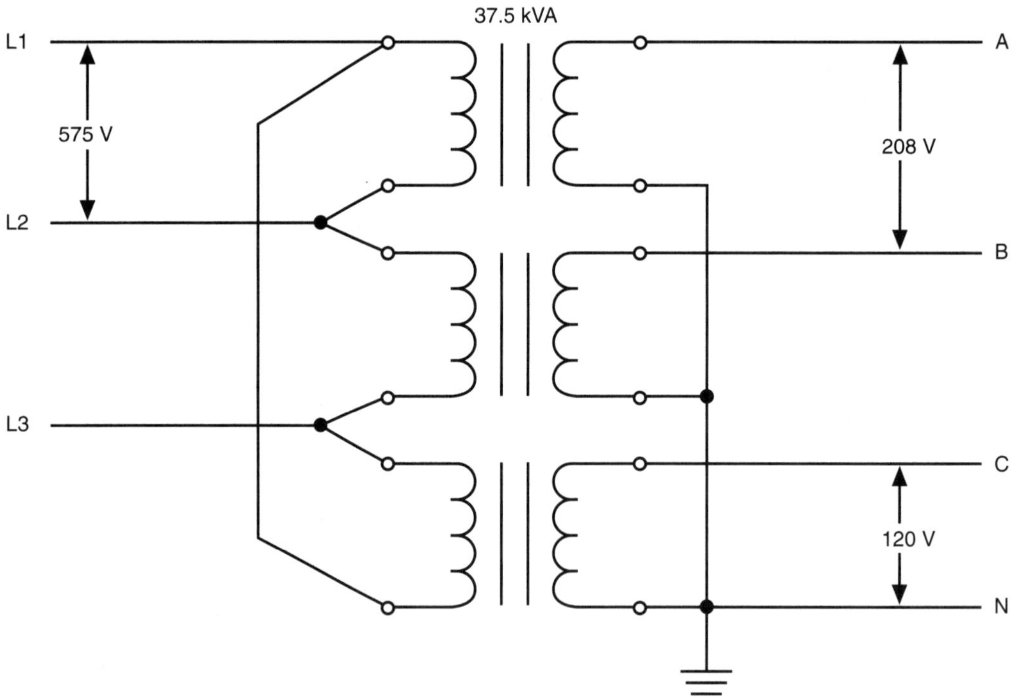

Fig. 15-26 A delta-to-wye three-phase transformation is often used to generate a three-phase, four-wire distribution.

Overrate computation: Ref. 450-3(b)(2)

125% of 104.1 amps = 130.1 amps

Standard OCPD ratings: Ref. 240-6(a)
NHS: 150 amps Ref. (NHS) 450-3(b)(2), Excp. No. 1

NE Code-maximum setting of the primary-circuit fuses determination:
Full-load current computation:

(37.5 kVA × 1,000) ÷ (1.732 × 575 volts) = 37.7 amps

Overrate computation: Ref. 450-3(b)(2)

250% of 37.7 amps = 94.1 amps

Standard OCPD ratings: Ref. 240-6(a)
NLS: 90 amps Ref. (NLS) 450-3(b)(2)

NE Code-minimum wire size of the secondary-circuit conductors:
Wire terminal temperature limitations: 75°C Ref. 110-14(c)(2), Excp. No. 1
Table-rated ampacity: 150 amps Ref. T310-16
Wire size: No. 1/0 AWG THHW-2 copper

NE Code-minimum wire size of the primary-circuit conductors:
Primary FLC computation:

(208 volts ÷ 575 volts) × 150 amps = 54.2 amps

Wire terminal temperature limitations: 60°C Ref. 110-14(c)(1), Excp. No. 1
Table-rated ampacity: 55 amps Ref. T310-16
Wire size: No. 6 AWG THHW-2 copper

The secondary-circuit feeder neutral conductor in this example was not sized. According to Sec. 220-22, the separately drived system neutral conductor would not have to be full sized as compared to the three No. 1/0 AWG copper secondary ungrounded phase conductors. The neutral conductor must be sized according to the maximum unbalanced load current of the panel or other load center supplied by the transformer.

Code challenge 15–18. If a 3-phase power transformer rated 66 kVA is used to derive a 3-phase, 4-wire, 208Y/120-volt supply from a 3-phase, 3-wire, 440-volt delta feeder circuit, determine:

1. The maximum setting of the secondary circuit breaker
2. The maximum rating of the primary-circuit time-delay fuses

3. The minimum wire size of the secondary-circuit conductors, using XHHW aluminum
4. The minimum wire size of the primary-circuit conductors, using XHHW aluminum

NE Code-maximum setting of the secondary circuit breaker determination:
Secondary FLC computation:

(__________ kVA × 1,000) ÷ (1.732 × __________ volts) = (A) __________ amps

Overrate computation: Ref. (B) __________________

(C) __________ % of __________ amps = __________ amps

Standard OCPD ratings: Ref. (D) __________________
NHS: (E) __________ amps Ref. (NHS) (F) __________________

NE Code-maximum rating determination of the primary-circuit fuses:
Primary FLC computation:

(__________ kVA × 1,000) ÷ (1.732 × __________ volts) = (G) __________ amps

Overrate computation: Ref. (H) __________________

(I) __________ % of __________ amps = __________ amps

Standard OCPD ratings: Ref. (J) __________________
NLS: (K) __________ amps Ref. (NLS) (L) __________________

NE Code-minimum wire size of the secondary-circuit conductors:
Wire terminal temperature limitations: (M) __________ °C
Ref. (N) __________________
Table-rated ampacity: (O) __________ amps Ref. (P) __________________
Wire size: (Q) __________ XHHW aluminum

NE Code-minimum wire size of the primary-circuit conductors:
Primary FLC computation:

(__________ volts ÷ __________ volts) × __________ amps = (R) __________ amps

Wire terminal temperature limitations: (S) __________ °C
Ref. (T) __________________
Table-rated ampacity: (U) __________ amps Ref. (V) __________________
Wire size: (W) __________ XHHW aluminum

This chapter covered the more common transformer bank configurations for power distribution. Chapter 16 will investigate other transformer configurations that are used in industry.

REVIEW QUESTIONS—PART ONE

1. Transformers are used as a way to reduce the voltage losses along the length of a circuit conductor.

   True or False

2. Transformers can be used to limit the magnitude of either short-circuit or ground-fault overcurrent available to a connected load such as a secondary panel, motor-control center, or other type of load center from a large switchgear or unit substation.

   True or False

3. The ____________ (primary or secondary) circuit of the transformer is the source of electrical power for both the transformer and its connected load. The ____________ circuit of the transformer is the source of power for the connected load.
4. The highest value of impedance reflected into the primary circuit of a transformer will occur when the secondary circuit is ____________ (open or shorted). The lowest value of impedance reflected into the primary circuit will occur when the secondary circuit is ____________ (open or shorted).
5. The open-secondary primary-circuit current is referred to as the transformer's ____________ current.
6. In the standard symbol for a transformer, the two parallel bars between the two winding symbols indicate a transformer with an ____________ ____________.
7. In the standard symbol for a transformer, the two open circles on the winding symbols indicate ____________ ____________.
8. On the standard symbol for a transformer, the capital letter H is used to identify the ____________ (higher or lower) -voltage winding. The capital letter X is used to identify the ____________-voltage winding. The ins of the windings are identified with ____________ (odd or even) numbers. The outs of the same windings are identified with ____________ numbers.
9. Solid dots, adjacent to the wire terminals on the winding symbol of a transformer, are used to identify the ____________ of the windings.
10. The principle of induction requires relative motion between a length of conductor and a concentrated magnetic field. What provides the relative motion in a transformer circuit? ____________
11. Because there is only one power rating for a transformer, the current ratio of primary to secondary current is ____________ (inversely or directly) proportional to the voltage ratio.
12. When the voltage ratio of a power transformer is 1-to-1, the transformer is being used for ____________ of the primary and secondary circuits.
13. Transformer circuits are set up in the same manner as motor circuits. Although good design would warrant a power rating greater than or equal to the full-load power requirements of the intended secondary loads, transformers are normally allowed to operate in an overload condition.

    True or False

14. In parallel to motor circuits, the overcurrent protection for a transformer circuit can be split with transformer overload protection installed in the primary circuit, and SC-GF overcurrent protection installed in the secondary circuit.

    True or False

15. If both the primary- and secondary-circuit voltage ratings are not over 600 volts nominal (600 volts or less), the allowed overrates of transformer overcurrent-protection devices are determined according to Sec. ____________.

16. If any circuit voltage rating is over 600 volts, the allowed overrates of transformer overcurrent-protection devices are determined according to Sec. ____________.

17. If a transformer circuit rated not over 600 volts is provided with combined primary overcurrent protection, the overrate of the OCPD can be computed at ____________ % of the primary FLC.

18. If a transformer circuit rated not over 600 volts is provided with split overcurrent protection, the overrate of the primary OCPD can be computed at ____________ % of the primary FLC. The overrate of the secondary OCPD can be computed at ____________ % of the secondary FLC.

19. If a transformer circuit rated over 600 volts is provided with combined primary overcurrent protection at a supervised location, the overrate of the fuses can be computed at ____________ % of the primary FLC. The overrate of a CB can be computed at ____________ % of the primary FLC.

20. If a transformer circuit rated over 600 volts primary, but not over 600 volts secondary, is provided with split overcurrent protection at a supervised location, the overrate of the primary fuses can be computed at ____________ % of the primary FLC. The overrate of the secondary OCPD (fuses or CB) can be computed at ____________ % of the secondary FLC.

21. If a transformer circuit rated over 600 volts primary, but not over 600 volts secondary, is provided with split overcurrent protection at an unsupervised location, the overrate of the primary fuses can be computed at ____________ % of the primary FLC. The overrate of the secondary OCPD (fuses or CB) can be computed at ____________ % of the secondary FLC.

REVIEW QUESTIONS—PART TWO

1. If a transformer circuit rated over 600 volts (both primary and secondary) is rated at 5% impedance and provided with split overcurrent protection, the overrate of the primary fuses can be computed at ____________ % of the primary FLC. The overrate of the secondary fuses can be computed at ____________ % of the secondary FLC.

2. If a transformer circuit rated over 600 volts (both primary and secondary) is rated at 10% impedance and provided with split overcurrent protection, the overrate of the primary fuses can be computed at ____________ % of the primary FLC. The overrate of the secondary fuses can be computed at ____________ % of the secondary FLC.

3. If a transformer circuit rated over 600 volts (both primary and secondary) is rated at 5% impedance and provided with split overcurrent protection, the overrate of the primary CB can be computed at ____________ % of the primary FLC. The overrate of the secondary CB can be computed at ____________ % of the secondary FLC.

4. If a transformer circuit rated over 600 volts (both primary and secondary) is rated at 10% impedance and provided with split overcurrent protection, the overrate of the primary CB can be computed at ____________ % of the primary FLC. The overrate of the secondary CB can be computed at ____________ % of the secondary FLC.

5. The greatest amount of short-circuit overcurrent available is at the shorted secondary terminals of a transformer. The interrupt rating of a fuse or circuit breaker connected to the load-side terminals has to be greater than or equal to this maximum overcurrent if the circuit is to be cleared in the event of a fault condition.

   True or False

6. The high-value overrates of transformer overcurrent protection as allowed by the NE Code are intended to protect the transformer windings. In order for the circuit conductors to be protected

against overcurrent, the table-rated ampacity listed according to wire size, should be determined according to the rating of the fuses or the setting of the circuit breaker providing the overload-overcurrent protection for the transformer.

True or False

7. When the transformer overcurrent protection is split into primary and secondary OPCDs, the setting of the secondary circuit breaker or the rating of the secondary fuses must be projected into the primary circuit according to the inverse of the voltage ratio to size the primary-circuit conductors to meet the requirements of Sec. ____________.
8. For single-phase transformer circuits, the line voltage is equal to the phase voltage, and the line current is equal to the phase current in both the primary and secondary circuits.

    True or False

9. The typical single-phase, 3-wire 120/240-volt service to a single-family dwelling is extended from a 2-phase transformer because 240 volts is available to loads connected between phase A and phase B.

    True or False

10. Two-phase distribution systems utilize two single-phase sources that are ____________ electrical degrees out of phase with each other. If the transformers have the same primary and secondary voltage and power ratings, the power rating of the transformer bank can be computed as ____________ times the product of either line or winding (phase) values of voltage and current in either the primary or secondary circuit.
11. The rated ampacity of the neutral conductor for 2-phase, 5-wire power distribution must be computed at a minimum of ____________ % of the maximum unbalanced load-current return for either of the single-phase transformers.
12. In a 3-phase transformer bank, if all three transformers have the same primary and secondary voltage and power ratings, then the total bank 3-phase power rating can be computed as ____________ times the product of the phase values of voltage and current in either the primary or secondary circuit.
13. In 3-phase transformer bank, if all three transformers have the same primary and secondary voltage and power ratings, then the total bank 3-phase power rating is equal to the product ____________ times the product of the line values of voltage and current in either the primary or secondary circuit.
14. In a 3-phase wye configuration, the line ____________ (voltage or current) and phase (winding) ____________ are equal. The line ____________ is 173% greater than the phase ____________.
15. In a 3-phase wye transformer bank there is a ____________ degree shift between any two of the three phase voltages. As a result the phase voltages add ____________ (vectorially or algebraically).
16. In comparison to the delta distribution, the 4-wire wye is the more common 3-phase power distribution system because of the dual voltage rating and the ease of establishing a ground-fault return path between the equipment-grounding system/building bonding grid and the fourth-wire service neutral conductor.

    True or False

17. The secondary circuit of a 3-phase wye-connected transformer can be either 3-wire or 4-wire. A 3-wire wye configuration, like a 3-wire delta-connected transformer, is a single-voltage system.

    True or False

18. When used as the service supply to a power panel within a building or other structure, the 3-phase, 4-wire wye offers the advantage of 3-phase power distribution, as it simultaneously permits lighting to be connected between any of the three phase conductors and the common neutral conductor.

    True or False

19. When a solidly grounded wye unit substation or switchgear is rated at 1,000 amps or more, with voltages ratings that do not exceed 600 volts phase-to-phase, but do exceed 150 volts phase-to-ground, the switchgear must be provided with additional ground-fault protection, other than the service fuses or circuit breaker.

    True or False

20. In a 3-phase delta configuration, the line ____________ (voltage or current) and phase (winding) ____________ are equal. The line ____________ is 173% greater than the phase ____________.
21. In a 3-phase delta transformer bank there is a ____________ degree shift between any two of the three phase currents. As a result the phase currents add ____________ (vectorially or algebraically).
22. A 3-phase, 3-wire floating-delta service is required to have a service grounding-electrode system established for the noncurrent-carrying metal parts of the distribution system (metal raceways, boxes, and other enclosures), even though there is no electrical connection from the delta secondary to it.

    True or False

23. One advantage of a 3-wire floating-delta power distribution is that the equipment will not single-phase (opening of one phase in a 3-phase supply) when any one phase faults to ground.

    True or False

24. One disadvantage of a 3-phase, 4-wire wye power distribution is that the equipment will single-phase (opening of one phase in a 3-phase supply) when any one phase faults to ground.

    True or False

25. A 3-phase motor will start if any one phase supply is open (in a single-phase condition), provided it drives only light to moderate torque loads.

    True or False

REVIEW PROBLEMS

1. If the full-load power rating of the secondary circuit of a transformer is given at 208 volts and 121 amps, determine the primary-circuit full-load current rating for a primary-circuit voltage rating of 440 volts.
2. If the secondary of a transformer used to step down 440 volts to 208 volts has a secondary full-load current of 70 amps, determine the rating of the primary current for this connected load.
3. If a power transformer has a nameplate-rated impedance of 3% and a secondary full-load current rating of 2,000 amps, determine the maximum value of short-circuit overcurrent available at the secondary load terminals.

4. A power transformer is used to step down 575 volts to 208 volts. If the secondary full-load current rating is equal to 106 amps, determine:
   a. The maximum rating of the time-delay fuses installed in the primary supply circuit for the combined overcurrent protection of the transformer
   b. The minimum rating of the primary-circuit supply conductors, using THHN aluminum
   c. The minimum rating of the secondary-circuit load conductors, using THHN aluminum
5. A power transformer is used to step down 575 volts to 240 volts. If the secondary full-load current rating is equal to 225 amps, determine:
   a. The maximum rating of the inverse-time circuit breaker installed in the secondary load circuit as the overload-overcurrent protection of the transformer
   b. The maximum rating of time-delay fuses installed in the primary supply circuit as the SC-GF overcurrent protection of the transformer
   c. The minimum rating of the secondary-circuit load conductors, using THHN copper
   d. The minimum rating of the primary-circuit supply conductors, using THHN copper
6. If a single-phase power transformer is rated 575/240 volts at 37.5 kVA, determine:
   a. The maximum rating of the inverse-time circuit breaker installed in the secondary load circuit as the overload-overcurrent protection of the transformer
   b. The maximum rating of time-delay fuses installed in the primary supply circuit as the SC-GF overcurrent protection of the transformer
   c. The minimum rating of the secondary-circuit load conductors, using RHW copper
   d. The minimum rating of the primary-circuit supply conductors, using RHW copper
7. If a single-phase power transformer is rated 480/208 volts at 30 kVA, determine:
   a. The maximum setting of the secondary circuit breaker
   b. The maximum rating of the primary-circuit time-delay fuses
   c. The minimum wire size of the secondary-circuit conductors, using XHHW c-c aluminum
   d. The minimum wire size of the primary-circuit conductors, using XHHW c-c aluminum
8. For a single-phase, 3-wire, 120/240-volt, 100-amp service supply to a dwelling structure through a transformer with a 7,200-volt primary, determine:
   a. The voltage ratio of the power transformer
   b. The kVA rating of the transformer
   c. The primary full-load current rating of the transformer
9. If a building will be supplied with a 225-amp, 120/240-volt, single-phase, 3-wire service, determine:
   a. The minimum kVA rating of the transformer based on the panel rating, if the NE Code-maximum setting of the secondary overcurrent-protection device is employed for an unsupervised installation
   b. The maximum rating of a trunk line fuse for a primary supply voltage of 7,200 volts
   c. The minimum wire size of a triplexed service drop using RHW c-c aluminum conductors
   d. The minimum wire size of RHW c-c aluminum conductors in conduit extended as the service-entrance to the service-disconnecting means
10. A 2-phase transformer has secondary voltage and full-load current ratings of 575 volts and 65 amps. Determine the kVA power rating of the transformer.
11. A 3-phase transformer has secondary voltage and full-load current ratings of 575 volts and 65 amps. Determine the kVA power rating of the transformer.
12. A 3-phase, 915-kVA transformer will be used to step down 4,160 volts to 440 volts. Determine both the primary and secondary full-load line currents.
13. A 3-phase, 1,219.4-kVA transformer will be used to step down 12,470 volts to 460 volts in a wye-to-wye wiring configuration. Determine:

a. The primary line voltage
b. The primary full-load line current
c. The primary phase voltage
d. The primary full-load phase current
e. The secondary line voltage
f. The secondary full-load line current
g. The secondary phase voltage
h. The secondary full-load phase current

14. A 3-phase, 4-wire 1,200-kVA unit substation will be installed for the power distribution within a building at an unsupervised location. For a primary supply voltage of 12,470Y/7,200 volts and a secondary service voltage of 460Y/265 volts, determine:
a. The maximum setting of the secondary inverse-time circuit breaker
b. The minimum wire size of the secondary conductors supplying the switchgear, using THHN copper paralleled five times
c. The maximum rating of the primary disconnect fuses
d. The minimum wire size of the bonding jumper connecting the metal enclosure of the switchgear to the neutral busbar at the transformer secondary

15. A 3-phase, 3-wire, 75-kVA wye-connected power transformer with voltage ratings of 480 volts and 6 volts will supply a process heating unit. If the secondary 6-volt circuit will consist of wye-connected copper-alloy busbars, and the control and overcurrent protection will be installed in the primary circuit only, determine:
a. The maximum rating of the primary disconnect fuses
b. The minimum wire size of the primary-circuit conductors using THHN copper
c. The secondary full-load current rating, based on the rating of the primary-circuit fuses

16. A 3-phase, 997.7-kVA transformer will be used to step down 13.8 kV to 480 volts in a delta-to-delta wiring configuration. Determine:
a. The primary line voltage
b. The primary full-load line current
c. The primary phase voltage
d. The primary full-load phase current
e. The secondary line voltage
f. The secondary full-load line current
g. The secondary phase voltage
h. The secondary full-load phase current

17. For a 3-phase, 1,200-amp switchgear supplied as a 3-wire delta with a primary supply voltage of 18.0 kV and a secondary service/power distribution voltage of 440 volts, determine:
a. The minimum kVA rating of the transformer within the unit substation if the NE Code-maximum setting of the secondary overcurrent-protection device is employed for an unsupervised installation
b. The minimum wire size of the secondary conductors supplying the switchgear, using THW copper paralleled five times
c. The maximum rating of the primary disconnect fuses

18. If a 3-phase power transformer rated 80 kVA is used to derive a 3-phase, 4-wire, 208Y/120-volt supply from a 3-phase, 3-wire, 575-volt delta feeder circuit, determine:
a. The maximum setting of the secondary circuit breaker
b. The maximum rating of the primary-circuit time-delay fuses
c. The minimum wire size of the secondary-circuit conductors, using THHW-2 aluminum
d. The minimum wire size of the primary-circuit conductors, using THHW-2 aluminum

Chapter 16

Transformers—Part II

As introduced in Chapter 15, transformers are used in power distribution systems when the relative line values of voltage and current are to be changed. The primary circuit is the source of power for both the transformer and its secondary load. The secondary circuit is the transformer source of power for the connected load.

The number of turns or wraps of wire in each respective winding dictates the operating voltage of that circuit. If the primary-circuit voltage is higher than the secondary voltage, then the transformer is wired in a step-down configuration. If the reverse is true, the transformer is wired in a step-up configuration. If the primary and secondary circuits have the same voltage ratings, the transformer is used primarily for isolation between the two circuits (primary and secondary).

The isolation may be for noise abatement, for limiting fault-overcurrent available to equipment, or for isolation from reference to earth ground when the primary system operates with one of its circuit (current-carrying) conductors grounded.

Regardless of the wiring configuration of the transformer windings, the current ratio will always be inversely proportional to the voltage ratio. In step-down configuration, the current is stepped up. In step-up configuration, the current is stepped down. In isolation, both voltage and current ratios are held at a 1-to-1 value.

A single-phase transformer many have one or more windings in either or both circuits for transformation at different voltage ratings. The power rating of the transformer is computed as the product of line voltage and line current in either circuit.

When the power distribution is a two-phase system, two single-phase transformers are banked. The two phases within this system do not interconnect or react with each other in terms of voltage or current. The power rating is computed at two times the line values of voltage and current. In both the single-phase and 2-phase power computations, the phase or winding values of voltage and current are equal to the line values of voltage and current.

When the power distribution is rated 3-phase, three single-phase transformers are banked. The three phases interact with each other in terms of voltage or current, according to whether the winding termination is in wye or delta. In the wye configuration, the values of line current and phase or winding current are equal. The value of line voltage is 173% (square root of 3) greater than the phase or winding voltage. In the delta configuration, the two trade places. The values of line and phase or winding voltage are equal. The value of line current is 173% greater than the phase or winding current.

The total power rating of the bank is the sum of the individual transformer ratings. If the power ratings are of equal value, then the product of 3 times the phase values of voltage and current of any one of the single-phase transformers equals the power rating of the bank.

If the power ratings of the transformers are not equal, or if two instead of the normal three transformers are used in the bank as will be investigated in this chapter with the open-delta configuration, then the power rating of the bank is computed in terms of line values: The power is computed at 1.732 times the product of line voltage and line current in either the wye or delta configuration. The line power rating is still equal to the sum of the individual transformer power ratings.

In terms of power distribution circuit conductors, single-phase transformers can be wired as either 2-wire or 3-wire systems. The 3-wire system is normally the typical service to the single-family dwelling or small business. The third wire is a mid-tap grounded neutral conductor.

The 2-phase distribution can be configured as either 4-wire or 5-wire. In the 5-wire distribution, two single-phase mid-tap transformers, same as used in the single-phase, 3-wire distribution, are wired one to each phase in the supply. The third wire of each secondary is connected as the common fifth-wire grounded neutral conductor for both phases.

The 3-phase distribution can be either 3-wire or 4-wire in either the wye or delta configuration. The 3-phase, 3-wire wire system, whether wye or delta, does not offer a fourth-wire grounded-return conductor common to all three windings; hence, it is a single-voltage system.

As discussed in Chapter 15, the 3-phase wye can be wired as a 4-wire dual-voltage distribution system. Both the line voltage and the lower phase (phase-to-neutral) voltages are available to the connected loads. The phase voltage can only supply single-phase loads. The line voltage can supply either single-phase, phase-to-phase loads, or 3-phase loads.

As will be discussed in this chapter, a delta system can also be wired as a 3-phase, 4-wire, dual-voltage distribution (high-leg delta). The difference between the two is that the wye distribution can be balanced in terms of neutral conductor load-return current. The delta distribution cannot.

In this chapter the corner-ground delta, the high-leg delta, the open-delta distributions, and autotransformers will be discussed.

As demonstrated in Chapter 15, if both circuit voltages are rated not over 600 volts, the overrate of the respective OCPDs (overcurrent-protection devices) is computed according to the percentages found in Sec. 450-3(b).If any circuit voltage is rated over 600 volts, the overrate percentages are found in Sec. 450-3(a) [either T450-3(a)(1), or T450-3(a)(2)b.].

The wire size of the respective circuit conductors must be based on the overload-overcurrent protection computations of the transformer circuits. If the overcurrent protection of the transformer is split between the primary and secondary circuits, Sec. 240-3 requires the rated ampacity of the circuit conductors to be based on the rating of the secondary fuse or circuit breaker. The rating of the secondary fuse or circuit breaker must be projected into the primary circuit by the inverse of the voltage ratio to determine the primary-circuit overload current rating of the conductors.

If the transformer is provided with combined overcurrent protection in the primary circuit only, the rated-ampacity of the primary-circuit conductors, according to wire size, must be greater than or equal to the rating of the primary OCPD. By the inverse of the voltage ratio, the rating of the primary OCPD must be projected into the secondary circuit to determine the secondary overload current rating of the conductors.

16.1 CORNER-GROUND DELTA

An alternative to the 3-phase, 3-wire floating delta discussed in Chapter 15, which does not offer a ground-fault return means to the power distribution system, is the corner-ground delta shown in Figure 16-1. In this configuration, which is also referred to as a "grounded B-phase delta," the B-phase conductor is referenced or tied to earth ground (through the service grounding-electrode system) such that either phase A or phase C will read phase-to-phase voltage to ground.

A lower, single-phase neutral voltage as with a 3-phase, 4-wire wye configuration is not available in the corner-ground delta power distribution. The grounding of the center phase provides a ground-fault return path should either phase A or C ungrounded supply conductor come in accidental contact with a grounded surface within the building or other structure.

Section 240-22 of the NE Code prohibits any type of overcurrent-protection device from being installed in series with a current-carrying conductor that has been intentionally grounded. By Exception No. 1, if a circuit breaker is used for the circuit disconnect and protection of a 3-phase transformer bank in a corner-ground (B-phase ground) configuration, the grounded current-carrying conductor can also be protected (opens all conductors simultaneously).

If the corner-ground delta is disconnected and protected by a fused knife switch, the center leg must be supplied through a fuse blank to meet the requirements of 240-22. As stipulated throughout the NE Code, the B phase has to be designated the center leg in the disconnect; hence, the B-phase ground delta.

In terms of overrate computations for the OCPDs or the circuit conductors, the corner-ground delta transformer installation is handled in the same manner as the floating delta, with one exception: The corner ground, which is the center leg or B-phase in the distribution, must also be sized for the ground-fault return requirements of Sec. 250-23(b). The wire size of this service conductor must at least equal the wire size of the

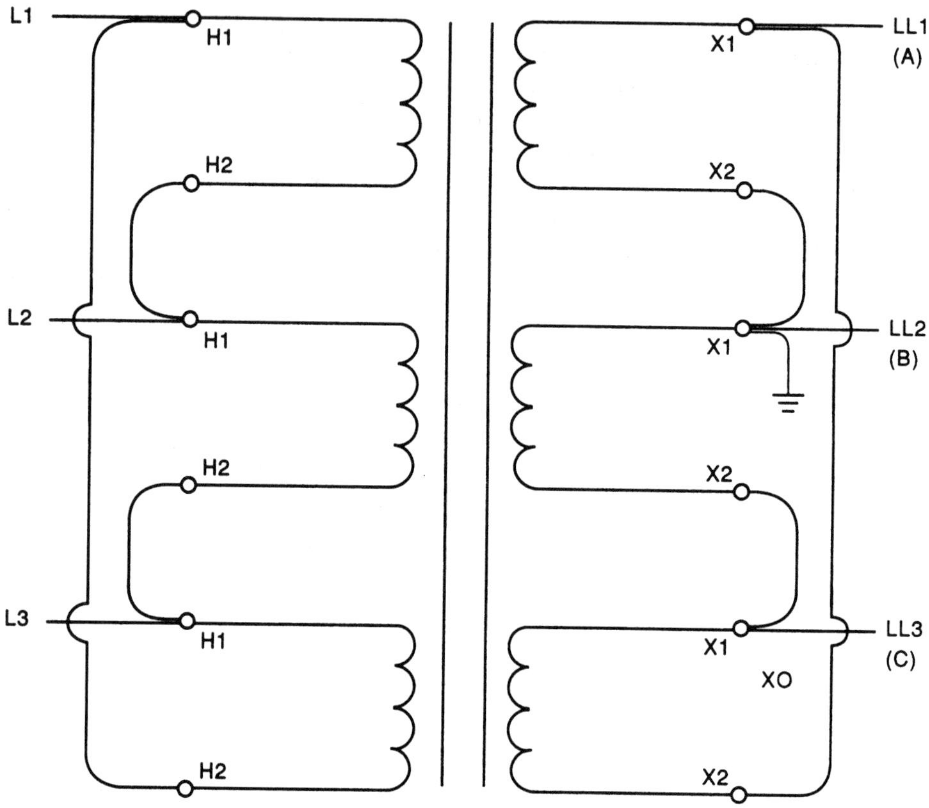

Fig. 16-1 In a grounded B-phase delta or corner-ground delta, the middle phase connection (LL2 or B-phase) is connected to a grounding-electrode system.

grounding-electrode conductor, which is based on the wire size of the ungrounded service conductors, phases A and C, in this type of service power distribution.

16.2 HIGH-LEG DELTA

Three-phase delta power distributions normally supply either 230, 460, or 575 volts phase-to-phase, with no lower phase-to-ground or neutral voltages available. The need of smaller transformers for lighting, receptacle, and/or control voltages (discussed in Chapter 15) can be avoided by changing the floating delta or corner-ground delta to a "high-leg delta" configuration.

Whether to use a 3-phase, 4-wire delta, or wye, and whether the transformation will require a bank of three single-phase transformers, or a single 3-phase transformer (common iron core), generally depends on the amount of power needed and the ratio of single-phase to 3-phase loads.

When the majority of the connected load is 3-phase motor loads, but lighting and receptacles must also be supplied, a high-leg delta transformation as shown in Figure

16-2, can be accomplished by using a lighting and power transformer that is center tapped to supply all the single-phase loads, and across the full winding, its part of the 3-phase loads. Two smaller single-phase transformers are wired in a delta connection with the lighting and power transformer to convey the required 3-phase power.

The lighting and power transformer of a high-leg delta normally has the same wiring configuration as the single-phase, 3-wire, 120/240-volt service supply for the single-family dwelling. The center tap between the two 120-volt windings is grounded to serve as the single-phase neutral service conductor for the larger transformer, and the ground-fault return conductor for the 3-phase system (all three ungrounded supply conductors).

The NE Code requires the center-tapped transformer to supply lines 1 and 3 (phases A and C) within the building or other structure. The third point of the closed-delta connection, which serves as the high-leg to ground, must be the center leg in the distribution, which is line 2 (or phase B).

Although many operating voltages are probably available, the more common high-leg delta voltage is the 240/120-volt, 3-phase, 4-wire distribution. When operating at 240-volts, the lower phase-to-neutral voltage (between either phase A or phase C and the neutral) will be half of the 240-volt winding voltage, or 120 volts. The high-leg conductor will operate at the square root of 3 times (173% greater) this lower phase-to-ground voltage, which is 208 volts.

Because of the higher voltage on this leg (or phase), Secs. 215-8 (feeders conduc-

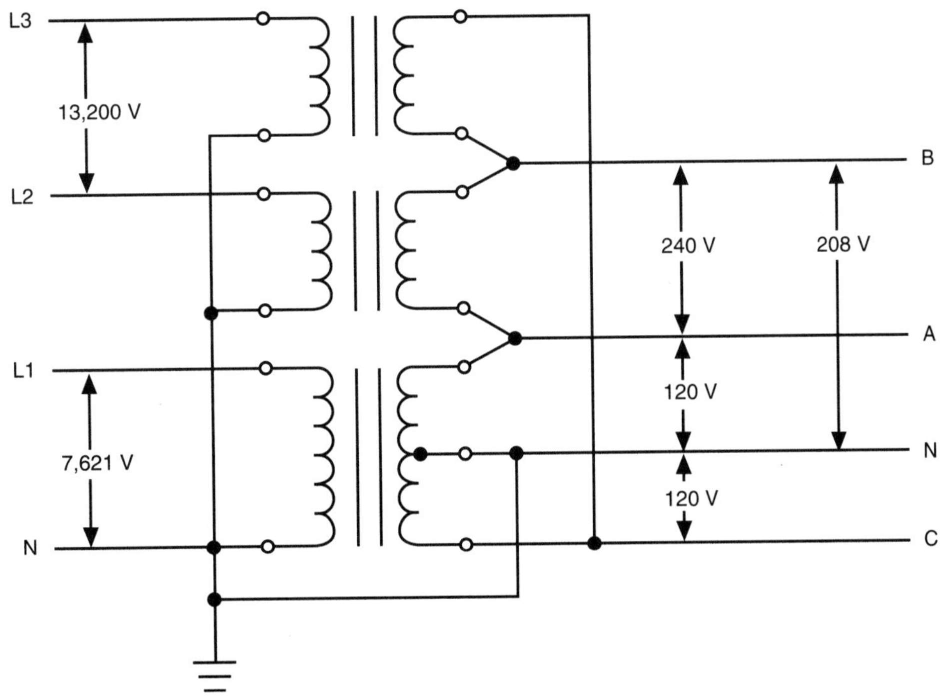

Fig. 16-2 In a high-leg delta three-phase transformation (older reference: red-leg delta), the center phase conductor has the higher voltage to ground.

tors), 230-56 (service conductors), and 384-3(e) (feeder supply and load-circuit conductors in a panel) of the NE Code require the center phase conductor (line 2 or phase B) to be identified with the color orange at all points of conductor (wire) termination.

The 3-phase, 4-wire, high-leg, delta configuration is not the more common power distribution for a building or other structure, simply because it cannot be balanced. The lighting and power transformer must be sized at the sum of the total single-phase kVA rating (again assuming the neutral loads are balanced, or its rating must be increased still further), plus one-third of the 3-phase kVA rating. Each of the other two power transformers must be sized at only one-third of the total 3-phase kVA load.

Once the transformers are sized, the circuit conductors and respective overcurrent-protection devices are sized in the same manner as for other power transformer installations, according to their respective currents.

Example 16–1. By evaluation of the intended loads in a building, the 3-phase loads have been assessed at 45 kVA. The single-phase loads have been assessed at 20 kVA. If the building service equipment will be supplied through a 3-phase, 4-wire, 240/120-volt, high-leg delta transformer bank from a 13.2-kV, 3-phase, 4-wire wye distribution, determine:

1. The minimum kVA rating of each transformer within the bank if the NE Code-maximum setting of the secondary overcurrent-protection device is employed and the transformer bank is located in an unsupervised location
2. The maximum rating of the secondary disconnect fuses
3. The minimum wire size of the secondary conductors supplying the panel, using THHN c-c aluminum conductors (assume a 70% service neutral load)
4. The maximum rating of the primary disconnect fuses

NE Code-minimum 3-phase power transformer kVA determination:

Underrate computation based on the secondary overcurrent protection allowance:

Ref. T450-3(a)(1)

$$45 \text{ kVA} \div 1.25 = 36 \text{ kVA}$$

Individual transformer kVA rating computation (for closed delta):

$$36 \text{ kVA} \div 3 = 12 \text{ kVA}$$

NE Code-minimum kVA rating determination of the lighting and power transformer:

Underrate of the single-phase kVA load computation based on the secondary overcurrent protection allowance:

Ref. T450-3(a)(1)

$$20 \text{ kVA} \div 1.25 = 16 \text{ kVA}$$

Combined transformer kVA rating computation:

12 kVA + 16 kVA = 28 kVA

NE Code-maximum rating determination of the secondary fuses:
Full-load current computation of 3-phase power (phases A, B, and C):

(12 kVA × 1,000) ÷ (1.732 × 240 volts) = 28.9 amps

Full-load current computation of single-phase lighting and power (phases A and C):

(16 kVA × 1,000) ÷ 240 volts = 66.7 amps

Total full-load current for lighting and power (phases A and C):

28.9 amps + 66.7 amps = 95.6 amps

NE Code-maximum rating determination of the high-leg power secondary fuse:
Overrate computation: Ref. T450-3(a)(1)

125% of 28.9 amps = 36.1 amps

Standard OCPD ratings: Ref. 240-6(a)
NHS: 40 amps Ref. (NHS) 450-3(a)(1), Excp. No. 1

NE Code-maximum rating determination of the lighting and power secondary fuses:
Overrate computation: Ref. T450-3(a)(1)

125% of 95.6 amps = 119.5 amps

Standard OCPD ratings: Ref. 240-6(a)
NHS: 125 amps Ref. (NHS) 450-3(a)(1), Excp. No. 1

NE Code-minimum wire size of the ungrounded high-leg secondary conductor:
Wire terminal temperature limitations: 60°C Ref. 110-14(c)(1), Excp. 1
Table-rated ampacity: 40 amps Ref. T310-16
Wire size: No. 6 AWG THHN c-c aluminum

NE Code-minimum wire size of the ungrounded secondary conductors (phases A and C):
Wire terminal temperature limitations: 75°C Ref. 110-14(c)(2), Excp. No. 1
Table-rated ampacity: 135 amps Ref. T310-16
Wire size: No. 2/0 AWG THHN c-c aluminum

NE Code-minimum wire size of the grounded service-neutral secondary conductor:
For load-return requirements: Ref. 220-22
FLC overrate computation: (secondary OCPD overrate) Ref. T450-3(a)(1)

125% of 66.7 amps = 83.4 amps

FLC derate computation: (given)

70% of 83.4 amps = 58.4 amps

Wire terminal temperature limitations: 60°C Ref. 110-14(c)(1), Excp. No. 1
Table-rated ampacity: 65 amps Ref. T310-16
Wire size: No. 3 AWG THHN c-c aluminum

For fault-return requirements: Ref. 250-23(b)
Wire size: No. 4 AWG THHN c-c aluminum Ref. T250-94

NE Code-minimum; compare for larger: Ref. 310-15(c)
Service neutral conductor wire size: No. 3 AWG THHN c-c aluminum

NE Code-maximum rating determination of the primary fuses:
Full-load current computation of 3-phase power (phases A, B, and C):

$$(240 \text{ volts} \div 13{,}200 \text{ volts}) \times 28.9 \text{ amps} = 0.53 \text{ amps}$$

Full-load current computation of single-phase lighting and power (phases A and C—wye configuration):

$$V_{phase} = (13.2 \text{ kV} \times 1{,}000) \div 1.732 = 7{,}621 \text{ volts}$$

$$(240 \text{ volts} \div 7{,}621 \text{ volts}) \times 66.7 \text{ amps} = 2.1 \text{ amps}$$

Total full-load current for lighting and power (phases A and C):

$$0.53 \text{ amps} + 2.10 \text{ amps} = 2.63 \text{ amps}$$

NE Code-maximum rating determination of the high-leg power primary fuse:
Overrate computation: Ref. T450-3(a)(1)

300% of 0.53 amps = 1.59 amps

Standard OCPD ratings: Ref. 240-6(a), Excp.
NHS: 3 amps Ref. (NHS) 450-3(a)(1), Excp. No. 1

NE Code-maximum rating determination of the lighting and power secondary fuses:
Overrate computation: Ref. T450-3(a)(1)

300% of 2.63 amps = 7.89 amps

Standard OCPD ratings: Ref. 240-6(a)
NHS: 10 amps Ref. (NHS) 450-3(a)(1), Excp. No. 1

For a supervised location, the single-phase and 3-phase kVA loads could have been underrated by 250% instead of just the 125%. Changing to a supervised location has no affect on the 300% overrate of the primary fuses.

The primary currents were computed by applying the turns ratio to the repsective single-phase and 3-phase currents because the primary circuit is connected in a wye configuration.

To compute the primary full-load currents of the larger power and lighting transformer using the power formulas, the 3-phase power formula would have to be solved for the 12 kVA in the same manner as for the high leg (phase B) using the higher line voltage of 13.2 kV. The single-phase power formula would have to be solved for the 16 kVA using the lower primary winding or phase voltage of 7,621 volts. The two currents would then have to be added to together in the same manner as with the secondary full-load current computations.

Code challenge 16–1. By evaluation of the intended loads in a building, the 3-phase loads have been assessed at 60 kVA. The single-phase loads have been assessed at 25 KVA. If the building service equipment will be supplied through a 3-phase, 4-wire, 240/120-volt, high-leg delta transformer bank from a 13.2-kV, 3-phase, 4-wire, wye distribution, determine:

1. The minimum kVA rating of each transformer within the bank if the NE Code-maximum setting of the secondary overcurrent protection device is employed and the transformer bank is located in an unsupervised location
2. The maximum rating of the secondary disconnect fuses
3. The minimum wire size of the secondary conductors supplying the panel, using XHHW aluminum conductors (assume a 70% service neutral load)
4. The maximum rating of the primary disconnect fuses

NE Code-minimum 3-phase power transformer kVA determination:

Underrate computation based on the secondary overcurrent protection allowance

Ref. (A) ______________

__________ kVA ÷ __________ = (B) __________ kVA

Individual transformer kVA rating computation (for closed delta):

__________ kVA ÷ __________ = (C) __________ kVA

NE Code-minimum kVA rating determination of the lighting and power transformer:

Underrate of the single-phase kVA load computation based on the secondary overcurrent protection allowance:

Ref. (D) ______________

__________ kVA ÷ __________ = (E) __________ kVA

Combined transformer kVA rating computation:

________ kVA + ________ kVA = (F) ________ kVA

NE Code-maximum rating determination of the secondary fuses:

Full-load current computation of 3-phase power (phases A, B, and C):

(________ kVA × 1,000) ÷ (1.732 × ________ volts) = (G) ________ amps

Full-load current computation of single-phase lighting and power (phases A and C):

(________ kVA × 1,000) ÷ ________ volts = (H) ________ amps

Total full-load current for lighting and power (phases A and C):

________ amps + ________ amps = (I) ________ amps

NE Code-maximum rating determination of the high-leg power secondary fuse:

Overrate computation: Ref. (J) ________

(K) ________ % of ________ amps = ________ amps

Standard OCPD ratings: Ref. (L) ________

NHS: (M) ________ amps Ref. (NHS) (N) ________

NE Code-maximum rating determination of the lighting and power secondary fuses:

Overrate computation: Ref. (O) ________

(P) ________ % of ________ amps = ________ amps

Standard OCPD ratings: Ref. (Q) ________

NHS: (R) ________ amps Ref. (NHS) (S) ________

NE Code-minimum wire size of the ungrounded high-leg secondary conductor:

Wire terminal temperature limitations: (T) ________ °C

Ref. (U) ________

Table-rated ampacity: (V) ________ amps Ref. (W) ________

Wire size: (X) ________ XHHW aluminum

NE Code-minimum wire size of the ungrounded secondary conductors (phases A and C):

Wire terminal temperature limitations: (Y) ________ °C

Ref. (Z) ________

Table-rated ampacity: (1A) ________ amps Ref. (1B) ________

Wire size: (1C) ________ XHHW aluminum

NE Code-minimum wire size of the grounded service-neutral secondary conductor:

For load-return requirements: Ref. (1D) ________

FLC overrate computation: (secondary OCPD overrate) Ref. (1E) ________

_____________ % of _____________ amps = (1F) _____________ amps

FLC derate computation: (given)

_____________ % of _____________ amps = (1G) _____________ amps

Wire terminal temperature limitations: (1H) _____________ °C

Ref. (1I) _____________

Table-rated ampacity: (1J) _____________ amps

Ref. (1K) _____________

Wire size: (1L) _____________ XHHW aluminum

For fault-return requirements:

Ref. (1M) _____________

Wire size: (1N) _____________ XHHW aluminum

Ref. (1O) _____________

NE Code-minimum; compare for (1P) _____________ :

Ref. (1Q) _____________

Service neutral conductor wire size: (1R) _____________ XHHW aluminum

NE Code-maximum rating determination of the primary fuses:

Full-load current computation of power (phase B):

(_____________ volts ÷ _____________ volts) × _____________ amps = (1S) _____________ amps

Overrate computation:

Ref. (1T) _____________

(1U) _____________ %of _____________ amps = _____________ amps

Standard OCPD ratings:

Ref. (1V) _____________

NHS: (1W) _____________ amps

Ref. (NHS) (1X) _____________

Full-load current computation of lighting and power (phases A and C—wye configuration):

V_{phase} = (_____________ kV × 1,000) ÷ _____________ = (1Y) _____________ volts

(_____________ volts ÷ _____________ volts) × _____________ amps = (1Z) _____________ amps

Total full-load current for lighting and power (phases A and C):

_____________ amps + _____________ amps = (2A) _____________ amps

Overrate computation:

Ref. (2B) _____________

(2C) _____________ % of _____________ amps = _____________ amps

Standard OCPD ratings:

Ref. (2D) _____________

NHS: (2E) _____________ amps

Ref. (NHS) (2F) _____________

16.3 THREE-PHASE OPEN-DELTA POWER

The delta-to-delta transformer connections can be configured as a closed delta, where three transformers are used to step-down the voltage, or as shown in Figure 16-3, delta-to-delta can be configured as an open delta (normally referred to as a "V–V" delta) where only two transformers are used to step-down the voltage of 3-phase power.

The two transformers are wired in series in the same manner as a closed delta, with the H2 lead of one transformer winding being connected to the H1 lead of the other transformer winding. Supplying lines L1, L2, and L3 are connected to leads H1, H1–H2, and H2 respectively. An open-delta primary requires the secondary output to also be connected in an open-delta configuration. In the secondary, the X2 lead of the first transformer winding will be connected to the X1 lead of the second transformer winding. The output is taken as LL1, LL2, and LL3 (phases A, B, and C) from X1, X1–X2, and X2 respectively.

Open-delta transformer banks are cost effective on the initial electrical installation, since only two transformers are required, but the power transfer capability or capacity of an open-delta bank is limited. The kVA rating of the bank equals only 86.6% of the total nameplate kVA rating of the two transformers. Computed another way, the kVA rating of an open-delta bank is limited to only 57.7% of three transformers of the same kVA rating connected in a closed-delta bank.

Example 16–2. Compute the bank power rating of two 60-kVA transformers connected in an open-delta configuration for step down of 4,160 volts to 575 volts.

$$86.6\% \text{ of } (2 \times 60 \text{ kVA}) = 103.9 \text{ kVA}$$

or

$$57.7\% \text{ of } (3 \times 60 \text{ kVA}) = 103.9 \text{ kVA}$$

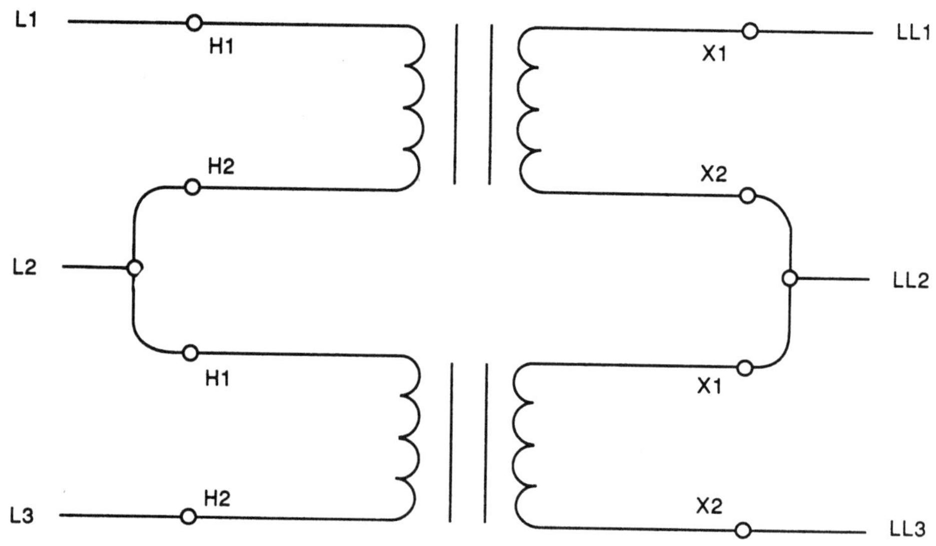

Fig. 16-3 In an open delta three-phase transformation, only two transformers are required. The kVA capacity of an open delta is restricted.

In the closed-delta bank of three transformers, the line current splits (vectorially) between the windings of the two transformers connected to that phase such that the phase or winding current (single-phase current) is only 57.7% of the 3-phase line current value.

In the open-delta, the 3-phase line current and the single-phase winding currents are equal value (single current path through one winding). This increases the single-phase power rating of the transformer.

Code challenge 16–2. Compute the bank power rating of two 37.5-kVA transformers connected in an open-delta configuration for step down of 2,300 volts to 440 volts.

(A) __________ % of (__________ × __________ kVA) = (B) __________ kVA

or

(C) __________ % of (__________ × __________ kVA) = (D) __________ kVA

When two transformers are to be banked in open-delta to step down the voltage for a connected load, the secondary full-load volt-amp rating must be divided by the 86.6% maximum rating to determine the required rating of the transformer bank. The bank rating, in turn is divided by 2 to determine each transformer's kVA rating.

As shown in Figure 16-4, to supply a closed-delta resistive bank load of 7,200 V, the closed-delta bank rating of three transformers equals the 7,200 VA. Each transformer is rated at 2,400 VA, which is one-third of the bank power rating.

In the open-delta configuration supplying the same load value (power rating) of resistive heating, the transformer bank has to have a power rating of 8,304 VA to supply the 7,200 VA load: 86.6% of 8,304 VA equals 7,200 VA.

In the open-delta configuration each transformer has to be rated at half the computed bank value: 50% of 8,304 VA equals 4,152 VA. Dividing 4,152 VA by the load-bank rating of 7,200 VA, each transformer in the open-delta has to be rated at 57.7% of the load bank (or equivalent closed-delta transformer bank) power rating of 7,200 VA.

As with 3-phase closed-delta and wye transformer configurations for transformers operating at 600 volts or less in both circuits, the transformer can be underrated, actually overloaded, by the 125% overrate of the secondary overload-overcurrent protection device rating or setting allowed by Sec. 450-3(b)(2). The overrate of the primary short-circuit, ground-fault overcurrent protection, which can be provided for the individual transformer circuit, or for two or more banks of 3-phase transformers by common feeder overcurrent protection, can be computed at a maximum of 250% of the lowest-rated primary FLC of any individual bank supplied.

For transformers used to step down voltages rated over 600 volts, the transformer can be underrated, actually overloaded, by the 125% (or 250%) overrate of the secondary overload-overcurrent protection device rating or setting allowed by T450-3(a)(1) (or T450-3(a)(2)b, supervised location). The overrate of the primary short-circuit, ground-fault overcurrent protection for these higher-voltage transformer banks depends on the type of protection (fuses or circuit breaker) and the nameplate-rated impedance (not more than 6% or not more than 10%).

As with all transformer installations, these higher overrates, which technically overload the transformer, would also overload the circuit conductors if the primary and sec-

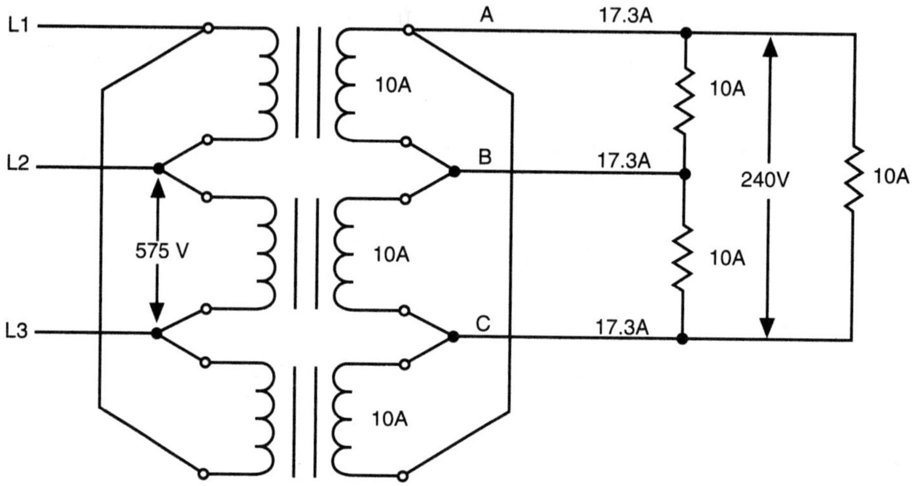

closed-delta bank rating: 3 x 10A x 240V = 7,200 VA

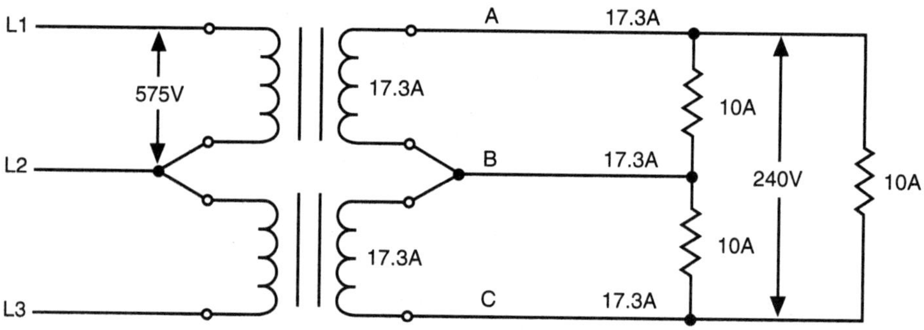

open-delta bank rating: 2 x 17.3A x 240V = 8,304 VA
86.6% of 8,304 VA = 7,200 VA
or, 7,200 VA / 0.866 = 8,304 VA

Fig. 16-4 In open delta, the transformer bank can only supply 86.6% of the total nameplate power rating.

ondary conductors were only sized at the full-load rating of their respective circuits. In accordance with Sec. 240-3, the circuit conductors must have a rated ampacity greater than or equal to the overload overpercentage applied to the secondary FLC—in other words, a rated-load ampacity greater than or equal to the rating of the secondary fuses or setting of the circuit breaker.

Example 16–3. A 3-phase, 3-wire, 240-volt, 200-amp, power panel will be supplied through an open-delta transformer bank from a 575-volt power distribution.

Determine:

1. The minimum kVA rating of each transformer within the bank if the NE Code-maximum setting of the secondary overcurrent-protection device is employed
2. The minimum wire size of the secondary conductors supplying the panel using XHHW-2 aluminum conductors
3. The maximum rating of the primary disconnect fuses
4. The minimum wire size of the primary conductors supplying the transformer using XHHW-2 aluminum conductors

NE Code-minimum transformer kVA determination:
Panel full-load kVA computation:

$$(1.732 \times 240 \text{ volts} \times 200 \text{ amps}) \div 1{,}000 = 83.1 \text{ kVA}$$

Underrate computation based on the secondary overcurrent protection allowance:
Ref. 450-3(b)(2)

$$83.1 \text{ kVA} \div 1.25 = 66.5 \text{ kVA}$$

Transformer bank full-load kVA overrate computation for V-V delta:

$$66.5 \text{ kVA} \div 0.866 = 76.8 \text{ kVA}$$

Individual transformer kVA rating computation:

$$76.8 \text{ kVA} \div 2 = 38.4 \text{ kVA}$$

NE Code-minimum wire size of the secondary conductors:
Wire terminal temperature limitations: 75° Ref. 110-14(c)(2), Excp. 1
Table-rated ampacity: 205 amps Ref. T310-16
Wire size: 250 kcmil XHHW-2 aluminum

NE Code-maximum rating determination of the primary fuses:
Primary FLC computation:

$$(76.8 \text{ kVA} \times 1{,}000) \div (1.732 \times 575 \text{ volts}) = 77.1 \text{ amps}$$

Overrate computation: Ref. 450-3(b)(2)

$$250\% \text{ of } 77.1 \text{ amps} = 192.8 \text{ amps}$$

Standard OCPD ratings: Ref. 240-6(a)
NLS: 175 amps Ref. (NLS) 450-3(b)(2)

E Code-minimum wire size of the primary conductors:
Primary FLC computation:

$$(240 \text{ volts} \div 575 \text{ volts}) \times 200 \text{ amps} = 83.5 \text{ amps}$$

Wire terminal temperature limitations: 60°C Ref. 110-14(c)(1), Excp. 1
Table-rated ampacity: 85 amps Ref. T310-16
Wire size: No. 1 AWG XHHW-2 aluminum

The 3-phase power formula had to be used to compute the primary-circuit full-load current in this example because the secondary full-load current never was computed. In comparison, the inverse of the voltage ratio was used to project the secondary OCPD rating into the primary circuit for determination of the rated ampacity of the primary-circuit supply conductors. In reference to Sec. 240-3, the primary-circuit conductors must be protected against overload overcurrent in accordance with their rated ampacity.

Code challenge 16–3. A 3-phase, 3-wire, 240-volt, 125-amp power panel will be supplied through an open-delta transformer bank from a 440-volt power distribution. **Determine:**

1. The minimum kVA rating of each transformer within the bank if the NE Code-maximum setting of the secondary overcurrent-protection device is employed.
2. The minimum wire size of the secondary conductors supplying the panel, using XHHW-2 copper conductors
3. The maximum rating of the primary disconnect fuses
4. The minimum wire size of the primary conductors supplying the transformer, using XHHW-2 copper conductors

NE Code-minimum transformer kVA determination:
Panel full-load kVA computation:

(1.732 × ____________ volts × ____________ amps) ÷ 1,000 = (A) ____________ kVA

Underrate computation based on the secondary overcurrent protection allowance
Ref. (B) ____________________

____________ kVA ÷ ____________ = (C) ____________ kVA

Transformer bank full-load kVA overrate computation for V-V delta:

____________ kVA ÷ ____________ = (D) ____________ kVA

Individual transformer kVA rating computation:

____________ kVA ÷ ____________ = (E) ____________ kVA

NE Code-minimum wire size of the secondary conductors:

Wire terminal temperature limitations: (F) ____________ °C

Ref. (G) ____________________

Table-rated ampacity: (H) ____________ amps Ref. (I) ____________________

Wire size: (J) ____________ XHHW-2 copper

NE Code-maximum rating determination of the primary fuses:

Primary FLC computation:

(__________ kVA× 1,000) ÷ (1.732 × __________ volts) = (K) __________ amps

Overrate computation: Ref. (L) ____________________

____________ % of ____________ amps = (M) ____________ amps

Standard OCPD ratings: Ref. (N) ____________________

NLS: (O) ____________ amps Ref. (NLS) (P) ____________________

NE Code-minimum wire size of the primary conductors:

Primary FLC computation:

(__________ volts ÷ __________ volts) × __________ amps = (Q) __________ amps

Wire terminal temperature limitations: (R) ____________ °C

Ref. (S) ____________________

Table-rated ampacity: (T) ____________ amps Ref. (U) ____________________

Wire size: (V) ____________ XHHW-2 copper

16.4 HIGH-LEG OPEN-DELTA POWER

As shown in Figure 16-5, the V–V or open-delta can also be used in a high-leg configuration to supply single-phase loads rated at either 120 or 240 volts, and 3-phase loads rated at 240 volts. The V–V connection is often used when the principle load is lighting and receptacle, but the motor loads are high-value horsepower ratings, which are often of intermittent use.

As with a high-leg closed-delta transformer bank, the high-leg open-delta bank requires two sizes of transformers. The larger (higher kVA rating) lighting and power transformer supplies both single-phase and 3-phase loads. The smaller power transformer only has to be large enough to supply the 3-phase loads.

In this configuration, the open-end of the smaller power transformer has to be designated the high leg. Both the single-phase, phase-to-neutral loads and 3-phase loads are supplied from the larger lighting and power transformer. All the operating voltages are the same as for the closed delta. The common center-tap conductor is grounded. This is the single-phase neutral or load-return conductor for phases A and C, and the ground-fault return conductor for all three phases.

In some instances, 208-volt single-phase loads are supplied between phase B and the neutral. The 3-phase neutral currents will not cancel, as in the 3-phase, 4-wire wye

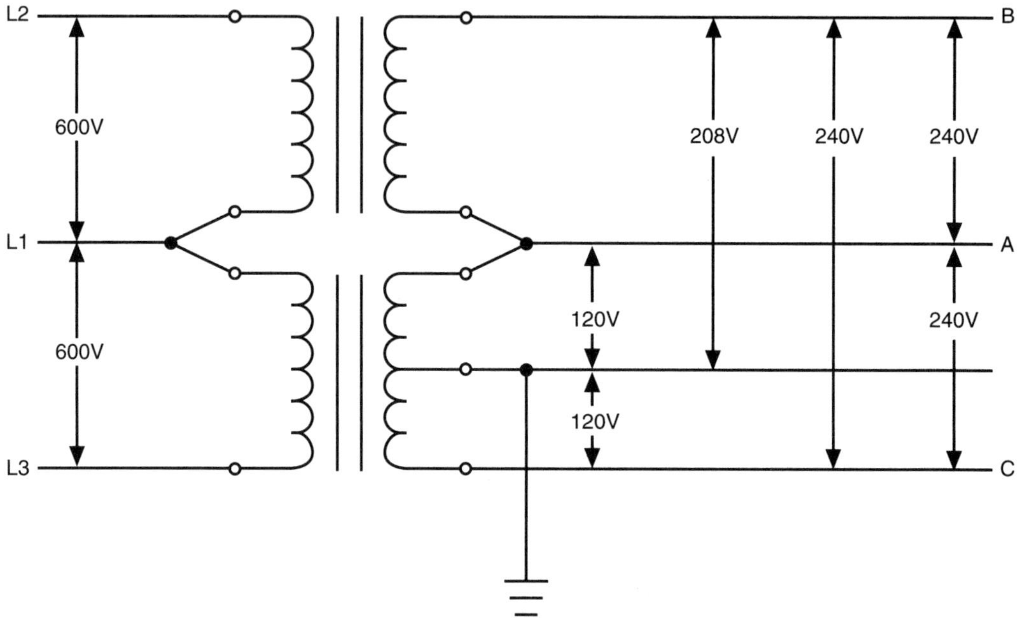

Fig. 16-5 Voltage relationships of a three-phase, four-wire high leg in open-delta transformation.

distribution because of the relative location, in terms of electrical degrees, of the center tap. This means the neutral would have to be sized for the algebraic sum (straight addition—not a phasor quantity) of the maximum unbalanced neutral current between phases A or C and all the B-phase, single-phase currents.

Used in this manner, the neutral conductor may turn out larger in wire size than any of the phase conductors. The phase B conductor, on the other hand, is normally the smaller ungrounded phase conductor (compared to phase A or C conductors), simply because of its limited loading.

The 3-phase 86.6% restriction of the overall open-delta transformer bank rating still applies. Instead of dividing this overrated or compensated load value by three as in a closed delta, each transformer in the open delta must carry half of the total 3-phase load value. In other words, the 3-phase kVA load value is divided by two.

For the single-phase loads, the rating of the lighting and power transformer must supply the trigonometric sum (vector or phasor quantity) of its share of the 3-phase loads added to the algebraic sum (straight addition) of all the single-phase loads.

Example 16–4. By evaluation of the intended loads in a building, the 3-phase loads have been assessed at 24 kVA. The single-phase loads have been assessed at 30 kVA. If the building service equipment will be supplied through a 3-phase, 4-wire, 240/120-volt, high-leg, open-delta transformer bank from a 18-kV, 3-phase, 3-wire delta distribution, assume an unsupervised location and determine:

1. The minimum kVA rating of each transformer within the bank if the NE Code-maximum setting of the secondary overcurrent-protection device is employed

2. The maximum rating of the secondary disconnect fuses
3. The minimum wire size of the secondary conductors supplying the panel, using THHN c-c aluminum conductors (assume a 70% service neutral load)
4. The maximum rating of the primary disconnect fuses

NE Code-minimum 3-phase power transformer kVA determination:

Underrate computation based on the secondary overcurrent protection allowance:

Ref. T450-3(a)(1)

$$24 \text{ kVA} \div 1.25 = 19.2 \text{ kVA}$$

Overrate computation (for open-delta):

$$19.2 \text{ kVA} \div 0.866 = 22.2 \text{ kVA}$$

Individual transformer kVA rating computation (for open-delta):

$$22.2 \text{ kVA} \div 2 = 11.1 \text{ kVA}$$

NE Code-minimum kVA rating determination of the lighting and power transformer:

Underrate of the single-phase kVA load computation based on the secondary overcurrent protection allowance:

Ref. T450-3(a)(1)

$$30 \text{ kVA} \div 1.25 = 24 \text{ kVA}$$

Combined transformer kVA rating computation:

$$11.1 \text{ kVA} + 24 \text{ kVA} = 35.1 \text{ kVA}$$

NE Code-maximum rating determination of the secondary fuses:

Full-load current computation of 3-phase power (phases A, B, and C):

$$(11.1 \text{ kVA} \times 1{,}000) \div (1.732 \times 240 \text{ volts}) = 26.7 \text{ amps}$$

Full-load current computation of single-phase lighting and power (phases A and C):

$$(24 \text{ kVA} \times 1{,}000) \div 240 \text{ volts} = 100 \text{ amps}$$

Total full-load current for lighting and power (phases A and C):

$$26.7 \text{ amps} + 100 \text{ amps} = 126.7 \text{ amps}$$

NE Code-maximum rating determination of the high-leg power secondary fuse:
Overrate computation: Ref. T450-3(a)(1)

125% of 26.7 amps = 33.4 amps

Standard OCPD ratings: Ref. 240-6(a)
NHS: 35 amps Ref. (NHS) 450-3(a)(1), Excp. No. 1

NE Code-maximum rating determination of the lighting and power secondary fuses:
Overrate computation: Ref. T450-3(a)(1)

125% of 126.7 amps = 158.4 amps

Standard OCPD ratings: Ref. 240-6(a)
NHS: 175 amps Ref. (NHS) 450-3(a)(1), Excp. No. 1

NE Code-minimum wire size of the ungrounded high-leg secondary conductor:
Wire terminal temperature limitations: 60°C Ref. 110-14(c)(1), Excp. No. 1
Table-rated ampacity: 40 amps Ref. T310-16
Wire size: No. 6 AWG THHN c-c aluminum

NE Code-minimum wire size of the ungrounded secondary conductors (phases A and C):
Wire terminal temperature limitations: 75°C Ref. 110-14(c)(2), Excp. No. 1
Table-rated ampacity: 180 amps Ref. T310-16
Wire size: No. 4/0 AWG THHN c-c aluminum

NE Code-minimum wire size of the grounded service-neutral secondary conductor:
For load-return requirements: Ref. 220-22
FLC overrate computation: (secondary OCPD overrate) Ref. T450-3(a)(1)

125% of 100 amps = 125 amps

FLC derate computation: (given)

70% of 125 amps = 87.5 amps

Wire terminal temperature limitations: 60°C Ref. 110-14(c)(1), Excp. No. 1
Table-rated ampacity: 100 amps Ref. T310-16
Wire size: No. 1/0 AWG THHN c-c aluminum

For fault-return requirements: Ref. 250-23(b)
Wire size: No. 2 AWG THHN c-c aluminum Ref. T250-94

NE Code-minimum; compare for larger: Ref. 310-15(c)

Service neutral conductor wire size: No. 1/0 AWG THHN c-c aluminum

NE Code-maximum rating determination of the primary fuses:

Full-load current computation of 3-phase power (phases A, B, and C):

$$(240 \text{ volts} \div 18{,}000 \text{ volts}) \times 26.7 \text{ amps} = 0.36 \text{ amps}$$

Full-load current computation of single-phase lighting and power (phases A and C):

$$(240 \text{ volts} \div 18{,}000 \text{ volts}) \times 100 \text{ amps} = 1.33 \text{ amps}$$

Total full-load current for lighting and power (phases A and C):

$$0.36 \text{ amps} + 1.33 \text{ amps} = 1.69 \text{ amps}$$

NE Code-maximum rating determination of the high-leg power primary fuse:

Overrate computation: Ref. T450-3(a)(1)

$$300\% \text{ of } 0.36 \text{ amps} = 1.08 \text{ amps}$$

Standard OCPD ratings: Ref. 240-6(a), Excp.

NHS: 3 amps Ref. (NHS) 450-3(a)(1), Excp. No. 1

NE Code-maximum rating determination of the lighting and power secondary fuses:

Overrate computation: Ref. T450-3(a)(1)

$$300\% \text{ of } 1.69 \text{ amps} = 5.07 \text{ amps}$$

Standard OCPD ratings: Ref. 240-6(a)

NHS: 6 amps Ref. (NHS) 450-3(a)(1), Excp. No. 1

In comparison to the primary full-load current ratings computed in Example 16–1, which was also terminated in a secondary 3-phase, 4-wire, high-leg configuration, but supplied by a wye primary, the secondary full-load currents of both the single-phase and 3-phase currents of the lighting and power transformer in this example were projected into the primary circuit with the ratio of the line voltages because of the delta-to-delta transformation. An open-delta connection must be a V–V connection—open-delta configuration in both the primary and secondary circuits.

Code challenge 16–4. By evaluation of the intended loads in a building, the 3-phase loads have been assessed at 48 kVA. The single-phase loads have been assessed at 30 kVA. If the building service equipment will be supplied through a 3-phase, 4-wire, 240/120-volt, high-leg delta transformer bank from a 13.2-kV, 3-phase, 3-wire, delta distribution, determine:

1. The minimum kVA rating of each transformer within the bank if the NE Code-maximum setting of the secondary overcurrent-protection device is employed. The transformer bank is located in an unsupervised location.
2. The maximum rating of the secondary disconnect fuses
3. The minimum wire size of the secondary conductors supplying the panel, using XHHW copper conductors (assume a 70% service neutral load)
4. The maximum rating of the primary disconnect fuses

NE Code-minimum 3-phase power transformer kVA determination:

Underrate computation based on the secondary overcurrent protection allowance:

Ref. (A) ____________________

____________ kVA ÷ ____________ = (B) ____________ kVA

Overrate computation (for open-delta):

____________ kVA ÷ ____________ = (C) ____________ kVA

Individual transformer kVA rating computation:

____________ kVA ÷ ____________ = (D) ____________ kVA

NE Code-minimum kVA rating determination of the lighting and power transformer:

Underrate of the single-phase kVA load computation based on the secondary overcurrent protection allowance:

Ref. (E) ____________________

____________ kVA ÷ ____________ = (F) ____________ kVA

Combined transformer kVA rating computation:

____________ kVA + ____________ kVA = (G) ____________ kVA

NE Code-maximum rating determination of the secondary fuses:

Full-load current computation of 3-phase power (phases A, B, and C):

(____________ kVA × 1,000) ÷ (1.732 × ____________ volts) = (H) ____________ amps

Full-load current computation of single-phase lighting and power (phases A and C):

(____________ kVA × 1,000) ÷ ____________ volts = (I) ____________ amps

Total full-load current for lighting and power (phases A and C):

____________ amps + ____________ amps = (J) ____________ amps

NE Code-maximum rating determination of the high-leg power secondary fuse:

Overrate computation: Ref. (K) ____________

(L) ____________ % of ____________ amps = ____________ amps

Standard OCPD ratings: Ref. (M) ____________

NHS: (N) ____________ amps Ref. (NHS) (O) ____________

NE Code-maximum rating determination of the lighting and power secondary fuses:

Overrate computation: Ref. (P) ____________

(Q) ____________ % of ____________ amps = ____________ amps

Standard OCPD ratings: Ref. (R) ____________

NHS: (S) ____________ amps Ref. (NHS) (T) ____________

NE Code-minimum wire size of the ungrounded high-leg secondary conductor:

Wire terminal temperature limitations: (U) ____________ °C

Ref. (V) ____________

Table-rated ampacity: (W) ____________ amps Ref. (X) ____________

Wire size: (Y) ____________ XHHW copper

E Code-minimum wire size of the ungrounded secondary conductors, phases A and C:

Wire terminal temperature limitations: (Z) ____________ °C

Ref. (1A) ____________

Table-rated ampacity: (1B) ____________ amps Ref. (1C) ____________

Wire size: (1D) ____________ XHHW copper

NE Code-minimum wire size of the grounded service-neutral secondary conductor:

For load-return requirements: Ref. (1E) ____________

FLC overrate computation: (secondary OCPD overrate) Ref. (1F) ____________

____________ % of ____________ amps = (1G) ____________ amps

FLC derate computation: (given)

____________ % of ____________ amps = (1H) ____________ amps

Wire terminal temperature limitations: (1I) ____________ °C

Ref. (1J) ____________

Table-rated ampacity: (1K) ____________ amps Ref. (1L) ____________

Wire size: (1M) ____________ XHHW copper

For fault-return requirements: Ref. (1N) ____________

Wire size: (1O) ____________ XHHW copper Ref. (1P) ____________

NE Code-minimum; compare for (1Q) ____________ : Ref. (1R) ________________

Service neutral conductor wire size: (1S) ____________ XHHW copper

NE Code-maximum rating determination of the primary fuses:

Full-load current computation of power (phase B):

(__________ volts ÷ __________ volts) × __________ amps = (1T) __________ amps

Overrate computation: Ref. (1U) ________________

(1V) ____________ % of ____________ amps = ____________ amps

Standard OCPD ratings: Ref. (1W) ________________

NHS: (1X) ____________ amps Ref. (NHS) (1Y) ________________

Full-load current computation of lighting and power (phases A and C):

(__________ volts ÷ __________ volts) × __________ amps = (1Z) __________ amps

Total full-load current for lighting and power (phases A and C):

____________ amps + ____________ amps = (2A) ____________ amps

Overrate computation: Ref. (2B) ________________

(2C) ____________ % of ____________ amps = ____________ amps

Standard OCPD ratings: Ref. (2D) ________________

NHS: (2E) ____________ amps Ref. (NHS) (2F) ________________

16.5 AUTOTRANSFORMER POWER

As shown in Figure 16-6, a single winding wrapped around a ferrous metal core, which could be described as an iron-core reactor, can be used as a transformer if the winding is tapped at different percentages of the winding's length. The winding of a transformer is rated in terms of turns per volt. If the supply of the secondary load circuit is tapped at the mid-point of the windings, only half of the 240-volt full-winding voltage or 120 volts would be applied to the connected load.

This type of transformer is an autotransformer. The autotransformer will obey all the laws concerning transformers (voltage, turns, and current ratios) except isolation. The primary and secondary circuits share a common connection at one end of the single winding. The taps are normally set at 50, 65, and 80 percent of the full winding voltage.

If the autotransformer is wired in a step-down configuration as shown in Figure 16-7, the higher voltage is connected across the full winding. The lower voltage is derived between one of the taps and the common connection with the full winding. The value or

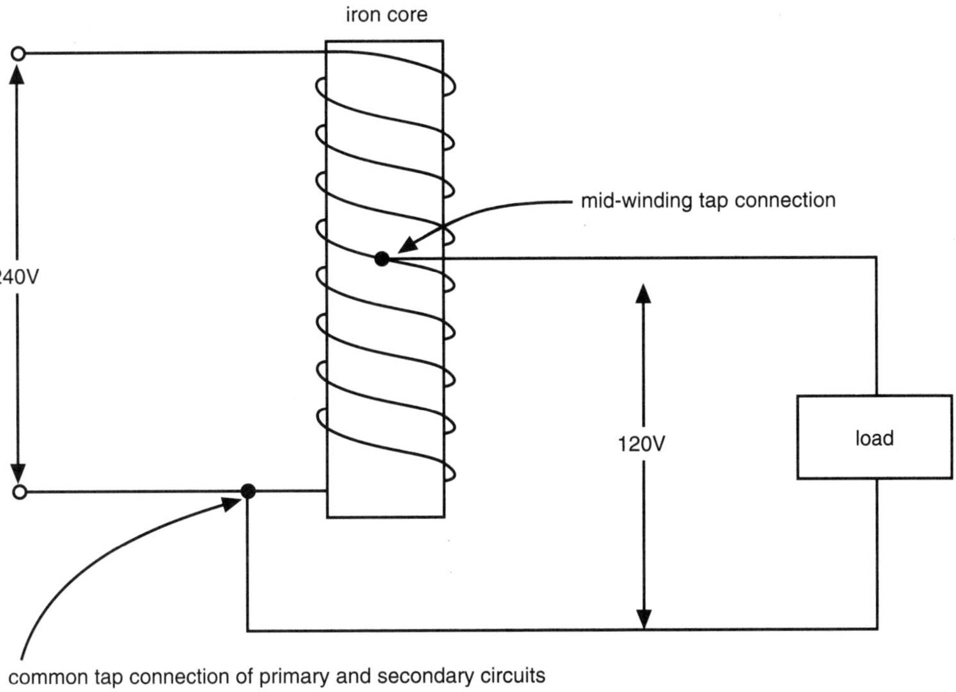

Fig. 16-6 Principle of voltage operation using an autotransformer.

rating of the applied primary voltage is multiplied by the percentage value of the tap to determine the derived secondary voltage value or rating.

If the autotransformer is wired in a step-up configuration as shown in Figure 16-8, the lower voltage is derived between one of the taps and the common winding terminal. The higher derived voltage is taken from the full-winding connections. The percentage value of the tap is divided into the primary (lower) voltage rating to determine the secondary (higher) voltage rating.

Autotransformers are often used with large ac induction motors for multiple-step, low-voltage reactance starting. The taps can be used as "step-voltages" from standstill, where torque is high-value and the in-rush starting current is at its maximum, to a run condition, where the established rotor field (magnetic field which produces a counter EMF in the stator circuit) reduces the stator current, and the torque balances out with the torque requirements of the driven load.

Another popular use of autotransformers is the adjustment or compensation for voltage drop. Compensation is required when voltage drop, due to length of conductor and magnitude of full-load current, reduces the supply voltage below the rated voltage of a motor or a resistance load. The 80% supply tap will boost the load voltage up to 125% of the supply voltage. The 50% supply tap will boost the load voltage up to 200% of the supply voltage.

As discussed with the 3-phase, 3-wire, wye power configuration, many industrial

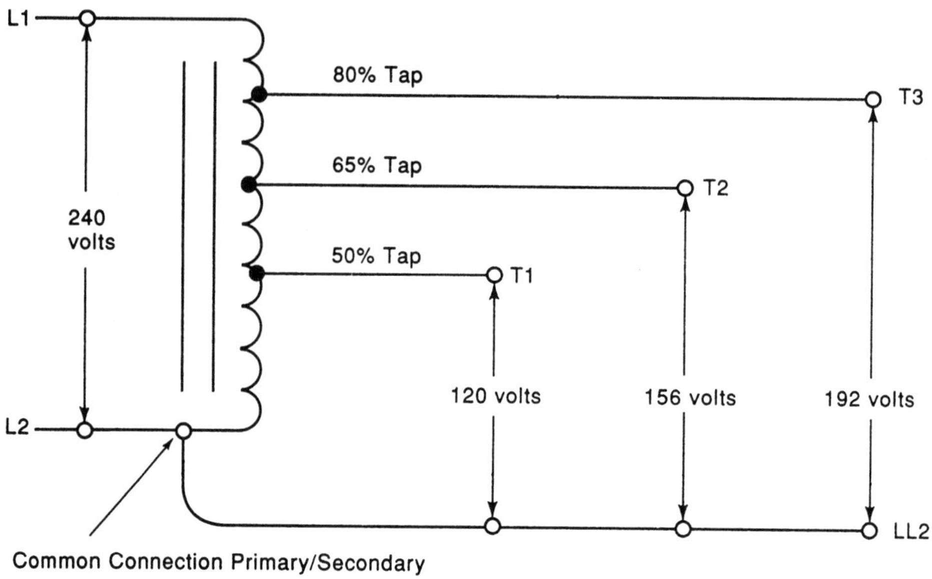

Fig. 16-7 The values of tap voltage from a step-down autotransformer are determined by multiplying the tap percentage values by the primary voltage rating.

plants required large amounts of process heating. When the plant power distribution is a 3-phase, 3-wire, 460-volt, single-voltage, delta system, a 3-phase, 3-wire, closed-delta bank of three individual autotransformers can be installed to derive a lower working voltage of 230 volts. The separately derived system as shown in Figure 16-9 can in turn supply load centers for 230-volt resistive heating banks.

The overrate of overcurrent protection devices for autotransformers as listed in Sec. 450-4 is somewhat misleading. The 125% overrate of the overload-overcurrent protection device appears to be restricted to the primary circuit only. Because this section reads "installed in series with the input (primary) circuit conductors only."

These restrictions are aimed at a conventional transformer (two separate windings) to where the lower-voltage secondary winding is purposely wired in series with the connected load and the series array is wired in parallel with the primary winding to either aid (boost) the supply voltage or oppose (buck) the supply voltage. An overcurrent protection device cannot be installed in series with the secondary shunt winding.

In a true autotransformer, provided only with taps to the partial winding, it would not be possible to wire an overcurrent protection device in series with the winding interconnected between points A and B as shown in Diagram 450-4 in the NE Code.

For an autotransformer bank installation, which simply drops (lowers) the supply voltage to a lower working voltage, Sec. 450-3(b)(2) will allow the overcurrent protection for the autotransformer to be split. As with the other distribution transformers we have discussed, the secondary fuses or circuit breaker can be sized at a maximum of 125% of the secondary full-load current rating of the transformer to provide full overcurrent pro-

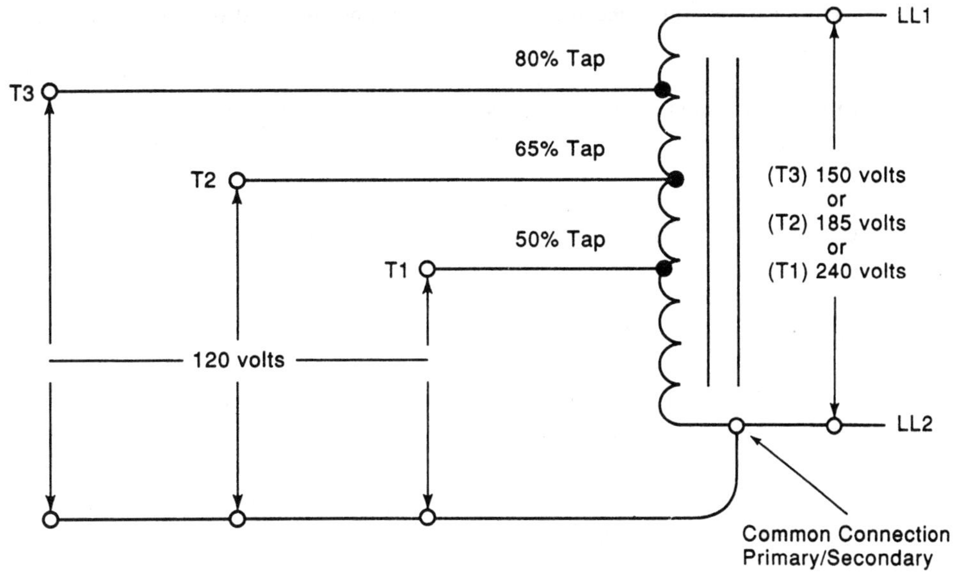

Fig. 16-8 The values of full-winding voltage from a step-up autotransformer are determined by dividing the tap percentage values into the primary voltage rating.

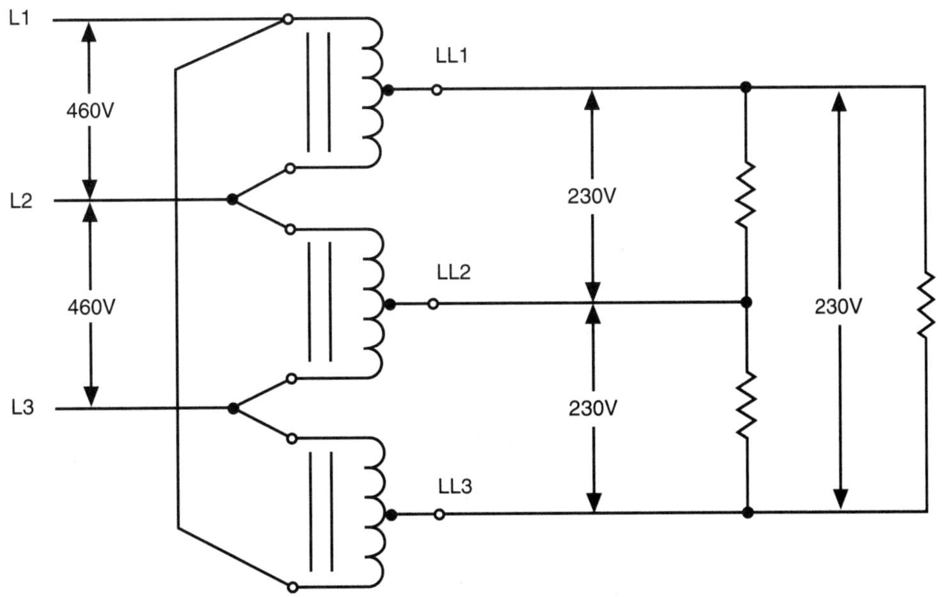

Fig. 16-9 A three-phase closed delta autotransformer bank supplying a closed-delta resistive heating bank.

tection for the connected load or load center and overload-overcurrent protection for the transformer.

The primary fuses or circuit breaker, set at a maximum of 250% of the primary FLC with the next lower standard size when the computation yields a nonstandard size or rating, provides the necessary short-circuit, ground-fault protection for the transformer.

Example 16–5. Three autotransformers with full-winding voltage ratings of 440 volts, and 50, 65, and 80 percent taps will be installed in a closed-delta configuration to step a 3-phase, 3-wire, 440-volt feeder circuit down to a 3-phase, 3-wire, 220-volt separately derived system to supply a process-heating control panel rated at 400 amps (change of voltage in Fig. 16-9). Determine:

1. The minimum kVA rating of each of the three autotransformers, if the NE Code-maximum setting of the secondary overcurrent-protection device is employed for the main circuit breaker in the process-heating control panel
2. The minimum wire size of the secondary conductors supplying the process-heating control panel, using THHN copper
3. The maximum rating of the primary disconnect fuses
4. The minimum wire size of the primary feeder conductors supplying the transformer bank, using THHN copper

NE Code-minimum transformer kVA determination:

Heating panel power rating computation:

$$(1.732 \times 220 \text{ volts} \times 400 \text{ amps}) \div 1{,}000 = 152.4 \text{ kVA}$$

Underrate computation based on the secondary overcurrent protection allowance:

Ref. 450-3(b)(2)

$$152.4 \text{ kVA} \div 1.25 = 121.9 \text{ kVA}$$

Individual transformer kVA rating computation (for closed-delta):

$$121.9 \text{ kVA} \div 3 = 40.6 \text{ kVA}$$

NE Code-minimum wire size of the secondary conductors:

Wire terminal temperature limitations: 75°C Ref. 110-14(c)(2), Excp. 1
Table-rated ampacity: 420 amps Ref. T310-16
Wire size: 600 kcmil THHN copper

NE Code-maximum rating determination of the primary fuses:

Full-load current computation:

$$(1{,}000 \times 121.9 \text{ kVA}) \div (1.732 \times 440 \text{ volts}) = 160 \text{ amps}$$

Overrate computation: Ref. 450-3(b)(2)

$$250\% \text{ of } 160 \text{ amps} = 400 \text{ amps}$$

Standard OCPD ratings: 400 amps Ref. 240-6(a)

NE Code-minimum wire size of the primary conductors:
Primary FLC computation:

$$(220 \text{ volts} \div 440 \text{ volts}) \times 400 \text{ amps} = 200 \text{ amps}$$

Wire terminal temperature limitations: 75°C Ref. 110-14(c)(2), Excp. 1
Table-rated ampacity: 200 amps Ref. T310-16
Wire size: No. 3/0 AWG THHN copper

If the computation of the primary full-load current in this example had not equaled to a standard rating of fuses, the next lower standard size would have been required by the lack of an exception to Sec. 450-3(b)(2).

Code challenge 16–5. Three autotransformers with full-winding voltage ratings of 600 volts, and 50, 65, and 80% taps will be installed in a closed-delta configuration to step a 3-phase, 3-wire, 600-volt feeder circuit down to a 3-phase, 3-wire, 480-volt separately derived system to supply a process-heating control panel rated at 225 amps. Determine:

1. The minimum kVA rating of each of the three autotransformers, if the NE Code-maximum setting of the secondary overcurrent-protection device is employed for the main circuit breaker in the process-heating control panel
2. The minimum wire size of the secondary conductors supplying the process-heating control panel, using THHN c-c aluminum
3. The maximum rating of the primary disconnect fuses
4. The minimum wire size of the primary feeder conductors supplying the transformer bank, using THHN c-c aluminum

NE Code-minimum transformer kVA determination:
Heating panel full-load kVA computation:

(1.732 × ____________ volts × ____________ amps) ÷ 1,000 = (A) ____________ kVA

Underrate computation based on the secondary overcurrent protection allowance:

Ref. (B) ________________

____________ kVA ÷ ____________ = (C) ____________ kVA

Individual transformer kVA rating computation:

_____________ kVA ÷ _____________ = (D) _____________ kVA

NE Code-minimum wire size of the secondary conductors:

Wire terminal temperature limitations: (E) _____________ °C

Ref. (F) ____________________

Table-rated ampacity: (G) _____________ amps Ref. (H) ____________________

Wire size: (I) _____________ THHN c-c aluminum

NE Code-maximum rating determination of the primary fuses:

Primary FLC computation:

(1,000 × ____________ kVA) ÷ (1.732 × ____________ volts) = (J) ____________ amps

Overrate computation: Ref. (K) ____________________

_____________ % of _____________ amps = (L) _____________ amps

Standard OCPD ratings: Ref. (M) ____________________

NLS: (N) _____________ amps Ref. (NLS) (O) ____________________

NE Code-minimum wire size of the primary conductors:

Primary FLC computation:

(__________ volts ÷ __________ volts) × __________ amps = (P) __________ amps

Wire terminal temperature limitations: (Q) _____________ °C

Ref. (R) ____________________

Table-rated ampacity: (S) _____________ amps Ref. (T) ____________________

Wire size: (U) _____________ THHN c-c aluminum

When two autotransformers are used in an open-delta configuration for 3-phase, 3-wire service distributions as shown in Figure 16-10, line 1 in the primary is connected to the H1 lead of one of the transformers. Line 2 is connected to the series connection of the H2 lead of this transformer to the H1 lead of the second transformer. Line 3 is connected to the H2 lead of the second transformer.

As with conventional transformers (primary and secondary windings isolated), the open-delta autotransformer configuration must be a "V–V" transformation. In other words, an open-delta secondary must have an open-delta primary. The draw back to this form of connection to the utility is the common connection of the center leg (line 2 or phase B) between the primary and secondary circuits, and the lack of reference for grounding lightning strikes or high-voltage crossovers (neutral stuff in a wye distribution).

The open-delta autotransformer configuration can overcome the requirement of the neutral stuff in the wye configuration by transmission as a corner-ground delta. In this type of 3-wire distribution, the center leg (line 2 or phase B) is redundantly grounded to

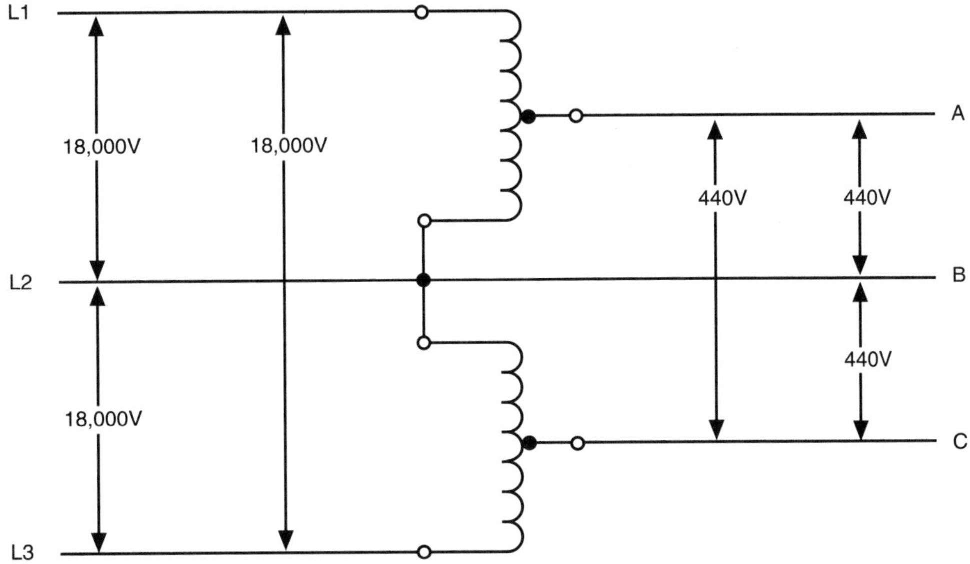

Fig. 16-10 Autotransformers used in a 3-phase open-delta configuration have a common connection between the primary and secondary circuits.

the earth through multiple grounding-electrode systems along the length of transmission and at the load transformer, which is the supply point for the service conductors in a building or other structure.

The power capability is limited by the use of the open delta. When three autotransformers are bank in closed delta, their kVA ratings add for a total bank rating. When two autotransformers are used in open-delta, their bank total is equal to only 86.6% of the sum of the two nameplate kVA ratings.

The overcurrent-protection requirements for autotransformer installations for the service supply (primary rated over 600 volts) are listed in Sec. 450-3(a) of the NE Code. If both the primary and secondary circuits are rated not greater than 600 volts, the overrate-protection requirments are listed in Sec. 450-3(b).

The rated-load ampacity of the circuit conductors is determined (again) by Sec. 240-3, based on the overrate percentages allowed for the secondary overload-overcurrent rating of the fuses or circuit breaker.

Example 16–6. For a 3-phase, 800-amp service panel supplied as a 3-wire delta from an open-delta autotransformer bank with a primary supply voltage of 18.0 kV and a secondary service/power distribution voltage of 440 volts (Fig. 16-10), determine:

1. The minimum kVA rating of the two autotransformers at an unsupervised location, if the NE Code-maximum setting of the secondary overcurrent protection device is employed for the main circuit breaker in the service power panel

2. The minimum wire size of the secondary conductors supplying the service panel, using THW copper paralleled three times
3. The maximum rating of the primary disconnect fuses

NE Code-minimum 3-phase power transformer kVA determination:
Service panel power rating computation:

$$(1.732 \times 440 \text{ volts} \times 800 \text{ amps}) \div 1{,}000 = 609.7 \text{ kVA}$$

Underrate computation based on the secondary overcurrent protection allowance:

Ref. T450-3(a)(1)

$$609.7 \text{ kVA} \div 1.25 = 487.8 \text{ kVA}$$

Overrate computation (for open-delta):

$$487.8 \text{ kVA} \div 0.866 = 563.2 \text{ kVA}$$

Individual transformer kVA rating computation (for open-delta):

$$563.2 \text{ kVA} \div 2 = 281.6 \text{ kVA}$$

NE Code-minimum wire size of the secondary conductors:
Parallel amperage computation: Ref. 310-4

$$800 \text{ amps} \div 3 = 266.7 \text{ amps}$$

Wire terminal temperature limitations: 75°C Ref. 110-14(c)(2), Excp. No. 1
Table-rated ampacity: 285 amps Ref. T310-16
Wire size: 300 kcmil THW copper

NE Code-maximum rating determination of the primary fuses:
Full-load current computation:

$$(563.2 \text{ kVA}) \div (1.732 \times 18.0 \text{ kV}) = 18.1 \text{ amps}$$

Overrate computation: Ref. T450-3(a)(1)

$$300\% \text{ of } 18.1 \text{ amps} = 54.2 \text{ amps}$$

Standard OCPD ratings: Ref. 240-6(a)
NHS: 60 amps Ref. (NHS) 450-3(a)(1), Excp. No. 1

The power rating of the transformer bank was not multiplied by 1,000 to compute the primary circuit full-load current in this example. Instead the power rating in kVA was divided by the product of the square root of 3 times the 18 kV.

Code challenge 16–6. For a 3-phase, 1,200-amp service panel supplied as a 3-wire delta from an open-delta autotransformer bank with a primary supply voltage of 18.0 kV and a secondary service/power distribution voltage of 575 volts (change of voltage in Fig. 16-10), determine:

1. The minimum kVA rating of the two autotransformer at an unsupervised location, if the NE Code-maximum setting of the secondary overcurrent-protection device is employed for the main circuit breaker in the service power panel
2. The minimum wire size of the secondary conductors supplying the service panel, using THW copper paralleled three times
3. The maximum rating of the primary disconnect fuses

NE Code-minimum 3-phase power transformer kVA determination:
Service panel power rating computation:

(1.732 × ____________ volts × ____________ amps) ÷ 1,000 = (A) ____________ kVA

Underrate computation based on the secondary overcurrent protection allowance:
Ref. (B) ____________________

____________ kVA ÷ ____________ = (C) ____________ kVA

Overrate computation (for open-delta):

____________ kVA ÷ ____________ = (D) ____________ kVA

Individual transformer kVA rating computation:

____________ kVA ÷ ____________ = (E) ____________ kVA

NE Code-minimum wire size of the secondary conductors:
Parallel amperage computation: Ref. (F) ____________________

(G) ____________ amps ÷ ____________ = ____________ amps

Wire terminal temperature limitations: (H) ____________ °C
Ref. (I) ____________________

Table-rated ampacity: (J) ____________ amps Ref. (K) ____________________
Wire size: (L) ____________ THW copper

NE Code-maximum rating determination of the primary fuses:
Full-load current computation:

(____________ kVA) ÷ (1.732 × ____________ kV) = (M) ____________ amps

Overrate computation: Ref. (N) ____________________

(O) ____________ % of ____________ amps = ____________ amps

Standard OCPD ratings: Ref. (P) ____________________

NHS: (Q) ____________ amps Ref. (NHS) (R) ____________________

16.6 BUCK OR BOOST AUTOTRANSFORMER POWER

A conventional or isolation transformer can be wired as a buck-or-boost autotransformer. Instead of tapping a single winding on an autotransformer, the low-voltage winding of an isolation transformer is wired in series with the connected load. In a "boost" configuration, the low-voltage secondary winding terminals are wired to aid the high-voltage primary winding. In a "buck" configuration, the terminals of the low-voltage secondary winding are wired to oppose the high-voltage primary winding.

A boost situation could arise when a 230-volt single-phase ac motor is installed on a 200-volt power supply. At the lower voltage rating, the motor would overheat, even under light-loading conditions.

As shown in Figure 16-11, an isolation transformer with a 200-volt primary and a 30-volt secondary can be wired to deliver 230 volts to the motor. The boost operation occurs when the 30-volt secondary winding is wired in additive to or the same polarity with

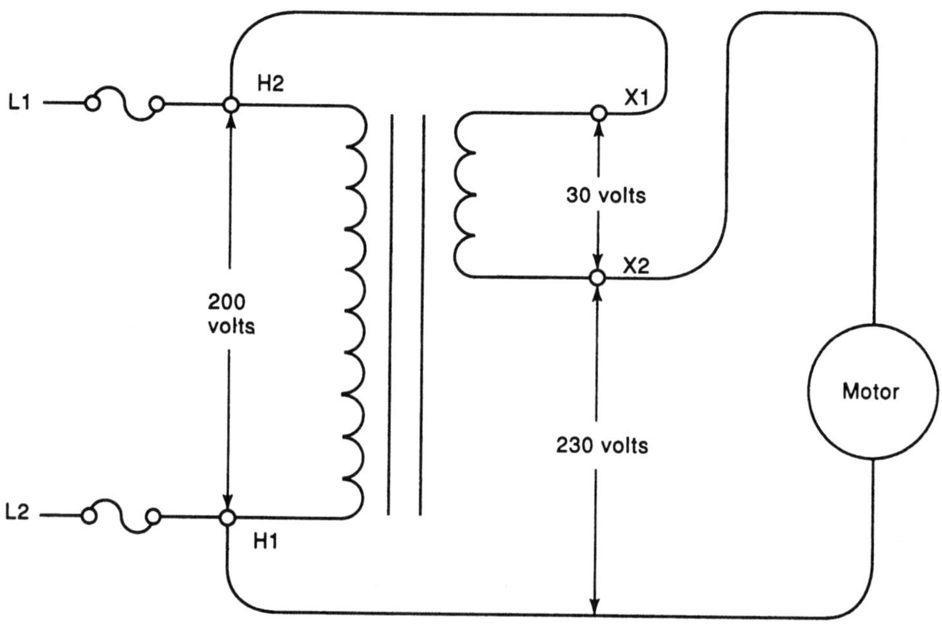

Fig. 16-11 An isolation transformer can be wired as an autotransformer in a *boost* configuration.

the 200-volt primary winding. The 30-volt secondary voltage is effectively added to the 200-volt primary voltage. The motor load receives 230 volts.

A typical application of this boost configuration is to correct for voltage drop on a distant motor load, such as a rural well pump, where the circuit conductors were not sized correctly to compensate for the voltage drop on initial installation, and replacement of the circuit conductors is cost prohibitive.

The circuit current in this boost application will increase the existing load current drawn from the branch-circuit supply, the extra current in the primary circuit, at the lower voltage, is converted to extra voltage at the original motor-load current in the secondary circuit.

Example 16–7. To compensate for excessive voltage drop, a conventional transformer will be wired in boost configuration to step-up the voltage supply at a remote pumping location. If the transformer has a higher-voltage winding rating of 200 volts and a lower-voltage winding rating of 30 volts (Fig. 16-11), and the single-phase ac, 230-volt pump motor rated at 1 hp is provided with running-overload protection set at 125% of the NE Code table-value motor FLC, determine:

1. The minimum VA rating of the boost-configured autotransformer based on the current rating of the running overload protection and the NE Code-maximum setting of the primary circuit overload-overcurrent device
2. The primary winding current, when the motor is operating at the allowed running-overload condition
3. The supply-circuit full-load current when the motor is operating at the allowed running-overload condition

NE Code-minimum autotransformer VA rating determination:

Table-value motor FLC: 8 amps Ref. T430-148

Maximum overrate of the primary-circuit overcurrent protection device:

125% Ref. 450-4

Power rating computation:

The transformer overload-protection overrate and the motor running overload protection overrate are equal, such that they cancel.

$$230 \text{ volts} \times 8 \text{ amps} = 1{,}840 \text{ VA}$$

Primary winding current computation in the maximum running-overload condition.

Running overload overrate computation: (given)

$$125\% \text{ of } 8 \text{ amps} = 10 \text{ amps}$$

Primary winding running-overload current computation:

$$(30 \text{ volts} \div 200 \text{ volts}) \times 10 \text{ amps} = 1.5 \text{ amps}$$

Supply-circuit full-load current in the maximum running-overload condition.

10 amps + 1.5 amps = 11.5 amps

In this example both the transformer, by Sec. 450-4(a), and the motor, by Sec. 430-32(a)(1), are allowed to operate in an overload up to 125% of the motor FLC, such that they cancel, the transformer power rating only has to equal the power rating of the motor at full load.

Code challenge 16–7. To compensate for excessive voltage drop, a conventional transformer will be wired in boost configuration to step up the voltage supply at a remote pumping location. If the transformer has a higher-voltage winding rating of 200 volts and a lower-voltage winding rating of 30 volts, and the single-phase ac, 230-volt pump motor rated at 3 hp is provided with running-overload protection set at 125% of the NE Code table-value motor FLC, determine:

1. The minimum VA rating of the boost-configured autotransformer based on the current rating of the running overload protection and the NE Code-maximum setting of the primary circuit overload-overcurrent device
2. The primary winding current, when the motor is operating at the allowed running-overload condition
3. The supply-circuit full-load current when the motor is operating at the allowed running-overload condition

NE Code-minimum autotransformer VA rating determination:
Table-value motor FLC: (A) ____________ amps Ref. (B) ____________________
Maximum overrate of the primary circuit overcurrent protection device:
Ref. (C) ____________ % Ref. (D) ____________________

Power rating computation: (the two overrate are equal, they cancel)

____________ volts × ____________ amps = (E) ____________ VA

Primary winding current computation in the maximum running-overload condition.
Running overload overrate computation:

____________ % of ____________ amps = (F) ____________ amps

Winding running overload current computation:

(__________ volts ÷ __________ volts)× __________ amps = (G) __________ amps

Supply-circuit full-load current in the maximum running-overload condition.

____________ amps + ____________ amps = (H) ____________ amps

A buck situation could arise when a 208-volt single-phase resistance load is installed on a 277-volt power source. At the higher voltage rating, the resistance load would overheat to the point of destruction or damage to the heating elements.

As shown in Figure 16-12, an isolation transformer with a 277-volt primary and a 69-volt secondary can be wired in a buck configuration to deliver 208 volts to the resistive load. The buck operation occurs when the secondary winding is wired in subtractive or reverse polarity to the primary winding. The 69-volt secondary voltage is effectively subtracted from the 277-volt primary voltage. The resistive load receives 208 volts.

Example 16-8. To prevent overvoltage from being applied to the fixed electric space heating bank in the central heating equipment of a building. A conventional transformer with a higher-voltage winding rating of 277 volts and a lower-voltage winding rating of 69 volts (Fig. 16-12) will be installed to buck the 277-volt supply voltage down to a 208-volt load voltage. If the connected load is a single-phase resistive heating bank rated at 10 kW, determine:

1. The minimum VA rating of the buck-configured autotransformer based on the full-load current rating of the heating bank and the NE Code-maximum setting of the primary-circuit overcurrent-protection device
2. The primary winding current when the heating bank is operating at the full-load condition
3. The supply-circuit full-load current when the heating bank is operating at the full-load condition

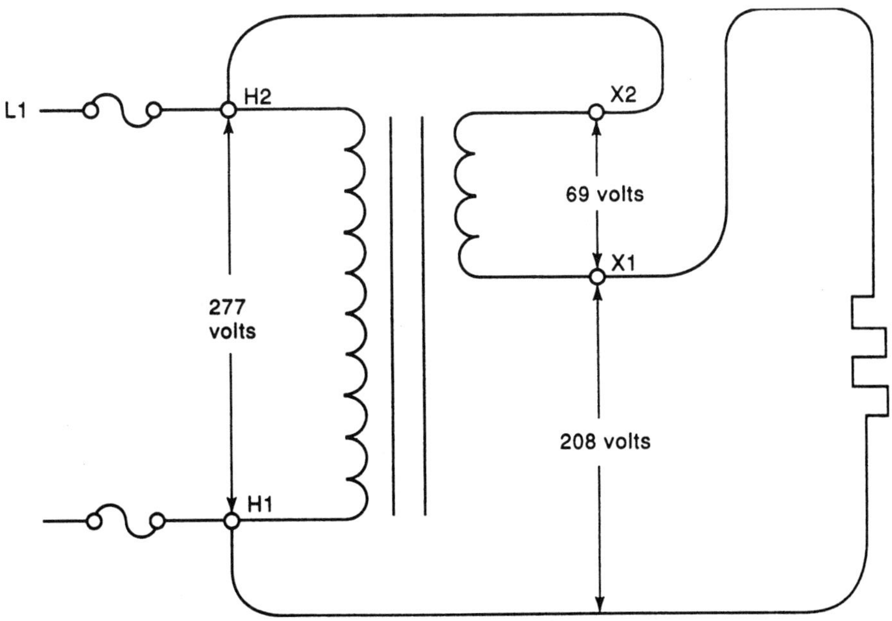

Fig. 16-12 An isolation transformer can be wired as an autotransformer in a *buck* configuration.

NE Code-minimum autotransformer VA rating determination:
Heating bank full-load current computation:

$$(10 \text{ kW} \times 1{,}000) \div 208 \text{ volts} = 48 \text{ amps}$$

Autotransformer power rating computation at full-load (secondary values):

$$69 \text{ volts} \times 48 \text{ amps} = 3{,}312 \text{ VA}$$

Underrate computation based on the primary overcurrent protection allowance:
Ref. 450-4

$$3{,}312 \text{ VA} \div 1.25 = 2{,}650 \text{ VA}$$

Primary winding current computation at full-load:

$$(69 \text{ volts} \div 277 \text{ volts}) \times 48 \text{ amps} = 12.0 \text{ amps}$$

Supply-circuit current computation at full-load:

$$48 \text{ amps} - 12 \text{ amps} = 36 \text{ amps}$$

The primary full-load current of 12 amps was subtracted from the secondary full-load current of 48 amps to equal a branch-circuit supply current of only 36 amps. The power ratings are equal: 277 volts × 36 amps = 208 volts × 48 amps.

Code challenge 16–8. To prevent overvoltage from being applied to the fixed electric space heating bank in the central heating equipment of a building. A conventional transformer with a higher-voltage winding rating of 277 volts and a lower-voltage winding rating of 37 volts will be installed to buck the 277-volt supply voltage down to a 240-volt load voltage. If the connected load is a single-phase resistive heating bank rated at 10 kW, determine:

1. The minimum VA rating of the buck-configured autotransformer based on the full-load current rating of the heating bank and the NE Code-maximum setting of the primary-circuit overcurrent-protection device
2. The primary winding current when the heating bank is operating at the full-load condition
3. The supply-circuit full-load current when the heating bank is operating at the full-load condition

NE Code-minimum autotransformer VA rating determination:
Heating bank full-load current computation:

(__________ kW × 1,000) ÷ __________ volts = (A) __________ amps

Autotransformer power rating computation at full-load (secondary values):

____________ volts × ____________ amps = (B) ____________ VA

Underrate computation based on the primary overcurrent protection allowance:
Ref. (C) ____________

____________ VA ÷ ____________ = (D) ____________ VA

Primary winding current computation at full-load:

(__________ volts ÷ __________ volts) × __________ amps = (E) __________ amps

Supply-circuit current computation at full-load:

____________ amps − ____________ amps = (F) ____________ amps

The fine print note (FPN) following Sec. 450-4(b) refers the student electrician to Sec. 210-9 for the NE Code-allowed uses of autotransformers. As a general rule, autotransformers cannot be used to supply individual branch circuits unless the common lead connection is to the grounded-circuit conductor in a two-wire branch circuit. Note exceptions.

There are other types of transformer system connections. The circuits shown and discussed in these two chapters are representative as to NE Code requirements of both transformer and circuit-conductor overcurrent protection.

REVIEW QUESTIONS

1. An alternate name for a 3-phase, 3-wire corner-ground delta is ____________ ____________ delta.
2. In the corner-ground delta wiring configuration the ____________ (A, B, or C) phase conductor is referenced or tied to earth ground through the service grounding-electrode system.
3. In the corner-ground delta wiring configuration, line voltage can be measured to earth ground from either of the other two phases that are operating ungrounded.

   True or False

4. In terms of fault-overcurrent protection, but not overload-overcurrent protection (must be separate), a fuse can be wired in series with the grounded phase conductor in the corner-ground delta distribution.

   True or False

5. In a high-leg delta transformation, a lighting and power transformer is center-tapped to supply all the single-phase loads, and across the full winding, its part of the 3-phase loads. Two smaller single-phase transformers are wired in a delta with the lighting and power transformer to convey the required 3-phase power.

   True or False

6. The lighting and power transformer of a high-leg delta normally has the same wiring configuration as the single-phase, 3-wire, 120/240-volt service supply for the single-family dwelling. The center-tap between the two 120-volt windings is grounded to serve as the single-phase neutral service conductor to the larger transformer, and the ground-fault return conductor for the 3-phase system (all three ungrounded supply conductors).

   True or False

7. The NE Code requires the center-tapped transformer to supply lines 1 and 3 (phases A and C) within the building or other structure. The third point of the closed-delta connection, which serves as the high-leg to ground, must be the center leg in the distribution, which is line 2 (or phase B).

   True or False

8. Because of the higher voltage reference to ground in a high-leg delta wiring configuration, the high-leg circuit conductor is required to be identified to the color ____________ .

9. The 3-phase, 4-wire, high-leg delta power distribution for a building or other structure is considered to be balanced when the single-phase, lower-voltage neutral load amperage is equal to the 3-phase, line-voltage load amperage.

   True or False

10. To compute the primary full-load current of the larger power and lighting transformer when a 3-phase power transformer bank is connected in a primary wye configuration and a secondary 3-phase, 4-wire, high-leg delta configuration, the 3-phase power formula must by solved using the total 3-phase power of the connected load and the primary line voltage. The single-phase power formula must by solved using the total single-phase power of the connected load and the primary phase voltage. The two currents are then added together.

    True or False

11. In an open-delta transformer configuration, the total power that can be transferred from the primary to the secondary circuit is equal to only ____________ % of the sum of the power ratings of the two transformers.

12. When two transformers are banked in open delta, the adjusted individual transformer power rating, when both circuits operate at 600 volts or less, can be underrated, actually overloaded, to ____________ % according to the rating or setting of the secondary OCPD.

13. When two transformers are banked in open delta, the adjusted individual transformer power rating, when either circuit operates at over 600 volts, can be underrated, actually overloaded, to ____________ % according to the rating or setting of the secondary OCPD when the transformer vault is installed in a supervised location.

14. When two transformers are banked in open delta, the adjusted individual transformer power rating, when either circuit operates at over 600 volts, can be underrated, actually overloaded, to ____________ % according to the rating or setting of the secondary OCPD when the transformer vault is installed in an unsupervised location.

15. Two single-phase transformers can be used to step down a 3-phase, 3-wire delta to a 3-phase, 4-wire high-leg delta in an open-delta configuration. Fortunately, the derate of the bank power rating for open delta does not apply.

    True or False

16. In a high-leg, V–V delta transformer configuration the open end (other end) of the smaller power transfomer has to be designated as the high leg.

    True or False

17. An ____________ consists of a single winding wrapped around a ferrous metal core that is tapped at different percentages of the winding length, such that one voltage can be applied to the full-winding and a second lower voltage can taken from a tap.

18. When an autotransformer is wired in a step-down configuration, the higher-voltage supply is connected across the ____________ (tapped or full) winding.

19. When an autotransformer is wired in a step-up configuration, the lower-voltage supply is connected across the ____________ (tapped or full) winding.

20. The drawback to using two autotransformers in an open-delta configuration for a 3-phase, 3-wire service distribution is the common connection of the center leg (line 2 or phase B) to the utility between the primary and secondary circuits, and the lack of reference for grounding lightning strikes or high-voltage crossovers.

    True or False

21. When a conventional transformer is wired as an autotransformer in a boost configuration, the secondary lower-voltage winding must be wired with additive polarity to increase the original line voltage being applied to the primary higher-voltage winding.

    True or False

22. When a conventional transformer is wired as an autotransformer in a buck configuration, the secondary lower-voltage winding must be wired with additive polarity to decrease the original line voltage being applied to the primary higher-voltage winding.

    True or False

23. When a conventional transformer is wired as an autotransformer in either a buck or boost configuration, the transformer overcurrent protection rating is limited to a maximum of ____________ % of the primary-circuit full-load current.

24. When a conventional transformer is wired as an autotransformer in either a buck or boost configuration, an overcurrent-protection device can be wired in series with the shunt winding, provided it is a fast-acting fuse (no time delay) and not an inverse-time circuit breaker.

    True or False

25. As long as the overcurrent-protection device is located in the primary supply circuit and the common-lead connection is with the grounded side of the circuit, an autotransformer wired in either a buck or boost configuration can be used to supply lower-voltage branch circuits in a building or other structure.

    True or False

REVIEW PROBLEMS

1. By evaluation of the intended loads in a building, the 3-phase loads have been assessed at 20 kVA. The single-phase loads have been assessed at 30 kVA. If the building service equipment will be supplied through a 3-phase, 4-wire, 240/120-volt, high-leg delta transformer bank from a 12.5 kV, 3-phase, 4-wire wye distribution, determine:
   a. The minimum kVA rating of each transformer within the bank if the NE Code-maximum setting of the secondary overcurrent-protection device is employed and the transformer bank is located in an unsupervised location

b. The maximum rating of the secondary disconnect fuses
c. The minimum wire size of the secondary conductors supplying the panel, using THHN c-c aluminum conductors (assume a 70% service neutral load)
d. The maximum rating of the primary disconnect fuses

2. Compute the bank power rating of two 150-kVA transformers installed in an open-delta configuration for step down of 2,300 volts to 440 volts.

3. A 3-phase, 3-wire, 480-volt, 150-amp, power panel will be supplied through a 3-phase, open-delta transformer bank from a 600-volt power distribution. Determine:
a. The minimum kVA rating of each transformer within the bank, if the NE Code-maximum setting of the secondary overcurrent-protection device is employed.
b. The minimum wire size of the secondary conductors supplying the panel, using XHHW-2 copper-clad aluminum conductors
c. The maximum rating of the primary disconnect fuses
d. The minimum wire size of the primary conductors supplying the transformer, using XHHW-2 copper-clad aluminum conductors

4. By evaluation of the intended loads in a building, the 3-phase loads have been assessed at 28 kVA. The single-phase loads have been assessed at 42 KVA. If the building service equipment will be supplied through a 3-phase, 4-wire, 240/120-volt, high-leg, open-delta transformer bank from a 13.2-kV, 3-phase, 3-wire delta distribution, determine:
a. The minimum kVA rating of each transformer within the bank if the NE Code-maximum setting of the secondary overcurrent-protection device is employed
b. The maximum rating of the secondary disconnect fuses
c. The minimum wire size of the secondary conductors supplying the panel, using THHN copper conductors (assume a 70% service neutral load)
d. The maximum rating of the primary disconnect fuses

5. Three autotransformers with full-winding voltage ratings of 480 volts and 50, 65, and 80% taps will be installed in a closed-delta configuration to step a 3-phase, 3-wire, 600-volt feeder circuit down to a 3-phase, 3-wire, 240-volt separately derived system to supply a process-heating control panel rated at 150 amps. Determine:
a. The minimum kVA rating of each of the three autotransformers, if the NE Code-maximum setting of the secondary overcurrent-protection device is employed for the main circuit breaker in the process-heating control panel
b. The minimum wire size of the secondary conductors supplying the process-heating control panel, using THHN aluminum
c. The maximum rating of the primary disconnect fuses
d. The minimum wire size of the primary feeder conductors supplying the transformer bank, using THHN aluminum

6. For a 3-phase, 600-amp service panel supplied as a 3-wire delta from an open-delta autotransformer bank with a primary supply voltage of 13.2 kV and a secondary service/power distribution voltage of 600 volts, determine:
a. The minimum kVA rating of the two autotransformers at an unsupervised location, if the NE Code-maximum setting of the secondary overcurrent-protection device is employed for the main circuit breaker in the service power panel
b. The minimum wire size of the secondary conductors supplying the service panel, using THW aluminum paralleled three times
c. The maximum rating of the primary disconnect fuses

7. To compensate for excessive voltage drop, a conventional transformer will be wired in boost configuration to step up the voltage supply at a remote pumping location. If the transformer has a higher-voltage winding rating of 175 volts and a lower-voltage winding rating of 25 volts, and the single-phase ac 200-volt pump motor rated at 1 hp is provided with running-overload protection set at 125% of the NE Code table-value motor FLC, determine:

a. The minimum VA rating of the boost-configured autotransformer based on the current rating of the running overload protection and the NE Code-maximum setting of the primary circuit overload-overcurrent device
b. The primary winding current, when the motor is operating at the allowed running-overload condition
c. The supply-circuit full-load current when the motor is operating at the allowed running-overload condition

8. To prevent overvoltage from being applied to the fixed electric space heating bank in the central heating equipment of a building. A conventional transformer with a higher-voltage winding rating of 575 volts and a lower-voltage winding rating of 95 volts will be installed to buck the 575-volt supply voltage down to a 480-volt load voltage. If the connected load is a single-phase resistive heating bank rated at 20 kW, determine:

a. The minimum VA rating of the buck-configured autotransformer based on the full-load current rating of the heating bank and the NE Code-maximum setting of the primary-circuit overcurrent-protection device
b. The primary winding current when the heating bank is operating at the full-load condition
c. The supply-circuit full-load current when the heating bank is operating at the full-load condition

Chapter 17

Feeder, Panel, and Service Load Computations

Once the size (rating) and number of both general-purpose (multioutlet) and specific-purpose (dedicated or individual) branch circuits have been determined for a building or structure, or a specific location within the building or structure, the feeder and service conductors must be sized (rated) and protected.

The feeder conductors may supply a motor control center, a lighting and appliance panel, a power panel, or some other type of load center. If the feeder supplies a motor control center, the conductors have to be sized in accordance with Sec. 430-24, and protected against overcurrent (short-circuit, ground-fault protection) in accordance with Sec. 430-62 or 430-63.

17.1 FEEDER OR SERVICE CONDUCTORS

Feeder or service load considerations, conductor ampacities, overcurrent protection, and installation requirements are governed by Art. 215 and Parts B, C, and D of Art. 220.

In general (Sec. 240-3), the rated-ampacity of feeder or service conductors must be greater than or equal to first the full-load current rating of the load served as determined by Sec. 220-10(a), and then greater than or equal to the rating of the feeder fuses or setting of the circuit breaker as determined by Sec. 220-10(b).

Normally, the load is computed in accordance with Part B of Art. 220 (standard method of load computation), and the rating of the feeder or service OCPD (fuse or circuit breaker) is selected from the standard listings in Sec. 240-6(a). If the computed load current does not correspond to a standard size, the next higher standard size of fuse or circuit breaker is selected.

The minimum wire size of the feeder or service conductors is determined according to the rating of the circuit OCPD and T310-16 [or other ampacity tables referenced by Sec. 310-15(a)]. The wire size selected should have a table-rated ampacity greater than or equal to the rating of the OCPD.

If prevailing conditions, such as higher ambient temperatures or more than three conductors in a single raceway exist for the feeder or service conductors, then the table-rated ampacity of the conductors must be derated for the existing prevailing conditions; Or the feeder or service OCPD rating can be overrated by the derating percentages of the prevailing conditions and the conductors sized for this larger current value. The rated-ampacity of the feeder or service conductors is also governed by the wire terminal temperature limitations imposed by Sec. 110-14(c).

If the length of the feeder or service conductors is in excess of 50 feet, the ampacity rating of the conductors should be checked for voltage drop. The feeder conductors can be sized for a maximum of 2% voltage drop [Sec. 215-2(b), FPN No. 2] according to the rating of the OCPD. The resulting wire size can be compared to the required wire size as determined by the prevailing conditions (if any exist). The larger of the two in comparison [Sec. 310-15(c)] would be the NE Code-minimum wire size of the feeder or service conductors.

Chapter 2 of this text discusses the rated ampacity of circuit conductors, whether branch-circuit, feeder-circuit, or in-coming service conductors, and how an increase in wire size is used to compensate for the loss of heat dissipation in all of these prevailing conditions.

Overcurrent protection for all conductors and supplied equipment (discussed in Chapter 7) must be sized (setting or rating) and installed according to the requirements of Art. 240 and Sec. 220-10(b). For the minimum rating or setting of the feeder or service overcurrent-protection device, the lighting loads (or other loads that are determined to be continuous) must be overrated to 125% of their full-load current rating. Loads determined to be noncontinuous are taken "as is" at 100% of their full-load current rating. The two computed currents are totaled to size the feeder or service overcurrent-protection device.

In the 1993 Edition of the NE Code, the wording of Sec. 220-10(b) would allow the rated ampacity of the feeder or service conductors to be computed according to the full-load current served without compensation of continuous load, due to the definition of ampacity in Art. 100. Conductor rated ampacity is still defined as the measure of current a conductor can convey continuously without overheating in the 1996 Edition of the NE Code, but the wording of Sec. 220-10(b) (use of "computed load") has been changed to require the 125% overrate for continuous load to be applied to both the OCPD rating and the rated ampacity of the circuit conductors.

There are some minor variations still allowed. Following the general rule of Sec. 240-3 stated in Sec. 240-3(c), the table-rated ampacity of the feeder or service conductors has to be greater than both the computed full-load current and the rating or setting of the feeder overcurrent-protection device rating. If the rating or setting of the feeder overcurrent-protection device selected is not in excess of 800 amps, Sec. 240-3(b) will allow the conductors to have a rated ampacity at least equal to the computed feeder or service full-load current and be protected against overcurrent by the next higher standard size of overcurrent-protection device.

As the feeder full-load current increases in value, the spacing between standard ratings or settings of overcurrent-protection devices listed in Sec. 240-6(a) increasingly widens. When the computed full-load current of the feeder or service exceeds 800 amps, Sec. 240-3(c) requires the rated ampacity of the circuit conductors, determined by wire size, to be greater than or equal to the rating of the feeder or service OCPD rating or setting.

17.2 PANELS AND OTHER LOAD CENTERS

Installation and use of panelboards (Fig. 17-1) and switchboards (Fig. 17-2), which are the more common load centers, are governed by the requirements listed in Art. 384 of the NE Code.

As defined in Art. 100, a switchboard is a large nonconductive board on which can be mounted on either front or back, or both, any combination of switches, overcurrent-protection devices, buses or conductor terminals, and a various array of monitoring or metering devices. Since switchboards are generally designed to be accessible from the rear as well as from the front, they are not intended to be installed in cabinets.

Modern switchboards (switchgear) are totally enclosed, which offers a large degree

Fig. 17-1 Panelboards are often surface-mounted power panels in an industrial setting.

Fig. 17-2 Switchboards or *switchgear* are located in a common room when large amounts of power are required in a manufacturing structure.

of protection against the contact of live parts and communication of fire to adjacent combustible materials. The enclosure has to be equipped with either doors or removable panels that allow access from both the front and the back of the switchboard. Switchboard busbars are required to be so arranged [Sec. 384-3(b)] so as to avoid inductive heating of either themselves or surrounding metals. Service busbars are usually isolated from the remainder of the switchboard by effective insulating barriers.

17.3 PANELS

Switchboards, especially at the branch-circuit supply level, have been replaced with panelboards in most installations. A panelboard is defined in Art. 100 as a busbar assembly with automatic overcurrent-protection devices installed in either cabinets or cutout boxes. The complete assembly is normally referred to as a "panel."

Cabinets, by definition (Art. 100), can be either flush or surface mounted, and have a removable front cover or trim to which a door or doors are hinged (Fig. 17-3). Cutout boxes, by definition (Art. 100), can only be surface mounted, and are manufactured with mounted hinged doors that do not extend beyond the outer dimensions of the box (Fig. 17-4).

Panels have live parts (busbars, wiring, etc.) exposed with the front trim removed. This allows for the installation and wiring of switches and/or overcurrent-protection de-

Fig. 17-3 Panelboards mounted in cabinets are normally flush (recessed) mounted.

vices. With the front trim or cover in place, the panel becomes "dead front." NE Code rules and regulations dealing specifically with panels are listed in Part C of Art. 384.

The busbars in a panel are usually installed vertically and parallel with the phase rotation (Fig 17-5). Phase rotation is read left to right as A–B for single-phase (3-wire, phase-to-phase with a common neutral) or 2-phase panels, and A–B–C for 3-phase panels. The panelboards will accomodate two vertical columns or rows of circuit breakers or fused switches installed side by side. The same slot in either column will be on the same phase. Phase rotation in either column, as required by Sec. 384-3(f), is read from top to bottom as A–B for single-phase panels (Fig. 17-6), A–B–C–D for 2-phase panels, and A–B–C for 3-phase panels.

A separate busbar for the neutral (or other grounded-circuit current-carrying) conductor is usually mounted in the lower section of the panel enclosure. Some newer design panels will have the neutral busbar mounted directly under the load side of the snap-in circuit breakers. This type of design aids in reducing the inductive heating effect of single conductors conveying ac current. It also aids in the reduction of nuisance tripping of ground-fault (GFCI) circuit breakers. The "neutral busbar" is insulated from the cabinet enclosure with the same type of insulating mating used for the phase busbars.

If the panel is installed and used as service equipment (Fig. 17-7), Sec. 384-3(c) requires the panel enclosure to be bonded (grounded) to the neutral busbar. Reference is made to Sec. 250-79(d), which requires the bonding jumper to be sized in accordance with T250-94. Table 250-94 is also used to size the grounding-electrode conductor for the

Fig. 17-4 Panelboards mounted in cut-out boxes can only be surface mounted.

in-coming service. The table is based on the wire size and conductor material of the service-entrance conductors.

Panels rated at 200 amps or less are normally furnished with a grounding screw suspended in the neutral busbar that is tagged with instructions to thread the screw into the panel enclosure if the panel is installed as service equipment, and to remove and discard the screw if the panel is not used as service equipment.

If the panel is used as service equipment, the equipment-grounding conductors and the grounded circuit or neutral conductors on the load side of the service-disconnecting means can be connected common to the neutral (grounded) busbar. Section 250-23(a) requires the neutral bus to serve as the central "tie point" for the service grounding-electrode system conductors, the equipment-grounding conductors on both the supply and load sides of the service-disconnecting means, and both the supply and load neutral or other grounded circuit (current-carrying) conductors within the distribution.

If the panel is supplied by a transformer as a separately derived system, the panel overcurrent protection must be located in the secondary circuit of the transformer. The panel, as required by Sec. 250-5(d), would have to be grounded with an established grounding-electrode system in accordance with Sec. 250-26. The panel enclosure, in terms of grounding, would have to be treated as service equipment, although in reference to Sec. 230-70, it does not have to be rated as service equipment.

Both the panel and the feeder ungrounded (phase) conductors of this separately derived system are required by Sec. 384-13 to have a rated ampacity at least equal to the

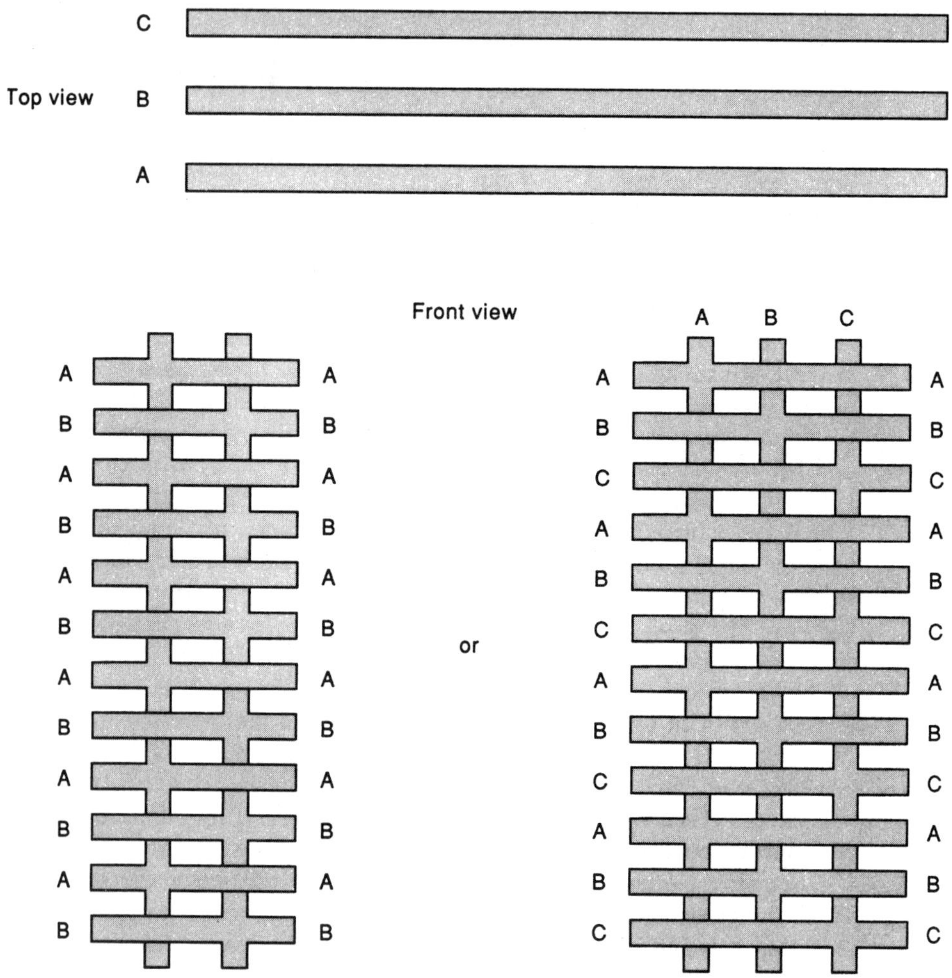

Fig. 17-5 Phase rotation or arrangement of busbars in panels, switchboards, or control centers is governed by Sec. 384-3(f).

full-load current requirements of the panel as computed in accordance with Art. 220. The feeder neutral conductor supplying the panel with main disconnecting means can be either full-size and equivalent to the size of the ungrounded phase or circuit conductors, or the grounded feeder neutral conductor can be reduced in wire size. The minimum wire size is determined according to where the equipment-grounding system and the grounding-electrode system will be interconnected (common tie point) with the feeder neutral conductor.

If the common tie point to the neutral conductor is located at the transformer enclosure, the feeder neutral conductor extended to the panel busbar must be isolated and insulated from equipment grounding [Sec. 250-23(a)]. The wire size can be reduced according to the load-return requirements of Sec. 220-22. The rated ampacity, determined according to wire size, must be at least equal to the maximum unbalanced load current as

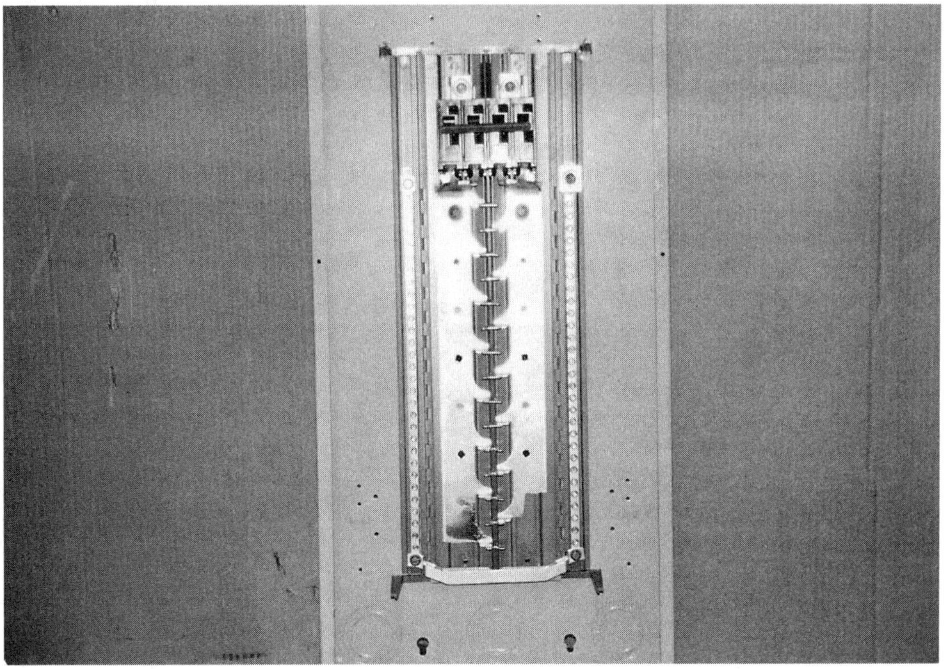

Fig. 17-6 Because of the busbar configuration, either the left or right slot on any given row is the same phase in a single-phase, 3-wire panel.

computed by Part B of Art. 220. The separate equipment-grounding conductor must be sized for the ground-fault return requirements of Sec. 250-79(e) by the rating of the feeder/panel OCPD and T250-95.

If the common tie point will be to the grounded busbar in the panel enclosure, the separate equipment-grounding conductor is omitted. The reduced feeder neutral conductor serves as both the load-return conductor sized by Sec. 220-22 and the ground-fault return conductor for the separately derived system as required by Sec. 250-26(a).

In parallel to Sec. 250-23(b) for the service neutral conductor, the feeder neutral conductor of this separately derived system is being used as the required bonding jumper or fault-return path between either the grounding-electrode system or the equipment-grounding system and the transformer windings.

For ground-fault return requirements, Sec. 250-26(a) will not allow the reduced separately derived system grounded feeder-neutral conductor to be smaller that the grounded-electrode conductor serving the same system. In accordance with Sec. 250-79(d) for the bonding jumper, and Secs. 250-26(b) and 250-81 for the grounding-electrode conductor, both are sized according to the wire size of the ungrounded phase conductors for the separately derived system by T250-94.

If the wire size of the ungrounded feeder conductors of the separately derived system exceeds the table cutoff of 1,100 kcmil copper or 1,750 kcmil aluminum (or c-c aluminum), the grounded feeder neutral conductor must be sized at a minimum of 12.5% of the largest conductor circular-mil area connected to any one phase winding of the trans-

Fig. 17-7 Panelboards mounted in cut-out boxes which serve as service equipment must be as service equipment. This panel would also have to be rated *raintight.*

former. If the required wire size for load-return and ground-fault return are not equal, Sec. 310-15(c) establishes the largest in comparison as NE Code minimum.

If the panel of the separately derived system is installed as a subfed panel from a fused-disconnect or circuit breaker enclosure, the common tie point of the grounding and grounded systems should be established with the grounded neutral conductor in the disconnect or CB enclosure in the same manner as just described for the panel enclosure.

As with the common tie point located in the transformer enclosure, the panel would be supplied with an isolated and insulated grounded feeder neutral conductor for load return, and with a separate equipment-grounding conductor for fault return.

In parallel to the requirements of Sec. 250-23(a) for service neutral conductors, load-side grounded neutral or other circuit conductors cannot be used for equipment grounding of the noncurrent-carrying metal parts that support the power distribution, for reference grounding of conductive parts within the building or other structure, or for connection to additional grounding electrodes located remote from the service equipment or the source of a separately derived system. Grounding and bonding are discussed in detail in Chapter 9 of this text.

Panels can be purchased or ordered through an electrical wholesale house at just about any desired voltage and/or current rating. The more common stock or "off the shelf" voltage ratings are 250, 480, and 600 volts. The more common current ratings are 60, 100, 125, 150, 200, 225, 300, 400, 600, and 800 amps.

Panels are normally identified by the manufacturer by the number of circuits they can contain. For instance, a 24-circuit panel is often used as a subpanel. This term of "circuits" can be misleading considering a three-pole circuit breaker supplies a "single" circuit. Section 384-15 lists the maximum number of overcurrent-protection devices allowed in a panel. The last paragraph in this section describes a two-pole circuit breaker as two overcurrent (protection) devices, and a three-pole circuit breaker as three overcurrent (protection) devices. If the NE Code term "overcurrent devices" is substituted for the term "circuits," the rating of the panel is easier to understand.

A 24-circuit panel offers 24 "slots" that can house (contain) 24 single-pole circuit breakers, or 12 double-pole breakers, or eight 3-pole (3-phase) circuit breakers. If the breakers are mismatched (combinations of single-pole, 2-pole, and 3-pole breakers), their combination of poles or slots cannot exceed 24.

Panels are normally labeled as "power panels." If a panel is to be labeled as a "lighting and appliance panel," Sec. 384-14 requires at least 10% of the contained branch-circuit overcurrent-protection devices for which a neutral or other grounded-circuit conductor is provided (two-wire branch circuits), to be rated at 30 amps or less.

The design of a lighting and appliance branch-circuit panel is restricted by Sec. 384-15 to a maximum of forty-two slots for overcurrent-protection devices not including the panel main. Power panels would not fall under this requirement.

Section 384-16(a) requires a lighting and appliance panel to have main overcurrent protection consisting of not more than two circuit breakers or two sets of main fuses on the supply side of its busbars. By Exception No. 1, the panel can be installed as "main lug only" if the supply feeder's overcurrent-protection device rating or setting does not exceed the rating of the panel.

17.4 CONDUCTOR IDENTIFICATION

Conductors are color identified in accordance with Sec. 310-12. The neutral or other intentionally grounded circuit (current-carrying) conductors must also meet the identifcation requirements of Sec. 200-6. The equipment-grounding conductors must also meet the identification requirements of Sec. 250-57.

Section 310-12(c) will allow the ungrounded (current-carrying) branch-circuit, feeder, or service conductors to be any color except white, natural gray, or green. Insulated conductors wire size No. 6 AWG and smaller are normally colored along their entire length at the point of manufacture. Insulated conductors wire size No. 4 AWG and larger are normally manufactured only in black. The larger conductors are allowed by Sec. 310-12(c) to be reidentified to other colors (if necessary) at the point of installation by paint, tape, or gum labels (single stripe or stripes or other regularly spaced series-identical marks), to any color other than white, natural gray, or green.

Section 310-12(c) will not allow the stripes of phase color tape to cover, hide, or obscure the manufacturer's marking installed on the conductor's insulation as required by Sec. 310-11.

Although not specifically required by the NE Code unless both voltages are present in a single building, structure, or occupancy [Sec. 210-4(d) and FPN], 3-phase distribution systems are normally color coded by phase according to the voltage rating. For 480-

volt systems with phase rotation A–B–C, the phase conductors are commonly identified as "B–O–Y," or brown–orange–yellow, respectively. For 208- or 240-volt 3-phase systems with phase rotation A–B–C, the phase conductors are commonly identified as "B–R–B," or black–red–blue, respectively.

When panels are supplied with a 3-phase, 4-wire, high-leg delta (also referred to as a red-leg delta), the B-phase (center phase) conductor is required by Secs. 384-3 (e) and (f) to be the high-leg or the phase with the higher voltage to ground. The high-leg conductor must either be the color orange along its entire length, or if a color other than orange (circuit conductors No. 4 AWG or larger), reidentified to orange with paint, tape, or tags at all points of splice or termination.

In reference to the last sentence of Sec. 215-8, if the delta feeder or branch circuit is 3-phase, 3-wire circuit and the neutral or other grounded-circuit conductor is not present, the high-leg or B phase is not NE Code-required to be colored orange. In this installation, it would be a judgment call on the part of the electrician.

If the 3-phase, 4-wire, 240/120-volt high-leg delta is derived from a 3-phase, 4-wire, 480/277-volt power distribution (parent voltage), the orange and red colors can be interchanged between the two color schemes just described. The 480-volt distribution can be color coded brown, red, and yellow. The high-leg, 240-volt distribution can be color coded black, orange, and blue to meet NE Code requirements.

A high-leg delta feeder or service is normally derived from a 3-phase transformer bank that has the secondary winding extended between derived phases A and C center tapped and grounded.

When separately derived systems are employed, Sec. 210-5(a) requires different voltage neutrals or other grounded-circuit conductors to be color identified. One system grounded-circuit conductor can be identified as either white or natural gray (light gray). The other voltage system grounded-circuit conductor must be white with a colored stripe (other than green) along its length.

Section 210-4(d) requires a plaque or other directory to be posted at all branch-circuit panels, motor control centers, and other load centers when dual-voltage systems are used in the same building or other structure.

Neutrals or other grounded-circuit conductors are required by Secs. 200-6 and 310-12(a) to have an outer finish of either white or natural gray along their entire length. Section 200-6(b) will allow grounded circuit conductors wire size No. 4 AWG and larger to be reidentified at the time of installation at all terminations.

Grounding-electrode conductors, equipment-grounding conductors, or bonding jumpers, which are intended to conduct only ground-fault overcurrent, are allowed by Sec. 250-57(b) to be either bare, covered, or insulated. When covered or insulated, grounding conductors are required to be green in color, or green with one or more yellow stripes. Section 310-12(b), Excp. No. 1 will allow grounding conductors wire size No. 4 AWG and larger to be permanently reidentified at all points of termination and accessibility at the time of installation.

"At all points of accessiblility" requires equipment-grounding conductors or bonding jumpers to be permanently identified at all junction and pull points. This differs from the ruling of Sec. 200-6(b), which requires reidentification of the larger grounded circuit or neutral conductors only at the point of termination. Some local authorities (Ref. 90-4) may require the neutral to be reidentified at all points of accessibility as well.

17.5 BRANCH-CIRCUIT LOAD REQUIREMENTS

A single panel can serve as a load center for several different types of loads. The panel can supply another panel, and the fused switch or circuit breaker installed on the load side of the first panel can serve as the feeder overcurrent protection for the other subfed panel (subpanel). The panel can supply a feeder, which in turn supplies a motor control center or some other load center. More commonly, the loads supplied by a panel are fed by individual branch circuits.

When branch circuits in a panel supply continuous loads, Sec. 384-16(c) restricts the full-load rating of the continuous-load current to a maximum of 80% of the rating of the branch circuit OCPD; or the rating branch-circuit overcurrent-protection device is required to be overrated to 125% of the total continuous full-load current it supplies. Either statement accomplishes the same end.

The minimum rated ampacity of the branch-circuit conductors supplying continuous loads is required by Secs. 220-3(b) and 240-3 to be greater than or equal to the rating of the fuse or circuit breaker protecting the circuit as determined by Sec. 240-6(a).

17.6 PANEL LOAD REQUIREMENTS

Section 384-13 requires the panel rating to be greater than or equal to the minimum feeder capacity (ampacity rating) as determined by Art. 220. To satisfy the requirements of Sec. 220-10(b) under Part B of Art. 220, the panel and its supply feeder conductors must have a capacity greater than or equal to the rating of the feeder overcurrent-protection device protecting the panel.

As discussed earlier for panel-supplied branch circuits, the computed rating of continuous-load current supplied by the main overcurrent protection of a panel, the overcurrent protection of a feeder supplying a panel, or the panel overcurrent protection for a feeder to another panel or load center is restricted by Sec. 384-16(c) to 80% of the rating of the OCPD. Section 384-16(c) parallels Sec. 220-3(a) for branch-circuit OCPD ratings, and Sec. 220-10(b) for feeder or service OCPD ratings.

When computing the rated load, all three sections require the continuous-load current to be overrated to 125% and added to the noncontinuous loads. If the computed load current does not correspond to a standard rating of fuses or setting of a circuit breaker as listed in Sec. 240-6(a), the next higher standard size must be used.

17.7 NONDWELLING LOAD CONSIDERATIONS

Part B of Art. 220 is referred to as the "standard method" of load computation and can be used for any type of load calculation either at the feeder, panel, or service level.

The rating or maximum load of the feeder, panel, or service is calculated under two different categories—either dwelling or nondwelling loads. Dwelling loads have special considerations because of the diversity of loading within a dwelling.

For nondwelling feeder, panel, or service loads, the connected branch-circuit or other feeder/panel loads fall under four basic categories:

1. Lighting loads
2. Convenience (general-purpose) receptacle loads
3. Specific appliance or other utilization equipment loads
4. Motor loads

As a rule, no derating demand factors are applied to industrial or commercial loads because of their continuous load nature.

17.8 LIGHTING LOADS

Lighting loads for commercial or industrial installations must be further divided into the five different types of lighting loads:

1. General-lighting (overhead lighting) loads
2. Additional lighting loads
3. Show-window lighting loads
4. Track-lighting loads
5. Outdoor signs and outline lighting loads

General-lighting loads represent overhead lighting provided for floor area in a structure. The required intensity of the lighting is normally determined by the work or other activity that will be carried on in that particular floor area. For example, the lighting in a storage area such as a warehouse would not have to be as bright as the lighting in an office area.

General-lighting loads do not include lighting used for athletic fields, parking lots, loading docks, or lighting used for walkways, passageways, ramps, or entrances located outside the building or structure. The NE Code-minimum load for lighting of this nature is computed at nameplate rating of the fixture or equipment without any NE Code demand derates. These lighting loads are computed in conjunction with the general-lighting load of the building or structure, and are referred to "additional lighting" loads.

General-lighting loads do not include "specialty lighting" used for show window or showcase lighting. Show window lighting loads must meet both the NE Code-minimum requirements of Sec. 220-12 and the total nameplate rating of the lighting fixtures or equipment [Sec. 220-10(a)].

"Show window" indicates the lighting fixtures or equipment are installed in a special opening of the building or structure walls. "Showcase" indicates the lighting fixtures or equipment are installed in special cases or counters free-standing within the building or structure. Show-window or showcase lighting is computed separate and in conjunction with the general-lighting load in a building or other structure.

General-lighting loads do not include track lighting. Track lighting is identified as "Lighting Track" by the NE Code, and is governed by Part R of Art. 410. Track-lighting loads must meet both the NE Code-minimum requirements of Sec. 410-102 and the total nameplate rating of the lighting fixtures (when known). Track lighting is computed separately and in conjunction with the general-lighting load in a building or other structure, unless the track lighting is to serve as the general-lighting and not specialty or accent lighting.

General-lighting loads do not include outdoor signs or building outline lighting that is used for decoration (accent) of the building and grounds. Article 600 governs the installation and load ratings of electric signs and outline lighting. Signs and outline lighting loads must meet both the NE Code-minimum requirements of Sec. 600-5 and the total nameplate rating of the sign and outline lighting (when known). Signs and outline lighting loads are computed in addition to the general-lighting load requirements of S & T 220-3(b).

17.9 GENERAL-LIGHTING LOADS

General-lighting loads have two NE Code ratings: The actual full-load current or volt-amp rating of the lighting fixtures as determined by Secs. 220-10(a) and (b), and the NE Code-minimum volt-amp load for general lighting as determined by Sec. and Table 220-3(b) and Sec. 220-10(b). The larger of the two loads in comparison must be used as the minimum general-lighting load.

Table 220-3(b) lists the minimum "volt-amps per square foot" (VA/sq. ft.) load allowance for the general-lighting load according to the type of occupancy, structure, or area served by a feeder, panel, or service. The volt-amps per square foot unit is multiplied times the overall floor area to determine the general-lighting volt-amp load.

As illustrated in Figure 17-8, Sec. 220-3(b) requires the general-lighting volt-amp load to be based on the outer floor dimensions of the structure or area served. This means the linear measurements along the outside walls of a building or other structure.

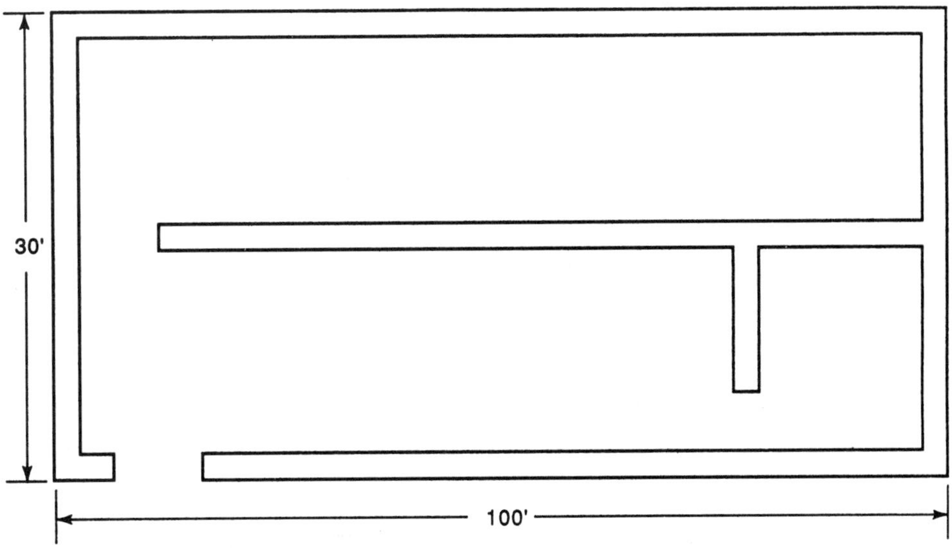

Fig. 17-8 The outer floor dimensions must be used in the general-lighting load computation of a building or other structure.

As illustrated in Figure 17-9, the overall area must be computed for each floor of a multistory building. Commercial or industrial store buildings may be two or three stories high. A commercial office building may be twenty or thirty stories high. The overall area must be computed by the outer floor dimensions of each floor level.

In many commercial store buildings the first floor may have more overall floor area than the other story floors. Different floor levels may have different types of occupancies. A two-story furniture store may use the first floor as a showroom of furniture displays, and the second floor as a storage area. The general-lighting load of the first floor would be computed as a store at 3 VA/sq. ft. The general-lighting load of the second floor would be computed as either a warehouse or just a storage space at ¼ VA/sq. ft.

The first floor of a three-story building on a school campus may be used as an auditorium. The upper two stories may be used as office space for faculty or staff. The general-lighting load of the first floor would be computed as an auditorium at 1 VA/sq. ft. The general-lighting load of the two upper floors would be computed as office area at 3.5 VA/sq.ft.

A single-story structure that uses part of the floor area for storage, and part of the floor area for showroom space would be computed in the same manner. The floor area used for storage would be computed as a separate floor area (by the outer dimensions) removed from the total floor area of the building. The storage area general-lighting load would be computed separately from the remaining showroom area general-lighting load. The two volt-amp loads would then be added together to determine the total general-lighting volt-amp load.

The same procedure would be used if a single-story structure or a single floor in a mul-

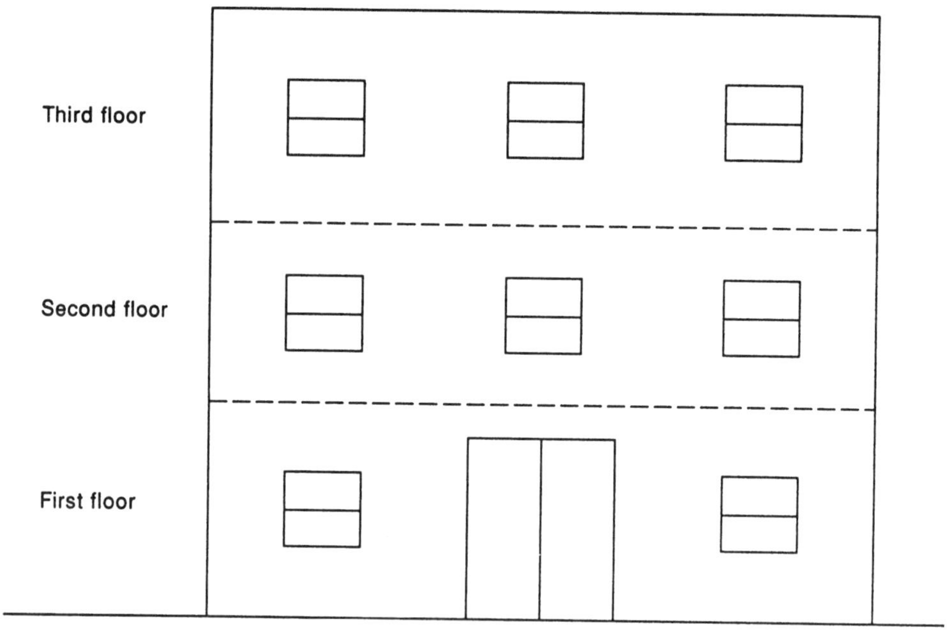

Fig. 17-9 In addition to outer floor dimensions, the general lighting load must be based on each floor level and type occupancy of that floor level.

tistory structure is divided up into showroom area and office area. In some buildings the overall floor area might be divided three ways into a showroom area, an office area, and a warehouse or storage area. The general-lighting load computation would then require three separate floor areas to be determined by their assigned outer dimensions. Each area would have a different value for the volt-amp unit-load assignment. The three separate volt-amp load computations would then be totaled for an overall general-lighting volt-amp load.

At the service level, a showroom/office main building may be annexed by a large warehouse structure. The panel in the main building may serve as the service equipment for the entire structure, with the warehouse loads supplied by another panel that is subfed (subpanel) from this service equipment panel.

To compute the service general-lighting load of the structure, the outer dimensions of the office floor area must be removed from the remaining showroom area of the main building. Three different general-lighting loads are then computed according to office area, showroom area, and warehouse area. These loads are then summed to determine the service general-lighting volt-amp load.

When the general-lighting schedule for a panel is known, the actual volt-amp lighting load must be determined by the nameplate rating and the number of units or fixtures installed. This actual volt-amp lighting load is compared to the minimum general-lighting volt-amp load as computed in accordance with Sec. and Table 220-3(b). The larger of the two volt-amp loads must be used for the NE Code-minimum feeder, panel, or service general-lighting load.

When fluorescent or other types of electric-discharge lighting are part or all of the general-lighting load, Sec. 210-22(b) requires the lighting load to be based on the volt-amp rating of the ballast, and not just on the wattage rating of the fixture tubes or bulbs.

Most lighting loads in nondwelling structures are considered by Sec. 210-22(c) to be continuous loads. Article 100 defines a continuous load as lighting (or other loads) that will draw current without interruption for three hours or more. When determining the rating or setting of the overcurrent-protection device (fuses or circuit breaker) for panels, feeders, or service conductors, Sec. 220-10(b) requires both the actual lighting volt-amp load [lighting schedule and Sec. 220-10(a)] and the calculated NE Code-minimum lighting volt-amp load [Sec. and Table 220-3(b)] to be computed at 125% for the comparison of the NE Code- minimum general-lighting load.

17.10 DERATE OF THE GENERAL-LIGHTING LOAD

Because of the diversity of loading in occupancies and/or structures such as hospitals, hotels, motels, and warehouses, S & T 220-11 permits the general-lighting load of a feeder, panel, or service to be derated to the percentages listed by type occupancy. This derate can be applied to both the NE Code-minimum general-lighting load as determined by S & T 220-3(b) and the actual general-lighting load (lighting schedule) as determined by Sec. 220-10(a).

The applicable derates of T220-11 as used on hospital general-lighting loads stem from the fact that the lighting for every patient room, every hallway, every clothes, utility, or storage closet, and every office, equipment room, dining room, or kitchen is not likely to all be on at any given time. To carry this idea a step further, if the case should arise that all of these lights are on, it is not likely that every motor, resistance heating element, or appliance (utilization equipment) for which the hospital service is also rated will also be turned on at that particular time. This is referred to as diversity of loading.

The applicable derates of T220-11 are restricted by the asterisk footnote. These derates cannot be applied to computed loads where the entire lighting is likely to be used at any given time. Hospital general-lighting loads that cannot be derated by S & T 220-11 include emergency and operating rooms (or facilities), special care areas like cardiac care, intensive care, and/or neonatal (baby) care units, as well as delivery rooms and dining halls.

In hotels or motels, the general-lighting loads that cannot be derated by S & T 220-11 include large conference halls or auditoriums, ballrooms, and central dining facilities.

17.11 ADDITIONAL LIGHTING LOADS

There is no NE Code-minimum volt-amp value or load assignment for outside lighting loads such as athletic fields, parking lots, walkways, hallways, or building security lighting. For this type of lighting, the additional lighting load of the feeder or service conductors is based on the total actual volt-amp or wattage lighting load. For the feeder, panel, or service overcurrent-protection device rating, the total actual load must be increased to 125% to compensate for continuous load.

This additional lighting load is computed independently of the general-lighting load and must be added to the NE Code-minimum general-lighting load to determine the total lighting load of the feeder, panel, or service.

17.12 SHOW-WINDOW LIGHTING LOADS

In commercial and industrial structures where show windows are installed, Sec. 220-12 requires a load assessment of "200 volt-amps per linear foot" of show window space. As shown in Figure 17-10, the linear footage must be measured horizontally along the longest base of the show window.

Some NE Code sources stipulate that the unit load of 200 volt-amps per linear foot takes into account the continuous load "nature" of show-window lighting, such that the 125% overrate required by Secs. 210-22(c) and 220-10(b) is not applicable to the feeder, panel, or service load computation. This statement cannot be validated by the wording of Sec. 220-12. In this text, the 200 volt-amps per linear foot of show-window lighting will be increased to 125% for the rating of the feeder, panel, or service overcurrent protection rating computations.

If the actual number of lighting outlets for show-window lighting is known, the lighting load can be based on Sec. 220-3(c)(7). Each outlet is assessed at 180 volt-amps. The product of the number of lighting outlets times 180 VA yields the show-window lighting load.

The 180 volt-amp load assessment per outlet is allowed only for the light-to medium-base lampholders. The heavy-duty mogul-base lampholders (larger screwshell) must be assessed at 600 volt-amps per outlet [Sec. 220-3(c)(4)]. If the show-window lighting will utilize heavy-duty lampholders, their wattage ratings are governed by Sec.

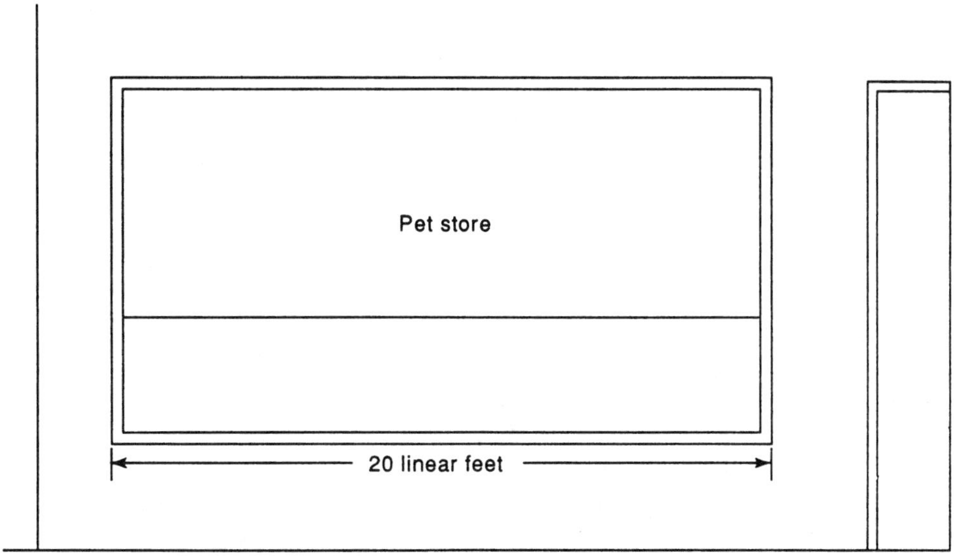

Fig. 17-10 The load assignment of 200 volt-amps per linear foot must be computed according to the *longest* horizontal base of the show window.

220-3(c)(4) and Sec. 410-53. Article 410 governs the rating, installation, and use of lighting fixtures.

If both the rating of each individual lighting fixture and the number of fixtures are known, the actual show-window lighting load can be computed by their product. Computed in this manner, the rating of the feeder, panel, or service overcurrent-protection device must be based on 125% of the total outlet show-window volt-amp lighting load to allow for continuous load [Sec. 220-10(b)].

In determining the feeder, panel, or service load, the highest or greatest (value of) volt-amp load computed by any of these methods should be used for the NE Code-minimum show-window lighting load.

Section 210-62 requires a receptacle outlet to be installed for every 12 linear feet of show-window space. Each receptacle outlet must be computed independently of the show-window lighting load at a value of 180 volt-amps [Sec. 220-3(c)(7)].

The show-window lighting volt-amp load must be computed independently of and in conjunction with the general-lighting load of a feeder, panel, or service.

17.13 TRACK-LIGHTING LOADS

Track lighting, as shown in Figure 17-11, is a form of specialty lighting that is often used to accent displays or display showrooms in nondwelling structures. Section 410-102 lists the NE Code-minimum volt-amp load that must be computed for track lighting. For each 2-foot section of track lighting, or for a section of track lighting that is less than two feet, the minimum load assignment is 150 volt-amps (changed in the 1996 NE Code—was 180 volt-amps).

Fig. 17-11 Track lighting is often used for accent lighting of displays and showrooms.

The length of a section, or the total length of several sections of track lighting is first divided by 2, then multiplied by 150 volt-amps to determine the feeder, panel, or service volt-amp load. If the measured length equals an odd number (such as 31 feet or 185 feet), 1 foot is added to the measured length to make it an even number before the division by 2.

As with all nondwelling lighting loads, Sec. 210-22(c) identifies track lighting as a continuous load. The computed volt-amp load according to length must be increased to 125% for the volt-amp load assignment for the feeder, panel, or service overcurrent-protection rating computations.

If the actual volt-amp load of track lighting can be determined from a panel schedule or nameplate of the intended track lighting fixtures. Both the actual and the NE Code-minimum volt-amp loads for track lighting must be computed and compared. The larger of the two in comparison must be taken as the NE Code-minimum track-lighting volt-amp load.

17.14 OUTDOOR SIGN LIGHTING LOADS

Article 600 governs the use of electric signs in or on a building or other structure. Electric signs are considered to be special equipment by the NE Code.

As indicated in Figure 17-12, a commercial store occupancy, located on the ground-floor level of a structure, is required by Sec. 600-5(a) to have at least one dedicated 20-amp branch circuit used to supply outside outlet(s) for a sign or for outline lighting (of the building) use.

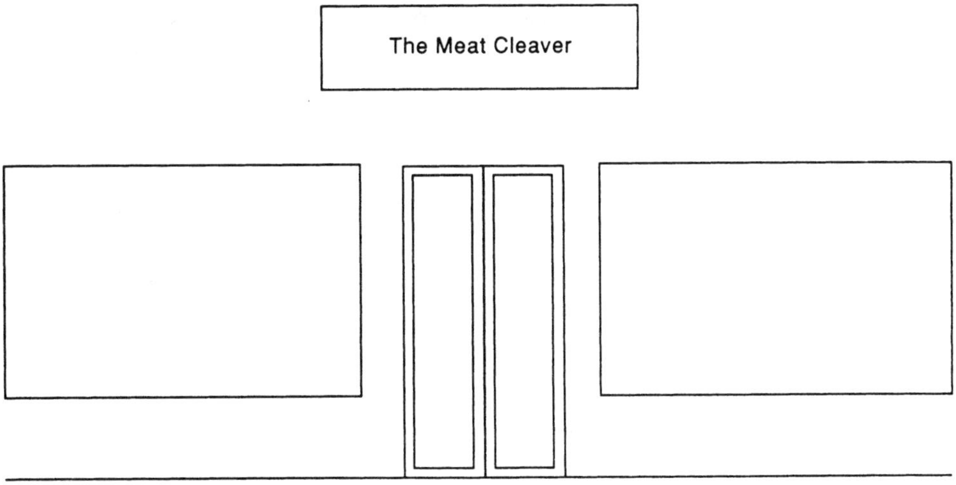

Fig. 17-12 All commercial stores with public street entrances must have a 20-amp branch circuit dedicated to outdoor signs and building outline lighting.

Section 600-5(a) requires the branch-circuit for the unknown sign or outline lighting load to be rated at a minimum of 20 amperes. If the circuit is rated at 120 volts, the volt-amp load rating (product of the two) is 2,400 volt-amps. For the 20-amp branch circuit, Sec. 600-5(b)(3) requires an NE Code-minimum load assessment of 1,200 volt-amps for the feeder, panel, or service load computations. As with other lighting loads, the 1,200 volt-amps will have to be overrated to 125% to compensate for the continuous-load requirements of an outdoor sign or building outline lighting.

As with any NE Code-assessed load computation, when the actual sign or building outline lighting volt-amp load is known, the larger of the two, in comparison to 1,200 volt-amps, must be used for the outdoor-sign/lighting portion of the feeder, panel, or service volt-amp load.

The general-lighting load, additional lighting loads such as parking lots or athletic fields, show-window lighting, track lighting, and outdoor sign or building outline lighting are separate lighting loads that may be connected to a feeder, panel, or service. The loads must be computed as separate lighting loads and, except for the general-lighting load, which is computed by the outer floor dimensions of each floor level, cannot be derated. The general-lighting load can only be derated when the type of occupancy served is listed in T220-11. In all lighting load computations, the NE Code required minimum must be compared to the actual lighting load (when known), and the larger volt-amp value must then be taken as the NE Code-minimum volt-amp load for that particular lighting load.

17.15 CONVENIENCE RECEPTACLE LOADS

Convenience (general-purpose or general-use) receptacles in all nondwelling feeder, panel, or service loads are required by Sec. 220-3(c)(7) to be computed at 180 volt-amps per outlet.

The overrate for continuous load should not be applied to this 180 volt-amp load assignment. If a convenience receptacle outlet will supply a continuous load, the actual volt-amp load assignment thus connected should be computed at 125% for the computed load of the feeder, panel, or service. The 180 volt-amp NE Code-minimum load assignment would be dropped.

General-purpose, general-use, or convenience receptacles (all three terms indicate either a single, duplex, or triplex receptacle outlet, with the duplex being the most common) are used for any type of portable, 120-volt equipment in a building or other structure. Their intended loads, other than the continuous loads as described in the previous paragraph, are normally not known at the time of feeder, panel, or service volt-amp load computation.

The majority of convenience receptacles installed in a building or other structure receive only intermittent use. In other words, most of the convenience receptacles do not supply a connected load simultaneously at any given time. Due to this diversity of loading among the convenience receptacles, the NE Code will allow the volt-amp load of convenience receptacles to be derated by either S & T 220-11, or S & T 220-13.

Section and Table 220-11 governs the derate of the general-lighting load in all dwelling units, hospitals, hotels and motels, and storage warehouses. The total volt-amp load for the convenience receptacles in these structures or occupancies can be added to the general-lighting volt-amp load, and the resulting volt-amp total derated by the "demand factor" percentages listed in T220-11.

To correctly size the feeder, panel, or service overcurrent protection, the general-lighting load should first be overrated to 125% to compensate for continuous load before it is combined with the convenience receptacle volt-amp load, and derated in accordance with S & T 220-11.

If the type of occupancy will not qualify the feeder, panel, or service convenience receptacle load to be derated by the demand percentages of T220-11, the nondwelling convenience receptacle load can be derated by or to the demand percentages of S & T 220-13. The derate percentage can only be applied to convenience receptacle loads that have a load assessment of 180 volt-amps as required by Sec. 220-3(c)(7), and to a computed receptacle load in excess of 10 kVA.

In lieu of the 180 volt-amp load assessment per convenience receptacle outlet required by Sec. 220-3(c)(7), the double asterisk (**) note of T220-3(b) will allow the 3½ VA/sq. ft. value as listed for the general-lighting load in banks and/or office buildings to be increased to 4½ VA/sq. ft. when the actual number of convenience receptacle outlets (to be installed) is not known. In other words, the convenience receptacle load would be included in the general-lighting load of the feeder, panel, or service, and no further allowance need be made. Of course, if the actual count or number of convenience receptacles is known, then the two loads (lighting and receptacle) must be computed independently, and then compared to the NE Code-minimums for each load.

17.16 PLUGMOLD LOADS

Plugmold (Fig. 17-13) is a type of surface raceway that contains single, 120-volt receptacles flush mounted along its length. The NE Code refers to this type of receptacle raceway as multioutlet assembly, which is governed by Art. 353. The volt-amp load of plug-

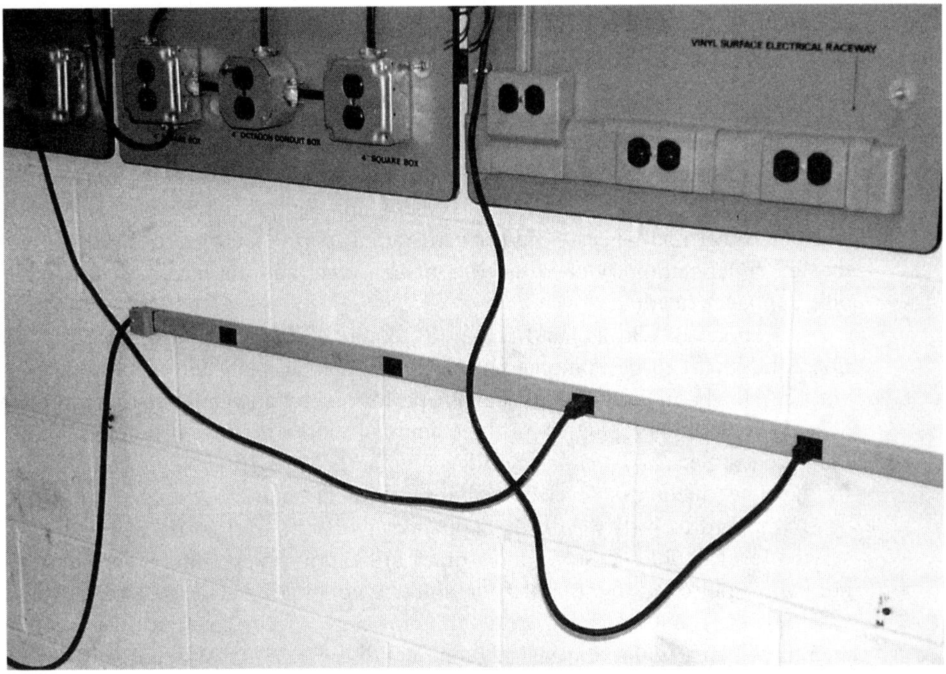

Fig. 17-13 Receptacle *plugmold* is often used when a concentration of outlets is required.

mold for feeder, panel, or service load computations is computed in one of two ways: light loading or heavy loading.

If the fixed assembly is installed for convenience and only light loading (such as long workbenches or studyhall countertops) will occur, Sec. 220-3(c), Excp. No. 1 will allow each 5-foot section of the multioutlet receptacle assembly to be counted as a single outlet rated at 180 volt-amps.

If the multioutlet assembly is to be heavily loaded, such as for lighting fixture displays in electrical wholesale houses or lab countertop spaces in health care facilities, where several lighting fixtures or appliances will be on simultaneously, Sec. 220-3(c), Excp. No. 1 requires each foot of the assembly to be computed at 180 volt-amps.

For heavy loading, the exception actually brings the load assignment of Sec. 220-3(c)(7) back into perspective, since the single receptacles assembled with the plugmold raceway are normally installed on 12-inch centers (1 foot apart).

17.17 SPECIFIC-APPLIANCE/UTILIZATION-EQUIPMENT LOADS

As defined in Art. 100, appliances are specific types of utilization equipment manufactured in standard sizes that are designed for specific chores, especially around the home. Utilization equipment is defined in a broader term as any type of equipment used to convert the supply of electrical energy into another form of energy to do useful work. Normally, any rule pertaining to an appliance per se applies to all utilization equipment in general.

This category in the breakdown of feeder, panel, or service loads is a catch-all for equipment loads, except heating and cooling loads or nondwelling kitchen equipment. Any type of utilization equipment such as water heaters, hair dryers, nondwelling electric clothes dryers or washers, dwelling-type cooking equipment used in a nondwelling structure, and other appliances, whether hard-wired (permanent) or cord- and plug-connected that are either stationary or fastened in place (fixed) and not portable come under the specific appliance heading.

Sections 220-3(c)(1) and 220-10(a) require each specific appliance (or other utilization equipment) to be computed at its nameplate rating in the feeder, panel, or service volt-amp load computations. If the task performed by a specific equipment or machine is deemed continuous (operates for three hours or more), then its volt-amp rating (except for motor loads, Sec. 220-14), in accordance with Sec. 220-10(b), must be increased to 125% for the panel, feeder, or service overcurrent-protection device rating computation.

It's hard to justify specific appliances as continuous loads at the feeder, panel, or service level. Appliances and most other utilization equipment, except motor loads, are normally controlled by a timer or some type of pilot device (thermostats, pressure switches, etc.) that restricts their continuous draw of current to a time value that is less than three hours under normal operating conditions. As a result, the appliance nameplate rating is taken at 100% for the feeder conductor, service conductor, panel, service equipment, and/or overcurrent-protection load calculations.

If the specific appliance is a combination load and/or multimotor equipment load, the panel and/or feeder volt-amp load is taken at nameplate rating [Secs. 430-25 and 430-54]. The 25% overrate by Sec. 220-10(b) for continuous-load does not apply to this type of equipment. The nameplate rating would also be used to size the service conductors, the service equipment, and the service or nonservice overcurrent-protection devices.

17.18 MOTOR LOADS

If the specific appliance or other utilization equipment is a motor load, such as a sump pump, an air handler, a water pump, or an air compressor, it can be counted as a specific appliance load at 100% of the motor full-load current or volt-amp rating provided the motor or motors are not the principal load of the feeder, panel, or service. When motors are included under the specific-appliance load, Sec. 220-14 by reference to specified sections in Art. 430 requires a 25% overrate of the largest motor volt-amp load to be added to the overall total of the feeder, panel, or service volt-amp load.

If one or more motors are the principal loads of a feeder, panel, or service, the rating of the overcurrent-protection device as determined in this chapter may not allow the motors to start under load conditions. (Refer to Chapters 12, 13, and 14 of this text or Sec. 430-63 in the NE Code.)

If motor loads are numerous but are not the principal load of a feeder, panel, or service, the motors may be computed as a separate load, independent or in conjunction with the specific appliance load, under Sec. 220-14. The motor volt-amp loads would be computed at 100% for the feeder, panel, or service load, with the 25% overrate of the largest motor (as before).

Very basically, whether one or several motor loads are to be installed as branch circuits supplied from a common feeder, panel, or service, Sec. 430-24 requires the total volt-amp load to include 25% of the largest motor volt-amp load.

When the motor VA or kVA load is not specified in the test, example, or list of connected loads for an intended installation (motor loads are normally specified in horsepower ratings), the motor volt-amp load can be determined by the horsepower rating (using NE Code-assigned values for full-load current), the voltage rating, and the number of phases of the motor in question by using the power formula. Section 430-6(a) requires the motor FLC (or FLA—motor full-load current) as used in the power formula to be taken according to the horsepower and voltage rating from either Table 430-147, 430-148, 430-149, or 430-150 for the motor in question.

With the power formula and the number of phases taken into account, the table-value motor FLC times the line voltage of the panel or service distribution will yield the motor's volt-amp load. Use of just the power formula does not take into consideration the efficiency or power factor ratings of connected loads, especially motors.

17.19 NONDWELLING COOKING/KITCHEN EQUIPMENT LOADS

Cooking equipment normally comes under the specific appliance heading for commercial and industrial feeder, panel, or service load computations. But in central cafeterias and dining halls for large institutions or office complexes, or for kitchens in fast-food vendor kitchens or public restuarants where the individual pieces of kitchen equipment are numerous, the overall (total) volt-amp load or kilovolt-amp load for such equipment can be derated in accordance with Sec. and Table 220-20.

Commercial cooking equipment, or household cooking equipment used in a nondwelling setting, cannot be derated according to the guidelines of S & T 220-19. Except for the household cooking equipment used in instructional programs (home economics in high-school curriculums, technical or vocational cooking programs, etc.), which are allowed by Note 5 of T220-19 to be derated as dwelling cooking equipment, all nondwelling installations must be derated by the guidelines of T220-20.

Section 220-20 stipulates that the kitchen equipment is not restricted just to cooking apparatus. Booster heaters, large vat blenders, deep fries, tray warmers, conveyor or tray line assemblies, and other kitchen equipment that is either thermostat controlled or only used intermittently can be derated according to T220-20. Equipment used for the heating and ventilation of the kitchen, including space heating units, hood exhaust fans, or air-conditioning equipment cannot be derated as kitchen cooking equipment.

No derate can be applied unless the total number of kitchen equipment units is three or more. If the count is less than three, the total nameplate rating of all the appliances must be taken at 100% when determining the feeder, panel, or service load.

Note in the second paragraph of Sec. 220-20 that if the two largest kitchen equipment loads constitute a greater percentage of the load than is allowed by the derate listed in T220-20, the total kitchen equipment load cannot be derated lower than 100% of these two largest loads.

When nondwelling cooking equipment is installed in a dwelling, the derates as listed in Sec. and Table 220-20 would not apply. If this situation should arise, it's unlikely that three or more pieces would be installed, so 100% of the nameplate rating would have to be considered.

17.20 HEATING AND COOLING LOADS

The fixed electric space heating for a structure is required by Sec. 220-15 to be computed at 100% of the total connected load (nameplate rating). If the space heating is accomplished by individual room units, the feeder, panel, or service heating load is computed at 100% of the total nameplate ratings of all the units supplied.

If the space heating is accomplished by a central electric furnace that is supplied by a single branch circuit, Sec. 424-3(b) requires the branch circuit to be overrated to 125% of the nameplate volt-amp (or wattage) rating. Section 220-15 will not allow the feeder rating to be less than the branch-circuit rating when this situation occurs.

Air-conditioning (refrigeration) equipment must be computed as a motor load (Sec. 440-3). For the feeder, panel, or service load, Secs. 220-14 and 440-35 require the air-conditioning equipment to be computed at nameplate rating. Air-conditioning equipment, when the nameplate rated-load or branch-circuit selection current of the individual hermetically sealed refrigerant motor compressor is known, must also be included when taking 25% of the largest motor load rating for the feeder, panel, or service load computations (Sec. 440-7). If the individual compressor rating is not known, the air-conditioning unit rating is omitted in comparison of largest motor load.

Heat pumps, which use the principles of air-conditioning in reverse for heating the home or other structure, also have emergency strip heating installed on the inside blower-coil assembly. The strip heaters are necessary or required when the outside temperature gets so low (cold) that the inside of the structure cannot be efficiently warmed with only heat exchange by the refrigeration equipment. The emergency strip heat also affords quick recovery when an immediate higher temperature within the structure is desired.

The outside motor-compressor unit and the inside heating unit are normally fed by separate branch circuits but function together when the strip heat is required. For feeder, panel, or service load computations, the heat pump compressor should be computed at 100% of the nameplate rating as part of the heating load.

Electric space heating equipment (Art. 424) and air-conditioning equipment (Art. 440) are dissimilar loads and except for heat pumps are normally not used simultaneously (noncoincidental loads). Section 220-21 will allow the omission of the smaller of the two loads in comparison for the computation of the feeder, panel, or service heating and cooling load.

If both are not present, then Sec. 220-21 has no bearing on either the heating or cooling load. As was stated above, except for the overrate to 125% when a single branch circuit supplies an electric furnace, either load would be taken at 100% for the feeder, panel, or service volt-amp load.

Under normal operation and climate conditions, it is not likely that either the electric space heating or the air-conditioning equipment will be on for three hours or more without cycling, so neither is computed as a continuous load in the feeder, panel, or service load computations. At the branch-circuit level, both are computed at 125% as required by other articles in the NE Code.

17.21 FUTURE GROWTH

Once all the feeder, panel, or service loads are determined (either NE Code minimum or actual, whichever is greater), the NE Code-minimum full-load current is determined using the power formula. The rating or setting of the overcurrent-protection device is determined in accordance with Secs. 230-42 (services), 240-3 (conductors), 384-13 (panels), and 240-6 (overcurrent-protection ratings).

Although Sec. 384-13 stipulates that the panel and its feeder conductors must have a rating at least equal to the computed load as determined by Art. 220, Sec. 90-8 recommends allowances of spare circuits and ampacity for future growth, modification, or convenience. This section does not specify a direct overrate of minimum load current because of the many different types of installations and their individual electrical requirements.

The accepted norm of the electrical trade is to allow 25% for future growth. For a 25% overrate, the computed full-load current must be divided by 75% or multiplied by 133⅓%. To overrate the full-load current by 125% is only allowing a 20% growth rate (1.25 is reciprocal to 0.80).

The overrated full-load current is then used to determining the actual feeder, panel, or service rating.

Very seldom will the computed load current before overrate for future growth equal a standard size or rating of a panel, other load center, or overcurrent-protection device. Sometimes just using the next higher standard size of panel (panels are manufactured in standard sizes of voltage and current ratings as well as number of phases) or other load-center allows for future growth "automatically" without having to overrate by 25%.

Following the requirements of Sec. 384-13, an "NE Code-legal" panel installation could result in a future addition of circuits that would overload the feeder conductors. For example, the panel and feeder full-load current is computed at 155 amps. The next higher standard setting or rating of overcurrent protection as listed in Sec. 240-6(a) is 175 amps. Section 240-3(b), will allow the feeder conductors, if they have a "table-rated" ampacity greater than or equal to 155 amps, to be protected at 175 amps.

If single insulated THWN aluminum conductors installed in conduit are used as the panel feeder, No. 3/0 AWG as listed in T310-16 can be used for the feeder conductors. The next higher standard size of manufactured panel is 200 amps. For a panel rated at 200 amps, the actual full-load current is only 155 amps, the feeder supply conductors are rated at 155 amps, and both the panel and the feeder conductors are protected at 175 amps. The requirements of Sec. 384-13 have been met. This is an NE Code-legal installation.

If this is a lighting and appliance panel, Sec. 384-16(a), Excp. No. 1 stipulates that the panel is not required to have a "main" or main overcurrent protection installed in the panel, provided the supply feeder conductors have overcurrent protection that is not rated in excess of the panel rating. Following our example, the feeder conductors, rated at 155 amps and protected at 175 amps are supplying a "main lug only" panel rated at 200 amps.

If an ammeter is left on the feeder conductors for a 24-hour period, the maximum current load should not exceed 155 amps. What happens if an electrician or department head elects to add a 15-amp branch-circuit load to this 200-amp panel that's only carrying 155 amps? The feeder conductors will be overloaded.

What happens if a 30-amp branch-circuit load is added to this 200-amp panel loaded at only 155 amps? The feeder conductors are now heavily overloaded. In time, the feeder overcurrent-protection device will open since it too is overloaded when the branch-circuits are at full-load capacity. Still the panel current rating was never approached.

The better electrical installation (although not specifically an NE Code requirement) would be to determine the actual full-load current requirements of the panel and feeder conductors, select the next higher size of manufactured panel, and size both the feeder conductors and the panel/feeder overcurrent-protection device for the rating of the panel. In the above example, if the panel had been sized and protected at the 200-amp rating, and the feeder conductors sized at 250 kcmil THWN aluminum (T310-16: rated ampacity 205 amps), the resulting overload situation could not have occurred unless the panel rating was also exceeded.

The idea of the feeder conductors offering overload protection to overcurrent-protection devices (which this type of installation would be) is both unsafe and unethical. Computing the size or rating of the panel and feeder conductors and their resulting overcurrent protection in the above manner ("better electrical installation") will avoid this situation and normally allow for future growth without the 25% overrate of full-load current being installed.

This same train of thought can be applied when determining the minimum rating of the service, the service conductors, the service-disconnecting means, and the service overcurrent protection for a building or other structure. The ampacity rating of the service conductors and service equipment should at least equal the maximum rating of overcurrent protection installed for said equipment.

17.22 NONDWELLING NEUTRAL CONSIDERATIONS

A panel schedule in the specifications of a structure, which lists both the branch-circuit volt-amp loads and to which phases these volt-amp loads are connected, is most helpful for the proper sizing of the panel and its supply conductors.

If the structure has small to moderate electrical power loads, a single panel may be the only service load. The panel may be a subfed panel from a fused knife switch or circuit breaker that serves as the building's service-disconnecting means and overcurrent protection, or the panel may contain main overcurrent protection and serve as the service equipment for the structure. With this type of installation, the panel schedule would also be used to determine the minimum service rating.

For larger buildings or structures, several panels may be supplied by central service equipment or switchgear. When this occurs, the panel schedules will aid in determining the minimum rating of the switchgear.

Either polyphase (2-phase or 3-phase) or single-phase, phase-to-phase loads (except dwelling electric cooking ranges or clothes dryers) do not contain or have a neutral component, and therefore do not contribute to the unbalanced neutral load of a panel or feeder.

For single-phase, 2-wire, phase-to-neutral loads, the neutral conductor carries the same or equal current as the ungrounded phase conductor supplying the load. Technically we are incorrect in referring to this grounded-circuit conductor as a neutral. A second phase conductor (another ungrounded-circuit conductor) is not present to offset the load current of the "neutral" conductor.

When a neutral conductor serves multiple-phase sources like 120/240-volt single-phase, 3-wire; 240/120-volt 2-phase, 5-wire; 240/120-volt 3-phase, 4-wire; 208Y/120-volt 3-phase, 4-wire; or 480Y/277-volt 3-phase, 4-wire, and the panel is fully loaded, the neutral conductor carries only the unbalanced (net) current load of all the phase-to-neutral loads of the system.

A 120/240-volt single-phase, 3-wire distribution is considered to be a multiphase source even though the generation of both 120-volt sources (phases A and B) are due to the action of the same single phase in the transformer primary. The two sources are taken from opposite ends of the secondary transformer winding, which in essence makes the two sources 180 degrees out of phase. As a result, the currents from the two sources returning along a common neutral conductor tend to cancel each other out for a minimum unbalanced neutral current load.

The maximum unbalanced neutral conductor load current is the maximum net computed current load between any ungrounded phase conductor and the neutral. Except for 2-phase, 5-wire systems, where the neutral conductor is sized at 140% of the maximum unbalanced neutral load current, Sec. 220-22 will allow the neutral conductor to be sized at an ampacity at least equal to the maximum unbalanced current load.

For larger feeders, panels, or services, Sec. 220-22 will allow a further reduction (demand factor) of the neutral current that is in excess of 200 amps. This further derate is only allowed when computing the panel or feeder neutral load for a 3-wire dc, a single-phase, 3-wire ac, a 3-phase, 4-wire ac, or a 2-phase, 5-wire ac distribution system. No derate or reduction of demand is allowed for the neutral load current, which consists of (supplies) electric-discharge lighting.

When computing the feeder, panel, or service neutral current for load-ampacity requirements and NE Code-minimum wire size, the power formula is solved for $A_{neutral}$. For 1-phase, 3-wire ac distributions, the power formula for neutral load current is:

$$A_{neutral} = \frac{\text{neutral power}_{VA}}{V_{line}}$$

V_{line} is the phase-to-phase voltage in the single-phase power equation. Many student electricians want to solve the power formula for the neutral current by using the phase-to-neutral voltage without allowing for the number of phases in polyphase neutral computations. This would make it appear that all the neutral current is drawn from only one phase or ungrounded source.

Using the power formula to solve for the neutral current in the same manner and with the same values (except for the actual volt-amp load) as is used for determining the phase conductor current (A_{line}) takes into account the multiple sources for the neutral current loads. This method of computation assumes the neutral loads are balanced between the available phases.

For 2-phase, 5-wire ac distributions, the power formula for neutral load current is:

$$A_{neutral} = \frac{\text{neutral power}_{VA}}{2 \times V_{line}}$$

For 3-phase, 4-wire ac distributions, the power formula for neutral load current is:

$$A_{neutral} = \frac{\text{neutral power}_{VA}}{1.732 \times V_{line}}$$

If the neutral volt-amp load has been computed in terms of kilovolt-amps, the term "neutral power_{kVA} × 1000 (volt-amps/kVA)" is substituted for "neutral power_{VA}" in all three power equations.

When the neutral conductor is a service conductor (service drop, service lateral, or service entrance), the neutral has a two-fold purpose. The service neutral serves as a current-carrying conductor that carries the unbalanced neutral current of the total connected load as does any feeder neutral conductor. The service conductor also serves as the ground-fault return conductor between the service equipment and the distribution power supply.

For ground-fault ampacity requirements, the neutral conductor must have an ampacity rating that is capable of handling the maximum ground-fault overcurrent available from the ungrounded phase conductors of the service. The reduced service neutral conductor, rated in accordance with Sec. 250-23(b), cannot be sized smaller than the NE Code-minimum wire size of the service grounding-electrode conductor of the intended service.

17.23 NONDWELLING VOLT-AMP LOAD COMPUTATIONS

Computation of the various loads within nondwelling structures as discussed and explained in the previous sections may be determined within the guidelines of the following example.

Example 17–1. A local plumber's supply building is supplied with a 120/240-volt, single-phase, 3-wire service. The building, which measures 60 by 120 feet, is divided into display/showroom and warehouse. The warehouse measures 60 by 80 feet and contains 27 general-use receptacles. The showroom occupies the rest of the building and contains the following loads:

14	Linear feet of show window
46	Convenience receptacles
1	5-kVA, 230-volt air conditioner
1	1-hp, 230-volt air handler
2	½-hp, 115-volt blower motors for gas heating units
1	3,500-watt, 230-volt, quick-recovery water heater

Determine:

1. The service rating (allow 25% overrate for future growth)
2. The wire size of the service-entrance conductors, phases A and B, using THW copper
3. The wire size of the reduced service neutral conductor, using THW copper
4. The wire size of the copper grounding-eletrode conductor run to the driven rod
5. The wire size of the copper bonding jumper ran to the interior metal water pipe

Volt-amp load computations: (worksheet) Ref. Art. 220-B

General lighting: Ref. S & T 220-3(b)

Warehouse (portion):

Floor area computation:

$$60 \times 80 = 4{,}800 \text{ square feet}$$

Load computation:

$$4{,}800 \text{ square feet} \times \tfrac{1}{4} \text{ volt-amps/square foot} = 1{,}200 \text{ volt-amps}$$

Showroom (portion):

Floor area computation:

$$60 \times 120 = 7{,}200 \text{ square feet}$$

$$7{,}200 - 4{,}800 = 2{,}400 \text{ square feet}$$

Load computation:

$$2{,}400 \text{ square feet} \times 3 \text{ volt-amps/square foot} = \underline{7{,}200 \text{ volt-amps}}$$

Total: 8,400 volt-amps

Show-window lighting: Ref. 220-12

$$14 \text{ linear feet} \times 200 \text{ volt-amps/linear foot} = 2{,}800 \text{ volt-amps}$$

Outside sign:

NE Code-minimum: 1,200 volt-amps Ref. 600-5(b)(3)

Lighting-load summation:

General lighting	8,400 volt-amps
Show window	2,800 volt-amps
Outside sign	1,200 volt-amps
Total:	12,400 volt-amps

Lighting overrate for continuous-load computation: Ref. 220-10(b)

125% of 12,400 volt-amps = 15,500 volt-amps

Convenience receptacles: Ref. 220-3(c)(7)

Warehouse: 27 × 180 volt-amps = 4,860 volt-amps
Showroom: 46 × 180 volt-amps = 8,280 volt-amps
Total: 13,140 volt-amps

Derate of convenience receptacle load: Ref. T220-13

First 10 kVA at 100%: 10,000 volt-amps
Remaining load computation:

13,140 VA–10,000 VA = 3,140 VA

Derate computation:

50% of 3,140 volt-amps = 1,570 volt-amps
Total: 11,570 volt-amps

Specific appliance: Ref. 220-10(a)

Water heater: 3,500 watts* = 3,500 volt-amps

Heating: Ref. 220-15

Blower motors: Ref. 220-14

½-hp motor FLC = 9.8 amps Ref. T430-148

Load computation:

2 × (120 volts × 9.8 amps) = 2,352 volt-amps

Cooling: Ref. 220-14

Air conditioner: Ref. 440-35

1,000 × 5 kVA = 5,000 volt-amps

Air handler:

1-hp motor FLC = 8.0 amps Ref. T430-148

Load computation:

240 volts × 8.0 amps = 1,920 volt-amps
Total: 6,920 volt-amps

*Consider watts equivalent to volt-amps.

Heating/Cooling: Ref. 220-21

Noncoincidental loads, compare for larger: 6,920 volt-amps

Largest motor overrate computation: Ref. 430-24

230-volt rating: 25% of 1,920 volt-amps = 480 volt-amps

115-volt rating: 25% of (120 volts × 9.8 amps) = 294 volt-amps

1. Service rating computations including 25% overrate for future growth:

Volt-amp load summation: Ref. 220-10(b)

Lighting	15,500 volt-amps
Convenience receptacles	11,570 volt-amps
Specific appliance	3,500 volt-amps
Heating/cooling	6,920 volt-amps
Largest motor: (240 volts)	480 volt-amps
Total:	37,970 volt-amps

Future growth overrate computation: Ref. 90-8

37,970 volt-amps ÷ 0.75 = 50,627 volt-amps

Service overcurrent protection rating determination:

FLC computation:

50,627 volt-amps ÷ 240 volts = 210.9 amps

Standard OCPD ratings: (NHS) 225 amps Ref. 240-6(a)

2. Wire size of the SE conductors, phases A and B determination:

Ref. 240-3

Wire terminal temperature limitations: 75°C Ref. 110-14(c)(2)

Table-rated ampacity: 230 amps Ref. T310-16

Wire size: No. 4/0 AWG THW copper

3. Wire size of the reduced SE neutral conductor determination:

Volt-amp load summation: Ref. 220-22

Lighting	15,500 volt-amps
Convenience receptacles	11,570 volt-amps
Heating	2,352 volt-amps
Largest motor: (120 volts)	294 volt-amps
Total:	29,716 volt-amps

Future growth overrate computation: Ref. 90-8

29,716 volt-amps ÷ 0.75 = 39,621 volt-amps

FLC computation:

39,621 volt-amps ÷ 240 volts = 165.1 amps

Overrate adjustment to align the neutral rated-load current with OCPD rating.

Adjustment computation:

$$\frac{\text{Service OCPD rating}}{\text{Phase FLC rating}} = \frac{225 \text{ amps}}{210.9 \text{ amps}} = 1.067$$

Overrate computation:

$$1.067 \times 165.1 \text{ amps} = 176.2 \text{ amps}$$

Wire size for load return:

Wire terminal temperature limitations: 75°C Ref. 110-14(c)(2)
Table-rated ampacity: 200 amps Ref. T310-16
Wire size: No. 3/0 AWG THW copper

Wire size for ground-fault return: Ref. 250-23(b)
Wire size of the ungrounded-SE conductors: No. 4/0 AWG THW copper
Neutral wire size: No. 2 AWG copper Ref. T250-94

NE Code-minimum: Ref. 310-15(c)
Compare for larger: No. 3/0 AWG THW copper

4. Copper grounding-electrode conductor run to the driven rod:
Ref. 250-81 Ref. 250-83(c)
Wire size: No. 6 AWG copper Ref. 250-94, Excp.(a)

5. Copper bonding jumper run to the interior metal water pipe:
Ref. 250-80(a)
Wire size of the ungrounded-SE conductors: No. 4/0 AWG THW copper
Jumper wire size: No. 2 AWG copper Ref. T250-94

The volt-amp load computations in this example were initially computed as a worksheet with any applicable derates or overrates applied with NE Code reference. This was done because some of the same load values are used for the load computations of both the service OCPD rating and the reduced neutral maximum unbalanced load-current ratings. Both the service rating and the wire size of the ungrounded service conductors are based on the rating of the service overcurrent-protection device rating (OCPD).

The NE Code-minimum service rating was computed as a summation of identified loads from the worksheet. In the comparison for the largest of the heating and cooling load, the blowers on the gas heating units were omitted in favor of the larger air-conditioning load. The largest motor load was taken as the 240-volt load, which was the air handler for the air-conditioning system.

The rated-load current of the reduced service-entrance neutral conductor was based on the summation of the lower-voltage loads only. Although the gas heater blower motors were omitted in the comparison of the heating and cooling loads in determining the service rating, they were not omitted in the determination of the neutral load, simply because the cooling load was rated at 240 volts.

Section 220-22 stipulates that the rated ampacity of the reduced service neutral conductor must be based on the maximum unbalanced neutral current load. Solving the single-phase power formula with the higher line voltage of 240 volts indicates that 165.1 amps of 120-volt neutral load current is sourced from phase A, and another 165.1 amps of 120-volt neutral load current is sourced from phase B for a total neutral load current of 330.2 amps.

The 165.1 amps corresponds to the computed phase (service) full-load of 210.9 amps. To honor the additional neutral loading allowed by the 225-amp service rating, the 165.1 amps must be overrated by the ratio of 225 amps to 210.9 amps. This computation is especially critical when the service rating is based on the next higher standard size of panel or other load center.

The grounding-electrode conductor and bonding jumper were sized according to the requirements of Art. 250. Grounding and bonding were discussed in Chapter 9 of this text.

Code challenge 17–1. A local pharmacy is supplied with a 120/240-volt, single-phase, 3-wire service. The building, which measures 80 by 140 feet, is divided into mercantile and storage area. The storage area measures 60 by 80 feet and contains 21 general-use receptacles. The mercantile area occupies the rest of the building, and contains the following loads:

30	Linear feet of show window
12	kVA of air conditioning
80	Convenience receptacles
1	¾-hp, 230-volt air handler
2	⅓-hp, 115-volt blower motors for gas heating units
1	30-gallon, 3,500-watt, 230-volt water heater

Determine:

1. The service rating (allow 20% overrate for future growth)
2. The wire size of the service-entrance conductors, phases A and B, using RHW copper
3. The wire size of the reduced service neutral conductor, using RHW copper
4. The wire size of the copper grounding-electrode conductor run to the metal underground water pipe
5. The wire size of the copper grounding-electrode conductor run to a ground-ring electrode

Volt-amp load computations: (worksheet) Ref. (A) ____________

General lighting: Ref. (B) ____________

Storage (portion):

Floor area computation:

____________ × ____________ = (C) ____________ square feet

Load computation:

__________ square feet × (D) __________ VA /square foot = __________ VA

Mercantile (portion):
Floor area computation:

________ × ________ = (E) ________ square feet

________ − ________ = (F) ________ square feet

Load computation:

__________ square feet × (G) __________ VA /square foot = __________ VA

Total: (H) __________ VA

Show-window lighting: Ref. (I) __________

__________ linear feet × __________ VA/linear foot = (J) __________ VA

Outside sign:
NE Code-minimum: (K) __________ volt-amps Ref. (L) __________

Lighting-load summation:

General lighting	__________ VA
Show window	__________ VA
Outside sign	__________ VA
Total:	(M) __________ VA

Lighting overrate for continuous-load computation: Ref. (N) __________

________ % of ________ volt-amps = (O) ________ volt-amps

Convenience receptacles: Ref. (P) __________

Storage:	(Q) __________ × __________ volt-amps =	__________ VA
Mercantile:	(R) __________ × __________ volt-amps =	__________ VA
Total:		(S) __________ VA

Derate of convenience receptacle load: Ref. (T) __________
First (U) __________ kVA at 100%: __________ VA
Remaining load computation:

________ VA − ________ VA = (V) ________ VA

Derate computation:
(W) __________ % of __________ volt-amps = __________ VA
Total: (X) __________ VA

Specific appliance: Ref. (Y) ____________

Water heater: __________ watts* = (Z) __________ volt-amps

Heating: Ref. (1A) ____________

Blower motors: Ref. (1B) ____________

⅓-hp motor FLC = (1C) __________ amps Ref. (1D) ____________

Load computation:

__________ × (__________ volts × __________ amps) = (1E) __________ VA

Cooling: Ref. (1F) ____________

Air conditioner: Ref. (1G) ____________

__________ × (1H) __________ kVA = __________ VA

Air handler: Ref. (1I) ____________

¾-hp motor FLC = (1J) __________ amps

Load computation:

__________ volts × __________ amps = (1K) __________ VA

Total: (1L) __________ VA

Heating/cooling: Ref. (1M) ____________

Noncoincidental loads, compare for (1N) __________: (1O) __________ VA

Largest motor overrate computation: Ref. (1P) ____________

230-volt rating:

__________ % of __________ volt-amps = (1Q) __________ volt-amps

115-volt rating:

__________ % of (__________ volt× __________ amps) = (1R) __________ volt-amps

1. Service rating computations including 25% overrate for future growth:

Volt-amp load summation: Ref. (1S) ____________

Lighting __________ VA

Convenience receptacles __________ VA

Specific appliance __________ VA

Heating/cooling __________ VA

Largest motor: (240 volts) __________ VA

Total: (1T) __________ VA

Future growth overrate computation: Ref. (1U) ____________

__________ volt-amps ÷ (1V) __________ = __________ volt-amps

*Consider watts equivalent to volt-amps.

Service overcurrent-protection rating determination:
FLC computation:

(1W) ____________ volt-amps ÷ ____________ volts = (1X) ____________ amps

Standard OCPD ratings: (NHS) (1Y) ____________ amps
Ref. (1Z) ____________________

2. Wire size of the SE conductors, phases A and B determination:
Ref. (2A) ____________________
Wire terminal temperature limitations: (2B) ____________ °C
Ref. (2C) ____________________

Table-rated ampacity: (2D) ____________ amps Ref. (2E) ____________________

Wire size: (2F) ____________ RHW copper

3. Wire size of the reduced SE neutral conductor determination:
Volt-amp load summation: Ref. (2G) ____________________

Lighting	____________	VA
Convenience receptacles	____________	VA
Heating/cooling	____________	VA
Largest motor: 120 volts)	____________	VA
Total:	(2H) ____________	VA

Future growth overrate computation: Ref. (2I) ____________________

____________ volt-amps ÷ (2J) ____________ = ____________ volt-amps

FLC computation:

(2K) ____________ volt-amps ÷ ____________ volts = (2L) ____________ amps

Overrate adjustment to align the neutral rated-load current with OCPD rating.
Adjustment computation:

$$\frac{\text{Service OCPD rating}}{\text{Phase FLC rating}} = \frac{\_\_\_\_\_\_\_\_\_\_\_\_ \text{ amps}}{\_\_\_\_\_\_\_\_\_\_\_\_ \text{ amps}} = \text{(2M)} \_\_\_\_\_\_\_\_\_\_\_\_$$

Overrate computation:

____________ × ____________ amps = (2N) ____________ amps

Wire size for load return:
Wire terminal temperature limitations: (2O) ____________ °C
Ref. (2P) ____________________
Table-rated ampacity: (2Q) ____________ amps Ref. (2R) ____________________
Wire size: (2S) ____________ RHW copper

Wire size for ground-fault return: Ref. (2T) ____________

Wire size of the ungrounded-SE conductors: (2U) ____________ RHW copper

Neutral wire size: (2V) ____________ copper Ref. (2W) ____________

NE Code-minimum: Ref. (2X) ____________

Compare for (2Y) ____________ : (2Z) ____________ RHW copper

4. Copper grounding-electrode conductor run to the metal underground water pipe.

Ref. (3A) ____________

Wire size of the ungrounded-SE conductors: (3B) ____________ RHW copper

Electrode-conductor wire size: (3C) ____________ copper

Ref. (3D) ____________

5. Copper grounding-electrode conductor run to a ground-ring electrode.

Ref. (3E) ____________

Wire size: (3F) ____________ copper Ref. (3G) ____________

To meet the requirements of Art. 600, the computed loads in both the example and the Code challenge problem had to include an outside sign circuit, even though an outside sign or other building outline lighting was not mentioned in the list of connected loads.

17.24 DWELLING LOAD CONSIDERATIONS

The volt-amp loads of a dwelling structure are computed different from the panel, feeder, or service volt-amp loads of a nondwelling structure. The difference stems from the diversity of loading.

The actual nameplate values or NE Code-minimum values of just about all electrical equipment in a dwelling has some type of derate listed in the National Electrical Code. The standard method of feeder, panel, or service calculation for dwelling structures is also governed by Part B of Art. 220.

As with nondwelling load computations, there are four principal loads in the standard method of dwelling structure load computation:

1. General-lighting and convenience receptacle load
2. Fixed-appliance (utilization equipment) and motor load
3. Heating and cooling load
4. Largest motor load

17.25 GENERAL-LIGHTING AND RECEPTACLE LOADS

Section 210-70(a) lists the required lighting outlets, and Sec. 210-52 lists the required convenience receptacle outlets that must be installed in all dwelling structures.

The requirements of S & T 220-3(b) list the convenience (duplex) receptacles installed in a dwelling as outlets intended for general illumination. The service loading of these receptacles is allowed for [single asterisk note of T220-3(b)] in the computation of the general-lighting load of the dwelling. No additional loading allowances are required.

This statement is true except for the 20-amp laundry receptacle branch circuit required by Sec. 220-4(c) and the minimum of two 20-amp, kitchen small-appliance receptacle branch circuits required by Sec. 220-4(b).

To determine the minimum general lighting and receptacle load of a dwelling, Sec. 220-3(b) requires a unit load assessment of 3 volt-amps per square foot [3 VA/sq. ft., T220-3(b)] to be multiplied by the outer floor dimensions of each floor level in a dwelling structure. This outer floor area computation is not to include areas contained in open porches, garages, attics, or underfloor crawl spaces that are not adaptable for future use.

"Future use" indicates future living space, where porches, carports, garages, attics, or basements can be converted at a future date over to bedrooms, dens, great rooms, or recreational rooms. If it is evident that these areas can or will be living quarters at a future date, then these areas should be included in the floor dimension computations for the general lighting and receptacle load.

The computations of Sec. and Table 220-3(b) are specified as the NE Code minimum because the actual general lighting and receptacle loads are not known at the time of electrical installation. Even if the actual number and load requirements of the lighting fixtures are listed on a specification sheet for the intended dwelling, there is no way of knowing what will be connected to the convenience receptacles.

Once the NE Code-minimum general lighting and receptacle volt-amp load for the dwelling structure is determined by the requirements of S & T 220-3(b), it can be derated by the guidelines listed in S & T 220-11. The FPN listed under Sec. 220-11 refers the student electrician to Sec. 220-16 for application of the demand factors in T220-11 to the kitchen small-appliance and laundry circuits in a dwelling.

Section 220-16(a) requires a load assessment of 1,500 volt-amps for each 20-amp, small-appliance branch circuit installed in the kitchen of a dwelling. Section 220-4(b) requires a minimum of two 20-amp, small-appliance branch circuits to be installed. If more than two are installed, Sec. 220-16(a) requires each one to be assessed at 1,500 volt-amps.

Section 220-4(c) requires at least one 20-amp laundry circuit to be installed in a dwelling. [Note the Exceptions for multifamily dwellings in Sec. 210-52(f).] Section 220-16(b) requires this laundry circuit to be assessed at 1,500 volt-amps.

The last sentence of both Secs. 220-16(a) and (b) allows the kitchen 20-amp small-appliance and laundry branch-circuit load assessments (respectively), which constitute a minimum load of 4,500 volt-amps [(2×1,500 VA) + 1,500 VA], to be derated with the general lighting and receptacle load under the guidelines of Sec. and Table 220-11.

After the volt-amp assessments for the kitchen small-appliance and laundry branch circuits are added to the computed volt-amp load of the general lighting and receptacle load, the total volt-amp load is derated by T220-11 as follows:

For a total volt-amp load of 120,000 volt-amps or less:

1. 3,000 volt-amps is subtracted from the total volt-amp load, because the first 3,000 volt-amps (small-appliance branch circuits) must be taken at 100%.

2. The remaining volt-amp load is derated to 35%.
3. The first 3,000 volt-amps is added back to the 35% volt-amp assessment to determine the overall volt-amp load.

For a total volt-amp load in excess of 120,000 volt-amps:

1. 120,000 volt-amps is subtracted from the total volt-amp load.
2. The remaining volt-amp load is derated to 25%.
3. 43,950 volt-amps is added back to the 25% volt-amp load assessment to determine the overall volt-amp load.

Unless the dwelling is just short of a multistory mansion, the summation of the general-lighting and convenience receptacle, kitchen small-appliance, and laundry volt-amp load will not be greater than 120,000 volt-amps. For single-family and two-family dwellings, all except 3,000 volt-amps of the combined volt-amp load is derated to 35% for feeder or service computations.

For larger multifamily dwellings, the summation of the general-lighting and convenience receptacle, kitchen small-appliance, and laundry volt-amp load will usually exceed 120,000 volt-amps, and all three derates of T220-11 are applied.

For volt-amp loads in excess of 120,000 volt-amps, the 43,950 volt-amps added back to the 25% derated, remaining volt-amp load is equal to 3,000 volt-amps, which must be taken at 100%, added to 117,000 volt-amps that is allowed to be derated to only 35%.

17.26 FIXED APPLIANCE AND MOTOR LOADS

Fixed appliances and motor loads in a dwelling would include equipment such as electric water heaters, garbage disposals, trash compactors, permanently installed dishwashers, attic exhaust and/or ventilating fans, kitchen hood-exhaust fans, bathroom exhaust fans, central vacuum systems, sump pumps, well pumps, swimming pool circulating or filtration pumps, electric garage-door openers, blowers on central heating and air-conditioning systems, as well as other fixed appliances or equipment installed in the home.

Once again, because of the diversity of loading, the NE Code will allow the total nameplate rating of all these appliances to be derated in accordance with Sec. 220-17. The derate can only be applied to appliances fastened in place, either permanently connected or cord- and plug-connected to the branch-circuit supply. Portable appliances are not counted since their loads are allowed for in the general-lighting, convenience receptacle, kitchen small-appliance, and laundry loads as derated by Sec. and Table 220-11.

The list does not include clothes washing machines, refrigerators or freezers, electric cooking ranges or other cooking equipment, electric clothes dryers, fixed electric space heating equipment, or air-conditioning equipment. The feeder, panel, or service volt-amp load requirements for these appliances or equipment are allowed for under other sections of Art. 220.

When four or more fixed appliances or motor loads (other than those specified above) are supplied by a common feeder, panel, or service for a dwelling structure (whether single-family or multifamily), Sec. 220-17 will allow the nameplate ratings of all these appliances to be added together for a total wattage or volt-amp rating. This total appliance volt-amp load is then derated to 75% for the feeder, panel, or service volt-amp load computations.

Whether the volt-amp loads are rated at 240 volts (phase-to-phase) or 120 volts (phase-to-neutral) has no bearing on the total volt-amp nameplate rating of the appliances or motor loads. Use of the power formula in solving for total service current will assume the phase-to-neutral loads are balanced between the two phases as two 120-volt sources (single-phase, 3-wire system).

If the dwelling service-entrance neutral conductor is to be reduced by the guidelines of Sec. 220-22, the fixed appliance and motor load for the neutral should only be summed and derated for the 120-volt loads.

In some dwelling feeder, panel, or service load calculations, the derate can be applied only to the service loads (240-volt and 120-volt loads together equals more than four) because the phase-to-neutral loads are counted as less than four. When this occurs, no derate can be applied to the lower-voltage loads. When the derate cannot be applied, the fixed appliances or motor loads are taken at nameplate rating (100%) for the volt-amp load calculations.

17.27 ELECTRIC CLOTHES DRYER LOADS

Clothes dryers are considered to be specific appliances by the National Electrical Code, and must be installed on individual branch circuits. The feeder, panel, or service volt-amp load of a dwelling electric clothes dryer is determined in accordance with S & T 220-18.

In terms of minimum loading, the branch circuit supplying an electric clothes dryer cannot be rated less than 5,000 volt-amps. If the clothes dryer has a nameplate rating greater than 5,000 volt-amps, Sec. 220-18 requires the dwelling feeder, panel, or service dryer volt-amp load to be computed at nameplate rating.

For multifamily dwellings, where individual dwelling units are furnished with laundry facilities fed by a common feeder, panel, or service, the total dryer volt-amp load can be derated according to the listed percentages in T220-18. The allowed derate is listed as a percentage according to the number of dryers supplied by the single feeder, panel, or service. Due to the diversity of loading, the greater the number of dryers, the lower is the listed percentage.

For a single-family or two-family dwelling, there is no derate allowed for the electric clothes dryer load. For multifamily dwellings, such as apartment complexes or condominiums, where the count of dryers is at least five, the percentage is taken of the total nameplate or NE Code-minimum volt-amp load (whichever is greater) of all the dryers supplied by a common (single) feeder, panel, or service.

These high-value derates of the total dryer volt-amp load are allowed because the number of dryers being used at the same time within a given multifamily dwelling structure will almost always be a lower value when compared to the number of dryers that are actually supplied in the same structure.

For reduced service neutral volt-amp load computations, the total dryer load, after any derates allowed by T220-18 have been applied, can be further derated to 70% of the service phase-to-phase volt-amp load value in accordance with Sec. 220-22.

17.28 ELECTRIC COOKING EQUIPMENT LOADS

Dwelling electric cooking equipment is also considered to be specific appliances by the National Electrical Code and must be supplied by individual branch circuits. NE Code rules governing the installation, supply, and protection for household cooking equipment, which includes free-standing ranges, surface or countertop (counter-mounted) cooking units, and wall-mounted ovens, are found in Arts. 210, 220, and 422.

Due to the diversity of loading, the nameplate full-load current ratings of these types of household cooking equipment are derated by Sec. and Table 220-19 for the load requirements of branch-circuit, feeder, panel, and service calculations.

Most branch-circuit, feeder, panel, or service computations are determined in volt-amp (VA) or kilovolt-amp (kVA) loads. Dwelling cooking equipment is normally rated in terms of kilowatts (kW) because the heating means employed is resistive. Section 220-19 allows kW loads to be treated equivalent to kVA loads for the purpose of total volt-amp load computations.

Table 220-19 consists of four columns. The first column to the left lists the number of ranges installed (supplied by) on any one (single) feeder, panel, or service. The second column, Column A, lists the "derated" demand or feeder load that can be used instead of the total nameplate ratings of all the ranges. The derated demand load corresponds to the number of ranges as listed in the first column.

The columns will be discussed and explained in order of Columns A, C, and B, with any applicable notes (Notes 1 through 4) discussed as they apply.

For a single free-standing range, counter-mounted cooking unit, or wall-mounted oven that has a nameplate rating of between 8¾ kW and 12 kW, the feeder, panel, or service demand load for the cooking unit can be derated by Column A of T220-19 to 8 kW.

When more than one range is supplied by a common feeder or service, the rating of the individual ranges determines the column or columns that are to be used. If none of the nameplate ratings exceed 12 kW, the total wattage of all the nameplate ratings can be disregarded. Instead, the value listed in Column A, which corresponds to the number of ranges supplied by a common (single) feeder, panel, or service, can be used as the total wattage load of all the ranges.

Except for the provisions of T220-19, Note 4, more than one range or other type of electric cooking equipment for dwelling structures cannot be installed on any one branch circuit. The derated demands as listed in T220-19 apply to any dwelling structure, which includes one- and two-family dwellings, multifamily dwellings (such as apartment and/or condominium complexes), public housing projects, and the efficiency apartments or kitchenettes contained in hotels or motels of resort or family vacation areas. The derated demand is restricted to the number of cooking appliances supplied by any one feeder, panel, or service.

As the total number of dwelling cooking appliances increases for any one feeder, panel, or service, the derated demand decreases. The demand factor for any number of ranges up to a count of 25, whose nameplate ratings are not greater than 12 kW, can be read directly from Column A for the number of ranges in the group.

When the count or number of ranges supplied by any one feeder or service exceeds 25, the value as listed in Column A must be computed. The demand load in Column A for a group of ranges whose count is between 26 and 40 must start with a base value of 15 kW. A product of 1 kW times the number of ranges contained in the group must be added to the base of 15 kW to determine the overall feeder or service demand volt-amp load.

When the total number or count of individual dwelling cooking equipment installed on any one feeder or service exceeds 40, the base as listed in Column A is increased to 25 kW, but the multiplier is decreased to ¾ kW.

When the nameplate rating of any type of dwelling cooking equipment exceeds 12 kW, the value as listed in Column A of T220-19 must be modified for the feeder or service derated volt-amp or kilowatt load.

For a single range or for several ranges of equal rating whose nameplate kW ratings are greater than 12 kW, Note 1 to T220-19 requires the value listed in Column A to be increased by 5% for each kW or major fraction thereof by which the actual nameplate rating exceeds 12 kW.

When the total number of ranges (or other type of cooking equipment) supplied by a single feeder, panel, or service exceeds 25 and their equal, individual nameplate ratings exceeds 12 kW, the calculation of the demand load is a two-step operation. The base value, which must be computed in accordance with Column A, must be overrated or increased according to the percentage value computed by Note 1.

When dwelling ranges or other dwelling cooking equipment with unequal nameplate kW ratings are supplied by a single feeder, panel, or service, and some or all of the individual nameplate ratings exceed 12 kW, the demand value as listed in Column A of T220-19 must be increased by the method described in Note 2.

To determine the percentage by which the table kW value must be increased, an "average kW rating" must be determined for all the ranges. Using a value of 12 kW for any range whose actual nameplate rating is less than 12 kW, all the nameplate ratings are added together. This total nameplate kW rating is then divided by the total number of ranges supplied to determine the average kW value.

The listed or computed value in Column A for the number of ranges supplied must be increased by 5% for each kW or major fraction thereof that the average kW value exceeds 12 kW.

If several ranges with unequal ratings are to be supplied by a common feeder, panel, or service, and none of the individual ranges have ratings greater than 12 kW, the demand load as listed in Column A of T220-19 would be taken as is for the load calculation.

When the rating of a single range or the rating of any individual household cooking unit supplied by a common feeder, panel, or service does not exceed 8¾ kW, Note 3 to T220-19 will allow the branch-circuit, feeder, panel, or service load to be derated to the percentages listed in either Column B or C (according to the individual nameplates ratings).

For a single range or several ranges rated 8¾ kW or less, but not less than 3½ kW, the total nameplate wattage load can be derated by the percentages listed in Column C of T220-19 for the number of ranges supplied by a common feeder, panel, or service (branch circuits are figured on an individual basis).

For a single range or several ranges rated less than 3½ kW, the total nameplate wattage load can be derated by the percentages listed in Column B of T220-19 for the number of ranges supplied by a common feeder, panel, or service.

The diversity of loading for ranges of lesser ratings is not as widespread as it is for ranges of greater ratings. For example, to boil water, cook bacon, and fry eggs on a four-burner surface unit rated at 3 kW would require a higher percentage of the unit's nameplate rating, as compared to the same task's being accomplished on a free-standing range rated at 7.6 kW. For this reason, the group percentages listed in Column B are higher than those listed in Column C.

If a common feeder, panel, or service supplies a mixture of household cooking appliances with ratings found in all three columns of T220-19 (the nameplate ratings range from 1 3/4 kW up to 27 kW), the total range load can be determined by the number of ranges and Column A, using Note 2 when and if necessary; or Column A can be used for the number of ranges for which it applies, Column B can be used for the number of ranges for which it applies, and Column C can be used for the number of ranges for which it applies.

Either method of load computation is acceptable by the NE Code (due to the wording of Note 3 to T220-19). Some NE Code sources recommend both computations, taking the lesser value of the dwelling cooking equipment volt-amp load for the feeder, panel, or service volt-amp computation.

17.29 HEATING AND COOLING LOADS

The installation of fixed electric space heating equipment in a dwelling is governed by the requirements of Art. 424. The installation of air-conditioning equipment, whether central cooling evaporator and condensor units or heat pumps, is governed by the requirements of Art. 440. Both fixed electric space heating and air-conditioning refrigeration equipment are considered to be utilization equipment by the NE Code that must be installed on specific (individual) branch circuits.

As outlined in Sec. 17.20 of this chapter, except for the 125% overrate for the feeder, panel, or service load when a single branch circuit supplies a central electric furnace, both heating and cooling loads are computed at 100% for the feeder, panel, or service volt-amp load computations. The two loads are considered dissimilar (noncoincidental) at the feeder level such that the smaller of the two is dropped (Sec. 220-21) for the overall volt-amp load computations.

17.30 MOTOR LOADS

The volt-amp loads of motor-driven equipment contained in a dwelling structure is computed at 100% and listed under the fixed-appliance load in the standard method of feeder, panel, or service load computations.

As with any feeder computation involving motor loads, Sec. 220-14, by reference to Sec. 430-24, requires 25% of the largest motor load to be added to the overall dwelling feeder, panel, or service volt-amp load.

If the air-conditioning equipment was omitted in the comparison of noncoincidental loads in Sec. 220-21, then the condensor unit should also be omitted for the determination of the largest motor load. If the dwelling either does not have fixed electric space heating, the heating of the dwelling structure is accomplished by means of a heat pump, or in comparing the noncoincidental loads (with fixed electric space heating) the condensor unit was the larger volt-amp load, Sec. 440-7 in the NE Code requires the rated-load current or branch-circuit selection current when specified on the condensor nameplate for the hermetically sealed refrigerant motor compressor (normally shortened to just motor compressor) to be considered in the largest motor load computation.

The motor-compressor rated-load or branch-circuit selection current can be converted to a volt-amp load by the power formula.

In the absence of motor-compressor rated-load or branch-circuit selection current specified on the condensor unit nameplate or given in a test problem or example, the heat pump or air-conditioning unit should not be considered for the largest motor load.

The heat pump or air-conditioner condensor unit is multimotor/combination-load equipment. Sections 430-25(b) and 440-35 both specify that the branch-circuit conductors should be sized according to the "minimum circuit capacity" specified on the equipment nameplate. In other words, the largest motor load within the equipment has already been adjusted for the 25% overrate (to 125%) by the manufacturer in the nameplate computation.

17.31 NEUTRAL CONSIDERATIONS

The feeder or service neutral can be "full size," which means equivalent in wire size to the ungrounded phase conductors of the dwelling feeder or service, or the reduced feeder or service neutral can be sized in accordance with Sec. 220-22.

When computing the volt-amp load for a reduced neutral conductor serving a dwelling structure, only the lower-voltage phase-to-neutral loads are considered. This is true except for the cooking equipment load, which is derated in accordance with S & T 220-19, and the electric clothes dryer load, which is derated in accordance with S & T 220-18. Section 220-22 will allow the already derated service load (240-volt load) for either of these multiwire branch-circuit appliance loads to be further derated to 70% for the reduced feeder or service neutral load computations.

The principal loads of the feeder or service are tabulated in the same manner for the reduced feeder or service neutral conductor as they are for the ungrounded feeder or service conductors of the dwelling. The general lighting and receptacle loads are normally all phase-to-neutral loads. The lower-voltage fixed-appliance and motor loads are tabulated for the allowable derate of Sec. 220-17. If any heating or cooling loads have the lower voltage rating, they must be computed and compared (noncoincidental loads) for the neutral feeder or service volt-amp load. If any motor loads are rated at the lower voltage, then 25% of the largest motor load must be added to the overall feeder or service neutral volt-amp load.

When the total dwelling feeder or service neutral volt-amp load is determined, the power formula is used with the higher phase-to-phase voltage rating to determine the reduced neutral current.

For a single-family dwelling, the service neutral conductor, or the main feeder neutral conductor (feeder circuit conveys the same load current as the service-entrance conductors) can be reduced in wire size (higher value of rated ampacity) according to the guidelines of Note 3 of T310-16. When either the ampacities of Note 3 or T310-16 are used to size the ungrounded circuit conductors, the feeder or service neutral conductor can be sized according to the guidelines of Sec. 220-22 for the load-return requirements.

The service neutral conductor must also meet the ground-fault return requirements of Sec. 250-23(b). For single-family dwellings, the service neutral must to be at least equal in wire size to the service grounding-electrode conductor as determined by T250-94.

17.32 SINGLE-FAMILY LOAD COMPUTATIONS

Due to the diversity of loading in a dwelling and the allowances made for it in the service volt-amp load computations by the standard method listed in Part B of Art. 220, continuous loads are not used as an applicable derate or overrate when determining the required load current of the feeder, panel, or service conductors.

Example 17–2. A single-family dwelling will be supplied with a 120/240-volt, single-phase, 3-wire service. The outer dimensions of the house are 36 by 75 feet. These dimensions include a double carport that measures 30 by 26 feet, which cannot be adapted for future living space.

Given the following loads:

Description	Rating	
Water heater	4.5 kW	240 V
Dryer	5.6 kW	240 V
Range	16.5 kW	240 V
Dishwasher	1.4 kVA	
Kitchen hood exhaust fan	500 VA	
Ceiling exhaust fan	750 VA	
Basement sump pump	960 VA	
Well pump	1.7 kVA	230 V
Five wall-insert electric heaters	4.5 kW	230 V each
Central air-conditioning:		
Outside unit:	7.2 kVA	230 V
Central air handler:	½ hp	230 V

Determine:

1. NE Code-minimum service rating
2. NE Code-minimum wire size of the service-entrance conductors, phases A and B, using THW copper

3. NE Code-minimum wire size of the reduced service-entrance neutral conductor, using THW copper
4. NE Code-minimum wire size of the service grounding-electrode conductor run to a driven-rod electrode

Volt-amp load computations: (worksheet) Ref. Art. 220, Part B

General lighting and receptacle load:

Floor area computation: dwelling structure Ref. 220-3(b)

36 feet × 75 feet = 2,700 square feet

Floor area computation: double carport

30 feet × 26 feet = 780 square feet

Net floor area computation:

2,700 square feet – 780 square feet = 1,920 square feet

Load computation: Ref. T220-3(b)

1,920 square feet × 3 VA/square foot = 5,760 volt-amps

Kitchen small-appliance load computation: Ref. 220-16(a)

2 × 1,500 volt-amps = 3,000 volt-amps

Laundry load: 1,500 volt-amps Ref. 220-16(b)

Summation of general lighting, receptacle and small-appliance load:

General lighting and receptacle	5,760 volt-amps
Kitchen small-appliance	3,000 volt-amps
Laundry	1,500 volt-amps
Total:	10,260 volt-amps

Derate of general lighting, receptacle, and small-appliance load:

Ref. S & T 220-11

First 3,000 volt-amps @ 100% = 3,000 volt-amps

Remaining load computation:

10,260 VA – 3,000 VA = 7,260 VA

Derate computation:

35% of 7,260 VA =	2,541 volt-amps
Total:	5,541 volt-amps

Fixed-appliance/motor load:

Decription	Rating
Water heater:	1,000 × 4.5 kW* = 4,500 volt-amps
Dishwasher:	1,000 × 1.4 kVA = 1,400 volt-amps
Kitchen hood exhaust fan	500 volt-amps
Ceiling exhaust fan	750 volt-amps
Basement sump pump	960 volt-amps
Well pump:	1.7 kVA × 1,000 = 1 ,700 volt-amps
Total:	9,810 volt-amps

Derate computation of fixed-appliance/motor load: Ref. 220-17

75% of 9,810 volt-amps = 7,358 volt-amps

Neutral fixed-appliance/motor load:

Decription	Rating
Dishwasher	1,400 volt-amps
Kitchen hood exhaust fan	500 volt-amps
Ceiling exhaust fan	750 volt-amps
Basement sump pump	960 volt-amps
Total:	3,610 volt-amps

Derate computation of neutral fixed-appliance/motor load: Ref. 220-17

75% of 3,610 volt-amps = 2,708 volt-amps

Dryer load determination: Ref. S & T 220-18
NE Code-minimum rating: 5,000 volt-amps
Nameplate rating: 5,600 watts*

Compare for larger: Use 5,600 volt-amps

Derate computation of neutral dryer load: Ref. 220-22

70% of 5,600 volt-amps = 3,920 volt-amps

Range load determination: Ref. 220-19
Table base value: 8 kW Ref. T220-19, Col. A

*Consider watts equivalent to volt-amps.

Overrate percentage computation: Ref. T220-19, Note 1

16.5 kW − 12 kW = 4.5 kW (round up)

5 kW × 5 %/kW = 25%

100% + 25% = 125%

Load computation:

1,000 × 8 kW = 8,000 watts

125% of 8,000 watts* = 10,000 volt-amps

Derate computation of neutral range load: Ref. 220-22

70% of 10,000 volt-amps = 7,000 volt-amps

Heating load computations: Ref. 220-15

1,000 × 4.5 kW = 4,500 watts

5 × 4,500 watts* = 22,500 volt-amps

Cooling load computations: Ref. 220-14

Outside unit: Ref. 440-35

1,000 × 7.2 kVA = 7,200 volt-amps

Central air handler: Ref. T430-148

Motor FLC = 4.9 amps

240 volts × 4.9 amps = 1,176 volt-amps

Total: 8,376 volt-amps

Noncoincidental loads: heating/cooling Ref. 220-15

Compare for larger: 22,500 volt-amps

Largest motor load: Ref. 220-14 Ref. 430-24

240-volt rating: well pump

25% of 1,700 volt-amps = 425 volt-amps

120-volt rating: sump pump

25% of 960 volt-amps = 240 volt-amps

*Consider watts equivalent to volt-amps.

1. **NE Code-minimum service rating determination:**

Summation of volt-amp loads: Ref. 220-10(a)

General lighting, receptacle, and small-appliance	5,541 volt-amps
Fixed appliance/motor	7,358 volt-amps
Dryer	5,600 volt-amps
Range	10,000 volt-amps
Heating/cooling	22,500 volt-amps
Largest motor	425 volt-amps
Total:	51,424 volt-amps

Full-load current computation:

$$51{,}424 \text{ volt-amps} \div 240 \text{ volts} = 214.3 \text{ amps}$$

Standard OCPD ratings: (NHS) 225 amps Ref. 240-6(a)

2. **Wire size of the SE conductors, phases A and B determination:**

Ref. 240-3

Wire terminal temperature limitation: 75°C Ref. 110-14(c)(2)

Table-rated ampacity: 230 amps Ref. T310-16

Wire size: No. 4/0 AWG THW copper

3. **Wire size of the reduced service-entrance neutral conductor determination:**

Summation of volt-amp loads: Ref. 220-10(a) Ref. 220-22

General lighting, receptacle, and small appliance	5,541 volt-amps
Fixed appliance/motor	2,708 volt-amps
Dryer	3,920 volt-amps
Range	7,000 volt-amps
Largest motor	240 volt-amps
Total:	19,409 volt-amps

Neutral full-load current computation:

$$19{,}409 \text{ volt-amps} \div 240 \text{ volts} = 80.9 \text{ amps}$$

Overrate adjustment to align the neutral rated-load current with OCPD rating:

Adjustment computation:

$$\frac{\text{Service OCPD rating}}{\text{Phase FLC rating}} = \frac{225 \text{ amps}}{214.3 \text{ amps}} = 1.050$$

Overrate computation:

$$1.050 \times 80.9 \text{ amps} = 84.9 \text{ amps}$$

Wire size for load return:

Wire terminal temperature limitation: 60°C Ref. 110-14(c)(1), Excp. 1
Table-rated ampacity: 85 amps Ref. T310-16
Wire size: No. 3 AWG THW copper

Wire size for ground-fault return: Ref. 250-23(b)
Wire size of ungrounded-SE conductors: No. 4/0 AWG THW copper
Neutral wire size: No. 2 AWG copper Ref. T250-94

NE Code-minimum: Ref. 310-15(c)

Compare for larger: No. 2 AWG THW copper

4. Grounding-electrode conductor run to a driven-rod electrode.

Ref. 250-83(c) Ref. 250-94, Excp. a
Wire size: No. 6 AWG copper Ref. (copper) 250-92(a)

As with the nondwelling feeder, panel, or service volt-amp and resulting current load computations shown in Example 17-1, the standard method of load computation for the dwelling structure is found in Part B of Art. 220.

The same format, in terms of a worksheet, then pulling the computed values for various computations in terms of service rating, size of the ungrounded service conductors, the grounded service conductor, and other determinations involving the service is followed whether dwelling or nondwelling load computations. The major difference is in the overrate for continuous load. General lighting of a dwelling structure is not overrated to 125%, due to the diversity of loading within the structure.

Code challenge 17–2. A single-family dwelling is supplied with a 120/240-volt, single-phase, 3-wire service. The outer dimensions of the house are 34 by 80 feet. These dimensions include a double carport that measures 28 by 32 feet, which cannot be adapted for future living space.

Given the following loads:

Decription	Rating	
Water heater	5.5 kW	240 V
Dryer	4.8 kW	240 V
Range	20.8 kW	240 V
Dishwasher	1.2 kVA	
Kitchen hood exhaust fan	300 VA	
Attic ventilating fan	640 VA	
Basement sump pump	864 VA	
Swimming pool filter pump	1.4 kVA	230 V
4 Wall-insert electric heaters	5.0 kW	230 V
Central air conditioning:		
Outside unit:	8.2 kVA	230 V
Air handler:	¾ hp	230 V

Determine:

1. NE Code-minimum service rating
2. NE Code-minimum wire size of the service-entrance conductors, phases A and B, using THWN copper
3. NE Code-minimum wire size of the reduced service-entrance neutral conductor, using THWN copper
4. NE Code-minimum wire size of the service grounding-electrode system run to a driven-rod electrode

Volt-amp load computations: (worksheet) Ref. (A)____________________

General lighting and receptacle load:

Floor area computation: dwelling structure Ref. (B)____________________

__________ feet × __________ feet = (C) __________ square feet

Floor area computation: double carport

__________ feet × __________ feet = (D) __________ square feet

Net floor area computation:

__________ feet − __________ feet = (E) __________ square feet

Load computation: Ref. (F) ____________________

__________ square feet × __________ VA/square foot = (G) __________ VA

Kitchen small-appliance load computation: Ref. (H)____________________

__________ × (I) __________ volt-amps = __________ volt-amps

Laundry load: (J) __________ volt-amps Ref. (K)____________________

Summation of general lighting, receptacle and small-appliance load:

General lighting and receptacle	__________	VA
Kitchen small-appliance	__________	VA
Laundry	__________	VA
Total:	(L) __________	VA

Derate of general lighting, receptacle, and small-appliance load:

Ref. (M) ____________________

First (N) ___________ volt-amps @ 100% = ___________ VA
Remaining load computation:

___________ VA – ___________ VA = (O) ___________ VA

Derate computation:

(P) ___________ % of ___________ volt-amps = ___________ VA
Total: (Q) ___________ VA

Fixed-appliance/motor load:

Decription				Rating
Water heater:	_________ ×	_________ kW* =	_________	VA
Dishwasher:	_________ ×	_________ kVA =	_________	VA
Kitchen hood exhaust fan			_________	VA
Attic ventilating fan			_________	VA
Basement sump pump			_________	VA
Swimming pool filter pump	_________ ×	_________ kVA =	_________	VA
Total:		(R)	_________	VA

Derate computation of fixed-appliance/motor load: Ref. (S) ___________________

___________ % of ___________ volt-amps = (T) ___________ volt-amps

Neutral fixed-appliance/motor load:

Decription		Rating
Dishwasher	___________	VA
Kitchen hood exhaust fan	___________	VA
Attic ventilating fan	___________	VA
Basement sump pump	___________	VA
Total: (U)	___________	VA

Derate computation of neutral fixed-appliance/motor load:
Ref. (V)____________________

___________ % of ___________ volt-amps = (W) ___________ volt-amps

Dryer load determination: Ref. (X)____________________
NE Code-minimum rating: ___________ volt-amps
Nameplate rating: ___________ kW*

Compare for (Y) ___________ : Use (Z) ___________ volt-amps

*Consider watts equivalent to volt-amps.

Derate computation of neutral dryer load: Ref. (1A)____________________

____________ % of ____________ volt-amps = (1B) ____________ volt-amps

Range load determination: Ref. (1C ____________________

Table base value: (1D) ____________ kW Ref. (1E) ____________________

Overrate percentage computation: Ref. (1F) ____________________

____________ kW − (1G) ____________ kW = ____________ kW

(1H) ____________ kW × ____________ %/kW = ____________ %

____________ % − ____________ % = (1I) ____________ %

Load computation:

____________ × ____________ kW = (1J) ____________ watts

____________ % of ____________ watts* = (1K) ____________ volt-amps

Derate computation of neutral range load: Ref. (1L) ____________________

____________ % of ____________ volt-amps = (1M) ____________ volt-amps

Heating load computations: Ref. (1N)____________________

____________ × ____________ kW = (1O) ____________ watts

____________ × ____________ watts* = (1P) ____________ volt-amps

Cooling load computations: Ref. (1Q)____________________

Outside unit: Ref. (1R)____________________

____________ × ____________ kVA = ____________ VA

Central air handler: Ref. (1S) ____________________

Motor FLC = (1T) ____________ amps

____________ volts × ____________ amps = ____________ VA

Total: (1U) ____________ VA

Noncoincidental loads: heating/cooling Ref. (1V)____________________

Compare for (1W) ____________ : ____________ volt-amps

*Consider watts equivalent to volt-amps.

Largest motor load: Ref. (1X) ____________ Ref. (1Y) ____________

230-volt rating: (1Z) ____________

____________ % of ____________ volt-amps = (2A) ____________ volt-amps

115-volt rating: (2B) ____________

____________ % of ____________ volt-amps = (2C) ____________ volt-amps

1. NE Code-minimum service rating determination:

Summation of volt-amp loads: Ref. (2D) ____________

General lighting, receptacle, and small-appliance	____________	VA
Fixed appliance/motor	____________	VA
Dryer	____________	VA
Range	____________	VA
Heating/cooling	____________	VA
Largest motor	____________	VA
Total:	(2E) ____________	VA

Full-load current computation:

____________ volt-amps ÷ ____________ volts = (2F) ____________ amps

Standard OCPD ratings: (NHS) (2G) ____________ amps
Ref. (2H) ____________

2. Wire size of the SE conductors, phases A and B determination:

Ref. (2I) ____________

Wire terminal temperature limitation: (2J) ____________ °C

Ref. (2K) ____________

Table-rated ampacity: (2L) ____________ amps Ref. (2M) ____________

Wire size: (2N) ____________ THWN copper

3. Wire size of the reduced service-entrance neutral conductor determination:

Summation of volt-amp loads:

Ref. (2O) ____________ Ref. (2P) ____________

General lighting, receptacle, and small-appliance	____________	VA
Fixed appliance/motor	____________	VA
Dryer	____________	VA
Range	____________	VA
Largest motor	____________	VA
Total:	(2Q) ____________	VA

Neutral full-load current computation:

____________ volt-amps ÷ ____________ volts = (2R) ____________ amps

Overrate adjustment to align the neutral rated-load current with OCPD rating:
Adjustment computation:

$$\frac{\text{Service OCPD rating}}{\text{Phase FLC rating}} = \frac{\text{(2S) \_\_\_\_\_\_\_\_\_\_ amps}}{\text{(2T) \_\_\_\_\_\_\_\_\_\_ amps}} = \text{\_\_\_\_\_\_\_\_\_\_}$$

Overrate computation:

(2U) ____________ × ____________ amps = ____________ amps

Wire size for load return:
Wire terminal temperature limitation: (2V) ____________ °C
Ref. (2W) ____________________
Table-rated ampacity: (2X) ____________ amps Ref. (2Y) ____________________
Wire size: (2Z) ____________ THWN copper

Wire size for ground-fault return: Ref. (3A) ____________________
Wire size of ungrounded-SE conductors: (3B) ____________ THWN copper
Neutral wire size: (3C) ____________ copper Ref. (3D) ____________________

NE Code-minimum: Ref. (3E) ____________________

Compare for (3F) ____________ : (3G) ____________ THWN copper

4. Grounding-electrode conductor run to a driven-rod electrode.
Ref. (3H) ____________________ Ref. (3I) ____________________
Wire size: (3J) ____________ copper Ref. (copper) (3K) ____________________

17.33 DWELLING SERVICE CONDUCTORS

The service-entrance phase and neutral conductors for a single-family dwelling are sized in accordance with the type of insulation, the type of conductor material, the wire terminal temperature limitations [Sec. 110-14(c)], and the calculated load current (required ampacity) in Table 310-16, or in the Note 3 Table to Table 310-16. The service-entrance conductors can be in the form of SE multiconductor cable or single-insulated conductors contained in conduit.

If the service-entrance conductors are to pass through an attic where ambient temperatures are expected to rise above 86°F (30°C), then the ampacity correction factors at the bottom of T310-16 (or other ampacity tables as they are used) should be applied to derate the ampacity of the intended service-entrance conductors. If the service-entrance conductors are contained with more than three conductors in a raceway that passes through an attic of higher ambient temperature, then both derates or overrates must be applied to the intended service-entrance conductors.

The derate for higher ambient temperatures (one of the prevailing conditions discussed in Chapter 2 of this text) is often abused in single-family dwellings. Electricians tend to route SE cable through an attic space without derating the contained conductor ampacities for the hot summer days when it's hotter than 100°F in the shade. The diversity of loading helps this shortcoming.

The derate for more than three current-carrying conductors (another prevailing condition) is very seldom (if ever) used for the service-entrance of a single-family dwelling, because most of these services are single-phase 3-wire. If the service-disconnecting means is located outside the dwelling structure, the interior panel becomes a single-phase, 4-wire, subfed panel. The fourth wire is an equipment-grounding conductor. The requirements of Note 8 for more than three circuit (current-carrying) conductors in a common raceway are voided by Note 11: Equipment-grounding conductors or bonding jumpers are not counted in the provisions of Note 8.

17.34 MINIMUM SERVICE RATING—SINGLE-FAMILY DWELLING

The standard service calculation as discussed in this chapter for Part B of Art. 220 determined the minimum service load current of the dwelling according to its connected loads. Section 230-42(b) establishes the minimum rating for a single-phase, 3-wire service for a single-family dwelling at 100 amps. A 100-amp service is required when either the dwelling has an initial net computed service load of 10 kVA or more, or when the dwelling service panel contains six or more 2-wire branch circuits.

The term 3-wire indicates two ungrounded (hot or phase) conductors and a third grounded neutral conductor. In comparison, 2-wire branch circuits can be phase-to-phase or phase-to-neutral circuits. Phase-to-phase, 2-wire branch circuits supply loads connected between two ungrounded conductors (higher voltage as compared to the phase-to-neutral circuit). Phase-to-neutral, 2-wire branch circuits supply loads connected between an ungrounded phase conductor and the grounded neutral conductor.

If the single-family dwelling has neither the six 2-wire branch circuits, nor the 10-kVA net computed load, then the service is required to have a minimum rating of 60 amperes [Sec. 230-42(b)(3)].

For a structure to quailify as a dwelling (Art. 100), the one or more rooms must provide areas for eating, living, and sleeping, and have permanent provisions for cooking and sanitation. If the structure is not a dwelling, then its electrical service comes under the jurisdiction of Sec. 230-42(b)(3)—its minimum service rating is 60 amperes.

REVIEW QUESTIONS—PART ONE

1. In general (standard method of computation), feeder or service conductors must have a rated ampacity according to wire size that is at least equal to the computed load (connected full-load current).

True or False

2. In general, feeder or service conductors must be protected against overcurrent in accordance with their rated ampacity.

True or False

3. Section ____________ lists the standard ratings of fuses and circuit breakers.

4. As defined in the NE Code, the rated ampacity of a conductor, according to wire size, type of insulation, and type of conductor material, was determined under continuous load conditions.

   True or False

5. The rated ampacity of the feeder or service conductors rated between 0 and 2,000 volts is listed, according to wire size, type insulation, and type of conductor material, in Tables ____________ through ____________ .

6. The rating of continuous-load currents, such as store lighting, must be increased to ____________ % before being added to the noncontinuous load currents such as convenience receptacles to size the overcurrent protection device of a feeder or service.

7. Phase rotation in a panel, switchboard, or other load center is read from top to bottom, left to right, or front to back.

   True or False

8. The feeder neutral conductor can be smaller than the ungrounded phase conductors if it is sized to carry the maximum unbalanced current load (heaviest loaded phase) of the panel or other loadcenter.

   True or False

9. The service neutral conductor can be sized to carry the maximum unbalanced current load (heaviest loaded phase) of the panel or other loadcenter, but its wire size cannot be smaller than the required service grounding-electrode conductor as sized by T250-94 (for phase conductors smaller than 1,100 kcmil copper or 1,750 kcmil aluminum).

   True or False

10. If a panel or other loadcenter is rated and used as service equipment, the equipment-grounding conductors can share a common connection with the service and load-side neutral conductors in the service enclosure.

    True or False

11. If the panel is subfed by a feeder from the service equipment panel, equipment-grounding conductors must be connected to a separate auxiliary grounding bus. The feeder and load-side neutral conductors must be insulated from equipment grounding or bonding.

    True or False

12. If a panel is to be identified as a lighting and appliance panel, at least ____________ % of the contained branch-circuit overcurrent devices for which a neutral conductor is provided must be rated at 30 amps or less.

13. If the percentage specified in Question No. 12 is not met, the panel should be identified as a power panel.

    True or False

14. The ungrounded (phase) circuit conductors can be identified by any color except ____________ , ____________ ____________ , or ____________ .

15. The grounded (neutral) circuit conductors can be identified by either ____________ or ____________ ____________ in color.

16. Grounded-circuit conductor wire ____________ AWG or smaller must be identified at the point of manufacture to one of the colors specified in Question 15.

17. Grounded-circuit conductor wire ____________ AWG or larger can be color identified black at the point of manufacture and reidentified to one of the colors specified in Question 15 at the time of installation at all terminations.

18. Equipment-grounding conductors, grounding-electrode conductors, or bonding jumpers used in the grounding and bonding of the noncurrent-carrying metal parts of an electrical distribution or the grounded surfaces of the building or other structure can be bare.

    True or False

19. When equipment-grounding conductors, grounding-electrode conductors, or bonding jumpers are covered or insulated, they must be color-identified ____________ or green with one or more ____________ stripes.

20. Insulated grounding conductor wire ____________ AWG or smaller must be identified at the point of manufacture to one of the colors specified in Question 19.

21. Insulated grounding conductor wire ____________ AWG or larger can be color identified black at the point of manufacture and reidentified to one of the colors specified in Question 19 at the time of installation at all points of accessibility.

Questions 22 and 23: The general or overhead lighting load in a nondwelling structure consists of the actual and the NE Code-minimum lighting load.

22. The actual lighting load is computed in accordance with Sec. ____________ .

23. The NE Code-minimum lighting load is computed in accordance with S & T ____________ .

REVIEW QUESTIONS—PART TWO

1. Parking lot lighting for a store or office building must be overrated to ____________ % to compensate for continuous load.

2. Show-window lighting must be computed at a minimum of ____________ volt-amps per linear foot, measured along the ____________ horizontal base of the window.

3. Track lighting must be computed at a minimum of ____________ volt-amps for each ____________ -foot section of track.

4. A commercial store with ground-floor entrance is required to have a ____________ -amp, 120-volt branch circuit provided for the installation of a sign or other outline lighting. When the actual sign load is not known, the NE Code requires a minimum of ____________ volt-amps to be assigned to this circuit for load computations.

5. Nondwelling convenience (duplex) receptacles must be computed at a minimum of ____________ volt-amps per outlet.

6. Nondwelling convenience receptacles installed in plugmold (multioutlet assembly) must be computed at a minimum of ____________ volt-amps per ____________ -foot section of lighting loading.

7. Nondwelling convenience receptacles installed in plugmold must be computed at a minimum of ____________ volt-amps per ____________ -foot section of appliance (heavy) loading.

8. No load assignment is made for dwelling convenience receptacles. These receptacles (except for small appliance and laundry) are computed as part of the general lighting load in a dwelling.

True or False

9. Appliances, motors, and other utilization equipment loads are computed at nameplate or NE Code-minimum load values for nondwelling feeder, panel, or service volt-amp computations.

True or False

10. The NE Code requires a minimum of ____________ 20-amp, 120-volt branch circuits to be installed for the supply of small appliances in every kitchen of a dwelling.

11. For the standard method of load computation for either feeder, panel, or service ratings, each kitchen small-appliance circuit of Question 10 has a minimum load assignment of ____________ volt-amps.

12. When four or more fixed appliances or motor loads are installed in a dwelling structure, the total nameplate or NE Code-minimum rating can be derated to ____________ %.

13. When the motor loads are computed at NE Code-minimum values, ____________ % of the largest motor volt-amp load must be added to the total load computation.

14. The volt-amp load of an electric clothes dryer must be computed at nameplate rating or ____________ volt-amps, whichever is greater, in the feeder, panel, or service load computations.

15. The total electric clothes dryer load cannot be derated for either feeder, panel, or service load computations unless the count exceeds ____________ .

16. For the feeder or service neutral conductor unbalanced load computations, the computed electric clothes dryer load can be derated to ____________ %.

17. The feeder, panel, or service volt-amp load of dwelling electric cooking equipment can be assigned a value of only ____________ kW if the nameplate rating does not exceed 12 kW.

18. When the nameplate rating of an individual unit of electric cooking equipment does not exceed 8¾ kW, the nameplate rating can be derated to ____________ % for the feeder, panel, or service volt-amp load computations.

19. For the dwelling feeder or service neutral conductor unbalanced load computations, the computed electric cooking equipment load can be derated to ____________ %.

20. Fixed electric space heating equipment for a building or other structure must be computed at 100% for the feeder, panel, or service load computations.

True or False

21. When the fixed electric space heating equipment in a building or other structure consists of an electric furnace supplied by single branch circuit, the load rating must be computed at ____________ % for the feeder, panel, or service load computations.

22. Air-conditioning equipment installed as a unit at a building or other structure must be computed at 100% for the feeder, panel, or service load computations.

True or False

23. When the structure is heated by fixed electric space heating, the heating and air-conditioning load can be compared for the larger of the two volt-amp loads and the larger of the two dropped in the feeder, panel, or service volt-amp load computation.

True or False

24. The total nameplate rating of nondwelling cooking and other central kitchen equipment can be derated according to the number of pieces of equipment and the percentages listed in Table __________.

25. The NE Code recommends that spare breaker space and spare amperage be left in all feeder, panel, or service loadcenter computations to allow for future growth and modification.

True or False

REVIEW PROBLEMS

1. A local electrical supply house is supplied with a 120/240-volt, single-phase, 3-wire service. The building, which measures 72 by 110 feet, has a storage area of 40 by 40 feet, a grade-level front entrance, and the following loads:

25	Linear feet of show window
17	kVA of air-conditioning
90	Convenience receptacles
4	⅓-hp, 115-volt air handlers for gas heating units
2	17-gallon, 2,000-watt, 120-volt water heaters

Determine:

a. Service rating (allow 20% overrate for future growth)
b. Wire size of the service-entrance conductors, phases A and B, using RHW copper
c. Wire size of the reduced service-entrance neutral conductor, using RHW copper
d. Wire size of the copper grounding-electrode conductor run to a metal-rebar, concrete-encased footing
e. Wire size of the copper grounding-electrode conductor run to a driven-rod electrode

2. A single-family dwelling will be supplied with a 120/240-volt, single-phase, 3-wire service. The outer dimensions of the house are 40 by 66 feet. These dimensions include a double carport that measures 24 by 34 feet. If the carport cannot be adapted for future use as living space, and the dwelling will have the following loads:

Description	Rating	
Water heater	3.5 kW	240 V
Dryer	4.7 kW	240 V
Range	16 kW	240 V
Well pump	1.8 kVA	230 V
Dishwasher	1.5 kVA	
Kitchen exhaust fan	360 VA	
Ceiling exhaust fan	580 VA	
Basement sump pump	760 VA	
Five wall-insert electric heaters	4.2 kW	240 V
Central air-conditioning		
Outside unit	9.4 kVA	230 V
Central air handler	2.1 kVA	230 V

Determine:

a. NE Code-minimum service rating
b. NE Code-minimum wire size of the service-entrance conductors, phases A and B, using RHW copper
c. NE Code-minimum wire size of the reduced service-entrance neutral conductor, using RHW copper
d. NE Code-minimum wire size of the grounding-electrode conductor run to a metal underground water pipe
e. NE Code-minimum wire size of the grounding-electrode conductor run to a driven-rod electrode

Chapter 18

Math Help

This chapter is designed as a self-help reference for students who may be having problems with the math requirements of this text. Different math fundamentals are presented for discussion and explained by example.

18.1 MATHEMATICAL EQUATIONS AND OPERATIONS

When the product, sum, difference, or division of numbers or unknowns are set to equal a given amount, a mathematical equation is formed. The product, as determined by multiplying two or more numbers or unknowns, the sum, as determined by the addition of two or more numbers or unknowns, the difference, as determined by subtracting two numbers or unknowns, and the division, as determined by dividing a number or unknown by another number or unknown, are mathematical operations used to solve a given equation.

The numbers or integers in a given mathematical equation are represented as is. The unknowns in a given mathematical equation are represented by the letters in the alphabet. For general algebraic equations (mathematical equations used to solve for one or more unknowns), the beginning letters of the alphabet are used for the unknowns.

Example.

$$A + 13 = 22$$

$$A = 22 - 13$$

$$A = 9$$

For electrical equations, the symbol used for the unknown to be determined is normally taken from the word used to describe the unknown.

Example. Ohm's law stipulates that the applied voltage of a circuit is equal to the product of the circuit current times the circuit resistance. If V equals the applied voltage, I equals the circuit current, and R equals the circuit resistance, then

$$V = I \times R \text{ (also stated as IR)}$$

If I has an assigned value of 10 amps, and R has an assigned value of 12 ohms, then V equals 120 volts. In equation form,

$$V = I \times R$$

$$V = 10 \text{ amps} \times 12 \text{ ohms}$$

$$V = 120 \text{ volts}$$

As a rule, algebraic equations can only be solved for one unknown. There are mathematical ways to determine unknown algebraic values in terms of other unknown algebraic values, but these ways are beyond the scope of this text.

When an equation contains more than one operation, which may be addition, subtraction, multiplication, or division, order is imperative. Order indicates the operations are not of equal rank. A simple statement that may help the student electrician to remember operation order is "My Dear Aunt Sally."

"My" represents multiplication. "Dear" represents division. "Aunt" represents addition. And "Sally" represents subtraction. Multiplication and division actually have equal rank over either subtraction or addition. If the equation contains both multiplication and division, which operation is performed first will have no effect on the outcome.

If the equation contains both addition and subtraction operations in addition to either multiplication or subtraction, which operation is performed first will have an effect on the outcome.

In the equation $10 + 14 \times 6 - 16 \div 4 = R$, "My Dear Aunt Sally" would require the product of 14×6 to occur before their result could be added to 10. "My Dear Aunt Sally" would also require the division of 16 by 4 before the other terms of the equation are reduced by the subtraction operation.

$$14 \times 6 = 84 \text{ and } 16 \div 4 = 4$$

Substituting results for the computed terms,

$$10 + 84 - 4 = R$$

As long as the values are maintained on their respective sides of the equal marks, the value of the equation remains unchanged. The equation can be rewritten in reverse order as:

$$R = 10 + 84 - 4$$

Solving by addition $\quad R = 94 - 4$
Solving by subtraction $\quad R = 90$

18.2 GROUPING TERMS FOR MATHEMATICAL OPERATIONS

The order required by "My Dear Aunt Sally" is simplified by the use of grouping. Grouping establishes priority as to what operations will occur in what order in a multistep equation.

Grouping is accomplished by parentheses ( ), brackets [ ], and braces { }. If only one grouping occurs in the equation, parentheses are used to contain collective terms.

All operations within the parentheses must be completed before operations outside the parentheses can occur. Grouping of terms or numbers indicates that the result of the operations contained within the parentheses, brackets, or braces must be considered as a "single term" within the equation.

If the equation $10 + 14 \times 6 - 16 \div 4 = R$ is written as $(10 + 14) \times 6 - 16 \div 4 = R$, the value of R as computed in the previous example will not be the same. The grouping of 10 + 14 within the parentheses requires their sum to be multiplied by 6.

$$24 \times 6 - 16 \div 4 = R$$

Reversing the order of the equation,

$$R = 24 \times 6 - 16 \div 4$$

Solving according to "My Dear Aunt Sally,"

$$24 \times 6 = 144 \text{ and } 16 \div 4 = 4$$

Substituting results for computed terms and substracting,

$$R = 144 - 4$$

$$R = 140$$

If a group of operations and terms are to be contained within another group, parentheses are used to contain the inner group. Brackets are used to contain the outer group of operations and terms.

If the equation $10 + 14 \times 6 - 16 \div 4 = R$ is written as $[(10 + 14) \times 6 - 16] \div 4 = R$, the value of R as computed in the previous example will not be the same. The operations contained in the parentheses must be performed first. Then operations contained within the brackets are performed next. The operations within the open equation are performed last.

To remove the parentheses, 10 is added to 14 to determine a single term of 24. The equation becomes

$$[24 \times 6 - 16] \div 4 = R$$

Two operations are contained within the brackets. "My Dear Aunt Sally" must be used to decide the order of operations within the brackets. The product of 24×6 must be determined before 16 can be taken away from the bracketed terms. The equation becomes

$$[144 - 16] \div 4 = R$$

The bracketed terms can now be reduced by subtracting 16 from 144. The equation becomes

$$128 \div 4 = R$$

Reversing the order of the equation

$$R = 128 \div 4$$

$$R = 32$$

If severals operations are to be performed as "inner groups" within an equation, parentheses must contain the innermost group. Brackets must contain the parentheses and other operations and terms. Braces must contain the brackets and other operations and terms. The open equation must contain the braces and other operations and terms.

If the equation $10 + 14 \times 6 - 16 \div 4 = R$ is written as $10 + \{14 \times [6 - (16 \div 4)]\} = R$, the value of R as determined in the previous example will not be the same. The order of operation established by "My Dear Aunt Sally" must be performed first within the parentheses to determine a single value or term. The operations within the brackets must then be solved next in the same manner to determine a single value or term. The operations within the braces are solved next for a single term. Finally, the open equation is solved for a single value of R.

To reduce the parentheses to a single term, 16 must be divided by 4. The equation becomes

$$10 + \{14 \times [6 - 4]\} = R$$

To reduce the brackets to a single term, 4 must be subtracted from 6. The equation becomes

$$10 + \{14 \times 2\} = R$$

To reduces the braces to a single term, 14 must be multiplied by 2. The equation becomes

$$10 + 28 = R$$

Reversing the order of the equation

$$R = 10 + 28$$

$$R = 38$$

Order of operations in an equation is of utmost importance if the correct value of an unknown is to be determined.

18.3 FRACTIONS

In a division operation the dividend reduced by the divisor yields a quotient.

$$\text{dividend} \div \text{divisor} = \text{quotient}$$

$$144 \div 12 = 12$$

As long as the dividend is larger than the divisor, the quotient will be a whole number or a mixed number that is greater than 1. (A mixed number consists of a whole number and a fraction representing a single term.) If the dividend is equal to the divisor, the quotient is equal to 1.

If the divisor is larger than the dividend, the quotient will have a value that is less than 1. This quotient is said to be a fraction.

Fractions consists of a numerator divided by a denominator. If the numerator is smaller than the denominator, the fraction is said to be a common fraction. If the numerator is larger than the denominator, the fraction is said to be an improper fraction.

Fractions are actually ratios of two numbers or ratios of two terms. If a fraction or ratio is stated as a mathematical equation that equals another fraction or ratio, the equal is said to be a proportion.

The division of 3 by 7 will yield a quotient less than 1. Written as a fraction, the division or ratio is simply left as $3/7$.

$$\text{The fraction can be written as: } \frac{3}{7}.$$

If the fraction is stated as a ratio, the ratio symbol of a colon (:) can be used in place of the divide sign, and the fraction is written as 3:7 (pronounced "3 to 7"). Ratios are often used to show the relationship of the number of windings in the primary of a transformer to the number of windings in the secondary of the same transformer.

When two fractions are to be added together, a common denominator must be found that is a multiple of both denominators. If the two denominators are both prime numbers, their product is the lowest common denominator (LCD). A prime number can only be divided by either 1 or the number itself and the quotient or result equals a whole number (without a fraction). Examples include 1, 2, 3, 5, 7, 11, 13, and 17.

Factoring is a process by which a given denominator is broken down into its prime factors. In other words, the prime factors are divided into the larger number until the quotient is equal to the prime number 1.

Example. The prime factors of the number 24 are determined as follows:

Division Operation	Prime Factors
$24 \div 2 = 12$	2
$12 \div 2 = 6$	2
$6 \div 2 = 3$	2
$3 \div 3 = 1$	3

The prime factors of the number 24 are:

$$24 = 2 \times 2 \times 2 \times 3$$

When two denominators are to be compared for a common denominator, the two denominators should be factored. The LCD will contain all the prime factors required for both denominators.

Example. The lowest common denominator for two denominators valued at 18 and 21 is determined as follows:

Division Operation		Prime Factors
For 18:	$18 \div 2 = 9$	2
	$9 \div 3 = 3$	3
	$3/3 \div 3 = 1$	3
For 21:	$21 \div 3 = 7$	3
	$7 \div 7 = 1$	7

LCD for denominators 18 and 21: $2 \times 3 \times 3 \times 7 = 126$

Note that the prime factors of both denominators are contained in the lowest common denominator. Comparing this LCD to the product of the two denominators, $18 \times 21 = 378$, the value of the LCD is much lower.

If 18 and 21 are denominators for 7 and 13, respectively, then the sum of the two fractions is determined as follows:

$$S = \frac{7}{18} + \frac{13}{21}$$

$$S = \frac{(7 \times 7) + (6 \times 13)}{126}$$

$$S = \frac{49 + 78}{126}$$

$$S = \frac{127}{126}$$

The sum of the common fractions $7/18$ and $13/21$ is equal to the improper fraction $127/126$.

To determine the multiple of the numerator 7 in the fraction of $7/18$, the LCD of 126 was reduced (divided) by 18 (7's denominator). In other words:

$$\frac{7 \times 7}{7 \times 18} = \frac{49}{126}$$

The fraction $7/7$ is equal to 1. Any term can be multiplied by the number 1 without changing its value.

To determine the multiple of the numerator 13 in the fraction $13/21$, the LCD of 126 was reduced by 21 (13's denominator). In other words:

$$\frac{6\times 13}{6\times 21}=\frac{78}{126}$$

If two fractions are to be subtracted, the same procedure must be followed. If both fractions have the same or equal denominators, the numerators are simply subtracted to determine the difference of the two fractions. If the two fractions have unequal or different denominators, then a common denominator (preferably the LCD) has to be determined. The numerators are a multiple of their respective denominators divided into the common denominator.

18.4 DECIMAL FRACTIONS

The addition or subtraction of fractions is made much easier if both fractions are changed to decimal fractions. Decimal fractions are fractions (either common or improper) whose denominator is either the number 10 or a multiple of 10 (100, 1000, 10,000, and so on, are all multiples of 10).

A decimal fraction is written without its denominator (10 or multiples of 10). Instead, the decimal fraction's denominator is implied with a decimal point in the numerator. If the fraction is an improper fraction (division equals a value greater than 1), the numerator is written as a mixed number. The decimal separates the decimal fraction from the whole numbers in the quotient. If the fraction is a common fraction (ratio or quotient is less than one), the decimal appears to the left of the quotient. To aid in clarity a zero is normally written to the left of the decimal of a common fraction to indicate the decimal value.

With the aid of a calculator or by long division, any fraction can be converted to a decimal fraction simply by dividing the numerator by the denominator.

Example. The fraction $7/8$ is a common fraction whose value is less than 1. If 7 is divided by 8, the quotient or result is stated as a decimal fraction with a value of .875. To add clarity to this term, it is normally written as 0.875.

Example. The fraction $18/12$ is an improper fraction whose value is greater than 1. If 18 is divided by 12, the quotient is stated as a mixed number that contains a whole number and a decimal fraction. The quotient or answer is 1.5.

In an earlier example, the fractions $7/18$ and $13/21$ were added together for a total value of $127/126$. The addition could have been performed with decimal fractions in the following manner:

$$\frac{7}{18}=0.389,\ \frac{13}{21}=0.619$$

$$0.389+0.619=1.008$$

To verify,

$$\frac{127}{126}=1.008$$

Two fractions can be subtracted in the same manner.

Example. Subtract $\frac{3}{16}$ from $\frac{11}{18}$.

$$D\ (\text{difference}) = \frac{11}{18} - \frac{3}{16}$$

The LCD must first be determined:

		Factors			Factors
For 16:	$16 \div 2 = 8$	2	For 18:	$18 \div 2 = 9$	2
	$8 \div 2 = 4$	2		$9 \div 3 = 3$	3
	$4 \div 2 = 2$	2		$3 \div 3 = 1$	3
	$2 \div 2 = 1$	2			

The lowest common denominator (LCD) must contain the product of all prime factors of both denominators:

$$\text{LCD: } 2 \times 2 \times 2 \times 2 \times 3 \times 3 = 144$$

$$D = \frac{(11 \times 8) - (3 \times 9)}{144}$$

$$D = \frac{88 - 27}{144}$$

$$D = \frac{61}{144}$$

The difference of $\frac{3}{16}$ subtracted from $\frac{11}{18}$ can also be found by converting the two fractions to decimals:

$$\frac{11}{18} = 0.6111, \quad \frac{3}{16} = 0.1875$$

$$D = 0.6111 - 0.1875$$

$$D = 0.4236$$

To verify, $$\frac{61}{144} = 0.4236$$

When fractions are converted to decimal fractions, mathematical operations (addition, subtraction, multiplication, and division) are made easier. When decimal fractions are added to or subtracted from whole numbers or mixed numbers (mixed numbers contain both whole numbers and decimal fractions), it is imperative that the location of the decimal be maintained throughout the operations.

Example. If the decimal fraction 0.73 is to be added to 1.41, the sum would be determined as follows:

$$\begin{array}{r} 1.41 \\ +\,0.73 \\ \hline 2.14 \end{array}$$

Example. If the decimal fraction 0.89 is to be subtracted from 2.64, the difference would be determined as follows:

$$\begin{array}{r} 2.64 \\ -\,0.89 \\ \hline 1.75 \end{array}$$

When a whole number, mixed number, or decimal fraction is multiplied by a decimal fraction, the product will always be less than the original whole number, mixed number, or decimal fraction.

Example. The multiplication of 0.63 times 86 will equal a number that is less than 86.

$$0.63 \times 86 = 54.18$$

Example. The multiplication of 0.45 times 8.75 will equal a number that is less than 8.75.

$$0.45 \times 8.75 = 3.94$$

Example. The multiplication of 0.67 times 0.89 will equal a decimal fraction that is less than 0.89.

$$0.67 \times 0.89 = 0.60$$

When a whole number, mixed number, or decimal fraction is multiplied by a mixed number (contains a whole number and a decimal fraction in a single term), the product will always be greater than the original whole number, mixed number, or decimal fraction.

Example. The multiplication of 86 times 2.25 will equal a number larger than 86.

$$86 \times 2.25 = 194$$

Example. The multiplication of 14.26 times 1.75 will equal a number larger than 14.26.

$$14.26 \times 1.75 = 24.96$$

Example. The multiplication of 0.75 times 1.55 will equal a number that is larger than 0.75.

$$0.75 \times 1.55 = 1.16$$

When a whole number, mixed number, or decimal fraction is divided by a decimal fraction, the quotient will always be larger than the original whole number, mixed number, or decimal fraction.

Example. The division of 74 by 0.65 will equal a quotient that is larger than 74.

$$74 \div 0.65 = 114$$

Example. The division of 17.8 by 0.4 will equal a quotient that is larger than 17.8.

$$17.8 \div 0.4 = 44.5$$

Example. The division of 0.85 by 0.468 will equal a quotient that is larger than 0.85.

$$0.85 \div 0.468 = 1.82$$

When a whole number, mixed number, or decimal fraction is divided by a mixed number, the quotient will always be smaller than the original whole number, mixed number, or decimal fraction.

Example. The division of 10 by 1.4 will equal a quotient that is smaller than 10.

$$10 \div 1.4 = 7$$

Example. The division of 17.85 by 6.8 will equal a quotient that is less than 17.85.

$$17.85 \div 6.8 = 2.6$$

Example. The division of 0.75 by 1.25 will equal a quotient that is less than 0.75.

$$0.75 \div 1.25 = 0.60$$

18.5 ROUNDING OFF

The accuracy of a product of either whole numbers or mixed numbers and either decimal fractions or mixed numbers is determined by the term with the least amount of numbers to the right of the decimal point. A whole number would be less accurate than a fraction or mixed number carried to the one-hundreths place (two numbers to the right of the decimal point).

The multiplication of a mixed number accurate to tenths (one number to the right of the decimal) times a decimal fraction accurate to hundreths will yield a product that is accurate only to tenths.

Example. The product of 14.3 times .75 is only accurate to one place to the right of the decimal point.

$$14.3 \times 0.75 = 10.725$$

Rounded to an accuracy of tenths: 10.725 = 10.7

The division of a mixed number accurate to thousandths (three numbers to the right of the decimal point) by a decimal fraction accurate to hundredths will yield a quotient that is only accurate to hundredths.

Example. The division of 67.894 by 0.67 is only accurate to two places to the right of the decimal point.

$$67.894 \div 0.67 = 101.3343$$

Rounded to an accuracy of hundredths: 101.3343 = 101.33

To reflect the accuracy of the quotient or product in a division or multiplication operation, the answer has to be rounded off. The accuracy of the answer is determined by the term with the least amount of numbers to the right of the decimal point.

If the accuracy of the answer is to be stated in tenths (one place to the right of the decimal point), but the answer was computed to ten-thousandths (four places to the right of the decimal point) accuracy, the value of the number in the hundredths place (two places to the right of the decimal point) will determine how the mixed number or decimal is rounded to tenths.

If the hundredths digit or number is 5 or greater, the tenths value will be rounded to the next higher whole number.

Example. If the quotient in a division operation was computed to be 8.4657, but only tenths was the accuracy of the divisor, the 4 in the tenths position would be rounded to 5 because the 6 in the hundreths position is larger than 5.

8.4657 is rounded to 8.5 for tenths accuracy.

If the hundredths digit is less than 5, the tenths value is left as is and the rest of the decimal fraction is dropped.

Example. If the product of a multiplication operation was computed to be 17.84699, but only tenths was the accuracy of the multiplier, the 8 in the tenths position would be left as is because the value of the hundredths position is less than 5.

17.84699 is rounded to 17.8 for tenths accuracy.

When a mixed number or fraction is to be rounded for accuracy, the value of the first number to the right of the place to be rounded to determines when the number will be left as is or when the number will be raised to the next whole number. If the first number to the right is 5 or greater, the number to the left is raised by one whole number. If the first number to the right is less than five, the number to the left is left as is in the rounding off.

Example. The following numbers are rounded to different degrees of accuracy:

$17.4 \times 0.87889 = 15.29269$	Rounded to tenths: 15.3
$0.694 \times 0.78 = 0.541$	Rounded to hundredths: 0.54
$178.947 \times 4 = 715.788$	Rounded to a whole number: 716
$16.405 \div 18.1 = 0.906$	Rounded to tenths: 0.9
$36.21 \div 8.8496 = 4.0917$	Rounded to hundredths: 4.09
$0.875 \div 0.5 = 1.750$	Rounded to tenths: 1.8

18.6 PERCENTAGES

A decimal fraction can be expressed as a percentage by moving the decimal point two places to the right and adding the symbol (%) for percentage. A percentage can be expressed as a decimal fraction by moving the decimal point two places to the left and dropping the percentage symbol (%).

Example.

0.75 is equal to 75%
0.6743 is equal to 67.43%
2.25 is equal to 225%
75% is equal to 0.75
250% is equal to 2.50
175% is equal to 1.75
1400% is equal to 14.00

In many of the NE Code computations within this text, the student electrician is required to take a percentage of either amperage, voltage, or volt-amp loads. The percentage computations are made easier if the percentage is first converted to a decimal fraction.

Example. When an over current-protection device supplies a continuous load, the load must be restricted to 80% of the fuse or circuit breaker's rating. If a 30-amp fuse supplies a continuous load, the maximum allowed current load is determined as follows:

$$\begin{aligned} A_{\text{load}} &= 80\% \text{ of } 30 \text{ amps} \\ &= 0.80 \times 30 \text{ amps} \\ &= 24 \text{ amps} \end{aligned}$$

When the amperage of a continuous load is given, and the rating of the overcurrent-protection device must be found, the percentage must be divided into the current rating of the continuous load.

Example. Determine the rating of the circuit breaker used to protect a 48-amp continuous load.

$$\begin{aligned} A_{\text{load}} &= 48 \text{ amps} \div 80\% \\ &= 48 \text{ amps} \div 0.80 \\ &= 60 \text{ amps} \end{aligned}$$

When determining the branch-circuit or feeder overcurrent protection of a motor, high overpercentages of the motor's full-load current (FLC) are permitted by the NE Code to allow for the large in-rush starting currents of motors. The overpercentages are converted to mixed numbers and multiplied by the motor's FLC.

Example. If a motor's FLC is listed as 80 amps and the motor has separate overload-overcurrent protection, the branch-circuit overcurrent protection can be rated up to 250% of the motor FLC when protected by an inverse-time circuit breaker.

$$\begin{aligned} A_{O/C} &= 250\% \text{ of } 80 \text{ amps} \\ &= 2.50 \times 80 \text{ amps} \\ &= 200 \text{ amps} \end{aligned}$$

When a percentage of a volt-amp load is to be added to the volt-amp load, the percentage can be added to 100%, and the volt-amp load then multiplied by their sum to determine the required value.

Example. The derated range load of a single dwelling cooking unit is 8 kW if the nameplate rating of the range is not greater than 12 kW. Note 1 to T220-19 requires the value of 8 kW to be increased 5% for each kW or major fraction thereof, when the range nameplate rating exceeds 12 kW. If the nameplate rating is 18 kW, the range derated load is determined as follows:

Percentage: $18 \text{ kW} - 12 \text{ kW} = 6 \text{ kW}$

$$6 \text{ kW} \times 5 \text{ \%/kW} = 30\%$$

$$100\% + 30\% = 130\%$$

Range kVA rating: $130\% \text{ of } 8 \text{ kW} = 1.30 \times 8 \text{ kW} = 10.4 \text{ kW}$

18.7 EXPONENTS (POWERS) OF NUMBERS

When a number or term appears several times in a multiplication or divison operation, the equation can be more conveniently written by expressing the repeated number or term as an exponential. The multiplication of $6 \times 6 \times 6 \times 6$ can be written as 6^4. This is read as "6 to the fourth power." The number or term itself is written one time as the "base" of the exponential, and the count of how many times the number or term appears in the equation is written as the "exponent" of the number or term in superscript beside the number or term.

A term, such as (R + 4), appearing several times in a given equation can be written as an exponential in the same manner:

$$(R + 4) \times (R + 4) \times (R + 4) = (R + 4)^3$$

If a number or term appears twice in a given equation, the number can be converted to an exponential with an exponent of 2. The number is said to be squared.

$$6 \times 6 = 6^1 \times 6^1 = 6^{(1 + 1)} = 6^2$$

If the number or term appears three times in a given equation, the number is said to be "cubed."

$$(V + 8) \times (V + 8) \times (V + 8) = 21$$
$$(V + 8)^1 \times (V+8)^1 \times (V + 8)^1 = 21$$
$$(V + 8)^{(1 + 1 + 1)} = 21$$
$$(V + 8)^3 = 21$$

If a number or term has an exponent larger than three, the exponential has no common name. The number or term is simply read as "raised to the fourth power," "raised to the fifth power," etc.

Example. 14^4 is read as "14 raised to the fourth power." $(Z + 7)^5$ is read as "Z plus seven raised to the fifth power."

18.8 RULES OF EXPONENTIALS

1. Any number or term written as a single factor is considered to be a base with an exponential of 1.

Example.
$$6 = 6^1$$

2. Any number or term written as a base with an exponential of 0 is equal to 1.

Example.
$$46^0 = 1$$
$$Y^0 = 1$$
$$(C + 14)^0 = 1$$

3. When the product (multiplication) of numbers or terms with "like bases" is taken, the exponents are added together.

Example.
$$Z = (R + 3) \times (R + 3)^2$$
$$Z = (R + 3)^{(1 + 2)}$$
$$Z = (R + 3)^3$$

4. When the quotient (division) of numbers or terms with "like bases" is taken, the exponents are subtracted. The exponent of the divisor is subtracted from the exponent of the dividend.

Example.
$$V = \frac{36^3}{36^2}$$
$$V = 36^{(3-2)}$$
$$V = 36$$

5. Any number or term written as a base with a negative value exponential is computed as the reciprocal of that number or term.

Example. $$6^{-2} = \frac{1}{6^2}$$

6. When an exponential term is raised to a power, the powers are multiplied for the result.

Example.
$$L = (13^3)^3$$
$$L = 13^{(3 \times 3)}$$
$$L = 13^9$$

Example.
$$476 = [(B + 4)^3]^{1/2}$$
$$476 = (B + 4)^{(3 \times 1/2)}$$
$$476 = (B + 4)^{3/2}$$

7. When a base is written with a fractional exponent (value less than one), the answer sought is referred to as the "root" of the base in the exponential term.

Example.
$$A = 25^{1/2}$$
$$A = (5^2)^{1/2}$$
$$A = 5^{(2 \times 1/2)}$$
$$A = 5^1$$
$$A = 5$$

Remember, the rules of exponentials are two-way streets: Exponentials in a given equation can be expanded as the product of the number or term times itself the number of times listed in the exponent. The number 4^3 is equal to 64 because 4^3 is equal to $4 \times 4 \times 4$. In evaluating this statement, $4 \times 4 = 16$, and $16 \times 4 = 64$.

Example. Expand the exponential term of 2^5 (two to the fifth power).
$$2^5 = 2^1 \times 2^1 \times 2^1 \times 2^1 \times 2^1 = ?$$
$$2^5 = 2 \times \{[(2 \times 2) \times 2] \times 2\}$$
$$2^5 = 2 \times \{[4 \times 2] \times 2\}$$
$$2^5 = 2 \times \{8 \times 2\}$$
$$2^5 = 2 \times 16$$
$$2^5 = 32$$

Example. Expand the exponential term $(R + 7)^3$ (R plus seven cubed).
$$(R + 7)^3 = (R + 7) \times (R + 7) \times (R + 7)$$

In the development of this text, the student electrician is not required to raise the value of a number or term past the second power of the number.

18.9 POWERS OF 10

When the base of an exponential is the number 10, a special benefit is obtained: The value of the exponential determines how many zeros must be added after the 1 in the results of expanding the exponential term.

Remember that any number written with a zero exponent has a defined value of 1. Also remember that a number written without an exponent is assumed to have an exponent of 1. With this in mind:

$$10^0 = 1$$

$$10^1 = 10$$

$$10^2 = 100$$

$$10^3 = 1000$$

Notice that the value of the exponential determines the required number of zeros that must be placed after the 1. This is very convenient when a student electrician must write large numbers that are the powers of 10. The number 1,000,000 can be written as 10^6. This tells the reader that the number "one million" is being indicated by the number 1 followed by 6 zeros.

As was stated when discussing decimal fractions, a whole number is written with the decimal to the right of the whole number. If the whole number is a fixed number that contains both a whole number and a decimal fraction, the whole number is written to the left of the decimal point, and the decimal fraction is written to the right of the decimal point. When just a whole number is written, the decimal is understood to be to the right of the last digit in the whole number.

Example. The whole number 84 can be written as 84.0. The decimal is to the right of the last digit.

Example. The fraction ¼ is written as the decimal fraction 0.25. The zero to the left of the decimal is just a "place holder" that indicates there is no whole number. The decimal fraction is written to the right of the decimal.

Example. The mixed number 6⅜ is written as the decimal mixed number of 6.375. The whole number appears to the left of the decimal point. The decimal fraction appears to the right of the decimal point.

Using the powers of 10, the decimal is assumed to be to the right side of the 1. The exponential, which tells how many zeros to add to the right of the 1, actually determines how many places the decimal point must be moved to the right from the one.

Example. When the number 10^3 is written, it indicates the number 1 with the decimal place moved three places to the right:

$$10^3 = 1{,}000$$

In using the powers of 10, when 10 is raised to a positive power, the number indicated is the number formed by a 1 followed by as many zeros as are indicated in the exponential. These zeros are created because the decimal point is moved to the right by that many places.

What effect does a negative exponent have on the powers of 10? It was shown in discussing the rules of exponents that a negative exponential value required the reciprocal of the expanded base to be taken. The same holds true with the powers of 10.

$$\text{The term } 10^{-1} \text{ requires the base to be written as: } \frac{1}{10^1}$$

A negative exponent has the opposite effect of a positive exponent on the movement of the decimal. A negative exponent adds zeros as it simultaneously moves the decimal point to the left from the understood location on the right side of the 1.

$$10^0 = 1.0$$

$$10^{-1} = \frac{1}{10} = 0.1$$

$$10^{-2} = \frac{1}{10 \times 10} = 0.01$$

$$10^{-3} = \frac{1}{10 \times 10 \times 10} = 0.001$$

Notice that even though the decimal point was moved two decimal places to the left in expanding 10^{-2}, only one zero was added. Notice also that when 10^{-3} was expanded, only two zeros were added. The reason for this is that the decimal is always located to the right of the one's position before the expansion is started. Therefore, when 10 is raised to a negative power, the number formed is a 1 preceded with a decimal point and a count of zeros that is one less than the value of the exponent.

18.10 SCIENTIFIC NOTATION

Using the powers of 10, any number can be written in terms of scientific notation. This is a shorthand method of handling large numbers. Consider the number 8,000. This number is actually the product of 8 multiplied by 1,000.

$$8{,}000 = 8 \times 1{,}000 = 8 \times 10^3$$

Since 1,000 is equal to 10^3 in powers of 10, 8,000 can be written as the product of 8 $\times 10^3$. This is scientific notation.

Scientific notation involves the product of two key factors: One, the whole or mixed number has a value that is greater than or equal to 1 but less than 10 ($1 \geq N < 10$). The other is a power of 10.

The pattern for expressing scientific notation is given as follows. The coefficient (whole or mixed number between 1 and 10) is determined from the given number by moving the decimal point either left or right until it lies immediately to the right of the first nonzero digit.

The exponent (in the powers of 10) is determined by giving 10 an exponent equivalent to the number of places the decimal point had to be moved to create the coefficient. Note that if the decimal point is moved to the left, the exponent of 10 is positive. If the decimal point is moved to the right, the exponent of 10 is negative. A negative exponent indicates that the scientific notation represents a decimal fraction that is less than 1.

Example. Convert the following numbers to scientific notation.

$$475 = 4.75 \times 100 = 4.75 \times 10^2$$

$$9{,}376 = 9.376 \times 1000 = 9.376 \times 10^3$$

$$0.32 = 3.2 \times 0.1 = 3.2 \times 10^{-1}$$

$$0.000089 = 8.9 \times 0.00001 = 8.9 \times 10^{-5}$$

$$1{,}760{,}000{,}000 = 1.76 \times 1{,}000{,}000{,}000 = 1.76 \times 10^9$$

18.11 SIGNIFICANT DIGITS

As was discussed under "rounding off" (Section 18.5), the accuracy of a mixed number or decimal fraction, which was the result of a multiplication or division operation, is only as accurate as the number with the least amount of digits to the right of the decimal point. These numbers, which represent the decimal fraction portion of the mixed number, are also referred to as significant digits.

When multiplying by longhand, the product of two numbers will contain as many decimal places as the sum of decimal places contained in both the multiplicand and the multiplier. The student electrician, using a product with so many decimal places may think the product is more accurate than it really is. The correct number of decimal places is maintained if the product is rounded off to the same number of significant digits as is contained in either the multiplicand or multiplier that has the least number of significant digits to the right of the decimal point.

When using scientific notation, digits contained in a given number or term may be significant until the large or small number is rewritten in scientific notation. After the conversion, many of the digits may become insignificant.

Example. Convert the number 16,879.476 to scientific notation.

$$16{,}879.476 = 1.6879476 \times 10{,}000 = 1.6879476 \times 10^4$$

Rounded, $16{,}879.476 = 1.688 \times 10^4$

If this number represents a measurement in feet, the decimal of 0.476 feet is equivalent to 5.7 inches, which is insignificant if we are dealing with almost 17,000 feet. From the old days of slide rules, when calculators and computers were expensive, innovative toys, the slide rule was considered accurate to only three places or thousandths. At best, the accuracy of a multiplication or division operation should not be carried more than four places.

In the electrical industry and our profession, analog meter movements normally cannot be read past a two-place accuracy. The third place or thousandths we normally "guesstimate" by eyesight. In NE Code computations, when the volt-amp loads consists of thousands or millions of watts, the difference of a few watts or a fraction of a watt is insignificant to the overall total volt-amp load.

18.12 UNIT PREFIXES (SCIENTIFIC NOTATION)

When an electrician has been in the field packing a pouch for a while, he or she will often hear terms such as kilowatts, megawatts, milliamps, k-ohms, or kVA. These are shorthand references to the powers of 10 that aid in referencing extra large or extra small quantities measured in the use and computation of electricity.

The terms kilo, mega, and milli are prefixes adopted from the SI system of measurement." "SI" stands for Systems International, indicating that these assigned prefixes are used all over the world. The student electrician should be familiar with the following table of prefixes (refer to Table 16-1). Note that only those most commonly used with Code reference material are the only ones shown.

The student electrician should commit these prefix values to memory, since many NE Code computations involved prefixed units of measurement.

TABLE 16-1 UNIT PREFIXES

Prefix	Symbol	Prefix Value
Mega	M	10^6
Kilo	k	10^3
Basic unit	none	10^0
Milli	m	10^{-13}
Micro	μ	10^{-6}

For familiarization, the following examples are offered:

Example. A dwelling cooking range with a nameplate rating of 12 kW is rated at 12,000 watts.

$$12 \text{ kW} = 12 \times 10^3 \text{ W} = 12{,}000 \text{ watts}$$

Example. A transformer with a nameplate rating of 37.5 kVA is rated at 37,500 volt-amps.

$$37.5 \text{ kVA} = 37.5 \times 10^3 \text{ VA} = 37{,}500 \text{ volt-amps}$$

Example. A single-family dwelling with a computed service load of 10 kVA is required to have a minimum of a 100-amp service. A load of 10 kVA is equal to 10,000 volt-amps.

$$10 \text{ kVA} = 10 \times 10^3 \text{ VA} = 10{,}000 \text{ volt-amps}$$

Example. A GFCI receptacle normally has a trip setting of 5 milliamps (5 mA), which is set to trip when the unbalanced current reaches a difference of 5-thousandths of 1 amp.

$$5 \text{ mA} = 5 \times 10^{-3} \text{ A} = 0.005 \text{ amps}$$

Example. A driven-rod electrode is required to have a maximum of 25 ohms resistance to ground. 25 ohms is a basic unit of resistance without prefix. The number is read as is.

$$25 \text{ ohms} = 25 \text{ ohms}$$

Example. An open circuit (open switch or broken wire) is considered to have a minimum of 50 megaohms resistance. This means an open circuit has a measured resistance of at least 50 million ohms.

$$50 \text{ M-ohms} = 50 \times 10^{6} \text{ ohms} = 50{,}000{,}000 \text{ ohms}$$

18.13 ROOT INDEXES

In discussing the rules of exponentials (Section 18.8), it was stated that when a base is written with a fraction exponent (value less than 1), the answer sought is referred to as the root of the base in exponential form.

To find the square root of 36, the student electrician must ask the question: What number multiplied by itself will equal the number 36? When the square root of 36 is sought, the operation is written as $36^{1/2}$. Although not used in this text, to find the square root of 36, 36 can be written under the radical sign [a check mark written to the left of the number 36 with the tail of the check mark extending as a straight line over the number (36)]. To state the operation in equation form:

$$36^{1/2} = (6^1 \times 6^1)^{1/2} = 6^{[(1+1) \times 1/2]} = 6$$

The square root of a number is sought when the number is raised to the ½ power. The cube root of a number is sought when the number is raised to the ⅓ power. Their are no common names or references to numbers that are raised to fractional powers smaller than ⅓.

To determine the cube root of a number, the student electrician is required to determine a number that when multiplied by itself three times (cubed) will equal the number in question.

Example. The cube root of the number 27 is equal to 3:

$$27^{1/3} = (3^1 \times 3^1 \times 3^1)^{1/3} = 3^{[(1+1+1) \times 1/3]} = 3$$

When the root or fractional exponent of a number exceeds one-third, determination of the root factor becomes a task.

Example. Determine the fourth root of the number 625.

$$625^{1/4} = (5^1 \times 5^1 \times 5^1 \times 5^1)^{1/4} = 5^{[(1+1+1+1) \times 1/4]} = 5$$

Fractional exponents are simple with the aid of a calculator which offers Y^X (Y to the X) or some other exponential function. In longhand, determining the root value (fractional exponents) of a number becomes an involved process that is beyond the scope of this text.

18.14 ESTIMATING ROOT VALUES

The student electrician will not be expected to go beyond the square root of a number when dealing with fractional exponents in this text.

When making NE Code computations without a calculator, the square root of a number can be estimated. Estimation requires the student electrician to determine the value of the perfect square (whole number without a fraction), which is the next value below the number in question, and the value of the perfect square, which is the next value above the number in question.

Example. What is the square root of the number 56?

The first lower perfect square value less than 56 is 49. The square root of 49 is 7. The first higher perfect square value greater than 56 is 64. The square root of 64 is 8.

The square root of 56 is a mixed number whose value lies between the whole numbers 7 and 8. The span or difference between 49 and 64 is 15. The span between 49 and 56 is 7. The ratio or fraction of 7 divided by 15 is approximately one-half. One-half is the decimal fraction of 0.5.

The square root of 56 is approximately equal to the number 7 plus the decimal fraction 0.5:

$$7.5^2 = 7.5 \times 7.5 = 56.25$$

18.15 SOLVING EQUATIONS FOR UNKNOWN VALUES

Within the scope of this text, the student electrician is expected to solve an algebraic equation with one unknown. The power formula and the voltage-drop formula are used often in this text, and are offered at this point for discussion.

The power formula for a 3-phase distribution system stipulates that the power delivered to the connected load is equal to the product of the square root of three times the line (system) voltage times the full-load current.

$$\text{Power}_{\text{volt-amps}} = 3^{1/2} \times V_{\text{line}} \times A_{\text{full-load}}$$

This formula assumes a unity power factor with no allowance for equipment efficiency. Unity power factor means the connected load is purely resistive, and the power consumed by the load is converted to heat.

If the full-load currents of several loads are given in a NE Code question, the power consumed by the load can be determined by summing the full-load currents, and plugging this total current value into the power formula.

Example. If a 3-phase branch circuit rated at 208 volts has the following connected current loads: a 7-amp lighting load, a 10-amp motor load, and a 4-amp heating load. The power consumed by the branch circuit can be determined as follows.

The total value of current in the branch circuit must first be found before it is plugged into the power formula:

$$A_{\text{full-load}} = 7 \text{ amps} + 10 \text{ amps} + 4 \text{ amps} = 21 \text{ amps}$$

The total branch-circuit power is

$$\begin{aligned} \text{Power}_{\text{volt-amps}} &= 3^{1/2} \times V_{\text{line}} \times A_{\text{line}} \\ &= 1.732 \times 208 \text{ volts} \times 21 \text{ amps} \\ &= 7{,}565 \text{ volt-amps} \end{aligned}$$

When a branch-circuit current rating is to be determined, the individual connected loads are normally given, and the required line current must be computed using the power formula.

Example. A branch circuit rated at 240 volts, 3-phase has the following connected power loads: 2,640, 2,115, and 1,480 volt-amps. The required current rating of the branch-circuit can be determined as follows:

The total power or volt-amp load (abbreviated to just VA) for the branch circuit must be found before it is plugged into the power formula:

$$\text{Power}_{\text{VA}} \text{ Total} = 2{,}640 \text{ VA} + 2{,}115 \text{ VA} + 1{,}480 \text{ VA} = 6{,}235 \text{ VA}$$

The required branch-circuit current is

$$\text{Power}_{\text{VA}} = 3^{1/2} \times V_{\text{line}} \times A_{\text{full-load}}$$

such that

$$A_{\text{full-load}} = \frac{\text{Power}_{\text{VA}}}{3^{1/2} \times V_{\text{line}}}$$

$$A_{\text{full-load}} = \frac{6{,}235 \text{ volt-amps}}{(1.732 \times 240 \text{ volts})}$$

$$A_{\text{full-load}} = \frac{6{,}235 \text{ volt-amps}}{415.7 \text{ volts}}$$

$$A_{\text{full-load}} = 15 \text{ amps}$$

The power formula is often made harder than it really is because student electricians tend to take shortcuts in their mathematical operations.

The voltage-drop formula has more plug-in values than the power formula. The voltage-drop formula must take into account the length of the conductors, the resistance of the conductors assigned by wire size, the full-load current of the connected load, and whether the distribution system is single-phase or 3-phase.

The voltage-drop formula is stated as:

$$VD = \frac{2 \times L \times A \times R \times (0.866)}{1{,}000}$$

Where: VD = actual voltage drop

L = one-way length of the conductor in question

A = measured full-load current

R = resistance of the conductor as listed according to wire size in Table 8, Chapter 9

The factor of 0.866 is used only when the formula is used for the voltage drop of a 3-phase circuit.

The voltage-drop formula as used in this text is worked in one of two ways. The formula can be solved for the actual voltage drop if the one-way length, the full-load current, and the resistance of the wire size in question are known (solving for only one unknown) Or the wire size required to restrict the voltage drop to some maximum value can be determined by solving the voltage-drop formula for a maximum value of R. The formula can be solved for the maximum value of R only if the maximum voltage drop, the one way length, and the full-load current are known.

Example. A No. 4/0 AWG copper conductor is used to supply a 3-phase load rated at 200 amps. Determine the actual voltage drop if the load is located 2,000 feet from its source.

$$VD = \frac{2 \times 2{,}000 \text{ feet} \times 200 \text{ amps} \times 0.0608 \text{ ohms} \times 0.866}{1{,}000 \text{ feet}}$$

The ohmic value of R, which is the wire resistance of No. 4/0 AWG copper, was determined by wire size in Table 8, Chapter 9 to be 0.0608 ohms.

By Ohm's law the product of a circuit's resistance and its delivered current is equal to the circuit's voltage.

$$VD = 42{,}122.2 \text{ volts} \div 1{,}000 = 42.1 \text{ volts}$$

The value of ohms per 1,000 feet as found in Table 8, Chapter 9 is allowed for by the value of 1,000 feet in the voltage-drop formula's denominator. This is the reason the ohmic value found in Table 8 is used as is in the formula.

When the actual voltage drop of a circuit is to be restricted to some maximum value. The maximum resistance value, which will yield a minimum wire size in Table 8, is solved for. Given the original voltage-drop formula,

$$VD = \frac{2 \times L \times A \times R \times 0.866}{1{,}000}$$

Instead of plugging in the given values for the unknowns in the formula, the formula can be restated in terms of solving for R. By tradition, the value to be solved for is listed by itself (coefficient of one) on the left side of the equals mark. All other terms are collected according to the operation to be performed on the right side of the equals mark.

Rewritten to position the R value on the left side:

$$\frac{2 \times L \times A \times 0.866 \times R}{1,000} = VD$$

Voltage-drop formula solved for R:

$$R_{max} = \frac{1,000 \times VD_{max}}{2 \times L \times A \times 0.866}$$

The formula was modified to indicate maximum values of voltage drop and resulting resistance of the wire. VD_{max} is a computed percentage of the supply voltage. The percentage to use depends on the type of electrical distribution circuit: For service and or feeder conductors, or the combination of the two, the NE Code allows a minimum of 2%. For the branch-circuit conductors the maximum voltage drop allowed is increased to 3%.

Example. Determine the required wire size of copper feeder conductors used to supply a load of 400 amps. The load is rated at 480 volts, 3-phase. The load is located 1,300 feet from the point of supply. The maximum voltage drop allowed is 2%.

$$R_{max} = \frac{1,000 \times (2\% \text{ of } 480 \text{ volts})}{2 \times 1,300 \text{ feet} \times 400 \text{ amps} \times 0.866}$$

$$R_{max} = \frac{9,600 \text{ volts}}{900,640 \text{ amps} \cdot \text{feet}}$$

$$R_{max} = 0.0107 \text{ ohms (per thousand feet)}$$

To find the minimum wire size, the value of R listed in Table 8, Chapter 9 must be less than or equal to the computed R value of 0.0107. The first lower value found is 0.0103, which corresponds to a 1,250 kcmil copper conductor.

Regardless of the formula used in NE Code computations, the same maneuvers outlined in these two examples can be used to solve for the required unknown.

Chapter 19

Electrical Formulas

In working problems involving electricity, standard formulas aid the student electrician in obtaining quick results or quick answers for what may seem to be a hard question. Knowing which formula to use and which value is to be solved for are key factors.

19.1 OHM'S LAW

By definition, Ohm's law states that the current consumed by a circuit is governed by both the total resistance of the circuit and the voltage applied to the circuit.

$$\text{current} = \text{voltage} \div \text{resistance}$$

Current is the actual flow of electrons in a closed electrical circuit. An open circuit, which constitutes a resistance of 50 million ohms or more, is so restrictive that virtually no current flows. Current is measured in amperes, which is commonly shortened to just amps. Current is represented as a capital I in dc computations or as a capital A in ac computations.

Voltage is referred to as the electromotive force of the circuit. Voltage is also referred to as the potential, or more correctly, the potential difference of the circuit. Voltage, measured in volts, is the pressure of the circuit that causes current to flow. Voltage is represented as a capital E in dc computations or as a capital V in ac computations.

Resistance is the restriction to current flow in an electrical circuit. True resistance dissipates heat when voltage is applied to and current flows through a restriction (resistance). Resistance is represented by a capital R. By formula,

For dc circuits: $I = E \div R$

For ac circuits: $A = V \div R$

For NE Code computations, whether I or A is used to represent current or whether E or V is used to represent voltage is not a crucial factor in solving these formulas.

Example. If 120 volts dc is applied to a total circuit resistance of 15 ohms, find the dc current.

$$I = 120 \text{ volts} \div 15 \text{ ohms} = 8 \text{ amps}$$

As a rule, the formula for Ohm's law can only be solved for one unknown. If the current is to be determined, then the values of both resistance and voltage must be known. If the resistance is to be determined, then both current and voltage must be known. And, if the voltage is to be determined, then both current and resistance must be known.

If the voltage is to be determined, the formula can be rewritten in terms of voltage:

For dc circuits: $E = I \times R$

For ac circuits: $V = A \times R$

Example. If 42 amps of ac current passes through a true resistance of 5 ohms, determine the ac voltage applied.

$$V = 42 \text{ amps} \times 5 \text{ ohms} = 210 \text{ volts}$$

If the resistance is to be determined, the formula can be rewritten in terms of resistance:

For dc circuits: $R = E \div I$

For ac circuits: $R = V \div A$

Example. If the amperage (current) of an ac circuit is measured at 12 amps, determine the circuit resistance with 460 volts applied.

$$R = 460 \text{ volts} \div 12 \text{ amps} = 38.3 \text{ ohms}$$

19.2 RESISTANCE—SERIES CIRCUITS

A resistive series circuit consists of one or more resistances which are wired in a series arrangement. As shown in Figure 19-1, the following statements apply for a series circuit:

1. The series resistances allow only one path of current flow.
2. The current flow is restricted by the total resistance of the circuit.

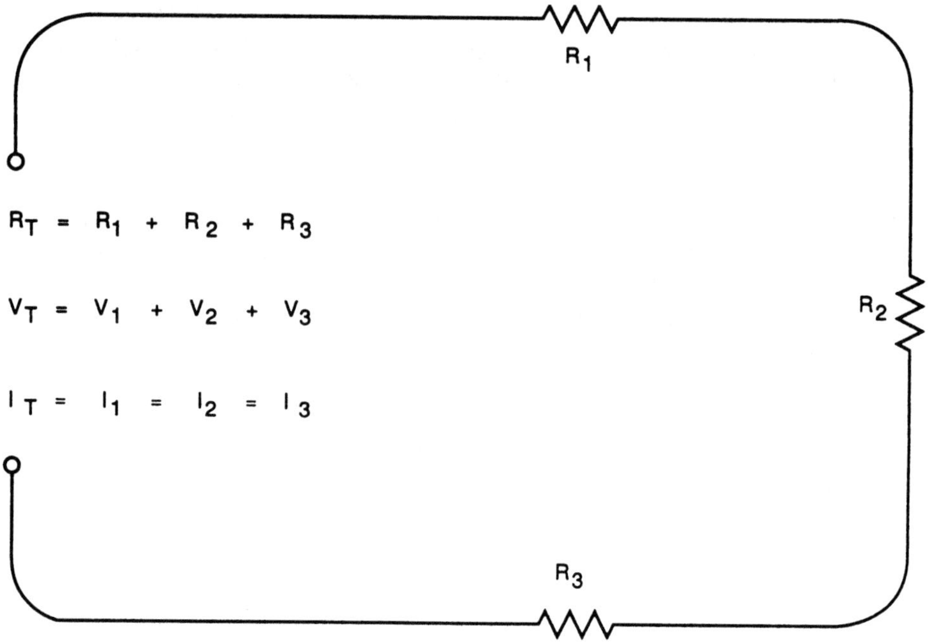

Fig. 19-1 Resistance: Series circuits.

3. The circuit voltage will divide as individual voltage drops across the series resistances.
4. The voltage drops will be proportional to the ratio of the individual resistances to the total resistance.
5. The sum of the voltage drops must equal the circuit voltage.

In equation form, if series resistances are labeled as R_1, R_2, R_3, . . . R_n, then

$$R_{total} = R_1 + R_2 + R_3 + \ . \ . \ . + R_n$$

If the voltage drops across resistors R_1, R_2, R_3, . . . R_n are labeled as E_1, E_2, E_3, . . . E_n, respectively, then

$$E_{total\ (circuit)} = E_1 + E_2 + E_3 + \ . \ . \ . + E_n$$

If the currents following through the series resistors R_1, R_2, R_3, ... R_n are labeled as I_1, I_2, I_3, . . . In, respectively, then

$$I_{total\ (circuit)} = I_1 = I_2 = I_3 = \ . \ . \ . = I_n$$

Example. The series resistances R_1, R_2, and R_3 in the circuit of Figure 19-1, have values of 20, 40, and 60 ohms, respectively. If 240 volts is applied to the circuit, determine the following: R_T, I_T, V_1, V_2, and V_3.

For the total circuit resistance:

$$R_T = 20 \text{ ohms} + 40 \text{ ohms} + 60 \text{ ohms} = 120 \text{ ohms}$$

For the total circuit current:

$$I_T = 240 \text{ volts} \div 120 \text{ ohms} = 2 \text{ amps}$$

For the voltage drops across the individual resistances:

$$V_1 = 2 \text{ amps} \times 20 \text{ ohms} = 40 \text{ volts}$$

$$V_2 = 2 \text{ amps} \times 40 \text{ ohms} = 80 \text{ volts}$$

$$V_3 = 2 \text{ amps} \times 60 \text{ ohms} = 120 \text{ volts}$$

For a quick check, the sum of the voltage drops must equal the circuit voltage:

$$V_T = V_1 + V_2 + V_3$$

$$240 \text{ volts} = 40 \text{ volts} + 80 \text{ volts} + 120 \text{ volts}$$

19.3 RESISTANCE—PARALLEL CIRCUITS

In a parallel resistance circuit, all the resistances have the same supply and return points. As shown in Figure 19-2, the following statements apply for a parallel circuit:

1. Parallel resistances offer more than one current path.
2. The circuit current (I_T) is determined by dividing the circuit resistance (R_T) into the applied voltage (V_T).
3. The circuit voltage is common or equal to the voltage applied to each parallel path.

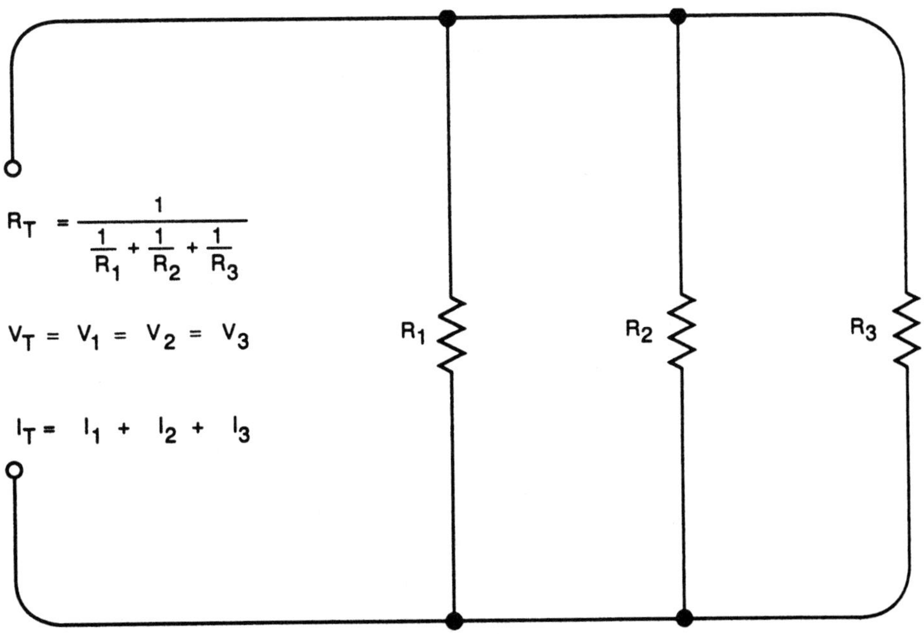

Fig. 19-2 Resistance: Parallel circuits.

4. Circuit current will divide inversely proportional to the magnitude of the resistance in each parallel path. The larger the value of resistance, the lower the value of current drawn by that parallel path.
5. The circuit resistance (R_T) will be less than the least value of resistance in any one parallel path.
6. If parallel circuits have equal values of resistance, the currents drawn by each path will be of equal value.

In equation form, if resistances R_1, R_2, R_3, . . . R_n constitute parallel branches in a parallel circuit, then

$$\frac{1}{R_{Total}} = \frac{1}{R_1} + \frac{1}{R_2} + \frac{1}{R_3} + \cdots + \frac{1}{R_n}$$

Or

$$R_{Total} = \frac{1}{\frac{1}{R_1} + \frac{1}{R_2} + \frac{1}{R_3} + \cdots + \frac{1}{R_n}}$$

$$V_T = V_1 = V_2 = V_3 = \cdots = V_n$$

$$I_T = I_1 + I_2 + I_3 + \cdots + I_n$$

Example. As shown in Figure 19-2, R_1, R_2, and R_3 are three parallel resistances in a circuit with values of 20, 40, and 60 ohms, respectively. If the applied circuit voltage is 240 volts ac, determine: R_T, V_1, V_2, V_3, I_T, I_1, I_2, and I_3.

$$R_T = \frac{1}{\frac{1}{20\text{ ohms}} + \frac{1}{40\text{ ohms}} + \frac{1}{60\text{ ohms}}}$$

$$= \frac{1}{(0.050 + 0.025 + 0.017)}$$

$$= \frac{1}{0.092}$$

$$= 10.9\text{ ohms}$$

$$V_T = V_1 = V_2 = V_3 = 240\text{ volts}$$

$$I_T = 240\text{ volts} \div 10.9\text{ ohms} = 22\text{ amps}$$

$$I_1 = 240\text{ volts} \div 20\text{ ohms} = 12\text{ amps}$$

$$I_2 = 240\text{ volts} \div 40\text{ ohms} = 6\text{ amps}$$

$$I_3 = 240\text{ volts} \div 60\text{ ohms} = 4\text{ amps}$$

For a quick check, I_T must be equal to the sum of I_1, I_2, and I_3:

$$I_T = I_1 + I_2 + I_3$$

$$22\text{ amps} = 12\text{ amps} + 6\text{ amps} + 4\text{ amps}$$

19.4 EQUAL PARALLEL RESISTANCES

As shown in Figure 19-3, when parallel resistances have equal ohmic values, the total resistance of the circuit is equal to the ohmic value of any branch divided by the number of parallel branches.

$$R_{Total} = R_{any\ branch} \div \text{number of branches}$$

The rules governing the circuit voltages remain unchanged:

$$V_T = V_1 = V_2 = V_3 = \ldots = V_n$$

The rules governing the circuit currents remain unchanged:

$$I_T = I_1 + I_2 + I_3 + \ldots + I_n$$

Example. If three parallel branch circuits R_1, R_2, and R_3 in Figure 19-3 each have an ohmic value of 60 ohms. The total circuit resistance is equal to 20 ohms.

$$R_{Total} = 60 \text{ ohms} \div 3 = 20 \text{ ohms}$$

For a quick check:

$$R_{Total} = \frac{1}{\frac{1}{60 \text{ ohms}} + \frac{1}{60 \text{ ohms}} + \frac{1}{60 \text{ ohms}}}$$

$$= \frac{1}{(0.0167 + 0.0167 + 0.0167)}$$

$$= \frac{1}{0.050}$$

$$= 20 \text{ ohms}$$

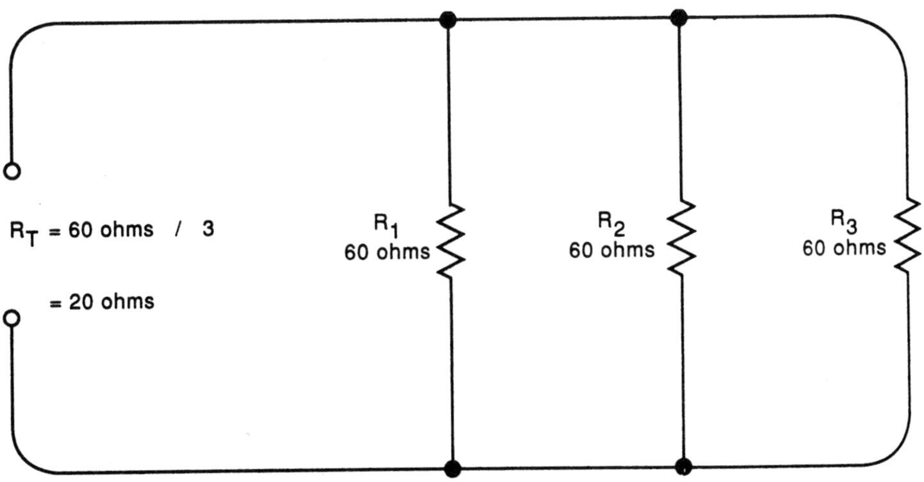

Fig. 19-3 *Equal* parallel resistances yield a circuit resistance that is equal to the resistance of any branch divided by the total number of branches.

19.5 TWO PARALLEL RESISTANCES

As shown in Figure 19-4, when two resistances of any value (equal or unequal) constitute the only two branches in a parallel circuit, the total circuit resistance can be determined by taking their product over their sum.

$$R_{Total} = \frac{R_1 \times R_2}{(R_1 + R_2)}$$

The rules governing the circuit voltages remain unchanged:

$$V_T = V_1 = V_2 = V_3 = \ . \ . \ . \ = V_n$$

The rules governing the circuit currents remain unchanged:

$$I_T = I_1 + I_2 + I_3 + \ . \ . \ . \ + I_n$$

Example. If R_1 equaling 60 ohms and R_2 equaling 180 ohms are the only two parallel branches in the resistance circuit of Figure 19-4, then the total circuit resistance R_T is equal to 45 ohms.

$$R_T = \frac{60 \times 180}{(60 + 180)}$$

$$= \frac{10,800}{240}$$

$$= 45 \text{ ohms}$$

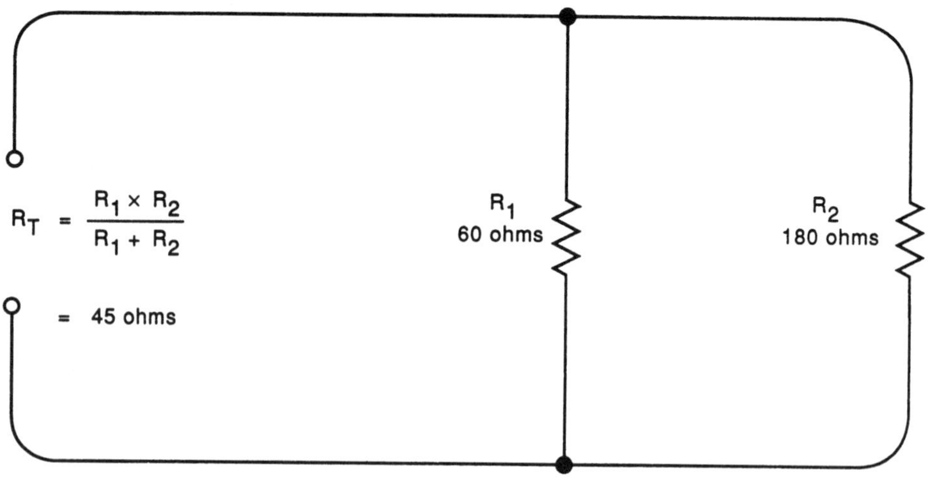

Fig. 19-4 The circuit resistance of two parallel branch circuits can be computed by taking the product over sum of the two resistances.

19.6 COMBINATION RESISTANCE CIRCUITS

When a resistance circuit contains both series elements and parallel elements of resistance as shown in Figure 19-5, it is referred to as a combination circuit. In this type of circuit, the resistances have to be determined in a few or several substeps to arrive at the total circuit resistance.

In combination circuits, voltage and current rules are obeyed according to the type of circuit seen. If part of the circuit is a series circuit, then the series rules are obeyed: Current is held constant according to total resistance; voltage divides as voltage drops according to the resistances. If part of the circuit is a parallel circuit, then the parallel rules are obeyed: Voltage is held constant on each parallel branch; current divides according to the resistance of each parallel branch.

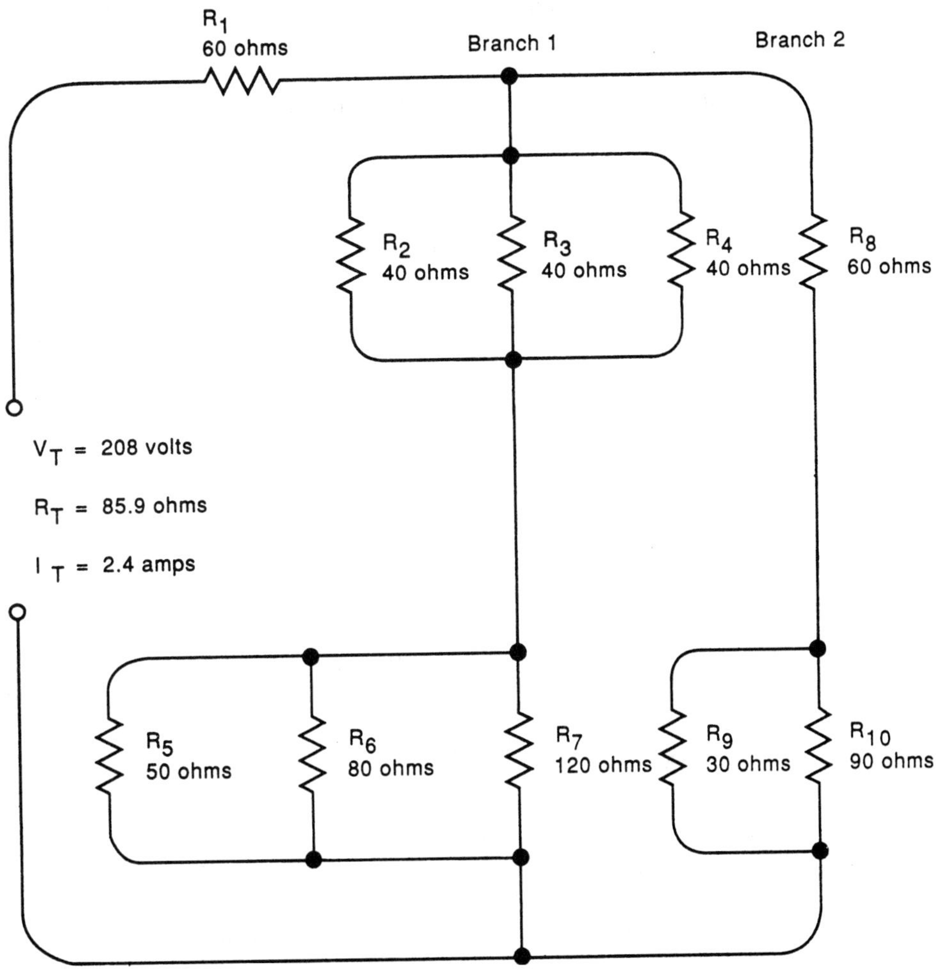

Fig. 19-5 In a combination resistance circuit, the resistances must be computed according to type of circuit (series or parallel) to determine circuit resistance.

Combination circuits offer the advantages of both series and parallel resistance circuits: Different voltage values can be obtained from the dropping point of a series circuit. Different current values can be obtained from the selected branches of a parallel circuit.

Example. In Figure 19-5, the resistances R_1 through R_{10} are configured as shown. For 208 volts applied and the resistance values given:

$R_1 = 60$ ohms	$R_6 = 80$ ohms
$R_2 = 40$ ohms	$R_7 = 120$ ohms
$R_3 = 40$ ohms	$R_8 = 60$ ohms
$R_4 = 40$ ohms	$R_9 = 30$ ohms
$R_5 = 50$ ohms	$R_{10} = 90$ ohms

Find R_T and I_T.

The value of R_T must be determined before the value of I_T can be determined.

Step 1: The net resistance of parallel circuit R_2, R_3, and R_4.

$$R_{(2\text{-}3\text{-}4)} = R_{\text{any branch}} \div \text{number of branches}$$

$$= 40 \text{ ohms} \div 3$$

$$= 13.3 \text{ ohms}$$

Step 2: The net resistance of parallel circuit R_5, R_6, and R_7.

$$R_{5\text{-}6\text{-}7} = \frac{1}{\dfrac{1}{R_5} + \dfrac{1}{R_6} + \dfrac{1}{R_7}}$$

$$= \frac{1}{\dfrac{1}{50 \text{ ohms}} + \dfrac{1}{80 \text{ ohms}} + \dfrac{1}{120 \text{ ohms}}}$$

$$= \frac{1}{(0.0200 + 0.0125 + 0.0083)}$$

$$= \frac{1}{0.0408}$$

$$= 24.5 \text{ ohms}$$

Step 3: The net resistance of branch 1.

$$R_{\text{branch 1}} = R_{(2\text{-}3\text{-}4)} + R_{(5\text{-}6\text{-}7)}$$

$$= 13.3 \text{ ohms} + 24.5 \text{ ohms}$$

$$= 37.8 \text{ ohms}$$

Step 4: The net resistance of parallel circuit R_9 and R_{10}.

$$R_{9-10} = \frac{R_9 \times R_{10}}{(R_9 \times R_{10})}$$

$$= \frac{30 \text{ ohms} \times 90 \text{ ohms}}{(30 + 90) \text{ ohms}}$$

$$= 2,700 \text{ ohms}^2 / 120 \text{ ohms}$$

$$= 22.5 \text{ ohms}$$

Step 5: The net resistance of branch 2.

$$R_{\text{branch 2}} = R_8 + R_{(9\text{-}10)}$$

$$= 60 \text{ ohms} + 22.5 \text{ ohms}$$

$$= 82.5 \text{ ohms}$$

Step 6: The net resistance of parallel branch circuits 1 and 2.

$$R_{\text{branch 1-2}} = \frac{R_{\text{branch 1}} \times R_{\text{branch 2}}}{(R_{\text{branch 1}} \times R_{\text{branch 2}})}$$

$$= \frac{37.8 \text{ ohms}^2 \times 82.5 \text{ ohms}}{(37.8 + 82.5) \text{ ohms}}$$

$$= 3118.5 \text{ ohms}/120.3 \text{ ohms}$$

$$= 25.9 \text{ ohms}$$

Step 7: The circuit resistance R_{total}.

$$R_{\text{total}} = R_1 + R_{\text{branch 1-2}}$$

$$= 60 \text{ ohms} + 25.9 \text{ ohms}$$

$$= 85.9 \text{ ohms}$$

The circuit current I_{total} is equal to the circuit voltage divided by the circuit resistance.

$$I_{\text{total}} = V_{\text{circuit}} \div R_{\text{total}}$$

$$= 208 \text{ volts} \div 85.9 \text{ ohms}$$

$$= 2.4 \text{ amps}$$

19.7 REACTANCE

True resistance in an electrical circuit is dissipated as heat in the resistive elements of the circuit. All circuits have another element that offers a type of resistance to the flow of electrical current—reactance.

Reactance does not effectively dissipate electrical resistance as heat. Instead, reac-

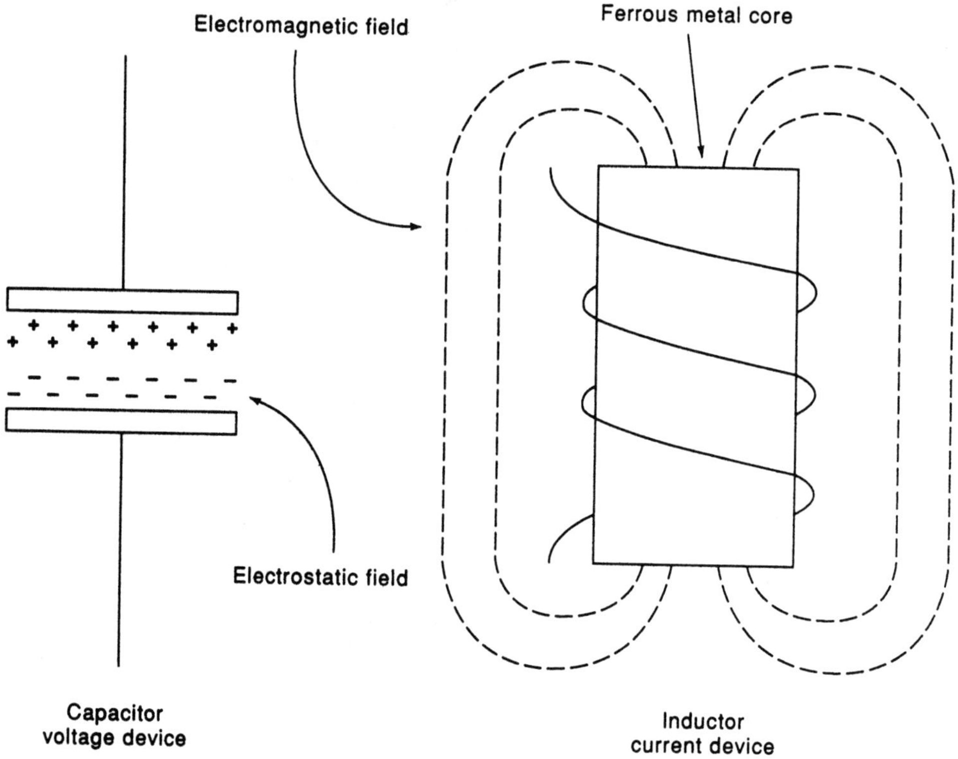

Fig. 19-6 A capacitor stores electrical energy in an electrostatic field. An inductor stores electrical energy in an electromagnetic field.

tance tends to store electrical energy in the circuit element. The stored electrical energy is released back into the circuit (such as a battery that can be charged and discharged) when circuit conditions, which include both voltage applied and current drawn, change.

As shown in Figure 19-6, the two storing elements of an electrical circuit are inductors (a wound or spiral-wrapped coil of continuous wire) and capacitors (two conductive plates with an insulating, nonconducting plate between them).

A capacitor is labeled as a voltage device because it stores electrical energy in an electrostatic field between at least two conductive plates. An inductor is labeled as a current device because it stores electrical energy in an electromagnetic field that surrounds the coil of wire.

19.8 CAPACITANCE

When more than one capacitor is connected in an electrical circuit, the way of connection, whether in series or parallel, will determine the net capacitive reactance of the circuit. Capacitors add in parallel as resistors add in series (more total capacitance), and capacitors divide in series as resistors divide in parallel (less circuit capacitance than the least value of individual capacitance in the circuit).

The value of a capacitor is labeled or expressed in farads (f). The value of a farad is so large in terms of the power stored on the capacitor that the value of the capcitor is normally reduced to microfarads.

When capacitors are wired in series as shown in Figure 19-7, the total capacitance of the circuit will be less than the least value of capacitance in the circuit. By formula:

$$\frac{1}{C_{\text{Total}}} = \frac{1}{C_1} + \frac{1}{C_2} + \frac{1}{C_3} + \cdots + \frac{1}{C_n}$$

Or

$$C_{\text{Total}} = \frac{1}{\frac{1}{C_1} + \frac{1}{C_2} + \frac{1}{C_3} + \cdots + \frac{1}{C_n}}$$

Example. If the three capacitors in Figure 19-7 having values of 10, 20, and 30 microfarads, respectively, are wired in a series circuit, the circuit capacitance is determined as follows:

$$\frac{1}{C_{\text{Total}}} = \frac{1}{10} + \frac{1}{20} + \frac{1}{30}$$

$$= 0.100 + 0.050 + 0.033$$

$$C_{\text{Total}} = \frac{1}{0.183} = 5.45 \text{ microfarads}$$

When capacitors are wired in parallel as shown in Figure 19-8, the circuit capacitance is equal to the sum of the individual capacitance values. By formula:

$$C_{\text{Total}} = C_1 + C_2 + C_3 + \ . \ . \ . \ + C_n$$

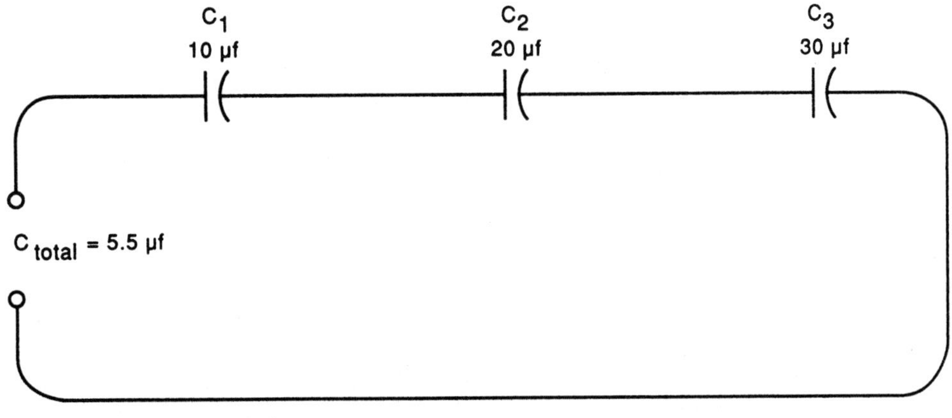

Fig. 19-7 Capacitors divide in a series circuit as resistors divide in a parallel circuit.

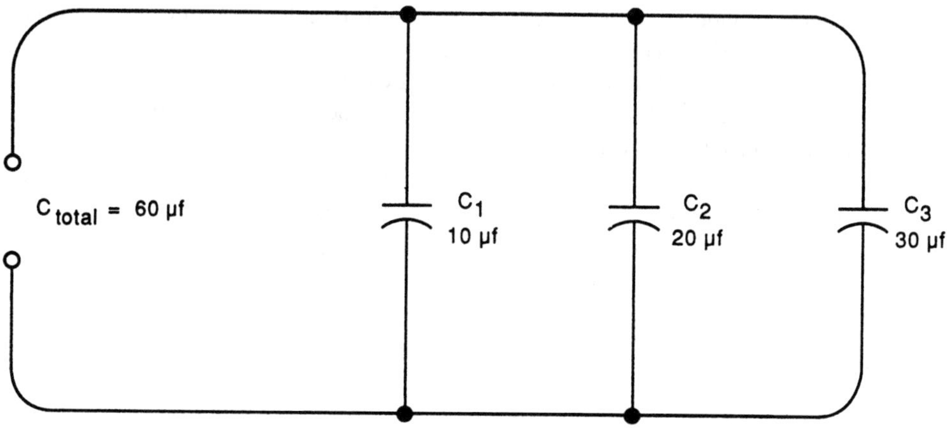

Fig. 19-8 Capacitors add in parallel as resistors add in series.

Example. If the three capacitors in Figure 19-8, having values of 10, 20, and 30 microfarads, respectively, are wired parallel in a circuit, the total circuit capacitance is determined as follows:

$$C_{Total} = 10 + 20 + 30 = 60 \text{ microfarads}$$

19.9 INDUCTANCE

When more than one inductor is connected in an electrical circuit, the way of connection, whether in series or parallel, will determine the net inductive reactance of the circuit. Inductors behave as resistors in terms of inductive reactance: In series, the value of the individual inductors will add, such that the sum of the inductances will equal the total circuit inductance. In parallel, the inductors will divide, such that the total circuit inductance will be less than the least value of inductance in the circuit.

The value of inductance in a circuit is labeled henrys (H). If the value of inductance is less than 1, the value may be stated as millihenrys or microhenrys.

When inductors are wired in series as shown in Figure 19-9, the value of the circuit inductance is equal to the sum of the values of individual inductances. By formula:

$$L_{Total} = L_1 + L_2 + L_3 + \ . \ . \ . \ + L_n$$

Example. If three inductive coils with values of 2, 4, and 8 millihenrys, respectively, are wired in series as shown in Figure 19-9, the total circuit inductance is determined as follows:

$$L_{Total} = 2 + 4 + 8 = 14 \text{ millihenrys}$$

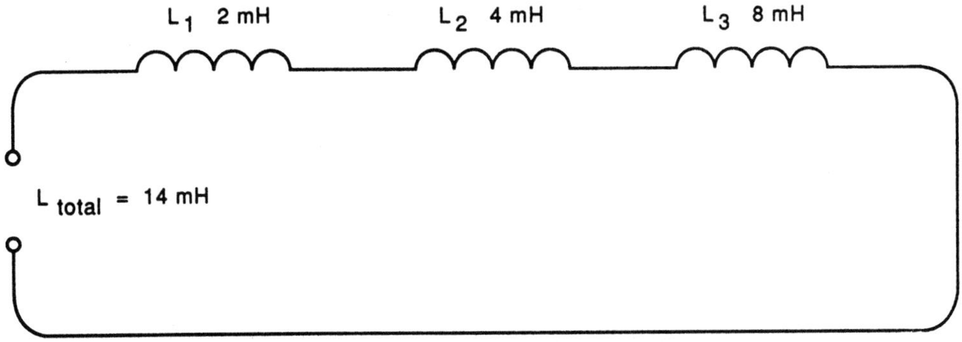

Fig. 19-9 When inductors are wired in series, the circuit inductance is equal to the sum of the individual inductances.

When inductors are wired in parallel as shown in Figure 19-10, the inductance will divide such that the total circuit inductance is less than the least value of inductance in the circuit. By formula:

$$\frac{1}{L_{Total}} = \frac{1}{L_1} + \frac{1}{L_2} + \frac{1}{L_3} + \cdots + \frac{1}{L_n}$$

Or

$$L_{Total} = \frac{1}{\frac{1}{L_1} + \frac{1}{L_2} + \frac{1}{L_3} + \cdots + \frac{1}{L_n}}$$

Example. If three inductive coils, with values of 2, 4, and 8 millihenrys, respectively, are wired in parallel as shown in Figure 19-10, the total circuit inductance is determined as follows:

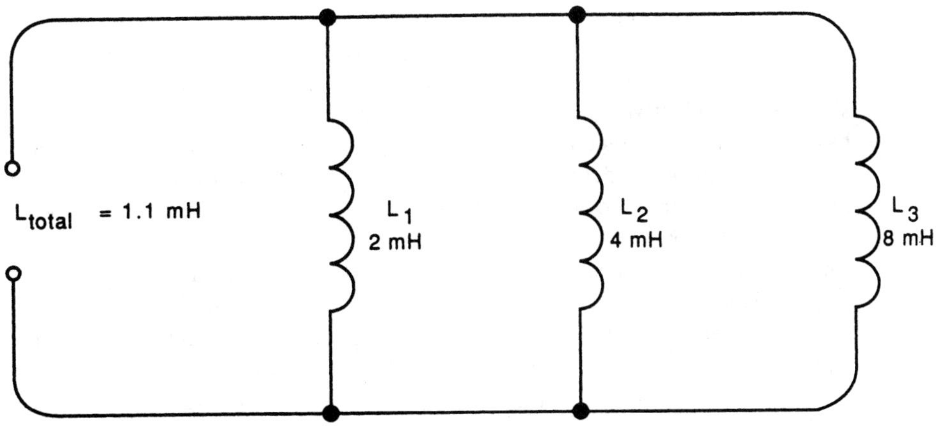

Fig. 19-10 When inductors are wired in parallel, the circuit inductance will be less than the least value of individual inductance.

$$\frac{1}{L_{Total}} = \frac{1}{2} + \frac{1}{4} + \frac{1}{8}$$

$$= 0.500 + 0.250 + 0.125$$

$$L_{Total} = \frac{1}{0.875} = 1.1 \text{ millihenrys}$$

19.10 IMPEDANCE

Both capacitors and inductors are frequency conscious, meaning their operation or rate of electrical energy storage depends on the rate of change of voltage and current magnitude in the power supply. Capacitors and inductors are found in both ac and dc circuits.

In a dc circuit, a capacitor will appear as a short circuit when initial power is applied, because no voltage drop can be seen across its terminals. When the capacitor establishes its full charge, it transcends into an open circuit as it blocks the flow of dc current. The full circuit voltage (ideally) is established as voltage drop across its terminals. Capacitors are also referred to as condensors.

In a dc circuit, an inductor will appear as an open circuit when initial power is applied. The full circuit voltage can be read across its terminals. As current flows in the coils, the electromagnetic field is established around it, the inductor transcends into a short circuit (ideally) because no voltage drop can be read across its terminals. There will be some voltage drop across the inductor, because any conductive material (coil of wire) will have some resistance to the flow of electrical current. Inductors are also referred to as coils or chokes.

In an ac circuit, the pronounced effect of either capacitive reactance (capacitance) or inductive reactance (inductance) is seen in the power factor of the circuit. An inductive ac circuit will have a lagging power factor. A capacitive ac circuit will have a leading power factor. A true resistive ac circuit will have a unity power factor.

Unity indicates the ac current and voltage that produced the consumed power are in phase. A unity power factor has a value of 1. A unity power factor indicates that all power being delivered to the circuit is being consumed as heat.

A lagging power factor indicates the ac circuit is imbalanced with a mixture of resistance and inductive reactance. The power factor has a value less than 1. When a circuit has a lagging power factor, the line current in the circuit lags behind the line voltage in reference to electrical degrees. The stored energy in the coil is continuously charged and discharged back into the circuit as the alternating current changes its polarity.

A leading power factor indicates the ac circuit is imbalanced with a mixture of resistance and capacitive reactance. The power factor also has a value less than 1. When a circuit has a leading power factor, the line current in the circuit leads the line voltage in reference to electrical degrees. The stored energy in the capacitor is continuously charged and discharged back into the circuit as the alternating current changes its polarity.

The set goal of any power distribution system is to have a unity power factor. By the loads served (ballasts, motors, and transformers), the typical power factor is lagging, indicating the inductive loads. Lagging power factors are normally corrected by adding capacitors to individual circuits or to the overall power distribution system.

When both capacitors and inductors are installed in an electrical circuit, they tend to interact or cancel the reactive effect of each other. As a result of this cancelling effect, a net reactance is determined for the circuit. This net reactance is added as a vector sum to the circuit resistance. This vector addition yields the circuit impedance.

Vector sum indicates that the two oppositions to current flow in the circuit, resistance and net reactance, cannot be added directly together. These are phasor quantities that must be added by the Pythagorean theorem.

The reactive power of either a capacitor or an inductor is 90° out of phase (ideally) to the resistance of the circuit. Reactive power is the result of either the inductive reactance or capacitve reactance in the circuit's impedance.

Inductive reactance is referenced by the term X_L · X_L is found by the following formula:

$$X_L = 2 \times 3.1416 \times \text{frequency} \times \text{inductance}$$

Where:

X_L = inductive reactance of the circuit.

3.1416 = fixed value of π (2 π is a value used to find the circumference of a circle)

frequency = frequency of the ac supply circuit (normally either 50 or 60 Hertz [Hz])

inductance = value of the coil or choke's magnetic field measured in henrys

Example. Find the value of inductive reactance of a 10-millihenry coil supplied by a 60 Hz ac circuit.

$$X_L = 2 \times 3.1416 \times 60 \times (10 \times 10^{-3}) = 3.7 \text{ ohms}$$

Capacitive reactance is referenced by the term X_C · X_C is found by the following formula:

$$X_C = \frac{1}{2 \times 3.1416 \times \text{frequency} \times \text{capacitance}}$$

Where:

X_C = capacitive reactance of the circuit.

3.1416 = fixed value of π (2 π is a value used to find the circumference of a circle)

frequency = frequency of the ac supply circuit (normally either 50 or 60 Hertz [Hz])

capacitance = measure of the capacitor or condensor's electrostatic field in farads

Example. Find the value of capacitve reactance for a 10-microfarad capacitor supplied by a 60 Hz ac circuit.

$$X_C = \frac{1}{2 \times 3.1416 \times 60 \times (10 \times 10^{-6})}$$

$$= \frac{1}{0.0038} = 265.2 \text{ ohms}$$

19.11 SERIES IMPEDANCE

When either capacitive or inductive reactance is combined in a series circuit with resistance, the resistance and reactance values are combined for the total circuit value by the Pythagorean theorem.

If only an inductor is placed in series with a resistor as shown in Figure 19-11, the current flowing to the inductor will be out of phase with the current flowing in the resistor. The inductive current will lag the resistive current which is in phase with the supply or line current.

The angle produced by the vector addition of the two currents is referred to as the phase angle. The trigonometric function cosine taken of the phase angle is equal to the power factor of the circuit.

Example. If a 30-ohm resistor is placed in series with a 40-ohm inductor coil in the ac circuit of Figure 19-11, the circuit impedance (Z_T) and power factor (pf) of the circuit are determined as follows:

$$Z_T = [(30)^2 + (40)^2]^{1/2}$$

$$= (2500)^{1/2} = 50 \text{ ohms}$$

$$\text{pf} = 30 \text{ ohms} \div 50 \text{ ohms} = 0.6$$

$$\cos^{-1} \text{ of } 0.6 = 53.1^\circ$$

If only a capacitor is placed in series with a resistor as shown in Figure 19-12, the current flowing to the capacitor will be out of phase with the current flowing in the resistor. The capacitive current will lead the resistive current which is in phase with the supply or line current.

The angle produced by the vector addition of the two currents is referred to as the phase angle. The trigonometric function cosine taken of the phase angle is equal to the power factor of the circuit.

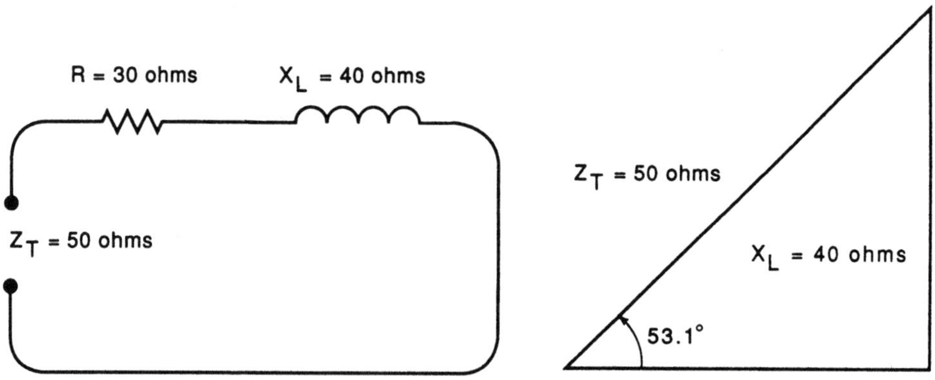

Fig. 19-11 In a series resistance/inductance ac circuit, the circuit impedance is a *vector sum* of both resistance and inductive reactance.

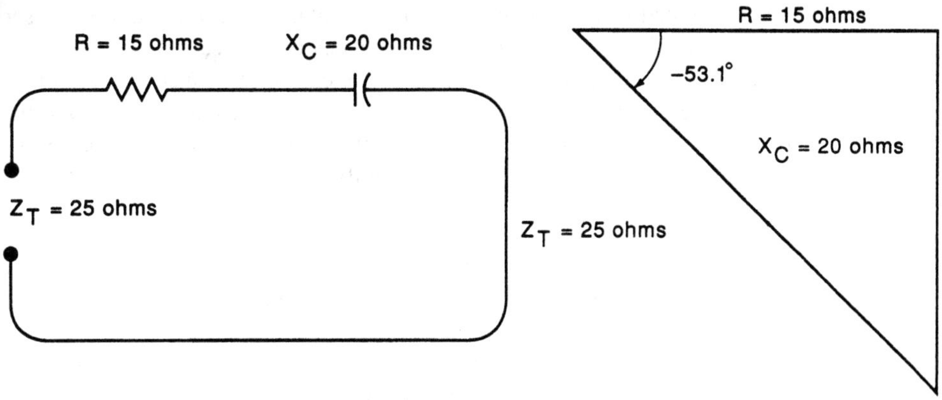

Fig. 19-12 In a series resistance/capacitance ac circuit, the circuit impedance is a *vector sum* of both resistance and capacitive reactance.

Example. If a 15-ohm resistor is placed in series with a 20 ohm capacitor in the series ac circuit of Figure 19-12, the circuit impedance (Z_T) and the power factor (pf) are determined as follows:

$$Z_T = [(15)^2 + (20)^2]^{1/2}$$
$$= (625)^{1/2} = 25 \text{ ohms}$$
$$\text{pf} = 15 \text{ ohms} \div 25 \text{ ohms} = 0.6$$
$$\cos^{-1} \text{ of } 0.6 = -53.1^\circ$$

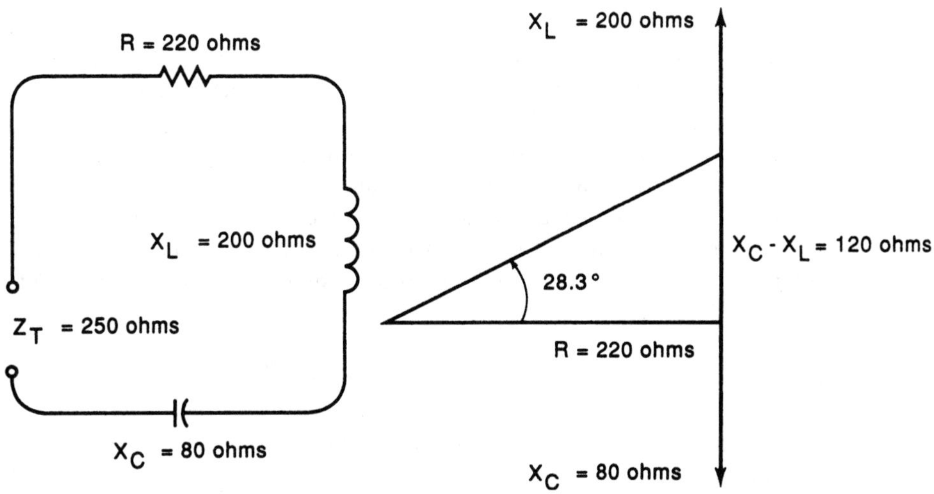

Fig. 19-13 When both capacitance and inductance are contained with resistance in a series ac circuit, the two reactances subtract for a *net reactance.*

When capacitors and inductors are both contained with resistance in a series ac circuit as shown in Figure 19-13, the two reactances tend to cancel each other out. The net reactance (after cancellation) combines with the circuit resistance as a vector sum to form the circuit impedance. If the capacitive and inductive reactances are equal, the net reactive power is equal to zero. The impedance of the circuit is equal to the resistance of the circuit. The power factor of the circuit is at unity, and equal to one.

Example. A 220-ohm resistor is in series with a 200-ohm inductor and an 80-ohm capacitor in the circuit of Figure 19-13. The circuit impedance (Z_T) and power factor (pf) are determined as follows:

$$Z_T = [R^2 + (X_L - X_C)^2]^{1/2}$$

$$= [(220)^2 + (200 - 80)^2]^{1/2}$$

$$= (62{,}800)^{1/2} = 250.6 \text{ ohms}$$

$$\text{pf} = 220 \text{ ohms} \div 250.6 \text{ ohms} = 0.88$$

$$\cos^{-1} \text{ of } 0.88 = 28.6^{\circ} \text{ (lagging)}$$

19.12 PARALLEL IMPEDANCE

In a parallel circuit, the sum of the reciprocals of parallel resistances is equal to the reciprocal of the total circuit resistance. If the circuit also contains inductance and capacitance in parallel with the resistance as shown in Figure 19-14, the circuit impedance is easiest found by determining the currents in the individual branch circuits.

Because it is a parallel branch circuit, the same supply voltage is applied to each branch. By Ohm's law, the respective branch-circuit currents are computed by dividing the supply voltage by the corresponding resistance or reactance. The computed currents add as a vector sum (using the Pythagorean theorem) in the same manner as the resistances and reactances add in a series combination circuit.

The currents in either the inductive branch or the capacitive branch are referred to as quadrature currents. The difference between the capacitive quadrature current and the inductive quadrature current is the net quadrature current. The net quadrature current combines with the resistive current as a vector sum (Pythagorean theorem) to form the line current of the circuit.

$$I_{line} = [I_R{}^2 + (I_L - I_C)^2]^{1/2}$$

The line voltage divided by the the line current in a parallel combination circuit is equal to the circuit impedance.

$$Z_{Total} = V_{line} \div I_{line}$$

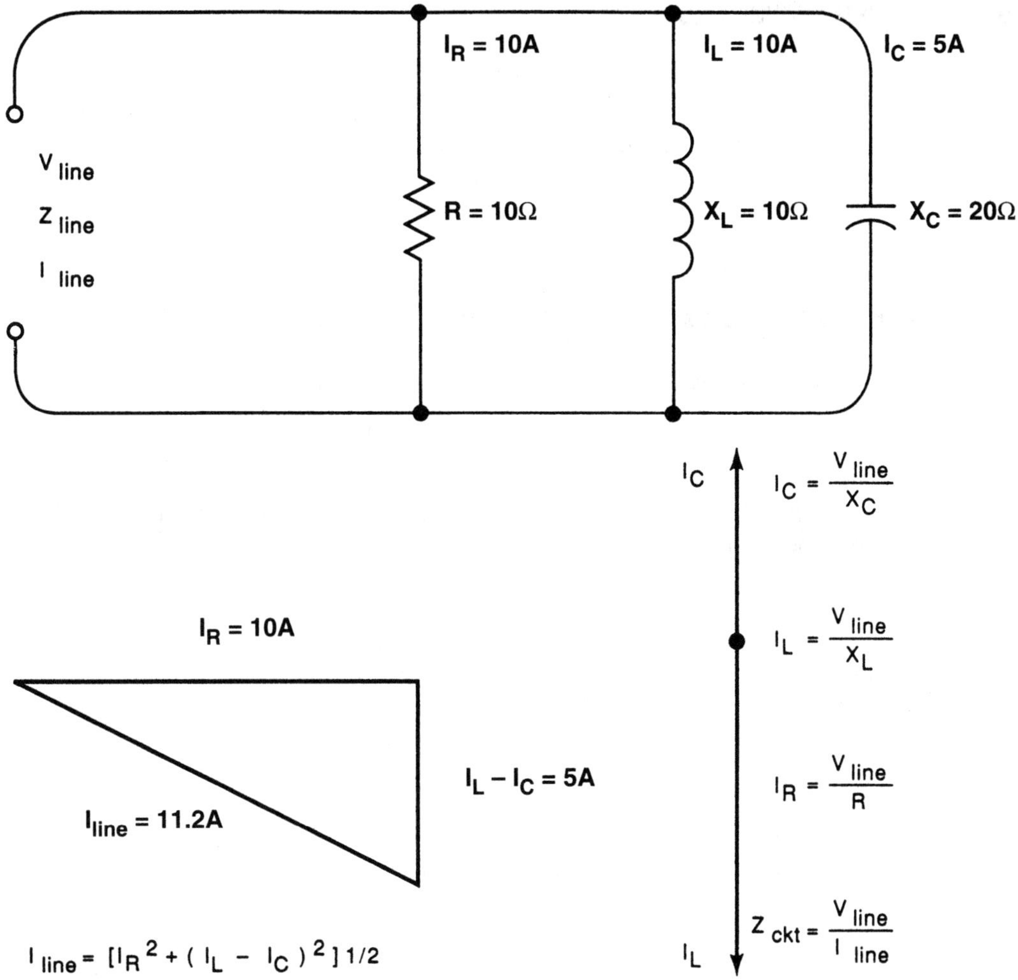

Fig. 19-14 The circuit impedance of a parallel ac resistance/reactance circuit is determined by dividing the line voltage by the line current.

Example. If a 10-ohm resistor is in parallel with both a 10-ohm inductor and a 20-ohm capacitor in the 100-volt ac circuit of Figure 19-14, the circuit impedance is determined as follows:

$$I_R = 100 \text{ volts} \div 10 \text{ ohms} = 10 \text{ amps}$$

$$I_L = 100 \text{ volts} \div 10 \text{ ohms} = 10 \text{ amps}$$

$$I_C = 100 \text{ volts} \div 20 \text{ ohms} = 5 \text{ amps}$$

$$I_{line} = [(10 \text{ amps})^2 + (10 \text{ amps} - 5 \text{ amps})^2]^{1/2}$$

$$= (125 \text{ amps}^2)^{1/2} = 11.2 \text{ amps}$$

$$Z_{ckt} = 100 \text{ volts} \div 11.2 \text{ amps} = 8.9 \text{ ohms}$$

19.13 POWER FORMULAS

The power consumed by a circuit, or the power consumed by an individual component within a circuit, is a product of the voltage applied times the measured current. Normally for NE Code purposes, the power applied to the circuit or system is simply the product of the line voltage times the line current. This is true only for dc circuits and ac circuits where no allowance is made for the power factor of the circuit.

19.14 TRUE POWER: DC AND SINGLE-PHASE AC CIRCUITS

In dc circuits and single-phase ac circuits with unity power factor:

$$\text{Power}_{\text{watts}} = V_{\text{line}} \times A_{\text{line}}$$

Example. If a dc circuit draws 30 amps with 240 volts applied, determine the wattage of the circuit.

$$\text{Power}_{\text{watts}} = 240 \text{ volts} \times 30 \text{ amps} = 7{,}200 \text{ watts}$$

To convert watts to kilowatts, divide the product of the line voltage and line current by 1,000.

$$\text{Power}_{\text{kilowatts}} = \frac{V_{\text{line}} \times A_{\text{line}}}{1{,}000}$$

Example. If a single-phase ac circuit with unity power factor draws 70 amps at 240 volts, determine the kilowatt power consumed by the circuit.

$$\text{Power}_{\text{kilowatts}} = \frac{240 \text{ volts} \times 70 \text{ amps}}{1{,}000} = 16.8 \text{ kW}$$

To solve for A_{line}, the power equation is divided by V_{line} such that

$$A_{\text{line}} = \text{Power}_{\text{watts}} \div V_{\text{Line}}$$

Or

$$A_{\text{line}} = \frac{\text{Power}_{\text{kilowatts}} \times 1{,}000}{V_{\text{line}}}$$

Example. If a circuit with unity power factor consumes 9.4 kW, determine the line current drawn when 120 volts is applied.

$$A_{\text{line}} = \frac{9.4 \text{ kW} \times 1{,}000}{120 \text{ volts}}$$

$$= \frac{9{,}400 \text{ watts}}{120 \text{ volts}} = 78.3 \text{ amps}$$

The power formula for true power (unity power factor) can be solved with resistance and either the line voltage or the line current, if Ohm's law (V = IR or AR) is substituted for the other component of either line current or line voltage.

To solve for true power when circuit (line) voltage and circuit resistance are known:

$$Power_{watts} = V_{line} \times A_{line}$$

And

$$A_{line} = V_{line} \div R_{ckt}$$

Then

$$Power_{watts} = V_{line} \times (V_{line} \div R_{ckt})$$

And

$$Power_{watts} = V_{line}^{\ 2} \div R_{ckt}$$

Example. Determine the power consumed by a dc circuit if the circuit voltage is 250 volts and the circuit resistance is 50 ohms.

$$Power_{watts} = (250 \text{ volts})^2 \div 50 \text{ ohms}$$
$$= 62{,}500 \text{ volts}^2 \div 50 \text{ ohms} = 1{,}250 \text{ watts}$$

To solve for power when circuit (line) current and circuit resistance are known:

$$Power_{watts} = V_{line} \times I_{line}$$

And

$$V_{line} = I_{line} \times R_{ckt}$$

Then

$$Power_{watts} = (I_{line} \times R_{ckt}) \times I_{line}$$

And

$$Power_{watts} = I_{line}^{\ 2} \times R_{ckt}$$

Example. If the line current drawn by a 40-ohm resistive circuit is 30 amperes, determine the power consumed by this circuit.

$$Power_{watts} = (30 \text{ amps})^2 \times 40 \text{ ohms} = 36{,}000 \text{ watts}$$

19.15 TRUE POWER: TWO-PHASE AC CIRCUITS

The true power formula must be modified for a 2-phase, 4- or 5-wire ac distribution system by a factor of 2:

$$Power_{watts} = 2 \times V_{line} \times I_{line}$$

Or

$$\text{Power}_{\text{watts}} = 2 \times V_{\text{line}}^{\,2} \div R_{\text{ckt}}$$

Or

$$\text{Power}_{\text{watts}} = 2 \times I_{\text{line}}^{\,2} \times R_{\text{ckt}}$$

Or

$$\text{Power}_{\text{kilowatts}} = 2 \times V_{\text{line}} \times I_{\text{line}} \div 1000$$

Example. Determine the true power of a 2-phase ac circuit if the power factor is unity, the line voltage is 240 volts, and the line current is 70 amps.

$$\text{Power}_{\text{watts}} = 2 \times 240 \text{ volts} \times 70 \text{ amps} = 33{,}600 \text{ watts}$$

Example. Determine the true power of a 2-phase ac circuit with unity power factor if the circuit resistance is 10 ohms and the circuit voltage is 240 volts.

$$\text{Power}_{\text{watts}} = 2 \times (240 \text{ volts})^2 \div 10 \text{ ohms} = 11{,}520 \text{ watts}$$

Example. Determine the true power of a 2-phase ac circuit with unity power factor if the circuit resistance is 24 ohms and the line current is 74 amps.

$$\text{Power}_{\text{watts}} = 2 \times 74 \text{ amps}^2 \times 24 \text{ ohms} = 262{,}848 \text{ watts}$$

Example. Determine the true power in kilowatts of a 2-phase ac circuit with unity power factor if the circuit voltage is 240 volts and the circuit current is 400 amps.

$$\text{Power}_{\text{kilowatts}} = (2 \times 240 \text{ volts} \times 400 \text{ amps}) \div 1000 = 192 \text{ kW}$$

19.16 TRUE POWER: THREE-PHASE AC CIRCUITS

The true power formula for a 3-phase ac circuit must be modified by the square root of 3 (1.732):

$$\text{Power}_{\text{watts}} = 1.732 \times V_{\text{line}} \times I_{\text{line}}$$

Or

$$\text{Power}_{\text{watts}} = 1.732 \times V_{\text{line}}^{\,2} \div R_{\text{ckt}}$$

Or

$$\text{Power}_{\text{watts}} = 1.732 \times I_{\text{line}}^{\,2} \times R_{\text{ckt}}$$

Or

$$\text{Power}_{\text{kilowatts}} = 1.732 \times V_{\text{line}} \times I_{\text{line}} \div 1000$$

Example. Determine the power of a 3-phase resistive circuit if the line voltage is 208 volts and the line current is 40 amps.

$$\text{Power}_{\text{watts}} = 1.732 \times 208 \text{ volts} \times 40 \text{ amps} = 14{,}410 \text{ watts}$$

Example. Determine the true power of a 3-phase resistive circuit if the circuit voltage is 465 volts and the circuit resistance is 17 ohms.

$$\text{Power}_{\text{watts}} = 1.732 \times (465 \text{ volts})^2 \div 17 \text{ ohms} = 22{,}030 \text{ watts}$$

Example. Determine the true power of a 3-phase circuit with unity power factor if the circuit current is 68 amps and the circuit resistance is 14 ohms.

$$\text{Power}_{\text{watts}} = 1.732 \times (68 \text{ amps})^2 \times 14 \text{ ohms} = 112{,}123 \text{ watts}$$

Example. Determine the resistive power in kilowatts of a 3-phase circuit which is rated at 480 volts and draws 314 amps.

$$\text{Power}_{\text{kilowatts}} = 1.732 \times 480 \text{ volts} \times 314 \text{ amps} \div 1000 = 261 \text{ kW}$$

19.17 APPARENT POWER

The true power of an ac circuit is the power applied to the resistance of the circuit. The reactive power of an ac circuit is the power applied to the reactance (either capacitive or inductive) of the circuit. The apparent power of an ac circuit is the power applied to the impedance of the circuit.

The apparent power of the circuit is the power of the product of the line voltage and line current as measured by metering equipment (voltmeter and ammeter) with the given load connected.

Apparent power is not used with dc circuits, because the reactance of inductors and capacitors goes into a static (steady) state when true dc power is applied. If the dc source offers a pulsating dc (contains ac ripple), then the reactive elements will act in the same manner as in an ac circuit.

The power formulas used for the apparent power of ac or pulsating dc circuits are the same as was stated for true power except the unit name is changed to volt-amps or kilovolt-amps. The substitution of Ohm's law to solve for apparent power is the same as was demonstrated for true power except the circuit impedance, and not just the circuit resistance is used in the Ohm's law formula. Computing the apparent power of an electrical circuit takes into account that the power factor may not be at unity.

For single-phase ac and pulsating dc, the apparent power formula is:

$$\text{Power}_{\text{volt-amps}} = V_{\text{line}} \times A_{\text{line}}$$

Or

$$\text{Power}_{\text{kilovolt-amps}} = V_{\text{line}} \times A_{\text{line}} \div 1000$$

Example. The maximum apparent power of a 30-amp, 120-volt branch circuit would be determined as follows:

$$\text{Power}_{\text{volt-amps}} = 120 \text{ volts} \times 30 \text{ amps} = 3{,}600 \text{ VA}$$

Example. The apparent kVA power consumed by a 240-volt motor circuit drawing 64 amps is

$$\text{Power}_{\text{kilovolt-amps}} = 240 \text{ volts} \times 64 \text{ amps} \div 1000 = 15.36 \text{ kVA}$$

For 2-phase ac circuits, the apparent power formula is:

$$\text{Power}_{\text{volt-amps}} = 2 \times V_{\text{line}} \times A_{\text{line}}$$

Or

$$\text{Power}_{\text{kilovolt-amps}} = 2 \times V_{\text{line}} \times A_{\text{line}} \div 1000$$

Example. The maximum apparent power of a 2-phase ac circuit that is rated at 240 volts and draws 80 amps is

$$\text{Power}_{\text{volt-amps}} = 2 \times 240 \text{ volts} \times 80 \text{ amps} = 38{,}400 \text{ VA}$$

Example. The apparaent kVA power consumed by a 2-phase, 5-wire feeder rated at 400 amps, 240 volts is

$$\text{Power}_{\text{kilovolt-amps}} = 2 \times 240 \text{ volts} \times 400 \text{ amps} \div 1000 = 192 \text{ kVA}$$

For 3-phase ac circuits, the apparent power formula is:

$$\text{Power}_{\text{volt-amps}} = 1.732 \times V_{\text{line}} \times A_{\text{line}}$$

Or

$$\text{Power}_{\text{kilovolt-amps}} = 1.732 \times V_{\text{line}} \times A_{\text{line}} \div 1000$$

Example. The maximum apparent power rating of a 3-phase, 208-volt circuit rated at 30 amps is

$$\text{Power}_{\text{volt-amps}} = 1.732 \times 208 \text{ volts} \times 30 \text{ amps} = 10{,}808 \text{ VA}$$

Example. The apparent kVA power consumed by a 480Y/277-volt, 3-phase, 4-wire service panel rated at 800 amps is

$$\text{Power}_{\text{kilovolt-amps}} = 1.732 \times 480 \text{ volts} \times 800 \text{ amps} \div 1000 = 665 \text{ kVA}$$

19.18 POWER FACTOR (pf)

As shown in Figure 19-15, the power factor of an electrical circuit is a ratio of the resistive power or wattage consumed by the load to the apparent power delivered to the load. Wattage, also referred to as true power, is determined by using a wattage meter within the ac circuit. Wattage is represented by a value of either watts (W) or kilowatts (kW). Apparent power, which is determined by measuring the power applied to the circuit with a voltmeter and an ammeter and taking a product of their readings, is represented by a value of either volt-amps (VA) or kilovolt-amps (kVA). By formula:

$$pf = \frac{\text{watts}}{\text{volt-amps}} = \frac{W}{VA}$$

Or

$$pf = \frac{\text{kilowatts}}{\text{kilovolt-amps}} = \frac{kW}{kVA}$$

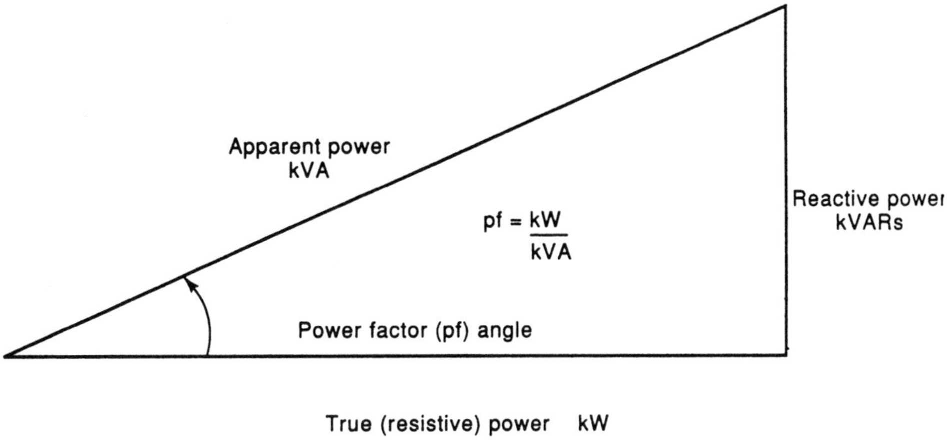

Fig. 19-15 The power factor (pf) of an ac circuit is equal to the true power of the circuit divided by the apparent power of the circuit.

Example. If the wattage consumed by a power circuit is equal to 1,400 watts and the apparent power delivered to this circuit is 1,800 volt-amps, the power factor of the circuit is

$$pf = 1{,}400 \text{ watts} \div 1{,}800 \text{ volt-amps} = 0.78$$

Example. If the wattage consumed by a power circuit is 165 kW, and the power delivered to the circuit is 185 kVA, the power factor of the circuit is determined as

$$pf = 165 \text{ kW} \div 185 \text{ kVA} = 0.89$$

The power factor of a circuit is normally expressed as a decimal fraction. To express the power factor as a percentage, the decimal is moved two places to the right and the symbol (%) is added for percentage.

Example. A power factor of 0.94 is equivalent to a 94% power factor.

A power factor expressed as a percentage can be expressed as a decimal fraction by dropping the percentage sign (%) and moving the decimal two places to the left. A zero should precede the decimal (on the left) to identify the decimal fraction.

Example. A power factor of 78% is equivalent to a 0.78 power factor.

As outlined in Figure 19-16, the power factor of an electrical circuit can never be greater than 1 (unity), because the true power (kilowatts) consumed by the circuit can never be greater than the apparent power (kilovolt-amps) delivered to the circuit.

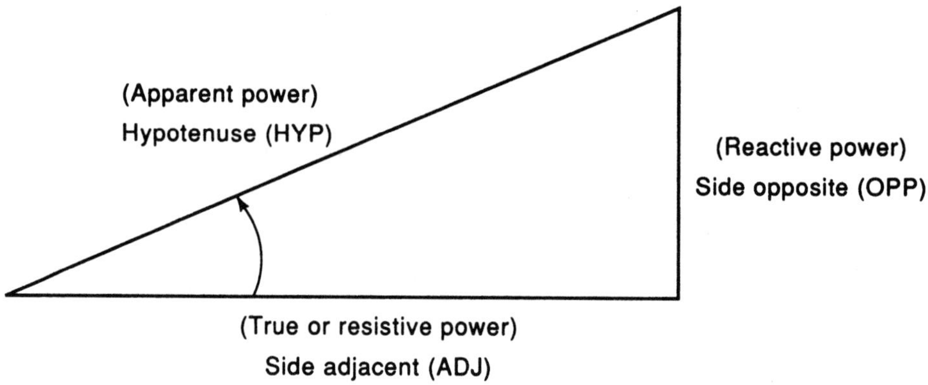

$$\text{Sine} = \frac{\text{OPP}}{\text{HYP}} \qquad \text{Cosine} = \frac{\text{ADJ}}{\text{HYP}} \qquad \text{Tangent} = \frac{\text{OPP}}{\text{ADJ}}$$

$$\text{Cosecant} = \frac{\text{HYP}}{\text{OPP}} \qquad \text{Secant} = \frac{\text{HYP}}{\text{ADJ}} \qquad \text{Cotangent} = \frac{\text{ADJ}}{\text{OPP}}$$

Fig. 19-16 The power triangle is a function of trigonometry.

If the kilowatt power measured by a wattmeter is equal to the product of the line voltage and line current as measured by a voltmeter and an ammeter then

$$\text{kilowatts} = \text{kilovolt-amps}$$

And

$$\text{pf} = \frac{\text{kilowatts}}{\text{kilovolt-amps}}$$

By substitution

$$\text{pf} = \frac{\text{kilovolt-amps}}{\text{kilovolt-amps}}$$

$$\text{pf} = 1$$

Example. The wattmeter reading of a three-phase power circuit is 28.1 kW. If the measured voltage applied to the circuit is 208 volts, and the measured current is 78 amps, determine the circuit's power factor.

$$\text{Power}_{\text{kVA}} = 1.732 \times 208 \text{ volts} \times 78 \text{ amps} \div 1000 = 28.1 \text{ kVA}$$

$$\text{pf} = 28.1 \text{ kW} \div 28.1 \text{ kVA} = 1$$

19.19 HORSEPOWER RATINGS

The horsepower rating of a motor is related by formula to the true power (wattage) of a motor. Ideally, a motor that draws 746 watts is rated at 1 horsepower. If a motor draws more than 746 watts of true power, then the motor is rated at more than 1 horespower. A motor rated as 15 horsepower would draw a total of (15 × 746 watts) 11,190 watts.

The difference between the wattage drawn by a motor and the apparent power supplied to the motor (measurement of supply voltage and current) is the losses of power due to both the efficiency and power factor of the motor.

19.20 EFFICIENCY

As with any type of machinery or mechanical conversion means, the efficiency of a motor is the ratio of output power to input power. By formula:

$$\text{motor efficiency percentage} = \frac{\text{output power}}{\text{input power}} \times 100$$

The efficiency of a motor takes into consideration the copper losses of the motor windings and the iron losses of the motor's rotor and stator.

The power factor of a motor takes into consideration the inductive losses (inductive reactance) of the motor. The construction of the motor and how the motor is wound are both contributing factors to the inductive losses of a motor.

The horsepower rating, which is derived from the true power rating of a motor, must also take into account the phase rating of a motor. This applies only to ac motors, since a dc motor will not have a phase rating.

The power factor (pf) is also another term that does not apply in the wattage and/or horsepower rating of a dc motor (unless its pulsating dc).

19.21 HORSEPOWER FORMULA: DC MOTORS

$$\text{Horsepower} = \frac{\text{volts} \times \text{amperes} \times \text{efficiency}}{\text{746 watts (per horsepower)}}$$

Example. A 500-volt dc compound motor draws a full-load current of 155 amps. If the motor's efficiency is rated at 96%, determine the horsepower rating of the dc motor.

$$\text{Horsepower} = \frac{\text{500 volts} \times \text{155 amps} \times 0.96}{\text{746 watts}} = \text{100 hp}$$

19.22 HORSEPOWER FORMULA: SINGLE-PHASE AC MOTORS

$$\text{Horsepower} = \frac{\text{volts} \times \text{amperes} \times \text{efficiency} \times \text{power factor}}{\text{746 watts (per horsepower)}}$$

Example. A single-phase ac motor rated at 240 volts, has a rated efficiency of 78% and a power factor of 86%. If the motor draws 23 amps, determine its horsepower rating.

$$\text{Horsepower} = \frac{240 \text{ volts} \times 23 \text{ amps} \times 0.78 \times 0.86}{746 \text{ watts}} = 5 \text{ hp}$$

19.23 HORSEPOWER FORMULA: TWO-PHASE AC MOTORS

$$\text{Horsepower} = \frac{2 \times \text{volts} \times \text{amperes} \times \text{efficiency} \times \text{power factor}}{746 \text{ watts (per horsepower)}}$$

Example. A 2-phase ac induction motor rated at 575 volts has an efficiency rating of 68% and a power factor rating of 80%. If the motor draws a full-load current of 24 amps, determine its horsepower rating.

$$\text{Horsepower} = \frac{2 \times 575 \text{ volts} \times 24 \text{ amps} \times 0.68 \times 0.80}{746 \text{ watts}} = 20 \text{ hp}$$

19.24 HORSEPOWER FORMULA: THREE-PHASE AC MOTORS

$$\text{Horsepower} = \frac{1.732 \times \text{volts} \times \text{amperes} \times \text{efficiency} \times \text{power factor}}{746 \text{ watts (per horsepower)}}$$

Example. A 3-phase induction motor rated at 480 volts has an efficiency rating of 86% and a power factor rating of 94%. If the motor draws a full-load current of 111 amps, determine its horsepower rating.

$$\text{Horsepower} = \frac{1.732 \times 480 \text{ volts} \times 111 \text{ amps} \times 0.86 \times 0.94}{746 \text{ watts}} = 100 \text{ hp}$$

19.25 MOTOR APPARENT POWER: NE CODE VALUES

In most NE Code computations, the apparent power of a given motor must be determined from the horsepower rating and the NE Code-given table values of full-load current. In this type calculation, the horsepower formula must be solved for the motor's VA or kVA load.

Example. The apparent power of a ¾-hp, 120-volt, single-phase ac motor with a full-load current rating of 13.8 amps would be determined as follows:

$$120 \text{ volts} \times 13.8 \text{ amps} = 1{,}656 \text{ volt-amps}$$

Example. The apparent power of a 10-hp, 240-volt, single-phase ac motor with a full-load current rating of 50 amps would be determined as follows:

$$240 \text{ volts} \times 50 \text{ amps} = 12{,}000 \text{ volt-amps}$$

Example. The apparent power of a 25-hp, 480-volt, 2-phase ac motor with a full-load current rating of 29 amps would be determined as follows:

$$2 \times 480 \text{ volts} \times 29 \text{ amps} = 27{,}840 \text{ volt-amps}$$

Example. The apparent power of a 75 hp, 240-volt, 3-phase ac motor with a full-load current rating of 192 amps would be determined as follows:

$$1.732 \times 240 \text{ volts} \times 192 \text{ amps} = 79{,}811 \text{ volt-amps}$$

19.26 VOLTAGE-DROP FORMULAS

For electrical equipment to function correctly, the rated nominal voltage must be supplied to its (their) terminals. Two voltage-drop formulas are available for use to the student electrician: One formula is based on the circular-mil area of the conductor according to given wire size. This formula uses an assigned resistance for either copper or aluminum (or copper-clad aluminum). The other formula is based only on the resistance of the conductor (based on the resistance of a 1,000-foot length) according to wire size and whether the conductor is copper or aluminum (or copper-clad aluminum).

Either formula must take into consideration the number of phases in the electrical circuit. If the circuit in question is either dc, single-phase ac, or 2-phase ac, either formula is taken as is for the voltage-drop computation. If the circuit is 3-phase, the voltage drop computation must be reduced by multiplying the equation by a value of 0.866.

Either voltage-drop formula computes voltage drop for only two conductors: one supply conductor and one return conductor. In a 3-phase circuit, a third conductor is used to supply current to the same load. Multiplying either formula by 0.866, which is equal to the square root of 3 divided by 2, allows for the reduction in voltage drop by the use of the third conductor.

The voltage-drop formulas will be used in two ways. One way will demonstrate substituting given values to find the actual voltage drop. The other way will demonstrate substituting actual values to find the minimum wire size allowed for a maximum voltage drop.

The fine print notes (FPNs) to Secs. 210-19(a) and 215-2(b) in the NE Code recommend voltage-drop considerations for service, feeder, and/or branch circuits. The maximum recommended percentage from point of supply to point of utilization (load) is 5%.

As a rule, service and feeder conductors will conductor heavier or larger current values as compared to branch circuits. As taught in this text, service or feeder conductors are computed at a maximum of 2% voltage drop. Branch-circuit conductors are computed at a maximum of 3% voltage drop.

Although inductive reactance is a principal load in the electrical world, it is considered as negligible using these voltage-drop formulas (power factor equal or near a value of 1). Instead of impedance, dc resistance is used where resistance values are required.

19.27 VOLTAGE-DROP: CIRCULAR-MIL AREA COMPUTATIONS

When using this voltage-drop formula, the resistance value is fixed for all wire sizes according to type of conductor material. The wire size is adjusted according to the circular-mil area of the conductor. The circular-mil voltage-drop formula is given as

$$VD = \frac{2 \times L \times A \times R \times 0.866}{CM}$$

Where: VD = either the maximum (computed) or actual voltage drop
L = one-way length of the circuit conductors in feet
A = full-load current in amps
R = fixed resistance of the conductor according to material: for copper, 12 ohms (per CM ft); for aluminum or copper-clad aluminum, 18 ohms (per CM ft)
CM = circular-mil area of the conductor as listed according to wire size in Table 8, Chapter 9 of the NEC

0.866 is used for 3-phase circuits only.

Example. Determine the actual voltage drop of a single-phase, 2-wire, 240-volt feeder for a full-load current of 250 amps. The feeder, which consists of two 250 kcmil THW copper conductors, has a one-way length of 900 feet.

$$VD = \frac{2 \times 900 \text{ feet} \times 250 \text{ amps} \times 12 \text{ ohms/CM-foot}}{250,000 \text{ CM}} = 21.6 \text{ volts}$$

When the wire size must be determined for a maximum allowed voltage drop, the formula is rewritten to solve for a maximum circular-mil area. The maximum value of voltage drop is computed as a percentage of the line voltage of the circuit in question. In formula, with the same assigned values as before:

$$CM_{max} = \frac{2 \times L \times A \times R \times 0.866}{VD_{max}}$$

Example. Determine the minimum wire size required to keep a 3-phase, 50-amp, 208-volt branch circuit that is 1,200 feet in one-way length under a 3% voltage drop. The branch circuit should consist of three THW aluminum conductors.

$$CM_{max} = \frac{2 \times 1,200 \text{ feet} \times 50 \text{ amps} \times 18 \text{ ohms/CM-foot} \times 0.866}{(0.03 \times 208 \text{ volts})}$$

$$CM_{max} = 299,769 \text{ circular mils}$$

NE Code Table 8, Chapter 9 (greater than or equal to 299,769 CM): 300 kcmil

19.28 VOLTAGE DROP: RESISTANCE COMPUTATIONS

When using this formula, the conductor wire size will change according to the given resistance (actual, as listed in Table 8, Chapter 9 of the NE Code, and maximum value as determined when a maximum voltage drop is specified). The value of 1,000 feet in the denominator allows for the "ohms per thousand (k) feet" as listed in the resistance heading of Table 8, Chapter 9 in the NE Code, which allows the R-value listed in the table, according to wire size and type of conductor material, to be used as is in the voltage-drop formula.

For resistance computations involving the actual voltage drop, the formula is:

$$VD = \frac{2 \times L \times R \times A \times 0.866}{1,000}$$

Where: VD = either maximum (computed) or actual voltage drop
L = one-way length of the circuit conductors in feet
A = circuit full-load current in amperes
R = assigned value of resistance as listed in Table 8, Chapter 9 of the NE Code according to conductor material and wire size

0.866 is used for 3-phase ac circuits only.

Example. Determine the actual voltage drop of a 3-phase, 3-wire, 575-volt feeder rated at 400 amps. The feeder circuit, which consist of 500 kcmil THHN copper conductors, has one-way length is 1,400 feet.

$$VD = \frac{2 \times 1,400 \text{ feet} \times 400 \text{ amps} \times 0.0258 \text{ ohms/k-feet} \times 0.866}{1,000 \text{ feet}}$$

$$VD = 25.0 \text{ volts}$$

When the wire size must be determined for a maximum allowed voltage drop, the formula is rewritten to solve for a maximum wire-resistance value. The maximum value of voltage drop is computed as a percentage of the line voltage of the circuit in question. In formula, with the same assigned values as before:

$$R_{max} = \frac{1,000 \times VD_{max}}{2 \times L \times A \times 0.866}$$

Example. Determine the minimum wire size needed for a 30-amp, 120-volt, 2-wire, single-phase branch circuit that is 600 feet long. Use THW copper.

$$R_{max} = \frac{1,000 \text{ feet} \times (0.03 \times 120 \text{ volts})}{2 \times 600 \text{ feet} \times 30 \text{ amps}}$$

$$R_{max} = 0.100 \text{ ohms}$$

NE Code Table 8, Chapter 9 ($R \leq 0.100$): No. 2/0 AWG THW copper

Journeyman's Guide to the NEC Appendix

ANSWERS TO CHAPTER CODE-CHALLENGE PROBLEMS

Chapter Two

2-1. (A) 85 amps; (B) T310-16
2-2. (A) 385 amps; (B) T310-16
2-3. (A) 125; (B) 240-6(a); (C) 130; (D) 240-3(b): (E) No. 3/0 AWG; (F) T310-16
2-4. (A) 60°C; (B) 110-14(c)(1), Excp. 1; (C) 55 amps; (D) No. 4 AWG; (E) T310-16
2-5. (A) 75°C; (B) 110-14(c)(2), Excp. 1: (C) 420 amps; (D) 600 kcmil; (E) T310-16
2-6. (A) 260; (B) T310-16; (C) 0.76; (D) T310-16, CF; (E) 197.6; (F) 75°C; (G) 110-14(c)(2), Excp. 1; (H) 230; (I) 197.6; (J) 310-15(c)
2-7. (A) 0.67; (B) T310-16, CF; (C) 89.6; (D) 100; (E) No. 3 AWG; (F) T310-16; (G) 60; (H) 110-14(c)(2), Excp. 1; (I) 70; (J) No. 4 AWG; (K) T310-16; (L) larger; (M) 310-15(c); (N) No. 3 AWG
2-8. (A) 25; (B) T310-16; (C) 70; (D) T310-16, N-8(a) tbl; (E) 17.5; (F) 60; (G) 110-14(c)(1), Excp. 1; (H) 25; (I) 17.5; (J) 310-15(c)
2-9. (A) 70; (B) T310-16, N-8(a) tbl; (C) 285.7; (D) 305; (E) 400 kcmil; (F) T310-16; (G) 75; (H) 110-14(c)(2), Excp. 1; (I) 205; (J) 250 kcmil; (K) T310-16; (L) larger; (M) 310-15(c); (N) 400 kcmil
2-10. (A) do not; (B) T310-16, N-11; (C) 8; (D) 60: (E) T310-16; (F) 0.70; (G) T310-

16, N-8(a) tbl; (H) 42; (I) 60; (J) 110-14(c)(1), Excp. 1; (K) 40; (L) 40; (M) 310-15(c)

2-11. (A) do not; (B) T310-16, N-10(a); (C) 8; (D) 70: (E) T310-16, N-8(a) tbl; (F) 214.3; (G) 230; (H) 250 kcmil; (I) T310-16; (J) 75; (K) 110-14(c)(2), Excp. 1; (L) 155; (M) No. 3/0 AWG; (N) T310-16; (O) larger; (P) 310-15(c); (Q) 250 kcmil

2-12. (A) does; (B) T310-16, N-10(c); (C) 12; (D) 20; (E) T310-16; (F) 50; (G) T310-16, N-8(a) tbl; (H) 10; (I) 60; (J) 110-14(c)(1), Excp. 1; (K) 20; (L) 10; (M) 310-15(c)

2-13. (A) 80; (B) 220-3(a); (C) 30; (D) T310-16; (E) 20; (F) T310-16, Note; (G) 16

2-14. (A) 125; (B) 220-10(b); (C) 200; (D) 75; (E) 110-14(c)(2), Excp. 1; (F) 205; (G) 250 kcmil; (H) T310-16

2-15. (A) 130; (B) T310-16; (C) 0.67; (D) T310-16, CF; (E) 80; (F) T310-16, N-8(a) tbl; (G) 87.1; (H) 69.7; (I) 60; (J) 110-14(c)(1), Excp. 1; (K) 110; (L) smaller; (M) 310-15(c); (N) 69.7

2-16. (A) 0.87; (B) T310-16, CF; (C) 70; (D) T310-16, N-8(a) tbl; (E) 132.2; (F) 188.8; (G) 195; (H) No. 2/0 AWG; (I) T310-16; (J) 75; (K) 110-14(c)(2), Excp. 1; (L) 115; (M) No. 2 AWG; (N) T310-16; (O) larger; (P) 310-15(c); (Q) No. 2/0 AWG

2-17. (A) 0.82; (B) T310-16, CF: (C) 70; (D) T310-16, N-8(a) tbl; (E) 125; (F) 220-10(b); (G) 122; (H) 174.2; (I) 195; (J) No. 2/0 AWG; (K) T310-16; (L) 60; (M) 110-14(c)(1), Excp. 1; (N) 125; (O) 130; (P) No. 1/0 AWG; (Q) T310-16; (R) larger; (S) 310-15(c); (T) No. 2/0 AWG

2-18. (A) 400; (B) 55; (C) 0.321; (D) 14.1; (E) Ch. 9, T8

2-19. (A) 0.02; (B) 480; (C) 0.866; (D) 0.0231; (E) 0.0214; (F) 600 kcmil; (G) Ch. 9, T8

2-20. (A) 0.96; (B) T310-16, CF; (C) 80; (D) T310-16, N-8(a) tbl; (E) 125; (F) 220-3(a); (G) 50; (H) 62.5; (I) 75; (J) No. 6 AWG; (K) T310-16; (L) 60; (M) 110-14(c)(1), Excp. 1; (N) 60; (O) 70; (P) No. 4 AWG; (Q) T310-16; (R) 210-19(a), FPN No. 4; (S) 0.03; (T) 208; (U) 0.866; (V) 0.0751; (W) 0.0608; (X) No. 4/0 AWG; (Y) Ch.9, T-8; (Z) larger; (1A) 310-15(c); (1B) No. 4/0 AWG

2-21. (A) 60; (B) 336-30(b); (C) copper; (D) 110-5; (E) 25; (F) T310-16

2-22. (A) 30; (B) T310-16; (C) 336-30(b), Excp.; (D) 0.76; (E) T310-16, CF; (F) 22.8; (G) 60; (H) 336-30(b); (I) 20; (J) T310-16, Note; (K) smaller; (L) 310-15(c); (M) 20

2-23. (A) 30; (B) T310-16; (C) 336-30(b), Excp.; (D) T310-16, N-11; (E) 16; (F) 50; (G) T310-16, N-8(a) tbl; (H) 15; (I) 60; (J) 336-30(b); (K) 20; (L) T310-16, Note; (M) smaller; (N) 310-15(c); (O) 15

2-24. (A) 25; (B) T310-16; (C) 0.71; (D) T310-16, CF; (E) 17.8; (F) 60; (G) 339-5; (H) 20; (I) T310-16, Note; (J) smaller; (K) 310-15(c); (L) 17.8

2-25. (A) 0.75; (B) T310-16, CF; (C) 166.7; (D) 180; (E) No. 4/0 AWG; (F) T310-16; (G) 75; (H) 110-14(c)(2), Excp. 1; (I) 135; (J) No. 2/0 AWG; (K) T310-16; (L) larger; (M) 310-15(c); (N) No. 4/0 AWG

2-26. (A) No. 1/0 AWG; (B) T310-16

2-27. (A) 60; (B) 338-4; (C) 336-30(b); (D) 40; (E) No. 8 AWG; (F) T310-16

Chapter Three

3-1. (A) App. C, T-C4; (B) 26; (C) 1½

3-2. (A) App. C, T-C8; (B) 3; (C) 2½

3-3. (A) App. C, T-C10; (B) 8; (C) 5

3-4. (A) App. C, T-C10; (B) 6; (C) 4
3-5. (A) App. C, T-C7; (B) 18; (C) 23; (D) 1¼
3-6. (A) App. C, T-C1A; (B) 8; (C) 2
3-7. (A) Ch. 9, T-5; (B) 0.0243
3-8. (A) Ch. 9, T-5; (B) 0.0353; (C) 0.026
3-9. (A) Ch. 9, T-5; (B) 0.0333; (C) 0.0507; (D) 0.0973; (E) 0.2624; (F) 2.3586; (G) Ch. 9, T-4; (H) 3.538; (I) 3
3-10. (A) Ch. 9, T-5A; (B) 0.0525; (C) 0.6939; (D) 6.9612; (E) Ch. 9, T-4; (F) 7.904; (G) 5
3-11. (A) Ch. 9, T-8; (B) 0.001; (C) 0.011; (D) 0.109; (E) 0.260; (F) 1.042
3-12. (A) Ch. 9, T-5; (B) Ch. 9, T-8; (C) 0.7073; (D) 0.3970; (E) 0.109; (F) 2.6279; (G) Ch. 9, T-4; (H) 3.000; (I) 3
3-13. (A) 0.87; (B) T310-16, CF; (C) 6; (D) T310-16, N-11; (E) 80; (F) T310-16, N-8(a) tbl; (G) 125; (H) 220-3(a); (I) 126.4; (J) 158.0; (K) 170; (L) No. 1/0 AWG; (M) T310-16; (N) 75; (O) 110-14(c)(2), Excp. 1; (P) 137.5; (Q) 150; (R) No. 1/0 AWG; (S) T310-16; (T) larger; (U) 310-15(c); (V) No. 1/0 AWG: (W) 345-7; (X) Ch. 9, T-5; (Y) Ch. 9, T-8; (Z) No. 1/0 AWG; (1A) 0.1825; (1B) 0.2223; (1C) 0.053; (1D) 1.2674; (1E) Ch. 9, T-4; (1F) 1.452; (1G) 2
3-14. (A) 1,000; (B) larger; (C) 318-10(a)(2); (D) Ch. 9, T-5; (E) 1.539; (F) 1.662; (G) 28.809; (H) 30; (I) T318-10
3-15. (A) 250; (B) 1,000; (C) 318-10(a)(2); (D) Ch. 9, T-5; (E) 0.4596; (F) 0.5281; (G) 0.5958; (H) 0.7073; (I) 23.0427; (J) 26.0; (K) 24; (L) T318-10, Col. 1
3-16. (A) 1,000; (B) larger; (C) 318-10(a)(3); (D) Ch. 9, T-5; (E) 1.31; (F) 1.539; (G) 11.396; (H) T318-10, Col. 2; (I) 1.1; (J) 12.5356; (K) 250; (L) 1,000; (M) 318-10(a)(3); (N) Ch. 9, T-5; (O) 0.5242; (P) 0.7901; (Q) 1.0496; (R) 15.7508; (S) 28.2864; (T) 32.5; (U) 30; (V) T318-10, Col. 2

Chapter Four

4-1. (A) 370-16(b); (B) 2; (C) 6; (D) 2; (E) 1; (F) 2; (G) 1; (H) T370-16(a); (I) 10; (J) 2⅛
4-2. (A) 370-16(b); (B) 2; (C) 4; (D) 6; (E) 6; (F) 2; (G) 1; (H) T370-16(a); (I) 11; (J) 1¼
4-3. (A) 370-16(b); (B) 2; (C) 4; (D) 2; (E) 1; (F) 2; (G) 1; (H) 1; (I) 2; (J) T370-16(a); (K) 8; (L) 3½
4-4. (A) 370-16(b); (B) 2; (C) 6; (D) 2; (E) 1; (F) 1; (G) 2; (H) T370-16(a); (I) 9; (J) 1½
4-5. (A) 370-16(b); (B) 4; (C) 4; (D) 2; (E) 1; (F) 1; (G) 1; (H) 2; (I) 4; (J) T370-16(a); (K) 13; (L) 2⅛
4-6. (A) 370-16(b); (B) 3; (C) 3; (D) 1; (E) 1; (F) 1; (G) 2; (H) T370-16(a); (I) 6; (J) 2⅛
4-7. (A) 370-16(b); (B) 12; (C) 12; (D) 12; (E) 12; (F) 4; (G) 1; (H) T370- 16(a); (I) 26; (J) 1½; (K) 1¼
4-8. (A) 370-16(b); (B) 4; (C) 4; (D) 8; (E) 8; (F) No. 10; (G) 1; (H) T370- 16(b); (I) 5; (J) 2.5; (K) 8; (L) 2.25; (M) 35.0; (N) T370-16(a); (O) 42; (P) 2⅛
4-9. (A) 370-16(b); (B) 1; (C) 2; (D) 2; (E) 2; (F) No. 12; (G) 1; (H) No. 12; (I) 1; (J) No. 12; (K) 1; (L) T370-16(b); (M) 5; (N) 2.25; (O) 8; (P) 2.00; (Q) 15.25; (R) T370-16(a); (S) 15.5; (T) 1½

Chapter Five

5-1. (A) 370-28(a)(1); (B) 8; (C) 32; (D) 300-10; (E) 2; (F) 6

5-2. (A) 370-28(a)(1); (B) 8; (C) 48; (D) 300-10; (E) 14; (F) 2; (G) 20; (H) 300-10; (I) 2; (J) 8

5-3. (A) 370-28(a)(2); (B) 6; (C) 21; (D) 370-28(a)(2); (E) 6; (F) 21; (G) 300-10; (H) 2; (I) 5½

5-4. (A) 370-28(a)(2); (B) 6; (C) 18; (D) 370-28(a)(2); (E) 6; (F) 18; (G) 300-10; (H) 2; (I) 5: (J) 370-28(a)(2); (K) 6; (L) 18; (M) 370-28(a)(2); (N) 18; (O) 12; (P) 180; (Q) 13.4

5-5. (A) 370-28(a)(2); (B) 6; (C) 43; (D) 370-28(a)(2); (E) 3; (F) 19; (G) 6; (H) 6; (I) 27: (J) larger: (K) 36; (L) 300-10; (M) 8

5-6. (A) 370-28(a)(2); (B) 6; (C) 2; (D) 27; (E) 300-10; (F) 6; (G) 370- 28(a)(2), Excp.; (H) 373-6(a); (I) 300-10; (J) 6; (K) larger: (L) 8

5-7. (A) 370-28(a)(2); (B) 6; (C) 2; (D) 30; (E) 370-28(a)(2); (F) 6; (G) 28; (H) 300-10; (I) 6

5-8. (A) 370-28(a)(2); (B) 4; (C) 28; (D) 5; (E) 30; (F) 12.5; (G) larger (H) 30; (I) 370-28(a)(1) & (2); (J) 4; (K) 30; (L) 3 1/2; (M) 21; (N) 5; (O) 33.5; (P) larger; (Q) 33.5; (R) 300-10; (S) 7

5-9. (A) Ch. 9, T-5; (B) 0.0139; (C) 0.026; (D) 0.0366; (E) 0.1855; (F) 2.627; (G) 362-5; (H) 13.135; (I) 4 by 4

5-10. (A) 374-5(a); (B) before; (C) after; (D) Ch. 9, T-5; (E) 0.9729; (F) 0.4596; (G) 374-5(a); (H) largest; (I) 2.9187; (J) 4 by 4; (K) 374-9(d); (L) T373-6(a); (M) one; (N) 8; (O) two; (P) 6; (Q) larger; (R) 8 by 8

Chapter Nine

9-1. (A) 75; (B) 110-14(c)(2); (C) 420; (D) 600 kcmil; (E) T310-16; (F) 250- 81(b); (G) No. 1/0 AWG; (H) T250-94

9-2. (A) 0.87; (B) T310-16, CF; (C) 459.7; (D) 480; (E) 900 kcmil; (F) T310-16; (G) 75; (H) 110-14(c)(2), Excp. 1; (I) 425; (J) 900 kcmil; (K) T310-16; (L) 310-15(c); (M) 900 kcmil; (N) 250-81(a); (O) No. 1/0 AWG; (P) T250-94

9-3. (A) 310-4; (B) 750; (C) 250-81(b); (D) No. 2/0 AWG; (E) T250-94

9-4. (A) Ch. 9, T-8; (B) 133,100; (C) 133.1; (D) 310-4; (E) 532.4; (F) 250- 83(b); (G) No. 1/0 AWG; (H) T250-94

9-5. (A) 250-23(b); (B) No. 2 AWG; (C) T250-94

9-6. (A) 310-4; (B) 1,400; (C) 250-23(b); (D) No. 4/0 AWG; (E) T250-94

9-7. (A) 250-23(b); (B) 218.8; (C) 250 kcmil; (D) Ch. 9, T-8

9-8. (A) 310-4; (B) 5,400; (C) 250-23(b); (D) 675; (E) 700 kcmil; (F) Ch. 9, T-8

9-9. (A) 310-4; (B) 1,250; (C) 250-23(b); (D) 31.3; (E) No. 4/0 AWG; (F) Ch. 9, T-8; (G) No. 1/0 AWG; (H) 310-4; (I) larger; (J) 310-15(c); (K) No. 1/0 AWG

9-10. (A) 250-79(d); (B) No. 1/0AWG; (C) T250-94

9-11. (A) 250-79(d); (B) copper; (C) No. 4 AWG; (D) T310-16, N-3 tbl; (E) No. 8 AWG; (F) T250-94

9-12. (A) 250-79(d); (B) No. 2 AWG; (C) T250-94

9-13. (A) 310-4; (B) 1,000; (C) 250-79(d); (D) No. 2/0 AWG; (E) T250-94

9-14. (A) 250-79(d); (B) 187.5; (C) No. 4/0 AWG; (D) Ch. 9, T-8

9-15. (A) aluminum; (B) 250-79(d); (C) 310-4; (D) 400; (E) 900; (F) T310-16; (G) 310-4; (H) 4,500; (I) 250-79(d); (J) 562.5; (K) 600; (L) Ch. 9, T-8

9-16. (A) 250-79(e); (B) No. 1/0 AWG; (C) T250-95

9-17. (A) 250-79(e); (B) 90; (C) 250-95; (D) No. 8 AWG; (E) T250-95

9-18. (A) 250-79(e); (B) 800; (C) 250-95; (D) No. 3/0 AWG; (E) T250-95

9-19. (A) 60; (B) 110-14(c)(1), Excp. 1; (C) No. 3 AWG; (D) T310-16; (E) 3; (F) 210-19(a), FPN No. 2; (G) 208; (H) 0.866; (I) 0.0643; (J) 0.0608; (K) No. 4/0 AWG; (L) Ch. 9, T-8; (M) 250-79(e); (N) No. 8 AWG; (O) T250-95; (P) 250-95; (Q) Ch. 9, T-8; (R) 52,620; (S) 211,600; (T) 16,510; (U) 4.021; (V) 66,387; (W) Ch. 9, T-8; (X) No. 1 AWG

Chapter Eleven

11-1. (A) black; (B) white; (C) black; (D) red; (E) white; (F) white; (G) red; (H) red; (I) white; (J) white; (K) red; (L) white

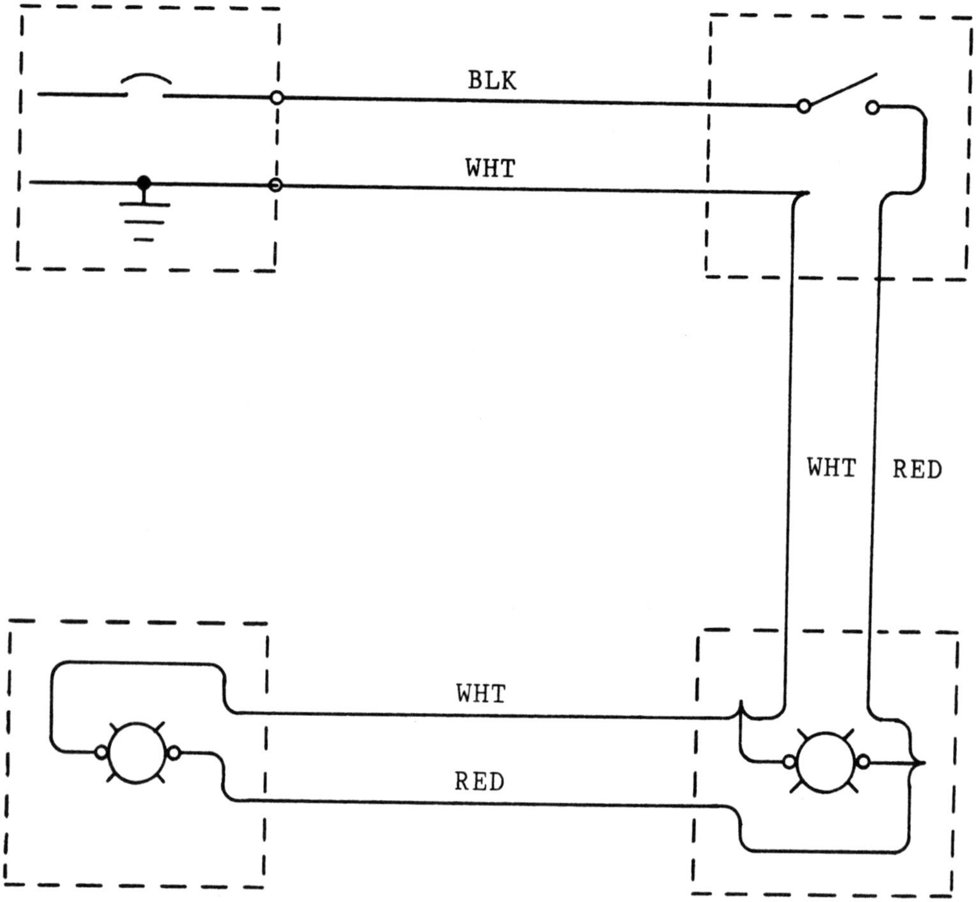

11-2. (A) black; (B) white; (C) black; (D) black; (E) white; (F) grounded; (G) white; (H) red; (I) ungrounded; (J) black; (K) red; (L) black; (M) white; (N) white; (O) black; (P) ungrounded; (Q) white; (R) grounded

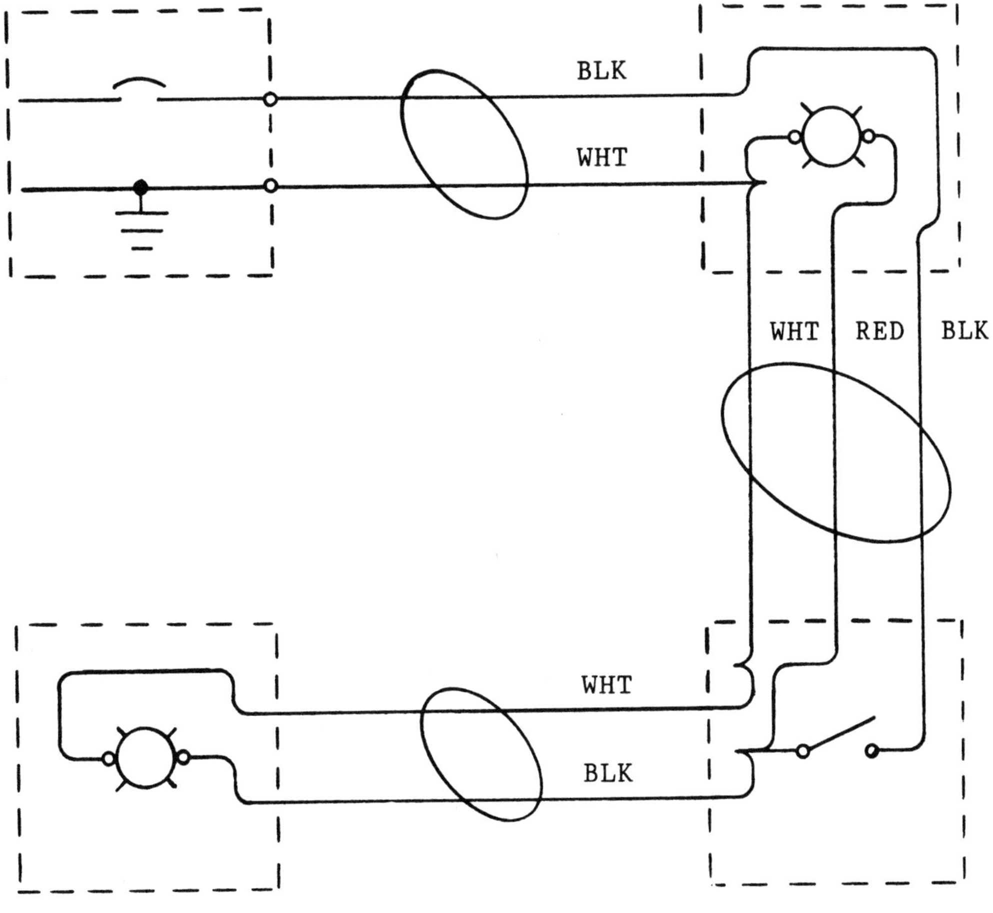

11-3. (A) black; (B) white; (C) black; (D) common; (E) white; (F) white; (G) brown; (H) switched; (I) white; (J) brown; (K) red; (L) red; (M) common; (N) brown; (O) switched

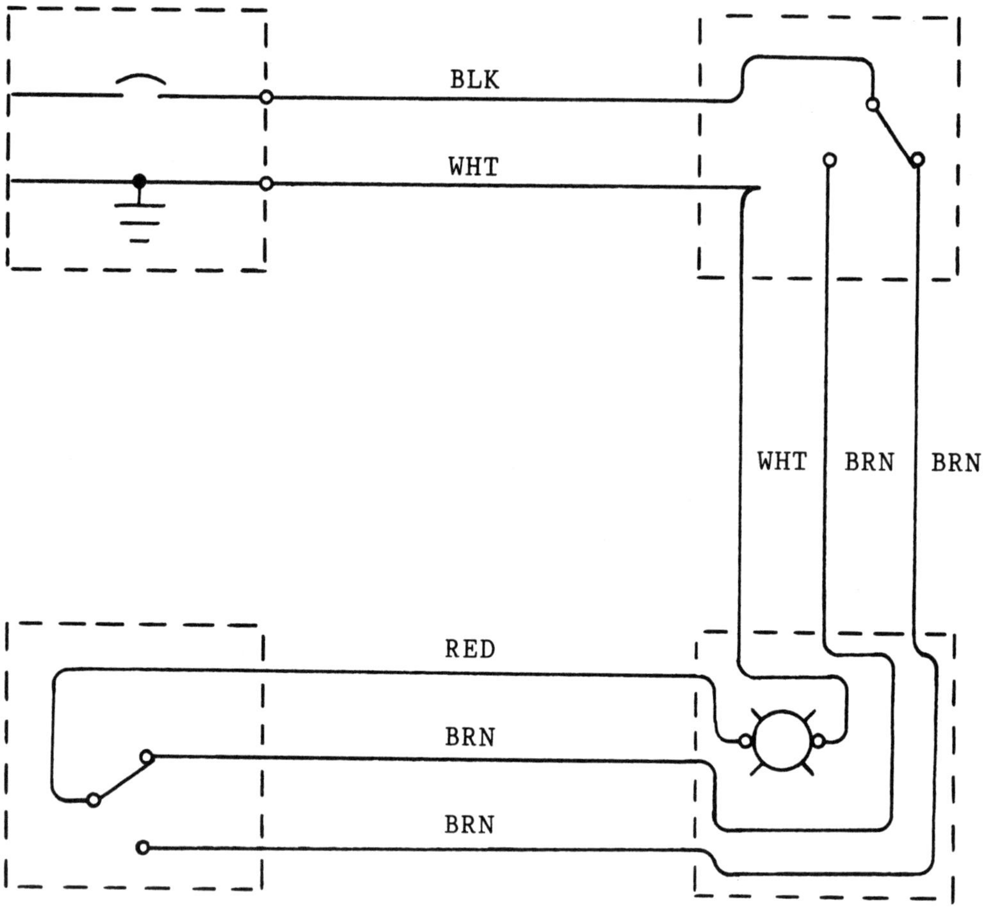

11-4. (A) black; (B) white; (C) black; (D) white-hot feed; (E) black; (F) white; (G) white; (H) red; (I) common; (J) black; (K) red; (L) switched; (M) white-hot feed; (N) common; (O) black; (P) red; (Q) switched; (R) white; (S) grounded; (T) black; (U) ungrounded; (V) red; (W) ungrounded

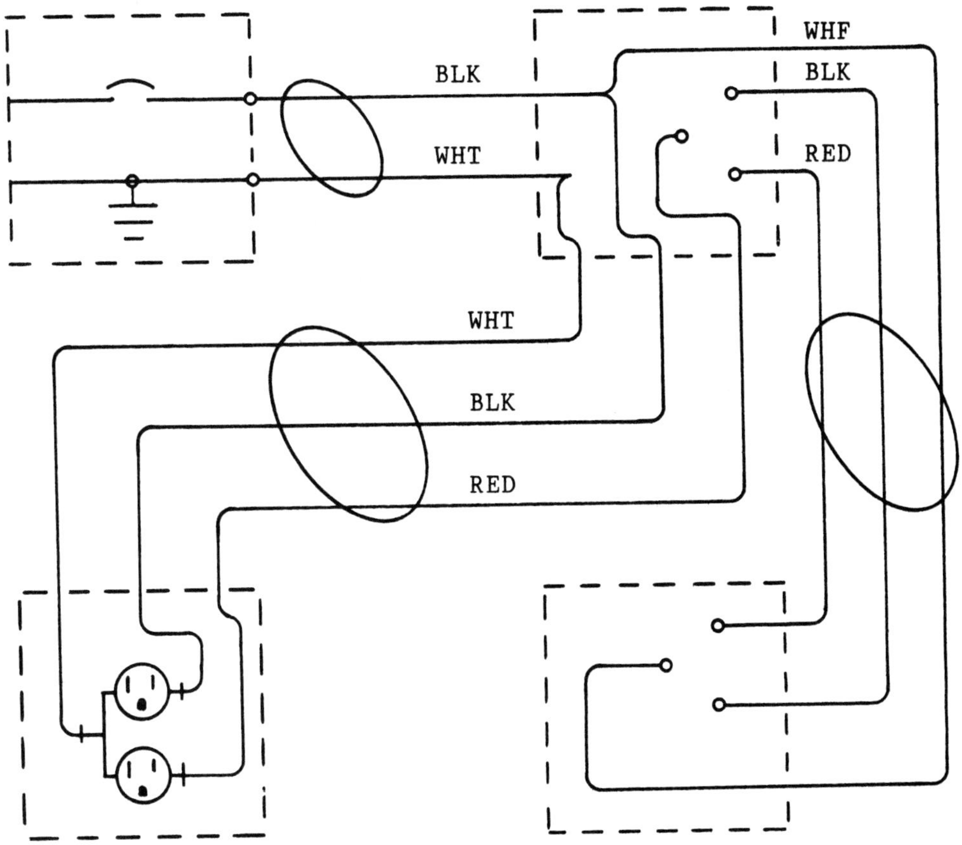

11-5. (A) black; (B) white; (C) black; (D) black; (E) connection; (F) white; (G) white; (H) connection; (I) red; (J) connection; (K) blue; (L) switched; (M) blue; (N) switched; (O) red; (P) common; (Q) blue; (R) switched; (S) red; (T) common; (U) blue; (V) switched; (W) red; (X) ungrounded; (Y) red; (Z) white; (1A) grounded; (1B) white; (1C) red; (1D) ungrounded; (1E) white; (1F) grounded

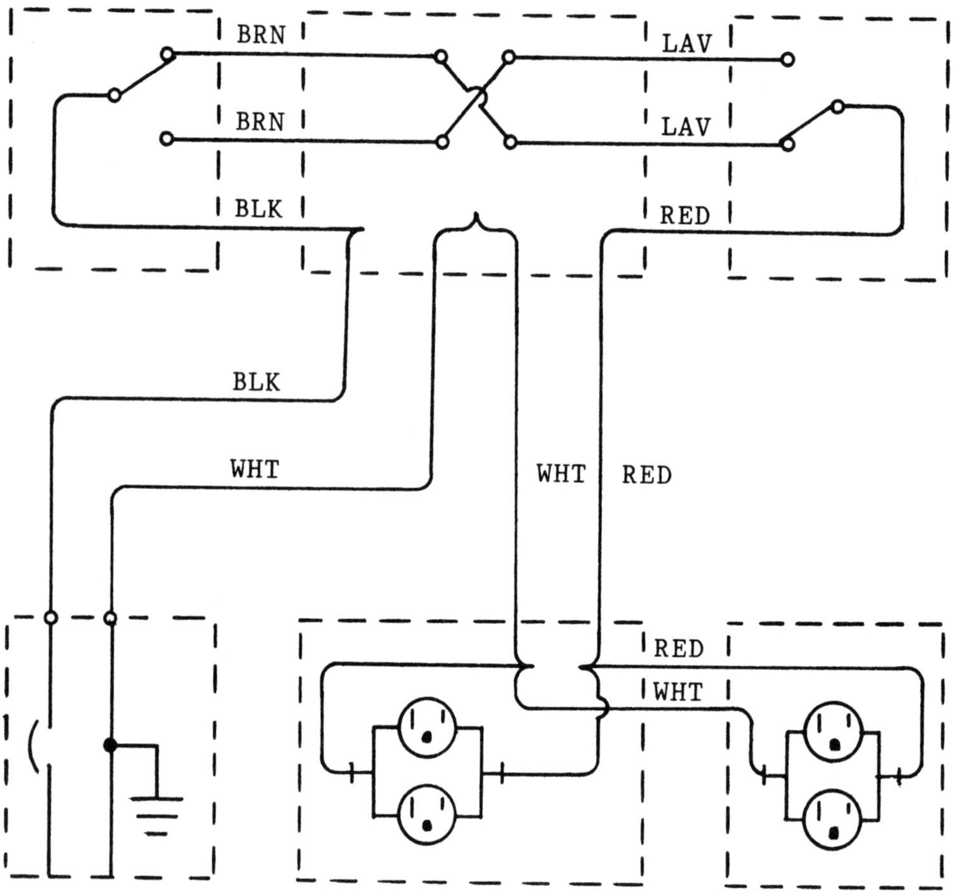

11-6. (A) ungrounded; (B) circuit breaker; (C) grounded; (D) grounded; (E) ungrounded; (F) common; (G) ungrounded; (H) switched; (I) grounded; (J) grounded; (K) connection; (L) grounded; (M) connection; (N) traveler; (O) switched; (P) traveler; (Q) switched; (R) return; (S) blue; (T) ungrounded; (U) ungrounded; (V) return; (W) blue; (X) common; (Y) traveler; (Z) switched; (1A) ungrounded; (1B) ungrounded; (1C) grounded; (1D) grounded

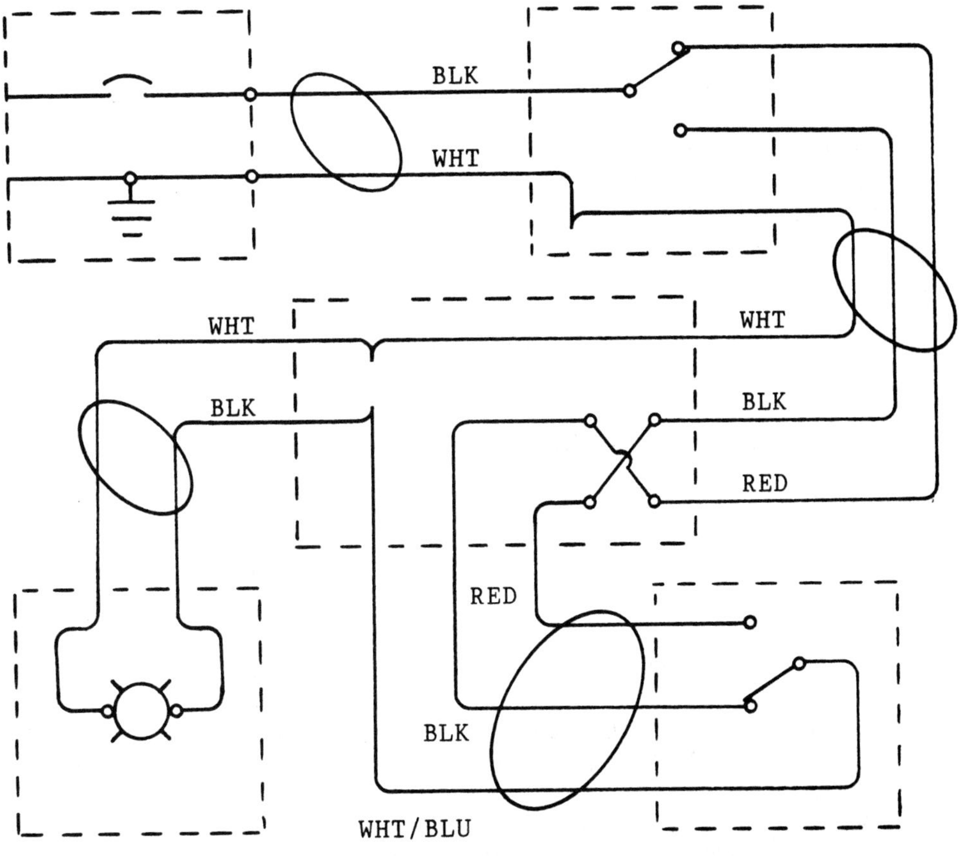

Chapter Twelve

12-1. (A) T430-150; (B) 28

12-2. (A) T430-150; (B) 65; (C) 430-110(a); (D) 115; (E) 74.8; (F) 100

12-3. (A) 440-12(a); (B) 75; (C) 440-12(a)(1); (D) 115; (E) 86.3; (F) 240-6(a); (G) 90; (H) 440-12(a)(2); (I) T430-150; (J) 88; (K) T430-151(b); (L) 404; (M) larger; (N) 30

12-4. (A) 1,000; (B) 1.732; (C) 12.6; (D) 440-2, Def.; (E) 30; (F) T430-150; (G) 11; (H) 21; (I) 165; (J) T430-151(b); (K) 91.5; (L) 169; (M) 2; (N) 2; (O) 2; (P) 136.6; (Q) 430-110(c)(2); (R) 440-12(b)(2); (S) 115; (T) 157; (U) 200; (V) 430-110(c)(1); (W) 440-12(b)(1)a.; (X) 150; (Y) T430-150; (Z) 2; (1A) 2; (1B) 2; (1C) 863.6; (1D) 430-110(c)(1); (1E) 440-12(b)(1)b.; (1F) 100; (1G) larger; (1F) 150

12-5. (A) T430-150; (B) 22; (C) 455-8(c), Excp. 1; (D) 250; (E) 240-6(a); (F) 60

12-6. (A) T430-150; (B) 21; (C) 455-8(c), Excp. 1; (D) 250; (E) 455-8(c), Excp. 1; (F) 100.6; (G) 240-6(a); (H) 110

12-7. (A) T430-151(b); (B) 73; (C) 455-8(c), Excp. 2; (D) 200; (E) 146; (F) T430-151(a); (G) 168; (H) 5

12-8. (A) T430-151(b); (B) 7 1/2; (C) T430-150; (D) 3; (E) 455-8(c), Excp. 2; (F) 127; (G) 9.6; (H) 146.2; (I) 455-8(c), Excp. 2; (J) 200; (K) 292.4; (L) T430-151(a); (M) 10

Chapter Thirteen

13-1. (A) 430-6(a); (B) 51; (C) 430-32(a)(1); (D) 63.8

13-2. (A) 430-6(a); (B) 270; (C) 430-34; (D) 378

13-3. (A) 38; (B) 430-6(a); (C) 430-32(a)(1); (D) 47.5; (E) 430-6(a); (F) 40; (G) T430-148; (H) 430-22(a); (I) 50; (J) 60; (K) 110-14(c)(1), Excp. 1; (L) 55; (M) T310-16; (N) No. 6 AWG

13-4. (A) 27; (B) 440-52(a)(1); (C) 430-32(a)(1); (D) 33.8; (E) 440-32; (F) 30; (G) 440-32; (H) 37.5; (I) 60; (J) 110-14(c)(1), Excp. 1; (K) 40; (L) T310-16; (M) No. 8 AWG

13-5. (A) 55; (B) 430-6(a); (C) 430-34; (D) 77; (E) 430-6(a); (F) 59; (G) T430-149; (H) 240-3; (I) 82.6; (J) 60; (K) 110-14(c)(1), Excp. 1; (L) 85; (M) T310-16; (N) No. 1 AWG

13-6. (A) T430-150; (B) 14; (C) 21; (D) 430-24; (E) 56; (F) 126; (G) 25; (H) 187.3; (I) 75; (J) 110-14(c)(2); (K) 205; (L) T310-16; (M) 250 kcmil

13-7. (A) 430-6(a); (B) 31; (C) 430-55; (D) 430-32(a)(1); (E) 38.8; (F) 240-6(a); (G) 40; (H) 430-34

13-8. (A) T430-150: (B) 27; (C) 430-52(c)(1); (D) T430-152; (E) 47.3; (F) 240-6(a); (G) 50; (H) 430-52(c)(1), Excp. 1; (I) 430-52(c)(1), Excp. 2(b); (J) 60.8; (K) 240-6(a); (L) 60; (M) 430-52(c)(1), Excp. 2(b); (N) 430-32(a)(1); (O) 430-22(a); (P) 33.8; (Q) 60; (R) 110-14(c)(1), Excp. 1; (S) 40; (T) T310-16; (U) No. 8 AWG ; (V) 430-34; (W) 240-3; (X) 37.8; (Y) 60; (Z) 110-14(c)(1), Excp. 1; (1A) 40; (1B) T310-16; (1C) No. 8 AWG

13-9. (A) T430-148; (B) 46; (C) 430-52(c)(1); (D) T430-152; (E) 138; (F) 240-6(a); (G) 150; (H) 430-52(c)(1), Excp. 1; (I) 430-52(c)(1), Excp. 2(a); (J) 184; (K) 240-6(a); (L) 175; (M) 430-52(c)(1), Excp. 2(a)

13-10. (A) T430-149; (B) 87; (C) 430-52(c)(1); (D) T430-152; (E) 217.5; (F) 240-6(a); (G) 225; (H) 430-52(c)(1), Excp. 1; (I) 430-52(c)(1), Excp. 2(c)(1); (J) 348; (K) 240-6(a); (L) 300; (M) 430-52(c)91), Excp. 2(c)(1)

13-11. (A) 130; (B) 430-6(a); (C) T430-150; (D) 126; (E) 430-6(a); (F) 430-52(c)(3); (G) T430-152; (H) 1,430; (I) 240-6(a); (J), 1,600; (K) 430-52(c)(3), Excp. 1; (L) 430-52(c)(3), Excp. 1; (M) 2,210; (N) 240-6(a); (O) 2,000; (P) 430-52(c)(3), Excp. 1; (Q) 430-52(c)(1); (R) T430-152; (S) 227.5; (T) 240-6(a); (U) 250; (V) 430-52(c)(1), Excp. 1; (W) 430-52(c)(1), Excp. 2(b); (X) 292.5; (Y) 240-6(a); (Z) 250; (1A) 430-52(c)(1), Excp. 2(b); (1B) 430-32(a)(1); (1C) 157.5; (1D) 430-34; (1E)

176.4; (1F) 430-32(a)(1); (1G) 430-22(a); (1H) 162.5; (1I) 75; (1J) 110-14(c)(1), Excp. 1; (1K) 175; (1L) T310-16; (1M) No. 2/0 AWG; (1N) 430-34; (1O) 240-3; (1P) 182; (1Q) 75; (1R) 110-14(c)(2), Excp. 1; (1S) 200; (1T) T310-16; (1U) No. 3/0 AWG

13-12. (A) 59; (B) 440-22(a); (C) 440-32; (D); 54; (E) 440-52(a)(1); (F) 440-22(a); (G) 103.3; (H) 240-6(a); (I) 110; (J) 440-22(a); (K) 132.8; (L) 240-6(a); (M) 125; (N) 440-22(a); (O) 430-32(a)(1); (P) 67.5; (Q) 440-52(a)(1); (R) 75.6; (S) 430-32(a)(1); (T) 440-32; (U) 73.8; (V) 60; (W) 110-14(c)(1), Excp. 1; (X) 75; (Y) T310-16; (Z) No. 2 AWG; (1A) 440-52(a)(1); (1B) 240-3; (1C) 82.6; (1D) 60; (1E) 110-14(c)(1), Excp. 1; (1F) 85; (1G) T310-16; (1H) No. 1 AWG

13-13. (A) T430-150; (B) 80; (C) 104; (D) 430-62(a); (E) T430-152; (F) 140; (G) 182; (H) 430-62(a); (I) 2; (J) 550; (K) 240-6(a); (L) 500; (M) 430-62(a); (N) 430-24; (O) 25; (P) 498; (Q) T310-16, CF; (R) 607.3; (S) 615; (T) 1,750 kcmil; (U) T310-16; (V) 75; (W) 110-14(c)(1), Excp. 1; (X) 520; (Y) 1,500 kcmil; (Z) T310-16; (1A) 310-15(c); (1B) larger; (1C) 1,750 kcmil

13-14. (A) T430-150; (B) 77; (C) 96; (D) 430-62(a); (E) T430-152; (F) 192.5; (G) 144; (H) 430-62(a); (I) 2; (J) 538.5; (K) 240-6(a); (L) 500; (M) 430-62(a)

13-15. (A) T430-150; (B) 74.8; (C) 88; (D) 90; (E) 440-22(a); (F) 430-52(c)(3); (G) T430-52; (H) 598.4; (I) 240-6(a); (J) 600; (K) 430-52(c)(3), Excp. 1; (L) 430-52(c)(3), Excp. 1; (M) 972.4; (N) 240-6(a); (O) 800; (P) 430-52(c)(3), Excp. 1; (Q) 430-52(c)(1); (R) T430-152; (S) 130.9; (T) 240-6(a); (U) 150; (V) 430-52(c)(1), Excp. 1; (W) 430-52(c)(1), Excp. 2(b); (X) 168.3; (Y) 240-6(a); (Z) 150; (1A) 430-52(c)(1), Excp. 2(b); (1B) 430-52(c)(1); (1C) T430-152; (1D) 132; (1E) 240-6(a); (1F) 150; (1G) 430-52 (c)(1), Excp. 1; (1H) 430-52(c)(1), Excp. 2(a); (1I) 352; (1J) 240-6(a); (1K) 350; (1L) 430-52(c)(1), Excp. 2(a); (1M) 440-22(a); (1N) 157.5; (1O) 240-6(a); (1P) 175; (1Q) 440-22(a); (1R) 202.5; (1S) 240-6(a); (1T) 200; (1U) 440-22(a); (1V) 430-62(a); (1W) T430-152; (1X) 440-22(a); (1Y) 250; (1Z) 150; (2A) 175; (2B) 430-62(a); (2C) 187; (2D) 2; (2E) 527.8; (2F) 240-6(a); (2G) 500; (2H) 430-62(a)

13-16. (A) 114; (B) T430-150; (C) 430-52(c)(1), Excp. 2(b); (D) 256.5; (E) 240-6(a); (F) 250; (G) 430-52(c)(1), Excp. 2(b); (H) 430-62(a); (I) 430-52(c)(1), Excp. 2(c)(2); (J) 300; (K) 220-10(b); (L) 125; (M) 388.3; (N) 240-6(a); (O) 350; (P) 430-62(a); (Q) 430-24; (R) 430-24; (S) 240-3; (T) 140; (U) 220-10(b); (V) 125; (W) 205.9; (X) 75; (Y) 110-14(c)(2), Excp. 1; (Z) 230; (1A) T310-16; (1B) No. 4/0 AWG

13-17. (A) 13.2; (B) T430-150; (C) 52; (D) 96; (E) 76; (F) 440-7, Excp.; (G) 430-63; (H) 430-62(a); (I) T430-152; (J) 91; (K) 168; (L) 440-22(a); (M) 133; (N) 430-62(a); (O) 3; (P) 3; (Q) 688; (R) 220-10(b); (S) 125; (T) 107.0; (U) 430-63; (V) 795; (W) 240-6(a); (X) 700; (Y) 430-63

13-18. (A) 8.14; (B) T430-150; (C) 4.8; (D); 40; (E) 52; (F) 440-35; (G) 60; (H) T310-16; (I) 110; (J) 440-22(c); (K) 150; (L) 430-63; (M) 430-62(a); (N) T430-152; (O) 175; (P) 175; (Q) 175; (R) 150; (S) 440-22(b); (T) 430-62(b); (U) 150; (V) 2; (W) 4; (X) 442.8; (Y) 220-10(b); (Z) 125; (1A) 45.6; (1B) 430-63; (1C) 488.4; (1D) 240-6(a); (1E) 450; (1F) 430-63; (1G) 430-24; (1H) 110; (1I) 2; (1J) 4; (1K) 25; (1L) 220-10(b); (1M) 464.2; (1N) 75; (1O) 110-14(c)(2), Excp. 1; (1P) 485; (1Q) T310-16; (1R) 1,250 kcmil

Chapter Fourteen

14-1. (A) 65; (B) T430-150; (C) 430-22(a); (D) 125; (E) 60; (F) 110-14(c)(1), Excp. 1; (G) 85; (H) T310-16; (I) No. 3 AWG; (J) 430-52(c)(1); (K) T430-152; (L) 250; (M) 240-6(a); (N) 175; (O) 430-52(c)(1), Excp. 1; (P) 430-52(c)(1), Excp. 2(c)(1); (Q) 400; (R) 240-6(a); (S) 250; (T) 430-52(c)(1), Excp. 2(c)(1); (U) No. 2 AWG; (V) 60; (W) 110-14(c)(1), Excp. 1; (X) 95; (Y) T310-16; (Z) 430-28; (1A), 1,000; (1B) have

14-2. (A) 192; (B) T430-150; (C) 430-22(a); (D) 125; (E) 75; (F) 110-14(c)(2), Excp. 1; (G) 255; (H) T310-16; (I) 250 kcmil; (J) 430-52(c)(1); (K) T430-150; (L) 175; (M) 240-6(a); (N) 350; (O) 430-52(c)(1), Excp. 1; (P) 430-52(c)(1), Excp. 2(b); (Q) 225; (R) 240-6(a); (S) 400; (T) 430-52(c)(1), Excp. 2(b); (U) 250 kcmil; (V) 75; (W) 110-14(c)(2), (X) 255; (Y) T310-16; (Z) 1,500 kcmil; (1A) 75; (1B) 110-14(c)(2); (1C) 625; (1D) T310-16; (1E) 430-28; (1F) 300; (1G) have

14-3. (A) yes; (B) T430-150; (C) 2.4; (D) 3.5; (E) no; (F) no; (G) 3; (H) 2; (I) 14.2; (J) no; (K) have; (L) 430-53(a)

14-4. (A) T430-150; (B) 7.6; (C) 11; (D) 26.2; (E) no; (F) 5; (G) 430-52(c)(1), Excp. 2(c)(1); (H) 400; (I) 240-6(a); (J) 30; (K) no; (L) have; (M) 430-53(b)

14-5. (A) 60; (B) 110-14(c)(1), Excp. 1; (C) 85; (D) T310-16; (E) 90; (F) 250-6(a); (G) have not; (H) cannot; (I) 430-53(c)(4)

14-6. (A) T430-148; (B) 3.2; (C) 7.6; (D) 440-22(b)(1); (E) 225; (F) 58.6; (G) 240-6(a); (H) 50; (I) 440-22(b)(1); (J) 440-34; (K) 25; (L) 44.6; (M) 60; (N) 110-14(c)(1), Excp. 1; (O) 55; (P) T310-16; (Q) No. 6 AWG

14-7. (A) 88; (B) T430-150; (C) 430-33; (D) T430-152; (E) 250; (F) 240-6(a); (G) 200; (H) 430-33; (I) T430-22(a), Excp.; (J) 150; (K) 75; (L) 110-14(c)(2); (M) 150; (N) T310-16; (O) No. 1/0 AWG

14-8. (A) 430-52(c)(1); (B) 58; (C) 430-52(c)(4); (D) T430-152; (E) 250; (F) 240-6(a); (G) 150; (H) 430-53(c)(1), Excp. 1; (I) 64; (J) 430-22(a); (K) 125; (L) 60; (M) 110-14(c)(1), Excp. 1; (N) 85; (O) T310-16; (P) No. 3 AWG

14-9. (A) T430-23(c); (B) 65; (C) 75; (D) 110-14(c)(2), Excp. 1; (E) 175; (F) T310-16; (G) No. 2/0 AWG

14-10. (A) 180; (B) 430-6(a); (C) T430-150; (D) 172; (E) 430-6(a); (F) 430-52(c)(1); (G) T430-152; (H) 540; (I) 240-6(a); (J) 600; (K) 430-52(c)(1), Excp. 1; (L) 430-52(c)(1), Excp. 2(a); (M) 720; (N) 240-6(a); (O) 700; (P) 430-52(c)(1), Excp. 2(a); (Q) 430-32(a)(1); (R) 58; (S) 430-32(a)(1); (T) 124.8; (U) 430-34; (V) 139.7; (W) 430-32(a)(1); (X) 430-22(a); (Y) 225; (Z) 75; (1A) 110-14(c)(1), Excp. 1; (1B) 230; (1C) T310-16; (1D) No. 4/0 AWG; (1E) 430-34; (1F) 240-3; (1G) 252; (1H) 75; (1I) 110-14(c)(2), Excp. 1; (1J) 255; (1K) T310-16; (1L) 250 kcmil; (1M) 430-22(a); (1N) 99.8; (1O) 430-32(a)(1); (1P) 430-22(a); (1Q) 124.8; (1R) 75; (1S) 110-14(c)(2), Excp. 1; (1T) 130; (1U) T310-16; (1V) No. 1 AWG; (1W) 430-34; (1X) 240-3; (1Y) 139.7; (1Z) 75; (2A) 110-14(c)(2), Excp. 1; (2B) 150; (2C) T310-16; (2D) No. 1/0 AWG

14-11. (A) 83; (B) T430-147; (C) 430-52(c)(1); (D) T430-152; (E) 150; (F) 240-6(a); (G) 125; (H) 430-52(c)(1), Excp. 1; (I) 430-39; (J) 430-22(a); (K) 125; (L) 75; (M) 110-14(c)(1), Excp. 1; (N) 115; (O) T310-16; (P) No. 2 AWG; (Q) 430-22(a), Excp. 2(a); (R) 190; (S) 75; (T) 110-14(c)(1), Excp. 1; (U) 175; (V) T310-16; (W) No. 2/0 AWG

14-12. (A) 42; (B) T430-150; (C) 455-7, Excp. 1; (D) 250; (E) 240-6(a); (F) 110; (G) 455-7, Excp. 2; (H) 455-6, Excp.; (I) 250; (J) 75; (K) 110-14(c)(2), Excp. 1; (L) 115; (M) T310-16; (N) No. 2/0 AWG; (O) 430-22(a); (P) 125; (Q) 60; (R) 110-14(c)91), Excp. 1; (S) 55; (T) 110-14(c)(1), Excp. 1; (U) No. 4 AWG

14-13. (A) 17; (B) T430-150; (C) 455-7, Excp. 1; (D) 250; (E) 106.3; (F) 240-6(a); (G) 110; (H) 455-7, Excp. 2; (I) 455-6, Excp.; (J) 250; (K) 106.3; (L) 75; (M) 110-14(c)(2), Excp. 1; (N) 115; (O) T310-16; (P) No. 2 AWG; (Q) 430-22(a); (R) 125; (S) 60; (T) 110-14(c)(1), Excp. 1; (U) 25; (V) T310-16; (W) No. 12 AWG

Chapter Fifteen

15-1. (A) 230; (B) 480; (C) 78.6

15-2. (A) 230; (B) 480; (C) 32.1

15-3. (A) 1; (B) 0.06; (C) 12,500

15-4. (A) 50.1; (B) 450-3(b)(1); (C) 125; (D) 240-6(a); (E) 70; (F) 450-3(b)(1), Excp. 1; (G) 60; (H) 110-14(c)(1); (I) 70; (J) T310-16; (K) No. 4 AWG; (L) 167.7; (M) 75; (N) 110-14(c)(2); (O) 175; (P) T310-16; (Q) No. 2/0 AWG

15-5. (A) 450-3(b)(2); (B) 125; (C) 240-6(a); (D) 100; (E) 450-3(b)(2), Excp. 1; (F) 41.9; (G) 450-3(b)(2); (H) 125; (I) 240-6(a); (J) 60; (K) 450-3(b)(2); (L) 60; (M) 110-14(c)(1), Excp. 1; (N) 100; (O) T310-16; (P) No. 1/0 AWG; (Q) 56.7; (R) 60; (S) 110-14(c)(1), Excp. 1; (T) 65; (U) T310-16; (V) No. 3 AWG

15-6. (A) 166.7; (B) 450-3(b)(2); (C) 125; (D) 240-6(a); (E) 225; (F) 450-3(b)(2), Excp.; (G) 83.4; (H) 450-3(b)(2); (I) 250; (J) 240-6(a); (K) 200; (L) 450-3(b)(2); (M) 75; (N) 110-14(c)(2); (O) 230; (P) T310-16; (Q) 300 kcmil; (R) 112.5; (S) 75; (T) 110-14(c)(2); (U) 120; (V) T310-16; (W) No. 1/0 AWG

15-7. (A) 156.3; (B) 450-3(b)(2); (C) 125; (D) 240-6(a); (E) 200; (F) 450-3(b)(2), Excp. 1; (G) 65.2; (H) 450-3(b)(2); (I) 250; (J) 240-6(a); (K) 150; (L) 450-3(b)(2); (M) 75; (N) 110-14(c)(2); (O) 205; (P) T310-16; (Q) 250 kcmil; (R) 83.5; (S) 60; (T) 110-14(c)(1), Excp. 1; (U) 85; (V) T310-16; (W) No. 1 AWG

15-8. (A) 31.3:1; (B) 1,000; (C) 34.5; (D) 4.79

15-9. (A) T450-3(a)(1); (B) 1.25; (C) 240; (D) 57.6; (E) 8; (F) T450-3(a)(1); (G) 300; (H) 240-6(a); (I) 25; (J) 450-3(a)(1), Excp.; (K) 75; (L) 110-14(c)(2); (M) 310; (N) T310-17; (O) No. 3/0 AWG; (P) 75; (Q) 110-14(c)(2); (R) 310; (S) T310-16; (T) 350 kcmil

15-10. (A) 2; (B) 25.4

15-11. (A) 1.732; (B) 52.8

15-12. (A) 80.2; (B) 2,510

15-13. (A) 12,500; (B) 1.732; (C) 115.4; (D) 7,217; (E) 115.4; (F) 208; (G) 6,940; (H) 120; (I) 6,940

15-14. (A) 1,600; (B) T450-3(a)(1); (C) 2,000; (D) 2,000; (E) 240-6(a), (F) 310-4; (G) 6; (H) 75; (I) 110-14(c)(2), Excp. 1; (J) 335; (K) T310-16; (L) 400 kcml; (M) 26.7; (N) T450-3(a)(1); (O) 80.1; (P) 240-6(a); (Q) 90; (R) 450-3(a)(1), Excp. 1; (S) 310-4; (T) 2,400; (U) 250-79(d); (V) 300

15-15. (A) 82.8; (B) 450-3(b)(1); (C) 125; (D) 240-6(a); (E) 110; (F) 450-3(b)(1), Excp. 1; (G) 75; (H) 110-14(c)(2), Excp. 1; (I) 115; (J) T310-16; (K) No. 2 AWG; (L) 4,217

15-16. (A) 18,000; (B) 4.82; (C) 18,000; (D) 2.78; (E) 480; (F) 180.4; (G) 480; (H) 104.2

15-17. (A) T450-3(a)(1); (B) 1,920; (C) 1,463; (D) 310-4; (E) 300; (F) 75; (G) 110-14(c)(2); (H) 310; (I) T310-16; (J) 500 kcmil; (K) 46.9; (L) T450-3(a)(1); (M) 300; (N) 240-6(a); (O) 150; (P) 450-3(a)91, Excp. 1

15-18. (A) 183.2; (B) 450-3(b)(2); (C) 125; (D) 240-6(a); (E) 250; (F) 450-3(b)(2), Excp. 1; (G) 86.6; (H) 450-3(b)(2); (I) 250; (J) 240-6(a); (K) 200; (L) 450-3(b)(2); (M) 75; (N) 110-14(c)(2); (O) 255; (P) T310-16; (Q) 350 kcmil; (R) 118.1; (S) 75; (T) 110-14(c)(2); (U) 120; (V) T310-16; (W) No. 1/0 AWG

Chapter Sixteen

16-1. (A) T450-3(a)(1); (B) 48; (C) 16; (D) T450-3(a)(1); (E) 20; (F) 36; (G) 38.5; (H) 83.3; (I) 121.8; (J) T450-3(a)(1); (K) 125; (L) 240-6(a); (M) 50; (N) 450-3(a)(1), Excp. 1; (O) T450-3(a)(1); (P) 125; (Q) 240-6(a); (R) 175; (S) 450-3(a)(1), Excp. 1; (T) 60; (U) 110-14(c)(1), Excp. 1; (V) 55; (W) T310-16; (X) No. 4 AWG; (Y) 75; (Z) 110-14(c)(2), Excp. 1; (1A) 180; (1B) T310-16; (1C) No. 4/0 AWG; (1D) 220-22; (1E) T450-3(a)(1); (1F) 104.1; (1G) 72.9; (1H) 60; (1I) 110-14(c)(1), Excp. 1; (1J) 75; (1K) T310-16; (1L) No. 2 AWG; (1M) 250-23(b); (1N) No. 2 AWG; (1O) T250-94; (1P) larger; (1Q) 310-15(c); (1R) No. 2 AWG; (1S) 0.70; (1T) T450-3(a)(1); (1U) 300; (1V) 240-6(a), Excp.; (1W) 3; (1X) 450-3(a)(1), Excp. 1; (1Y) 7,621; (1Z) 2.62; (2A) 3.32; (2B) T450-3(a)(1), Excp. 1; (2C) 300; (2D) 240-6(a); (2E) 10; (2F) 450-3(a)(1), Excp. 1

16-2. (A) 86.6; (B) 65.0; (C) 57.7; (D) 64.9

16-3. (A) 52.0; (B) 450-3(b)(2); (C) 41.6; (D) 48; (E) 24; (F) 75; (G) 110-14(c)(2), Excp. 1; (H) 135; (I) T310-16; (J) No. 1 AWG; (K) 31.5; (L) 450-3(b)(2); (M) 78.7; (N) 240-6(a); (O) 70; (P) 450-3(b)(2); (Q) 68.2; (R) 60; (S) 110-14(c)(1), Excp. 1; (T) 70; (U) T310-16; (V) No. 4 AWG

16-4. (A) T450-3(a)(1); (B) 38.4; (C) 44.3; (D) 22.2; (E) T450-3(a)(1); (F) 24; (G) 46.2; (H) 53.4; (I) 100; (J) 153.4; (K) T450-3(a)(1); (L) 125; (M) 240-6(a); (N) 70; (O) 450-3(a)(1), Excp. 1; (P) T450-3(a)(1); (Q) 125; (R) 240-6(a); (S) 200; (T) 450-3(a)(1), Excp. 1; (U) 60; (V) 110-14(c)(1). Excp. 1; (W) 70; (X) T310-16; (Y) No. 4 AWG; (Z) 75; (1A) 110-14(c)(2), Excp. 1; (1B) 200; (1C) T310-16; (1D) No. 3/0 AWG; (1E) 220-22; (1F) T450-3(a)(1); (1G) 125; (1H) 87.5; (1I) 60; (1J) 110-14(c)(91), Excp. 1; (1K) 95; (1L) T310-16; (1M) No. 2 AWG; (1N) 250-23(b); (1O) No. 4 AWG; (1P) T250-94; (1Q) larger; (1R) 310-15(c); (1S) No. 2 AWG; (1T) 0.97; (1U) T450-3(a)(1); (1V) 300; (1W) 240-6(a); (1X) 3; (1Y) 450-3(a)(1), Excp. 1; (1Z) 1.82; (2A) 2.79; (2B) T450-3(a)(1); (2C) 300; (2D) 240-6(a), Excp. (2E) 10; (2F) 450-3(a)(1), Excp. 1

16-5. (A) 187.1; (B) 450-3(b)(2); (C) 150.0; (D) 50; (E) 75; (F) 110-14(c)(2), Excp. 1; (G) 230; (H) T310-16; (I) 300 kcmil; (J) 144.3; (K) 450-3(b)(2); (L) 360.9; (M) 240-6(a); (N) 350; (O) 450-3(b)(2); (P) 180; (Q) 75; (R) 110-14(c)(2), Excp. 1; (S) 205; (T) 310-16; (U) 250 kcmil

16-6. (A) 1,195; (B) T450-3(a)(1); (C) 956; (D) 1,104; (E) 552; (F) 310-4; (G) 1,200; (H) 75; (I) 110-14(c)(2); (J) 420; (K) T310-16; (L) 600 kcmil; (M) 35.4; (N) T450-3(a)(1); (O) 300; (P) 240-6(a); (Q) 110; (R) 450-3(a)(1), Excp. 1

16-7. (A) 17; (B) T430-148; (C) 125; (D) 450-4; (E) 510; (F) 21.3; (G) 3.20; (H) 24.5

16-8. (A) 41.7; (B) 1,543; (C) 450-4; (D) 1,234; (E) 5.57; (F) 36.1

Chapter Seventeen

17-1. (A) Art. 220, Pt. B; (B) 220-3(b); (C) 4,800; (D) 1/4; (E) 11,200; (F) 6,400; (G) 3; (H) 20,400; (I) 220-12; (J) 6,000; (K) 1,200; (L) 600-5(b)(3); (M) 27,600; (N) 220-10(b); (O) 34,500; (P) 220-3(c)(7); (Q) 21; (R) 80; (S) 18,180; (T) T220-13; (U) 10; (V) 8,180; (W) 50; (X) 14,090; (Y) 220-10(a); (Z) 3,500; (1A) 220-15; (1B) 220-14; (1C) 7.2; (1D) T430-148; (1E) 1,728; (1F) 220-14; (1G) 440-35; (1H) 12; (1I) T430-148; (1J) 6.9; (1K) 1,656; (1L) 13,565; (1M) 220-21; (1N) larger; (1O) 13,656; (1P) 220-14; (1Q) 414; (1R) 216; (1S) 220-10(b); (1T) 66,160; (1U) 90-8; (1V) 0.80; (1W) 82,700; (1X) 344.6; (1Y) 350; (1Z) 240-6(a); (2A) 240-3; (2B) 75; (2C) 110-14(c)(2); (2D) 380; (2E) T310-16; (2F) 500 kcmil; (2G) 220-22; (2H) 50,534; (2I) 90-8; (2J) 0.80; (2K) 63,168; (2L) 263.1; (2M) 1.02; (2N) 268.4; (2O) 75; (2P) 110-14(c)(2); (2Q) 285; (2R) T310-16; (2S) 300 kcmil; (2T) 250-23(b); (2U) 500 kcmil; (2V) No. 1/0 AWG; (2W) T250-94; (2X) 310-15(c); (2Y) larger; (2Z) 300 kcmil; (3A) 250-81(a); (3B) 500 kcmil; (3C) No. 1/0 AWG; (3D) T250-94; (3E) 250-81(d); (3F) No. 2 AWG; (3G) 250-94, Excp. c.

17-2. (A) Art. 220, Pt. B; (B) 220-3(b); (C) 2,720; (D) 896; (E) 1,824; (F) T220-3(b); (G) 5,472; (H) 220-16(a); (I) 1,500; (J) 1,500; (K) 220-16(b); (L) 9,972; (M) T220-11; (N) 3,000; (O) 6,972; (P) 35; (Q) 5,440; (R) 9,904; (S) 220-17; (T) 7,428; (U) 3,004; (V) 220-17; (W) 2,253; (X) 220-18; (Y) larger; (Z) 5,000; (1A) 220-22; (1B) 3,500; (1C) 220-19; (1D) 8; (1E) T220-19, Col. A; (1F) T220-19, N-1; (1G) 12; (1H) 9; (1I) 145; (1J) 8,000; (1K) 11,600; (1L) 220-22; (1M) 8,120; (1N) 220-15; (1O) 5,000; (1P) 20,000; (1Q) 220-14; (1R) 440-35; (1S) T430-148; (1T) 6.9; (1U) 9,856; (1V) 220-21; (1W) larger; (1X) 220-14; (1Y) 430-24; (1Z) filter pump; (2A) 350; (2B) sump pump; (2C) 216; (2D) 220-10(a); (2E) 49,818; (2F) 207.5; (2G) 225; (2H) 240-6(a); (2I) 240-3; (2J) 75; (2K) 110-14(c)(2); (2L) 230; (2M) T310-16; (2N) No. 4/0 AWG; (2O) 220-10(a); (2P) 220-22; (2Q) 19,529; (2R) 81.4; (2S) 225; (2T) 207.5; (2U) 1.084; (2V) 60; (2W) 110-14(c)(1), Excp. 1; (2X) 95; (2Y) T310-16; (2Z) No. 2 AWG; (3A) 250-23(b); (3B) No. 4/0 AWG; (3C) No. 2 AWG; (3D) T250-94; (3E) 310-15(c); (3F) larger; (3G) No. 2 AWG; (3H) 250-83(c); (3I) 250-94, Excp. a.; (3J) No. 6 AWG; (3K) 250-92(a)

Index

A

B

F

G

H

I

J

K

L

M

N

O

P

R

S

T

U

V

W